ASTRONOMY AND ASTROPHYSICS LIBRARY

For further volumes:
http://www.springer.com/series/848

Rudolf Kippenhahn
Alfred Weigert
Achim Weiss

Stellar Structure and Evolution

Second Edition

Rudolf Kippenhahn
Göttingen
Germany

Alfred Weigert
Universität Hamburg
Hamburg
Germany

Achim Weiss
Max-Planck-Institut für Astrophysik
Garching
Germany

ISSN 0941-7834
ISBN 978-3-642-44524-8 ISBN 978-3-642-30304-3 (eBook)
DOI 10.1007/978-3-642-30304-3
Springer Heidelberg New York Dordrecht London

Softcover reprint of the hardcover 2nd edition 2012

Printed on acid-free paper

Springer is part of Springer Science+Business Media (www.springer.com)

To Our Wives

Preface to the First Edition

The attempt to understand the physics of the structure of stars and their change in time – their evolution – has been bothering many physicists and astronomers ever since the last century. This long chain of successful research is well documented not only by numerous papers in the corresponding journals but also by a series of books. Some of them are so excellently written that despite their age they can still be recommended and not only as documents of the state of the art at that time. A few outstanding examples are the books of Emden (1907), Eddington (1926), Chandrasekhar (1939), and Schwarzschild (1958). But our science has rapidly expanded in the last few decades, and new aspects have emerged which could not even be anticipated, say, 30 years ago and which today have to be carefully explored.

This does not mean, however, that our ambition is to present a complete account of the latest and most refined numerical results. This can well be left to the large and growing number of excellent review articles. This book is intended rather to be a textbook that will help students and teachers to understand these results as far as possible and present them in a simple and clear manner. We know how difficult this is since we ourselves have tried for the largest part of our scientific career to understand "how the stars work" – and then to make others believe it. In these attempts we have found that often enough a simplified analytical example can be more helpful than the discussion of an exceptionally beautiful numerical solution. Therefore we do not hesitate to include many simple considerations and estimates, if necessary, even at the expense of rigour and the latest results. The reader should also note that the list of references given in this book is not intended to represent a table of honour for the (known and unknown) heroes of the theory of stellar structure; it is merely designed to help the beginner to find a few first paths in the literature jungle and presents those papers from which we have more or less randomly chosen the numbers for figures and numerical examples (There are others of at least the same quality!).

The choice of topics for a book such as this is difficult and certainly subject to personal preferences. Completeness is neither possible nor desirable. Still, one may wonder why we did not include, for example, binary stars, although we are obviously interested in their evolution. The reason is that here one would have had to include the physics of essentially non-spherical objects (such as disks), while we concentrate mainly on spherical configurations; even in the brief description of rotation the emphasis is on small deviations from spherical symmetry.

This book would never have been completed without the kind and competent help of many friends and colleagues. We mention particularly Wolfgang Duschl and Peter Schneider who read critically through the whole manuscript; Norman Baker, Gerhard Börner, Mounib El Eid, Wolfgang Hillebrandt, Helmuth Kahler, Ewald Müller, Henk Spruit, Joachim Wambsganß, and many others read through particular chapters and gave us their valuable advice. In fact it would probably be simpler to give a complete list of those of our colleagues who have *not* contributed than of those who helped us.

In addition we have to thank many secretaries at our institutes; several have left their jobs (for other reasons!) during the five years in which we kept them busy. Most of this work was done by Cornelia Rickl and Petra Berkemeyer in Munich and Christa Leppien and Heinke Heise in Hamburg, while Gisela Wimmersberger prepared all the graphs. We are grateful to them all.

Finally we wish to thank Springer-Verlag for their enthusiastic cooperation.

Munich and Hamburg Rudolf Kippenhahn
December 1989 Alfred Weigert

Preface to the Second Edition

Twenty years after its first publication, this textbook is still a major reference for scientists and students interested in or working on problems of stellar structure and evolution. But with the incredible growth of computational power, the computation of stellar models has to large extent become a standard tool for astrophysics. While the early computations were restricted to single choices for mass, compositions and possibly evolutionary stage, by now models for the whole parameter space exist. The first edition of this book was restricted to a few examples for low- and intermediate-mass star evolution and lacked the broader view now being possible. There are even semi-automatic stellar evolution codes that may be used remotely via the Internet.

However, stellar evolution programs should not be used without a thorough understanding of the stellar physics. Therefore, a textbook concentrating on the foundations of the theory and explaining in detail specific phases and events in the life of a star is very much needed to allow scientifically solid modelling of stars. This is the reason why this book deserved a second edition.

Much to our regret, A. Weigert passed away two years after publication of the first edition. He left a gap that cannot be filled. Given the above mentioned need for a second edition and the requirement to add up-to-date stellar models, it was decided to have A. Weiss join R. Kippenhahn in preparing the new edition.

The two authors of this book came to discriminate between the *eternal truth* and the *mutable* parts. The latter ones refer to the current state of modelling and knowledge obtained from numerical models and their comparison to observations. Such chapters were updated, extended, or added. As far as possible, the stellar models shown were specifically calculated for this purpose, with the present, much evolved version of the original code by Kippenhahn, Weigert, and Hofmeister. The numerical results are therefore much more homogeneous and consistent than in the first edition.

The *eternal truth* concerns the aforementioned basic physics and their understanding. These parts of the book have been left almost untouched, since the authors (and those readers who were consulted) did not see any reason to change them.

The authors are indebted to many friends and colleagues who gave their advice or comments, with respect to both necessary changes and the new text passages.

The support of Santi Cassisi, Jørgen Christensen-Dalsgaard, Wolfgang Hillebrandt, Thomas Janka, Ralf Klessen, Ewald Müller, Hans Ritter, Maurizio Salaris, and Helmut Schlattl was essential for us.

We are also very grateful to all those colleagues who very generously provided their own data to help filling gaps that we could not fill with our own models. They were (again in alphabetical order) Leandro Althaus, Isabelle Baraffe, Raphael Hirschi, Marco Limongi, Marcelo Miller Bertolami, Aldo Serenelli, and Lionel Siess. Needless to say, their data also came with much wanted and helpful advice and sometimes fruitful scientific discussions about details of the models.

Norbert Grüner's help in the difficult task of generating a useful index is acknowledged, too.

Last, but not least, we thank Mrs. Rosmarie Mayr-Ihbe, who designed, corrected, and improved the many figures that we added to this second edition.

Garching
February 2012

Achim Weiss

Contents

Part I
The Basic Equations

Chapter 1
Coordinates, Mass Distribution, and Gravitational Field in Spherical Stars

1.1 Eulerian Description

For gaseous, non-rotating, single stars without strong magnetic fields, the only forces acting on a mass element come from pressure and gravity. This results in a spherically symmetric configuration. All functions will then be constant on concentric spheres, and we need only one spatial variable to describe them. It seems natural to use the distance r from the stellar centre as the spatial coordinate, which varies from $r = 0$ at the centre to the total radius $r = R$ at the surface of the star. In addition, the evolution in time t requires a dependence of all functions on t. If we thus take r and t as independent variables, we have a typical "Eulerian" treatment in the sense of classical hydrodynamics. Then all other variables are considered to depend on these two, for example, the density $\varrho = \varrho(r, t)$.

In order to provide a convenient description of the mass distribution inside the star, in particular of its effect on the gravitational field, we define the function[1] $m(r, t)$ as the mass contained in a sphere of radius r at the time t (Fig. 1.1). Then m varies with respect to r and t according to

$$dm = 4\pi r^2 \varrho dr - 4\pi r^2 \varrho v \, dt. \tag{1.1}$$

The first term on the right is obviously the mass contained in the spherical shell of thickness dr (Fig. 1.1), and it gives the variation of $m(r, t)$ due to a variation of r at constant t, i.e.

$$\frac{\partial m}{\partial r} = 4\pi r^2 \varrho. \tag{1.2}$$

[1] In most textbooks our function $m(r, t)$ is denoted by M_r.

R. Kippenhahn et al., *Stellar Structure and Evolution*, Astronomy and Astrophysics Library, DOI 10.1007/978-3-642-30304-3_1, © Springer-Verlag Berlin Heidelberg 2012

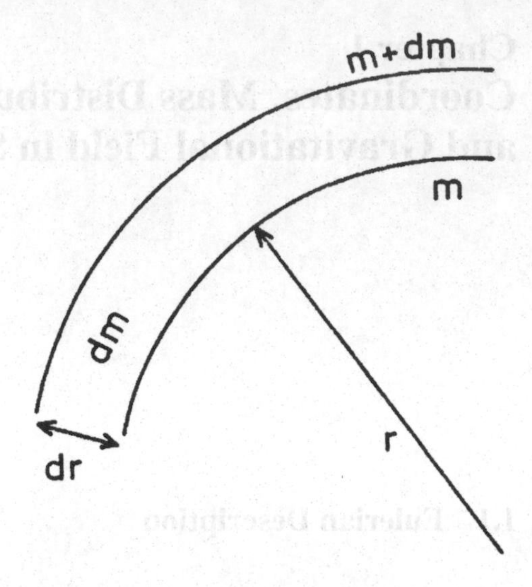

Fig. 1.1 The variation of m with r at a fixed moment $t = t_0$. The quantities dm and dr are connected by (1.2)

Since it is preferable to describe the mass distribution in the star by $m(r, t)$ (instead of ϱ), (1.2) will be taken as the first of our basic equations in the Eulerian description.

The last term in (1.1) gives the (spherically symmetric) mass flow out of the sphere of (constant) radius r due to a radial velocity v in the outward direction in the time interval dt:

$$\frac{\partial m}{\partial t} = -4\pi r^2 \varrho v \ . \tag{1.3}$$

The partial derivatives in the last two equations indicate as usual that the other independent variable (t or r) is held constant.

Differentiating (1.2) with respect to t and (1.3) with respect to r and equating the two resulting expressions gives

$$\frac{\partial \varrho}{\partial t} = -\frac{1}{r^2} \frac{\partial(\varrho r^2 v)}{\partial r} \ . \tag{1.4}$$

This is the well-known continuity equation of hydrodynamics, $\partial\varrho/\partial t = -\nabla \cdot (\varrho v)$, for the special case of spherical symmetry.

1.2 Lagrangian Description

It will turn out that, in the spherically symmetric case, it is often more useful to take a Lagrangian coordinate instead of r, i.e. one which is connected to the mass elements. The spatial coordinate of a given mass element then does not vary in time. We choose for this coordinate the above defined m: to any mass element, the value

m (which is the mass contained in a concentric sphere *at a given moment* t_0) is assigned once and for all (see Fig. 1.1).

The new independent variables are then m and t, and all other variables are considered to depend on them, for example, $\varrho(m,t)$. This also includes the radial distance r of our mass element from the centre, which is now described by the function $r = r(m,t)$. Since there is certainly no singularity of ϱ at the centre, we have here $m = 0$, while the star's total mass $m = M$ is reached at the surface (i.e. where $r = R$). This already shows one advantage of the new description for the (normal) case of stars with constant total mass: while the radius R varies strongly in time, the star always extends over the same interval of the independent variable $m : 0 \le m \le M$. Although real stars do lose mass, for example, by stellar winds or due to gravitational interaction in binary systems, over short timescales the assumption of constant mass is justified nevertheless. In any case, the range of m never changes by more than a factor of a few.

As just indicated, there will certainly be no problem concerning a unique one-to-one transformation between the two coordinates r and m. We then easily find the connection between the partial derivatives in the two cases from well-known formulae. For any function depending on two variables, one of which is substituted by a new one ($r, t \to m, t$), the partial derivatives with respect to the new variables are given by

$$\frac{\partial}{\partial m} = \frac{\partial}{\partial r} \cdot \frac{\partial r}{\partial m},$$
$$\left(\frac{\partial}{\partial t}\right)_m = \frac{\partial}{\partial r} \cdot \left(\frac{\partial r}{\partial t}\right)_m + \left(\frac{\partial}{\partial t}\right)_r . \tag{1.5}$$

Subscripts indicate which of the spatial variables (m or r) is considered constant.

Let us apply the first of (1.5) to m. The left-hand side is then simply $\partial m/\partial m = 1$, and the first factor on the right-hand side is equal to $4\pi r^2 \varrho$, according to (1.2). So we can solve for the last factor and obtain

$$\frac{\partial r}{\partial m} = \frac{1}{4\pi r^2 \varrho}. \tag{1.6}$$

This is a differential equation describing the spatial behaviour of the function $r(m,t)$. It replaces (1.2) in the Lagrangian description and shall be the new first basic equation of our problem.

Introducing (1.6) into the first equation (1.5) gives the general recipe for the transformation between the two operators:

$$\frac{\partial}{\partial m} = \frac{1}{4\pi r^2 \varrho} \frac{\partial}{\partial r}. \tag{1.7}$$

The second equation (1.5) reveals the main reason for the choice of the Lagrangian description. Its left-hand side gives the so-called substantial time derivative of hydrodynamics. It describes the change of a function in time when following a given mass element, for example, the change of a physical property of this mass element. The conservation laws for time-dependent spherical stars give very simple equations only in terms of this substantial time derivative. In terms of a *local* time derivative, $(\partial/\partial t)_r$, the description would become much more complicated since the "convective" terms with the velocity $(\partial r/\partial t)_m$ [corresponding to the first term on the right-hand side of the second equation (1.5)] would appear explicitly.

1.3 The Gravitational Field

It follows from elementary potential theory that, inside a spherically symmetric body, the absolute value g of the gravitational acceleration at a given distance r from the centre does not depend on the mass elements outside of r. It depends only on r and the mass within the concentric sphere of radius r, which we have called m:

$$g = \frac{Gm}{r^2}, \tag{1.8}$$

where $G = 6.673\times10^{-8}\,\mathrm{dyn\,cm^2\,g^{-2}}$ is the gravitational constant. So the gravitating mass appears only in the form of our variable m.

Generally, the gravitational field inside the star can be described by a gravitational potential Φ, which is a solution of the *Poisson equation*

$$\nabla^2\Phi = 4\pi G\varrho, \tag{1.9}$$

where ∇^2 denotes the Laplace operator. For spherical symmetry this reduces to

$$\frac{1}{r^2}\frac{\partial}{\partial r}\left(r^2\frac{\partial\Phi}{\partial r}\right) = 4\pi G\varrho. \tag{1.10}$$

The vector of the gravitational acceleration points towards the stellar centre and may in spherical coordinates be written as $\boldsymbol{g} = (-g, 0, 0)$ with $0 < g = |\boldsymbol{g}|$. It is obtained from Φ by the vector relation $\boldsymbol{g} = -\nabla\Phi$, where in our spherically symmetric case, only the radial component is non-vanishing:

$$g = \frac{\partial\Phi}{\partial r}. \tag{1.11}$$

With (1.8), (1.11) becomes

$$\frac{\partial\Phi}{\partial r} = \frac{Gm}{r^2}, \tag{1.12}$$

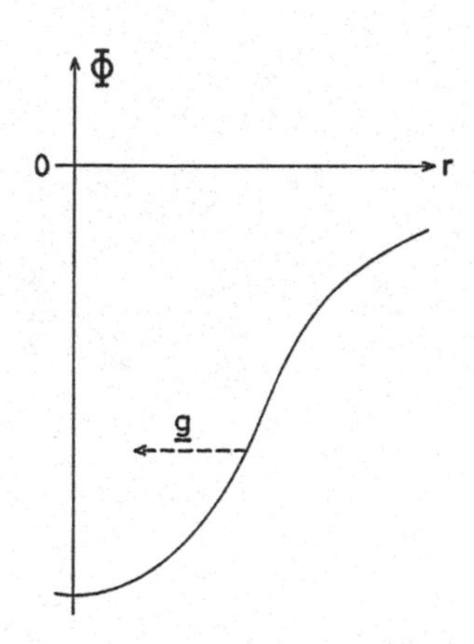

Fig. 1.2 Gravitational potential and vector of gravitational acceleration (*dashed*) in a spherically symmetric star

which is indeed a solution of (1.10), as is easily verified by substitution. The potential then becomes

$$\Phi = \int_0^r \frac{Gm}{r^2} dr + \text{constant.} \tag{1.13}$$

Unless otherwise mentioned we will fix the free constant of integration in such a way that Φ vanishes for $r \to \infty$. Φ has a minimum at the stellar centre. Figure 1.2 shows schematically the function $\Phi(r, t)$ at a given time.

Chapter 2
Conservation of Momentum

Conservation of momentum provides the next basic differential equation of the stellar-structure problem. We will derive this in several steps of gradually increasing generality. The first assumes mechanical equilibrium (Sect. 2.1), the equation of motion for spherical symmetry follows in Sect. 2.4, while in Sect. 2.5 even the assumption of spherical symmetry is dropped. In Sect. 2.6 we briefly discuss general relativistic effects in the case of hydrostatic equilibrium.

2.1 Hydrostatic Equilibrium

Most stars are obviously in such long-lasting phases of their evolution that no changes can be observed at all. Then the stellar matter cannot be accelerated noticeably, which means that all forces acting on a given mass element of the star compensate each other. This mechanical equilibrium in a star is called "hydrostatic equilibrium", since the same condition also governs the pressure stratification, say, in a basin of water. With our assumptions (gaseous stars without rotation, magnetic fields, or close companions), the only forces are due to gravity and to the pressure gradient.

For a given moment of time, we consider a thin spherical mass shell with (an infinitesimal) thickness dr at a radius r inside the star. Per unit area of the shell, the mass is $\varrho\, dr$, and the weight of the shell is $-g\varrho\, dr$. This weight is the gravitational force acting towards the centre (as indicated by the minus sign).

In order to prevent the mass elements of the shell from being accelerated in this direction, they must experience a net force due to pressure of the same absolute value, but acting outwards. This means that the shell must feel a larger pressure P_{i} at its interior (lower) boundary than the pressure P_{e} at its outer (upper) boundary (see Fig. 2.1). The total net force per unit area acting on the shell due to this pressure difference is

$$P_{\mathrm{i}} - P_{\mathrm{e}} = -\frac{\partial P}{\partial r}\, dr. \tag{2.1}$$

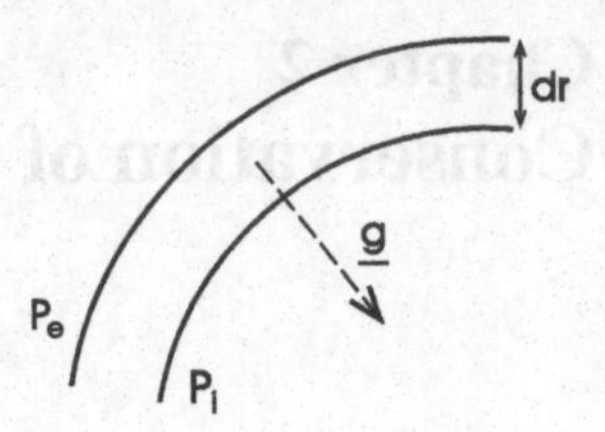

Fig. 2.1 Pressure at the upper and lower border of a mass shell of thickness *dr*, and the vector of gravitational acceleration (*dashed*) acting at one point on the shell

(The right-hand side of this equation is in fact a positive quantity, since P decreases with increasing r.) The sum of the forces arising from pressure and gravity has to be zero,

$$\frac{\partial P}{\partial r} + g\varrho = 0, \tag{2.2}$$

which gives the condition of hydrostatic equilibrium as

$$\frac{\partial P}{\partial r} = -g\varrho. \tag{2.3}$$

This shows the balance of the forces from pressure (left-hand side) and gravity (right-hand side), both per unit volume of the thin shell. Equation (1.8) gives $g = Gm/r^2$ so that (2.3) finally becomes

$$\frac{\partial P}{\partial r} = -\frac{Gm}{r^2}\varrho. \tag{2.4}$$

This hydrostatic equation is the second basic equation describing the stellar-structure problem in the Eulerian form (r as an independent variable).

If we take m as the independent variable instead of r, we obtain the hydrostatic condition by multiplying (2.4) with $\partial r/\partial m = (4\pi r^2 \varrho)^{-1}$, according to (1.5) and (1.6):

$$\frac{\partial P}{\partial m} = -\frac{Gm}{4\pi r^4}. \tag{2.5}$$

This is the second of our basic equations in the Lagrangian form.

2.2 The Role of Density and Simple Solutions

We have dealt up to now with the distribution of matter, the gravitational field, and the pressure stratification in the star. This purely mechanical problem yielded two differential equations, for example, with m as independent variable (a choice not affecting the discussion),

$$\frac{\partial r}{\partial m} = \frac{1}{4\pi r^2 \varrho}, \qquad \frac{\partial P}{\partial m} = -\frac{Gm}{4\pi r^4}. \tag{2.6}$$

Let us see whether solutions can be obtained at this stage for the problem as stated so far.

We have only two differential equations for three unknown functions, namely r, P, and ϱ. Obviously we can solve this mechanical problem only if we can express one of them in terms of the others, for example, the density ϱ as a function of P. In general, this will not be the case. But there are some exceptional situations where ϱ is a well-known function of P and r or P and m. We can then treat the equations as ordinary differential equations, since they do not contain the time explicitly.

If such integrations are to be carried out starting from the centre, the difficulty occurs that (2.6) are singular there, since $r \rightarrow 0$ for $m \rightarrow 0$, though one can easily overcome this problem by the standard procedure of expansion in powers of m, as given later in (11.3) and (11.6).

A rather artificial example that can be solved by quadrature is $\varrho = \varrho(m)$, in particular ϱ = constant in the homogeneous gaseous sphere.

Physically more realistic are solutions obtained for the so-called *barotropic* case, for which the density is a function of the gas pressure only: $\varrho = \varrho(P)$. A simple example would be a perfect[1] gas at constant temperature. After assuming a value P_c for the central pressure, both equations (2.6) have to be solved simultaneously, since $\varrho(P)$ in the first of them is not known before P is evaluated.

As we will see later (for instance, in Sects. 19.3 and 19.8), there are also cases for which no choice of P_c yields a surface of zero pressure at finite values of r. In the theory of stellar structure there is even a use for these types of solution.

Among the barotropic solutions is a wide class of models for gaseous spheres called *polytropes*. These important solutions will later be discussed extensively (Chap. 19). Barotropic solutions also describe white dwarfs, i.e. stars that really exist (Sect. 37.1).

But in general the density is not a function of pressure only but depends also on the temperature T. For a given chemical composition of the gas, its thermodynamic behaviour yields an equation of state of the form $\varrho = \varrho(P, T)$. A well-known case is that of a perfect gas, where

$$\varrho = \frac{\mu}{\Re} \frac{P}{T} \tag{2.7}$$

with the gas constant $\Re = 8.315 \times 10^7 \,\mathrm{erg\,K^{-1}\,g^{-1}}$ (which we define per g instead of per mole), while μ is the (dimensionless) mean molecular weight, i.e. the average number of atomic mass units per particle; in the case of ionized hydrogen, $\mu = 0.5$ (see Sect. 4.2).

[1] Throughout this book we will use the terms *perfect* and *ideal* gas synonymously, as they describe the same physical concept.

Once the temperature appears in the equation of state and cannot be eliminated by means of additional conditions, it then becomes much more difficult to determine the internal structure of a self-gravitating gaseous sphere. The mechanical structure is then also determined by the temperature distribution, which in turn is coupled to the transport and generation of energy in the star. This requires new equations, with which we shall deal in Chaps. 4 and 5.

2.3 Simple Estimates of Central Values P_c, T_c

The hydrostatic condition (2.5) together with an equation of state for a perfect gas (2.7) enables us to estimate the pressure and the temperature in the interior of a star of given mass and radius.

Let us replace the left-hand side of (2.5) by an average pressure gradient $(P_0 - P_c)/M$, where $P_0(=0)$ and P_c are the pressures at the surface and at the centre. On the right-hand side of (2.5) we replace m and r by rough mean values $M/2$ and $R/2$, and we obtain

$$P_c \approx \frac{2GM^2}{\pi R^4}. \tag{2.8}$$

From the equation of state for a perfect gas, and with the mean density

$$\bar{\varrho} = \frac{3M}{4\pi R^3}, \tag{2.9}$$

we find with (2.8) that

$$\begin{aligned} T_c &= \frac{P_c}{\varrho_c}\frac{\mu}{\Re} = P_c \frac{\mu}{\Re}\frac{\bar{\varrho}}{\varrho_c}\frac{4\pi R^3}{3M} \\ &\approx \frac{8}{3}\frac{\mu}{\Re}\frac{Gm}{R}\frac{\bar{\varrho}}{\varrho_c}. \end{aligned} \tag{2.10}$$

Since in most stars the density increases monotonically from the surface to the centre, we have $\bar{\varrho}/\varrho_c < 1$ (Numerical solutions show that $\bar{\varrho}/\varrho_c \approx 0.03 \ldots 0.01$.). Therefore (2.10) yields

$$T_c \lesssim \frac{8}{3}\frac{G\mu}{\Re}\frac{M}{R}. \tag{2.11}$$

With the mass and the radius of the Sun ($M_\odot = 1.989 \times 10^{33}$ g, $R_\odot = 6.96 \times 10^{10}$ cm) and with $\mu = 0.5$, we find that

$$P_c \approx 7 \times 10^{15}\,\mathrm{dyn/cm^2}, \quad T_c < 3 \times 10^7\,\mathrm{K}. \tag{2.12}$$

Modern numerical solutions (Chap. 29) give $P_c = 2.4 \times 10^{17}$ dyn/cm^2, $T_c = 1.6 \times 10^7$ K.

So we can expect to encounter enormous pressures and very high temperatures in the central regions of the stars. Moreover, our assumption of a perfect gas turns out to be fully justified for these values of P and T.

2.4 The Equation of Motion for Spherical Symmetry

Our equation of hydrostatic equilibrium (2.5) is a special case of conservation of momentum. If the (spherical) star undergoes accelerated radial motions, we have to consider the inertia of the mass elements, which introduces an additional term. We confine ourselves here to the Lagrangian description (m, t as independent variables), which is especially convenient for spherical symmetry.

We go back to the derivation of the hydrostatic equation in Sect. 2.1 and again consider a thin shell of mass dm at the distance r from the centre (Fig. 1.1). Owing to the pressure gradient, this shell experiences a force per unit area f_P given by (2.1), the right-hand side of which is easily rewritten in terms of $\partial P/\partial m$ according to (1.7):

$$f_P = -\frac{\partial P}{\partial m}\, dm. \tag{2.13}$$

The gravitational force per unit area acting on the mass shell is, with the use of (1.8),

$$f_g = -\frac{g\, dm}{4\pi r^2} = -\frac{Gm}{r^2}\frac{dm}{4\pi r^2}. \tag{2.14}$$

If the sum of the two forces is not equal to zero, the mass shell will be accelerated according to

$$\frac{dm}{4\pi r^2}\frac{\partial^2 r}{\partial t^2} = f_P + f_g. \tag{2.15}$$

This gives with (2.13) and (2.14) the equation of motion as

$$\frac{1}{4\pi r^2}\frac{\partial^2 r}{\partial t^2} = -\frac{\partial P}{\partial m} - \frac{Gm}{4\pi r^4}. \tag{2.16}$$

The signs in (2.16) are such that the pressure gradient alone would produce an outward acceleration (since $\partial P/\partial m < 0$), while the gravity alone would produce an inward acceleration.

Equation (2.16) would give exactly the equation of hydrostatic equilibrium (2.5) if the second time derivative of r vanished, i.e. if all mass elements were at rest or moved radially at constant velocity. Moreover, the term on the left-hand side is certainly unimportant if its absolute value is small compared to the absolute values of any term on the right, i.e. if the two terms on the right-hand side cancel each other nearly to zero. Then the hydrostatic condition is a very good approximation, and the configuration moves through neighbouring near-equilibrium states. In this sense we are allowed to apply the simpler hydrostatic equation to a much wider class of solutions than those fulfilling the strict requirement $\partial^2 r/\partial t^2 = 0$. To illustrate this further we assume a deviation from hydrostatic equilibrium such that, for example, in (2.16), the pressure term suddenly "disappears". The inertial term on the left would then have to compensate the gravitational term on the right. We now define a characteristic time-scale τ_{ff} for the ensuing collapse of the star by setting $|\partial^2 r/\partial t^2| = R/\tau_{ff}^2$. Then we obtain from (2.16) $R/\tau_{ff}^2 \approx g$, or

$$\tau_{ff} \approx \left(\frac{R}{g}\right)^{1/2}. \tag{2.17}$$

This is some kind of a mean value for the free-fall time over a distance of order R following the sudden disappearance of the pressure. We can correspondingly determine a timescale τ_{expl} for the explosion of our star for the case that gravity were suddenly to disappear: $R/\tau_{expl}^2 = P/\varrho R$, where we have replaced $\partial P/\partial r$ by P/R after writing $4\pi r^2(\partial P/\partial m) = (\partial P/\partial r)/\varrho$ (P and ϱ are here average values over the entire star). We then find that

$$\tau_{expl} \approx R\left(\frac{\varrho}{P}\right)^{1/2}. \tag{2.18}$$

Since $(P/\varrho)^{1/2}$ is of the order of the mean velocity of sound in the stellar interior, one can see that τ_{expl} is of the order of the time a sound wave needs to travel from the centre to the surface.

If our model is near hydrostatic equilibrium, then the two terms on the right side of (2.16) have about equal absolute value and $\tau_{ff} \approx \tau_{expl}$. We then call this timescale the *hydrostatic timescale* τ_{hydr}, since it gives the typical time in which a (dynamically stable) star reacts on a slight perturbation of hydrostatic equilibrium. With $g \approx GM/R^2$, we obtain from (2.17) up to factors of order 1 that

$$\tau_{hydr} \approx \left(\frac{R^3}{GM}\right)^{1/2} \approx \frac{1}{2}(G\bar{\varrho})^{-1/2}. \tag{2.19}$$

In the case of the Sun we find the surprisingly small value $\tau_{hydr} \approx 27\,\mathrm{min}$. Even in the case of a red giant ($M \approx M_\odot, R \approx 100 R_\odot$), one has only $\tau_{hydr} \approx 18\,\mathrm{days}$, while for a white dwarf ($M \approx M_\odot, R \approx R_\odot/50$), the hydrostatic timescale is extremely short: $\tau_{hydr} \approx 4.5\,\mathrm{s}$. In most phases of their life the stars change slowly on a timescale that is very long compared to τ_{hydr}. Then they are very close to hydrostatic equilibrium and the inertial terms in (2.16) can be ignored.

2.5 The Non-spherical Case

Up to now we have dealt with spherically symmetric configurations only. It is easy to see how the equations would have to be modified for more general cases without this symmetry.

After rewriting (2.16) for the independent variable r, we easily identify it as a special case of the Eulerian equation of motion of hydrodynamics

$$\varrho \frac{d\boldsymbol{v}}{dt} = -\nabla P - \varrho \nabla \Phi, \tag{2.20}$$

where $\boldsymbol{v}$ is the velocity vector, and the substantial time derivative on the left is defined by the operator

$$\frac{d}{dt} = \frac{\partial}{\partial t} + \boldsymbol{v} \cdot \nabla. \tag{2.21}$$

The general form of (1.4) has already been shown to be the continuity equation of hydrodynamics

$$\frac{\partial \varrho}{\partial t} = -\nabla \cdot (\varrho \boldsymbol{v}), \tag{2.22}$$

and, as described in Sect. 1.3, the gravitational potential Φ is connected with an arbitrary distribution of the density by the Poisson equation (1.9):

$$\nabla^2 \Phi = 4\pi G \varrho. \tag{2.23}$$

We see in fact that the stellar-structure equations discussed up to now are just special cases of normal textbook hydrodynamics.

2.6 Hydrostatic Equilibrium in General Relativity

To help with subsequent work (Chap. 38), we briefly refer to the change of the equation of hydrostatic equilibrium due to effects of general relativity. For details see, for example, Zeldovich and Novikov (1971).

Very strong gravitational fields, as in the case of neutron stars, are described by the Einstein field equations

$$R_{ik} - \frac{1}{2} g_{ik} R = \frac{\kappa}{c^2} T_{ik}, \quad \kappa = \frac{8\pi G}{c^2}, \tag{2.24}$$

where R_{ik} is the Ricci tensor, g_{ik} is the metric tensor and the scalar R is the Riemann curvature. T_{ik} is the energy-momentum tensor, which for a perfect gas has as the only non-vanishing components $T_{00} = \varrho c^2$, $T_{11} = T_{22} = T_{33} = P$ (ϱ includes the energy density, P = pressure). We are interested in static (time-independent), spherically symmetric mass distributions. Then the line element ds, i.e. the distance between two neighbouring events, is given in spherical coordinates (r, ϑ, φ) by the general form

$$ds^2 = e^{\nu} c^2 dt^2 - e^{\lambda} dr^2 - r^2(d\vartheta^2 + \sin^2\vartheta \, d\varphi^2) \tag{2.25}$$

with $\nu = \nu(r), \lambda = \lambda(r)$. With these expressions for T_{ik} and ds, the field equations (2.24) can be reduced to three ordinary differential equations:

$$\frac{\kappa P}{c^2} = e^{-\lambda}\left(\frac{\nu'}{r} + \frac{1}{r^2}\right) - \frac{1}{r^2}, \tag{2.26}$$

$$\frac{\kappa P}{c^2} = \frac{1}{2} e^{-\lambda}\left(\nu'' + \frac{1}{2}\nu'^2 + \frac{\nu' - \lambda'}{r} - \frac{\nu'\lambda'}{2}\right), \tag{2.27}$$

$$\kappa \varrho = e^{-\lambda}\left(\frac{\lambda'}{r} - \frac{1}{r^2}\right) + \frac{1}{r^2}, \tag{2.28}$$

where primes denote derivatives with respect to r. After multiplication with $4\pi r^2$, (2.28) can be integrated giving

$$\kappa m = 4\pi r(1 - e^{-\lambda}). \tag{2.29}$$

Here m denotes the *gravitational mass* inside r defined by

$$m = \int_0^r 4\pi r^2 \varrho \, dr. \tag{2.30}$$

For $r = R, m$ becomes the gravitational mass M of the star. It is the mass a distant observer would measure by its gravitational effects, for example, on orbiting planets. It is not, however, the mass which we naïvely identify with the baryon number times the atomic mass unit: M contains not only the rest mass, but the whole energy (divided by c^2). This includes the internal and the gravitational energy, the latter being negative and reducing the gravitational mass (just as the binding energy of a nucleus results in a mass defect; see Chap. 18). The seemingly familiar form of (2.30) is treacherous. First of all, $\varrho = \varrho_0 + U/c^2$ contains the whole energy density U as well as the rest-mass density ϱ_0, and the changed metric would give the spherical volume element as $e^{\lambda/2} 4\pi r^2 \, dr$ instead of the usual form $4\pi r^2 \, dr$ [over which (2.30) is integrated].

Differentiation of (2.26) with respect to r gives $P' = P'(\lambda, \lambda', \nu', \nu'', r)$. When $\lambda, \lambda', \nu', \nu''$ are eliminated by (2.26), (2.27) and (2.29), one arrives at the

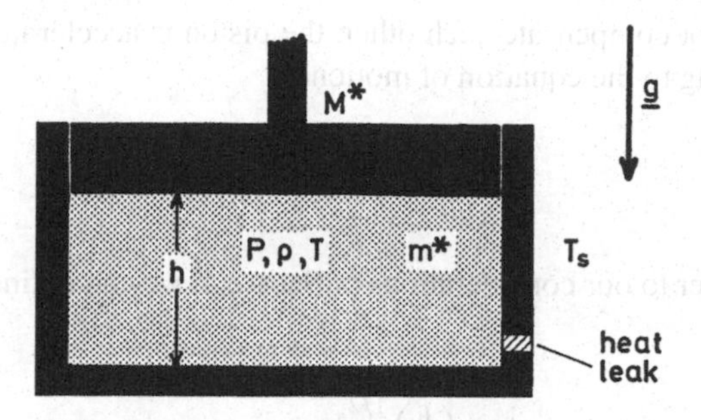

Fig. 2.2 The piston model. Gas of mass m^* (with pressure P, density ϱ, temperature T) is held in a container with a movable piston of mass M^*. The gravitational acceleration g acts on the piston. The container is embedded in a medium of temperature T_s; a possible heat leak is indicated (*dashed*) in the right wall of the container. In Chap. 2, only the mechanical properties of the model are discussed

Tolman-Oppenheimer-Volkoff (TOV) equation for hydrostatic equilibrium in general relativity:

$$\frac{dP}{dr} = -\frac{Gm}{r^2}\varrho\left(1 + \frac{P}{\varrho c^2}\right)\left(1 + \frac{4\pi r^3 P}{mc^2}\right)\left(1 - \frac{2Gm}{rc^2}\right)^{-1} . \tag{2.31}$$

Obviously this reverts to the usual form (2.4) for $c^2 \to \infty$.

For gravitational fields that are not too large (small deviations from Newtonian mechanics), one can expand the product of the parentheses in (2.31) and retain only terms linear in $1/c^2$. This gives the so-called *post-Newtonian approximation:*

$$\frac{dP}{dr} = -\frac{Gm}{r^2}\varrho\left(1 + \frac{P}{\varrho c^2} + \frac{4\pi r^3 P}{mc^2} + \frac{2Gm}{rc^2}\right). \tag{2.32}$$

2.7 The Piston Model

From time to time we shall make use of a simple mechanical model which in some respects mimics the behaviour of stars, and which is shown in Fig. 2.2. A piston of mass M^* encloses a gas of mass m^* in a box. $G^* = gM^*$ is the weight of the piston in a gravitational field described by the gravitational acceleration g. A is the cross-sectional area of the piston and h its height above the bottom. Then $V = Ah$ is the volume of the gas, while its density is $\varrho = m^*/V$.

In the case of hydrostatic equilibrium, the gas pressure P adjusts in such a way that the weight per unit area is balanced by the pressure:

$$G^* = PA. \tag{2.33}$$

If the forces do not compensate each other, the piston is accelerated in the vertical direction according to the equation of motion

$$M^* \frac{d^2 h}{dt^2} = -G^* + PA. \tag{2.34}$$

In a similar manner to our considerations of Sect. 2.4, we can define two timescales τ_{ff} and τ_{expl}:

$$\tau_{\mathrm{ff}} \approx \left(\frac{h}{g}\right)^{1/2}, \tag{2.35}$$

$$\tau_{\mathrm{expl}} \approx h \left(\frac{\varrho}{P}\right)^{1/2} \left(\frac{M^*}{m^*}\right)^{1/2}. \tag{2.36}$$

In the limit of hydrostatic equilibrium both timescales are the same, and we then call $\tau_{\mathrm{ff}} = \tau_{\mathrm{expl}}$ the hydrostatic timescale τ_{hydr}.

Chapter 3
The Virial Theorem

3.1 Stars in Hydrostatic Equilibrium

While the virial theorem generally plays a relatively minor role in physics, it is of vital importance for the understanding of stars. It connects two important energy reservoirs of a star and allows predictions and interpretations of certain evolutionary phases.

If we multiply (2.5) by $4\pi r^3$ and integrate over dm in the interval $[0, M]$, i.e. from centre to surface, we obtain on the left-hand side an integral which can be simplified by partial integration:

$$\int_0^M 4\pi r^3 \frac{\partial P}{\partial m} dm = \left[4\pi r^3 P\right]_0^M - \int_0^M 12\pi r^2 \frac{\partial r}{\partial m} P\, dm , \tag{3.1}$$

where the term in brackets vanishes, since $r = 0$ at the centre and $P = 0$ at the surface. With (1.6) the integrand of the last term in (3.1) is reduced to $3P/\varrho$. Therefore, after multiplication by $4\pi r^3$ and integration, (2.5) gives

$$\int_0^M \frac{Gm}{r} dm = 3 \int_0^M \frac{P}{\varrho} dm . \tag{3.2}$$

Both sides of (3.2) have the dimensions of energy and can be easily interpreted. We define the *gravitational energy* E_g by

$$E_g := -\int_0^M \frac{Gm}{r} dm . \tag{3.3}$$

Consider a unit mass at the position r. Its potential energy due to the gravitational field of the mass m inside r is $-Gm/r$. Therefore E_g is the potential energy of all mass elements dm of the star (normalized to zero at infinity). The energy $-E_g(> 0)$

R. Kippenhahn et al., *Stellar Structure and Evolution*, Astronomy and Astrophysics Library, DOI 10.1007/978-3-642-30304-3_3, © Springer-Verlag Berlin Heidelberg 2012

is necessary to expand all mass shells into infinity, and it is released when the stellar configuration forms out of an infinitely distributed medium.

We see that E_g varies if the configuration undergoes expansion or contraction: if all mass shells inside the configuration expand or contract simultaneously, then E_g increases or decreases, respectively. And the same must be true for the integral on the right of (3.2). Note that these radial motions must be slow compared to τ_{hydr} in order that hydrostatic equilibrium is always maintained, otherwise (3.2) would not hold.

In order to understand the meaning of the term on the right of (3.2) we first assume a perfect gas. Then

$$\frac{P}{\varrho} = \frac{\Re}{\mu} T = (c_P - c_v)T = (\gamma - 1)c_v T, \tag{3.4}$$

where c_P, c_v are the specific heats per unit mass (and we make use of $\Re/\mu = c_P - c_v$ and replace c_P/c_v by γ). For a monatomic gas $\gamma = 5/3$, and we have

$$\frac{P}{\varrho} = \frac{2}{3} u, \tag{3.5}$$

where $u = c_v T$ is the internal energy per unit mass of the perfect gas. Therefore (3.2) can be written as

$$E_g = -2E_i \tag{3.6}$$

with the total internal energy of the star

$$E_i := \int_0^M u \, dm \, . \tag{3.7}$$

Equation (3.6) is the *virial theorem* for a perfect monatomic gas. For a general equation of state we define a quantity ζ by

$$\zeta u := 3 \frac{P}{\varrho} \, . \tag{3.8}$$

For a perfect gas $\zeta = 3(\gamma - 1)$, in the monatomic case $\gamma = 5/3$, and therefore $\zeta = 2$. For a pure photon gas, $P = aT^4/3$, and $u\varrho = aT^4$ (a = radiation density constant), giving $\zeta = 1$. If ζ is constant throughout the star, (3.2) leads to the more general virial theorem:

$$\zeta E_i + E_g = 0 \, . \tag{3.9}$$

We now define the *total energy* W of our configuration,

$$W = E_i + E_g \, , \tag{3.10}$$

where for a gravitationally bound system $W < 0$, and with (3.9) we find that

$$W = (1 - \zeta) E_{\rm i} = \frac{\zeta - 1}{\zeta} E_{\rm g} \,. \tag{3.11}$$

In the case of $\zeta = 1$ ($\gamma = 4/3$) the total energy vanishes.

But in general W, $E_{\rm g}$, and $E_{\rm i}$ are coupled. A change of the total energy of the configuration is then connected with a change of its internal energy and with expansion or shrinking. A gas of finite temperature must radiate and W must decrease. Let L be the *luminosity* of the star, i.e. the total energy loss per unit time by radiation; then conservation of energy demands that $(dW/dt) + L = 0$, so that with (3.11) we obtain

$$L = (\zeta - 1)\frac{dE_{\rm i}}{dt} = -\frac{\zeta - 1}{\zeta}\frac{dE_{\rm g}}{dt} \,. \tag{3.12}$$

We have seen that $\dot{E}_{\rm g} < 0$ for contraction of all mass shells (where the dot denotes a derivative with respect to time t). For a perfect gas (3.12) gives $L = -\dot{E}_{\rm g}/2 = \dot{E}_{\rm i}$, which means that half of the energy liberated by the contraction is radiated away and the other half is used to heat the star ($L > 0$, $\dot{E}_{\rm i} > 0$). The surprising fact that a star heats up while losing energy can be described by saying that the star has a negative specific heat (cf. the gravothermal specific heat defined in Sect. 25.3.4).

We have to keep in mind that it is the luminosity that causes the shrinking: a configuration in hydrostatic equilibrium has a finite temperature and therefore radiates into the (cold) universe.

3.2 The Virial Theorem of the Piston Model

Let us consider the situation for the piston model of Sect. 2.7 for the case of a perfect gas. Assuming $M^* \gg m^*$, we define $E_{\rm g} := +G^* h$, where the free additional constant is chosen such that $E_{\rm g} = 0$ for $h = 0$. Hydrostatic equilibrium (2.33) with $m^* = Ah\varrho$ and (3.4) demands that

$$hG^* = \frac{P}{\varrho} m^* = (\gamma - 1) c_v T\, m^* \,. \tag{3.13}$$

The internal energy $E_{\rm i}$ of the gas is $E_{\rm i} = c_v T m^*$, and we find that

$$E_{\rm g} = (\gamma - 1) E_{\rm i} \,, \tag{3.14}$$

which is the virial theorem for the piston model. Differentiating with respect to time, with $\gamma = 5/3$, results in

$$\frac{dE_\mathrm{g}}{dt} = \frac{2}{3}\frac{dE_\mathrm{i}}{dt} . \tag{3.15}$$

Hence we see that in contrast to the situation in stars, a reduction of E_g is connected with *cooling* of the gas. Indeed the piston can only sink if the gas cools.

This different behaviour comes from the fact that the gravitational field is assumed to be constant here. In order to demonstrate this we now assume the weight G^* to be a function of h and differentiate (3.13) with respect to h :

$$G^*(1 + G_h^*) = (\gamma - 1)\frac{dE_\mathrm{i}}{dh} \tag{3.16}$$

with $G_h^* := (d \ln G^*/d \ln h)$. Indeed, if $G_h^* = 0$ (constant gravity), we see that E_i increases with h. If, however, G^* decreases sufficiently with increasing h (such that $G_h^* < -1$), then E_i increases with decreasing h, corresponding to the behaviour of stars. In fact in an expanding star each mass shell also loses weight with increasing r.

3.3 The Kelvin–Helmholtz Timescale

Returning now to consider stars, since according to (3.12) L is of the order of $|dE_\mathrm{g}/dt|$, we can define a characteristic time-scale

$$\tau_\mathrm{KH} := \frac{|E_\mathrm{g}|}{L} \approx \frac{E_\mathrm{i}}{L} \tag{3.17}$$

called the *Kelvin–Helmholtz timescale* (after the two physicists who estimated this as the evolutionary timescale for a contracting or cooling star).

A rough estimate for $|E_\mathrm{g}|$ is

$$|E_\mathrm{g}| \approx \frac{G\bar{m}^2}{\bar{r}} \approx \frac{GM^2}{2R} , \tag{3.18}$$

where quantities with a bar indicate mean values for m and r (which we have replaced by $M/2$ and $R/2$). Then we have

$$\tau_\mathrm{KH} \approx \frac{GM^2}{2RL} . \tag{3.19}$$

For the Sun, with $L = 3.827 \times 10^{33}$ erg/s, we find $\tau_\mathrm{KH} \approx 1.6 \times 10^7$ years. In the early days of astrophysics the source of stellar energy was still uncertain, and it was suggested, among other proposals, that the Sun "lived" from its gravitational energy E_g. Our estimate shows that this can work only for some 10^7 years, after which time it would have contracted to a very condensed body. As it became obvious that the Sun has been radiating in roughly the same way for some 10^9 years, the contraction hypothesis had to be abandoned. But there are phases in a stellar life when E_g is the main or even the only stellar energy source (Chap. 28); then the star evolves on the timescale τ_{KH}. A more detailed discussion of the evolution of a star in time appears in Sect. 4.5.

3.4 The Virial Theorem for Non-vanishing Surface Pressure

One often needs the virial theorem for gaseous spheres imbedded in a medium of finite pressure. In this case, at the surface ($m = M$), $P = P_0 > 0$ instead of $P = 0$. Consequently the first term on the right of (3.1) does not vanish at the surface, and (3.2) is modified to

$$\int_0^M \frac{Gm}{r} dm = 3 \int_0^M \frac{P}{\varrho} dm - 4\pi R^3 P_0 \ . \tag{3.20}$$

Correspondingly we find, rather than (3.9), that

$$\zeta E_i + E_g = 4\pi R^3 P_0 \ . \tag{3.21}$$

Chapter 4
Conservation of Energy

Since we do not wish to interrupt the derivation of the energy equation for stars with lengthy formalisms, we first provide a few thermodynamic relations which will be used extensively later on.

4.1 Thermodynamic Relations

The first law of thermodynamics relates the heat dq added per unit mass,

$$dq = du + Pdv, \tag{4.1}$$

to the internal energy u and the specific volume $v = 1/\varrho$ (both also defined per unit mass).

We now assume rather general equations of state, $\varrho = \varrho(P, T)$ and $u = u(\varrho, T)$. Usually they will also depend on the chemical composition, but here this is assumed to be fixed. With the derivatives defined as

$$\alpha := \left(\frac{\partial \ln \varrho}{\partial \ln P}\right)_T = -\frac{P}{v}\left(\frac{\partial v}{\partial P}\right)_T, \tag{4.2}$$

$$\delta := -\left(\frac{\partial \ln \varrho}{\partial \ln T}\right)_P = \frac{T}{v}\left(\frac{\partial v}{\partial T}\right)_P, \tag{4.3}$$

the equation of state can be written in the form $d\varrho/\varrho = \alpha dP/P - \delta dT/T$.

We also need the specific heats:

$$c_P := \left(\frac{dq}{dT}\right)_P = \left(\frac{\partial u}{\partial T}\right)_P + P\left(\frac{\partial v}{\partial T}\right)_P, \tag{4.4}$$

$$c_v := \left(\frac{dq}{dT}\right)_v = \left(\frac{\partial u}{\partial T}\right)_v. \tag{4.5}$$

R. Kippenhahn et al., *Stellar Structure and Evolution*, Astronomy and Astrophysics Library, DOI 10.1007/978-3-642-30304-3_4, © Springer-Verlag Berlin Heidelberg 2012

With

$$du = \left(\frac{\partial u}{\partial v}\right)_T dv + \left(\frac{\partial u}{\partial T}\right)_v dT \tag{4.6}$$

and with (4.1) we find the change $ds = dq/T$ of the specific entropy to be

$$ds = \frac{dq}{T} = \frac{1}{T}\left[\left(\frac{\partial u}{\partial v}\right)_T + P\right] dv + \frac{1}{T}\left(\frac{\partial u}{\partial T}\right)_v dT. \tag{4.7}$$

Since ds is a total differential form, $\partial^2 s/\partial T \partial v = \partial^2 s/\partial v \partial T$ and

$$\frac{\partial}{\partial T}\left[\frac{1}{T}\left(\frac{\partial u}{\partial v}\right)_T + \frac{P}{T}\right] = \frac{1}{T}\frac{\partial^2 u}{\partial T \partial v}, \tag{4.8}$$

which after the differentiation on the left is carried out gives

$$\left(\frac{\partial u}{\partial v}\right)_T = T\left(\frac{\partial P}{\partial T}\right)_v - P. \tag{4.9}$$

Next we derive an expression for $(\partial u/\partial T)_P$, taking P, T as independent variables. From (4.6) it follows that

$$\frac{du}{dT} = \left(\frac{\partial u}{\partial T}\right)_v + \left(\frac{\partial u}{\partial v}\right)_T \frac{dv}{dT}, \tag{4.10}$$

and therefore

$$\begin{aligned}\left(\frac{\partial u}{\partial T}\right)_P &= \left(\frac{\partial u}{\partial T}\right)_v + \left(\frac{\partial u}{\partial v}\right)_T \left(\frac{\partial v}{\partial T}\right)_P \\ &= \left(\frac{\partial u}{\partial T}\right)_v + \left(\frac{\partial v}{\partial T}\right)_P \left[T\left(\frac{\partial P}{\partial T}\right)_v - P\right],\end{aligned} \tag{4.11}$$

where we have made use of (4.9). From the definitions (4.4), (4.5) and from (4.11) we write

$$\begin{aligned}c_P - c_v &= P\left(\frac{\partial v}{\partial T}\right)_P + \left(\frac{\partial u}{\partial T}\right)_P - \left(\frac{\partial u}{\partial T}\right)_v \\ &= \left(\frac{\partial v}{\partial T}\right)_P \left(\frac{\partial P}{\partial T}\right)_v T.\end{aligned} \tag{4.12}$$

On the other hand, the definitions (4.2) and (4.3) for α and δ imply that

$$\left(\frac{\partial P}{\partial T}\right)_v = -\frac{\left(\frac{\partial v}{\partial T}\right)_P}{\left(\frac{\partial v}{\partial P}\right)_T} = \frac{P\delta}{T\alpha}, \tag{4.13}$$

and therefore

$$c_P - c_v = T\left(\frac{\partial v}{\partial T}\right)_P \frac{P\delta}{T\alpha} = \frac{P\delta^2}{\varrho T\alpha}, \tag{4.14}$$

where we have made use of $T(\partial v/\partial T)_P = v\delta = \delta/\varrho$; hence we arrive at the basic relation

$$c_P - c_v = \frac{P\delta^2}{\varrho T\alpha}. \tag{4.15}$$

For a perfect gas this equation reduces to the well-known relation $c_P - c_v = \Re/\mu$ [see (4.33)].

We have now derived all the tools for rewriting (4.1) in terms of T and P. The first step is to write it in the form

$$\begin{aligned} dq = du + Pdv &= \left(\frac{\partial u}{\partial T}\right)_v dT + \left[\left(\frac{\partial u}{\partial v}\right)_T + P\right] dv \\ &= \left(\frac{\partial u}{\partial T}\right)_v dT + T\left(\frac{\partial P}{\partial T}\right)_v dv \end{aligned} \tag{4.16}$$

by making use of (4.9), and then with (4.5) and (4.13) we have

$$\begin{aligned} dq &= c_v dT - \frac{T}{\varrho}\left(\frac{\partial P}{\partial T}\right)_v \frac{d\varrho}{\varrho} = c_v dT - \frac{P\delta}{\varrho\alpha}\frac{d\varrho}{\varrho} \\ &= c_v dT - \frac{P\delta}{\varrho\alpha}\left(\alpha\frac{dP}{P} - \delta\frac{dT}{T}\right) = \left(c_v + \frac{P\delta^2}{\varrho T\alpha}\right) dT - \frac{\delta}{\varrho} dP. \end{aligned} \tag{4.17}$$

The terms in parentheses in the last expression are, according to (4.15), simply c_P and therefore

$$dq = c_P dT - \frac{\delta}{\varrho} dP. \tag{4.18}$$

Next we define the adiabatic temperature gradient ∇_{ad}, a quantity often used in astrophysics, by

$$\nabla_{ad} := \left(\frac{\partial \ln T}{\partial \ln P}\right)_s, \tag{4.19}$$

where the subscript s indicates that the definition is valid for constant entropy. Since for adiabatic changes the entropy has to remain constant, i.e. $ds = dq/T = 0$, we can easily derive an expression for ∇_{ad} from (4.18), i.e.

$$0 = dq = c_P dT - \frac{\delta}{\varrho} dP \tag{4.20}$$

or $(dT/dP)_s = \delta/\varrho c_P$ and

$$\nabla_{\rm ad} \equiv \left(\frac{P}{T}\frac{dT}{dP}\right)_s = \frac{P\delta}{T\varrho c_P} . \tag{4.21}$$

4.2 The Perfect Gas and the Mean Molecular Weight

For a perfect gas consisting of n particles per unit volume that all have the molecular weight μ, the equation of state is

$$P = nkT = \frac{\Re}{\mu}\varrho T , \tag{4.22}$$

with $\varrho = n\mu m_{\rm u}$($k = 1.38 \times 10^{-16}$ erg K^{-1} = Boltzmann constant; $\Re = k/m_{\rm u} = 8.31 \times 10^7$ erg K^{-1} g^{-1} = universal gas constant; $m_{\rm u} = 1$ amu $= 1.66053 \times 10^{-24}$ g = the atomic mass unit). Note that we here use the gas constant with a dimension (energy per K and per *unit mass*) different from that in thermodynamic text books (energy per K and per mole). This has the consequence that here the molecular weight μ is dimensionless (instead of having the dimension mass per mole); it is simply the particle mass divided by 1 amu.

In the deep interiors of stars the gases are fully ionized, i.e. for each hydrogen nucleus, there also exists a free electron, while for each helium nucleus, there are two free electrons. We therefore have a mixture of two gases, that of the nuclei (which in itself can consist of more than one component) and that of the free electrons. The mixture can be treated similarly to a one-component gas, if all single components obey the perfect gas equation.

We consider a mixture of fully ionized nuclei. The chemical composition can be described by specifying all X_i, the weight fractions of nuclei of type i, which have molecular weight μ_i and charge number Z_i. If we have n_i nuclei per volume and a "partial density" ϱ_i, then obviously $X_i = \varrho_i/\varrho$ and

$$n_i = \frac{\varrho_i}{\mu_i m_{\rm u}} = \frac{\varrho}{m_{\rm u}}\frac{X_i}{\mu_i} . \tag{4.23}$$

(Here and in the following, we neglect the mass of the electrons compared to that of the ions.) The total pressure P of the mixture is the sum of the partial pressures

$$P = P_{\rm e} + \sum_i P_i = \left(n_{\rm e} + \sum_i n_i\right) kT . \tag{4.24}$$

Here $P_{\rm e}$ is the pressure of the free electrons, while P_i is the partial pressure due to the nuclei of type i. The contribution of one completely ionized atom of element i

to the total number of particles (nucleus plus Z_i free electrons) is $1 + Z_i$; therefore

$$n = n_e + \sum_i n_i = \sum_i (1 + Z_i) n_i . \tag{4.25}$$

With this and (4.23), (4.24) becomes

$$P = nkT = \Re \sum_i \frac{X_i(1 + Z_i)}{\mu_i} \varrho T , \tag{4.26}$$

which can be written simply in the form (4.22) with the *mean molecular weight*

$$\mu = \left(\sum_i \frac{X_i(1 + Z_i)}{\mu_i} \right)^{-1} . \tag{4.27}$$

By introducing the mean molecular weight, we are able to treat a mixture of perfect gases as a uniform perfect gas. We just have to replace the molecular weight in (4.22) by the mean molecular weight. In the case of pure (fully ionized) hydrogen with $X_H = 1$, $\mu_H = 1$, $Z_H = 1$, we have $\mu = 1/2$, while for a fully ionized helium gas ($X_{He} = 1$, $\mu_{He} = 4$, $Z_{He} = 2$), we find $\mu = 4/3$.

Equation (4.27) can be easily modified for the partial gas consisting of the ions only, or equivalently, for the case of a *neutral* gas where all the electrons are still in the atom. In (4.25) we just have to replace $1 + Z_i$ by 1 and we find

$$\mu_0 = \left(\sum_i \frac{X_i}{\mu_i} \right)^{-1} . \tag{4.28}$$

Here we have dealt with the cases of full ionization and of no ionization at all. In Chap. 14 we will deal with the case of partial ionization.

At this point we also define the mean molecular weight per free electron μ_e, a quantity which we shall need later. For a fully ionized gas each nucleus i contributes Z_i free electrons and we have

$$\mu_e = \left(\sum_i X_i Z_i / \mu_i \right)^{-1} . \tag{4.29}$$

Since for all (not too rare) elements heavier than helium $\mu_i / Z_i \approx 2$ is a good approximation, we find

$$\mu_e = \left(X + \frac{1}{2} Y + \frac{1}{2}(1 - X - Y) \right)^{-1} = \frac{2}{1 + X} , \tag{4.30}$$

where we have followed the custom of using $X := X_{\rm H}, Y := X_{\rm He}$ for the weight fractions of hydrogen and helium. Then $1 - X - Y$ is the mass fraction of the elements heavier than helium.

4.3 Thermodynamic Quantities for the Perfect, Monatomic Gas

If the gas is monatomic, the internal energy per gram is the kinetic energy of the translational motion of the particles only

$$u = \frac{3}{2} kT \frac{n}{\varrho}. \tag{4.31}$$

From (4.2) and (4.3) we find

$$\alpha = \delta = 1, \tag{4.32}$$

and from (4.15)

$$c_P - c_v = \frac{P}{\varrho T} = \frac{\Re}{\mu} \tag{4.33}$$

and therefore with (4.5)

$$c_v = \left(\frac{\partial u}{\partial T}\right)_\varrho = \frac{3}{2} k \frac{n}{\varrho} = \frac{3}{2} \frac{\Re}{\mu} \tag{4.34}$$

and with (4.33)

$$c_P = \frac{5}{2} \frac{\Re}{\mu}. \tag{4.35}$$

Equation (4.21) therefore yields

$$\nabla_{\rm ad} = \frac{\Re}{\mu c_P} = \frac{c_P - c_v}{c_P} = \frac{2}{5}. \tag{4.36}$$

Sometimes also the quantity

$$\gamma_{\rm ad} := \left(\frac{\partial \ln P}{\partial \ln \varrho}\right)_s \tag{4.37}$$

for adiabatic changes is needed. If we differentiate the equation of state (4.22), we find

$$\frac{dP}{P} = \frac{d\varrho}{\varrho} + \frac{dT}{T} \tag{4.38}$$

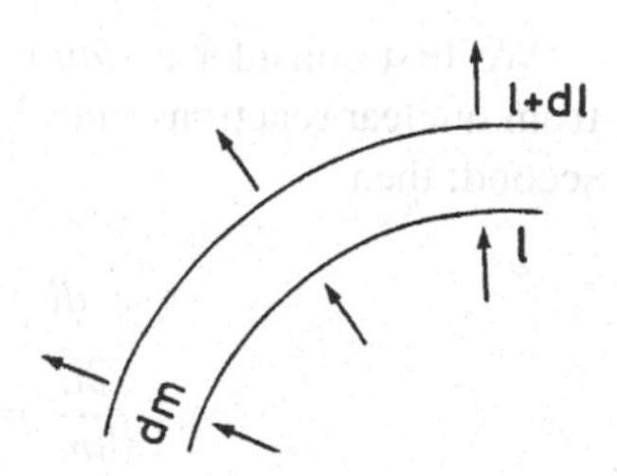

Fig. 4.1 Energy flux through a mass shell

which holds for all variations of the variables in the perfect gas equation, including the adiabatic variation. For these we obtain from (4.36)

$$\frac{dT}{T} = \nabla_{\rm ad} \frac{dP}{P} = \left(1 - \frac{c_v}{c_P}\right) \frac{dP}{P}. \tag{4.39}$$

Eliminating dT/T from (4.38) and (4.39) gives

$$\left(\frac{d\varrho}{\varrho}\right)_{\rm ad} = \frac{c_v}{c_P} \left(\frac{dP}{P}\right)_{\rm ad} \tag{4.40}$$

or

$$\gamma_{\rm ad} = \left(\frac{d \ln P}{d \ln \varrho}\right)_s = \frac{c_P}{c_v}. \tag{4.41}$$

4.4 Energy Conservation in Stars

By $l(r)$ we define[1] the net energy per second passing outward through a sphere of radius r. The function l is zero at $r = 0$, since there can be no infinite energy source at the centre, while l reaches the total luminosity L of the star at the surface. In between, l can be a complicated function, depending on the distribution of the sources and sinks of energy.

The function l comprises the energies transported by radiation, conduction, and convection, transport mechanisms with which we shall deal in Chaps. 5 and 7. Not included is a possible energy flux by neutrinos, which normally have negligible interaction with the stellar matter (see below). Included in l are only those fluxes which require a temperature gradient.

Consider a spherical mass shell of radius r, thickness dr, and mass dm, as indicated in Fig. 4.1. The energy per second entering the shell at the inner surface is l, while $l + dl$ is the energy per second leaving it through the outer surface. The surplus power dl can be provided by nuclear reactions, by cooling, or by compression or expansion of the mass shell.

[1]In many textbooks our function l is denoted by L_r.

We first consider a *stationary* case in which dl is due to the release of energy from nuclear reactions only. Let ε be the nuclear energy released per unit mass per second; then

$$dl = 4\pi r^2 \varrho\varepsilon\, dr = \varepsilon\, dm\,, \qquad \text{or} \tag{4.42}$$

$$\frac{\partial l}{\partial m} = \varepsilon. \tag{4.43}$$

In general ε depends on temperature and density and on the abundance of the different nuclear species that react, described in detail in Chap. 18.

If we relax the condition of time independence, then dl can become non-zero even if there are no nuclear reactions. A *non-stationary* shell can change its internal energy, and it can exchange mechanical work ($P\, dV$) with the neighbouring shells. Instead of (4.43) we write

$$dq = \left(\varepsilon - \frac{\partial l}{\partial m}\right) dt, \tag{4.44}$$

where dq is the heat per unit mass added to the shell in the time interval dt. Replacing dq by the first law of thermodynamics (4.1) we obtain

$$\begin{aligned}\frac{\partial l}{\partial m} &= \varepsilon - \frac{\partial u}{\partial t} - P\frac{\partial v}{\partial t}\\ &= \varepsilon - \frac{\partial u}{\partial t} + \frac{P}{\varrho^2}\frac{\partial \varrho}{\partial t}\end{aligned} \tag{4.45}$$

This can be rewritten in terms of P and T, with the help of (4.18), as

$$\frac{\partial l}{\partial m} = \varepsilon - c_P\frac{\partial T}{\partial t} + \frac{\delta}{\varrho}\frac{\partial P}{\partial t}\,, \tag{4.46}$$

where δ is defined in (4.3). This is the third of the basic equations of stellar structure. One often combines the terms containing the time derivatives in a source function

$$\begin{aligned}\varepsilon_g &:= -T\frac{\partial s}{\partial t}\\ &= -c_P\frac{\partial T}{\partial t} + \frac{\delta}{\varrho}\frac{\partial P}{\partial t}\\ &= -c_P T\left(\frac{1}{T}\frac{\partial T}{\partial t} - \frac{\nabla_{\rm ad}}{P}\frac{\partial P}{\partial t}\right),\end{aligned} \tag{4.47}$$

where use is made of the fact that $ds = dq/T$ and of (4.21).

Let us now turn to the problem of *neutrino losses.* These can be formed in appreciable amounts in a star either as a by-product of nuclear energy generation or by other reactions. Stellar material is normally transparent to neutrinos and therefore they can easily "tunnel" the energy they have to the surface. This is the reason we have excluded the energy flux due to neutrinos from l. The only mass elements affected by the neutrinos are at the place of their creation, where they act as an energy sink; hence ε_ν is used to represent the energy taken per unit mass per second from the stellar material in the form of neutrinos. In general, the energy lost by neutrinos in nuclear reactions is already taken into account in the net energy Q released in each reaction (see Sect. 18.3). By definition, $\varepsilon_\nu > 0$. Obviously the complete energy equation is then

$$\frac{\partial l}{\partial m} = \varepsilon - \varepsilon_\nu + \varepsilon_g . \tag{4.48}$$

As mentioned at the beginning of Sect. 4.4, the boundary values of l are $l = 0$ at the centre and $l = L$ at the surface. In between, l is not necessarily monotonic, since the right-hand side of (4.48) may be positive or negative; l can even become larger than L, or negative. For instance, the surface luminosity L of an expanding star can be smaller than the energy produced in the central core by nuclear reactions ($\varepsilon > 0$), since part of it is used to expand the star ($\varepsilon_g < 0$); and strong neutrino losses can make $l < 0$ in certain parts of the stellar interior (see Sect. 33.5).

The energy per second carried away from the star by neutrinos is often called the *neutrino luminosity:*

$$L_\nu := \int_0^M \varepsilon_\nu \, dm. \tag{4.49}$$

4.5 Global and Local Energy Conservation

In Chap. 3 we considered gravitational energy (E_g) and internal energy (E_i), but ignored nuclear and neutrino energies, as well as the kinetic energy E_{kin} of radial motion. We now define the total energy of the star as $W = E_{kin} + E_g + E_i + E_n$, where E_n is the nuclear energy content of the whole star. Obviously the energy equation is

$$\frac{d}{dt}(E_{kin} + E_g + E_i + E_n) + L + L_\nu = 0 , \tag{4.50}$$

and, of course, this must also be obtained from the *local* energy equation (4.48) by integration over m. Clearly, the integration of $\partial l/\partial m$ gives L, the integration of $-\varepsilon_\nu$ gives $-L_\nu$, while the integral over ε gives $-dE_n/dt$. Integration over ε_g, however, needs some consideration.

Let us write ε_g as in (4.45):

$$\varepsilon_g = -\frac{\partial u}{\partial t} + \frac{P}{\varrho^2}\frac{\partial \varrho}{\partial t} . \tag{4.51}$$

Then integration over $-\partial u/\partial t$ gives $-dE_{\rm i}/dt$. In order to deal with the last term in (4.51) we use (3.2, 3.3) and find that

$$E_{\rm g} = -3\int_0^M \frac{P}{\varrho} dm , \tag{4.52}$$

which we differentiate with respect to time (indicated by dots):

$$\dot{E}_{\rm g} = -3\int_0^M \frac{\dot{P}}{\varrho} dm + 3\int_0^M \frac{P}{\varrho^2}\dot{\varrho}\, dm . \tag{4.53}$$

We first treat hydrostatic equilibrium ($dE_{\rm kin}/dt = 0$). Then differentiation of (2.5) gives

$$\frac{\partial \dot{P}}{\partial m} = 4\frac{Gm}{4\pi r^4}\frac{\dot{r}}{r} \tag{4.54}$$

We multiply this by $4\pi r^3$ and integrate over m:

$$\int_0^M 4\pi r^3 \frac{\partial \dot{P}}{\partial m} dm = 4\int_0^M \frac{Gm}{r}\frac{\dot{r}}{r} dm = 4\,\dot{E}_{\rm g} . \tag{4.55}$$

Partial integration of the left-hand side gives

$$[4\pi r^3 \dot{P}]_0^M - 3\int_0^M 4\pi r^2 \frac{\partial r}{\partial m}\dot{P}\, dm , \tag{4.56}$$

where the term in brackets vanishes at both ends of the interval, since either $r = 0$ or $P = 0$ independent of time. If we replace $\partial r/\partial m$ by $1/4\pi r^2 \varrho$ we find from (4.55) that

$$-3\int_0^M \frac{\dot{P}}{\varrho} dm = 4\dot{E}_{\rm g} . \tag{4.57}$$

Introducing this into the right-hand side of (4.53) gives

$$\dot{E}_{\rm g} = -\int_0^M \frac{P}{\varrho^2}\dot{\varrho}\, dm , \tag{4.58}$$

and therefore the integration of the last term of (4.51) gives $\dot{E}_{\rm g}$ so that the equation (4.50) without $\dot{E}_{\rm kin}$ is now recovered.

If, instead of hydrostatic equilibrium, we had used the full equation of motion (2.16), after multiplication with $4\pi r^2 \dot{r}$ and integration over m, we would have obtained the full equation (4.50) *with* the term $\dot{E}_{\rm kin}$.

4.6 Timescales

Consider a star balancing its energy loss L essentially by release of nuclear energy. If L remains constant this can go on for a *nuclear timescale* τ_n defined by

$$\tau_n := \frac{E_n}{L} . \tag{4.59}$$

Note that E_n means the nuclear energy reservoir from which energy can be released under the given circumstances, i.e. the corresponding reactions must be possible. The most important reaction is the fusion of ^1H into ^4He. This "hydrogen burning" releases $Q = 6.3 \times 10^{18}\,\mathrm{erg\,g^{-1}}$, and, if the Sun consisted completely of hydrogen, E_n would be $QM_\odot = 1.25 \times 10^{52}$ erg. With $L_\odot = 4 \times 10^{33}$ erg/s, (4.59) gives $\tau_n = 3 \times 10^{18}$ s, or 10^{11} years. A comparison with the earlier estimates of τ_{hydr} (Sect. 2.4) and τ_{KH} (Sect. 3.3) shows that

$$\tau_n \gg \tau_{KH} \gg \tau_{hydr} , \tag{4.60}$$

which is not only true for the Sun, but for all stars that survive by hydrogen and helium burning. We emphasize this point, since under these circumstances the equation of energy conservation (4.46) can be simplified. As an illustration, we assume that the star changes its properties considerably within the timescale τ (which may be either small or large compared to τ_{KH}). This change may, for instance, be due to exhaustion of nuclear fuel or artificial "squeezing" of the star from the exterior. We now give rough estimates for the four terms in (4.46), assuming a perfect gas:

$$\left|\frac{\partial l}{\partial m}\right| \approx \frac{L}{M} \approx \frac{E_i}{\tau_{KH} M} , \tag{4.61}$$

$$\varepsilon \approx \frac{L}{M} = \frac{E_n}{M\tau_n} \approx \frac{E_i}{\tau_{KH} M} , \tag{4.62}$$

$$\left|c_P \frac{\partial T}{\partial t}\right| \approx \frac{c_P T}{\tau} \approx \frac{E_i}{\tau M} , \tag{4.63}$$

$$\left|\frac{\delta}{\varrho}\frac{\partial P}{\partial t}\right| \approx \frac{\Re}{\mu}\frac{T}{\tau} \approx \frac{c_P T}{\tau} \approx \frac{E_i}{\tau M} . \tag{4.64}$$

In the case $\tau \gg \tau_{KH}$, the terms in (4.63) and (4.64) are small compared to those in (4.61) and (4.62); therefore the time derivatives in the energy equation (4.46) can be neglected ($|\varepsilon_g| \ll \varepsilon$), and the energy equation is $\partial l/\partial m = \varepsilon$, as in (4.43). This occurs if, for instance, the consumption of hydrogen and helium steers the evolution, i.e. $\tau = \tau_n$ ($\gg \tau_{KH}$), and represents a considerable simplification for calculating models which are said to be in *complete equilibrium* (i.e. mechanical and thermal equilibrium).

In the case $\tau \ll \tau_{KH}$, the right-hand sides of (4.63) and (4.64) are large compared to those of (4.61) and (4.62). Therefore in (4.46) the last two terms containing the time derivatives must (at least very nearly) cancel each other, which means that $dq/dt \approx 0$, or the change is nearly adiabatic. Note that a relatively small deviation from the strict adiabatic change can still be of the order ε, and therefore ε_g cannot be neglected in the energy equation. An example for this case is a star pulsating with the timescale $\tau = \tau_{hydr} \ll \tau_{KH}$ (see Chaps. 40 and 41). The variable luminosity of a pulsating star, for instance, is not due to changes of ε but of ε_g.

Here we have assumed the simplest case, namely that the star changes more or less uniformly. The situation can be much more complicated if, for example, only parts of the star are affected and local timescales have to be considered which may be quite different.

Chapter 5
Transport of Energy by Radiation and Conduction

The energy the star radiates away so profusely from its surface is generally replenished from reservoirs situated in the very hot central region. This requires an effective transfer of energy through the stellar material, which is possible owing to the existence of a non-vanishing temperature gradient in the star. Depending on the local physical situation, the transfer can occur mainly via radiation, conduction, and convection. In any case, certain "particles" (photons, atoms, electrons, "blobs" of matter) are exchanged between hotter and cooler parts, and their mean free path together with the temperature gradient of the surroundings will play a decisive role. The equation for the energy transport, written as a condition for the temperature gradient necessary for the required energy flow, will supply our next basic equation for the stellar structure.

5.1 Radiative Transport of Energy

5.1.1 Basic Estimates

Rough estimates show important features of the radiative transfer in stellar interiors and justify an enormous simplification of the formalism.

Let us first estimate the mean free path ℓ_{ph} of a photon at an "average" point inside a star like the Sun:

$$\ell_{ph} = \frac{1}{\kappa \varrho} , \tag{5.1}$$

where κ is a mean absorption coefficient, i.e. a radiative cross section per unit mass averaged over frequency. Typical values for stellar material are of order $\kappa \approx 1\,\mathrm{cm}^2\,\mathrm{g}^{-1}$; for the ionized hydrogen in stellar interiors, a lower limit is certainly the value for electron scattering, $\kappa \approx 0.4\,\mathrm{cm}^2\,\mathrm{g}^{-1}$ (see Chap. 17). Using this and the

R. Kippenhahn et al., *Stellar Structure and Evolution*, Astronomy and Astrophysics Library, DOI 10.1007/978-3-642-30304-3_5, © Springer-Verlag Berlin Heidelberg 2012

mean density of matter in the Sun, $\bar{\varrho}_\odot = 3M_\odot/4\pi R_\odot^3 = 1.4\,\mathrm{g\,cm^{-3}}$, we obtain a mean free path of only

$$\ell_{\rm ph} \approx 2\,\mathrm{cm}\,, \tag{5.2}$$

i.e. stellar matter is very opaque.

The typical temperature gradient in the star can be roughly estimated by averaging between centre ($T_{\rm c} \approx 10^7$ K) and surface ($T_0 \approx 10^4$ K):

$$\frac{\Delta T}{\Delta r} \approx \frac{T_{\rm c} - T_0}{R_\odot} \approx 1.4 \times 10^{-4}\,\mathrm{K\,cm^{-1}}\,. \tag{5.3}$$

The radiation field at a given point is emitted from a small, nearly isothermal surrounding, the differences of temperature being only of order $\Delta T = \ell_{\rm ph}(dT/dr) \approx 3\times10^{-4}$ K. Since the energy density of radiation is $u \sim T^4$, the relative anisotropy of the radiation at a point with $T = 10^7$ K is $4\Delta T/T \sim 10^{-10}$. The situation in stellar interiors must obviously be very close to thermal equilibrium, and the radiation very close to that of a black body. Nevertheless, the small remaining anisotropy can easily be the carrier of the stars' huge luminosity: this fraction of 10^{-10} of the flux emitted from 1 cm^2 of a black body of $T = 10^7$ K is still 10^3 times larger than the flux at the solar surface ($6 \times 10^{10}\,\mathrm{erg\,cm^{-2}\,s^{-1}}$). Radiative transport of energy occurs via the non-vanishing net flux, i.e. via the surplus of the outwards-going radiation (emitted from somewhat hotter material below) over the inwards-going radiation (emitted from less-hot material above).

5.1.2 Diffusion of Radiative Energy

The above estimates have shown that for radiative transport in stars, the mean free path $\ell_{\rm ph}$ of the "transporting particles" (photons) is very small compared to the characteristic length R (stellar radius) over which the transport extends: $\ell_{\rm ph}/R_\odot \approx 3 \times 10^{-11}$. In this case, the transport can be treated as a diffusion process, which yields an enormous simplification of the formalism. We derive the corresponding equation by analogy to those for particle diffusion. A more rigorous derivation can be found in any textbook about radiation transport, for instance, in Chaps. 2 and 8 of Weiss et al. (2004).

The diffusive flux j of particles (per unit area and time) between places of different particle density n is given by

$$\boldsymbol{j} = -D\,\nabla n\,, \tag{5.4}$$

where D is the coefficient of diffusion,

$$D = \frac{1}{3} v\,\ell_{\rm p}\,, \tag{5.5}$$

determined by the average values of mean velocity v and mean free path ℓ_p of the particles.

In order to obtain the corresponding diffusive flux of radiative energy $\boldsymbol{F}$, we replace n by the energy density of radiation U,

$$U = aT^4 , \tag{5.6}$$

v by the velocity of light c, and ℓ_p by ℓ_{ph} according to (5.1).

In (5.6), $a = 7.57 \times 10^{-15}\,\mathrm{erg\,cm^{-3}\,K^{-4}}$ is the *radiation density constant.* Owing to the spherical symmetry of the problem, $\boldsymbol{F}$ has only a radial component $F_r = |\boldsymbol{F}| = F$ and ∇U reduces to the derivative in the radial direction

$$\frac{\partial U}{\partial r} = 4\,a\,T^3 \frac{\partial T}{\partial r} . \tag{5.7}$$

Then (5.4) and (5.5) give immediately that

$$F = -\frac{4ac}{3} \frac{T^3}{\kappa \varrho} \frac{\partial T}{\partial r} . \tag{5.8}$$

Note that this can be interpreted formally as an equation for heat conduction by writing

$$\boldsymbol{F} = -k_{rad} \nabla T , \tag{5.9}$$

where

$$k_{rad} = \frac{4ac}{3} \frac{T^3}{\kappa \varrho} \tag{5.10}$$

represents the coefficient of conduction for this radiative transport.

We solve (5.8) for the gradient of the temperature and replace F by the usual local luminosity $l = 4\pi r^2 F$; then

$$\frac{\partial T}{\partial r} = -\frac{3}{16\pi ac} \frac{\kappa \varrho l}{r^2 T^3} . \tag{5.11}$$

After transformation to the independent variable m (as in Sect. 2.1), the basic equation for radiative transport of energy is obtained in the form

$$\frac{\partial T}{\partial m} = -\frac{3}{64\pi^2 ac} \frac{\kappa l}{r^4 T^3} . \tag{5.12}$$

Of course, this neat and simple equation becomes invalid when one approaches the surface of the star. Because of the decreasing density, the mean free path of the photons will there become comparable with (and finally larger than) the remaining distance to the surface; hence the whole diffusion approximation breaks down, and one has to solve the far more complicated full set of transport equations for radiation in the stellar atmosphere (These equations indeed yield our simple diffusion approximation as the proper limiting case for large optical depths.). Fortunately, however, we have then left the stellar-interior regime with which this book deals, and we happily leave the complicated remainder to those of our colleagues who feel the call to treat the problem of stellar atmospheres.

5.1.3 The Rosseland Mean for κ_ν

The above equations are independent of the frequency ν; F and l are quantities integrated over all frequencies, so that the quantity κ must represent a "proper mean" over ν. We shall now prescribe a method for this averaging.

In general the absorption coefficient depends on the frequency ν. Let us denote this by adding a subscript ν to all quantities that thus become frequency dependent: $\kappa_\nu, \ell_\nu, D_\nu, U_\nu$, etc.

For the diffusive energy flux $\boldsymbol{F}_\nu$ of radiation in the interval $[\nu, \nu + d\nu]$, we write now, as in Sect. 5.1.2,

$$\boldsymbol{F}_\nu = -D_\nu \nabla U_\nu, \qquad \text{with} \tag{5.13}$$

$$D_\nu = \frac{1}{3} c\, \ell_\nu = \frac{c}{3\kappa_\nu \varrho}, \tag{5.14}$$

while the energy density in the same interval is given by

$$U_\nu = \frac{4\pi}{c} B(\nu, T) = \frac{8\pi h}{c^3} \frac{\nu^3}{e^{h\nu/kT} - 1}. \tag{5.15}$$

$B(\nu, T)$ denotes here the Planck function for the *intensity* of black-body radiation (differing from the usual formula for the energy density simply by the factor $4\pi/c$). For simplicity, we will not always write the arguments ν and T explicitly in the following formulae. From (5.15) we have

$$\nabla U_\nu = \frac{4\pi}{c} \frac{\partial B}{\partial T} \nabla T, \tag{5.16}$$

which together with (5.14) is inserted into (5.13), the latter then being integrated over all frequencies to obtain the total flux $\boldsymbol{F}$:

$$\boldsymbol{F} = -\left[\frac{4\pi}{3\varrho} \int_0^\infty \frac{1}{\kappa_\nu} \frac{\partial B}{\partial T} d\nu \right] \nabla T. \tag{5.17}$$

We have thus regained (5.9), but with

$$k_{\text{rad}} = \frac{4\pi}{3\varrho} \int_0^\infty \frac{1}{\kappa_\nu} \frac{\partial B}{\partial T} d\nu. \tag{5.18}$$

Equating this expression for $k_{\rm rad}$ with that in the averaged form of (5.10), we have immediately the proper formula for averaging the absorption coefficient:

$$\frac{1}{\kappa} = \frac{\pi}{acT^3} \int_0^\infty \frac{1}{\kappa_\nu} \frac{\partial B}{\partial T} d\nu . \tag{5.19}$$

This is the so-called *Rosseland mean* (after Sven Rosseland).

Since

$$\int_0^\infty \frac{\partial B}{\partial T} d\nu = \frac{acT^3}{\pi} , \tag{5.20}$$

the Rosseland mean is formally the harmonic mean of κ_ν with the weighting function $\partial B/\partial T$, and it can simply be calculated, once the function κ_ν is known from atomic physics.

In order to see the physical interpretation of the Rosseland mean, we rewrite (5.13) with the help of (5.14)–(5.16):

$$\boldsymbol{F}_\nu = -\left(\frac{1}{\kappa_\nu} \frac{\partial B(\nu, T)}{\partial T}\right) \frac{4\pi}{3\varrho} \nabla T . \tag{5.21}$$

This shows that, for a given point in the star (ϱ and ∇T given), the integrand in (5.19) is at all frequencies proportional to the net flux $\boldsymbol{F}_\nu$ of energy. The Rosseland mean therefore favours the frequency ranges of maximum energy flux. One could say that an average *transparency* is evaluated rather than an *opacity*–which is plausible, since it is to be used in an equation describing the transfer of energy rather than its blocking.

One can also easily evaluate the frequency where the weighting function $\partial B/\partial T$ has its maximum. From (5.15) one finds that, for given a temperature, $\partial B/\partial T \sim x^4 e^x (e^x - 1)^{-2}$ with the usual definition $x = h\nu/kT$. Differentiation with respect to x shows that the maximum of $\partial B/\partial T$ is close to $x = 4$.

The way we have defined the Rosseland mean κ, which is a kind of weighted harmonic mean value, has the uncomfortable consequence that the opacity κ of a mixture of two gases having the opacities κ_1, κ_2 is not the sum of the opacities: $\kappa \neq \kappa_1 + \kappa_2$.

Therefore, in order to find κ for a mixture containing the weight fractions X of hydrogen and Y of helium, the mean opacities of the two single gases are of no use. Rather one has to add the frequency-dependent opacities $\kappa_\nu = X\kappa_{\nu \rm H} + Y\kappa_{\nu \rm He}$ before calculating the Rosseland mean. For any new abundance ratio X/Y the averaging over the frequency has to be carried out separately.

In the above we have characterized the energy flux due to the diffusion of photons by $\boldsymbol{F}$. Since in the following we shall encounter other mechanisms for energy transport, from now, on we shall specify this radiative flux by the vector $\boldsymbol{F}_{\rm rad}$. Correspondingly we shall use $\kappa_{\rm rad}$ instead of κ, etc.

5.2 Conductive Transport of Energy

In heat conduction, energy transfer occurs via collisions during the random thermal motion of the particles (electrons and nuclei in completely ionized matter, otherwise atoms or molecules). A basic estimate similar to that in Sect. 5.1.1 shows that in "ordinary" stellar matter (i.e. in a non-degenerate gas), conduction has no chance of taking over an appreciable part of the total energy transport. Although the collisional cross sections of these charged particles are rather small at the high temperatures in stellar interiors ($10^{-18} \cdots 10^{-20}\,\mathrm{cm}^2$ per particle), the large density ($\bar{\varrho} = 1.4\,\mathrm{g\,cm}^{-3}$ in the Sun) results in mean free paths several orders of magnitude less than those for photons; and the velocity of the particles is only a few per cent of c. Therefore the coefficient of diffusion (5.5) is much smaller than that for photons.

The situation becomes quite different, however, for the cores of evolved stars (see Chap. 33), where the electron gas is highly degenerate. The density can be as large as $10^6\,\mathrm{g\,cm}^{-3}$. But degeneracy makes the electrons much faster, since they are pushed up close to the Fermi energy; and degeneracy increases the mean free path considerably, since the quantum cells of phase space are filled up such that collisions in which the momentum is changed become rather improbable. Then the coefficient of diffusion (which is proportional to the product of mean free path and particle velocity) is large, and heat conduction can become so efficient that it short-circuits the radiative transfer (see Sect. 17.6).

The energy flux $\boldsymbol{F}_{\mathrm{cd}}$ due to heat conduction may be written as

$$\boldsymbol{F}_{\mathrm{cd}} = -k_{\mathrm{cd}} \nabla T \ . \tag{5.22}$$

The sum of the conductive flux $\boldsymbol{F}_{\mathrm{cd}}$ and the radiative flux $\boldsymbol{F}_{\mathrm{rad}}$ as defined in (5.9) is

$$\boldsymbol{F} = \boldsymbol{F}_{\mathrm{rad}} + \boldsymbol{F}_{\mathrm{cd}} = -(k_{\mathrm{rad}} + k_{\mathrm{cd}}) \nabla T \ , \tag{5.23}$$

which shows immediately the benefit of writing the radiative flux in (5.9) formally as an equation of heat conduction. On the other hand, we can just as well write the conductive coefficient k_{cd} formally in analogy to (5.10) as

$$k_{\mathrm{cd}} = \frac{4ac}{3} \frac{T^3}{\kappa_{\mathrm{cd}} \varrho} \ , \tag{5.24}$$

hence defining the "conductive opacity" κ_{cd}. Then (5.23) becomes

$$\boldsymbol{F} = -\frac{4ac}{3} \frac{T^3}{\varrho} \left(\frac{1}{\kappa_{\mathrm{rad}}} + \frac{1}{\kappa_{\mathrm{cd}}} \right) \nabla T \ , \tag{5.25}$$

which shows that we arrive formally at the same type of equation (5.11) as in the pure radiative case, if we replace $1/\kappa$ there by $1/\kappa_{\mathrm{rad}} + 1/\kappa_{\mathrm{cd}}$. Again the result is plausible, since the mechanism of transport that provides the largest flux will dominate the sum, i.e. the mechanism for which the stellar matter has the highest "transparency".

Equation (5.12), which, if we define κ properly, holds for radiative and conductive energy transport, can be rewritten in a form which will be convenient for the following sections.

Assuming hydrostatic equilibrium, we divide (5.12) by (2.5) and obtain

$$\frac{(\partial T/\partial m)}{(\partial P/\partial m)} = \frac{3}{16\pi acG}\frac{\kappa l}{mT^3} . \tag{5.26}$$

We call the ratio of the derivatives on the left $(dT/dP)_{\text{rad}}$, and we mean by this the variation of T in the star with depth, where the depth is expressed by the pressure, which increases monotonically inwards. In this sense, in a star which is in hydrostatic equilibrium and transports the energy by radiation (and conduction), $(dT/dP)_{\text{rad}}$ is a gradient describing the temperature variation with depth. If we use the customary abbreviation

$$\nabla_{\text{rad}} := \left(\frac{d \ln T}{d \ln P}\right)_{\text{rad}} , \tag{5.27}$$

(5.26) can be written in the form

$$\nabla_{\text{rad}} = \frac{3}{16\pi acG}\frac{\kappa l P}{mT^4} , \tag{5.28}$$

in which conduction effects are now included. Note the difference in definition and meaning of ∇_{rad} and of ∇_{ad} introduced in (4.21), which concerns not only their (in general different) numerical values. As just explained, ∇_{rad} means a spatial derivative (connecting P and T in two neighbouring mass shells), while ∇_{ad} describes the thermal variation of one and the same mass element during its adiabatic compression. Only in special cases $(d \ln T/d \ln P)$ and ∇_{ad} will have the same value, and we then speak of an "adiabatic stratification".

We will use ∇_{rad} also in connection with more general cases (other modes of energy transport like convection as in Chap. 7, deviation from hydrostatic equilibrium). It then means the gradient to which a radiative, hydrostatic layer would adjust at a corresponding point (same values of P, T, l, m), or simply an abbreviation for the expression on the right-hand side of (5.28), which is valid only for hydrostatic equilibrium and as long as an effective κ as in (5.25) can be defined.

5.3 The Thermal Adjustment Time of a Star

We can write (5.12), which holds for radiative and conductive energy transport, in the form

$$l = -\sigma^* \frac{\partial T}{\partial m} , \qquad \sigma^* = \frac{64\pi^6 ac T^3 r^4}{3\kappa} . \tag{5.29}$$

Now, combining this with (4.45) and replacing the internal energy u by its value $c_v T$ for the perfect gas, it follows that

$$\frac{\partial}{\partial m}\left(\sigma^* \frac{\partial T}{\partial m}\right) - c_v \frac{\partial T}{\partial t} = -\left[\varepsilon + \frac{P}{\varrho^2}\frac{\partial \varrho}{\partial t}\right] . \qquad (5.30)$$

If we put the right-hand side equal to zero, then (5.30) has the form of the equation of heat transfer with variable conductivity σ^*. Indeed variation of the temperature with time along a rod of conductivity σ and specific heat c is governed by the equation

$$\frac{\partial}{\partial x}\left(\sigma \frac{\partial T}{\partial x}\right) = c \frac{\partial T}{\partial t} , \qquad (5.31)$$

where x is the spatial coordinate along the rod (see Landau and Lifshitz, vol. 6, 1987) . There exists a vast amount of mathematical theory associated with this equation, especially for the case where σ is constant. For example, one can define an initial-value problem with given $T = T(x)$ at $t = 0$. How, then, does this initial temperature profile evolve in time? There are classical methods for determining $T = T(x,t)$ for $t > 0$. One of the basic results is that one can start with an exciting temperature profile $T(x)$, for instance, one which resembles the skyline of Manhattan or the panorama of the Alps, and after some time, the temperature profile always looks like the landscape of Nebraska: $T(x,t)$ approaches the limit solution $T =$ constant after sufficient time.

One can easily estimate the timescale over which (5.31) demands considerable changes of an initially given temperature profile, *the timescale of thermal adjustment*, by replacing in (5.31) ∂T by ΔT, ∂x by a characteristic length d, and ∂t by τ_{adj}:

$$\tau_{\text{adj}} = \frac{c}{\sigma} d^2 , \qquad (5.32)$$

where d is a characteristic length over which the (initially given) temperature variation changes. Obviously, only temperature profiles with variations over small distances can change rapidly in time.

The inhomogeneous term on the right of (5.30) is a source term. It takes into account that energy can be added everywhere by nuclear reactions or by compression. In the case of the rod it would correspond to extra heat sources adding heat at different values of x. Similarly to (5.32) we can derive a characteristic time for a star:

$$\tau_{\text{adj}} = \frac{c_v M^2}{\overline{\sigma}^*} , \qquad (5.33)$$

where we have replaced the operator $\partial/\partial m$ by $1/M$ and introduced a mean value $\overline{\sigma}^*$, which we can estimate from (5.29). We find for the luminosity L of the star

$L \approx \overline{\sigma}^* \overline{T}/M$, where $\overline{T}$ is a mean temperature of the star. Therefore, for a rough estimate, we have from (5.33) that

$$\tau_{\text{adj}} \approx \frac{c_v \overline{T} M}{L} = \frac{E_{\text{i}}}{L} = \tau_{\text{KH}} . \tag{5.34}$$

This means that the Kelvin–Helmholtz timescale as defined in (3.17) can be considered a characteristic time of thermal adjustment of a star or – in other words – the time it takes a thermal fluctuation to travel from centre to surface.

In spite of the indicated equivalence of τ_{adj} and τ_{KH}, it is often advisable to consider τ_{adj} separately, in particular if it is to be applied to parts of a star only. For example, we will encounter evolved stars with isothermal cores of very high conductivity (Chap. 33). The luminosity there is zero so that formally the corresponding τ_{KH} becomes infinite. The decisive timescale that in fact enforces the isothermal situation is the very small τ_{adj}. The difference can be characterized as follows: how much energy may be transported after a temperature perturbation is often much more important than how much energy is flowing in the unperturbed configuration.

5.4 Thermal Properties of the Piston Model

We now investigate the thermal properties of the piston model discussed in Sects. 2.7 and 3.2 by first assuming that the gas of mass m^* in the container is thermally isolated from the surroundings. If the piston is moved, the gas changes adiabatically, i.e.

$$dQ = m^* du + P dV = 0 , \tag{5.35}$$

dQ being the heat added to the total mass of the gas. For a perfect gas the energy per unit mass is $u = c_v T$, and for adiabatic conditions, with $V = Ah$, this leads to

$$dQ = c_v m^* dT + PA\, dh = 0 . \tag{5.36}$$

We now relax the adiabatic condition in three ways. First, we allow a small leak through which heat (but no gas) can escape from the interior (gas at temperature T) to the surroundings (at temperature T_{s}) see Fig. 2.2. The corresponding heat flow will be $\chi(T - T_{\text{s}})$, where χ is a measure of the heat conduction at the leak indicated in Fig. 2.2. Second, in order to make the gas more similar to stellar matter, we assume the release of nuclear energy with a rate ε. Third, we assume that a radiative energy flux F penetrates the gas and that the energy $\kappa F m^*$ is absorbed per second. The energy balance of the gas in the stationary case then can be expressed by

$$\varepsilon m^* + \kappa F m^* = \chi(T - T_{\text{s}}) . \tag{5.37}$$

In general the heat dQ added to the gas within the time interval dt is

$$dQ = [\varepsilon m^* + \kappa F m^* - \chi(T - T_s)]dt , \tag{5.38}$$

and, if we compare (5.36) and (5.38), we find that

$$c_v m^* \frac{dT}{dt} + PA \frac{dh}{dt} = \varepsilon m^* + \kappa m^* F - \chi(T - T_s) . \tag{5.39}$$

This is the equation of energy conservation of the gas.

If we assume $\varepsilon = \kappa = 0$, then (5.39) has only one time-independent solution: $T = T_s$. What is the timescale of this adjustment of T?

The two time derivatives on the left-hand side of (5.39) give the same estimate for τ; indeed a change of h occurs only as a consequence of, and together with, the change of T. For our rough estimate we can therefore replace the left-hand side of (5.39) by $c_v \Delta T m^*/T$ where $\Delta T = |T - T_s|$:

$$c_v m^* \Delta T/\tau \approx \chi |T - T_s| . \tag{5.40}$$

For the timescale by which ΔT decays we obtain

$$\tau_{adj} \approx c_v m^*/\chi , \tag{5.41}$$

which is the time it takes the gas to adjust its temperature to that of the surroundings. This timescale for our piston model plays a role similar to the Kelvin–Helmholtz timescale in stars. For sufficiently small χ (sufficiently large τ_{adj}), we have $\tau_{hydr} \ll \tau_{adj}$, similar to the situation in stars, where $\tau_{hydr} \ll \tau_{KH}$.

Chapter 6
Stability Against Local, Non-spherical Perturbations

We have based our treatment on the assumption of strict spherical symmetry, meaning that all functions and variables (including velocities) are constant on concentric spheres. In reality there will arise small fluctuations on such a sphere, for example, simply from the thermal motion of the gas particles. Such local perturbations of the average state may be ignored if they do not grow. But in a star sometimes small perturbations may grow and give rise to macroscopic local (non-spherical) motions that are also statistically distributed over the sphere. In the basic equations the assumption of spherical symmetry can still be kept if we interpret the variables as proper average values over a concentric sphere.

However, these motions have to be considered carefully because they can have a strong influence on the stellar structure. They not only mix the stellar material but also transport energy: hot gas bubbles rise, while cooler material sinks down, i.e. energy transport is by convection, something which is known to play an important role in the earth's atmosphere.

Whether convection occurs in a certain region of a star obviously depends on the question whether the small perturbations always present will grow or stay small: a question of *stability*. We shall derive criteria which tell us whether stellar material at a certain depth is stable or not. Depending on the physical conditions one can make different simplifying assumptions which lead to different stability problems. The following dynamical problem covers most of the "normal" cases in stars.

6.1 Dynamical Instability

The kind of stability we are discussing here is based on the assumption that the moving mass elements have no time to exchange appreciable amounts of heat with the surroundings and therefore move adiabatically. This type of stability (or instability) is called *dynamical*. We will soon learn that there are other types of instability.

R. Kippenhahn et al., *Stellar Structure and Evolution*, Astronomy and Astrophysics Library, DOI 10.1007/978-3-642-30304-3_6, © Springer-Verlag Berlin Heidelberg 2012

First, we consider the possibility that the physical quantities (temperature, density, etc.) may not be exactly constant on the surface of a concentric sphere but rather may show certain fluctuations. In the global problem of stellar structure, one then has only to interpret the previously used functions as proper averages. For the local description, we shall simply represent a fluctuation by a mass "element" (subscript e) in which the functions have constant, but somewhat different, values than in the average "surroundings" (subscript s). For any quantity A we define the difference DA between element and surroundings[1] as

$$DA := A_e - A_s \; . \tag{6.1}$$

One can easily imagine an initial fluctuation of temperature, for example, a slightly hotter element with $DT > 0$. Normally one could then also expect an excess of pressure DP. However, the element will expand immediately until pressure balance with the surroundings is restored, and since this expansion occurs with the velocity of sound, it is usually much more rapid than any other motion of the element. Therefore we can assume here (and in the following) that the element always remains in pressure balance with its surroundings:

$$DP = 0 \; . \tag{6.2}$$

Consequently the assumed $DT > 0$ requires that, for a perfect gas with $\varrho \sim P/T$, $D\varrho < 0$, i.e. the element is lighter than the surrounding material, and the buoyancy forces will lift it upwards: temperature fluctuations are obviously accompanied by local motions of elements in a radial direction.

So, we can also take a radial shift $\Delta r > 0$ of the element as the initial perturbation for testing the stability of a layer. Consider an element that was in complete equilibrium with the surroundings at its original position r but has now been lifted to $r + \Delta r$ (cf. Fig. 6.1). In general its density will differ from that of its new surroundings by

$$D\varrho = \left[\left(\frac{d\varrho}{dr} \right)_e - \left(\frac{d\varrho}{dr} \right)_s \right] \Delta r \; , \tag{6.3}$$

$(d\varrho/dr)_e$ determining the change of the element's density while it rises by dr; the other derivative is the spatial gradient in the surroundings.

A finite $D\varrho$ gives the radial component $K_r = -gD\varrho$ of a buoyancy force $\boldsymbol{K}$ (per unit of volume), where g again is the absolute value of the acceleration of gravity. If $D\varrho < 0$, the element is lighter and $K_r > 0$, i.e. $\boldsymbol{K}$ is directed upwards. This situation is obviously unstable, since the element is lifted further, the original perturbation being increased.

---

[1] Note that we use the subscript s, which is different from s used for the specific entropy in other parts of this book.

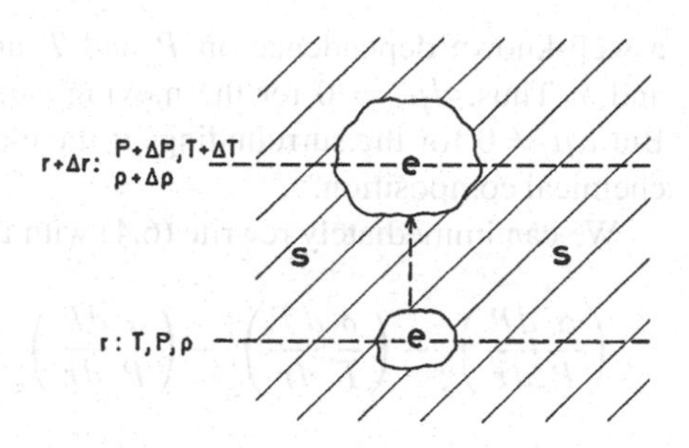

Fig. 6.1 In order to test the stability of a "surrounding" layer (s), a test "element" (e) is lifted from level r to $r + \Delta r$

If on the other hand $D\varrho > 0$, then $K_r < 0$, i.e. $\boldsymbol{K}$ is directed downwards. The element, which is heavier than its new surroundings, is drawn back to its original position, the perturbation is removed, and the layer is stable. As the *condition for stability* we obtain with $D\varrho > 0$ from (6.3) the result

$$\left(\frac{d\varrho}{dr}\right)_{\mathrm{e}} - \left(\frac{d\varrho}{dr}\right)_{\mathrm{s}} > 0 \,. \tag{6.4}$$

Unfortunately this criterion is highly impractical, since it requires knowledge of density gradients that do not appear in the basic equations. It is therefore preferable to turn to temperature gradients as used in the equations of radiative and conductive transport. In order to evaluate $(d\varrho/dr)_{\mathrm{e}}$ correctly, we would have to take into account the possible energy exchange between the element and its surroundings. For simplicity let us here assume that no such exchange of energy occurs, i.e. that the element rises *adiabatically*. This is very close to reality for the deep interior of a star (see Chap. 7).

In order to transform the gradients of ϱ into those of T, we write the equation of state $\varrho = \varrho(P, T, \mu)$ in the following differential form:

$$\frac{d\varrho}{\varrho} = \alpha\frac{dP}{P} - \delta\frac{dT}{T} + \varphi\frac{d\mu}{\mu} \,, \tag{6.5}$$

where α and δ have already been defined in (4.2) and (4.3). But here, we have made allowance also for a possible variation of the chemical composition, which is characterized by the molecular weight μ. We therefore have

$$\alpha := \left(\frac{\partial \ln \varrho}{\partial \ln P}\right), \quad \delta := -\left(\frac{\partial \ln \varrho}{\partial \ln T}\right), \quad \varphi := \left(\frac{\partial \ln \varrho}{\partial \ln \mu}\right), \tag{6.6}$$

where the three partial derivatives correspond to constant values of T, μ; P, μ; and P, T, respectively, and for a perfect gas with $\varrho \sim P\mu/T$, one has $\alpha = \delta = \varphi = 1$. In this description $d\mu$ shall represent only the change of μ due to the change of chemical composition, i.e. the variation of the concentrations of different nuclei in the deep interior. Of course, μ can also change in the outer regions for constant composition if the degree of ionization changes. This effect, however, has a well-known dependence on P and T and is supposed to be incorporated in α and δ. Thus, $d\mu = 0$ for the moving element that carries its composition along. But $d\mu \neq 0$ for the surroundings if the element passes through layers of different chemical composition.

We can immediately rewrite (6.4) with the help of (6.5) in the form

$$\left(\frac{\alpha}{P}\frac{dP}{dr}\right)_{\rm e} - \left(\frac{\delta}{T}\frac{dT}{dr}\right)_{\rm e} - \left(\frac{\alpha}{P}\frac{dP}{dr}\right)_{\rm s} + \left(\frac{\delta}{T}\frac{dT}{dr}\right)_{\rm s} - \left(\frac{\varphi}{\mu}\frac{d\mu}{dr}\right)_{\rm s} > 0\,. \quad (6.7)$$

The two terms containing the pressure gradient cancel each other owing to (6.2), and the other terms are usually multiplied by the so-called *scale height of pressure* H_P:

$$H_P := -\frac{dr}{d\ln P} = -P\frac{dr}{dP}\,. \quad (6.8)$$

With (2.3), the condition for hydrostatic equilibrium, we find $H_P = P/\varrho g$, i.e. $H_P > 0$, since P decreases with increasing r. H_P has the dimension of length, being the length characteristic of the radial variation of P. In the solar photo-sphere ($g = 2.7 \times 10^4\,\mathrm{cm\,s^{-2}}$, $P = 1.0 \times 10^5\,\mathrm{dyn\,cm^{-2}}$, $\varrho = 2.6 \times 10^{-7}\,\mathrm{g\,cm^{-3}}$), one finds $H_P = 1.4 \times 10^7$ cm, while at $r = R_\odot/2$ ($g = 9.8 \times 10^4\,\mathrm{cm\,s^{-2}}$, $P = 7.3 \times 10^{14}\,\mathrm{dyn\,cm^{-2}}$, $\varrho = 1.4\,\mathrm{g\,cm^{-3}}$), H_P is much bigger, at 5.5×10^9 cm. If one approaches the stellar centre–where $g = 0$, while P remains finite–then $H_P \to \infty$.

Multiplication of (6.7) by H_P yields as a condition for stability

$$\left(\frac{d\ln T}{d\ln P}\right)_{\rm s} < \left(\frac{d\ln T}{d\ln P}\right)_{\rm e} + \frac{\varphi}{\delta}\left(\frac{d\ln\mu}{d\ln P}\right)_{\rm s}\,. \quad (6.9)$$

Similar to the previously defined quantities $\nabla_{\rm rad}$ and $\nabla_{\rm ad}$, we define three new derivatives:

$$\nabla := \left(\frac{d\ln T}{d\ln P}\right)_{\rm s}, \quad \nabla_{\rm e} := \left(\frac{d\ln T}{d\ln P}\right)_{\rm e}, \quad \nabla_\mu := \left(\frac{d\ln\mu}{d\ln P}\right)_{\rm s}\,. \quad (6.10)$$

Here the subscripts s indicate that the derivatives are to be taken in the surrounding material. In both cases they are spatial derivatives in which the variations of T and μ with depth are considered and P is taken as a measure of depth. The quantity $\nabla_{\rm e}$ describes the variation of T in the element during its motion, where the position of the element is measured by P. In this sense $\nabla_{\rm e}$ and $\nabla_{\rm ad}$ are similar, since both describe the temperature variation of a gas undergoing pressure variations; on the other hand, $\nabla_{\rm rad}$ and ∇_μ describe the spatial variation of T and μ in the surroundings.

With the definitions (6.10) the condition (6.9) for stability becomes

$$\nabla < \nabla_{\rm e} + \frac{\varphi}{\delta}\nabla_\mu\,. \quad (6.11)$$

In (5.27) and (5.28) we defined $\nabla_{\rm rad}$, which describes the temperature gradient for the case that the energy is transported by radiation (or conduction) only. Therefore in a layer that indeed transports all energy by radiation the actual gradient ∇ is equal to $\nabla_{\rm rad}$. Let us test such a layer for its stability and assume the elements change adiabatically: $\nabla_{\rm e} = \nabla_{\rm ad}$; the radiation layer is stable if

$$\nabla_{\rm rad} < \nabla_{\rm ad} + \frac{\varphi}{\delta}\nabla_\mu , \tag{6.12}$$

a form known as the *Ledoux criterion* (named after Paul Ledoux) for dynamical stability. In a region with homogeneous chemical composition, $\nabla_\mu = 0$, and one has then simply the famous *Schwarzschild criterion* for dynamical stability (named after Karl Schwarzschild):

$$\nabla_{\rm rad} < \nabla_{\rm ad} . \tag{6.13}$$

If in the criteria (6.12) and (6.13) the left-hand side is larger than the right, the layer is dynamically unstable. If they are equal, one speaks of marginal stability. The difference between the two criteria obviously plays a role only in regions where the chemical composition varies radially. We will see that such regions occur in the interior of evolving stars, where heavier elements are usually produced below the lighter ones, such that the molecular weight μ increases inwards (as the pressure does) and $\nabla_\mu > 0$. Then the last term in inequality (6.12) obviously has a stabilizing effect (φ and δ are both positive). This is plausible since the element carries its heavier material upwards into lighter surroundings and gravity will tend to draw it back to its original place.

If these criteria favour stability, then no convective motions will occur, and the whole flux will indeed be carried by radiation, i.e. the actual gradient at such a place is equal to the radiative one: $\nabla = \nabla_{\rm rad}$. If they favour instability, then small perturbations will increase to finite amplitude until the whole region boils with convective motions that carry part of the flux–and the actual gradient has to be determined in a manner described in Chap. 7. This instability can be caused either by the fact that $\nabla_{\rm rad}$ has become too high (large flux or very opaque matter), or else by a depression of $\nabla_{\rm ad}$; both cases occur in stars. And, finally, in a twilight zone, where one of the two criteria (6.12) and (6.13) says stability and the other one says instability, strange things may happen (see, for instance, Sects. 6.3 and 30.4.2).

Note that (6.12) and (6.13) are strictly local criteria, which means good and bad news. They are very practical since they can be evaluated easily for any given place by using the local values of P, T, ϱ only, without bothering about other parts of the star. And in most cases this will give satisfactory answers. In critical cases, however, this may not be sufficient. Strictly speaking, convective motions are not only dependent on the local forces (which are solely regarded by the criteria), but must be coupled (by momentum transfer, inertia, the equation of continuity) to their neighbouring layers. And in extreme cases the reaction of the whole star against a local perturbation should be taken into account. An obvious example is the precise determination of the border of a convective zone, where elements that were accelerated elsewhere "shoot over" until their motion is braked. We will come back later to such problems when they arise (see Sect. 30.4.1).

We can immediately derive a qualitative relation between the different gradients. They are best visualized in a diagram such as Fig. 6.2, where $\ln T$ is plotted against $\ln P$ (decreasing outwards) for an unstable layer violating the Schwarzschild criterion. In such a diagram, an adiabatic change follows a line with slope $\nabla_{\rm ad}$, the changes in a rising element are given by a line with slope $\nabla_{\rm e}$, while the stratifications in the surroundings and in a radiative layer are shown by lines with slopes ∇ and $\nabla_{\rm rad}$, respectively.

Suppose we have convection in a chemically homogeneous layer ($\nabla_\mu = 0$). The criterion (6.11) must be violated, i.e. $\nabla > \nabla_{\rm e}$. If some part of the flux is carried by convection, then the actual gradient $\nabla < \nabla_{\rm rad}$, since only a part of the total flux is left for radiative transfer. Consider a rising element that has started from a point with P_0, T_0. In Fig. 6.2 this element moves downwards to the left along the line with slope $\nabla_{\rm e}$. Since $\nabla > \nabla_{\rm e}$, the element (although cooling) will obviously have an increasing temperature excess over its new surroundings (the temperature of which changes with ∇). Therefore it will radiate energy into its surroundings, which means that the element cools more than adiabatically: $\nabla_{\rm e} > \nabla_{\rm ad}$. Combining these inequalities, we arrive at the relation illustrated in Fig. 6.2:

$$\nabla_{\rm rad} > \nabla > \nabla_{\rm e} > \nabla_{\rm ad} \, . \tag{6.14}$$

The fact that $\nabla_{\rm e}$ must always be between $\nabla_{\rm ad}$ and ∇ of the surroundings shows that the criteria (6.12) and (6.13) are also to be used in near-surface regions, where the rising elements lose much of their energy by radiation.

6.2 Oscillation of a Displaced Element

In a dynamically stable layer a displaced mass element is pushed back by buoyancy. When coming back to its original position, it has gained momentum and will overshoot and therefore start to oscillate. In the following we shall discuss this oscillation.

Consider a mass element lifted from its normal (equilibrium) position in the radial direction by an amount Δr (see Fig. 6.1). There it has an excess of density $D\varrho$ over its new surroundings given by (6.3), which for balance of pressure ($DP = 0$) and with (6.5) and the definitions (6.6), (6.8), (6.10) can easily be written as

$$D\varrho = \frac{\varrho\delta}{H_P}\left[\nabla_{\rm e} - \nabla + \frac{\varphi}{\delta}\nabla_\mu\right]\Delta r \, . \tag{6.15}$$

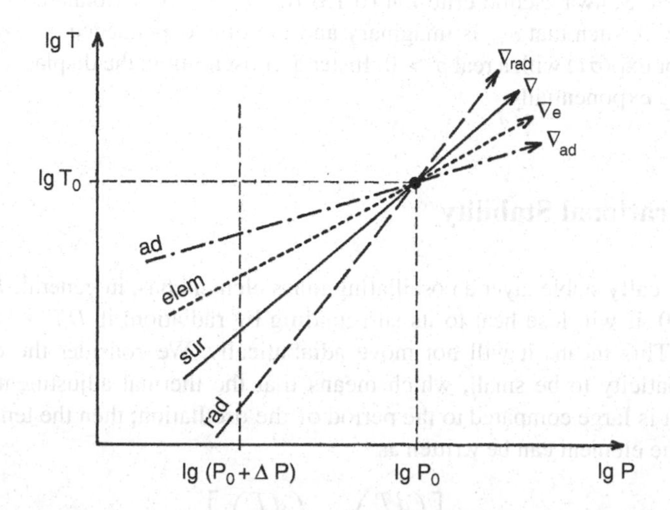

Fig. 6.2 Temperature-pressure diagram with a schematic sketch of the different gradients $\nabla(\equiv \partial \ln T/\partial \ln P)$ in a convective layer. Starting at a common point with P_0 and T_0, the different types of changes (adiabatic, in a rising element, in the surroundings, for radiative stratification) lead to different temperatures at a slightly higher point with $P_0 + \Delta P$ ($< P_0$, since P decreases outwards)

In the presence of gravity g, the resulting buoyancy force per unit volume is $K_r = -gD\varrho$, producing an acceleration of the element of

$$\frac{\partial^2(\Delta r)}{\partial t^2} = -\frac{g\delta}{H_P}\left[\nabla_e - \nabla + \frac{\varphi}{\delta}\nabla_\mu\right]\Delta r\,. \tag{6.16}$$

Suppose now that the element, after an original displacement Δr_0, moves adiabatically ($\nabla_e = \nabla_{ad}$) through a dynamically stable layer ($D\varrho/\Delta r > 0$). The element is accelerated back towards its equilibrium position around which it then oscillates according to the solution of (6.16):

$$\Delta r = \Delta r_0\, e^{i\omega t}\,. \tag{6.17}$$

The frequency $\omega = \omega_{ad}$ of this adiabatic oscillation is the so-called *Brunt-Väisälä frequency* given by

$$\omega_{ad}^2 = \frac{g\delta}{H_P}\left(\nabla_{ad} - \nabla + \frac{\varphi}{\delta}\nabla_\mu\right)\,. \tag{6.18}$$

(It plays, e.g. a role in the discussion of non-radial oscillations of a star, see Chap. 42.) The corresponding period is $\tau_{ad} = 2\pi/\omega_{ad}$.

We see immediately what happens in an unstable layer. If the Ledoux criterion (6.12) [or the Schwarzschild criterion (6.13) for $\nabla_\mu = 0$] is violated, then (6.18) gives $\omega_{\rm ad}^2 < 0$, such that $\omega_{\rm ad}$ is imaginary and the time dependence of Δr is given by the factor $\exp(\sigma t)$ with a real $\sigma > 0$. Instead of oscillating, the displaced element moves away exponentially.

6.3 Vibrational Stability

In a dynamically stable layer an oscillating mass element has, in general, $DT \neq 0$. If $DT > 0$, it will lose heat to its surrounding by radiation; if $DT < 0$, it will gain heat. This means it will not move adiabatically. We consider the deviation from adiabaticity to be small, which means that the thermal adjustment time of the element is large compared to the period of the oscillation; then the temperature excess of the element can be written as

$$DT = \left[\left(\frac{dT}{dr}\right)_{\rm e} - \left(\frac{dT}{dr}\right)_{\rm s}\right]\Delta r$$
$$= -\frac{T}{H_P}(\nabla_{\rm e} - \nabla)\Delta r \,. \tag{6.19}$$

Dynamical stability means that $D\varrho/\Delta r > 0$ and therefore (6.11) is fulfilled. If the layer is chemically homogeneous, then $\nabla_\mu = 0$, and (6.11) becomes $\nabla_{\rm e} - \nabla > 0$, such that (6.19) gives $DT < 0$ for $\Delta r > 0$. Above its equilibrium position the element is cooler than the surroundings and receives energy by radiation. This reduces $\nabla_{\rm e} - \nabla$, $D\varrho$, and the restoring force, such that the element is less accelerated back towards the equilibrium position. The result will be an oscillation with slowly decreasing amplitude. Formally this radiative damping shows up as a small positive imaginary part of ω in (6.17) after the exchange of heat with the surroundings is included in (6.16). The oscillatory part (real part of ω) is still very close to the adiabatic value (6.18).

If the stable layer is inhomogeneous with $\nabla\mu > 0$, it can be that with (6.11) $\nabla_{\rm e} - \nabla > 0$ also (*both* criteria for stability are fulfilled), i.e. we find again that $DT < 0$ for $\Delta r > 0$ and radiative damping as before. However, we can also imagine a situation with $\nabla_{\rm e} - \nabla < 0$ in spite of (6.11) for large enough ∇_μ. Then $DT > 0$ for $\Delta r > 0$ according to (6.19), and the lifted element, being hotter than its surroundings, will now *lose* energy by radiation. This increases $\nabla_{\rm e} - \nabla$, $D\varrho$, and the restoring force, and the element will oscillate with slowly increasing amplitude. This is an *over-stability,* or *vibrational instability.* The difficulties in this strange situation are obvious [it being the above mentioned twilight zone between the two criteria (6.12) and (6.13)]. The growing oscillation may lead to a chemical mixing of elements and surroundings and thus decrease, or eventually even destroy, the stabilizing gradient ∇_μ. But then again, it is not clear whether in such critical situations a local analysis suffices at all. The reaction of other layers of the star might provide enough damping to suppress the over-stability.

With these considerations it follows that we have to distinguish between *dynamical stability* and *vibrational stability*. The first applies to purely adiabatic behaviour of the moving mass, while the second takes heat exchange into account. A layer with a temperature gradient ∇ such that the Ledoux criterion is fulfilled but the Schwarzschild criterion is not, i.e.

$$\nabla_{ad} < \nabla < \nabla_{ad} + \frac{\varphi}{\delta}\nabla_{\mu} , \tag{6.20}$$

is dynamically stable but vibrationally unstable.

A dynamical instability grows on a timescale given by $(H_P/g)^{1/2}$, while in the case of a vibrational instability, the growth of amplitude is governed by the time it takes a mass element to adjust thermally to its surrounding, i.e. by the fraction of the total energy of the moving element lost by radiation per unit time. In the following we shall estimate this timescale τ_{adj}.

6.4 The Thermal Adjustment Time

Let us consider a mass element with $DT > 0$, i.e. one that will radiate into the surroundings. Superposed onto the radial energy flux $\boldsymbol{F}$, carrying energy from the stellar interior to the surface, there will be a local, non-radial flux $\boldsymbol{f}$, carrying the surplus energy of the element to its surroundings. According to (5.9) and (5.10), the absolute value f of the radiative flux from the element due to its excess temperature will be

$$f = \frac{4acT^3}{3\kappa\varrho}\left|\frac{\partial T}{\partial n}\right| , \tag{6.21}$$

where $\partial/\partial n$ indicates the differentiation perpendicular to the surface of the element. Suppose our element to be a roughly spherical "blob" with diameter d. We will approximate the temperature gradient in the normal direction by $\partial T/\partial n \approx 2DT/d$. The radiative loss λ per unit time from the whole surface V of the blob is then

$$\lambda = Sf = \frac{8acT^3}{3\kappa\varrho}DT\frac{S}{d} . \tag{6.22}$$

The quantity λ is a sort of "luminosity" of the blob, and it determines the rate by which the thermal energy of the blob of volume V changes:

$$\varrho V c_P \frac{\partial T}{\partial t} = -\lambda . \tag{6.23}$$

Here we can replace $\partial T/\partial t$ by $\partial(DT)/\partial t$, since the temperature of the (large) surroundings scarcely changes, owing to radiative losses of the blob. Furthermore, let $V/S \approx d/6$ (as for a sphere); then one obtains from (6.22) and (6.23) that

$$\frac{\partial(DT)}{\partial t} = -\frac{DT}{\tau_{\rm adj}}, \tag{6.24}$$

with the timescale for thermal adjustment

$$\tau_{\rm adj} = \frac{\kappa \varrho^2 c_P d^2}{16acT^3} = \frac{\varrho V c_P DT}{\lambda}. \tag{6.25}$$

The second equation follows from a comparison of (6.22)–(6.24). We see that $\tau_{\rm adj}$ is roughly the excess thermal energy divided by the luminosity, i.e. an equivalent to the Kelvin–Helmholtz timescale for a star (3.17). For sufficiently large elements that are far enough from a region of marginal stability, one has $\tau_{\rm adj} \gg 1/\omega_{\rm ad}$, which means that the radiative losses give only a small deviation from adiabatic oscillations, as discussed in Sect. 6.2.

6.5 Secular Instability

Even a small exchange of heat between a displaced mass element and its surroundings can lead to another kind of instability, which is called *thermal* or *secular instability.* We first discuss this qualitatively with an experiment which can easily be carried out with water and kitchen equipment.

In a glass jar containing cold fresh water we carefully pour over a layer of warm salty water. The salt increases the specific weight of the upper layer, but the warmth shall be enough to reduce (despite the salt content) its specific weight to below that of the underlying fresh water. If, owing to a perturbation, a blob of salty water is pushed downwards, buoyancy will push it back, i.e. the two layers are then *dynamically stable.*

But the buoyancy acts as a restoring force only as long as the element stays warm during its excursion into the cold layers. On the timescale by which it loses its excess temperature, the buoyancy diminishes and the element moves downwards because of its salt content. Indeed if one watches the two layers for some time, one can see (especially if the salty water is coloured) that small blobs of salty water slowly sink, a phenomenon called *salt-fingers.* It is an instability controlled by the heat leakage of the element. This is *secular instability.* It can not only occur in glass jars, but also in stars!

Consider a blob of stellar matter situated in surroundings of somewhat different, but homogeneous, composition, i.e. $D\mu \neq 0$, but $\nabla_\mu = 0$ (Such a situation can occur, for example, if two homogeneous layers of different compositions are above each other and a blob from one layer is displaced into the other.). The blob is supposed to be in mechanical equilibrium with its surroundings, i.e. $DP = D\varrho = 0$. This requires, however, a temperature difference according to (6.5):

$$\delta \frac{DT}{T} = \varphi \frac{D\mu}{\mu} . \tag{6.26}$$

For $D\mu > 0$, for example, the blob is hotter and therefore radiates towards the surroundings; the loss of energy under pressure balance ($DP = 0$) leads to an increased density and the blob sinks until again $D\varrho = 0$. Equation (6.26) is still valid and, since $D\mu$, is unchanged, $DT > 0$ as before, and so on. Obviously the blob will slowly sink (or rise for $D\mu < 0$) with a velocity v_μ such that DT always remains constant according to (6.26).

Owing to radiation, the temperature of the blob changes at the rate–DT/τ_{adj} [see (6.24)]. While sinking or rising it changes also because of the adiabatic compression (or expansion) that occurs as a result of the change of pressure, even in the absence of energy exchange. The rate of change of DT can then immediately be written as

$$\frac{1}{T}\frac{\partial}{\partial t}(DT) = \left(\nabla_{ad} \frac{\partial \ln P}{\partial t} - \frac{DT}{T \tau_{adj}} \right) - \nabla \frac{\partial \ln P}{\partial t} . \tag{6.27}$$

The rate of change of pressure is simply linked to the velocity v_μ by

$$\frac{\partial \ln P}{\partial t} = -\frac{v_\mu}{H_P} . \tag{6.28}$$

Using this and (6.26), together with the condition $\partial(DT)/\partial t = 0$ [which follows from (6.26), since $D\mu$ does not vary if the element moves in a chemically homogeneous region], we can solve (6.26)–(6.28) for the velocity and obtain

$$v_\mu = -\frac{H_P}{(\nabla_{ad} - \nabla)\tau_{adj}} \frac{\varphi}{\delta} \frac{D\mu}{\mu} . \tag{6.29}$$

In this case of thermal instability, therefore, the blob sinks ($v_\mu < 0$ for $D\mu > 0$) through a dynamically stable surrounding ($\nabla_{ad} > 0$) with the adjustment timescale for radiative losses.

The idea of blobs finding themselves in strange surroundings ($D\mu > 0$) is not far-fetched. Secular instabilities of the kind discussed here can occur in stars, for example, of about one solar mass. After hydrogen has been transformed to helium in their cores, their central region is cooled by neutrinos, which take away energy without interacting with the stellar matter. The temperature in these stars, therefore, is highest somewhere off-centre and decreases towards the stellar surface as well as towards the centre. If, then, helium "burning" is ignited in the region of maximum temperature, the newly formed carbon is in a shell surrounding the central core (Sects. 33.4 and 33.5). This carbon-enriched shell has a higher molecular weight than the regions below: carbon "fingers" will grow and sink inwards. In later evolutionary phases, other nuclear reactions, such as neon burning, may ignite off-centre, and heavier fingers of material may sink.

6.6 The Stability of the Piston Model

Our piston model (Sects. 2.7 and 5.4) shows a stability behaviour in many respects similar to that of the blobs.

We start with the two equations that together with the equation of state describe the time dependence of the piston model. These are (2.34) and (5.39), where we assume for the sake of simplicity that $\varepsilon = \kappa = 0$. The equilibrium state is given by $T = T_s$ and $G^* = PA$.

In order to investigate the stability we denote the equilibrium values by the subscript "0" and make small perturbations of the form

$$\begin{aligned} h(t) &= h_0(1 + x\mathrm{e}^{\mathrm{i}\omega t}) \\ P(t) &= P_0(1 + p\mathrm{e}^{\mathrm{i}\omega t}) \\ T(t) &= T_0(1 + \vartheta\mathrm{e}^{\mathrm{i}\omega t}) \end{aligned} \tag{6.30}$$

with $|x|, |p|, |\vartheta| \ll 1$. We therefore neglect quadratic and higher-order expressions in these quantities.

From mass conservation ϱh = constant and from the perfect gas equation $P \sim \varrho T$, we obtain

$$p = \vartheta - x \,. \tag{6.31}$$

We now introduce (6.30) into (2.34) and obtain after linearization and using $G^* = PA$

$$M^* h_0 \omega^2 x + P_0 A p = 0 \,, \tag{6.32}$$

which with $g = P_0 A / M^*$ and with (6.31) can be replaced by

$$\left(\frac{\omega^2 h_0}{g} - 1\right) x + \vartheta = 0 \,, \tag{6.33}$$

while the corresponding perturbation and linearization of (5.39) gives

$$\mathrm{i}\omega P_0 A h_0 x + (\mathrm{i}\omega c_v m^* T_0 + \chi T_0)\vartheta = 0 \,. \tag{6.34}$$

The two linear homogeneous equations (6.33) and (6.34) for x and ϑ can be solved if the determinant vanishes. This condition gives an algebraic equation of third order for the eigenvalue ω.

The problem becomes simple if we assume that the trapped gas changes adiabatically, i.e. if $\chi = 0$. Then (6.34), with $m^*/(Ah_0) = \varrho_0$ and with the perfect gas equation, yields

$$\frac{\Re}{\mu c_v} x + \vartheta = 0 , \tag{6.35}$$

and with $\Re/\mu = c_P - c_v$ (4.33) and $c_P/c_v = \gamma_{ad}$ (4.37) it follows that

$$(\gamma_{ad} - 1)x + \vartheta = 0 . \tag{6.36}$$

Setting the determinant of the equations (6.33) and (6.36) to zero gives the eigenvalue for the adiabatic motion:

$$\omega = \pm\omega_{ad}, \qquad \omega_{ad} = (\gamma_{ad} g/h_0)^{1/2} . \tag{6.37}$$

Since ω is real, the adiabatic motion is an oscillation with frequency ω and constant amplitude. Therefore in the language of Sect. 6.1 our perfect gas piston model is *dynamically stable.* Note that $1/\omega_{ad}$ is of the order of the hydrostatic timescale τ_{hydr} defined in Sect. 2.7.

How do non-adiabatic effects change the picture? With the χ term in (6.34) we have, instead of (6.36),

$$(\gamma_{ad} - 1)x + \left(1 + \frac{a}{\mathrm{i}\omega}\right)\vartheta = 0 , \tag{6.38}$$

with $a = \chi/(c_v m^*)$. Setting the determinant of (6.33) and (6.38) equal to zero now gives a *cubic equation* in ω. In general ω will be complex.

We assume χ to be small, so that the oscillation frequency must be close to the adiabatic value and we can put $\omega = \omega_{ad} + \xi$, with $|\xi| \ll |\omega_{ad}|$. If we neglect higher terms in ξ and χ, we find from the vanishing determinant of the system of homogeneous linear equations (6.33) and (6.38) and after some algebraic manipulation that

$$\mathrm{i}\xi = -\frac{\gamma_{ad} - 1}{2\gamma_{ad}} \frac{\chi}{c_v m^*} = -\frac{\gamma_{ad} - 1}{2\gamma_{ad}} \frac{1}{\tau_{adj}} < 0 , \tag{6.39}$$

where we have used (5.41). The (almost adiabatic) oscillation is therefore damped since the exponents of (6.30), $\mathrm{i}\omega = \mathrm{i}\omega_{ad} + \mathrm{i}\xi$, have a negative real part that decreases the amplitude on a timescale τ_{adj}. The piston model with a leak is *vibrationally stable.*

The cubic equation for ω must have a third root, which we find easily by assuming that it describes an evolution so slow that the inertia term in (2.34) can be neglected (This has to be checked later.). Then (6.33) has to be replaced by

$$\vartheta - x = 0 , \tag{6.40}$$

which according to (6.31) is equivalent to $p = 0$. Indeed if the evolution is so slow that there is always hydrostatic equilibrium, the pressure is given by the (constant) weight of the piston. We then have from (6.34) and (6.40)

$$i\omega = -\frac{\chi T}{P_0 A h_0 + c_v m^* T_0} = -\frac{\chi}{c_P m^*} = -\frac{1}{\gamma_{\mathrm{ad}} \tau_{\mathrm{adj}}} . \tag{6.41}$$

For the latter equation we have used the relation $P_0 A h_0 = \Re m^* T_0 / \mu$ and (5.41). The third root gives an exponential decay in time of the initial perturbation, the timescale being comparable with τ_{adj}. If χ is sufficiently small and the evolution slow, the assumption that the inertia term is negligible is justified.

Our result (6.41) means that any deviation from thermal equilibrium ($T - T_s \neq 0$) vanishes within the thermal adjustment time, i.e. the thermally adjusted piston model for $\varepsilon = \kappa = 0$ is *secularly stable.* We see that it shows the same limiting cases for the stability problem (dynamical, vibrational, and secular stability) as the blobs. In Sect. 41.1 we will consider the influence on the stability of the piston model of the (here neglected) terms in (5.39) due to ε and κ.

To summarize: if the trapped gas is changing adiabatically, the piston model is dynamically stable. If there is a leak, the oscillations are damped and the gas vibrationally stable. If the thermal evolution is so slow that hydrostatic equilibrium is always achieved, it is secularly stable, if κ and ϵ are zero.

Chapter 7
Transport of Energy by Convection

Convective transport of energy means an exchange of energy between hotter and cooler layers in a dynamically unstable region through the exchange of macroscopic mass elements ("blobs", "bubbles", "convective elements"), the hotter of which move upwards while the cooler ones descend. The moving mass elements will finally dissolve in their new surroundings and thereby deliver their excess (or deficiency) of heat. Owing to the high density in stellar interiors, convective transport can be very efficient. However, this energy transfer can operate only if it finds a sufficient driving mechanism in the form of the buoyancy forces.

A thorough theoretical treatment of convective motions and transport of energy is extremely difficult. It is the prototype of the many astrophysical problems in which the bottleneck preventing decisive progress is the difficulty involved in solving the well-known hydrodynamic equations. For simplifying assumptions, solutions are available that may even give reasonable approximations for certain convective flows in the laboratory (or in the kitchen). Unfortunately, convection in stars proceeds under rather malicious conditions: turbulent motion transports enormous fluxes of energy in a very compressible gas, which is stratified in density, pressure, temperature, and gravity over many powers of ten. Nevertheless, large efforts have been made over many years to solve this notorious problem, and they have partly arrived at promising results. Canuto (2008) summarizes the state of the art of models for the underlying Navier-Stokes equations, which in the field of oceanography and atmospheric sciences have had great success, and which aim at modelling the fluctuations around an average state. None of these so-called *Reynolds stress models*, however, has reached a stage where it could provide a procedure easy enough to be handled in everyday stellar-structure calculations, and at the same time would describe the full properties of convection accurately enough. On the other hand, full two- and three-dimensional hydrodynamical simulations have also made large progress, thanks to the impressive advances in supercomputer technology and efficient numerical algorithms (see the review by Kupka 2008). They give valuable hints to the true nature of convection and often serve as numerical experiments to test the dynamical methods. Nevertheless, these numerical simulations are still limited in their size and thus can follow convection in most cases only for a limit

R. Kippenhahn et al., *Stellar Structure and Evolution*, Astronomy and Astrophysics Library, DOI 10.1007/978-3-642-30304-3_7, © Springer-Verlag Berlin Heidelberg 2012

time and only for thin convection zones. But even if these restrictions can be foreseen to get relaxed with time, such full hydrodynamical simulations will never be used in full stellar evolution models, as they would unnecessarily follow the star's evolution on a dynamical timescale, which is so much shorter than the dominant nuclear one. Therefore, we limit ourselves exclusively to the description of the old so-called "mixing-length" theory. The reason for this is not that we believe it to be sufficient, but it does provide at least a simple method for treating convection locally, at any given point of a star. Moreover, empirical tests of the resulting stellar models show a surprisingly good agreement with observations. And, finally, even this poor approximation shows without any doubt that in the very deep interior of a star, a detailed theory is normally not necessary.

Note that in the following we are dealing only with convection in stars that are in hydrostatic equilibrium. We furthermore assume that the convection is time independent, which means that it is fully adjusted to the present state of the star. Otherwise, a convection theory for rapidly changing regions (time-dependent convection) has to be developed.

Equation (5.28) gives the gradient ∇_{rad} that would be maintained in a star if the whole luminosity l had to be transported outwards by radiation only. If convection contributes to the energy transport, the actual gradient ∇ will be different (namely smaller). It is the purpose of this section to estimate ∇ in the case of convection.

7.1 The Basic Picture

The mixing-length theory goes back to Ludwig Prandtl, who in 1925 modelled a simple picture of convection in complete analogy to molecular heat transfer: the transporting "particles" are macroscopic mass elements ("blobs") instead of molecules; their mean free path is the "mixing length" after which the blobs dissolve in their new surroundings. Prandtl's theory was adapted for stars by L. Biermann. There exist different variations and formulations of the mixing-length theory in the literature. Two widely used versions are those by Böhm-Vitense (1958) and Cox (see Weiss et al. 2004). We follow here the former one.

The total energy flux $l/4\pi r^2$ at a given point in the star consists of the radiative flux F_{rad} (in which the conductive flux may already be incorporated) plus the convective flux F_{con}. Their sum defines according to (5.28) the gradient ∇_{rad} that would be necessary to transport the whole flux by radiation:

$$F_{rad} + F_{con} = \frac{4acG}{3} \frac{T^4 m}{\kappa P r^2} \nabla_{rad}. \tag{7.1}$$

However, part of the flux is transported by convection. If the actual gradient of the stratification is ∇, then the radiative flux is obviously only

$$F_{rad} = \frac{4acG}{3} \frac{T^4 m}{\kappa P r^2} \nabla. \tag{7.2}$$

Note that ∇ is not yet known; in fact, we hope to obtain it as the result of this consideration. The first step is to derive an expression for F_{con}.

Consider a convective element (a blob) with an excess temperature DT over its surroundings. It moves radially with velocity v and remains in complete balance of pressure, that is, $DP = 0$ [see (6.2) and Fig. 6.1]. This gives a local flux of convective energy

$$F_{\mathrm{con}} = \varrho v c_P DT, \tag{7.3}$$

which we can take immediately as the correct equation for the average convective flux, if we consider vDT replaced by the proper mean over the whole concentric sphere. One should be aware that this "proper mean" comprises most of the difficulties for a strict treatment. We adopt the following simple model.

All elements may have started their motion as very small perturbations only, that is, with initial values that can be approximated by $DT_0 = 0$ and $v_0 = 0$. Because of differences in temperature gradients and buoyancy forces, DT and v increase as the element rises (or sinks) until, after moving over a distance ℓ_{m}, the element mixes with the surroundings and loses its identity. ℓ_{m} is called the *mixing length.* The elements passing at a given moment through a sphere of constant r will have different values of v and DT since they have started their motion at quite different distances, from zero to ℓ_{m}. We assume, therefore, that the "average" element has moved $\ell_{\mathrm{m}}/2$ when passing through the sphere. Then,

$$\begin{aligned}\frac{DT}{T} &= \frac{1}{T}\frac{\partial(DT)}{\partial r}\frac{\ell_{\mathrm{m}}}{2} \\ &= (\nabla - \nabla_{\mathrm{e}})\frac{\ell_{\mathrm{m}}}{2}\frac{1}{H_P}.\end{aligned} \tag{7.4}$$

The density difference [for $DP = D\mu = 0$, see (6.3) and (6.5)] is simply $D\varrho/\varrho = -\delta DT/T$ and the (radial) buoyancy force (per unit mass), $k_r = -g \cdot D\varrho/\varrho$. On average, half of this value may have acted on the element over the whole of its preceding motion ($\ell_{\mathrm{m}}/2$), such that the work done is

$$\frac{1}{2}k_r\frac{\ell_{\mathrm{m}}}{2} = g\delta(\nabla - \nabla_{\mathrm{e}})\frac{\ell_{\mathrm{m}}^2}{8H_P}. \tag{7.5}$$

Let us suppose that half of this work goes into the kinetic energy of the element ($v^2/2$ per unit mass), while the other half is transferred to the surroundings, which have to be "pushed aside". Then, we have for the average velocity v of the elements passing our sphere

$$v^2 = g\delta(\nabla - \nabla_{\mathrm{e}})\frac{\ell_{\mathrm{m}}^2}{8H_P}. \tag{7.6}$$

Inserting this and (7.4) into (7.3), we obtain for the average convective flux

$$F_{\rm con} = \varrho c_P T \sqrt{g\delta} \frac{\ell_{\rm m}^2}{4\sqrt{2}} H_P^{-3/2} (\nabla - \nabla_{\rm e})^{3/2} . \tag{7.7}$$

Finally, we shall consider the change of temperature $T_{\rm e}$ inside the element (diameter d, surface S, volume V) when it moves with velocity v. This change has two causes, one being the adiabatic expansion (or compression), and the other being the radiative exchange of energy with the surroundings. The total energy loss λ per unit time is given by (6.22); the corresponding temperature decrease per unit length over which the element rises is $\lambda/\varrho V c_P v$, and the total change per unit length is then

$$\left(\frac{dT}{dr}\right)_{\rm e} = \left(\frac{dT}{dr}\right)_{\rm ad} - \frac{\lambda}{\varrho V c_P v} . \tag{7.8}$$

Multiplying this by H_P/T, we have

$$\nabla_{\rm e} - \nabla_{\rm ad} = \frac{\lambda H_P}{\varrho V c_P v T} . \tag{7.9}$$

Here, λ may be replaced by (6.22), with the average DT given by (7.4). The resulting equation then contains a "form factor" $\ell_{\rm m} S/Vd$, which would be $6/\ell_{\rm m}$ for a sphere of diameter $\ell_{\rm m}$. In the literature, one often finds

$$\frac{\ell_{\rm m} S}{Vd} \approx \frac{9/2}{\ell_{\rm m}} , \tag{7.10}$$

which we will use in the following.

Equation (7.9), with the help of (6.22) and (7.10), then becomes

$$\frac{\nabla_{\rm e} - \nabla_{\rm ad}}{\nabla - \nabla_{\rm e}} = \frac{6acT^3}{\kappa \varrho^2 c_P \ell_{\rm m} v} . \tag{7.11}$$

Let us now summarize what we have achieved and describe what is still lacking. To start with the latter, we have obviously not yet used any physics that could *determine* the mixing length $\ell_{\rm m}$. Since we do not know a reasonable approach for this, we shall simply treat $\ell_{\rm m}$ as a free parameter and make (more or less) plausible assumptions for its value (This is typical for all versions of the mixing-length approach and in fact also for many others that seem to be less arbitrary at a first glance.). In any case, the heat transfer mainly operates via the largest possible elements, and they can scarcely move over much more than their own diameter before differential forces destroy their identity.

Now, however, the prospect looks quite favourable: we have obtained the five equations (7.1), (7.2), (7.6), (7.7) and (7.11), which we can solve for the five quantities $F_{\rm rad}$, $F_{\rm con}$, v, $\nabla_{\rm e}$, and ∇, if the usual local quantities ($P, T, \varrho, l, m, c_P, \nabla_{\rm ad}, \nabla_{\rm rad}$, and g) are given.

7.2 Dimensionless Equations

For a simpler treatment of the five equations obtained from the mixing-length theory, we define two dimensionless quantities:

$$U := \frac{3acT^3}{c_P \varrho^2 \kappa \ell_m^2} \sqrt{\frac{8H_P}{g\delta}} , \tag{7.12}$$

$$W := \nabla_{rad} - \nabla_{ad} . \tag{7.13}$$

The meaning of U will become clear later; that of W is obvious. Note that both can be calculated immediately for any point in the star when the usual variables and the mixing length ℓ_m are given.

If v is eliminated with the help of (7.6), then (7.11) becomes

$$\nabla_e - \nabla_{ad} = 2U\sqrt{\nabla - \nabla_e}. \tag{7.14}$$

Eliminating F_{rad}, F_{con} from (7.1), (7.2) and (7.7) and using (2.4) and (6.8), we arrive at

$$(\nabla - \nabla_e)^{3/2} = \frac{8}{9} U(\nabla_{rad} - \nabla) . \tag{7.15}$$

We have thus replaced the set of five equations by the two equations (7.14) and (7.15) for ∇ and ∇_e, and we will now even reduce them to one final equation.

Rewriting the left-hand side of (7.14) as $(\nabla - \nabla_{ad}) - (\nabla - \nabla_e)$, one sees immediately that this is a quadratic equation for $(\nabla - \nabla_e)^{1/2}$ with the solution

$$\sqrt{\nabla - \nabla_e} = -U + \xi , \tag{7.16}$$

where ξ is a new variable given by the positive root of

$$\xi^2 = \nabla - \nabla_{ad} + U^2 . \tag{7.17}$$

In (7.15), we insert (7.16) on the left-hand side, eliminate ∇ on the right-hand side with (7.17), and obtain

$$(\xi - U)^3 + \frac{8U}{9}(\xi^2 - U^2 - W) = 0 . \tag{7.18}$$

So we have arrived at a cubic equation for ξ that can be solved for any given set of parameters U and W. It turns out that (7.18) has only one real solution. The resulting ξ, together with (7.17), then gives the decisive quantity ∇, that is, the average temperature gradient to which the layer settles in the presence of convection.

Other characteristic quantities of the convection are then also easily calculable, for example, the velocity v from (7.6) and (7.14).

We note for completeness that the cubic equation (7.18) should be solved numerically and not by the analytical formulae for the solution of third order equations, because the individual terms appearing therein can be many magnitudes larger than the root of the formula.

7.3 Limiting Cases, Solutions, Discussion

For a given difference $W = \nabla_{\rm rad} - \nabla_{\rm ad}$, the convection depends decisively on the value of U. Let us write (7.2) as $F_{\rm rad} = \sigma_{\rm rad}\nabla$, and (7.7) as $F_{\rm con} = \sigma_{\rm con}(\nabla - \nabla_{\rm e})^{3/2}$. Then, U, defined in (7.12), is essentially the ratio of the "conductivities": $\sigma_{\rm rad}/\sigma_{\rm con}$.

The dimensionless quantity U can also be written in terms of the time $\tau_{\rm ff}$ it takes a mass element to fall freely over the distance H_P. With $\tau_{\rm ff} = (2H_P/g)^{1/2}$ and (6.25), we have

$$U \approx \frac{\tau_{\rm ff}}{\tau_{\rm adj}} \frac{d^2}{\ell_{\rm m}^2} , \tag{7.19}$$

where we have ignored a factor $3/(8\delta^{1/2})$, which is of order 1. One normally assumes that $\ell_{\rm m} \approx d$, and therefore, $U \approx \tau_{\rm ff}/\tau_{\rm adj}$.

The quantity U is also related to another dimensionless quantity Γ defined by

$$\Gamma := \frac{(\nabla - \nabla_{\rm e})^{1/2}}{2U} = \frac{\nabla - \nabla_{\rm e}}{\nabla_{\rm e} - \nabla_{\rm ad}} , \tag{7.20}$$

where we have made use of (7.14). Numerator and denominator have simple meanings as can easily be shown. For a roughly spherical convective element of radius $\ell_{\rm m}/2$, cross-section A, volume V, lifetime $\tau_l = \ell_{\rm m}/v$, and thermal energy $e_{\rm th} = \varrho V c_P T$, one finds from (7.3) and (7.4) that

$$\nabla - \nabla_{\rm e} = \frac{(F_{\rm con}A)\tau_l}{e_{\rm th}} \frac{4H_P}{3\ell_{\rm m}} \tag{7.21}$$

and from (7.9) that

$$\nabla_{\rm e} - \nabla_{\rm ad} = \frac{\lambda\tau_l}{e_{\rm th}} \frac{H_P}{\ell_{\rm m}} , \tag{7.22}$$

and therefore,

$$\Gamma = \frac{4}{3} \frac{F_{\rm con}A}{\lambda} \approx \frac{\text{energy transported}}{\text{energy lost}} . \tag{7.23}$$

For an average element, Γ gives the convective energy flowing through A relative to the radiative energy loss per second. It is a measure for the *efficiency of convection*. Large values of Γ (small U) are typical for very dense matter, where radiation losses are relatively unimportant compared to the convective flux. In regions of small density, however, the radiative losses can be so large that even very violent movements are ineffective for energy transport; the elements then lose nearly all of their excess heat through radiation to the surroundings, and cool down to $DT \approx 0$. In this case, Γ is very small (i.e. U is very large). The meaning of Γ can also be represented in terms of two typical timescales for the elements, namely, lifetime and adjustment time: in the second equation (6.25), replace DT by (7.4) and solve for $\nabla - \nabla_e$. This expression is then divided by (7.22) giving

$$\Gamma = \frac{\nabla - \nabla_e}{\nabla_e - \nabla_{ad}} = 2\frac{\tau_{adj}}{\tau_l} . \tag{7.24}$$

Let us consider the limiting cases for very large and very small U (or Γ). One should keep in mind that all gradients are finite; except for ∇_{rad}, they are all smaller than unity. And for the discussion in terms of Γ, one can easily rewrite (7.14) and (7.15) with the help of (7.20).

$U \to 0$ (or $\Gamma \to \infty$): Equation (7.14) gives $\nabla_e \to \nabla_{ad}$, and thus, (7.15) yields $\nabla \to \nabla_{ad}$. A negligible excess of ∇ over the adiabatic value is sufficient to transport the whole luminosity. This is the case in the very dense central part of a star. Here, we do not need to solve the mixing-length equations ($\nabla = \nabla_{ad}$ is known), and the uncertainties of this theory do not arise.

$U \to \infty$ (or $\Gamma \to 0$): In (7.15), the gradients on the left-hand side must be finite, and therefore on the right-hand side, $\nabla \to \nabla_{rad}$. Convection is ineffective and cannot transport a substantial fraction of the luminosity. Therefore, $F \to F_{rad}$, and the gradient ∇ is again known without further calculations. This is the case near the photosphere of a star.

The situation is difficult where the two limiting cases do not apply, for example, in the upper part of an outer convective envelope. There the equations of the mixing-length theory have to be solved, and they will yield a value for ∇ somewhere between ∇_{ad} and ∇_{rad}, the convection being said to be *superadiabatic*.

The following gives a more detailed discussion of the solutions of (7.18), which depend strongly on the (given) parameters U and W. We illustrate them in a diagram, where lg W is plotted over lg U (Fig. 7.1).

Instead of using the variable ξ, the solutions may be discussed in terms of the over-adiabaticity

$$x := \nabla - \nabla_{ad} = \xi^2 - U^2 , \tag{7.25}$$

which describes the gradient ∇ of the stratification relative to the (known) adiabatic gradient. With this definition, the cubic equation (7.18) is transformed to

$$\left[\sqrt{x + U^2} - U\right]^3 + \frac{8}{9}U(x - W) = 0 . \tag{7.26}$$

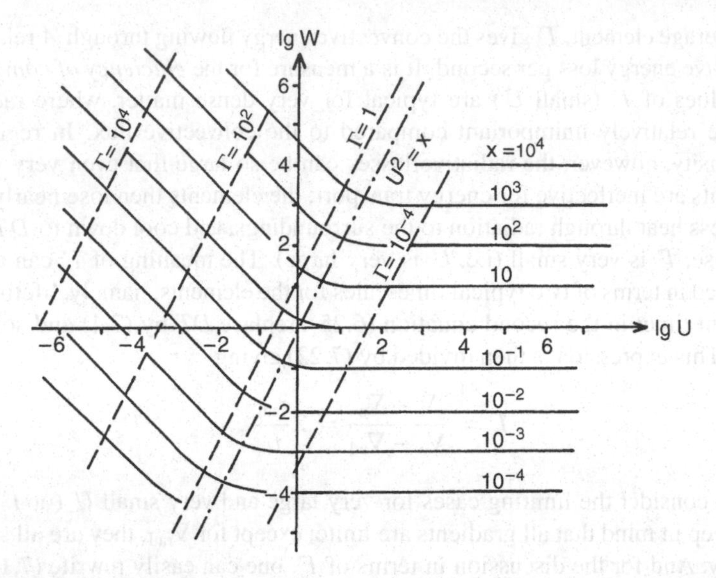

Fig. 7.1 The plane of the parameters U, W (on logarithmic scales) that determine the convection. The lines $x = \nabla - \nabla_{\text{ad}}$ = constant are *solid*; the line where $U^2 = x$ is *dot-dashed*. Some lines Γ = constant are *dashed*

1. $\underline{\Gamma = 1}$: Let us first derive the line which separates the regimes of effective convection (at small U) and ineffective convection (at large U). Equation (7.20) for $\Gamma = 1$ is introduced into (7.16), which gives $\xi = 3U$ such that from (7.25), we have $x = 8U^2$. Inserting this into (7.26), we find the condition for $\Gamma = 1$ to be
$$W = 17\,U^2 . \tag{7.27}$$
The corresponding straight line lg W = 2 lg U + 1.23 is shown by dashes in Fig. 7.1 (Lines for other values of Γ are obtained by a parallel shift.). We will now derive the lines on which x is constant. This is easily done by considering the following two limiting cases.
2. $\underline{U^2 \gg x}$: In (7.26), the term in square brackets on the left, divided by U, goes to zero, and one has
$$x = W . \tag{7.28}$$
Therefore, x = constant on straight lines parallel to the abscissa (right part of Fig. 7.1).
3. $\underline{U^2 \ll x}$: In (7.26), the term in square brackets goes to $x^{3/2} \gg Ux$, such that
$$x^{3/2} = \frac{8}{9}UW \tag{7.29}$$
and x = constant on the lines lg W = − lg U + lg(9/8) + (3/2) lg x (left part of Fig. 7.1).

Finally, we derive the equation for the border between the regimes $U^2 \gg x$ and $U^2 \ll x$.

4. $\underline{U^2 = x}$: With this condition, (7.26) gives

$$W = U^2 \left[\frac{9}{8}(\sqrt{2} - 1)^3 + 1 \right] . \tag{7.30}$$

The corresponding straight line $\lg W = 2 \lg U + 0.033$ (dot-dashed line in Fig. 7.1) is below and parallel to that for $\Gamma = 1$.

The meaning of the different regions in Fig. 7.1 is now quite clear. Below and left of a line of sufficiently small x (say, $x = 10^{-2}$), we have nearly $\nabla = \nabla_{ad}$; above that line, the convection is superadiabatic. Not too far to the right of the line $\Gamma = 1$, the efficiency is so small that $\nabla \approx \nabla_{rad}$.

For an estimate for the interior of a star, let us assume a perfect monatomic gas with $\delta = \mu = 1, c_P/\Re = 5/2$ and a mixing-length $\ell_m = H_P$. For an average point in a star like the Sun, we may take $r = R_\odot/2, m = M_\odot/2, T = 10^7$ K, $\kappa = 1\,\mathrm{cm^2\,g^{-1}}$ and $\varrho = 1\,\mathrm{g\,cm^{-3}}$. Then, we obtain $U \approx 10^{-8}$, which is so far to the left in Fig. 7.1 that, for reasonable values of $W = \nabla_{rad} - \nabla_{ad}$ (say between 1 and 10^2), $\nabla - \nabla_{ad} \approx 10^{-5} \ldots 10^{-4}$. For the central region of the Sun, ϱ and κ are larger by factors of 10^2 and 10, respectively. Then, $U \approx 10^{-13}$, and (for the same values of W) the difference $\nabla - \nabla_{ad}$ is even smaller by a factor 10^3 or more, that is, $< 10^{-7}$. The stratification of such convective zones is indeed very close to an adiabatic one, and we can simply set $\nabla = \nabla_{ad}$, independent of the uncertainties of the theory (The situation is difficult only near the interface between convective and radiative zones, where one should have a smooth transition between the two modes of transport.).

Convective elements in such dense layers are so effective ($\Gamma \approx 10^6 \ldots 10^9$) that they can transport the whole luminosity with surprisingly little effort. Compared with the surroundings, they only need very small excesses of the T gradient, $D(dT/dr) \approx 10^{-12} \ldots 10^{-10}\,\mathrm{K\,cm^{-1}}$, and an average temperature excess $DT \approx 10^{-2} \ldots 1$ K; their velocities are typically $v \approx 1 \ldots 100\,\mathrm{m\,s^{-1}}$ (which is $10^{-6} \ldots 10^{-4}$ times the velocity of sound), and their lifetime is between 1 and 10^2 days.

The Reynolds number decides whether the flow of an incompressible viscous fluid is turbulent or laminar (Landau, Lifshitz, vol. 6, 1987). It is defined as

$$Re = \frac{v \varrho \ell_m}{\eta} . \tag{7.31}$$

Here, η is the viscosity of the fluid and ℓ_m and v are the typical distance elements travel and their velocity. For high Reynolds numbers, the flow is turbulent. In spite of the small velocities of convective elements, the Reynolds number is $\gg 1$, since the flow extends over such a large distance ℓ_m. The situation is quite different for convection near the surface of the star, where the density is low. This gives small effectivity and positive lg U. Here, the cubic equation for ξ (or x) has to be solved for each point to find the proper ∇ for that place, and the results are affected by the uncertainties of the theory.

In any case, we use the resulting value of ∇ in the transport equation written in the form

$$\frac{dT}{dm} = -\frac{T}{P}\frac{Gm}{4\pi r^4}\nabla\,. \tag{7.32}$$

(Here, we have replaced dP/dm by the right-hand side of the hydrostatic equation since the theory is suitable only for hydrostatic equilibrium.) For convection in the very deep interior, $\nabla = \nabla_{ad}$, where ∇_{ad} is given by (4.21), while for envelope convection, we take ∇ as given by the solution of the mixing-length theory. And we can even take the same equation (7.32) for transport by radiation, if we set $\nabla = \nabla_{rad}$ (compare Sect. 5.2).

Aside from the more or less effective (and more or less well-determined) transport of energy, turbulent convection, if it occurs, has a side effect that is important for the life of the star: it mixes the stellar matter very thoroughly and rapidly compared to other relevant timescales, and thus, it contributes directly to the long-lasting chemical record of the star's history.

7.4 Extensions of the Mixing-Length Theory

The mixing-length theory, as described above, has many open and hidden assumptions. Most prominent is the mixing length itself, usually expressed as the "mixing length parameter" α_{MLT}, which is the mixing-length in units of the pressure scale height H_P. It is generally assumed that α_{MLT} is both constant within a star and does vary neither with stellar mass, composition, nor with evolutionary stage. Its value is not known better than that from general physical arguments, it should be of order 1. To determine a reasonable numerical value, a comparison of the effective temperature or radius of stellar models with observed stars is done, preferentially in the case of the Sun. This yields values for α_{MLT} between 1.5 and 2.0. Ludwig et al. (1999) have done a comparison with numerical simulations of convection and found only a weak dependence of the order of 20 % on stellar parameters for stars of solar metallicity.

But there are even more hidden parameters and assumptions. The theory contains, for example, several mean values entering the equations from (7.4) on. For (7.10), we assumed a certain geometrical form of the blobs to obtain the ratio of surface to volume. Different formulations of the mixing-length theory may make different assumptions for all this. Therefore, the mixing-length parameter may not be directly comparable between such different formulations.

The most basic limitation in all these variants of the theory is the assumption of one single size (and form) for the convective elements. The theory of turbulence, numerical simulations, laboratory experiments and astrophysical observations all show that this is certainly not the case. Instead, numerical simulations and helioseismology showed that convection often operates by extended funnel-like downdrafts and turbulent updrafts. Convective energy is thus realistically not transported in laminar flows of blobs of identical size and energy content, but by turbulent elements ("eddies") of all sizes. F_{con} can therefore not be calculated as in (7.3) but rather results from an integration over the full spectrum of convective eddies.

Canuto and Mazzitelli (1991) developed and presented an extension of the mixing-length theory, in which the full turbulent kinetic energy spectrum is taken into account. This "full spectrum turbulence" theory (FST) can be formulated in much the same way as the mixing-length theory, which is the limiting case for a δ-function like energy spectrum. It also results in a cubic equation to be solved. Canuto and Mazzitelli use the formulation of Cox and Giuli (Weiss et al. 2004, Chap. 14, and eq. 14.82); in this formulation, the cubic equation reads

$$\frac{9}{4}\Gamma'^3 + \Gamma'^2 + \Gamma' - \frac{1}{U^2}(\nabla_{\text{rad}} - \nabla_{\text{ad}}) = 0, \tag{7.33}$$

where we have replaced already some terms by quantities of our own formulation. Γ' is defined as

$$2\Gamma' + 1 = \left(1 + \frac{\nabla - \nabla_{\text{ad}}}{U^2}\right)^{1/2} \tag{7.34}$$

and corresponds to the convective efficiency.

After modelling the convective flux in the FST model, (7.33) is modified by multiplying the Γ'^3-term with a function $\Omega(\Gamma')$, which is the new turbulent convective flux relative to that of the mixing-length theory. Ω rises monotonically from 0 to 1 for Γ' going from $\Gamma' \approx 0$ to $\Gamma' \to \infty$ and can be approximated by an analytical fitting function.

As a rule of thumb convective fluxes in this theory are larger than predicted by the mixing-length theory in case of efficient convection, and superadiabatic regions are narrower but more superadiabatic. The temperature gradient in the solar convection zone predicted by the FST model agrees much better with that obtained in numerical simulations than it does in the mixing-length case. The numerical value for the mixing length parameter in this case is around 0.7.

Chapter 8
The Chemical Composition

8.1 Relative Mass Abundances

The chemical composition of stellar matter is obviously very important, since it directly influences such basic properties as absorption of radiation or generation of energy by nuclear reactions. These reactions in turn alter the chemical composition, which represents a long-lasting record of the nuclear history of the star.

The composition of stellar matter is extremely simple compared to that of terrestrial bodies. Because of the high temperatures and pressures, there are no chemical compounds in the stellar interior, and the atoms are for the most part completely ionized. It suffices then to count and keep track of the different types of nuclei.

We denote by X_i that fraction of a unit mass which consists of nuclei of type i. This requires that

$$\sum_i X_i = 1 . \tag{8.1}$$

The chemical composition of a star at time t is then described, if for the relevant nuclei the functions $X_i = X_i(m,t)$ are given in the interval $[0, M]$ of m.

The commonly used particle number per volume, n_i, of nuclei with mass m_i, is related to the mass abundances by

$$X_i = \frac{m_i n_i}{\varrho} . \tag{8.2}$$

Usually, one does not need to specify very many X_i because most elements are either too rare or play no relevant role, or their abundances remain constant in time. In fact, for many purposes, it is even sufficient to specify only the mass fractions of hydrogen, helium, and "the rest" with the notation

$$X \equiv X_{\mathrm{H}}, \quad Y \equiv X_{\mathrm{He}}, \quad Z \equiv 1 - X - Y . \tag{8.3}$$

R. Kippenhahn et al., *Stellar Structure and Evolution*, Astronomy and Astrophysics Library, DOI 10.1007/978-3-642-30304-3_8, © Springer-Verlag Berlin Heidelberg 2012

This requires additional conventions about the relative distribution of the elements in Z, collectively called "metals", in particular the amount of C, N, and O, which are important for hydrogen burning.

Young stars throughout, and most stars in their envelopes, contain an overwhelming amount of hydrogen and helium: $X = 0.65\dots0.75$, $Y = 0.30\dots0.25$, $Z = 0.05\dots0.0001$.

Of course, nuclear reactions will eventually change this simple picture drastically. For example, if many competing reactions occur simultaneously, or if one is interested in such aspects as isotopic ratios, one may have to specify a large number of different X_i. Only if inverse β decay, the big equalizer in late stages of evolution, has destroyed all elements does the composition then return to utmost simplicity–just neutrons (Chap. 38).

The advantages of the use of m instead of r as independent variable become particularly evident when we have to describe the chemical composition. If we took $X_i(r,t)$ instead, any expansion would immediately lead to a change of all the functions X_i, this holds, of course, for all functions depending on the chemical composition.

8.2 Variation of Composition with Time

8.2.1 Radiative Regions

In radiative regions, there is no exchange of matter between different mass shells, if we can neglect diffusion. Then, the X_i can change only if nuclear reactions create or destroy nuclei of type i in the mass element under consideration.

The frequency of a certain reaction is described by the *reaction rate* r_{lm}, that is, the number of reactions per unit volume and time that transform nuclei from type l into type m (see Chap. 18). The reaction itself will in most cases involve more than just one mother and one daughter nucleus, but for simplicity, we characterize it by one index only. In general, an element i can be affected simultaneously by many reactions, some of which create it (r_{ji}) and some of which destroy it (r_{ik}). These reaction rates give directly the change per second of n_i. Then, with (8.2), we have

$$\frac{\partial X_i}{\partial t} = \frac{m_i}{\varrho}\left[\sum_j r_{ji} - \sum_k r_{ik}\right], \quad i = 1\dots I \tag{8.4}$$

for any of the elements $1\cdots I$ which are involved in reactions (If more than one nucleus of type i is created or destroyed per reaction, the corresponding terms in the sums have simply to be normalized by the number of nuclei of type i involved.).

The reaction $p \rightarrow q$ in which one nucleus of type p is transformed may be connected with a release of energy e_{pq}. In the equation of energy conservation, we have used the energy generation rate ε per unit mass, which normally contains contributions from several different reactions. The ε are simply proportional to the reaction rates:

$$\varepsilon = \sum_{p,q} \varepsilon_{pq} = \frac{1}{\varrho} \sum_{p,q} r_{pq}\, e_{pq} . \tag{8.5}$$

Let us introduce the energy generated when one mass unit of type p nuclei is transformed into type q:

$$q_{pq} = \frac{e_{pq}}{m_p} . \tag{8.6}$$

For simple cases, it is convenient to rewrite (8.4) in terms of the ε, which already occur in the equation of energy conservation. If all reactions give a positive contribution to ε, then instead of (8.4), we can write

$$\frac{\partial X_i}{\partial t} = \sum_j \frac{\varepsilon_{ji}}{q_{ji}} - \sum_k \frac{\varepsilon_{ik}}{q_{ik}} . \tag{8.7}$$

If I different nuclei are simultaneously subject to nuclear transformations, equations (8.4) or (8.7) form a set of I differential equations, technically called a "nuclear reactions network". One of them could be replaced by the normalization (8.1), such that we need only $I - 1$ of them to complete the basic equations of our problem. Technically, however, this is not advisable, as (8.1) can then serve as an independent consistency check: if the set of differential equations is solved correctly, mass must be conserved.

Note that for simple cases, it may even suffice to consider just one of these equations. For example, if hydrogen burning is to be taken into account only by way of an overall generation rate ε_{H} (giving the sum over all single reactions), then the only equation needed is

$$\frac{\partial X}{\partial t} = -\frac{\varepsilon_{\mathrm{H}}}{q_{\mathrm{H}}} \tag{8.8}$$

with $\partial Y/\partial t = -\partial X/\partial t$, where q_{H} is the energy release per unit mass when hydrogen is converted into helium.

In Sect. 4.6, we defined the nuclear timescale for a certain burning, $\tau_{\mathrm{n}} = E_{\mathrm{n}}/L$. One can actually define a nuclear timescale for each type of nuclear burning since each nuclear energy reservoir is proportional to an integral of $X_i \cdot dm$ over the whole star, where X_i refers to the element consumed by the reactions; therefore, τ_{n} is equivalent to τ_{Xi}, the timescale for the exhaustion of the element i.

8.2.2 Diffusion

Certain microscopic effects can also change the chemical composition in a star. If gradients occur in the abundances of chemical elements, then *concentration diffusion* tends to smooth out the differences. Even in chemically homogeneous stellar layers, heavier atoms can migrate towards the regions of higher temperature, owing to the effect of *temperature diffusion.* Also, the pressure gradient in a stratified layer causes the heavier particles to diffuse towards the region of higher pressure, that is, *pressure diffusion*. The detailed statistical theory of diffusion is derived in Burgers (1969), Chapman and Cowling (1970), and Choudhuri (1998).

We start with the simplest case: concentration diffusion. Let c be the concentration of particles of a certain species, that is, the number density of particles of that type divided by the number density of all particles, and $\boldsymbol{j}_\mathrm{D}$ be the "flux of concentration"; then, Fick's first law states that

$$\boldsymbol{j}_\mathrm{D} = -D\nabla c\,, \tag{8.9}$$

where D is the diffusion coefficient (We will derive (8.9) later.). With $\boldsymbol{j}_\mathrm{D} = c\boldsymbol{v}_\mathrm{D}$, where $\boldsymbol{v}_\mathrm{D}$ is the diffusion velocity, one has

$$\boldsymbol{v}_\mathrm{D} = -\frac{D}{c}\nabla c\,. \tag{8.10}$$

With the continuity equation

$$\frac{\partial c}{\partial t} = -\nabla\cdot\boldsymbol{j}_\mathrm{D}, \tag{8.11}$$

we find that

$$\frac{\partial c}{\partial t} = \nabla\cdot(D\nabla c)\,, \tag{8.12}$$

and in the case of constant D that

$$\frac{\partial c}{\partial t} = D\nabla^2 c\,, \tag{8.13}$$

a rough estimate for the characteristic timescale is given by

$$\tau_\mathrm{D} \approx \frac{S^2}{D}\,, \tag{8.14}$$

where S is a characteristic length for the variation of c.

By generalizing (8.10) one can formally include the two other types of diffusion, i.e.

$$\boldsymbol{v}_\mathrm{D} = -\frac{1}{c}D(\nabla c + k_T\nabla\ln T + k_P\nabla\ln P)\,, \tag{8.15}$$

if the coefficients k_T and k_P are properly specified. In order to do that we first consider the combined effects of concentration and temperature diffusion.

We assume ∇T to be perpendicular to the x–y plane in a Cartesian coordinate system; then the flux of particles of a certain type in the $+z$ direction due to the statistical motion of the particles is determined by the density n and the mean velocity $\overline{v}$, both taken at $z = -\ell$, where ℓ is the mean free path of the particles of this type:

$$j^+ = \frac{1}{6} c(-\ell)\overline{v}(-\ell) , \tag{8.16}$$

where the numerical factor originates in averaging over $\cos^2$. This takes into account that the particles penetrating the x–y plane had their last encounter at $z = -\ell$.

If one expands n and $\overline{v}$ at $z = 0$ in (8.16) and in a corresponding expression for j^-, the fluxes in the $+z$ and $-z$ directions are

$$j^\pm = \frac{1}{6}\left(c(0) \mp \frac{\partial c}{\partial z}\ell\right)\left(\overline{v}(0) \mp \frac{\partial \overline{v}}{\partial z}\ell\right) , \tag{8.17}$$

and therefore there is a net flux

$$j = j^+ - j^- = -\frac{1}{3}\left(\frac{\partial c}{\partial z}\ell\overline{v} + \frac{\partial \overline{v}}{\partial z}\ell c\right) , \tag{8.18}$$

which in general does not vanish, i.e. we have obtained Fick's law.

We now consider the relative diffusion velocity $v_{D_1} - v_{D_2}$ resulting from the motion of two different types of particles (1, 2), with fluxes j_1, j_2 and concentrations c_1, c_2:

$$v_{D_1} - v_{D_2} = \frac{j_1}{c_1} - \frac{j_2}{c_2} . \tag{8.19}$$

With (8.18) we can replace the j_i by $\ell_i, \overline{v}_i$, and the gradients of c_i, while the velocity gradient–with the help of $\overline{v}_i = (3\Re T/\mu_i)^{1/2}$–can be replaced by the temperature gradient. Using the continuity equation (and after some algebra) an expression of the form

$$v_{D_1} - v_{D_2} = -\frac{D}{c_1 c_2}\left(\frac{\partial c_1}{\partial z} + k_T \frac{\partial \ln T}{\partial z}\right) \tag{8.20}$$

follows. The two terms in the brackets are responsible for concentration diffusion and temperature diffusion. In a mixture of two species ($i = 1, 2$) D and k_T have the form

$$D = \frac{1}{3}(c_2\ell_1\overline{v}_1 + c_1\ell_2\overline{v}_2) = \left(\frac{\Re T}{3}\right)^{1/2}\left(c_2\ell_1\mu_1^{-1/2} + c_1\ell_2\mu_2^{-1/2}\right) , \tag{8.21}$$

$$k_T = \frac{1}{2}\frac{\ell_1\sqrt{\mu_2} - \ell_2\sqrt{\mu_1}}{\ell_1 c_2\sqrt{\mu_2} + \ell_2 c_1\sqrt{\mu_1}} c_1 c_2 (c_2 - c_1) , \tag{8.22}$$

where ℓ_1 and ℓ_2 are the mean free paths of the two species (Landau and Lifshitz 1987, vol. 6). The absolute value k_T is of order 1 or less, and its sign is not immediately clear, though more detailed considerations indicate that $k_T > 0$ for a typical ionized hydrogen–helium mixture in stars.

From (8.21) it is obvious that D is of order

$$D \approx \left(\frac{\Re T}{3}\right)^{1/2} \ell \approx \frac{1}{3} v^* \ell \,, \tag{8.23}$$

where v^* and ℓ are some kind of averages of the statistical velocities and the mean free paths of both components. This expression for D can be used to estimate the timescale τ_D according to (8.14). As long as $|k_T| \approx 1$ this also gives the characteristic timescale for temperature diffusion.

Since $D > 0$, in the case of $k_T > 0$ for pure temperature diffusion, one has $\mathrm{sign}(v_D) = -\mathrm{sign}(\partial \ln T/\partial x)$. Let us now consider the case of a mixture of hydrogen and helium. Here $v_D = v_H - v_{He}$ is the z component of the diffusion velocity and $v_D > 0$ means that hydrogen diffuses in the direction of lower temperature, i.e. "upwards" in the star. For the central region of the Sun ($T \approx 10^7$ K, $\varrho \approx 100\,\mathrm{g\,cm^{-3}}$) one finds that $\ell \approx 10^{-8}$ cm and $D \approx 6\,\mathrm{cm^2\,s^{-1}}$, and with a characteristic length-scale $S \approx R_\odot \approx 10^{11}$ cm, the characteristic timescale τ_D (according to (8.14)) there becomes $\tau_D \approx 10^{13}$ years. Although τ_D is much larger than the age of the universe and therefore the effects of concentration and temperature diffusion seem to be astrophysically irrelevant for the Sun, diffusion does have enough influence on stellar evolution such that high-precision observations require models that include its effect. This will become evident in the case of the standard solar model (see Chap. 29). We will therefore briefly discuss the situation. If a layer is homogeneous, then there is no concentration diffusion, but the hydrogen particles diffuse towards the regions of lower temperature. This causes an outward increase of n_H which in turn triggers concentration diffusion acting against the temperature diffusion ($\mathrm{sign}(\partial c_H/\partial z) = -\mathrm{sign}(\partial T/\partial z)$) until both types of diffusion compensate each other.

We now turn to pressure diffusion, which is the cause of what is often called "sedimentation" or "gravitational settling". A statistical consideration similar to that used to make temperature diffusion plausible also shows that there is diffusion in isothermal layers with a non-vanishing pressure gradient. The reader is again referred to Chapman and Cowling (1970), or any of the other standard textbooks. In a way similar to that for k_T an expression for k_P in (8.15) can also be obtained.

We here confine ourselves to the discussion of the final outcome of this process of pressure diffusion, i.e. the state of final equilibrium for an isothermal layer in hydrostatic equilibrium in a gravitational field pointing towards the $-z$ direction. Let us assume that the material consists of two components ($i = 1, 2$) of perfect gases of different molecular weights μ_i and partial pressures P_i. Then there exist two pressure-scale heights $H_{P_i} = -dz/d \ln P_i$ with which (6.8) can be written in the form

$$H_{P_i} = \frac{P_i}{g\varrho_i} = \frac{\Re T}{g\mu_i} \,, \tag{8.24}$$

where $dP_i/dz = -g\varrho_i$ and $P_i = \Re\varrho_i T/\mu_i$ are used. The particle densities are proportional to the P_i, which are here approximately proportional to $\exp(-z/H_{Pi})$. Therefore the component with the higher μ_i falls off more sharply in the z direction than that with smaller μ_i, so that in a very simplified way, one can say that the heavier component has "moved below" the lighter one. This is the final state, which would be brought about by pressure diffusion alone even if the species were originally in a completely mixed state. Of course, in reality, the two other types of diffusion would also act and therefore influence the final state.

Estimates show that not only $|k_T|$ but also $|k_P|$ is of order one. Therefore it normally takes rather a long time before an appreciable separation occurs in stars. Although in general we will ignore the effect of diffusion in this book, it can be very relevant in certain special cases. Equation (8.12), using (8.15), can be formulated in terms of relative mass fractions X_i instead of particle concentrations and for the case of spherical symmetry as

$$\frac{\partial X_i}{\partial t} = -\frac{1}{\rho r^2}\frac{\partial}{\partial r}\left[r^2 X_i T^{5/2}\left(A_P(i)\frac{\partial \ln p}{\partial r} + A_T(i)\frac{\partial \ln T}{\partial r} + \sum_{k\neq e,\mathrm{He}^4} A_k(i)\frac{\partial \ln C_i}{\partial r}\right)\right]. \tag{8.25}$$

This formulation follows the one by Thoul et al. (1994), and the $T^{5/2}$ factor results from a convenient definition of the diffusion constants called A_P, A_T, and A_k here. In this description the concentration C_i is defined as c_i/c_e, i.e. as the usual particle concentration in units of the electron concentration. Note that the concentration diffusion is taken as a sum over all species, since the concentration of species i may also change due to the diffusion of all other elements. The sum actually has not to be taken over all species as mass and charge conservation reduce the number of independent A_k by two. Here we have taken out helium and electrons.

When diffusion is to be taken into account, proper evaluation of the diffusive constants D (or A in (8.25)) for the various types of diffusion is necessary. This involves correct treatment of the interaction forces between the particles and will be quite sophisticated. Two widely used sources for calculating the diffusion constants are Paquette et al. (1986) and Thoul et al. (1994), both using a method described in the book by Burgers (1969). This method is also sketched in Weiss et al. (2004). An improvement by applying quantum corrections was introduced by Schlattl and Salaris (2003). In general these diffusive speeds or constants are considered to be accurate to 20 %. There is an additional effect not discussed here: Coupling of the radiation field to partially ionized atoms results in a net upward force, counteracting the downward sedimentation. This sort of diffusion is called radiative levitation and can lead to strong variations in surface element abundances, in particular for those elements with rich energy level systems. A derivation of the relevant coefficients was given by Richer et al. (1998).

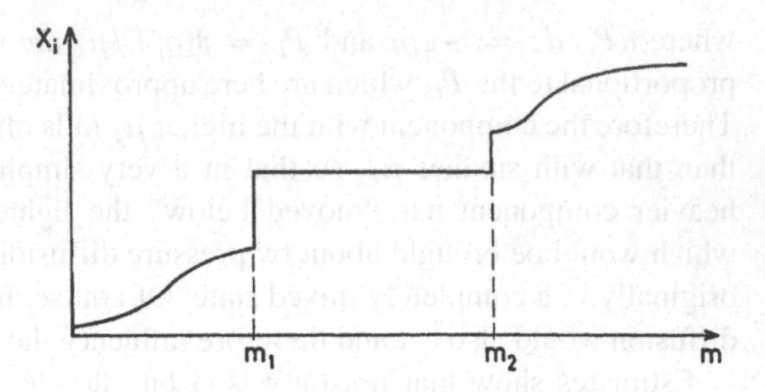

Fig. 8.1 The abundances X_i are smeared out owing to rapid mixing inside a convection zone extending from m_1 to m_2. At these borders, X_i can be discontinuous

8.2.3 *Convective Regions*

Here we deal with the much more important effect of mixing due to turbulent convective motion, a process that normally is very rapid compared to the extremely slow change of the chemical composition produced by nuclear reactions. Therefore we can assume that the composition in a convective region in most cases remains homogeneous,

$$\frac{\partial X_i}{\partial m} = 0 . \tag{8.26}$$

This requires a dispersion not only of the newly created nuclei, but of all elements inside a convective zone.

Suppose a convective zone extends between the mass values m_1 and m_2 (Fig. 8.1). Inside that interval all $X_i = \overline{X}_i$ are constant. At the boundaries one can generally have a discontinuity, such that the "outer" values X_{i1} and X_{i2} are different from the "inner" values–which are simply $\overline{X}_i$. But m_1 and m_2 can change in time, and hence one can easily see that the abundances in the convective zone vary with the rate

$$\frac{\partial \overline{X}_i}{\partial t} = \frac{1}{m_2 - m_1} \times \left(\int_{m_1}^{m_2} \frac{\partial X_i}{\partial t} dm + \frac{\partial m_2}{\partial t}(X_{i2} - \overline{X}_i) - \frac{\partial m_1}{\partial t}(X_{i1} - \overline{X}_i) \right) . \tag{8.27}$$

The X_{i1}, X_{i2} should here be taken as the value on the side that the corresponding boundary moves towards. The integral in the bracket describes the change due to nuclear reactions and can be replaced by an integral over the rates $-\varepsilon_i/q_i$, as in (8.8), where q_i is the energy released if a mass unit of the nucleus i is transformed. Without any nuclear reaction ($\partial X_i/\partial t = 0$) in the convective zone, its composition can still change if the boundaries move into a region of inhomogeneous composition, and this can have important consequences. For example, "ashes" of earlier nuclear burnings may be brought to the surface, fresh fuel may be carried into a zone of nuclear burning, or discontinuities can be produced that drastically influence the later evolution.

In cases of very fast nuclear reactions (or short nuclear timescales) the assumptions of instantaneous mixing is no longer correct. In such situations one normally treats convective mixing as a diffusive process with the diffusive velocity v_c estimated from the solution of the mixing length theory and using (7.6) and (7.16). In this case (8.25) can simply be extended by adding the additional term

$$D_c \frac{\partial X_i}{\partial r} = \left(\frac{1}{3} v_c \alpha_{\mathrm{MLT}} H_P \varrho r^2 \right) \frac{\partial X_i}{\partial r}, \tag{8.28}$$

where we used the estimate for D_c by Langer et al. (1985). Since usually D_c is by orders of magnitude larger than any of the diffusion constants in (8.25), the types of diffusion discussed in Sect. 8.2.2 can in fact be neglected in convective regions.

Chapter 9
Mass Loss

So far we have always assumed that stars have constant total mass. This is, however, not at all the case. The Sun is losing mass via the solar wind at a rate of about $10^{-14}\ M_\odot/\text{year}$. While this mass loss is so slow that it can savely be ignored, other stars may have mass loss rates of up to $10^{-8}\ M_\odot/\text{year}$ or even beyond. The highest mass loss rates for single stars are known for very massive stars ($M \gtrsim 50\,M_\odot$) and for stars of intermediate mass (around $5\,M_\odot$) in a very late stage of their evolution. In addition, stars in binary systems can lose (and gain) mass at any rate due to the gravitational interaction between the two components. Mass loss can therefore range from being totally irrelevant for the evolution of a star up to reducing the mass by up to 50 % or more.

For completeness, we add that the nuclear processes, which provide the overwhelming part of the radiation lost from the stellar surface, imply a conversion of matter to energy and therefore lead to a reduction of the stellar mass, too. For the Sun, this is of the same order as the solar wind, and can therefore be safely ignored. This is also true for all other stars either because this effect is very small, anyhow, or because stellar wind mass loss is much larger.

Evidence for mass loss and estimates of its size come from the direct detection of circumstellar matter and from spectral signatures, such as Doppler shifts and spectral line shapes. Wind velocities can range from a few to a few thousand km/s.

Physically, stellar winds result in many cases from the interaction of the photons emitted from the photosphere with atoms, molecules, or dust grains in the atmosphere. It is therefore a complicated radiation-hydrodynamics problem, which, in addition, may depend on chemical processes, too. An example for the latter are winds from very cool stars, which depend on the coupling of radiation to dust grains. Their formation is a complicated chemical process depending strongly on temperature and density in the stellar atmosphere, which may be subject to regular variations due to stellar pulsations. High mass loss rates are often associated with pulsations in extended stellar envelopes. In some cases, solid physical models exist, which describe the mechanisms for stellar winds. This is particularly true for winds from hot stars (so-called *radiation-driven winds*) and for *dust-driven winds* of cool stars with carbon-rich chemistry. For the observational evidence and for

R. Kippenhahn et al., *Stellar Structure and Evolution*, Astronomy and Astrophysics Library, DOI 10.1007/978-3-642-30304-3_9, © Springer-Verlag Berlin Heidelberg 2012

introductions to stellar wind theories, we refer the reader to the reviews "Winds from hot stars" by Kudritzki and Puls (2000), "Mass loss from cool stars" by Willson (2000), and "Dust driven winds" by Sedlmayer and Dominik (1994).

Since a full theoretical model for any stellar wind is not available, and would not be reasonable to be used in modelling stellar evolution, and since most information about stellar mass loss still results from observations, empirical mass loss formulations are used in the models. They have all been obtained from observations of some class of stars, and therefore differ from each other. Therefore, different mass loss formulae have to be used for different type of stars. None of them is very accurate, but in most cases, it suffices to have the correct order of magnitude of mass loss and its dependence on the global properties of the star. We now introduce a few such empirical mass loss formulations, which are widely used in stellar evolution calculations.

The most famous mass loss formula of all is that of Reimers (1975), obtained from red giants with heavy element abundances similar to those in the Sun. Reimers showed that the dependence of the mass loss rate on basic stellar parameters can be expressed by the simple fitting formula

$$\dot{M}_{\mathrm{R}} = -4 \times 10^{-13} \eta \frac{L}{gR} \cdot \frac{g_{\odot} R_{\odot}}{L_{\odot}}. \tag{9.1}$$

The unit of $\dot{M}$ is $M_{\odot}$/year. This formula reflects the intuitive expection that mass loss increases with luminosity L, and decreases with a deeper gravitational potential well $gR = GM/R$. The parameter η was introduced later to use Reimers' formula for other types of stars, too. It usually varies between 0.2 and 1.0 and is lower for metal-poor stars, indicating a weaker coupling of the photons to the gas if fewer heavy elements are present.

Reimers' formula has no strong theoretical justification, but seems to be a useful estimate for the order of magnitude of mass loss from cool stars. It has been modified from time to time to take into account a more detailed dependence on stellar parameters. One of the latest of such modifications, which is fitting better to recent mass loss determinations, is that of Schröder and Cuntz (2005), which is

$$\dot{M}_{\mathrm{SC}} = -8 \times 10^{-14} \frac{LR}{M} \frac{M_{\odot}}{L_{\odot} R_{\odot}} \left(\frac{T_{\mathrm{eff}}}{4{,}000\,\mathrm{K}} \right)^{3.5} \left(1 + \frac{g}{4{,}300 g_{\odot}} \right). \tag{9.2}$$

For very cool and luminous stars on the *asymptotic giant branch*, which experience an almost catastrophic mass loss event with mass loss rates up to $10^{-4} M_{\odot}$/year, a simple and useful formula has been derived by Blöcker (1995), based on observations and dust-driven wind theories. There are more sophisticated theoretical or empirical mass loss functions (see Sect. 34.6), but Blöcker's is in most cases sufficient for an estimate:

$$\dot{M}_{\mathrm{B}} = -4.83 \times 10^{-9} \dot{M}_{\mathrm{R}} (M_{\star}/M_{\odot})^{-2.1} (L/L_{\odot})^{2.7} \tag{9.3}$$

There are two variants of this formula, in which for $M_\star$ either the initial or the present mass of the star is used. Since (9.3) is only a rough estimate of the actual mass loss, this is acceptable.

Finally, we add a formula fitting empirical mass loss rates for hot stars of spectral type O and B, obtained by Lamers (1981):

$$\dot{M}_{\mathrm{L}} = -1.48 \times 10^{-5} \left(\frac{L}{1{,}000 L_\odot} \right)^{1.42} \left(\frac{R}{30 R_\odot} \right)^{0.61} \left(\frac{30 M_\odot}{M} \right)^{0.99} \tag{9.4}$$

A more physical discussion of mass loss from hot stars can be found in the mentioned review by Kudritzki and Puls (2000).

Obviously, all these formulae contain, in some form or other, the basic dependence on M, R, and L by Reimers. Sometimes a dependence on chemical composition is added. It is generally assumed that $\dot{M} \sim X_{\mathrm{res}}^{1/2}$, where X_{res} denotes the mass fraction of all elements other than hydrogen and helium.

Equations (9.1)–(9.4) already indicate that the main effect of mass loss is simply to reduce the total mass of a star. This has to be taken into account in stellar modelling and will be discussed in Sect. 12.5.

Part II
The Overall Problem

Chapter 10
The Differential Equations of Stellar Evolution

10.1 The Full Set of Equations

Collecting the basic differential equations for a spherically symmetric star in hydrostatic equilibrium derived in Chap. 1, we are then led by (1.6), (2.16), (4.47), (4.48), (7.32), and (8.4) to

$$\frac{\partial r}{\partial m} = \frac{1}{4\pi r^2 \varrho}\,, \tag{10.1}$$

$$\frac{\partial P}{\partial m} = -\frac{Gm}{4\pi r^4}\,, \tag{10.2}$$

$$\frac{\partial l}{\partial m} = \varepsilon_{\mathrm{n}} - \varepsilon_\nu - c_P \frac{\partial T}{\partial t} + \frac{\delta}{\varrho}\frac{\partial P}{\partial t}\,, \tag{10.3}$$

$$\frac{\partial T}{\partial m} = -\frac{GmT}{4\pi r^4 P}\nabla\,, \tag{10.4}$$

$$\frac{\partial X_i}{\partial t} = \frac{m_i}{\varrho}\left(\sum_j r_{ji} - \sum_k r_{ik}\right)\,, \quad i = 1, \ldots, I\,. \tag{10.5}$$

Equation (10.2) has an additional term $-\partial^2 r/\partial t^2 (4\pi r^2)^{-1}$ in case the assumption of hydrostatic equilibrium is not fulfilled. In (10.5) we have a set of I equations (one of which may be replaced by the normalization $\sum_i X_i = 1$) for the change of the mass fractions X_i of the relevant nuclei $i = 1, \ldots, I$ having masses m_i. Additional formulae regulate the mixing of the composition in convective regions, (8.27) or (8.28), or in case of diffusive processes (8.25). In (10.3), $\delta \equiv -(\partial \ln \varrho/\partial \ln T)_P$, and in (10.4), $\nabla \equiv d \ln T/d \ln P$. If the energy transport is due to radiation (and conduction), then ∇ has to be replaced by ∇_{rad}, which is given by (5.28):

R. Kippenhahn et al., *Stellar Structure and Evolution*, Astronomy and Astrophysics Library, DOI 10.1007/978-3-642-30304-3_10, © Springer-Verlag Berlin Heidelberg 2012

$$\nabla = \nabla_{\text{rad}} = \frac{3}{16\pi acG} \frac{\kappa l P}{m T^4} . \tag{10.6}$$

If the energy is carried by convection, then ∇ in (10.4) has to be replaced by a value obtained from a proper theory of convection; this may be ∇_{ad} in the deep interior or obtained from a solution of the cubic equation (7.26) for superadiabatic convection in the outer layers. Note that the expression on the right-hand side of (10.4) assumes hydrostatic equilibrium. This does not matter in the case of radiative transport, since the local adjustment time of the radiation field is very short, and the convection theory of Chap. 7 is valid only for stars in hydrostatic equilibrium. Otherwise another convection theory valid in rapidly changing regions would have to be used. Additional criteria such as (6.12) and (6.13) distinguish between radiative and convective transport.

In the system (10.1)–(10.5) one can distinguish certain subsystems, i.e. (10.1) and (10.2) give the mechanical part, being coupled to the thermo-energetic part only through the density ϱ–which usually also depends on T. If for some reason or other this dependence of ϱ on T is not present (or can be eliminated), then (10.1) and (10.2) can be solved regardless of the other equations to give the mechanical structure $r(m)$, $P(m)$. Equations (10.5) may be regarded as the chemical part. Under normal conditions (τ_{n} much larger than the other timescales; see Sect. 10.2) they can be decoupled from the spatial parts (10.1)–(10.4), which describe the structure of the star for a given time and given composition $X_i(m)$. This would be questionable, of course, if the chemical composition changed as rapidly as the other variables, and for changes of $X_i(m)$ more rapid than those of P, T, one would rather assume to have an "equilibrium composition" $X_i(P, T)$ at any time (see Chap. 36).

Equations (10.1)–(10.5) contain functions which describe properties of the stellar material such as $\varrho, \varepsilon_{\text{n}}, \varepsilon_\nu, \kappa, c_P, \nabla_{\text{ad}}, \delta$ and the reaction rates r_{ij}. We shall deal with these functions in Part III. Meanwhile we assume them to be known functions of P, T and the chemical composition described by the functions $X_i(m, t)$. We therefore have an equation of state

$$\varrho = \varrho(P, T, X_i) \tag{10.7}$$

and equations for the other thermodynamic properties of the stellar matter

$$c_P = c_P(P, T, X_i) , \tag{10.8}$$

$$\delta = \delta(P, T, X_i) , \tag{10.9}$$

$$\nabla_{\text{ad}} = \nabla_{\text{ad}}(P, T, X_i) , \tag{10.10}$$

as well as the Rosseland mean of the opacity (including conduction)

$$\kappa = \kappa(P, T, X_i) , \tag{10.11}$$

and the nuclear reaction rates and the energy production and energy loss via neutrinos:

$$r_{jk} = r_{jk}(P, T, X_i) , \tag{10.12}$$

$$\varepsilon_n = \varepsilon_n(P, T, X_i) , \tag{10.13}$$

$$\varepsilon_\nu = \varepsilon_\nu(P, T, X_i) . \tag{10.14}$$

In these equations, the arguments X_i stand for *all* types of nuclei ($i = 1, \ldots, I$).

It is now time to count the equations and the unknown variables. We consider the material functions on the right-hand sides of (10.1)–(10.5) to be replaced with the help of the corresponding equations (10.7)–(10.14), i.e. by functions of P, T, X_i. For I different types of nuclei being affected by reactions, (10.1)–(10.5) form a set of $4 + I$ differential equations for the $4 + I$ variables $r, P, T, l, X_1, \ldots, X_I$. We therefore have the same number of equations and unknown variables.

The independent variables are m and t. If we assume that the total mass of the star does not change in time (i.e. no gain nor loss of mass) and if we define the time at which evolution starts as $t = t_0$, then we are looking for solutions in the intervals

$$0 \leq m \leq M, \quad t \geq t_0 . \tag{10.15}$$

In the full problem we are confronted with a set of non-linear, partial differential equations. As usual, physically relevant solutions require the specification of boundary conditions (here at $m = 0, m = M$) and of initial values [e.g. $X_i(m, t_0)$]. The boundary conditions will be dealt with in Chap. 11. In order to see more clearly which initial values have to be specified we replace the two terms with time derivatives of P and T in (10.3) by one term containing the change of the entropy $s, -T\partial s/\partial t$, according to (4.47). Obviously the full problem requires specification of the functions $r(m, t_0), \dot{r}(m, t_0), s(m, t_0)$, and $X_i(m, t_0)$.

After proper initial values and boundary conditions are specified, together with the stellar mass M, the problem is to find solutions of the basic equations, i.e. the unknown variables as functions of m and t. A solution $r(m), P(m), \ldots, X_i(m)$ for a given time t in the interval $[0, M]$ is called a *stellar model*. But before we discuss in more detail how solutions of our set of differential equations can be obtained, we first discuss simplifications of the full problem.

10.2 Timescales and Simplifications

There are three types of time derivatives in our set of equations. To each of them belongs a certain characteristic timescale. In Sect. 2.4 the term $(\partial^2 r/\partial t^2)/4\pi r^2$ in (2.16)–the dynamical version of (10.2)–was used to derive $\tau_{\rm hydr}$. From the time derivatives in (10.3) we have derived $\tau_{\rm KH}$ in Sect. 3.3. The time derivatives in (10.5) define chemical timescales τ_{Xi} which were shown to be equivalent to τ_n [see (4.59)] at the end of Sect. 8.2.1.

In Sect. 2.4 we showed that the inertia term in (10.2) can be neglected if the evolution is slow compared to τ_{hydr}. Therefore, if the evolution of a star is governed by thermal adjustment or by nuclear reactions ($\tau_{KH} \gg \tau_{hydr}$ and $\tau_n \gg \tau_{hydr}$), the equation of hydrostatic equilibrium (10.2) is appropriate. The star then evolves along a sequence of states of hydrostatic equilibrium. As initial conditions, the functions $s(m, t_0)$ and $X_i(m, t_0)$ have to be specified in this approximation.

If the star evolves on the timescale $\tau_n \gg \tau_{KH}$, then according to the discussion in Sect. 4.4, the time derivatives in the energy equation can also be neglected and (10.3) is reduced to

$$\frac{\partial l}{\partial m} = \varepsilon_n - \varepsilon_\nu. \tag{10.16}$$

The star now evolves along a sequence of states in which it is not only in hydrostatic equilibrium but also thermally adjusted. We call this *complete* (mechanical and thermal) *equilibrium*. The only initial values to be given in this case are the $X_i(m, t_0)$.

In complete equilibrium the basic equations split into two parts: the "structure equations" (10.1), (10.2), (10.16) and (10.4) contain only spatial derivatives while the "chemical equations" (10.5) contain only time derivatives. Therefore, if at a certain time $t = t_0$ the $X_i(m, t_0)$ are given, the structure equations can be taken as a set of four *ordinary* differential equations describing the structure of the star at t_0.

Complete equilibrium is a good approximation for stars in many important evolutionary phases, for example, the stars on the main sequence. But even without complete equilibrium the full set of equations is usually split into two parts: the spatial part solved as a boundary value problem for a given chemical composition $X_i(m, t_0)$, and the time-dependent initial-value problem of the chemical changes. These two parts are solved in two different, alternating steps with different numerical schemes. This introduces a basic problem of inconsistency: consider the spatial problem to be solved at time t_0, with some chemical stratification given. Once the solutions for $r(m)$, $T(m)$, $P(m)$, and $l(m)$ have been found, some layers may be convective. Therefore, the chemical stratification for which the solution was determined may be altered by convective (or, more general, by any kind of) mixing, and the solution will not be consistent with the real chemical composition. The mixing is done only in the second step, after the spatial problem is solved, over a time step Δt, and after this step, the physical variables are again not a solution of the new $X_i(m, t_0 + \Delta t)$, etc. Of course one can control the severeness of this inconsistency by keeping Δt small, but one should be aware of its fundamental nature. Another problem arises if the structure variables are kept constant over the time step Δt. In nuclear burning regions, for example, temperature and density are usually rising with time. Therefore, they would be underestimated in the nuclear reactions if kept constant over Δt, and so would be the chemical changes due to nuclear burning. This effect leads to an overestimate of main-sequence lifetimes. Again, it can be minimized by using very small values for Δt, or by a clever prediction how T and ρ (and other quantities) may change during a time step.

Chapter 11
Boundary Conditions

As usual in mathematical physics, the boundary conditions constitute an important part of the whole problem, and their influence on the solutions is not easy to foresee. This is connected with the fact that the boundary conditions for the problem of stellar structure cannot be imposed at one end of the interval $[0, M]$ only but rather are split into some that are given at the centre and some near the surface of the star. The central conditions are simple, whereas the surface conditions implicate observable quantities and a completely different, much more complicated transport equation. It is therefore advisable to get some feeling about their influence on the stellar structure. We discuss these problems for the case of complete equilibrium.

11.1 Central Conditions

Two boundary conditions can be immediately written down for the centre, defined by $m = 0$. Since the density ϱ must go to a reasonable, finite, and non-vanishing value (there can be no singularity and no cavity in the centre), we must have $r = 0$. And since the energy sources also remain finite (positive or negative), l must vanish at the centre as well:

$$m = 0: \quad r = 0, \quad l = 0. \tag{11.1}$$

This was the simple part. Unfortunately nothing is a priori known about the central values of pressure P_c and temperature T_c, so the conditions (11.1) still allow a two-parameter set of solutions, obtained by outward integrations starting with arbitrary P_c, T_c, and $r = l = 0$.

It is useful to know the behaviour of the four functions r, l, P, T in the vicinity of the centre, $m \to 0$, for a given time $t = t_0$. The equation of continuity (10.1) may be written as

$$d(r^3) = \frac{3}{4\pi\varrho} dm, \tag{11.2}$$

R. Kippenhahn et al., *Stellar Structure and Evolution*, Astronomy and Astrophysics Library, DOI 10.1007/978-3-642-30304-3_11, © Springer-Verlag Berlin Heidelberg 2012

which can be integrated for constant $\varrho = \varrho_c$, i.e. for small enough values of m and r, giving

$$r = \left(\frac{3}{4\pi\varrho_c}\right)^{1/3} m^{1/3}. \tag{11.3}$$

This can be considered the first term in a series expansion of r around $m = 0$. A corresponding integration of the energy equation (10.3) yields

$$l = (\varepsilon_n - \varepsilon_\nu + \varepsilon_g)_c\, m. \tag{11.4}$$

In both cases we have used the proper boundary conditions (11.1) by taking the integration constants to be zero.

Eliminating r for small values of m by (11.3), we obtain from the hydrostatic equation (10.2)

$$\frac{dP}{dm} = -\frac{G}{4\pi}\left(\frac{4\pi\varrho_c}{3}\right)^{4/3} m^{-1/3}, \tag{11.5}$$

which can be integrated to yield

$$P - P_c = \frac{3G}{8\pi}\left(\frac{4\pi}{3}\varrho_c\right)^{4/3} m^{2/3}. \tag{11.6}$$

The pressure gradient must, of course, vanish at the centre, which can be seen by writing the hydrostatic equation (2.4) in the form

$$\frac{dP}{dr} \sim \frac{m}{r^2} \sim \frac{r^3}{r^2} \to 0 \tag{11.7}$$

for $r \to 0$.

The variation of temperature will first be considered in the radiative case, for which (5.12) requires that

$$\frac{dT}{dm} = -\frac{3}{64\pi^2 ac}\frac{\kappa l}{r^4 T^3}. \tag{11.8}$$

With $P \to P_c, T \to T_c$, κ tends to some well-defined value κ_c. Replacing $l(\sim m)$ by (11.4) and $r(\sim m^{1/3})$ by (11.3) now, we can integrate (11.8) for small values of m and obtain the first equation (11.9). In the case of (adiabatic) convection we start from (7.32) with $\nabla = \nabla_{ad}$ and replace r by (11.3). An integration for small values of m then gives the second equation (11.9):

$$T^4 - T_c^4 = -\frac{1}{2ac}\left(\frac{3}{4\pi}\right)^{2/3} \kappa_c(\varepsilon_n - \varepsilon_\nu + \varepsilon_g)_c \varrho_c^{4/3} m^{2/3} \text{ (radiative)},$$

$$\ln T - \ln T_c = -\left(\frac{\pi}{6}\right)^{1/3} G\frac{\nabla_{ad,c}\varrho_c^{4/3}}{P_c} m^{2/3} \text{ (convective)}. \tag{11.9}$$

11.2 Surface Conditions

The strict surface conditions are rather complicated and unwieldy. For rough estimates one might therefore prefer to use a crude approximation, provided that it is simple.

An extreme step in this direction would be to take the naïve "zero conditions"

$$m \to M: \quad P \to 0, \quad T \to 0. \tag{11.10}$$

These at least reflect correctly the fact that, in the outermost region of the star, P and T go to very small values compared to those in the interior. But, of course, in reality, there is a gradual and rather extended transition to the finite values of P, T of the diffuse interstellar medium.

The next step is to find a sphere that we can reasonably call the "surface" of the star and that defines the total stellar radius $r = R$. The theory of stellar atmospheres suggests the use of the *photosphere*, from where the bulk of the radiation is emitted into space, and which is found where the optical depth τ of the overlying layers,

$$\tau := \int_R^\infty \kappa \varrho \, dr = \bar{\kappa} \int_R^\infty \varrho \, dr, \tag{11.11}$$

is equal to 2/3. Here we have defined a mean opacity $\bar{\kappa}$, averaged over the stellar atmosphere. In hydrostatic equilibrium the pressure at this level is given by the weight of the matter above. We can well approximate the gravitational acceleration by the constant value $g_0 = GM/R^2$, since the bulk of the matter in these layers is anyway very close to the photosphere. Then

$$P_{r=R} = \int_R^\infty g \varrho \, dr = g_0 \int_R^\infty \varrho \, dr, \tag{11.12}$$

and if we eliminate here the integral over ϱ by that in the second equation (11.11), we find with $\tau = 2/3$ that

$$P_{r=R} = \frac{GM}{R^2} \frac{2}{3} \frac{1}{\bar{\kappa}}. \tag{11.13}$$

The temperature at the photosphere is equal to the *effective temperature* $T_{r=R} = T_{\text{eff}}$ of the star defined by

$$L = 4\pi R^2 \sigma \, T_{\text{eff}}^4. \tag{11.14}$$

Here $\sigma = ac/4$ is the Stefan-Boltzmann constant of radiation. T_{eff} is thus the temperature of that black body which yields the same surface flux of energy as the star.

In (11.11) we have replaced κ by an average value. As we usually have detailed knowledge about the opacity, an obvious improvement is to take into account the pressure (or density) and temperature dependency of κ. This requires knowledge about the temperature stratification in the atmosphere. One common approach is to use the *Eddington approximation*, which is

$$T^4(\tau) = \frac{3}{4}(L/4\pi R^2\sigma)\left(\tau + \frac{2}{3}\right), \tag{11.15}$$

which obviously results in $T = T_{\rm eff}$ for $\tau = 2/3$.

Equation (11.11) can be transformed into a differential equation for the radius, $dr/d\tau = -1/(\kappa\varrho)$, and with $dP/dr = -g\varrho$ we obtain

$$\frac{dP}{d\tau} = \frac{Gm}{r^2\kappa} \tag{11.16}$$

which is to be integrated from $\tau = 0$ to $\tau = 2/3$ with the boundary condition $P(\tau = 0) = 0$. The Eddington approximation (11.15) is used to determine $\kappa(P, T)$. For the gravitational acceleration on the right-hand side of (11.16) we can savely use GM/R^2. We thus obtain an improved value for $P_{r=R}$. This approach is called the *Eddington grey atmosphere* because of the use of the Eddington approximation and the Rosseland mean for the opacity, and it is indeed used in stellar evolution codes, which solve the stellar structure equations (10.1)–(10.5) from $m = 0$ to $m = M$.

The *photospheric conditions* (11.13) and (11.14) or (11.16) and (11.14) represent two relations between the surface values ($m \to M$) of the functions P, T, r, l. They are certainly a better approximation for the surface conditions than (11.10). Their severest defect is that they refer to a level where the assumption made for deriving the transport equation (5.12) (small mean free path of the photons) breaks down. At this level, one should use the more complicated transport equation for stellar atmospheres. Indeed such attempts have been made, and full stellar atmosphere models are connected to those of the stellar interior at a suitable optical depth. Examples are the work by Schlattl et al. (1997) for the solar case and VandenBerg et al. (2008) for low-mass stars.

Quite generally, the correct surface conditions can be formulated as follows: the interior solution should fit smoothly to a solution of the stellar-atmosphere problem. Let us put this into a more mathematical form.

The transition between interior and outer (atmospheric) solutions is made at a certain mass value $m_{\rm F}$, the "fitting mass", which should be far enough in to ensure that the interior equations are still valid there. On the other hand, $m_{\rm F}$ should still be close enough to M that, for simplicity, we can always use thermally adjusted outer solutions with constant $l = L$. The smaller $M - m_{\rm F}$, the less energy can be stored or released in these outer layers.

For the stellar-interior problem, we consider the mass M and the chemical composition to be given. The theory of stellar atmospheres tells us that for given M and $X_i(M)$, there is a two-parameter set of possible atmospheric solutions, the parameters being, for example, R and $T_{\rm eff}$, or R and L [which are connected by (11.14)]. Any one of these possible atmospheric solutions can be extended by integration downwards to $m_{\rm F}$ and may yield there the four "exterior" values $r = r_{\rm F}^{\rm ex}, P = P_{\rm F}^{\rm ex}, T = T_{\rm F}^{\rm ex}, l = l_{\rm F}^{\rm ex} = L$.

The outer boundary conditions now require for $m = m_F$ that one quartet $r_F^{ex}, \ldots, l_F^{ex}$ obtained from an outer solution has to match the corresponding values $r_F^{in}, \ldots, l_F^{in}$ of the interior solution, which extends from the centre to m_F:

$$r_F^{ex} = r_F^{in}, \quad P_F^{ex} = P_F^{in}, \quad T_F^{ex} = T_F^{in}, \quad l_F^{ex} = l_F^{in}. \tag{11.17}$$

These four simultaneous fits are in principle possible, since the solutions have enough degrees of freedom: the interior solution has two (we can vary the central values P_c and T_c), and the outer solution also has two (variation of R and L). The fact that both solutions have two degrees of freedom is reflected in the following alternative representation, which is often used in numerical computations. Imagine that many outer integrations are carried out for many pairs of parameters R and L. At $m = m_F$, they yield the four functions $r_F^{ex}(R, L), P_F^{ex}(R, L), T_F^{ex}(R, L), l_F^{ex}(R, L)$. The last one is very simple, namely $l_F^{ex} = L$. The first one is certainly well behaved, and we can invert it without complications, obtaining $R = R(r_F^{ex}, L)$. This is now used to replace the argument R in the functions P_F^{ex} and T_F^{ex}, which can then be considered known functions π and θ of r_F^{ex} and $l_F^{ex} = L$:

$$\begin{aligned} P_F^{ex}\big(R(r_F^{ex}, L), L\big) &:= \pi\big(r_F^{ex}, L\big), \\ T_F^{ex}\big(R(r_F^{ex}, L), L\big) &:= \theta\big(r_F^{ex}, L\big). \end{aligned} \tag{11.18}$$

For any given pair r_F^{ex}, L, the π and θ give the corresponding values of pressure and temperature for one outer solution. We now replace the variables $P_F^{ex}, \ldots, l_F^{ex} = L$ in (11.18) by $P_F^{in}, \ldots, l_F^{in}$, using the fit conditions (11.17):

$$P_F^{in} = \pi\big(r_F^{in}, L\big), \quad T_F^{in} = \theta\big(r_F^{in}, L\big). \tag{11.19}$$

These are the outer boundary conditions for the interior solution. Obviously, if these are fulfilled, there is always an outer solution that continuously matches the interior solution. We can now drop the distinction between the variables of the exterior and interior solutions at $m = m_F$ expressed in the superscripts "ex" and "in".

The fulfilment of the boundary conditions is illustrated in Fig. 11.1, where the functions π and θ (obtained from outer solutions) are sketched over the r_F-L plane. We have also indicated the surfaces $\tilde{\pi}(r_F, L)$ and $\tilde{\theta}(r_F, L)$, which give the corresponding functions of the *interior* solutions obtained by varying P_c and T_c. The intersection of the surfaces ($\pi = \tilde{\pi}$ and $\theta = \tilde{\theta}$) gives the matches of P_F or of T_F, respectively. We project the intersections into the r_F-L plane (dot-dashed lines), and where these projections intersect, we have the desired match of all four variables.

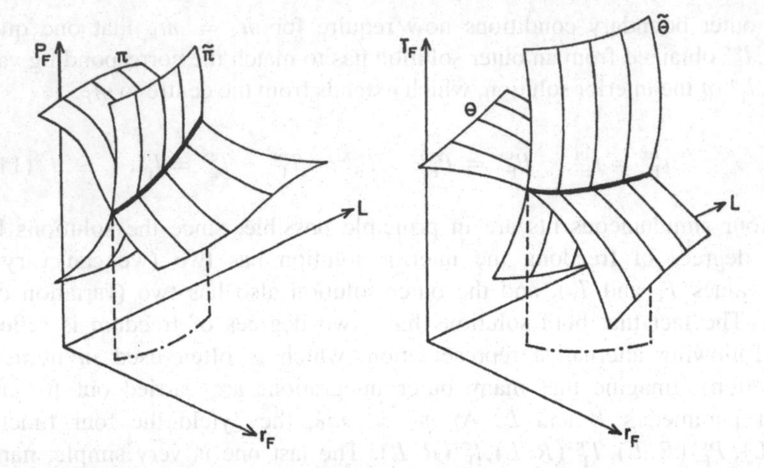

Fig. 11.1 The function values P_F (or T_F) at the fitting mass $m = M_F$ are plotted over r_F and L. The surface π (or θ) contains the values obtained by all possible integrations downwards from the photosphere. The surface $\tilde{\pi}$ (or $\tilde{\theta}$) contains the corresponding values obtained from all possible integrations outwards from the centre. The *heavy line* shows the intersection of π and $\tilde{\pi}$ (or θ and $\tilde{\theta}$), the *dot-dashed line* the projection of this intersection into the r_F-L plane (All surfaces are freely invented sketches)

11.3 Influence of the Surface Conditions and Properties of Envelope Solutions

We confine ourselves here to "normal" stars in complete (mechanical and thermal) equilibrium. For the outer envelope of such a star, it is characteristic that l and m vary very little over wide ranges of r (This is because ε is negligible and ϱ is very small; for example, only about 10 % of the solar mass lies outside $r = R_\odot/2$.). This allows the derivation of approximate solutions that demonstrate the influence of the outer layers on the interior solution.

11.3.1 Radiative Envelopes

Since m varies so little in the envelope, it seems advisable to take another independent variable, for which we may choose the pressure P, since it varies monotonically with m. The equation of radiative transport is derived from (5.12) and (2.5) as

$$\frac{\partial T}{\partial P} = \frac{3}{64\pi\sigma G}\frac{\kappa l}{T^3 m} \tag{11.20}$$

($\sigma = ac/4$). Let us approximate the dependence of κ on P and T by a power law of the form

$$\kappa = \kappa_0 P^a T^b , \tag{11.21}$$

with κ_0 = constant and exponents typically $a > 0, b < 0$. By proper choice of κ_0, a, and b we can represent reasonably (though, of course, not correctly) the run of κ over wide ranges of the envelope. Introducing (11.21) into (11.20) results in

$$\frac{T^{3-b}}{P^a}\frac{\partial T}{\partial P} = \frac{3\kappa_0}{64\pi\sigma G}\frac{l}{m}, \tag{11.22}$$

and now we take $l \approx L$ and $m \approx M$ (this, together with the approximation of κ, determines how far inwards we are allowed to extend our solution). Then the right-hand side is constant and (11.22) can be integrated by separation of the variables:

$$T^{4-b} = B(P^{1+a} + C), \tag{11.23}$$

where C is a constant of integration, while the positive constant B is given by

$$B = \frac{4-b}{1+a}\frac{3\kappa_0}{64\pi\sigma G}\frac{L}{M}. \tag{11.24}$$

For an illustrative example we now fix the exponents: $a = 1, b = -4.5$, which corresponds to the famous Kramers opacity for bound–free and free–free absorption in stellar material (see Chap. 17), and which is a good approximation for envelopes of moderate temperatures. Then (11.23) becomes

$$T^{8.5} = B(P^2 + C), \tag{11.25}$$

a solution for the envelope that will now be discussed. It is illustrated in Fig. 11.2, which gives $\lg T$ against $\lg P$, so that the slope of a solution is equal to the value of $\nabla \equiv d\ln T/d\ln P$. Differentiation of (11.25) gives the slope

$$\nabla = 0.235\frac{BP^2}{T^{8.5}}. \tag{11.26}$$

The multitude of possible solutions differ by their value of the integration constant C.

$\underline{C = 0}$: The solution (11.25) now gives

$$\frac{T^{8.5}}{BP^2} = 1, \tag{11.27}$$

for which (11.26) yields the slope $\nabla = 2/8.5 \approx 0.235$. This is smaller than the usual value of $\nabla_{ad} = 2/5$ (see Chap. 14), and therefore the solution is consistent with (the assumed) radiative transport, shown in Fig. 11.2 as the straight solid line $\lg T = (2\lg P + \lg B)/8.5$. Obviously $T \to 0$ for $P \to 0$, and this solution would reach the zero boundary condition if we were to extend it outwards over the photosphere.

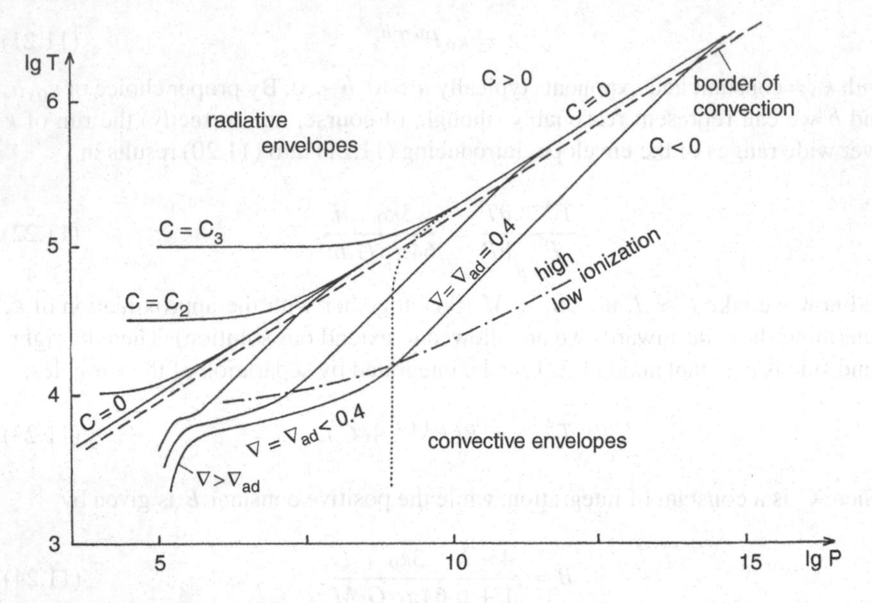

Fig. 11.2 A lg$T$–lg$P$ diagram for illustrating typical properties of envelope solutions as discussed in the text (see there for details)

<u>$C > 0$</u>: Since $B > 0$, (11.25) yields

$$\frac{T^{8.5}}{BP^2} > 1. \tag{11.28}$$

Comparing this with (11.26) and (11.27), we see that in Fig. 11.2, the solutions with $C > 0$ lie above that with $C = 0$ and that they have a smaller slope, $\nabla < 2/8.5$. The layers are therefore all the more radiative. For $P^2 \ll C$ equation (11.25) becomes $T^{8.5} \approx BC = \text{constant}$. This shows that towards the surface these solutions tend to a constant (and rather high) T. Three of them (for 3 different values $C_1 < C_2 < C_3$ of C) are illustrated by solid lines on the left of Fig. 11.2. On each line, one point corresponds to the photosphere with $T = T_{\text{eff}}$. Obviously we will find such radiative-envelope solutions below the photospheres with T_{eff} larger than some critical value (close to 10^4 K). Towards the interior, P will finally increase so far that $P^2 \gg C$ in (11.25) and the solution approximates closely that for $C = 0$. Since all solutions with $C > 0$ asymptotically approach the solution $C = 0$, the precise starting values at the surface do not greatly influence the solution in the deep interior.

<u>$C < 0$</u>: Equation (11.25) now gives

$$\frac{T^{8.5}}{BP^2} < 1, \tag{11.29}$$

which with (11.26) and (11.27) shows that these solutions lie below the curve for $C = 0$ and that their slope is larger, $\nabla > 2/8.5$. A discussion quite analogous to that for $C > 0$ shows immediately that these solutions have the structure indicated in Fig. 11.2 by the dotted line. They bend downwards from the line $C = 0$, become gradually steeper, and tend vertically to a finite P for $T \to 0$ (With a proper scaling of the coordinates the curves $C > 0$ and $C < 0$ are simply symmetric with respect to the line $C = 0$.). However, the assumption of radiative transport breaks down when convection sets in, which is the case for $\nabla = \nabla_{ad}$ (see Sect. 6.1). This is close to 0.4 in the interior of not too massive stars, while ionization effects near the surface can make it considerably smaller (see Chap. 14). This limit is derived by equating the right-hand side of (11.26) with ∇_{ad}:

$$T^{8.5} = \frac{0.235}{\nabla_{ad}} BP^2. \tag{11.30}$$

For constant ∇_{ad} this corresponds to a straight line given by $\lg T = (2 \lg P + \lg B + \lg(0.235/\nabla_{ad}))/8.5$. For $\nabla_{ad} = 0.4$ this lower border for radiative solutions is plotted in Fig. 11.2 (dashed line). Near the surface, ionization effects decrease ∇_{ad} considerably below 0.4, and therefore the border line should be curved *upwards* in its lowest part.

11.3.2 Convective Envelopes

The radiative solutions with $C < 0$ extending from the interior have to be terminated at the broken line in Fig. 11.2 given by (11.30), where convection sets in, and have to be replaced in the outer regions by solutions valid for convective transport. Three such convective solutions are shown as solid lines in the lower part of Fig. 11.2. In order to construct them we have to consider their slope $d \lg T/d \lg P$ $(= \nabla)$. As long as the solutions stay in regions of high enough density, convection is very effective (cf. Sect. 7.3) and the slope is equal to the adiabatic gradient ∇_{ad}.

We can start the convective solutions near the border of convection with a slope given by $\nabla = \nabla_{ad} = 0.4$. With decreasing temperature the curves come into regions where the most abundant elements (hydrogen and helium) are no longer completely ionized (see Chap. 14). For hydrogen this occurs around $T = 10^4$ K, depending somewhat on P (cf. the dependence of the Saha equation on the electron density). Partial ionization depresses ∇_{ad} appreciably below 0.4 such that the curves with a slope $\nabla = \nabla_{ad}$ are less steep and closely approach one another.

Finally the curves come into regions of such low density that convection is ineffective and the stratification is over-adiabatic, $\nabla > \nabla_{ad}$ (Chap. 7). Correspondingly the curves in Fig. 11.2 become rather steep until they reach the photospheric point. Unfortunately the precise slope ∇ in the over-adiabatic part can only be calculated from a convection theory, with all its uncertainties. Anyway, convective envelopes start at cool photospheres, and with decreasing T_{eff}, the convection gradually reaches deeper into the interior. Small variations (due to numerical or physical uncertainties) of T_{eff} or of the over-adiabatic part lead to curves that are widely separated in the interior.

11.3.3 Summary

Making a few simplifying assumptions, we have been able to derive convenient solutions for the temperature-pressure stratification of stellar envelopes, i.e. for the layers below the photosphere. In the case of radiative envelopes, the assumptions concerned κ, m, and l. An opacity law like (11.21) is certainly a poor approximation if one takes the same values of a, b, κ_0 for too wide a range, or for very different envelopes. The discussions can, however, be easily repeated for different values of a, b, κ_0 [e.g. $a = 0, b = 0, \kappa_0 = 0.2(1 + X_H)$, as in the case of electron scattering, Sect. 17.1] giving essentially similar results. The assumption l = constant certainly holds for $T < 10^6$ K, where nuclear burning is negligible, though the assumption m = constant = M breaks down much earlier. But, even if we stress these assumptions somewhat by extending the solutions too far inwards, we will still obtain the correct qualitative behaviour.

Radiative envelopes are found below all hot photospheres ($T > 9{,}000$ K). Towards the deep interior these solutions converge rapidly to the solution with $C = 0$. The interior is therefore relatively insensitive to details of the outer boundary conditions, in particular to the photospheric details.

Below cool atmospheres there are convective envelopes, which extend farther downwards the smaller T_{eff} is. This suggests that a minimum value of T_{eff} might exist where the whole star has become convective (cf. the Hayashi line, Chap. 24). The inward extension of the convective part depends rather sensitively on the precise position of the photosphere and the details of the over-adiabatic layer. Small changes in even the outer solution, which are otherwise rather unimportant, can exert a remarkable influence on the interior, and the same is true for the uncertainties in the treatment of superadiabatic convection.

11.3.4 The T−r Stratification

Sometimes it is useful to know how $T = T(r)$ increases below the photosphere. From the definition of $\nabla \equiv d \ln T / d \ln P$ we have $dT = T\nabla dP/P$, where we replace dP by using the hydrostatic equation in the form

$$dP = -\frac{Gm}{r^2}\varrho\, dr = Gm\varrho d\left(\frac{1}{r}\right) \tag{11.31}$$

and eliminate $T\varrho/P = \mu/\Re$ by means of the equation of state for a perfect gas. We then have

$$dT = \nabla \frac{G\mu}{\Re} m \, d\left(\frac{1}{r}\right). \tag{11.32}$$

For the outer envelope with low density we may approximate m by the surface value M, so that if ∇ is constant between points 1 and 2, we can integrate (11.32) to obtain

$$T_1 - T_2 = \nabla \frac{GM\mu}{\Re} \left(\frac{1}{r_1} - \frac{1}{r_2}\right). \tag{11.33}$$

Let the subscript 2 indicate the photosphere, i.e. $T_2 = T_{\rm eff}$ and $r_2 = R$. Now at any point $r = r_1$ in the envelope we have

$$T - T_{\rm eff} = f\left(\frac{R}{r} - 1\right), \quad f = \nabla \frac{g\mu}{\Re} \frac{M}{R}. \tag{11.34}$$

As a simple example we take $M = M_\odot$, $R = R_\odot$ and a solution with $C = 0$ (see Sect. 11.3.1), for which we found that $\nabla = 0.235$. With $\mu = 1$ we find that $f = 5.4 \times 10^6$ K. This large value of f provides for a very rapid increase of T below the photosphere. Within only 2 % of the radius, T has reached 10^5 K. And at $r \approx 0.8R$ (where $m \approx 0.99M$ still) the temperature exceeds 10^6 K, which also shows that the "average" T for all mass elements of the star is well above 10^6 K.

Chapter 12
Numerical Procedure

For realistic material functions no analytic solutions are possible, so that one depends all the more on numerical solutions of the basic differential equations. Consequently the activity and the number of results in this field has increased with the numerical capabilities. The growth of computing facilities by leaps and bounds since the 1960s may be illustrated by a remark of Schwarzschild (1958): "A person can perform more than twenty integration steps per day", so that "for a typical single integration consisting of, say, forty steps, less than two days are needed". The situation has changed drastically since those days when the scientist's need for meals and sleep was an essential factor in the total computing time for one model. Nowadays one asks rather for the number of solutions produced per second. And these modern solutions are enormously more refined (numerically and physically) than those produced 40 years ago. This progress has been possible because of the introduction of large and fast electronic computers and the simultaneous development of an adequate numerical procedure connected with the name of L.G. Henyey. His method for calculating models in hydrostatic equilibrium is now generally used and will be described later. For more details and for further references see Kippenhahn et al. (1967). If inertia terms with $\ddot{r} \neq 0$ become important, one needs a so-called "hydrodynamic" procedure (see Sect. 12.3).

12.1 The Shooting Method

It is not difficult to see that the appropriate choice of a numerical procedure is anything but a trivial matter. Consider the simplest case, the calculation of a model in complete equilibrium at a given time, for given mass M and given chemical composition $X_i(m)$. The "spatial problem" can then be separated and is described by the structure equations (10.1), (10.2), (10.4) and (10.16). The naïve attempt simply to integrate them from one boundary to the other would encounter the difficulty that the boundary conditions are split, one pair being given at the centre,

R. Kippenhahn et al., *Stellar Structure and Evolution*, Astronomy and Astrophysics Library, DOI 10.1007/978-3-642-30304-3_12,

the other at the surface. Moreover, a test calculation starting with trial values P_c, T_c at the centre has little chance of meeting the correct surface conditions. Outward integrations differing only a little near the centre have the tendency to diverge strongly when approaching the surface (see Sect. 11.3). The reason is that for radiative transport (10.4) with (10.6) contains the factor T^{-4}. For inward integrations starting with trial values R, L at the surface another divergence occurs near the centre owing to the singularity produced by the factor r^{-4} in (10.2).

A compromise between these two possibilities is a fitting procedure often used in earlier, non-automized computations. Outward and inward integrations were both carried to an intermediate fitting point, where they were fitted smoothly to each other by a gradual variation of the trial values P_c, T_c and R, L. The simultaneous fit of four variables (r, P, T, l) is, in principle, possible, since one can vary four free parameters (P_c, T_c, R, L) in the partial solutions. The fitting point is preferably chosen to be at the interface between physically different regions. For example, one takes the border between a convective central core and a radiative envelope, or between regions of different composition.

Fitting methods turned out to be unsuitable for calculating large series of complicated models. For these purposes they were generally replaced by the Henyey method. There are, however, certain applications where a fitting method is still unsurpassed, for example, if one wishes to find *all* possible solutions for given core and envelope parameters. Another application is the generation of the very first model for an evolutionary sequence, since the relaxation methods, which will be introduced in the next section, always need a trial model for finding a solution. For chemically homogeneous stars the shooting methods are well suited to construct such initial models.

12.2 The Henyey Method

This method is very practical, especially for solving boundary-value problems where the conditions are given at both ends of the interval. A trial solution for the whole interval is gradually improved upon in consecutive iterations until the required degree of accuracy is reached. In each iteration, corrections to *all* variables at *all* points are evaluated in such a way that the effect of each of them on the whole solution (including the boundaries) is taken into account. In a generalized Newton–Raphson method, corrections are obtained from linearized algebraic equations.

For spherical stars in hydrostatic equilibrium we have the partial differential equations (10.1)–(10.5) together with boundary conditions at the centre and at the surface. In addition the proper initial values have to be specified as well as the stellar mass M. The general structure of the system of equations suggests that one should treat two subsystems separately and alternately. First, the system (10.1)–(10.4) is solved for given $X_i(m)$, then (10.5) is applied to a small time step Δt, after which (10.1)–(10.4) is solved for the new values of $X_i(m)$, and so on. In modern language such an approach is called *operator splitting*. In this way one can construct a whole evolutionary sequence of models (But one should be aware of the fundamental inconsistency inherent to this approach, which was discussed in Chap. 10.). We now describe in detail the first of these two steps, the solution of the "spatial system".

If there is complete equilibrium ($\ddot{r} = \dot{P} = \dot{T} = 0$), the initial values to be given are the $X_i(m)$, so that we can treat them as known parameters for any point. According to (10.7)–(10.14) the material functions $\varepsilon, \kappa, \varrho, \ldots$ on the right-hand sides of (10.1), (10.2), (10.4) and (10.16) can be replaced by their dependencies upon P and T. Then we have to solve the four ordinary differential equations (10.1), (10.2), (10.4) and (10.16) for the four unknown variables r, P, T, l in the interval $[0, M]$ (where M is also thought to be given).

The case of hydrostatic equilibrium ($\ddot{r} = 0$) but thermal non-equilibrium ($\dot{P} \neq 0, \dot{T} \neq 0$) is almost equivalent, the only difference being the additional term ε_g in (10.3), which contains the partial derivatives $\dot{P}$ and $\dot{T}$. This requires as initial values for the earlier time $t_0 - \Delta t$ not only the $X_i(m)$ but also $T(m)$ and $P(m)$ (See the remarks on possible initial values in Chap. 10.). Assume that we take them from a "foregoing" solution, calling these given functions $P^*(m), T^*(m)$. At any point $m = m_j$, we denote the variables by P_j, T_j and replace the time derivatives $\dot{P}_j, \dot{T}_j$ by

$$\dot{P}_j = \frac{1}{\Delta t}(P_j - P_j^*), \quad \dot{T}_j = \frac{1}{\Delta t}(T_j - T_j^*) . \tag{12.1}$$

The given values of $\Delta t, P_j^*, T_j^*$ can now be considered known parameters. Then $\dot{P}_j, \dot{T}_j$ are functions of P_j, T_j only, as is the case with all material functions, and therefore we can also consider ε_g to be replaced by the function $\varepsilon_g(P, T)$, and the situation is as before with the complete equilibrium models: we again have the four ordinary differential equations (10.1)–(10.4) for the four unknown variables r, P, T, l, but with a somewhat different right-hand side of (10.3).

Let us write these four differential equations briefly as

$$\frac{dy_i}{dm} = f_i(y_1, \ldots, y_4), \quad i = 1, \ldots, 4 , \tag{12.2}$$

where we have used the abbreviations $y_1 = r, y_2 = P, y_3 = T, y_4 = l$. The next step is discretization, i.e. we proceed from the differential equations (12.2) to corresponding difference equations for a finite mass interval $[m^j, m^{j+1}]$. Let us denote the variables at both ends of this interval by upper indices, for example, $y_1^j, y_1^{j+1}, \ldots, y_4^j, y_4^{j+1}$. The functions f_i on the right-hand sides of (12.2) have to be taken for some average arguments we call $y_i^{j+1/2}$; they are a combination of y_i^j and y_i^{j+1}, for example, the arithmetic or the geometric mean. If we define the four functions

$$A_i^j := \frac{y_i^j - y_i^{j+1}}{m^j - m^{j+1}} - f_i\left(y_1^{j+1/2}, \ldots, y_4^{j+1/2}\right), \quad i = 1, \ldots, 4 , \tag{12.3}$$

then the difference equations replacing (12.2) for the mass interval between m_j and m_{j+1} are

$$A_i^j = 0, \quad i = 1, \ldots, 4 . \tag{12.4}$$

The difference equations (12.4) and (12.1) represent a linearization of the differential equations and are therefore an approximation, the accuracy of which has to be controlled. Obviously, the smaller Δt and $\Delta m^j = m^j - m^{j-1}$, the better the approximation. In practical circumstances the spatial discretization is not constant throughout the stellar model, but depends on the changes of the physical variables. A good approach is to choose Δm^j for each j such, that all variables change by less than a predefined upper limit between points j and $j-1$. That maximum change will differ between variables and has to be determined by numerical experiments reducing it to a limit from where on the numerical solution no longer depends on the Δm^j significantly. Apart from this basic control algorithm there are more advanced methods, which, for example, take into account not only the slope but the curvature of the functions $T(m)$, $P(m)$ (Wagenhuber and Weiss 1994). The advantage of this method is that it is sensitive to deviations from linear behaviour. It places many grid points where the variables are a strongly non-linear function of m, while it uses very few in the opposite case. Wagenhuber called this the *curvature method*, as opposed to the simpler *gradient method*.

It is possible to exclude the outermost envelope of the star from the iteration procedure, since time-consuming computations may be necessary for this part (e.g. partial ionization and superadiabatic convection). With sufficient computing power this is no longer a necessity, however. Another situation where this would be advisable is when fully realistic atmospheres are to be connected to the interior of the star, since the diffusion approximation (10.6) is not valid at $m = M$ but at some deeper layer where the optical depth $\tau \gg 1$. The lower boundary of such an atmosphere then provides the upper boundary of the interior model. As described in Sect. 11.2 the outer boundary conditions are imposed at a fitting mass m_{F}, which may have the special value $m = M$ and may have the upper index $j = 1$, and they are formulated by the two equations (11.18) that relate the variables $y_1^1, \ldots, y_4^1$ at $m^1 = m_{\mathrm{F}}$. These equations are specific choices and may differ. With the definitions

$$B_1 := y_2^1 - \pi\left(y_1^1, y_4^1\right), \quad B_2 := y_3^1 - \theta\left(y_1^1, y_4^1\right) , \tag{12.5}$$

equations (11.19) become

$$B_i = 0 , \quad i = 1, 2 . \tag{12.6}$$

As described in Sect. 11.2 the functions π, θ have to be derived by "downward" integrations starting with different trial values of R, L. In practice this may be greatly simplified if we content ourselves with a linear approximation for π and θ (i.e. taking the tangential planes instead of the complicated surfaces in Fig. 11.1). Then only three trial integrations suffice to determine all coefficients in B_1 and B_2.

In the innermost interval of m, between the central point $m^K (= 0)$ and m^{K-1}, we apply series expansions for all four variables as given by (11.3), (11.4), (11.6) and (11.9). These four equations are written as

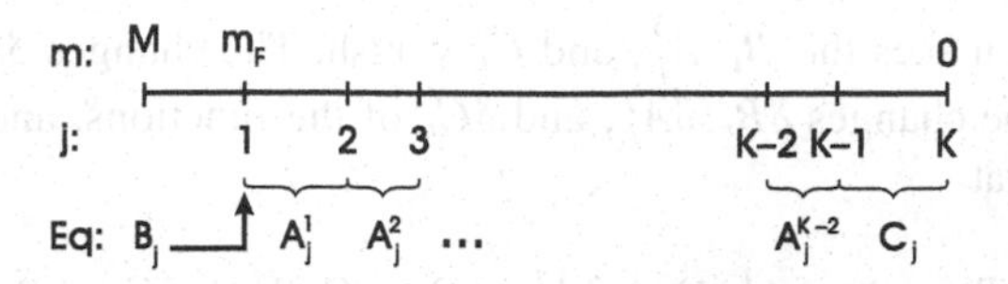

Fig. 12.1 Sketch of the mesh points in the interior solution, from the fitting mass $m = m_F$ (in this example $m_F < M$) to the centre ($m = 0$). It is also indicated which of the equations (12.4), (12.6) and (12.7) have to be fulfilled at m_F or between two adjacent mesh points

$$C_i\left(y_1^{K-1}, \ldots, y_4^{K-1}, y_2^K, y_3^K\right) = 0, \quad i = 1, \ldots 4, \tag{12.7}$$

which already incorporates the central boundary conditions $y_1^K = y_4^K = 0$ (i.e. $r = l = 0$ at the centre).

Consider now the whole interval of m, between $m^K = 0$ and the fitting mass $m^1 = m_F$, to be divided into $K - 1$ intervals (usually not equidistant) by K mesh points as sketched in Fig. 12.1. At these K mesh points we have $(4K - 2)$ unknown variables (since $y_1^K = y_4^K = 0$), and in order to have a solution, these unknowns have to fulfil the following equations: (12.6) for the outer boundary, (12.4) for each interval except the last one ($j = 1, \ldots, K-2$), and (12.7) for the central boundary; thus there are $2 + 4(K - 2) + 4 = 4K - 2$ equations, which may be written:

$$\begin{aligned} B_i &= 0, \quad i = 1, 2, \\ A_i^j &= 0, \quad i = 1, \ldots, 4, \quad j = 1, \ldots, K-2, \\ C_i &= 0, \quad i = 1, \ldots, 4. \end{aligned} \tag{12.8}$$

Suppose that we are looking for a solution for given values of $M, X_i(m)$, $P^*(m), T^*(m)$ (which all enter into these equations as parameters). And suppose, furthermore, that we have a first approximation to this solution, say, $(y_i^j)_1$ with $i = 1, \ldots, 4$, $j = 1, \ldots, K$ (This may be a rough first guess, e.g. obtained by an extrapolation of a foregoing solution or a solution for similar parameters. It may also be obtained from a shooting method.). Since the $(y_i^j)_1$ are only an approximation, they will not fulfil (12.8), i.e. when we use them as arguments in the functions A_i^j, B_i, and C_i, we find that

$$B_i(1) \neq 0, \quad A_i^j(1) \neq 0, \quad C_i(1) \neq 0, \tag{12.9}$$

where we indicate by (1) that the first approximation is used as arguments. Let us now look for corrections δy_i^j for all variables at all mesh points such that the second approximation

$$\left(y_i^j\right)_2 = \left(y_i^j\right)_1 + \delta y_i^j \tag{12.10}$$

of the arguments makes the B_i, A_i^j, and C_i vanish. The changes δy_i^j of the arguments produce the changes $\delta B_i, \delta A_i^j$, and δC_i of the functions, and we obviously have to require that

$$B_i(1) + \delta B_i = 0, \quad A_i^j(1) + \delta A_i^j = 0, \quad C_i(1) + \delta C_i = 0 \,. \tag{12.11}$$

For small enough corrections, we may expand the $\delta B_i, \ldots$ in terms of increasing powers of the corrections δy_i^j, and keep only the linear terms in this expansion; for example,

$$\delta B_1 \approx \frac{\partial B_1}{\partial y_1^1}\delta y_1^1 + \frac{\partial B_1}{\partial y_2^1}\delta y_2^1 + \frac{\partial B_1}{\partial y_3^1}\delta y_3^1 + \frac{\partial B_1}{\partial y_4^1}\delta y_4^1 \,. \tag{12.12}$$

For (12.5) the third term would vanish because in this special case B_1 is independent of y_3. With this linearization (12.11) can be written as

$$\frac{\partial B_i}{\partial y_1^1}\delta y_1^1 + \cdots + \frac{\partial B_i}{\partial y_4^1}\delta y_4^1 = -B_i \,, \qquad i = 1, 2 \,,$$

$$\frac{\partial A_i^j}{\partial y_1^j}\delta y_1^j + \cdots + \frac{\partial A_i^j}{\partial y_4^j}\delta y_4^j + \frac{\partial A_i^j}{\partial y_1^{j+1}}\delta y_1^{j+1} + \cdots + \frac{\partial A_i^j}{\partial y_4^{j+1}}\delta y_4^{j+1} = -A_i^j \,,$$
$$i = 1, \ldots, 4, \quad j = 1, \ldots, K-2 \,, \tag{12.13}$$

$$\frac{\partial C_i}{\partial y_1^{K-1}}\delta y_1^{K-1} + \cdots + \frac{\partial C_i}{\partial y_4^{K-1}}\delta y_4^{K-1} + \frac{\partial C_i}{\partial y_2^K}\delta y_2^K + \frac{\partial C_i}{\partial y_3^K}\delta y_3^K = -C_i \,, \qquad i = 1, \ldots, 4 \,.$$

(The B_i, A_i^j, C_i, and all derivatives have here to be evaluated using the first approximation as arguments.) This is a system of $2+4(K-2)+4 = 4K-2$ linear, inhomogeneous equations for the $4K-2$ unknown corrections δy_i^j ($i = 1, \ldots, 4$ and $j = 1, \ldots, K$; but $\delta y_1^K = \delta y_4^K = 0$ because of the central boundary conditions). Equation (12.13) may be written concisely in matrix form as

$$H \begin{pmatrix} \delta y_1^1 \\ \cdot \\ \cdot \\ \cdot \\ \delta y_3^K \end{pmatrix} = - \begin{pmatrix} B_1 \\ \cdot \\ \cdot \\ \cdot \\ C_4 \end{pmatrix}, \tag{12.14}$$

where the matrix H of the coefficients is called the *Henyey matrix*; its elements are the derivatives on the left-hand sides of (12.13).

Usually H has a non-vanishing determinant, det $H \neq 0$, and we can solve these linear equations, obtaining the wanted corrections δy_i^j. These are applied as shown in (12.10) to obtain a second, better approximation $(y_i^j)_2$. When using these

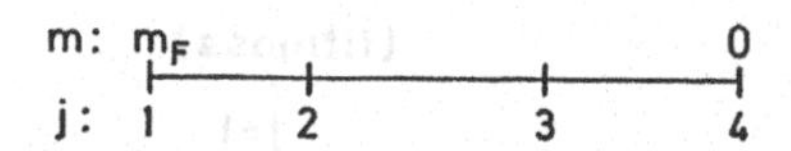

Fig. 12.2 Mesh points in the "three-layer model"

second approximations as arguments, we will generally still find $B_i \neq 0, A_i^j \neq 0$, and $C_i \neq 0$, i.e. equations (12.8) are not yet fulfilled. This is because the corrections were calculated from the *linearized* equations (12.13), while equations (12.8) are non-linear (Even if we had linear equations instead of (12.8), the solution might require several iterations, since the numerical solution of (12.13) has only limited accuracy.). Therefore in a second iteration step we calculate new corrections by the same procedure to obtain a third approximation

$$(y_i^j)_3 = (y_i^j)_2 + \delta y_i^j \ , \tag{12.15}$$

and so on. In consecutive iterations of this type, the approximate solution can be improved until either the absolute values of all corrections δy_i^j, or the absolute values of all right-hand sides in (12.13), drop below a chosen limit. Then we have approached the solution with the required accuracy.

If a time sequence of models is to be produced, one can now change the parameters appropriately for a new small time step Δt [by evaluating from (10.5) the change of the $X_i(m)$, and by redefining the just-calculated $P(m), T(m)$ as the new $P^*(m), T^*(m)$]. The new model for $t + \Delta t$ is then calculated by the Henyey method in the same manner as for the model for t.

Of course, there is no guarantee that the iteration procedure for improving the approximations really does converge. In fact often enough one finds divergence if the chosen approximation is too far from the solution; then the required corrections are so large that one cannot neglect the second-order terms when evaluating $\delta B_i, \delta A_i^j$, and δC_i in (12.11), and the linearized equations (12.14) therefore yield wrong corrections.

What happens, on the other hand, if we take a given precise solution as the "first approximation"? It fulfils (12.8) such that the right-hand sides of (12.14) vanish. Equation (12.14) is then a system of *homogeneous* linear equations, which for det $H \neq 0$ has only the trivial solution $\delta y_i^j = 0$: in this (normal) case, there is no other solution ("local uniqueness" as mentioned in Sect. 12.6). If, however, det $H = 0$, then we obtain solutions $\delta y_i^j \neq 0$, i.e. other solutions for the same parameters. In this somewhat pathological situation the "local uniqueness" of the solution is violated.

The Henyey matrix and its determinant are obviously important quantities. This concerns also their connection with the stability properties (see Sect. 12.6). It is worthwhile noting the general structure of H, which turns out to be very simple. This is most easily demonstrated by considering the simple "three-layer model", which has only four mesh points from centre to fitting mass (Fig. 12.2). One interval is adjacent to m_F, one to the centre, while the intermediate interval

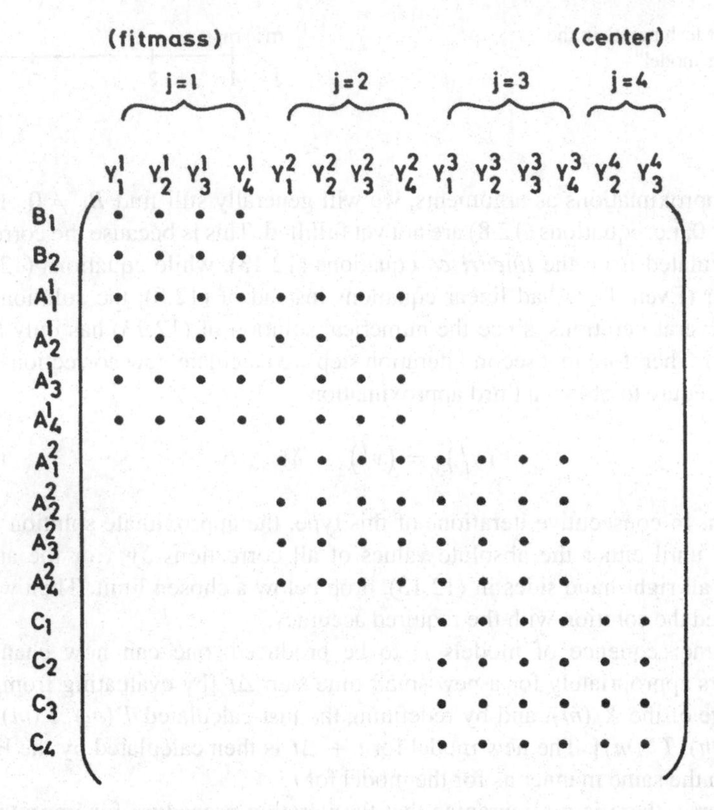

Fig. 12.3 Structure of the Henyey matrix H for the three-layer star sketched in Fig. 12.2. A *dot* in, for example, the column y_i^j and the row A_k^l represents the matrix element $\partial A_k^l/\partial y_i^j$. All matrix elements outside the *dotted area* are zero

borders on neither of these two boundaries, so that the full generality of possible cases is exhibited. Any further mesh point will only duplicate the situation of the intermediate interval. The Henyey matrix H for this three-layer star is indicated in Fig. 12.3, where a dot in a column under y_i^j and in a row denoted at the left-hand side by A_k^l represents a matrix element $\partial A_k^l/\partial y_i^j$. Some of these derivatives will be zero, since some basic equations do not depend on all variables [e.g. (10.16) does not contain $y_1 = r$]. Outside the dotted area there are only zero elements, because the first-order scheme (12.13) connects only neighbouring points. The Henyey matrix therefore has non-vanishing elements only in overlapping blocks along the main diagonal, so that this can be easily used for devising simple and well-behaved algorithms for computing det H and inverting the matrix through elimination processes. The most widely used method for solving such block matrices in stellar evolution codes is that by Henyey et al. (1964), which was described in all details by Kippenhahn et al. (1967). The basic idea is to express the corrections of the block matrix connecting points $(j, j + 1)$ in terms of the quantities of the next block $(j+1, j+2)$, and so on. At the end there is a final block (usually the innermost one), for which the corrections are determined by matrix inversion, and from which on then all the other corrections can be calculated by going backwards again. The Henyey method has K inversions of matrices of size 4×8 instead of straightforwardly inverting the Henyey matrix of size $K \times K$. It therefore grows only linearly–instead of quadratically–with increasing number of grid points.

12.3 Treatment of the First- and Second-Order Time Derivatives

When devising a numerical scheme for solving our partial differential equations one can choose many details more or less arbitrarily without greatly affecting the results. This concerns questions such as the prescription for averaging between spatial mesh points, and the definition of the variables; these can be, for example, the physical quantities themselves, their logarithms, or any other functions describing them properly.

Concerning the manner in which the time derivatives are approximated, one distinguishes between explicit and implicit schemes that are known to behave differently, in particular when one is dealing with second-order time derivatives. Forward integration in time, starting from given initial values, can require time steps of various length, and the results can be unstable with respect to small numerical errors. In Sect. 12.2 we encountered examples of both types of scheme:

An *explicit* scheme was indicated in the case of the chemical equations (10.5). Consider the time interval between t^n (at which all variables q^n are supposed to be known) and t^{n+1} (for which the variables q^{n+l} are to be calculated). We may use (10.5) simply in order to calculate time derivatives $\dot{X}_i^n$ of the chemical composition from the known reaction rates r_{ik}^n and densities ϱ^n. The composition for t^{n+l} is then evaluated as $X_i^{n+1} = X_i^n + \Delta t \dot{X}_i^n$ before the other variables for this time are derived. In fact the X_i^{n+1} are used as fixed parameters when calculating the solution at t^{n+1} by iteration. Such a procedure is relatively simple, and in general, the results can be sufficiently accurate if the time steps are kept small enough. However, there is no guarantee to prevent unphysical solutions in explicit methods. For example, if $\dot{X}_i^n$ is sufficiently negative even a small time step might lead to a negative X_i^{n+1}. To prevent this, an *implicit* treatment is indicated. If $\dot{X}_i$ depends on the chemical abundances itself, as is the case for the nuclear reactions (10.5), the abundance at t^{n+1} is used on the right-hand side, too. This constitutes a set of implicit equations, which need to be solved by inversion methods, but which are numerically stable. An easy way is by writing $X_i^{n+1} = X_i^n + \Delta t \dot{X}_i^n = X_i^n + \Delta X_i^n$ and linearizing the equations in the ΔX_i^n, neglecting all higher terms. The resulting system of equations is linear in ΔX_i^n and can be solved by one matrix inversion. However, the quality of the linearization again depends on the size of Δt. Such implicit schemes are generally used to solve networks of nuclear reactions, where the terms $\dot{X}_i^n$ may vary by many order of magnitudes.

In the set of structure equations (10.1)–(10.4) to be solved at time t_i^{n+1} for given X_i^{n+1} the energy equation (10.3) contains the time derivatives of $\dot{P}$ and T. With respect to these an *implicit* scheme was used in Sect. 12.2. According to (12.1) the $\dot{P}$ and $\dot{T}$ are replaced by $(P^{n+1} - P^n)/\Delta t$ and $(T^{n+1} - T^n)/\Delta t$, respectively. These time derivatives are therefore considered to depend also on the variables at time t^{n+1} and are evaluated together with them in the iteration procedure. In principle one could also have used an explicit method. For example, replace $\dot{P}$ and $\dot{T}$ in (10.3) by the time derivative of the entropy s and use this equation only in order to evaluate $\dot{s}^n$ at time t^n. Then, as in the case of the chemical composition, the solution for t^{n+1} is calculated for a given, fixed entropy $s^{n+1} = s^n + \Delta t \dot{s}^n$ from the other equations.

It is well known that, for differential equations that involve first-order derivatives in time and first- (or higher-) order spatial derivatives, implicit methods allow larger time steps for a given spacing in mass; for explicit difference schemes the time step has to be kept small to avoid numerical instability (For details see, for instance, Richtmyer and Morton 1967.).

Let us now turn to the so-called*hydrodynamical problem,* which arises when the inertial term in the equation of motion cannot be neglected. Then in addition to the first-order time derivatives in (10.3) there is a second-order time derivative in (10.2), as in (2.16). One usually introduces the radial velocity

$$v = \frac{\partial r}{\partial t} \tag{12.16}$$

of the mass elements as a new variable, with which (10.2) becomes

$$\frac{\partial P}{\partial m} = -\frac{Gm}{4\pi r^4} - \frac{1}{4\pi r^2}\frac{\partial v}{\partial t} \,. \tag{12.17}$$

When using (12.16) and (12.17) instead of (2.16) one has again to deal with first-order time derivatives only. These can be replaced by ratios of differences, and one can use an explicit or an implicit scheme as before, the explicit being simpler but demanding smaller time steps. However, this is not the only choice to be made. For example, within the framework of an explicit method, the different variables can be defined at different times (say, the radius values at $t^n, t^{n+1}, \ldots$, and the velocities at the intermediate times $t^{n-1/2}, t^{n+1/2}, \ldots$). Furthermore, one may devise a scheme which treats the mechanical equations explicitly but is implicit with respect to the time derivatives in the energy equation (10.3).

The presence of the second-order time derivatives changes the properties of the equations and the behaviour of the numerical procedure considerably. Whenever an explicit scheme is used, the time steps have to be kept small in order to fulfil the Courant condition, according to which the time step Δt must not exceed $\Delta r / v_s$, where Δr is the thickness of the smallest mass shell and v_s is the local velocity of sound.

12.4 Treatment of the Diffusion Equation

The diffusion equation (8.25) contains first-order derivatives in time and second-order derivatives in space for the N chemical species. It may be supplemented by the nuclear reactions of (10.5), and by the additional term for diffusive convective mixing (8.28) to achieve a consistent treatment of "burning and mixing", but these terms do not change the nature of the equations further.

The left-hand side of (8.25) can again be written as $(X_i(t + \Delta t) - X_i(t))/\Delta t$, and $X_i(t + \Delta t)$ is the quantity to be determined. As with the nuclear reactions (10.5) discussed in 12.3, an implicit scheme is to be preferred for sake of numerical stability, implying that on the right-hand side $X_i(t + \Delta t)$ is used, too. This constitutes at each grid point a set of N implicit equations, which can be solved either through linearization or iteration. However, in contrast to the situation we found for the nuclear network, these sets of equations are now coupled between grid points due to the spatial derivatives of the diffusion equation.

These second-order spatial derivatives of, for example, $\ln T$, are calculated in two steps. First, the first-order derivative for grid point j is approximated in the standard way by

$$\frac{\Delta \ln T^j}{\Delta r^j} = \frac{\ln T^j - \ln T^{j-1}}{r^j - r^{j-1}} \tag{12.18}$$

and similarly for $j + 1$. Then the second-order derivative at grid point j can be calculated from

$$\left.\frac{\partial^2 \ln T}{\partial r^2}\right|_j \approx \left(\frac{\Delta \ln T^{j+1}}{\Delta r^{j+1}} - \frac{\Delta \ln T^j}{\Delta r^j}\right) / \left(\bar{r}^{j+1} - \bar{r}^j\right), \tag{12.19}$$

where $\bar{r}^j$ is a suitable mean value for r in the interval $(j, j + 1)$. In the simplest case it is the arithmetic mean and thus the denominator in (12.19) reduces to $(r^{j+1} - r^{j-1})/2$. All other quantities in (8.25) appearing in front of the first-order derivatives, such as $A_T(i)$, also have to be taken as mean quantities for the second derivative in analogy to (12.19). We note that the spatial derivatives are defined here at each grid point, contrary to the system of equations (12.3), where the derivatives were defined for the shell between j and $j + 1$. One may imagine that the shells now are centred at a grid point, extending halfway to the neighbouring ones. The advantage of this definition is that the diffusion equations are defined at the same location as the nuclear network equations.

In this way, the discretized equations for the N elements at the $M - 2$ grid points from $j = 2, \ldots, M - 1$ contain values of the X_i at three grid points $j - 1, j, j + 1$. As for the structure equations, they are solved by iterating for $X_i(t + \Delta t)$, starting with the initial trial values $X_i(t)$, which are already known. The iteration method can again be the standard Newton–Raphson method, which requires first-order derivatives of all quantities appearing in (8.25). The complete system of equations is similar to (12.13) with the exception that three instead of two neighbouring grid points are connected. It is therefore obvious that the Henyey method will be applicable again, the only difference being that the block matrices are now of dimension $N \times 3N$.

The missing two equations to complete the system for the N elements result from the boundary conditions at $j = 1$ and $j = M$, which follow from mass conservation. We follow here the formulation by Schlattl (1999), where also more technical details concerning the solution of (8.25) can be found.

Mass conservation leads to

$$\sum_{j=1}^{M} \left(X_i^j(t + \Delta t) - X_i^j(t) \right) \Delta m^j = 0, \qquad 1 \leq i \leq N \tag{12.20}$$

where j again denotes the grid point $(1 \leq j \leq M)$ and i the element. Since (8.25) is formulated in Eulerian space, the mass intervals Δm^j have to be defined appropriately, for example, by

$$\Delta m^j = \begin{cases} \frac{1}{2}(m^1 - m^2) & j = 1 \\ \bar{m}^j - \bar{m}^{j+1} & 2 \leq j \leq M - 1 \\ \frac{1}{2} m^{M-1} & j = M. \end{cases} \tag{12.21}$$

Note that mean values for m^j are used in the second line. This way M mass intervals are created.

As an example we formulate the expression for the $\ln T$ term in (8.25), abbreviating $r^2 X_i(t + \Delta t) T^{5/2} A_T$ by K_T. With $\Delta r^j = \Delta m^j / (4\pi \varrho^j (r^j)^2)$ the boundary conditions translate into expressions like

$$\frac{1}{\varrho r^2} \frac{\partial}{\partial r} \left(K_T \frac{\partial \ln T}{\partial r} \right)_{r=R} \approx \frac{2}{\varrho^{j=1} R^2} \frac{\bar{K}_i^{j=2} (\Delta \ln T / \Delta r)^{j=2}}{r^{j=1} - r^{j=2}} \tag{12.22}$$

and

$$\frac{1}{\varrho r^2} \frac{\partial}{\partial r} \left(K_T \frac{\partial \ln T}{\partial r} \right)_{r=0} \approx \frac{24}{\varrho^{j=M} (r^{j=M-1})^2} \frac{\bar{K}_T^{j=M} (\Delta \ln T / \Delta r)^{j=M}}{r^{j=M-1}} \tag{12.23}$$

To simplify reading we have written suffixes indicating grid numbers j explicitly. In (12.23), the linear expansion (11.3) of m at the centre was used to compute $\Delta r^{j=M}$ from $\Delta m^{j=M}$, which involves m^{M-1}.

We finally add that Schlattl (1999) justifies the Eulerian formulation for the diffusion equations, as opposed to our otherwise preferred Lagrangian one, with the necessity for very dense spatial resolution in situations of shallow convective layers.

12.5 Treatment of Mass Loss

The mass loss formulae (9.1)–(9.4) describe only how the stellar mass reduces with time due to stellar winds. Therefore, the treatment in stellar evolution calculations is very simple. Over a time step Δt, during which the chemical composition changes as described in Sect. 12.3, the stellar mass will change according to

$$M(t + \Delta t) = M(t) - \Delta M(t) = M(t) - \dot{M}(t)\Delta t, \quad (12.24)$$

where $\dot{M}(t)$ is the mass loss rate evaluated according to (9.1) or any other similar prescription, using the stellar parameters at time t.

In terms of the mass grid established in Sect. 12.2 a simple removal of all grid points i with $m_i \geq M(t) - \Delta M(t)$ can be done. Such a procedure, of course, ignores all effects of accelerating matter and moving it out of the star's gravitational potential. To treat this correctly, however, a hydrodynamical method with an open outer boundary would be needed, which in most cases is not necessary. Consider the energy spent to remove mass from the stellar surface to infinity. This is, according to (1.13), GM/R per mass unit of the stellar wind. Multiplying with the mass loss rate we obtain the result that $(\dot{M}GM)/R$ erg/s are needed. For the Sun this amounts to 1.2×10^{27} erg/s, which is only 10^{-7} of the solar luminosity, and can therefore be safely ignored. For a very evolved red giant with very strong mass loss the energy spent for expelling mass can reach values up to 0.001 or even 1 % of the stellar luminosity.

While the simple removal of grid points is correct in terms of mass distribution and chemical composition, it is not taking into account thermal effects. Imaging a mass layer that was deep inside the stellar envelope now suddenly being the outermost one, since all overlying layers were expelled. It will be hotter than the surface layers have been before and temperature and pressure will not be that of a photosphere. Thermal relaxation will therefore set in. While the Sun is losing mass continuously, its surface temperature is constant. This is because the timescale for mass loss, $\tau_{\mathrm{ML}} \approx M/\dot{M}$ is of the order of 10^{14} years and therefore much longer than even the nuclear timescale. As long as $\tau_{\mathrm{ML}} \gg \tau_{\mathrm{KH}}$ the outermost layers will quickly expand and restore the previous photospheric conditions. The adjustment, of course, vanishes with increasing depth. Numerical schemes are therefore trying to take this into account: while grid points are removed due to mass loss, the thermal structure of the star remains almost unperturbed. In the opposite case, when $\tau_{\mathrm{ML}} \lesssim \tau_{\mathrm{KH}}$, the layers uncovered by mass loss indeed have no time to change their temperature (pressure can be adjusted, since τ_{hydro} is still much shorter). This, however, may happen only in binary systems during extreme mass transfer episodes. In such cases, a hydrodynamical treatment of the complete system is indicated, anyhow.

12.6 Existence and Uniqueness

As every numerical scheme, the Henyey method sometimes does not converge easily to a solution, and there are cases when it seems to oscillate between two solutions. While in most cases this is a purely numerical issue, resulting for example from insufficiently accurate derivatives, one wonders whether there could also be deeper mathematical reasons. This relates to questions about the existence and uniqueness of the solution. It is closely connected to the determinant of the Henyey matrix, as $\det \mathrm{H} = 0$ obviously does not allow an inversion for determining the δy_i and $\det \mathrm{H} \approx 0$ will lead to numerical problems during the inversion.

An old problem is whether, for stars in complete equilibrium and of given "parameters" (stellar mass M and chemical composition X_i), there exists one, and only one, solution of the basic equations of stellar structure. From simple considerations concerning uncomplicated cases, answers to this question were given in the 1920s by Heinrich Vogt und Henry Norris Russell; however, there is no mathematical basis for this so-called Vogt–Russell theorem, and when by numerical experiments multiple solutions for the same parameters were found to exist it had to be abandoned. The conditions under which uniqueness is violated, and why, have therefore been investigated. A linearized treatment (concerning "local" uniqueness) is easier to understand, whereas non-linear results refer to the "global" behaviour of the solutions and require a more involved mathematical apparatus. Relevant work concerning these issues was done by Kähler (1972, 1975, 1978). For another representation, particularly of the linear problem, see Paczyński (1972).

The mathematical discussion is usually restricted to models in complete or at least hydrostatic equilibrium and analyses the behaviour of solutions under (infinitesimally) small changes of the parameters. Mathematical conditions can be formulated when a solution is locally unique, which can be translated into the statement that the evolution–considered as being a change of parameters (chemical composition and/or entropy) with time–follows a unique sequence of solutions. However, there is no general statement about when such conditions are fulfilled. The condition for having a locally unique solution is equivalent to $\det \mathrm{H} \neq 0$. But even if this condition is fulfilled, there still might be multiple, well-separated solutions. If one of them is unstable, the star switches to the stable one when perturbed. This is related to the general stability of stars.

Behind the mathematical question there is thus also interest concerning the predicted evolution of stars. For example, after learning that often more than one solution exists, that solutions can disappear, or that new solutions appear in pairs, one might begin to wonder whether the star really "knows" how to evolve. But we should keep in mind that normally the star will be brought into one particular state (corresponding to a certain solution) according to its history. And if the equations indicate that the evolution approaches a "critical point", then this means in general only that the approximation used breaks down. For example, if an evolutionary sequence calculated for complete equilibrium comes to a critical point beyond which continuation is not possible, then the difficulties are normally removed by allowing for thermal non-equilibrium. Correspondingly if hydrostatic models that are not in thermal equilibrium evolve to a critical point, the difficulties are usually removed after the introduction of inertia terms. An example would be the reaction of a star when reaching the *Schönberg–Chandrasekhar limit* (Sect. 30.5), where two existing solutions of complete equilibrium merge. The star easily switches from one to the other by leaving thermal equilibrium.

Broadly speaking, it was found that indeed several solutions for the same set of parameters (stellar mass M and chemical composition X_i) exist but that they are widely separated and a star's evolution proceeds along a well-defined sequence of locally unique solutions.

Part III
Properties of Stellar Matter

In addition to the basic variables (m, r, P, T, l) in terms of which we have formulated the problem, the differential equations of stellar structure (Sect. 10.1) also contain quantities such as density, nuclear energy generation, or opacity. These describe properties of stellar matter for given values of P and T and for a given chemical composition as indicated in (10.7)–(10.14) and are quantities that certainly do not depend on m, r, or l at the given point in the star. They could just as well describe the properties of matter in a laboratory for the same values of P, T, and chemical composition. We can therefore deal with them without specifying the star or the position in it for which we want to use them. In this chapter we shall discuss these "material functions", and we start by specifying the dependence of the density ϱ on P, T, and the chemical composition. This is described by an *equation of state,* which is especially simple if we have a perfect gas. We already discussed this case in Sect. 4.2. But radiation and ionization also influence the pressure and the internal energy. We therefore have to include them.

Chapter 13
The Perfect Gas with Radiation

13.1 Radiation Pressure

The pressure in a star is not only given by that of the gas because the photons in the stellar interior can contribute considerably to the pressure, and therefore our discussion of the perfect gas of Sect. 4.2 has to be extended. Since the radiation is practically that of a black body (see Sect. 5.1.1), its pressure P_{rad} is given by

$$P_{\mathrm{rad}} = \frac{1}{3}U = \frac{a}{3}T^4 , \tag{13.1}$$

where U is the energy density and a is the radiation density constant $a = 7.56464 \times 10^{-15}$ erg cm^{-3} K^{-4}. Then the total pressure P consists of the gas pressure P_{gas} and radiation pressure P_{rad}:

$$P = P_{\mathrm{gas}} + P_{\mathrm{rad}} = \frac{\Re}{\mu}\varrho T + \frac{a}{3}T^4 , \tag{13.2}$$

where on the right we have assumed that the gas is perfect. We now define a measure for the importance of the radiation pressure by

$$\beta := \frac{P_{\mathrm{gas}}}{P} , \quad 1 - \beta = \frac{P_{\mathrm{rad}}}{P} . \tag{13.3}$$

For $\beta = 1$ the radiation pressure is zero, while $\beta = 0$ means that the gas pressure is zero. The definition (13.3) can also be used if the gas is not perfect.

Two other relations which can be derived by differentiation of (13.3) are sometimes useful:

R. Kippenhahn et al., *Stellar Structure and Evolution*, Astronomy and Astrophysics Library, DOI 10.1007/978-3-642-30304-3_13, © Springer-Verlag Berlin Heidelberg 2012

$$\left(\frac{\partial \beta}{\partial T}\right)_P = -\left[\frac{\partial(1-\beta)}{\partial T}\right]_P = -\frac{4}{T}(1-\beta)\,, \tag{13.4}$$

$$\left(\frac{\partial \beta}{\partial P}\right)_T = -\left[\frac{\partial(1-\beta)}{\partial P}\right]_T = \frac{1}{P}(1-\beta)\,. \tag{13.5}$$

13.2 Thermodynamic Quantities

From (13.2) we obtain

$$\varrho = \frac{\mu}{\Re}\frac{1}{T}\left(P - \frac{a}{3}T^4\right)\,, \tag{13.6}$$

and with the definitions (6.6) with (13.4), and (13.5) we find that

$$\alpha = \frac{1}{\beta}\,, \qquad \delta = \frac{4-3\beta}{\beta}\,, \quad \varphi = 1\,. \tag{13.7}$$

Indeed, if the radiation pressure can be neglected ($\beta = 1$), we find $\alpha = \delta = 1$, as should be expected for a perfect monatomic gas.

If the gas components are monatomic, then the internal energy per unit mass is

$$u = \frac{3}{2}kT\frac{n}{\varrho} + \frac{aT^4}{\varrho} = \frac{3}{2}\frac{\Re}{\mu}T + \frac{aT^4}{\varrho} = \frac{\Re T}{\mu}\left[\frac{3}{2} + \frac{3(1-\beta)}{\beta}\right]\,, \tag{13.8}$$

so that according to the definition (4.4) of c_P we have

$$c_P = \left(\frac{\partial u}{\partial T}\right)_P + P\left(\frac{\partial v}{\partial T}\right)_P = \left(\frac{\partial u}{\partial T}\right)_P - \frac{P}{\varrho^2}\left(\frac{\partial \varrho}{\partial T}\right)_P\,. \tag{13.9}$$

Using (13.8), after some algebraic manipulations involving (13.4), we obtain

$$\left(\frac{\partial u}{\partial T}\right)_P = \frac{\Re}{\mu}\left[\frac{3}{2} + \frac{3(4+\beta)(1-\beta)}{\beta^2}\right]\,. \tag{13.10}$$

From the definition of δ with (13.7)–(13.9) we write

$$c_P = \frac{\Re}{\mu}\left[\frac{3}{2} + \frac{3(4+\beta)(1-\beta)}{\beta^2} + \frac{4-3\beta}{\beta^2}\right]\,, \tag{13.11}$$

and then the relation (4.21) may be applied in order to determine the adiabatic gradient ∇_{ad} for the perfect gas plus radiation:

$$\nabla_{\mathrm{ad}} = \frac{\Re\delta}{\beta\mu c_P} = \left(1 + \frac{(1-\beta)(4+\beta)}{\beta^2}\right) \Big/ \left(\frac{5}{2} + \frac{4(1-\beta)(4+\beta)}{\beta^2}\right)\,. \tag{13.12}$$

For $\beta \rightarrow 1$, (13.11) and (13.12) give the well-known values for the perfect monatomic gas: $c_P = 5\Re/(2\mu)$ and $\nabla_{ad} = 2/5$, while for $\beta \rightarrow 0$, one has $\nabla_{ad} \rightarrow 1/4$ and c_P becomes infinite.

Sometimes the derivative

$$\frac{1}{\gamma_{ad}} := \left(\frac{d \ln \varrho}{d \ln P}\right)_{ad} \tag{13.13}$$

is required (4.37, 4.41). If in the definition

$$\frac{d\varrho}{\varrho} = \alpha \frac{dP}{P} - \delta \frac{dT}{T} \tag{13.14}$$

of α and δ the adiabatic condition $PdT/(TdP) = \nabla_{ad}$ is introduced, one finds

$$\gamma_{ad} = \frac{1}{\alpha - \delta \nabla_{ad}} . \tag{13.15}$$

In the case of a perfect gas with radiation pressure we have to introduce the expressions (13.7), while for the limit $\beta = 1$, we find

$$\gamma_{ad} = \frac{1}{1 - \nabla_{ad}} . \tag{13.16}$$

For a monatomic gas without radiation pressure ($\beta = 1$) one has $\nabla_{ad} = 0.4$ and therefore $\gamma_{ad} = 5/3$, whereas in the limit $\beta \rightarrow 0$–after α, δ, and ∇_{ad} are inserted from (13.7) and (13.12)–we find for a gas dominated by radiation pressure that

$$\gamma_{ad} \rightarrow \frac{4}{3} , \qquad \nabla_{ad} \rightarrow \frac{1}{4} . \tag{13.17}$$

Instead of γ_{ad}, ∇_{ad}, one often uses the "adiabatic exponents" introduced by Chandrasekhar, which are defined by

$$\Gamma_1 := \left(\frac{d \ln P}{d \ln \varrho}\right)_{ad} = \gamma_{ad} , \tag{13.18}$$

$$\frac{\Gamma_2}{\Gamma_2 - 1} := \left(\frac{d \ln P}{d \ln T}\right)_{ad} = \frac{1}{\nabla_{ad}} , \tag{13.19}$$

$$\Gamma_3 := \left(\frac{d \ln T}{d \ln \varrho}\right)_{ad} + 1 , \tag{13.20}$$

and obey the relation

$$\frac{\Gamma_1}{\Gamma_3 - 1} = \frac{\Gamma_2}{\Gamma_2 - 1} . \tag{13.21}$$

Chapter 14
Ionization

In Sect. 4.2 and Chap. 13 we assumed complete ionization of all atoms. This is a good approximation in the very deep interior, where T and P are sufficiently large, but the degree of ionization certainly becomes smaller if one approaches the stellar surface, where T and P are small. In the atmosphere of the Sun, for instance, hydrogen and helium atoms are neutral. When a gas is partially ionized the mean molecular weight and thermodynamic properties such as c_P depend on the degree of ionization. It is the aim of this section to show how this can be calculated and how it influences the properties of the stellar gas.

14.1 The Boltzmann and Saha Formulae

We consider the atoms of a chemical element in a certain state of ionization, contained in a unit volume of gas in thermodynamic equilibrium. They are distributed over many states of excitation, which we denote by subscript s, and these different states can be degenerate such that the state of number s consists in reality of g_s substates. The number g_s is the *statistical weight*. Consider in particular the atoms of a certain element in state s and in the ground state $s = 0$, separated by the energy difference ψ_s, and the transition between both, say, by emission and absorption of photons. In equilibrium, the rate of such upward transitions is equal to that of downward transitions. This gives as the ratio between the numbers of atoms in the two states:

$$\frac{n_s}{n_0} = \frac{g_s}{g_0} \mathrm{e}^{-\psi_s/kT} . \tag{14.1}$$

Equation (14.1) is the well-known *Boltzmann formula*, which governs the distribution of particles over states of different energy.

Instead of referring to the atoms in the ground state, we want to compare the atoms of state s with the number n of *all* atoms of that element:

R. Kippenhahn et al., *Stellar Structure and Evolution*, Astronomy and Astrophysics Library, DOI 10.1007/978-3-642-30304-3_14, © Springer-Verlag Berlin Heidelberg 2012

$$n = \sum_s n_s \ . \tag{14.2}$$

From (14.1), multiplication by g_0 and summation over all states leads to

$$g_0 \frac{n}{n_0} = g_0 \sum_{s=0}^{\infty} \frac{n_s}{n_0} = g_0 + g_1 \, \mathrm{e}^{-\psi_1/kT} + g_2 \, \mathrm{e}^{-\psi_2/kT} + \ldots := u_\mathrm{p} \ , \tag{14.3}$$

where $u_\mathrm{p} = u_\mathrm{p}(T)$ is the so-called *partition function*. From (14.1) and (14.3) we obtain the Boltzmann formula in the form

$$\frac{n_s}{n} = \frac{g_s}{u_\mathrm{p}} \, \mathrm{e}^{-\psi_s/kT} \ . \tag{14.4}$$

We can also use the Boltzmann formula to determine the degree of ionization, but there are differences between excitation and ionization that require attention. Excitation concerns ions and bound electrons distributed over *discrete* states only. In the case of ionization the upper state consists of two separate particles, the ion and the electron; and the free electron has a *continuous* manifold of states. After ionization, say by absorption, the electron "thrown out" can have an arbitrary amount of kinetic energy, and recombination can occur with electrons of arbitrary kinetic energy.

We say an atom is in the rth state of ionization if it has already lost r electrons. The energy necessary to take away the next electron from the ground state is χ_r. After ionization this electron is in general not at rest, but has a momentum relative to the atom of absolute value p_e. Then $p_\mathrm{e}^2/(2m_\mathrm{e})$ is its kinetic energy; therefore relative to its original bound state the free electron has the energy $\chi_r + p_\mathrm{e}^2/(2m_\mathrm{e})$, while the state of ionization of the atom is now $r + 1$.

Let us consider as the lower state an r-times ionized ion in the ground state. The upper state may be that of the $(r + 1)$ times ionized ion plus the free electron with momentum in the interval $[p_\mathrm{e}, p_\mathrm{e} + dp_\mathrm{e}]$. The number densities of ions in these two states are n_r and dn_{r+1}. The statistical weight of the upper state is the product of g_{r+1} of the ion and of $dg(p_\mathrm{e})$, the statistical weight of the free electron. Transitions upwards and downwards occur between the two states with equal rates. In the case of thermodynamic equilibrium the Boltzmann formula (14.1) applies and gives

$$\frac{dn_{r+1}}{n_r} = \frac{g_{r+1} dg(p_\mathrm{e})}{g_r} \exp\left(- \frac{\chi_r + p_\mathrm{e}^2/(2m_\mathrm{e})}{kT} \right) . \tag{14.5}$$

What is the statistical weight $dg(p_\mathrm{e})$ of the electron in the momentum interval $[p_\mathrm{e}, p_\mathrm{e} + dp_\mathrm{e}]$? The Pauli principle of quantum mechanics tells us that in phase space a cell of volume $dq_1 \, dq_2 \, dq_3 \, dp_1 \, dp_2 \, dp_3 = dV \, d^3p$ can contain up to $2dV \, d^3p/h^3$ electrons, namely up to two electrons per quantum cell of volume h^3. Here the q's and the p's are the space and momentum variables of the (six-dimensional) phase space, while dV and d^3p are the (three-dimensional) "volumes" and h is the *Planck constant* ($h = 6.62620 \times 10^{-27}$ erg s). Then

$$dg(p_e) = \frac{2\,dV\,d^3p_e}{h^3} \; . \tag{14.6}$$

If the electron density in (three-dimensional) space is n_e then per electron the volume $dV = 1/n_e$ is available, while the volume in (three-dimensional) momentum space containing all points belonging to the interval $[p_e, p_e + dp_e]$ is $d^3p_e = 4\pi p_e^2 dp_e$, since all these points are on a spherical shell of radius p_e and thickness dp_e. We then have

$$dg(p_e) = \frac{8\pi p_e^2 dp_e}{n_e h^3} \tag{14.7}$$

and (14.5) yields

$$\frac{dn_{r+1}}{n_r} = \frac{g_{r+1}}{g_r}\frac{8\pi p_e^2 dp_e}{n_e h^3}\exp\left(-\frac{\chi_r + p_e^2/(2m_e)}{kT}\right) . \tag{14.8}$$

All upper states (ions of degree $r+1$ in the ground state and free electrons of all momenta) are then obtained by integration over p_e:

$$\frac{n_{r+1}}{n_r} = \frac{g_{r+1}}{g_r}\frac{8\pi}{n_e h^3}\,e^{-\chi_r/kT}\int_0^\infty p_e^2 \exp\left(-\frac{p_e^2}{2m_e kT}\right) dp_e \; . \tag{14.9}$$

Since for $a > 0$

$$\int_0^\infty x^2 e^{-a^2x^2} dx = \frac{\sqrt{\pi}}{4a^3} , \tag{14.10}$$

we obtain

$$\frac{n_{r+1}}{n_r} n_e = \frac{g_{r+1}}{g_r} f_r(T) , \quad \text{with} \quad f_r(T) = 2\frac{(2\pi m_e kT)^{3/2}}{h^3} e^{-\chi_r/kT} \; . \tag{14.11}$$

This is the *Saha equation* (named after the physicist Meghnad Saha) though it is still not yet in its final form, since we have considered only the ground states. Therefore, in order to be more precise, we now use the quantities $n_{r+1,0}, n_{r,0}, g_{r+1,0}, g_{r,0}$, where the second subscript indicates the ground state for which these quantities are defined. By $n_{r+1}, n_r, g_{r+1}, g_r$ we from now on mean number densities of ions and statistical weights for *all* states of excitation. A particular state of excitation is indicated by a second subscript such that $n_{i,k}$ is the number density of atoms in the stage i of ionization and in state k of excitation and $g_{i,k}$ is the corresponding statistical weight. The Saha equation (14.11) is then written more precisely as

$$\frac{n_{r+1,0}}{n_{r,0}} n_e = \frac{g_{r+1,0}}{g_{r,0}} f_r(T) \; . \tag{14.12}$$

The number density of ions in the ionization state r (in *all* states of excitation) is

$$n_r = \sum_s n_{r,s} , \tag{14.13}$$

which corresponds to (14.2), and we now write the Boltzmann formula (14.1) for ions of state r as

$$\frac{n_{r,s}}{n_{r,0}} = \frac{g_{r,s}}{g_{r,0}} \mathrm{e}^{-\psi_{r,s}/kT} , \tag{14.14}$$

where $\psi_{r,s}$ is the excitation energy of state s; then (14.13) can be written in the form

$$\begin{aligned} \frac{g_{r,0}}{n_{r,0}} n_r &= g_{r,0} \sum_s \frac{n_{r,s}}{n_{r,0}} \\ &= g_{r,0} + g_{r,1}\mathrm{e}^{-\psi_{r,1}/kT} + g_{r,2}\mathrm{e}^{-\psi_{r,2}/kT} + \ldots := u_r , \end{aligned} \tag{14.15}$$

where $u_r = u_r(T)$ is the partition function for the ion in state r. With the help of $n_r \, g_{r,0} = n_{r,0} \, u_r$, which follows from (14.15), the Saha equation can be written for all stages of excitation as

$$\frac{n_{r+1}}{n_r} n_\mathrm{e} = \frac{u_{r+1}}{u_r} f_r(T) , \tag{14.16}$$

where $f_r(T)$ is given in (14.11). With $P_\mathrm{e} = n_\mathrm{e} kT$ one has

$$\frac{n_{r+1}}{n_r} P_\mathrm{e} = \frac{u_{r+1}}{u_r} 2 \frac{(2\pi m_\mathrm{e})^{3/2}}{h^3} (kT)^{5/2} \, \mathrm{e}^{-\chi_r/kT} . \tag{14.17}$$

14.2 Ionization of Hydrogen

In order to see the consequences of the Saha equation we shall apply it to a pure hydrogen gas. We define the *degree of ionization* x by

$$x = \frac{n_1}{n_0 + n_1} , \tag{14.18}$$

i.e. $n_1/n_0 = x/(1-x)$. If the gas is neutral, then $x = 0$; if it is completely ionized, $x = 1$. Also the left-hand side of (14.17) can be replaced by $x P_\mathrm{e}/(1-x)$, and if $n = n_0 + n_1$ is the total number of hydrogen atoms, then we can relate the partial pressure of the electrons to the total gas pressure:

$$P_\mathrm{e} = n_\mathrm{e} kT = (n + n_\mathrm{e})kT \frac{n_\mathrm{e}}{n + n_\mathrm{e}} = P_\mathrm{gas} \frac{n_\mathrm{e}}{n + n_\mathrm{e}} . \tag{14.19}$$

For each ionized atom there is just one electron ($n_e = n_1$); therefore

$$P_e = \frac{x}{1+x} P_{gas} \tag{14.20}$$

and (14.17) can be written in the form

$$\frac{x^2}{1-x^2} = K_H , \quad \text{with} \quad K_H = \frac{u_1}{u_0} \frac{2}{P_{gas}} \frac{(2\pi m_e)^{3/2}}{h^3} (kT)^{5/2} e^{-\chi_H/kT} . \tag{14.21}$$

Here $\chi_H = 13.6$ eV is the ionization energy of hydrogen. Now with (14.21) we have come up with a quadratic equation for the degree of ionization that can be solved if T and P_{gas} are given. If radiation pressure is important, it is sufficient to give T and the total pressure P, and then P_{gas} can be obtained from (13.2).

In order to compute the degree of ionization, the partition function has to be known. For this we need the statistical weights of the different states of excitation, which are given by quantum mechanics. Since the higher states contribute little to the partition function, we may approximate it by the weight of the ground state, $u_0 \approx g_{0,0} = 2$, while for ionized hydrogen, $u_1 = 1$ (see, for instance, Cox, 2000, pp. 2–34).

We now give some numerical examples. In the solar photosphere we have in cgs units $P_{gas} = 1.01 \times 10^5, T = 5,779$ K, and we obtain $x = 5 \times 10^{-5}$, while in a deeper layer with $P_{gas} = 3.35 \times 10^{12}, T = 7.17 \times 10^5$ K, hydrogen is almost completely ionized: $x = 0.985$.

Since in (14.21) K_H increases with T and decreases with P_{gas}, and since the left-hand side increases with x, one can see that the degree of ionization increases with temperature and decreases with the gas pressure. This can be easily understood: with increasing temperature, the collisions become more violent, the photons more energetic, and the processes of "kicking off" the electrons from the atoms more frequent. If, on the other hand, the temperature is kept constant but the pressure increases, then the probability grows that the ion meets an electron and recombines.

In Chap. 4 we have defined the mean molecular weight μ for a mixture of gases and have seen that it is different for ionized and non-ionized gases. Therefore mean molecular weights depend on the degree of ionization.

In order to determine μ for the hydrogen gas having the degree of ionization x, we define the number E of free electrons per atom (neutral or ionized), which is here simply

$$E = \frac{n_e}{n} = x . \tag{14.22}$$

Remember that μm_u, $\mu_0 m_u$, and $\mu_e m_u$ are defined as the average particle masses per free particle, per nucleus, and per free electron, respectively. This means that the density can be written as

$$\varrho = (n + n_e)\mu m_u = n\mu_0 m_u = n_e \mu_e m_u . \tag{14.23}$$

Using (14.22) and $n = n_0 + n_1$, we solve (14.23) for the mean molecular weight and find

$$\mu = \frac{\varrho}{m_\mathrm{u} n} \frac{1}{1+E} = \frac{\mu_0}{1+E} = \mu_\mathrm{e} \frac{E}{1+E} , \tag{14.24}$$

where we have neither replaced μ_0 by its value 1 for hydrogen nor E by x, since (14.24) also holds for a mixture of gases.

14.3 Thermodynamical Quantities for a Pure Hydrogen Gas

Many thermodynamic properties depend on the degree of ionization. We here indicate roughly how the formulae can be derived for the relatively simple case of the pure hydrogen gas. This is not because of its importance, but rather because the treatment is quite analogous to that in the much more involved case of mixtures. The gas is supposed to be perfect, since partial ionization usually occurs only in the stellar envelope, where effects of degeneracy can be neglected.

In Sect. 4.1 we defined the quantity $\delta = -(\partial \ln\varrho / \partial \ln T)_P$. In the case of pure hydrogen obeying the perfect-gas equation we have $\delta = 1$ for $x = 0$ and $x = 1$, since μ is constant in both cases (Remember that we wished to incorporate in α and δ the changes of μ due to partial ionization, while φ should be reserved for changes of μ due to changing chemical composition.). For partial ionization, x varies with T, and therefore δ is given by a complicated expression. From the perfect-gas equation $\varrho \sim \mu P/T$ and (14.24) with μ_0 = constant we find

$$\delta = 1 + \frac{1}{1+E} \left(\frac{\partial E}{\partial \ln T} \right)_P , \tag{14.25}$$

which also holds for a mixture of gases. For pure hydrogen $E = x$ and we need the derivative of x, which can be obtained by differentiation of the Saha equation (14.21). This gives

$$\delta = 1 + \frac{1}{2} x(1-x) \left(\frac{5}{2} + \frac{\chi_\mathrm{H}}{kT} \right) . \tag{14.26}$$

While the mean molecular weight as given by (14.24) depends only on the degree of ionization, δ depends also on T, and if in addition radiation pressure is taken into account, one has to add terms proportional to $(1-\beta)/\beta$ to the right-hand sides of (14.25) and (14.26).

The definition (4.4) of c_P together with $P = \Re\varrho T/\mu$ gives

$$c_P = \left(\frac{\partial u}{\partial T} \right)_P + \frac{\Re}{\mu} \delta . \tag{14.27}$$

So we need the internal energy per mass unit

$$u = \frac{3}{2}\frac{\Re}{\mu_0}(1+E)T + u_{\text{ion}} , \tag{14.28}$$

where the first term gives the kinetic energy of ions and electrons, and the second term u_{ion} means the energy that has been used for ionization and that again becomes available if the ions recombine. Again (14.27) and (14.28) also hold for mixtures. For pure hydrogen, $E = x$ and $u_{\text{ion}} = x\chi_{\text{H}}/(\mu_0 m_{\text{u}}) = x\chi_{\text{H}}/m_{\text{u}}$, and after lengthy manipulations, one gets

$$c_P\frac{\mu_0}{\Re} = \frac{5}{2}(1+x) + \frac{\Phi_{\text{H}}^2}{G(x)} , \tag{14.29}$$

with the abbreviations

$$\Phi_{\text{H}} := \frac{5}{2} + \frac{\chi_{\text{H}}}{kT} \quad \text{and} \quad G(x) := \frac{1}{x(1-x)} + \frac{1}{x(1+x)} = \frac{2}{x(1-x^2)} . \tag{14.30}$$

If radiation plays a role, it appears not only in the equation for the pressure, but also in the internal energy. The result for c_P is that in (14.29) the factor 5/2 has to be replaced by $5/2 + 4(1-\beta)(4+\beta)/\beta^2$.

We can now easily derive an expression for ∇_{ad}:

$$\nabla_{\text{ad}} = \frac{P\delta}{T\varrho c_P} = \frac{2 + x(1-x)\Phi_{\text{H}}}{5 + x(1-x)\Phi_{\text{H}}^2} . \tag{14.31}$$

14.4 Hydrogen–Helium Mixtures

As a next step in the general problem we consider a gas of hydrogen and helium with weight fractions X, Y respectively. This is important for stellar envelopes and shows the difficulties which arise if mixtures are treated. We now have six types of particles: neutral and ionized hydrogen; neutral, ionized, and double ionized helium; and electrons. There are three types of ionization energy: χ_{H}^0 for hydrogen and $\chi_{\text{He}}^0, \chi_{\text{He}}^1$ for neutral and single ionized helium ($\chi_{\text{H}}^0 = 13.598$ eV, $\chi_{\text{He}}^0 = 24.587$ eV, $\chi_{\text{He}}^1 = 54.418$ eV). Each ionized hydrogen atom contributes the energy χ_{H}^0 to the internal energy, each helium atom in the first stage of ionization the energy χ_{He}^0 and each helium atom completely stripped of its two electrons the energy $\chi_{\text{He}}^0 + \chi_{\text{He}}^1$. By $x_{\text{H}}^0, x_{\text{H}}^1, x_{\text{He}}^0, x_{\text{He}}^1, x_{\text{He}}^2$ we define degrees of ionization, i.e. x_i^r gives the number of atoms of type i in ionization state r (= r electrons lost) divided by the total number of atoms of type i (irrespective of their state of ionization):

$$x_{\mathrm{H}}^0 = \frac{n_{\mathrm{H}}^0}{n_{\mathrm{H}}}\,, \quad x_{\mathrm{H}}^1 = \frac{n_{\mathrm{H}}^1}{n_{\mathrm{H}}}\,, \quad x_{\mathrm{He}}^0 = \frac{n_{\mathrm{He}}^0}{n_{\mathrm{He}}}\,,$$

$$x_{\mathrm{He}}^1 = \frac{n_{\mathrm{He}}^1}{n_{\mathrm{He}}}\,, \quad x_{\mathrm{He}}^2 = \frac{n_{\mathrm{He}}^2}{n_{\mathrm{He}}}\,, \tag{14.32}$$

with $n_{\mathrm{H}} = n_{\mathrm{H}}^0 + n_{\mathrm{H}}^1$ and $n_{\mathrm{He}} = n_{\mathrm{He}}^0 + n_{\mathrm{He}}^1 + n_{\mathrm{He}}^2$, where the n_i^r are number densities of ions of type i in ionization state r. Note that the degrees of ionization x_{H}^0 and x_{H}^1 correspond to $1 - x$ and x in Sect. 14.2.

The contribution of the ionization energy to the internal energy per unit mass [cf. (14.28)] is

$$u_{\mathrm{ion}} = \frac{1}{m_{\mathrm{u}}} \left\{ X x_{\mathrm{H}}^1 \chi_{\mathrm{H}}^0 + \frac{1}{4} Y \left[x_{\mathrm{He}}^1 \chi_{\mathrm{He}}^0 + x_{\mathrm{He}}^2 \left(\chi_{\mathrm{He}}^0 + \chi_{\mathrm{He}}^1 \right) \right] \right\}\,, \tag{14.33}$$

since X/m_{u}, $Y/(4m_{\mathrm{u}})$ are the numbers of hydrogen and helium atoms (neutral and ionized) per unit mass. Correspondingly we have for the number E of electrons per atom (irrespective of ionization state and chemical type)

$$E = \left[X x_{\mathrm{H}}^1 + \frac{1}{4} Y \left(x_{\mathrm{He}}^1 + 2 x_{\mathrm{He}}^2 \right) \right] \mu_0\,. \tag{14.34}$$

We now have three Saha equations:

$$\frac{x_{\mathrm{H}}^1}{x_{\mathrm{H}}^0} \frac{E}{E+1} = K_{\mathrm{H}}^0\,, \quad \frac{x_{\mathrm{He}}^1}{x_{\mathrm{He}}^0} \frac{E}{E+1} = K_{\mathrm{He}}^0\,, \quad \frac{x_{\mathrm{He}}^2}{x_{\mathrm{He}}^1} \frac{E}{E+1} = K_{\mathrm{He}}^1\,, \tag{14.35}$$

with

$$K_i^r = \frac{u_{r+1}}{u_r} \frac{2}{P_{\mathrm{gas}}} \frac{(2\pi m_{\mathrm{e}})^{3/2} (kT)^{5/2}}{h^3} \mathrm{e}^{-\chi_i^r/kT} \tag{14.36}$$

for i = H, He, and by definition

$$x_{\mathrm{H}}^0 + x_{\mathrm{H}}^1 = 1\,, \quad x_{\mathrm{He}}^0 + x_{\mathrm{He}}^1 + x_{\mathrm{He}}^2 = 1\,. \tag{14.37}$$

We now consider X, Y, P_{gas}, and T to be given. Then (14.34), (14.35) and (14.37) are six equations for the six unknown quantities $x_{\mathrm{H}}^0, x_{\mathrm{H}}^1, x_{\mathrm{He}}^0, x_{\mathrm{He}}^1, x_{\mathrm{He}}^2, E$. The equations (14.35) are coupled to each other via E, which, for instance, means that the degree of ionization of hydrogen also depends on the degree of ionization of helium. But this is to be expected, since a hydrogen ion can also recombine with free electrons that originally came from helium, since it has no prejudices concerning the origin of a captured electron.

The coupling of the three Saha equations (14.35) makes an analytical treatment impossible: the degrees of ionization have to be obtained numerically. In general,

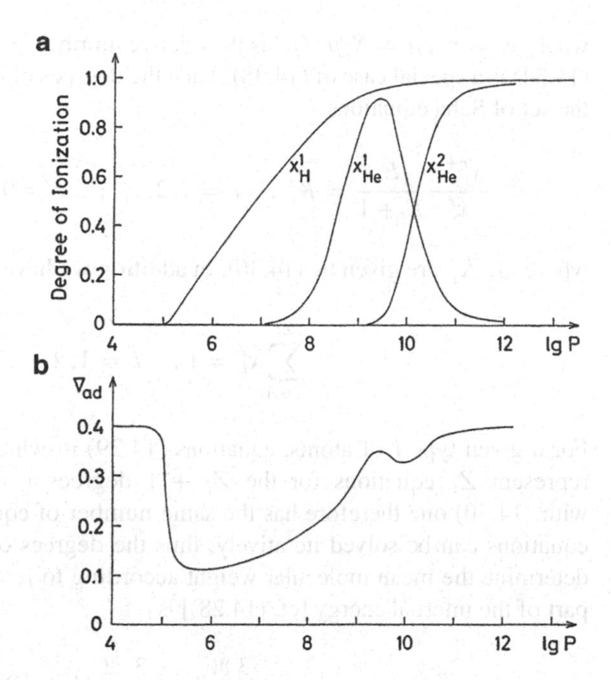

Fig. 14.1 Ionization in the outer layers of the Sun. (**a**) Degrees of ionization of hydrogen and helium. (**b**) The influence of ionization on ∇_{ad}

this is done by an iteration procedure, starting with a trial value of E, which is then gradually improved.

In Fig. 14.1 we give the degrees of ionization and ∇_{ad} for the outer layers of the Sun. One can see that the regions of partial ionization of H and He are almost separated. This is because the ionization energies $\chi^0_H, \chi^1_{He}, \chi^2_{He}$ differ from each other appreciably. The second helium ionization does not start until the hydrogen is almost completely ionized. Therefore one may, for an approximative treatment, solve at most two of equations (14.35) simultaneously, which simplifies the situation. Each of the three ionization layers produces a lowering of ∇_{ad} where influences of hydrogen and first helium ionization overlap.

14.5 The General Case

If X_i is the weight fraction of the chemical element i with charge number Z_i and molecular weight μ_i, and if x_i^r are the degrees of ionization (the numbers of atoms of type i in ionization state r in units of the total number of atoms of type i), then

$$E = \sum_i \nu_i \sum_{r=0}^{Z_i} x_i^r = \sum_i \frac{\mu_0}{\mu_i} X_i \sum_{r=0}^{Z_i} x_i^r r \ , \tag{14.38}$$

where $\nu_i = n_i/n = X_i\mu_0/\mu_i$ is the relative number of particles of type i. Equation (14.34) is a special case of (14.38). Then the degrees of ionization are obtained from the set of Saha equations

$$\frac{x_i^{r+1}}{x_i^r}\frac{E}{E+1} = K_i^r\ , \quad i = 1,2,\ldots, \quad r = 0,1,\ldots Z_i\ , \tag{14.39}$$

where the K_i^r are given by (14.36). In addition we have the relations

$$\sum_{r=0}^{Z_i} x_i^r = 1\ , \quad i = 1,2,\ldots\ . \tag{14.40}$$

For a given type i of atoms, equations (14.39) in which E is replaced by (14.38) represent Z_i equations for the $Z_i + 1$ degrees r of ionization, and together with (14.40) one therefore has the same number of equations as of variables. The equations can be solved iteratively; thus the degrees of ionization can be used to determine the mean molecular weight according to $\mu = \mu_0/(1+E)$. The kinetic part of the internal energy [cf. (14.28)] is

$$u_{\text{kin}} = \frac{3}{2}\frac{\Re}{\mu}T = \frac{3}{2}\frac{\Re}{\mu_0}(1+E)T\ , \tag{14.41}$$

while the ionization energy per mass unit is

$$u_{\text{ion}} = \sum_i \frac{X_i}{\mu_i m_{\text{u}}} \sum_{r=0}^{Z_i} x_i^r \sum_{s=0}^{r-1} \chi_i^s\ , \tag{14.42}$$

which is the general form of (14.33).

For the determination of δ and c_P according to (14.25) and (14.27) we need derivatives of the degrees of ionization: $(\partial x_i^r/\partial \ln T)_P$. They can be computed numerically by evaluating the x_i^r for neighbouring arguments, though one has to be careful if the radiation pressure is not negligible. The derivatives of the x_i^r are needed for constant *total* pressure P, whereas the argument for evaluating the degrees of ionization is the *gas* pressure. One therefore has to choose the neighbouring arguments P_{gas} and T such that $P = P_{\text{gas}} + P_{\text{rad}}$ = constant. The general theory of ionization and, in particular, the influence on the thermodynamic functions for arbitrary mixtures are given in Baker and Kippenhahn (1962, Appendix A).

In modern stellar evolution calculations the equation of state is no longer computed online, but rather pre-calculated tables are used. These tables result from sophisticated models of the properties of stellar matter, which are too complicated to be integrated into the stellar evolution programs. Ionization is only one of the many physical effects treated in such models for many chemical elements, their number amounting to up to 20.

14.6 Limitation of the Saha Formula

In the derivation of the Saha formula we have assumed thermodynamic equilibrium. This is certainly fulfilled in the interior of stars, and the Saha formula is even a sufficient approximation for many atmospheres as long as one can assume so-called LTE (local thermodynamic equilibrium), which is the case when collisions dominate over radiative processes. One cannot apply it for non-LTE, as, for example, in the solar corona.

But even in the deep interior of a star, where local thermodynamic equilibrium is certainly a very good approximation, the naïve application of the Saha formula gives wrong results. For instance let us apply it to the centre of the Sun ($P_c \approx P_{gas} = 2.32 \times 10^{17}$ dyn/cm^2, $T_c = 1.57 \times 10^7$ K) and assume for simplicity pure hydrogen ($X = 1$); then (14.21) gives for the degree of ionization $x_H = 0.80$. This would mean that 20 % of the hydrogen atoms are neutral. Indeed, for sufficiently high temperatures, the exponential in the Saha formula can be replaced by 1, and x_H^1 decreases inwards with K_H if $\nabla \equiv d \ln T / d \ln P_{gas} < 2/5$, as can be seen from (14.21).

The solution of this paradox has to do with the decrease of the ionization energy with increasing density. Let us consider ions at a distance d from each other: their electrostatic potentials have to be superimposed in order to obtain their total potential (Fig. 14.2). Obviously the higher quantum states of the ions are strongly disturbed, and the ionization energy is reduced for high density. This should be taken into account in the Saha formula, which would then give a higher degree of ionization. Furthermore, the neighbouring ions allow only a finite number of bound states. This has the consequence that in the partition function as given by (14.15) one has to sum over a finite number of excited states only.

In order to estimate roughly at which density these effects become important, we consider a pure hydrogen gas. If the mean distance between two atoms is d, then there will be no bound states if the orbital radius a of the electron is comparable with, or larger than, $d/2$. With

$$a = a_0 \nu^2 , \quad d \approx \left(\frac{3}{4\pi n_H} \right)^{1/3} , \tag{14.43}$$

where $a_0 = 5.3 \times 10^{-9}$ cm is the Bohr radius, ν the quantum, number and n_H the number density of the atoms, we obtain from the condition $a < d/2$ (which must be fulfilled for a bound state) that

$$\nu^2 < \left(\frac{3}{4\pi n_H} \right)^{1/3} \frac{1}{2a_0} . \tag{14.44}$$

This allows a rough estimate of the principal quantum number of the highest bound state. In the centre of the Sun, with $\varrho_c \approx 150$ g/cm^3, we have $n_H \approx \varrho_c / m_u \approx 10^{26}$ cm^{-3}, and therefore $\nu^2 < 0.13$, which means that even the ground state of hydrogen does not exist. Therefore all hydrogen atoms will be ionized.

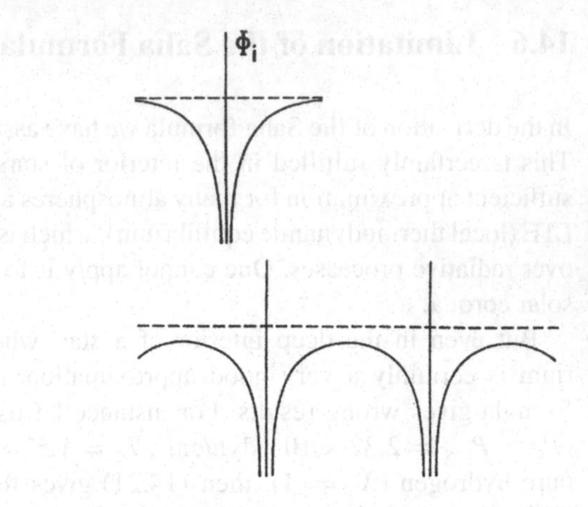

Fig. 14.2 Sketch of the electrostatic potential of an isolated ion (*above*) and the superposition of the potentials of neighbouring ions (*below*)

For this so-called *pressure ionization*, no perfect theory is at hand. The picture we have used above is a static one, since it does not take into account that the ions move relative to each other. It also ignores that at high densities electrons can tunnel from a bound state of one ion into a bound state of another ion in the neighbourhood. In the specialized computations of the equation of state for an astrophysical plasma, more elaborate models are used to solve this problem. An example is the hydrogen–helium equation of state by Saumon et al. (1995).

For simplified stellar-model calculations one may use the Saha formula for the outer layers of the stars and then switch to complete ionization when the Saha formula gives degrees of ionization which decrease again towards deeper layers. This switching normally does not produce a noticeable discontinuity in the run of ionization, since the maximum often occurs close to complete ionization.

If we assume that pressure ionization can be neglected as long as $d > 10a_0$, then the Saha formula would be valid only for densities:

$$\varrho = \mu_0 m_\mathrm{u} n_\mathrm{ion} < \frac{3\mu_0 m_\mathrm{u}}{4\pi (10a_0)^3} = 2.66 \times 10^{-3} \mu_0 \,\mathrm{g\,cm^{-3}} \,. \tag{14.45}$$

Chapter 15
The Degenerate Electron Gas

15.1 Consequences of the Pauli Principle

We consider a gas of sufficiently high density in the volume dV so that it is practically fully pressure ionized (Sect. 14.6). Here we shall deal with the free electrons, of number density n_e. If the velocity distribution of the electrons is given by Boltzmann statistics, then their mean kinetic energy is $3kT/2$. In momentum space p_x, p_y, p_z each electron of a given volume dV in local space is represented by a point, and these points form a "cloud" which is spherically symmetric around the origin. If p is the absolute value of the momentum ($p^2 = p_x^2 + p_y^2 + p_z^2$), then the number of electrons in the spherical shell $[p, p+dp]$ is, according to the Boltzmann distribution function,

$$f(p)dpdV = n_e \frac{4\pi p^2}{(2\pi m_e kT)^{3/2}} \exp\left(-\frac{p^2}{2m_e kT}\right) dp\, dV. \tag{15.1}$$

Consider a reduction of T with n_e = constant. Then the maximum of the distribution function, which is at $p_{\max} = (2m_e kT)^{1/2}$, tends to smaller values of p, and the maximum of $f(p)$ becomes higher, since n_e is given by $\int_0^\infty f(p)dp$. This is indicated in Fig. 15.1 by the thin curves. But with this classical picture we can come into contradiction with quantum mechanics, since electrons are fermions, for which Pauli's exclusion principle holds: each quantum cell of the six-dimensional phase space (x, y, z, p_x, p_y, p_z) cannot contain more than two electrons (here x, y, z are the space coordinates of the electrons with $dV = dxdydz$). The volume of such a quantum cell is $dp_x dp_y dp_z dV = h^3$, where h is Planck's constant. Therefore in the shell $[p, p+dp]$ of momentum space there are $4\pi p^2 dpdV/h^3$ quantum cells, which can contain not more than $8\pi p^2 dpdV/h^3$ electrons. Quantum mechanics therefore demands that

$$f(p)dpdV \le 8\pi p^2 dpdV/h^3 , \tag{15.2}$$

R. Kippenhahn et al., *Stellar Structure and Evolution*, Astronomy and Astrophysics Library, DOI 10.1007/978-3-642-30304-3_15, © Springer-Verlag Berlin Heidelberg 2012

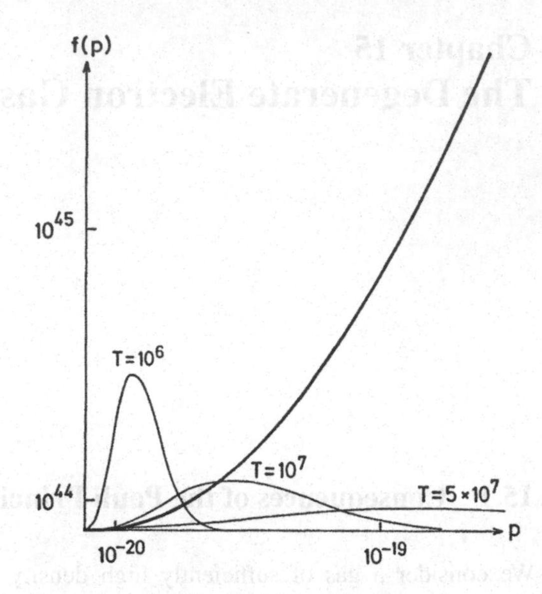

Fig. 15.1 For an electron gas with $n_e = 10^{28}\,\mathrm{cm}^{-3}$ (corresponding to a density of $\varrho = 1.66 \times 10^4\,\mathrm{g\,cm}^{-3}$ for $\mu_e = 1$), the Boltzmann distribution function $f(p)$ is shown by *thin lines* over the absolute value of the momentum p (both in cgs units) for three different temperatures (in K). The *heavy line* shows the parabola that gives an upper bound to the distribution function owing to the Pauli principle (Note that the coordinates are not logarithmic but linear as in Figs. 15.2 and 15.5)

as indicated by the heavy parabola in Fig. 15.1, giving an upper bound for $f(p)$. One can immediately see that the Boltzmann distribution for n_e = constant is in contradiction with quantum mechanics for sufficiently low temperatures. The same holds for T = constant and sufficiently high density, since the Boltzmann distribution is proportional to n_e. We therefore have to include quantum-mechanical effects if the temperature of the gas is too low or if the electron density is too high, in order to avoid the distribution function exceeding its upper bound. One then says that the electrons become *degenerate*.

We first consider an electron gas of temperature zero, i.e. all the electrons have the lowest energy possible.

15.2 The Completely Degenerate Electron Gas

The state in which all electrons have the lowest energy without violating Pauli's principle is that in which all phase cells up to a certain momentum p_F are occupied by two electrons, all other phase cells above p_F being empty:

$$f(p) = \frac{8\pi p^2}{h^3} \quad \text{for} \quad p \le p_F\,,$$
$$f(p) = 0 \quad \text{for} \quad p > p_F\,. \tag{15.3}$$

This distribution function is shown in Fig. 15.2, and the total number of electrons in the volume dV is given by

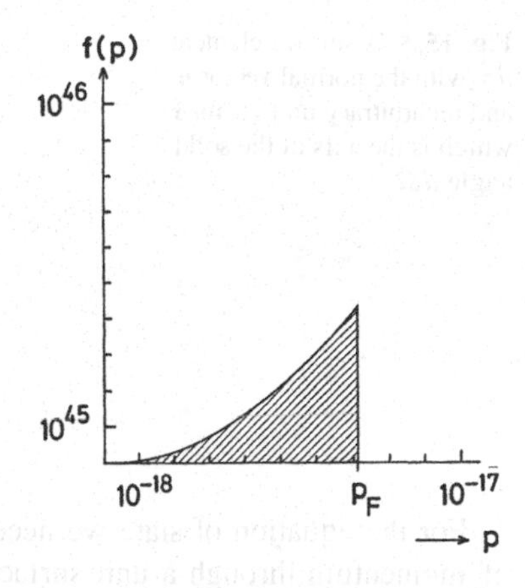

Fig. 15.2 The distribution function $f(p)$ against the momentum p (both in cgs units) in the case of a completely degenerate electron gas with $T = 0\,\mathrm{K}$ and $n_e = 10^{28}\,\mathrm{cm}^{-3}$ (cf. Fig. 15.1)

$$n_e dV = dV \int_0^{p_F} \frac{8\pi p^2 dp}{h^3} = \frac{8\pi}{3h^3} p_F^3 dV . \tag{15.4}$$

If therefore the electron density is given, (15.4) gives the *Fermi momentum* $p_F \sim n_e^{1/3}$. Further, if the electrons are non-relativistic, then $E_F = p_F^2/2m_e \sim n_e^{2/3}$ is the *Fermi energy*, and, although the temperature of our electron gas is zero, the electrons have finite energies up to E_F. But there are no electrons of higher energy. If the electron density is sufficiently large, then according to (15.4) p_F can become so high that the velocities of the fastest electrons may become comparable with c, the velocity of light. We therefore write the relations between velocity v, energy E_{tot}, and momentum p of the electrons in the form given by special relativity (see, for instance, Landau and Lifshitz, vol. 2, 1976):

$$p = \frac{m_e v}{\sqrt{1 - v^2/c^2}} , \tag{15.5}$$

$$E_{tot} = \frac{m_e c^2}{\sqrt{1 - v^2/c^2}} = m_e c^2 \sqrt{1 + \frac{p^2}{m_e^2 c^2}} , \tag{15.6}$$

where m_e is the *rest mass* of the electron. From (15.5) and (15.6) it follows that

$$\frac{1}{c}\frac{\partial E_{tot}}{\partial p} = \frac{p/(m_e c)}{[1 + p^2/(m_e^2 c^2)]^{1/2}} = \frac{v}{c} . \tag{15.7}$$

In the following we have to distinguish between the *total energy* E_{tot} as given by (15.6) and the *kinetic energy* E:

$$E = E_{tot} - m_e c^2 . \tag{15.8}$$

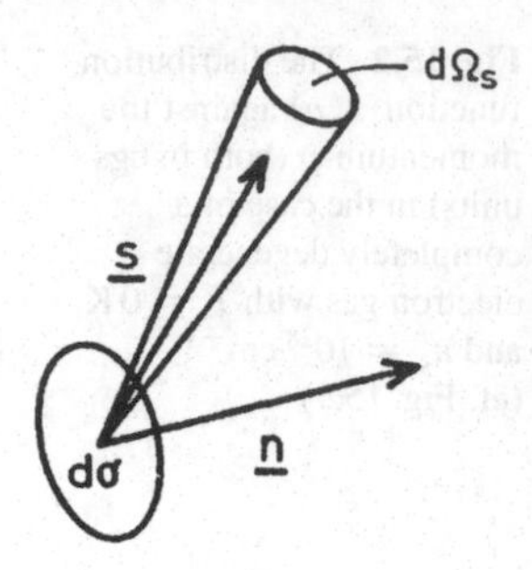

Fig. 15.3 A surface element $d\sigma$ with the normal vector $\boldsymbol{n}$ and an arbitrary unit vector $\boldsymbol{s}$ which is the axis of the solid angle $d\Omega_s$

For the equation of state we need the pressure, which by definition is the flux of momentum through a unit surface per second. We consider a surface element $d\sigma$ having a normal vector $\boldsymbol{n}$, as indicated in Fig. 15.3. An arbitrary unit vector $\boldsymbol{s}$, together with $\boldsymbol{n}$, defines an angle ϑ.

Let us determine the number of electrons per second that go through $d\sigma$ into a small solid angle $d\Omega_s$ around the direction $\boldsymbol{s}$. We restrict ourselves to electrons for which the absolute value of their momentum lies between p and $p + dp$. At the location of the surface element there are $f(p)dpd\Omega_s/(4\pi)$ electrons per unit volume that have the right momentum (i.e. the right value of p and the right direction). Therefore $f(p)dpd\Omega_s v(p)\cos\vartheta d\sigma/(4\pi)$ electrons per second move through the surface element $d\sigma$ into the solid-angle element $d\Omega_s$. Here $v(p)$ is the velocity that according to (15.5) belongs to the momentum p. The factor $\cos\vartheta$ arises, since the electrons moving into the solid-angle element see only a projection of $d\sigma$. Each electron carries a momentum of absolute value p and of direction $\boldsymbol{s}$. The component in direction $\boldsymbol{n}$ is therefore $p\cos\vartheta$. We obtain the total flux of momentum in direction $\boldsymbol{n}$ by integration over all directions $\boldsymbol{s}$ of a hemisphere and over all absolute values p; hence the pressure P_e of the electrons is

$$P_e = \int_{2\pi}\int_0^\infty f(p)v(p)p\cos^2\vartheta dpd\Omega_s/(4\pi) = \frac{8\pi}{3h^3}\int_0^{p_F} p^3 v(p)dp, \qquad (15.9)$$

where we have replaced $f(p)$ by (15.3) and taken the value $4\pi/3$ for the integration of $\cos^2\vartheta$ over a hemisphere. It is obvious that the orientation of $d\sigma$ does not enter into the expression for P_e: the electron pressure is isotropic because f is spherically symmetric in momentum space.

With (15.5) we obtain from (15.9) that

$$\begin{aligned} P_e &= \frac{8\pi c}{3h^3}\int_0^{p_F} p^3 \frac{p/(m_e c)}{[1 + p^2/(m_e^2 c^2)]^{1/2}} dp \\ &= \frac{8\pi c^5 m_e^4}{3h^3}\int_0^x \frac{\xi^4 d\xi}{(1+\xi^2)^{1/2}}, \qquad (15.10) \end{aligned}$$

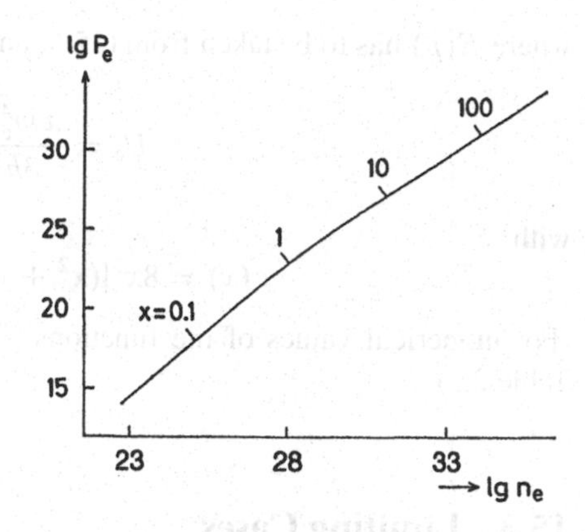

Fig. 15.4 The equation of state for the fully degenerate electron gas. On logarithmic scales the pressure P_e (in dyn cm^{-2}) is plotted against the number density n_e (in cm^{-3}). The relativity parameter $x = p_F/m_e c$ increases along the curve from the *lower left* to the *upper right*; values of x are indicated above the curve

where we have introduced new variables:

$$\xi = p/(m_e c) , \qquad x = p_F/(m_e c) . \tag{15.11}$$

The integral is

$$\int_0^x \frac{\xi^4 d\xi}{(1+\xi^2)^{1/2}} = \frac{1}{8}[x(2x^2-3)(1+x^2)^{1/2} + 3\mathrm{arcsinh}(x)] \tag{15.12}$$

(where arcsinh is the inverse function of sinh); therefore

$$P_e = \frac{\pi m_e^4 c^5}{3h^3} f(x) , \tag{15.13}$$

with

$$f(x) = x(2x^2-3)(x^2+1)^{1/2} + 3\mathrm{arcsinh}(x) \equiv x(2x^2-3)(x^2+1)^{1/2} + 3\ln[x+(1+x^2)^{1/2}] . \tag{15.14}$$

We now write (15.4) in the form

$$n_e = \frac{\varrho}{\mu_e m_u} = \frac{8\pi m_e^3 c^3}{3h^3} x^3 . \tag{15.15}$$

Equations (15.13)–(15.15) define the function $P_e(n_e)$, which is plotted in Fig. 15.4 for the fully degenerate electron gas. Before discussing this and deriving an equation of state $P_e = P_e(\varrho)$, we give an expression for the internal energy U_e of the electron gas per volume:

$$U_e = \int_0^{p_F} f(p)E(p)dp = \frac{8\pi}{h^3}\int_0^{p_F} E(p)p^2 dp , \tag{15.16}$$

where $E(p)$ has to be taken from (15.6) and (15.8). One obtains

$$U_e = \frac{\pi m_e^4 c^5}{3h^3} g(x) , \tag{15.17}$$

with

$$g(x) = 8x^3[(x^2+1)^{1/2} - 1] - f(x) . \tag{15.18}$$

(For numerical values of the functions $f(x)$ and $g(x)$ see Chandrasekhar 1939, Table 23.)

15.3 Limiting Cases

The parameter x as defined in (15.11) is a measure of the importance of relativistic effects for electrons with the highest momentum. With (15.5) we can write x in the form

$$x = \frac{p_F}{m_e c} = \frac{v_F/c}{(1 - v_F^2/c^2)^{1/2}} \quad \text{or} \quad \frac{v_F^2}{c^2} = \frac{x^2}{1+x^2} , \tag{15.19}$$

where v_F is the velocity of the electrons with $p = p_F$. If $x \ll 1$, then $v_F/c \ll 1$ and all electrons move much slower than the velocity of light (non-relativistic case). On the other hand if $x \gg 1$, then v_F/c is very close to one: the bigger x, the more electrons with velocities near v_F become relativistic, and for very high values of x almost all electrons are relativistic.

The functions $f(x)$ and $g(x)$ as defined in (15.14) and (15.18) have the following asymptotic behaviour:

$$x \to 0 : \ f(x) \to \frac{8}{5} x^5 , \quad g(x) \to \frac{12}{5} x^5 . \tag{15.20}$$

$$x \to \infty : \ f(x) \to 2x^4 , \quad g(x) \to 6x^4 . \tag{15.21}$$

We first consider the case $x \ll 1$, where relativistic effects can be ignored, for which (15.13) yields

$$P_e = \frac{8\pi m_e^4 c^5}{15h^3} x^5 , \tag{15.22}$$

and together with (15.15) we obtain the equation of state for a *completely degenerate non-relativistic* electron gas:

$$\begin{aligned} P_e &= \frac{1}{20}\left(\frac{3}{\pi}\right)^{2/3} \frac{h^2}{m_e} n_e^{5/3} = \frac{1}{20}\left(\frac{3}{\pi}\right)^{2/3} \frac{h^2}{m_e m_u^{5/3}} \left(\frac{\varrho}{\mu_e}\right)^{5/3} \\ &= 1.0036 \times 10^{13} \left(\frac{\varrho}{\mu_e}\right)^{5/3} \quad \text{(cgs)} \end{aligned} \tag{15.23}$$

where we have used $\varrho = n_e \mu_e m_u$. The internal energy U_e of the electrons per unit volume and the electron pressure are related by

$$P_e = \frac{2}{3} U_e , \tag{15.24}$$

which can be obtained from (15.17), (15.20) and (15.22).

For the *extreme relativistic* case ($x \gg 1$) of a *completely degenerate* electron gas, one has according to (15.13) and (15.21)

$$P_e = \frac{2\pi m_e^4 c^5}{3h^3} x^4 , \tag{15.25}$$

and therefore

$$\begin{aligned} P_e &= \left(\frac{3}{\pi}\right)^{1/3} \frac{hc}{8} n_e^{4/3} = \left(\frac{3}{\pi}\right)^{1/3} \frac{hc}{8 m_u^{4/3}} \left(\frac{\varrho}{\mu_e}\right)^{4/3} \\ &= 1.2435 \times 10^{15} \left(\frac{\varrho}{\mu_e}\right)^{4/3} \text{(cgs)} , \end{aligned} \tag{15.26}$$

while (15.17), (15.21) and (15.25) give

$$P_e = \frac{1}{3} U_e . \tag{15.27}$$

15.4 Partial Degeneracy of the Electron Gas

For a finite temperature, not all electrons will be densely packed in momentum space in the cells of lowest possible momentum. Indeed, if the temperature is sufficiently high, we expect them to have a Boltzmann distribution. Further, there must be a smooth transition from the completely degenerate state (as discussed in Sects. 15.2 and 15.3) to the non-degenerate case.

The most probable occupation of the phase cells of the shell $[p, p + dp]$ in momentum space is determined by Fermi–Dirac statistics (see Landau, Lifshitz, vol. 5, 1980):

$$f(p) dp dV = \frac{8\pi p^2 dp dV}{h^3} \frac{1}{1 + e^{E/kT - \psi}} \tag{15.28}$$

(where the so-called degeneracy parameter ψ will be discussed later). The first factor gives again the maximally allowed occupations for this shell; see (15.2). However, for $p \le p_F$, there are fewer electrons in the shell than in the case of complete degeneracy: the second factor is smaller than one; it is a "filling factor", telling us what fraction of the cells is occupied. This factor depends on the temperature and the kinetic energy E of a particle with momentum p as defined in Sect. 15.2.

With (15.28) n_e, P_e, and U_e become

$$n_e = \frac{8\pi}{h^3}\int_0^\infty \frac{p^2 dp}{1+e^{E/kT-\psi}} \tag{15.29}$$

$$P_e = \frac{8\pi}{3h^3}\int_0^\infty p^3 v(p)\frac{dp}{1+e^{E/kT-\psi}}, \tag{15.30}$$

$$U_e = \frac{8\pi}{h^3}\int_0^\infty \frac{E p^2 dp}{1+e^{E/kT-\psi}}. \tag{15.31}$$

We first deal only with the *non-relativistic case* for which $E = p^2/(2m_e)$, and the electron density n_e is given by

$$n_e = \frac{8\pi}{h^3}\int_0^\infty \frac{p^2 dp}{1+e^{p^2/2m_e kT-\psi}} = \frac{8\pi}{h^3}(2m_e kT)^{3/2} a(\psi), \tag{15.32}$$

with

$$a(\psi) = \int_0^\infty \frac{\eta^2}{1+e^{(\eta^2-\psi)}} d\eta, \tag{15.33}$$

where we have used the variable $\eta = p/(2m_e kT)^{1/2}$.

We conclude from (15.32) that the *degeneracy parameter* ψ is a function of $n_e/T^{3/2}$ only:

$$\psi = \psi\left(\frac{n_e}{T^{3/2}}\right). \tag{15.34}$$

We now discuss limiting cases for ψ, beginning with large negative values for ψ (again non-relativistic). In this case $a(\psi)$ in (15.33) can be made arbitrarily small, and from (15.32) we infer that for a given electron density this is the case for high temperatures. We know that then $f(p)$ must become the Boltzmann distribution. Comparing (15.1) with (15.28) [where in the denominator the 1 can be neglected against $\exp(E/kT - \psi)$], we see that

$$e^\psi = \frac{h^3 n_e}{2(2\pi m_e kT)^{3/2}}. \tag{15.35}$$

Here we have replaced $E/(kT)$ by its non-relativistic value $p^2/(2m_e kT)$. Indeed in this limit ψ is a function of $n_e/T^{3/2}$, as concluded for the general case.

We now want to consider the case $\psi \to \infty$ (again non-relativistic) and introduce an energy E_0 by $\psi = E_0/(kT)$. We then have for large enough ψ

$$\frac{1}{1+e^{E/kT-\psi}} = \frac{1}{1+e^{\psi(E/E_0-1)}} \approx \begin{cases} 1 & \text{for} \quad E < E_0 \\ 0 & \text{for} \quad E > E_0 \end{cases}. \tag{15.36}$$

The transition of the numerical value of expression (15.36) from one to zero near E_0 becomes all the more steep, the larger the value of ψ. In the limiting case $\psi \to \infty$ it becomes a discontinuity, and comparison of (15.36) with (15.3) shows that E_0 is the Fermi energy $E_F = p_F^2/(2m_e)$. One immediately sees that $\psi \to \infty$ corresponds to the case of complete degeneracy, where the distribution function is given by (15.3).

We now deal with the (non-relativistic) case where the numerical value of ψ is moderate. In (15.32) we replace the variable p by E. With $m_e dE = p dp$ and $p = (2m_e E)^{1/2}$, we have

$$n_e = \frac{4\pi}{h^3}(2m_e)^{3/2} \int_0^\infty \frac{E^{1/2} dE}{1 + e^{E/kT - \psi}}, \tag{15.37}$$

and defining the so-called Fermi–Dirac integrals $F_\nu(\psi)$ by

$$F_\nu(\psi) := \int_0^\infty \frac{u^\nu}{e^{(u-\psi)} + 1} du, \tag{15.38}$$

we find that

$$n_e = \frac{\varrho}{\mu_e m_u} = \frac{4\pi}{h^3}(2m_e kT)^{3/2} F_{1/2}(\psi), \tag{15.39}$$

which again manifests the relation (15.34) and which, by inversion of (15.38), allows to determine ψ for given n_e and T.

The distribution function for partial (non-relativistic) degeneracy as given by (15.28) is shown in Fig. 15.5 for $T = 1.9\times10^7$ K and $\psi = 10$ [$F_{1/2}(10) = 21.34$, see Table 15.1]. One can see that for small values of p the function $f(p)$ is close to the Pauli parabola, but in contrast to the case $T = 0$ it is smooth near p_F. The higher the temperature the smoother the transition around p_F, until finally $f(p)$ resembles a Boltzmann distribution. The electron pressure P_e is given in (15.30). Now (in the non-relativistic case), we have $p^3 v(p) dp = m_e^4 v^4 dv = m_e^3 v^3 dE = m_e^{3/2} 2^{3/2} E^{3/2} dE$ and

$$P_e = \frac{8\pi}{3h^3}(2m_e)^{3/2} \int_0^\infty \frac{E^{3/2} dE}{1 + e^{E/kT - \psi}}. \tag{15.40}$$

With $y = E/(kT)$ the integral becomes one of the type defined in (15.38):

$$P_e = \frac{8\pi}{3h^3}(2m_e kT)^{3/2} kT\, F_{3/2}(\psi). \tag{15.41}$$

For the internal energy U_e per unit volume we have from (15.34) with the non-relativistic relation $p^2 = 2m_e E$:

$$U_e = \frac{4\pi}{h^3}(2m_e kT)^{3/2} kT\, F_{3/2}(\psi) = \frac{3}{2} P_e, \tag{15.42}$$

in agreement with (15.24).

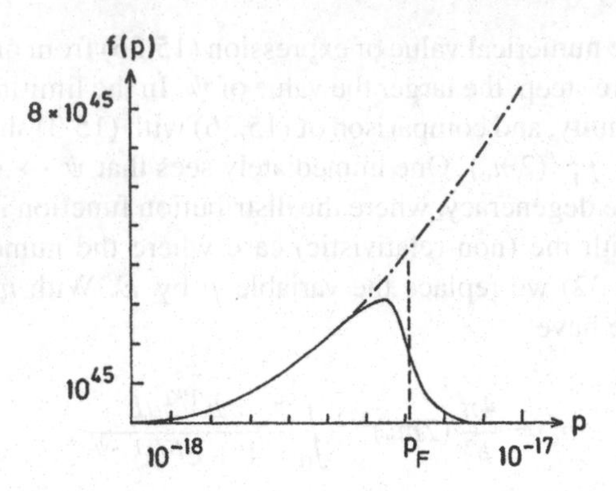

Fig. 15.5 The *solid line* gives the distribution function ($f(p)$ and p in cgs) for a partially degenerate electron gas with $n_e = 10^{28}\,\mathrm{cm}^{-3}$ and $T = 1.9 \times 10^7$ K, which corresponds to a degeneracy parameter $\psi = 10$ (cf. the case of complete degeneracy of Fig. 15.2). The *dot-dashed line* shows the further increase of the parabola that defines an upper bound for the distribution function

Again, (15.39) and (15.41) define an equation of state for the electron gas. If T and n_e are given, then (15.39) gives ψ (since $F_{1/2}(\psi)$ has a unique inverse function) and P_e can be determined. Numerical values for some of the functions F_ν are given in Table 15.1 to allow a quick estimate of the equation of state.

Without proof we give an expansion of the integrals F_ν for large positive values of ψ, i.e. for strong degeneracy:

$$F_\nu(\psi) = \frac{\psi^{\nu+1}}{\nu+1}\{1 + 2[c_2(\nu+1)\nu\psi^{-2} + c_4(\nu+1)\nu(\nu-1)(\nu-2)\psi^{-4} + \ldots]\}\,, \tag{15.43}$$

with $c_2 = \pi^2/12, c_4 = 7\pi^4/720$. We therefore have for $\psi \gg 1$ that $F_{1/2}(\psi) \approx 2\psi^{3/2}/3, F_{3/2}(\psi) \approx 2\psi^{5/2}/5$. If we introduce these expressions into (15.39) and (15.41) and eliminate ψ, we come to the relation (15.23) for non-relativistic strong degeneracy.

On the other hand for $\psi \to -\infty$ (the electrons behave almost like a perfect gas) we can make the approximation

$$F_\nu(\psi) = \int_0^\infty \frac{y^\nu dy}{1 + e^{(y-\psi)}} \approx e^\psi \int_0^\infty y^\nu e^{-y} dy. \tag{15.44}$$

For $\nu = 1/2$ and $\nu = 3/2$ integration gives $F_{1/2}(\psi) \approx \sqrt{\pi}\,e^\psi/2, F_{3/2}(\psi) \approx 3\sqrt{\pi}e^\psi/4$. If we introduce these approximations into (15.39) and (15.41) and eliminate ψ, we recover $P_e = n_e kT$, which is the equation of state for the perfect (non-degenerate) electron gas.

Table 15.1 Numerical values for Fermi–Dirac functions $F_{1/2}, F_{3/2}, F_2, and F_3$ (after Gong et al. 2001, using the computer program for numerical integration provided there)

Ψ	$F_{3/2}(\Psi)$	$F_{1/2}(\Psi)$	$F_2(\Psi)$	$F_3(\Psi)$
−4.00	0.024269	0.016128	0.036548	0.109768
−3.50	0.039931	0.026481	0.060169	0.180844
−3.00	0.065612	0.043366	0.098963	0.297802
−2.50	0.107581	0.070724	0.162525	0.490023
−2.00	0.175801	0.114588	0.266265	0.805319
−1.50	0.285772	0.183802	0.434567	1.320880
−1.00	0.460849	0.290501	0.705130	2.159840
−0.50	0.734659	0.449793	1.134368	3.515199
0.00	1.152804	0.678094	1.803085	5.682197
0.50	1.772794	0.990209	2.820969	9.098521
1.00	2.661683	1.396375	4.328331	14.389356
1.50	3.891976	1.900833	6.494369	22.412444
2.00	5.537254	2.502458	9.512668	34.298283
2.50	7.668804	3.196599	13.595529	51.482510
3.00	10.353715	3.976985	18.968568	75.729812
3.50	13.654202	4.837066	25.866374	109.150502
4.00	17.627703	5.770727	34.529354	154.211461
4.50	22.327332	6.772574	45.201594	213.743156
5.00	27.802446	7.837976	58.129472	290.944038
5.50	34.099195	8.962995	73.560777	389.383271
6.00	41.261003	10.144285	91.744165	513.002403
6.50	49.328972	11.378986	112.928816	666.116392
7.00	58.342217	12.664638	137.364234	853.414231
7.50	68.338129	13.999097	165.300117	1,079.959324
8.00	79.352594	15.380486	196.986283	1,351.189722
8.50	91.420172	16.807137	232.672619	1,672.918257
9.00	104.574241	18.277560	272.609060	2, 051.332632
9.50	118.847118	19.790412	317.045564	2, 492.995468
10.00	134.270160	21.344471	366.232105	3, 004.844342
10.50	150.873848	22.938625	420.418670	3, 594.191796
11.00	168.687863	24.571846	479.855250	4, 268.725360
11.50	187.741147	26.243190	544.791837	5, 036.507549
12.00	208.061959	27.951777	615.478430	5, 905.975874
12.50	229.677920	29.696791	692.165026	6, 885.942840
13.00	252.616059	31.477465	775.101624	7, 985.595952
13.50	276.902852	33.293083	864.538223	9, 214.497712
14.00	302.564251	35.142971	960.724822	10, 582.585620
14.50	329.625717	37.026492	1, 063.911422	12, 100.172179
15.00	358.112248	38.943047	1, 174.348023	13, 777.944887
15.50	388.048402	40.892064	1, 292.284623	15, 626.966247
16.00	419.458325	42.873005	1, 417.971224	17, 658.673757
16.50	452.365762	44.885355	1, 551.657824	19, 884.879918
17.00	486.794087	46.928625	1, 693.594425	22, 317.772230
17.50	522.766312	49.002348	1, 844.031026	24, 969.913193
18.00	560.305110	51.106078	2, 003.217626	27, 854.240307
18.50	599.432825	53.239389	2, 171.404227	30, 984.066072
19.00	640.171486	55.401871	2, 348.840828	34, 373.077988
19.50	682.542825	57.593132	2, 535.777429	38, 035.338556
20.00	726.568284	59.812795	2, 732.464029	41, 985.285274

For the non-relativistic case we have derived the tools to deal with partial degeneracy. For the *extreme relativistic case* similar approximations are possible, since in the integrals (15.29) and (15.30) p can be replaced by E/c, and v by c. Then the same procedure which led to (15.39) and (15.41) now yields

$$n_e = 8\pi \left(\frac{kT}{hc}\right)^3 F_2(\psi) , \tag{15.45}$$

$$P_e = \frac{8\pi}{3h^3c^3} (kT)^4 F_3(\psi) , \tag{15.46}$$

where F_2 and F_3 are defined by (15.38). For strong degeneracy ($\psi \to \infty$) the first term of the expansion (15.43) is introduced into (15.45) and (15.46), and elimination of ψ gives the already derived equation of state (15.26) for a completely degenerate, relativistic electron gas.

No analytical approach is known for the case of partial degeneracy if the electron gas is only moderately relativistic, because the relation between E and p cannot be approximated by a simpler expression and in the integrals (15.29) and (15.30) the full relation (15.6) has to be taken; hence the problem has to be treated numerically. However, to do this efficiently, no general integration scheme should be used for the many integrations needed to calculate the equation of state at the many mesh points of a stellar model, but instead optimized methods adapted to the special form of the integrals should be used. The integrals can, for instance, be determined by using Laguerre polynomials as an approximation of the integrand (Kippenhahn and Thomas 1964). This method was extended to higher accuracy by Pichon (1989), who also discusses alternative schemes of numerical integration. A very efficient and convenient method was developed by Gong et al. (2001), who also provide a ready-to-use computer code, available from the publisher of that paper, which conveniently also computes the derivatives of the Fermi integrals, needed for the Henyey method of Chap. 12. This code was also used to compute the values of Table 15.1. Alternatively, Blinnikov et al. (1996) have extended the approach of expansions to more general cases of degeneracy and relativism than discussed here. These authors also provide a computer code on request.

Chapter 16
The Equation of State of Stellar Matter

In Chap. 15 we dealt with degeneracy of arbitrary degree for the electron gas. We now discuss the combined effect of *all* components of stellar matter, starting with the ion gas.

16.1 The Ion Gas

In the non-degenerate case, electron pressure $P_e = n_e kT$ and ion pressure $P_{ion} = n_{ion} kT$ are of the same order of magnitude; they are even equal in the case of ionized hydrogen with $n_e = n_{ion}$. For sufficiently low temperature or sufficiently high density the ions can become degenerate, too. If they are Fermi particles such as protons, they will behave in phase space like the electrons, so that, for P_{ion} and n_{ion}, relations such as (15.29)–(15.31) hold if the mass of the ions m_{ion} is used instead of m_e, and ψ is now the degeneracy parameter for the ions. Again the transition between perfect-gas behaviour and degeneracy is roughly at $\psi = 0$. We write (15.39) in the form

$$\frac{n_j}{T^{3/2}} = \text{constant}\,(m_j)^{3/2} F_{1/2}(\psi), \tag{16.1}$$

where n_j and m_j refer to either electrons or ions. Suppose that the electron gas has a certain value of $\psi = \psi^*$ for $n_e = n_e^*$. An ion gas of the same temperature has the same degeneracy parameter $\psi = \psi^*$ for $n_{ion} = (m_{ion}/m_e)^{3/2} n_e^* \approx 8 \times 10^4 n_e^*$. Therefore the ions require much higher densities to become degenerate. For the interior of normal stars one can assume that even if the electrons are degenerate the ions still obey Boltzmann statistics; thus, because of the Pauli principle, the degenerate electrons have much higher momentum than the non-degenerate ions, and the electron pressure is much larger than the pressure of the ions: $P = P_{ion} + P_e \approx P_e$.

R. Kippenhahn et al., *Stellar Structure and Evolution*, Astronomy and Astrophysics Library, DOI 10.1007/978-3-642-30304-3_16, © Springer-Verlag Berlin Heidelberg 2012

Even when the ion gas does not contribute noticeably to the pressure, it provides the main contribution to the mass density ϱ. This has already been taken into account by relating n_e to $\varrho = n_e \mu_e m_u$, for example in (15.39). Furthermore, the ions can influence the thermodynamic properties of the plasma considerably.

One should be aware that, for certain types of stars, the treatment of the ions is not as simple as described here, since they can be subject to rather complicated interactions, for example, those indicated in Sects. 16.4 and 16.5.

16.2 The Equation of State

For normal stellar matter, the equation of state is then given by

$$P = P_{\text{ion}} + P_e + P_{\text{rad}} = \frac{\Re}{\mu_0}\varrho T + \frac{8\pi}{3h^3}\int_0^\infty p^3 v(p) \frac{dp}{e^{E/kT-\psi}+1} + \frac{a}{3}T^4 , \quad (16.2)$$

$$\varrho = \frac{4\pi}{h^3}(2m_e)^{3/2} m_u \mu_e \int_0^\infty \frac{E^{1/2} dE}{e^{E/kT-\psi}+1} , \quad (16.3)$$

where $v(p) = \partial E/\partial p$ according to (15.7) and where E is given by (15.8). If the electron gas is highly degenerate, then also $P_{\text{rad}} \ll P_e$ and $P \approx P_e$.

For given ϱ and T and chemical composition (μ_0), (16.3) can be used to determine ψ. Then ϱ, ψ, and T determine P via (16.2). The equation of state $P = P(\varrho, T)$ for all degrees of degeneracy, including relativistic effects, is therefore given here in implicit form.

An expression similar to (16.2) can be obtained for the internal energy u per unit mass:

$$u = \frac{U_{\text{ion}} + U_e + U_{\text{rad}}}{\varrho} = \frac{3}{2}\frac{\Re}{\mu_0}T + \frac{8\pi}{h^3 \varrho}\int_0^\infty \frac{p^2 E(p) dp}{e^{E/kT-\psi}+1} + \frac{aT^4}{\varrho} , \quad (16.4)$$

where the U are the energies per unit volume, and the first term on the right corresponds to the (perfect monatomic) ion gas.

Figure 16.1 shows the $\lg \varrho$–$\lg T$ plane for the ranges relevant for the interiors of most stars. In different regions, different effects dominate the total pressure, for example, in some places the electron degeneracy and in others the radiation pressure. We will derive rough borders between these different regimes.

Let us first consider the lines ψ = constant for given μ_e in this diagram. In the non-relativistic regime, (15.39) shows that ψ is constant for $T \sim \varrho^{2/3}$, i.e. on straight lines of slope 2/3 in the $\lg \varrho$–$\lg T$ plane. In the relativistic regime ψ = constant for $T \sim \varrho^{1/3}$ according to (15.45), i.e. on straight lines with slope 1/3.

We have already seen that the perfect-gas approximation $P_{\text{gas}} = \Re \varrho T/\mu$ becomes valid for large negative values of ψ. For large positive values of ψ complete degeneracy is a good approximation for the electron gas, and $P \approx P_e$ for

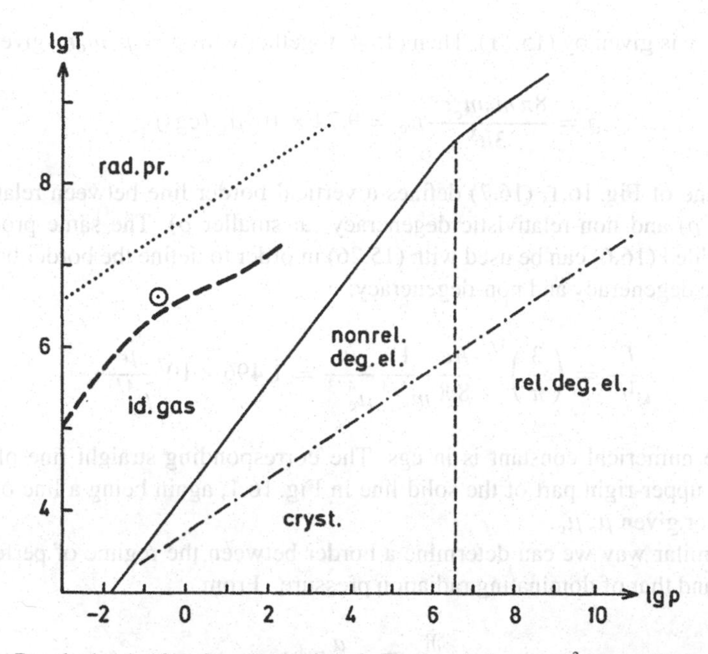

Fig. 16.1 Rough sketch of regions in the lg ϱ–lg T plane (ϱ in g cm^{-3}, T in K), in which the equation of state is dominated by radiation pressure (above the *dotted line* given here by $P_{\mathrm{rad}} = P_{\mathrm{gas}}$ for $\mu = 0.5$), and by the degenerate electron gas (below the *solid line* given here by (16.6) and (16.8) for $\mu_{\mathrm{e}} = 2$), which can be relativistic (right of the *vertical broken line* given by (16.7) for $\mu_{\mathrm{e}} = 2$) or non-relativistic (left of the *vertical broken line*). The *dot-dashed line* indicates the melting temperature as given by (16.26) for $\mu_0 = 4$. By comparing with (14.45) one can see that the Saha formula is valid almost nowhere in the plotted domain. The *heavy dashed curve* on the left corresponds to a model of the present Sun

the non-relativistic case is given by (15.23). We can define the border between the two regimes by the condition that both approximations yield the same value for the pressure:

$$\frac{\Re}{\mu}\varrho T = \frac{1}{20}\left(\frac{3}{\pi}\right)^{2/3}\frac{h^2}{m_{\mathrm{e}}}\left(\frac{\varrho}{\mu_{\mathrm{e}} m_{\mathrm{u}}}\right)^{5/3}. \tag{16.5}$$

Equation (16.5) is equivalent to

$$\frac{T}{\varrho^{2/3}} = \frac{1}{20}\left(\frac{3}{\pi}\right)^{2/3}\frac{h^2}{m_{\mathrm{e}}\Re m_{\mathrm{u}}^{5/3}}\frac{\mu}{\mu_{\mathrm{e}}^{5/3}} = 1.207\times 10^5\frac{\mu}{\mu_{\mathrm{e}}^{5/3}}, \tag{16.6}$$

where the numerical constant is in cgs units. Equation (16.6) gives a straight line with slope 2/3 in Fig. 16.1 (lower left part of the solid line), which is obviously a line of ψ = constant for given μ, μ_{e}. To the left of it the electrons behave almost like a perfect gas; to the right they are degenerate and dominate the pressure.

We now ask where relativistic effects become important. The transition between the non-relativistic and relativistic cases occurs around $x \approx 1$, where the relativity parameter x is given by (15.11). Then (15.4) together with $\varrho = \mu_e m_u n_e$ gives

$$\varrho = \frac{8\pi m_u m_e^3 c^3}{3h^3} \mu_e = 9.74 \times 10^5 \mu_e \text{ (cgs)} . \tag{16.7}$$

In the plane of Fig. 16.1, (16.7) defines a vertical border line between relativistic (at larger ϱ) and non-relativistic degeneracy (at smaller ϱ). The same procedure which yielded (16.6) can be used with (15.26) in order to define the border between relativistic degeneracy and non-degeneracy:

$$\frac{T}{\varrho^{1/3}} = \left(\frac{3}{\pi}\right)^{1/3} \frac{hc}{8\Re} \frac{1}{m_u^{4/3}} \frac{\mu}{\mu_e^{4/3}} = 1.496 \times 10^7 \frac{\mu}{\mu_e^{4/3}} , \tag{16.8}$$

where the numerical constant is in cgs. The corresponding straight line of slope 1/3 is the upper-right part of the solid line in Fig. 16.1, again being a line of $\psi =$ constant for given μ, μ_e.

In a similar way we can determine a border between the regime of perfect gas pressure and that of dominating radiation pressure . From

$$\frac{\Re}{\mu} \varrho T = \frac{a}{3} T^4 \tag{16.9}$$

we find

$$\frac{T}{\varrho^{1/3}} = \left(\frac{3\Re}{a\mu}\right)^{1/3} = \frac{3.2 \times 10^7}{\mu^{1/3}} , \tag{16.10}$$

where the constant is in cgs. This line of slope 1/3 is dotted in Fig. 16.1.

In Fig. 16.1 it is indicated how T grows with increasing density in the Sun. As one can see, the interior regions of the Sun avoid the area in the diagram where radiation pressure is important, as well as that of degeneracy. However, we will have to deal with other cases in which the equation of state is more complicated. This concerns highly evolved stars, but also unevolved stars of very low mass (For a review see Van Horn 1986.).

16.3 Thermodynamic Quantities

With the implicit form (16.2) and (16.3) and with the expression (16.4) for the internal energy we are in principle able to determine δ, c_P, and ∇_{ad}. Since in general no analytic methods are known one can try to determine the thermodynamic quantities numerically. Here we just give them for some limiting cases for which analytic expressions can be derived. For the sake of simplicity we neglect the effects of radiation and we suppose the ions to be a perfect gas.

In the cases of complete degeneracy of a non-relativistic or an extremely relativistic electron gas, it is obvious from equations (15.23) and (15.26) that the quantities α, δ as defined in (4.2) and (4.3) are $\alpha = 3/5, \delta = 0$, or $\alpha = 3/4, \delta = 0$ respectively.

We define the ratio η of ion pressure to total pressure

$$\eta := \frac{P_{\text{ion}}}{P_{\text{ion}} + P_{\text{e}}} . \tag{16.11}$$

For strong non-relativistic degeneracy (15.39), (15.41), and (15.43) for $\psi \gg 1$, imply that

$$P_{\text{e}} \approx \frac{4}{15} B_1 (\psi k T)^{5/2} , \quad B_1 = \frac{4\pi}{h^3} (2m_{\text{e}})^{3/2} ,$$

$$\varrho \approx \frac{2}{3} \mu_{\text{e}} m_{\text{u}} B_1 (\psi k T)^{3/2} , \tag{16.12}$$

which together with $P_{\text{ion}} = \Re \varrho T / \mu_0 = k \varrho T / (m_{\text{u}} \mu_0)$ and (16.11) result in

$$\eta \approx \frac{5}{2} \frac{\mu_{\text{e}}}{\mu_0} \frac{1}{\psi} . \tag{16.13}$$

The larger ψ (the stronger the degeneracy), the smaller η, and therefore the smaller the contribution of the ion gas to the total pressure.

The value of δ can be obtained from the relation

$$\delta = -\left(\frac{\partial \ln \varrho}{\partial \ln T}\right)_P = -\left(\frac{\partial \ln \varrho}{\partial \ln T}\right)_\psi + \frac{\left(\frac{\partial \ln \varrho}{\partial \ln \psi}\right)_T \left(\frac{\partial \ln P}{\partial \ln T}\right)_\psi}{\left(\frac{\partial \ln P}{\partial \ln \psi}\right)_T} , \tag{16.14}$$

which follows from the total differentials of the functions $\varrho = \varrho(\psi, T), P = P(\psi, T)$. For $P = P_{\text{e}}$ the partial derivatives can be taken from (16.12), and (16.14) gives $\delta = 0$. For a small but non-vanishing contribution P_{ion} we write according to (16.11) the total pressure $P = P_{\text{e}}/(1 - \eta) \approx (1 + \eta) P_{\text{e}}$. If we then use the expressions (16.12) and (16.13), we obtain for the non-relativistic case

$$\delta \approx \frac{3}{5} \eta \approx \frac{3}{2} \frac{\mu_{\text{e}}}{\mu_0} \frac{1}{\psi} . \tag{16.15}$$

For the extremely relativistic electron gas we find from (15.45) and (15.46), with the lowest terms of the expansion (15.43), that

$$P_{\text{e}} = \frac{B_2}{4} (\psi k T)^4 , \quad B_2 = \frac{8\pi}{3c^3 h^3} ,$$

$$\varrho = \mu_{\text{e}} m_{\text{u}} B_2 (\psi k T)^3 , \tag{16.16}$$

and in the same way we obtained (16.13) and (16.15) we now get

$$\eta \approx 4\frac{\mu_e}{\mu_0}\frac{1}{\psi}, \quad \delta = \frac{3}{4}\eta = \frac{3\mu_e}{\mu_0}\frac{1}{\psi}. \tag{16.17}$$

In order to derive c_P we need the internal energy u. Let us again neglect the radiation field here; then u contains a component u_e of the (degenerate) electron gas and a component u_{ion} of the (perfect) ion gas: $u = u_e + u_{ion}$. In the non-relativistic case, (15.42) gave $U_e = 3P_e/2$ for the internal energy U_e per unit volume of the electron gas, independent of ψ. A corresponding relation $U_{ion} = 3P_{ion}/2$ holds for the non-degenerate ions, and therefore

$$u = \frac{U}{\varrho} = \frac{3}{2}\frac{P_{ion} + P_e}{\varrho} = \frac{3}{2}\frac{P}{\varrho}. \tag{16.18}$$

This gives the derivative

$$\left(\frac{\partial u}{\partial T}\right)_P = -\frac{3}{2}\frac{P}{\varrho T}\left(\frac{\partial \ln \varrho}{\partial \ln T}\right)_P = \frac{3}{2}\frac{P\delta}{\varrho T}, \tag{16.19}$$

which is used in the definition (4.4) of c_P:

$$c_P = \left(\frac{\partial u}{\partial T}\right)_P - \frac{P}{\varrho^2}\left(\frac{\partial \varrho}{\partial T}\right)_P = \left(\frac{\partial u}{\partial T}\right)_P + \frac{P\delta}{\varrho T} = \frac{5}{2}\frac{P\delta}{\varrho T}. \tag{16.20}$$

Then (4.21) gives $\nabla_{ad} = 2/5$, the same value we obtained for the perfect gas with $\beta = 1$ [see (13.12)]. Since we have derived it without making use of the degree of degeneracy, the numerical value 2/5 for ∇_{ad} is independent of ψ, but holds only for non-relativistic degeneracy.

In the extreme relativistic case, (15.27) shows that $U_e = 3P_e$, while again $U_{ion} = 3P_{ion}/2$ for the non-degenerate ions. The total energy density is then

$$u = u_e + u_{ion} = 3\frac{P_e}{\varrho} + \frac{3}{2}\frac{P_{ion}}{\varrho} = 3\frac{P}{\varrho} - \frac{3}{2}\frac{P_{ion}}{\varrho} = 3\frac{P}{\varrho} - \frac{3}{2}\frac{\Re}{\mu_0}T; \tag{16.21}$$

the specific heat is

$$c_P = -\frac{4P}{\varrho^2}\left(\frac{\partial \varrho}{\partial T}\right)_P - \frac{3}{2}\frac{\Re}{\mu_0} = \frac{4P}{\varrho T}\delta - \frac{3}{2}\frac{\Re}{\mu_0}, \tag{16.22}$$

so that we can now determine ∇_{ad}:

$$\nabla_{ad} = \frac{P\delta}{\varrho T c_P} = \frac{1}{4 - \frac{3}{2}\frac{\Re}{\mu_0}\frac{\varrho T}{P\delta}}. \tag{16.23}$$

From (16.16) and (16.17) we find that

$$P \approx P_e = \frac{B_2}{4}(\psi kT)^4 , \quad \varrho = B_2 \mu_e m_u (\psi kT)^3 , \quad \delta = 3\frac{\mu_e}{\mu_0}\frac{1}{\psi} , \tag{16.24}$$

and therefore $3\Re \varrho T/\mu_0 = 4P\delta$, which with (16.23) gives $\nabla_{ad} = 1/2$. This is the value for the fully degenerate, extreme relativistic case.

16.4 Crystallization

Up to now we have treated the ions as a perfect gas, which means we have neglected their interaction. However, this no longer suffices for high densities and particularly low temperatures, in which case the Coulomb interaction of the ions must be considered: instead of moving freely, the ions tend to form a rigid lattice, which minimizes their total energy. This occurs when the thermal energy $3kT/2$ becomes comparable with the Coulomb energy per ion of charge $-Ze$. If we define a volume V_{ion} per ion by $n_{ion}V_{ion} = 1$ (where n_{ion} is the number density of ions) and a mean separation r_{ion} between the ions, we have $V_{ion} = 4\pi r_{ion}^3/3$. Then the ratio

$$\Gamma_C := \frac{(Ze)^2}{r_{ion} kT} = 2.7 \times 10^{-3} \frac{Z^2 n_{ion}^{1/3}}{T} \tag{16.25}$$

is a measure for the importance of this effect, the numerical constant having units of cgs. $\Gamma_C \ll 1$ would mean that the electrostatic energy plays a minor role and the ions have a Boltzmann distribution, while $\Gamma_C \gg 1$ indicates that the kinetic energy of the ions is negligible and that they try to form a conglomerate that has a lower energy, i.e. they form a crystal.

More detailed considerations (see, for instance, Shapiro and Teukolsky 1983) indicate that $\Gamma_C \approx 170$ is a critical value for the transition between the two types of behaviour of the ion gas. With this value for Γ_C and using the relation $\varrho = \mu_0 m_u n_{ion}$ we obtain the critical temperature T_m (melting temperature):

$$T_m \approx \frac{Z^2 e^2}{\Gamma_c k}\left(\frac{4\pi\varrho}{3\mu_0 m_u}\right)^{1/3} = 1.3 \times 10^3 Z^2 \mu_0^{-1/3} \varrho^{1/3} , \tag{16.26}$$

where the numerical constant is in cgs units. The corresponding straight line is plotted (dot-dashed) in Fig. 16.1.

In the interior of evolved stars we have high densities, but the temperature is well above the melting temperature. The situation is different in cooling white dwarfs, where the temperature becomes smaller with time, while the density remains virtually unchanged. We will come back to this in Chap. 37, which deals with white dwarfs.

16.5 Neutronization

If in a plasma the electrons have sufficient energy, they can combine with the protons to form neutrons. If m_n and m_p are the masses of neutron and proton, then the electron must have the total energy $E_{tot} > E^* = c^2(m_n - m_p)$. At low densities the neutron will decay within 11 min back into a proton–electron pair, where the electron has the total energy E^* and a kinetic energy $E^*_{kin} = E^* - m_e c^2$; however, the situation can be different if the gas is completely degenerate and the phase space is filled up to the (kinetic) Fermi energy E_F. If the Fermi energy E_F exceeds E^*_{kin}, the electrons released do not have enough energy to find an empty cell in phase space, and the neutrons cannot decay, i.e. the Fermi sea of electrons has stabilized the neutrons.

In order to estimate under which conditions this occurs we write the relation (15.6) between E and p in the form

$$p = \frac{1}{c}(E^2 - m_e^2 c^4)^{1/2} . \tag{16.27}$$

If we put $E = E_{kin} + m_e c^2 = E_F + m_e c^2 = c^2(m_n - m_p) = 1.294 \times 10^6$ eV, we can determine the corresponding Fermi momentum p_F from (16.27) and obtain $x = p_F/(m_e c) \approx 2.2$. Then, according to (15.15) and taking $\varrho = \mu_e m_u n_e$ with $\mu_e = 2$, we find $\varrho \approx 2.4 \times 10^7 \text{ g cm}^{-3}$. Therefore, if a proton–electron gas is compressed to a density above this value, then the gas undergoes a transition into a neutron gas ("neutronization").

For stellar matter the situation is more complicated, since at sufficiently high densities the plasma contains heavier nuclei, and not just protons. The nuclei capture electrons (inverse β decay) and become neutron-rich isotopes. This requires much higher electron energies than those just estimated, since the neutrons in the nucleus are degenerate and the new ones have to be raised above the Fermi energy. Correspondingly higher plasma densities are required to provide the electrons with the necessary energy. If the nuclei become too neutron rich they start to break up, releasing free neutrons. The density at which this "neutron drip" starts is of the order of several $10^{11} \text{ g cm}^{-3}$, but the exact value depends on the nuclear model one is using in detailed calculations. Hillebrandt (1991) gives $\varrho_{drip} \approx 3 \times 10^{11} \text{ g cm}^{-3}$, Pethick and Ravenhall (1991) estimate $\varrho_{drip} \approx 3.5 \times 10^{11} \text{ g cm}^{-3}$.

Let us briefly consider the effect on the equation of state. Up to ϱ_{drip} the total pressure $P \approx P_e$ is provided by relativistic electrons. With further increases of ϱ, the number density n_e increases by less than an amount proportional to ϱ, owing to the capture of some electrons. Therefore the pressure rises by less than $\varrho^{4/3}$. Consequently $\gamma_{ad} \equiv (d \ln P / d \ln \varrho)_{ad}$ is reduced below 4/3, which can be seen in Fig. 16.2, where the slope of the curve $P = P(\varrho)$ is suddenly reduced for $\log \varrho \gtrsim 11.7$. At still higher ϱ the increasing number of free neutrons contribute gradually more to P.

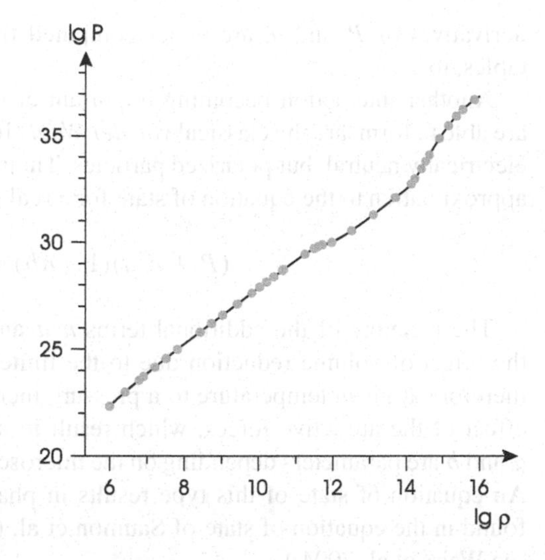

Fig. 16.2 The equation of state for very high densities. On logarithmic scales the pressure P_e (in dyn cm^{-2}) is plotted against the density ϱ (in g cm^{-3}). The *grey* symbols refer to experimental or theoretical data points from various sources. See Haensel et al. (2007), p. 15, for more details. Figure adapted from their Fig. 1.3

With increasing ϱ the neutrons become increasingly degenerate–as a perfect Fermi gas they would give the slope 5/3. But then interaction between neutrons becomes important, and the details of the equation of state are very uncertain, for example, depending on rather badly known properties of the particles. For more details see Sects. 37.2 and 38.1 and Shapiro and Teukolsky (1983).

16.6 Real Gas Effects

Although the stellar plasma can to a great extent be treated as a perfect gas, the assumptions for a perfect gas are not truly fulfilled: there are interaction forces, such as the Coulomb force, acting between the constituents, and atoms and ions cannot always be considered to be pointlike. Therefore an accurate equation of state has to include such effects.

We already encountered *pressure ionization* in Sect. 14.6, which is a consequence of the spatially overlapping energy levels, which leads to interacting ionic potentials. We noted that there is no good theory to treat this in a simple way, but that one has to modify the Saha equation somehow to avoid its wrong behaviour at high pressure. A good theory for pressure ionization has to work with quantum-mechanical atomic models. The effect of pressure ionization is increasingly important for cool, dense stars of low mass ($M < 1\,M_\odot$) and gas planets. Saumon et al. (1995) have developed an equation for state for dense gases, which, due to the complexity of the problem, is limited to hydrogen-helium mixtures. This and other modern equations of state are provided in tabular form, for example, as tables of $P(\rho, T)$ and $u(\rho, T)$ for various chemical mixtures. The thermodynamic quantities, which are or use derivatives of P and u, are either computed from the tables, or are provided as tables, too.

Another interaction becoming important at low temperatures, when molecules are able to form, are the classical *van der Waals* forces, which are attractive forces of electrically neutral, but polarized particles. Their consideration leads in the simplest approximation to the equation of state for a real gas

$$(P + n^2 a)(1 - nb) = nkT. \tag{16.28}$$

The meaning of the additional terms $n^2 a$ and nb is easy to understand: nb is the effect of volume reduction due to the finite size of the particles, which leads therefore at given temperature to a pressure increase. The second term, $n^2 a$, is the effect of the attractive forces, which result in a reduction of the gas pressure P. a and b are parameters depending on the microscopic properties of the gas particles. An equation of state of this type results in phase transitions, which indeed were found in the equation of state of Saumon et al. (1995) (For a derivation of (16.28) see Weiss et al. 2004.).

In Sect. 16.4, we discussed crystallization, which is due to the electrostatic interaction between ions at high densities, in the limit of $\Gamma_C \gg 1$. At the other extreme, when $\Gamma_C \ll 1$, the gas is close to being a perfect one, but not quite so. Consider an ionized gas, which consists of positively charged ions and unbound electrons. As long as these are not degenerate, they can move freely and will feel the Coulomb forces in the plasma. In particular, ions will attract electrons and it is plausible that clouds of electrons gather around ions such that from a sufficiently large distance the ion electron cloud will appear as being electrically neutral. This picture of *electron shielding (in the weak limit)* requires low particle densities, because the inter-ion distances must be larger than the typical electron cloud size. The physical effect is usually treated within the *Debye–Hückel theory* (Landau and Lifshitz 1980, Chap. 78; Weiss et al. 2004, Chap. 17.15), which we will encounter in detail in Sect. 18.4. Here it suffices to state that based on a shielded Coulomb potential around the ions,

$$\Phi(r) = \frac{Ze}{r} \cdot e^{-r/r_D}, \tag{16.29}$$

where r_D is the Debye-radius (18.50), the resulting attractive electrostatic forces lead to a reduction of the gas pressure according to

$$P = nkT\left[1 - 3.2 \times 10^7 \frac{\varrho^{1/2}}{T^{3/2}} \mu \zeta^{3/2}\right], \tag{16.30}$$

where

$$\zeta = \sum_i \frac{Z_i(Z_i + 1)}{A_i} X_i \tag{16.31}$$

is, as in (18.47), the mass weighted average of free electrons times ionic charge Z_i of all ion species i.

For the centre of the Sun, $\mu \approx 0.8$, $\zeta \approx 1.7$, $T \approx 15 \times 10^6$ K, and $\varrho \approx 140\,\mathrm{g/cm^{-3}}$, and the correction to P is 1.6 %. Although this effect appears to be small, it has turned out that equations of state with an accuracy of this order are needed for modern solar and stellar models.

These are the most important non-ideal effects that modify the equation of state. In addition there are even more interaction forces of quantum nature (such as spin–spin interaction), which may in some situations become important. Some of these effects are considered in equation of states published by specialized groups. The most important ones are the OPAL and MHD equations of state (Rogers et al. 1996 and Mihalas et al. 1988, and later improvements), both available in tabular form and for a variety of chemical mixtures. They are widely used in current stellar evolution calculations, and have helped to improve the solar model considerably. More on this issue can be found in Weiss et al. (2004), Chap. 15-A.

Chapter 17
Opacity

In this chapter we deal with the material function $\kappa(\varrho, T)$. While for the equation of state it was possible to use certain approximations (for instance, that of a perfect gas) without introducing too much error, this is almost impossible for the opacity. Although there are similar approximations (such as those for electron scattering or free–free transitions) they never hold for the whole star and are used only in simplifying approaches. Therefore, nowadays, when solving the stellar-structure equations, one uses numerical opacity tables for different chemical mixtures, which give $\kappa(\varrho, T)$ in the full range of ϱ and T.

In the following we describe the basic processes that contribute to the opacity and give approximate analytic formulae without deriving them from quantum mechanics. The reader who wants to learn more of the methods by which opacities are computed is referred to Weiss et al. (2004) and to the original papers quoted there.

17.1 Electron Scattering

If an electromagnetic wave passes an electron, the electric field makes the electron oscillate. The oscillating electron represents a classical dipole that radiates in other directions, i.e. the electron scatters part of the energy of the incoming waves. The weakening of the original radiation due to scattering is equivalent to that by absorption, and we can describe it by way of a cross section at frequency ν per unit mass (which we called κ_ν in Sect. 5.1). This can be calculated classically giving the result

$$\kappa_\nu = \frac{8\pi}{3}\frac{r_e^2}{\mu_e m_u} = 0.20\,(1+X), \tag{17.1}$$

where r_e is the classical electron radius, X the mass fraction of hydrogen, and the constant is in $cm^2\,g^{-1}$. The term $\mu_e m_u$ arises because κ_ν is taken per unit mass; and μ_e is replaced by (4.30). Since κ_ν does not depend on the frequency, we immediately obtain the Rosseland mean for electron scattering:

R. Kippenhahn et al., *Stellar Structure and Evolution*, Astronomy and Astrophysics Library, DOI 10.1007/978-3-642-30304-3_17, © Springer-Verlag Berlin Heidelberg 2012

$$\kappa_{sc} = 0.20\,(1 + X)\,\mathrm{cm^2\,g^{-1}} \,. \tag{17.2}$$

The "Thomson scattering" just described neglects the exchange of momentum between electron and radiation. If this becomes important, then κ_ν will be reduced compared to the value given in (17.1), though this effect plays a role only at temperatures sufficiently high for the scattered photons to be very energetic. In fact during the scattering process the electron must obtain such a large momentum that its velocity is comparable to c, say $v \gtrsim 0.1c$ for (17.2) to become a bad approximation. The momentum of the photon is $h\nu/c$, which after scattering is partly transferred to the electron, $m_e v \sim h\nu/c$. Therefore relativistic corrections ("Compton scattering") become important if the average energy of the photons is $h\nu \gtrsim 0.1 m_e c^2$. For $h\nu$ we take the frequency at which the Planck function has a maximum; then according to Wien's law this is at $h\nu = 4.965\,kT$, and the full Compton scattering cross section has to be taken into account if $T > 0.1 m_e c^2/(4.965k)$, or roughly $T > 10^8$ K. In fact even at $T = 10^8$ K Compton scattering reduces the opacity by only 20 % of that given by (17.2).

17.2 Absorption Due to Free–Free Transitions

If during its thermal motion a free electron passes an ion, the two charged particles form a system which can absorb and emit radiation. This mechanism is only effective as long as electron and ion are sufficiently close. Now, the mean thermal velocity of the electrons is $v \sim T^{1/2}$, and the time during which they form a system able to absorb or emit is proportional to $1/v \sim T^{-1/2}$; therefore, if in a mass element the numbers of electrons and ions are fixed, the number of systems temporarily able to absorb is proportional to $T^{-1/2}$.

The absorption properties of such a system have been derived classically by Kramers, who calculated that the absorption coefficient per system is proportional to $Z^2\nu^{-3}$, where Z is the charge number of the ion. We therefore expect the absorption coefficient κ_ν of a given mixture of (fully ionized) matter to be

$$\kappa_\nu \sim Z^2 \varrho T^{-1/2} \nu^{-3} \,. \tag{17.3}$$

Here the factor ϱ appears because for a given mass element the probability that two particles are accidentally close together is proportional to the density.

For the determination of the Rosseland mean κ of this absorption coefficient we make use of a simple theorem which can be easily proved by carrying out the integration (5.19): a factor ν^α contained in κ_ν gives a factor T^α in κ. With this and with (17.3) we find

$$\kappa_{ff} \sim \varrho T^{-7/2} \,. \tag{17.4}$$

All opacities of the form (17.4) are called *Kramers opacities* and give only a classical approximation. One normally multiplies the Kramers formula (17.4) by a correction factor g, the so-called *Gaunt factor*, in order to take care of the quantum-mechanical correction (see, for instance, Weiss et al. 2004). In (17.4) we have still omitted the factor Z^2 which appears in (17.3). In general, one has a mixture of different ions, and therefore one has to add the contributions of the different chemical species. The (weighted) sum over the values of Z^2 is taken into the constant of proportionality in (17.4), which then depends on the chemical composition. For a fully ionized mixture a good approximation is given by

$$\kappa_{\mathrm{ff}} = 3.8 \times 10^{22}(1+X)[(X+Y)+B]\varrho T^{-7/2} , \tag{17.5}$$

with the numerical constant in cgs. The mass fractions of H and He are X and Y, respectively. Here the factor $1+X$ arises, since κ_{ff} must be proportional to the electron density–which is proportional to $(1+X)\varrho$. The term $(X+Y)$ in the brackets can be understood in the following way: there are X/m_{u} hydrogen ions and $Y/(4m_{\mathrm{u}})$ helium ions. The former have the charge number 1, the latter the charge number 2. But since $\kappa_\nu \sim Z^2$ [see (17.3)], when adding the contributions of H and He to the total absorption coefficient, we obtain the factor $X/m_{\mathrm{u}} + 4Y/(4m_{\mathrm{u}}) = (X+Y)m_{\mathrm{u}}$. Correspondingly the term B gives the contribution of the heavier elements:

$$B = \sum \frac{X_i Z_i^2}{A_i} , \tag{17.6}$$

where the summation extends over all elements higher than helium and A_i is the atomic mass number.

17.3 Bound–Free Transitions

We first consider a (neutral) hydrogen atom in its ground state, with an ionization energy of χ_0, i.e. a photon of energy $h\nu > \chi_0$ can ionize the atom. Energy conservation then demands that

$$h\nu = \chi_0 + \frac{1}{2} m_{\mathrm{e}} v^2 , \tag{17.7}$$

where v is the velocity of the electron released (relative to the ion, which is assumed to be at rest before and after ionization).

If we define an absorption coefficient a_ν per ion ($a_\nu = \kappa_\nu \varrho / n_{\mathrm{ion}}$), we expect $a_\nu = 0$ for $\nu < \chi_0/h$ and $a_\nu > 0$ for $\nu \geq \chi_0/h$. Classical considerations similar to those which lead to the Kramers dependence (17.3) of κ_ν for free–free transitions give $a_\nu \sim \nu^{-3}$ for $\nu \geq \chi_0/h$. Quantum-mechanical corrections can again be taken into account by a Gaunt factor (see, for instance, Weiss et al. 2004). The absorption coefficient of the hydrogen atom in its ground state has a frequency dependence as given in Fig. 17.1a. But if we have neutral hydrogen atoms in different stages of excitation, the situation is different: an atom in the first excited stage has an

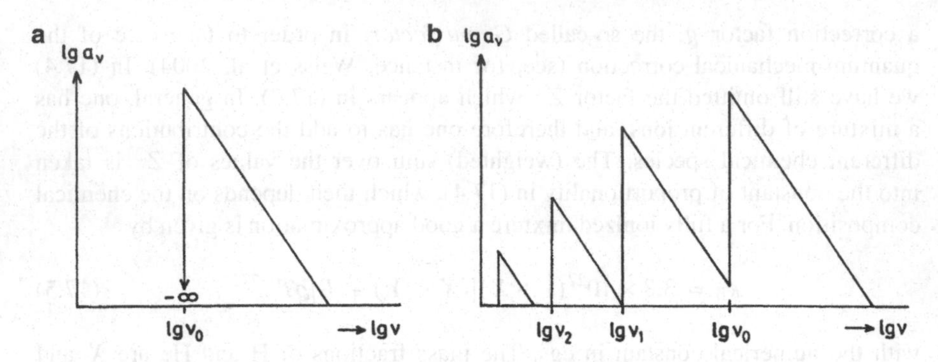

Fig. 17.1 (**a**) The absorption coefficient a_ν of a hydrogen atom in the ground state as a function of the frequency ν; $\nu_0 = \chi_0/h$ (**b**) The absorption coefficient of a mixture of hydrogen atoms in different stages of excitation

absorption coefficient $a_\nu = 0$ for $h\nu < \chi_1$, where χ_1 is the energy necessary to ionize a hydrogen atom from the first excited state, while $a_\nu \sim \nu^{-3}$ for $h\nu \geq \chi_1$. The absorption coefficient κ_ν for a mixture of hydrogen atoms in different states of excitation is a superposition of the a_ν for different stages of excitation. The resulting κ_ν is a sawtooth function, as indicated in Fig. 17.1b. In order to obtain κ_ν for a certain value of the temperature T, one has to determine the relative numbers of atoms in the different stages of excitation by the Boltzmann formula; then their absorption coefficients a_ν, weighted with their relative abundances, are to be summed. To obtain the Rosseland mean one has to carry out the integration (5.19).

If there are ions of different chemical species with different degrees of ionization, one has to sum the functions a_ν for all species in all stages of excitation and all degrees of ionization before carrying out the Rosseland integration. An important source of opacity are bound–free transitions of neutral hydrogen atoms, in which case the opacity must be proportional to the number of neutral hydrogen atoms and κ can be written in the form

$$\kappa_{\mathrm{bf}} = X(1-x)\tilde{\kappa}(T) . \tag{17.8}$$

Here $\tilde{\kappa}(T)$ is obtained by Rosseland integration over (weighted) sums of functions a_ν for the different stages of excitation, while x is the degree of ionization as defined in Sect. 14.2. The function $\tilde{\kappa}(T)$ is plotted in Fig. 17.2.

17.4 Bound–Bound Transitions

For absorption by an electron bound to an ion, more than just the bound–free transitions discussed in Sect. 17.3 contribute to the opacity. If, after absorption of a photon from a directed beam, the electron does not leave the atom but jumps to a

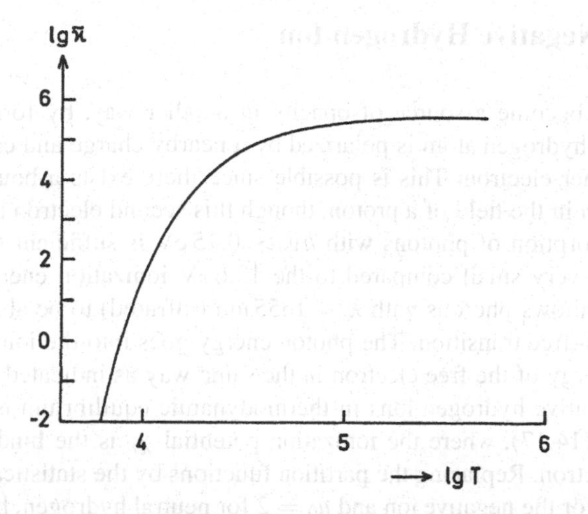

Fig. 17.2 The function $\widetilde{\kappa}(T)$ of (17.8), where $\widetilde{\kappa}$ is in $cm^2\ g^{-1}$ and T in K

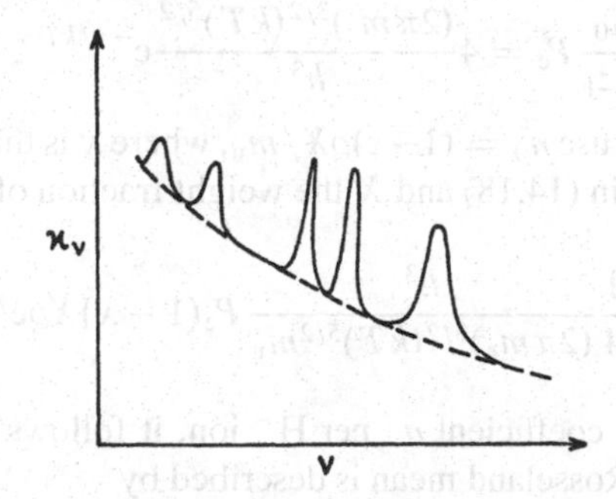

Fig. 17.3 Bound–bound transitions contributing to the opacity κ_ν

higher bound state, the energy will later on be re-emitted in an arbitrary direction, so that the intensity of the directed beam is weakened. This mechanism is effective only at certain frequencies, and one would expect that absorption in a few lines gives only a small contribution to the overall opacity; however, the absorption lines in stars are strongly broadened by collisions, and as one can see in Fig. 17.3, they can occupy considerable regions of the spectrum. Bound–bound absorption can become a major contribution to the (Rosseland mean) opacity if $T < 10^6$ K. It can then increase the total opacity by a factor 2, while for higher temperatures (say $T \approx 10^7$ K) the contribution of bound–bound transitions to the total opacity is much smaller (10 %). Calculation of the absorption coefficients due to bound-bound transitions obviously requires detailed knowledge about the energy levels of all atoms and ions, all mechanisms that lead to line broadening, and of all the transition probabilities. In addition, occupation levels and ionization levels have to be known, which links the calculation of opacities closely to that of the equation of state. Such calculations have again to be done in separate calculations by specialists in atomic physics.

17.5 The Negative Hydrogen Ion

Hydrogen can become a source of opacity in another way, by forming negative ions: a neutral hydrogen atom is polarized by a nearby charge and can then attract and bind another electron. This is possible since there exists a bound state for a second electron in the field of a proton, though this second electron is only loosely bound–the absorption of photons with $h\nu > 0.75\,\text{eV}$ is sufficient for its release. This energy is very small compared to the 13.6 eV ionization energy for neutral hydrogen and allows photons with $\lambda < 1655\,\text{nm}$ (infrared) to be absorbed, giving rise to a bound–free transition. The photon energy goes into the ionization energy and kinetic energy of the free electron in the same way as indicated in (17.7). The number of negative hydrogen ions in thermodynamic equilibrium is given by the Saha formula (14.17), where the ionization potential χ_r is the binding energy of the second electron. Replacing the partition functions by the statistical weights, we have $u_{-1} = 1$ for the negative ion and $u_0 = 2$ for neutral hydrogen; hence the Saha equation gives

$$\frac{n_0}{n_{-1}} P_\mathrm{e} = 4 \frac{(2\pi m_\mathrm{e})^{3/2} (kT)^{5/2}}{h^3} \mathrm{e}^{-\chi/kT} , \tag{17.9}$$

with $\chi = 0.75\,\text{eV}$. If we use $n_0 = (1-x)\varrho X/m_\mathrm{u}$, where x is the degree of ionization of hydrogen as defined in (14.18) and X the weight fraction of hydrogen, we find

$$n_{-1} = \frac{1}{4} \frac{h^3}{(2\pi m_\mathrm{e})^{3/2} (kT)^{5/2} m_\mathrm{u}} P_\mathrm{e} (1-x) X \varrho \mathrm{e}^{\chi/kT} . \tag{17.10}$$

Now, for an absorption coefficient a_ν per H^- ion, it follows that $\kappa_\nu = a_\nu n_{-1}/\varrho$, which implies that the Rosseland mean is described by

$$\kappa_{\mathrm{H}^-} = \frac{1}{4} \frac{h^3}{(2\pi m_\mathrm{e})^{3/2} (kT)^{5/2} m_\mathrm{u}} P_\mathrm{e} (1-x) X \, a(T) \mathrm{e}^{\chi/kT} , \tag{17.11}$$

where $a = a(T)$ is obtained from a_ν by Rosseland integration (5.19). The opacity κ_{H^-} is proportional to n_{-1}, which in turn is proportional to $n_0 n_\mathrm{e}$ (or $n_0 P_\mathrm{e}$), since the H^- ions are formed from *neutral* hydrogen atoms and free electrons.

For a completely neutral, pure hydrogen gas there would be no free electrons and therefore no H^- ions. If now the temperature is increased and the hydrogen becomes slightly ionized, giving $n_\mathrm{e} \sim X$, the free electrons can combine with neutral hydrogen atoms. One therefore would expect an increase of κ as long as $1 - x$ is not too small.

The situation is different in the case of a more realistic mixture of stellar material. Heavier elements have lower ionization potentials (a few eV) and provide electrons even at relatively low temperatures; hence, although there is only a small mass fraction of heavier elements, they determine the electron density at low temperatures where hydrogen is neutral. When the elements heavier than helium are singly ionized (say from 3,000 K to 5,000 K) one has

$$n_e = \varrho\,[xX + (1 - X - Y)/A]/m_u\,, \tag{17.12}$$

where $\varrho(1 - X - Y)/(Am_u)$ is the number density of atoms of higher elements ("metals") of mean mass number A. Even if the metals constitute only a small percentage in weight (and number), they still determine the opacity as long as $1 - X - Y > xXA$ (which becomes very small for low temperatures where x is small). The metal content can therefore be of great influence on κ for the surface layers and thus the outer boundary conditions of stars.

17.6 Conduction

Electrons, like all particles, can transport heat by conduction. Their contribution to the total energy transport can normally be neglected compared to that of photons, since the conductivity is proportional to the mean free path ℓ, and in normal (non-degenerate) stellar material $\ell_{\text{photon}} \gg \ell_{\text{particle}}$.

However, conduction by electrons becomes important in the dense degenerate regions in the very interior of evolved stars, as well as in white dwarfs. The reason is that in the case of degeneracy, all quantum cells in phase space below p_F are filled up, and electrons, when approaching ions and other electrons, have difficulty exchanging their momentum. This is equivalent to saying that "encounters" are rare or that the mean free path is large. In Sect. 5.2 we saw that the contribution to conduction can be formally taken into account in the equation of radiative transport by defining a "conductive opacity" κ_{cd}, as in (5.24). If κ_{rad} is the Rosseland mean of the (radiative) opacity, then conduction reduces the "total" opacity κ, as can be seen from (5.25):

$$\frac{1}{\kappa} = \frac{1}{\kappa_{\text{rad}}} + \frac{1}{\kappa_{\text{cd}}}\,. \tag{17.13}$$

The thermal conductivity of the electron component of a gas is mainly determined by collisions between electrons and ions, but electron–electron collisions can also be important. Analytic formulae can be found in Weiss et al. (2004), while tables of the thermal conductivity due to electrons in stellar material have been computed first by Hubbard and Lampe (1969). They list the conductivities of a pure hydrogen gas, a mixture of pure helium and pure carbon, a solar composition, and a mixture typical for the core of an evolved star. More recent work is published by Itoh and co-workers (Itoh et al. 1983) and Potekhin et al. (1999) for variable chemical mixtures.

The later source, which provides conductive opacities for any ion charge, was used to plot Fig. 17.4, which shows the dependence of the conductive opacity on density for a given temperature. For extremely strong degeneracy, κ_{cd} is proportional to $\varrho^{-2}T^2$.

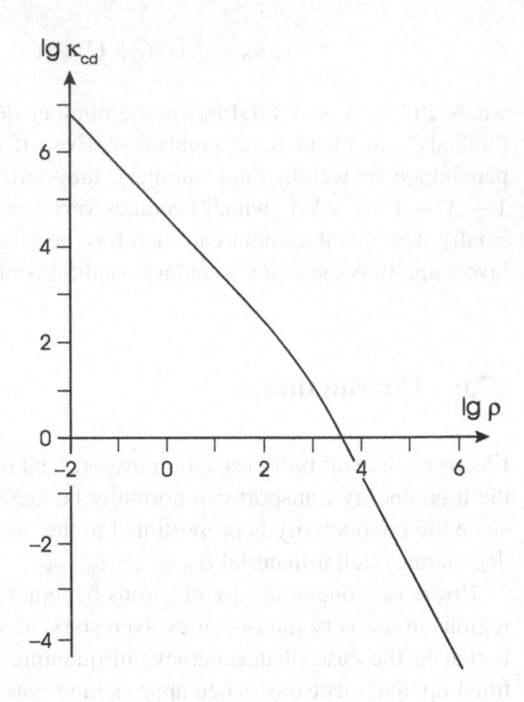

Fig. 17.4 The "conductive opacity" κ_{cd} (in $\mathrm{cm}^2\,\mathrm{g}^{-1}$) of a hydrogen gas at $T = 10^7$ K against the density ϱ (in $\mathrm{g\,cm}^{-3}$) (Data from Potekhin et al. 1999)

17.7 Molecular Opacities

For temperatures below $\approx 10,000$ K the formation of molecules in the envelopes of cool stars becomes increasingly important. Due to their rich system of energy levels, corresponding to the various states of rotational and vibrational excitation, they are important absorbers. They contribute significantly to the opacity below $\approx 5,000$ K and begin to dominate it for $T \lesssim 3,000$ K. The importance of any absorber for the Rosseland mean opacity depends primarily on its absorption properties and not so much on its abundance. This is even more true for molecules and is the reason why Ti, which is three orders of magnitudes less abundant than oxygen, dominates–along with the water molecule–the opacity in the form of TiO as long as there is enough oxygen available for its formation. This is normally the case, unless there is more carbon than oxygen, in which case the oxygen is bound in CO molecules. In that case, other carbon molecules, such as C_2, CN, or C_2H_2, dominate.

Obviously, molecular opacities depend on atomic abundances, on the formation and stability of the various molecules, and finally on their energy level spectrum. This problem is sufficiently complicated that it can again be treated only in separate calculations including atomic and molecular physics, thermodynamics, and chemical processes. The results are again made available in tabular form for the stellar modelling. The largest sets of such tables has been provided by Alexander and Ferguson (Alexander and Ferguson 1994; Ferguson et al. 2005). The calculations consider more than 30 elements, over 50 molecules, and some 800 million atomic and molecular lines. In addition, absorption by dust grains is also

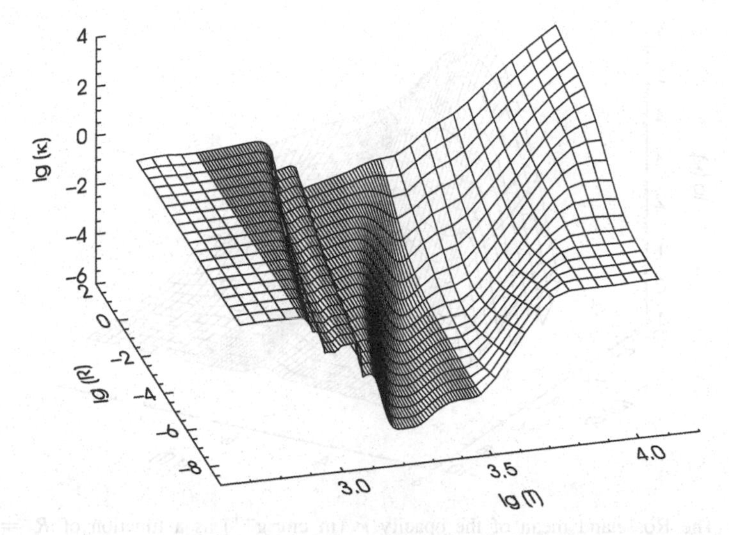

Fig. 17.5 The Rosseland mean of the opacity κ (in $\mathrm{cm^2\,g^{-1}}$) as a function of $R = \varrho/T_6^3$ (in $\mathrm{g\,cm^{-3}}$, since $T_6 = T/10^6\,\mathrm{K}$) and T (in K) for a mixture of $X = 0.70$, $Y = 0.29$, $Z = 0.01$, using data for atomic, molecular, and dust opacity from Ferguson et al. (2005)

included. They dominate below 1500 K, temperatures which are usually found in stellar atmospheres only.

Figure 17.5 shows the total Rosseland opacity for a mixture with 70 % hydrogen and 1 % of metals. The varying density of the grid lines reflects the density of the (R, T) points computed for the table [In this and the following, similar figures, the quantity $R = \varrho/T_6^3$ (with $T_6 = T/10^6\,\mathrm{K}$) has been used as this has become customary in the opacity community. This quantity is roughly constant in large parts of main-sequence stars. This R must not be confused with the stellar radius and is used with this meaning in this chapter only.]. In regions of many different opacity sources the opacities were calculated at many temperatures and densities. At higher temperatures, atomic absorption dominates; the steep rise to the right is mainly caused by the $\mathrm{H^-}$ ion. The "shoulder" around $\lg T = 3.4$ and low R is caused by the first formation of molecules, such as CO, NO, and H_2. The first sharp rise at lower temperatures after we passed the minimum around $\lg T = 3.3$ is due to formation of TiO and H_2O, which is followed by a slight decrease in κ once temperatures are too low to allow many excited states in the molecules. The various maxima at even lower temperatures are caused by different grains appearing and disappearing. For example, the one around $\lg T = 3.1$ is due to Al_2O_3 and $CaTiO_3$. Solid silicates and iron grains form at even lower temperatures. Each of these features is also present at higher densities, but then already occurring at higher temperatures.

Such low temperatures, where molecular or even dust absorption dominates, are usually found in stars only in convective envelopes. However, the outermost parts of these envelopes are highly superadiabatic (Chap. 7), such that $\nabla \lesssim \nabla_{\mathrm{rad}}$, and therefore the opacities determine the temperature stratification even in this case.

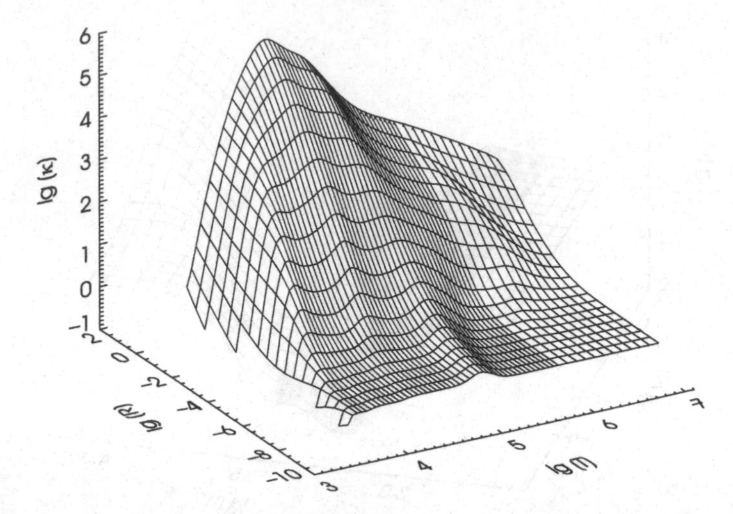

Fig. 17.6 The Rosseland mean of the opacity κ (in $\mathrm{cm^2\,g^{-1}}$) as a function of $R = \varrho/T_6^3$ (in $\mathrm{g\,cm^{-3}}$, since $T_6 = T/10^6\,\mathrm{K}$) and T (in K) for a mixture with a hydrogen and helium content $X = 0.70$, $Y = 0.29$. These are opacities calculated by the OPAL project at Lawrence Livermore National Laboratory (Rogers and Iglesias 1992; Iglesias and Rogers 1996). The dominant absorption mechanisms at different parts of the model are discussed in the text. The continuation towards higher temperatures is shown in Fig. 17.7

17.8 Opacity Tables

In view of the complexity of modern opacity calculations, the basic considerations of Sects. 17.1–17.6 are not sufficient for calculating accurate stellar models. Instead, specialized groups have published extensive tables of opacities for different chemical mixtures over a wide range of temperatures and densities. Each group, however, may specialize on one specific aspect. The *Opacity Project* (Mendoza et al. 2007) and the Livermore *OPAL* group concentrate on atomic absorption important for higher temperatures (Fig. 17.6); the Wichita group (Alexander and Ferguson) on molecular and dust absorption for temperatures below 10^4 K, and finally Itoh, Pothekin, and others on electron conduction. These various sources then have to be combined to opacity tables covering the whole stellar structure. Indeed, the low-and high-temperature opacities, which are shown in Figs. 17.5 and 17.6, agree very well in the overlapping temperature range. The conductive opacities can finally be added by use of (17.13).

In Fig. 17.7 we give a graphical representation of such a combined opacity table for a mixture with a metal fraction of 0.01 and a hydrogen content of 0.70. Figures 17.6 and 17.5 show the corresponding individual parts for the same mixture.

Indeed one sees that, over the whole range of arguments, $\kappa(R, T)$ is a rather complicated function. In order to give a feeling for the parts of the plotted surface that are relevant to stars, we discuss a model of the present Sun, which is plotted in Fig. 17.7 (thick solid line). We are using only the $T - \varrho$ structure; the chemical

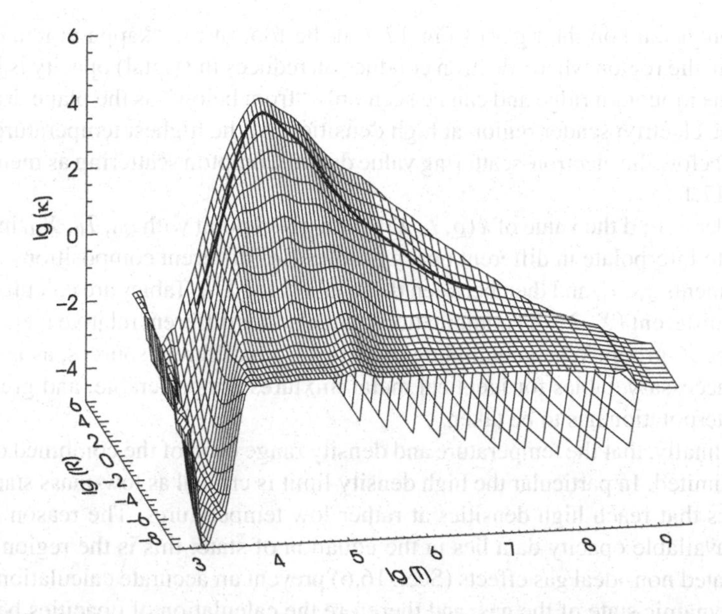

Fig. 17.7 Combination of the opacity κ shown in Figs. 17.6 and 17.5, and of electron conduction opacities as in Fig. 17.4. The latter is a steeply declining function of ϱ (or R) and can be seen "from below" in the back of the figure. Electron scattering provides the flat region in the right foreground. The *thick solid line* represents the $T - \varrho$ structure of a solar model (for details see text)

composition of the Sun is in fact quite different from that of the table. For example, in the solar centre, $X \approx 0.34$, and at the surface, the composition is $X = 0.737$, $Y = 0.245$, and $Z = 0.018$. Nevertheless, the main features are still visible. Note also, that R indeed is rather constant throughout the solar interior, although ϱ varies by eight orders of magnitude.

The model starts with the photospheric values $\lg T = 3.76$, $\lg \varrho = -6.58$ (in cgs), or $\lg R = 0.14$. The corresponding point lies on the left end of the thick solid line in Fig. 17.7 and on the rising slope on the right of Fig. 17.5, where molecular opacities are still contributing. On moving deeper into the Sun the opacity sharply increases owing to the onset of hydrogen ionization, which provides the electrons for H^- formation as described in Sect. 17.5, and the opacity rises by several powers of 10 until it reaches a maximum value. This occurs when an appreciable amount of hydrogen becomes ionized and is not available for H^- formation, because the factor $1 - x$ in (17.11) reduces the opacity. In the regions below, bound–free transitions become the leading opacity source and still further inwards free–free transitions take over. There a simple power law seems to be a good approximation, as indicated in (17.4). Note that in the logarithmic representation the opacity surface for a power law is just a plane. Equation (17.4) therefore corresponds to a tangential plane which osculates the opacity surface. The line for the interior remains in the domain of free–free transitions. The region of dominant electron scattering is the horizontal plateau in the foreground on the right of Fig. 17.7 at the foot of the "kappa mountain". In this figure the region where electron conduction reduces the (total) opacity is hidden behind the mountain ridge and can be seen only "from below" as the plane dropping below the electron scatter region at high densities. At the highest temperatures, κ is reduced below the electron-scattering value due to Compton scattering as mentioned in Sect. 17.1.

In order to find the value of $\kappa(\varrho, T, X_i)$ for a given point with ϱ_0, T_0, X_{i0} in a star, one has to interpolate in different opacity tables (for different compositions X_i) for the arguments ϱ_0, T_0 and then between these tables for X_{i0}. Tables are calculated not only for different (X, Y, Z) combinations, but also for different relative metal ratios within the Z-group. When combining opacities from different sources, as is almost always necessary, tables for identical metal mixtures are preferable, and great care in the interpolation has to be taken.

Note finally, that the temperature and density range even of the combined opacity table is limited. In particular the high density limit is critical as low-mass stars have structures that reach high densities at rather low temperatures. The reason for the lack of available opacity data lies in the equation of state: this is the region where complicated non-ideal gas effects (Sect. 16.6) prevent an accurate calculation of the thermodynamic state of the gas, and therefore the calculation of opacities becomes impossible. In practical stellar evolution calculations, such a situation asks for the creativity of the modeller to somehow supplement the missing data.

Chapter 18
Nuclear Energy Production

We shall limit ourselves here to a very rough summary of the most important features of nuclear reactions in stars. This will suffice completely for the consideration of the main band of stellar structures, while the study of particular aspects of nuclear astrophysics anyway requires the consultation of specialized literature (see Clayton 1968, or Iliadis 2007). For example, we will only deal with energy production of equilibrium nuclear burning, i.e. we will neglect the effects occurring when the timescale of a rapidly changing star becomes comparable to that of an important nuclear reaction. On the other hand, we will also briefly touch on such topics as electron screening or neutrino production, about which a certain minimum of information seems to be indispensible for general discussions.

We begin with a few historical comments. That thermonuclear reactions can provide the energy source for the stars was first shown by R. Atkinson and F. Houtermans in 1929, after G. Gamow discovered the tunnel effect. Later, two important discoveries were published almost simultaneously in 1938: H. Bethe and Ch. Critchfield described the *pp* chain, and C.F. von Weizsäcker and Bethe independently found the CNO cycle. The reactions of helium burning were then described in 1952 by E.E. Salpeter. Finally, a classic paper summarized the state of the art in 1957, “Synthesis of the Elements in Stars” (Burbidge et al. 1957).

18.1 Basic Considerations

Most observed stars (including the Sun) live on so-called thermonuclear fusion. In such nuclear reactions, induced by the thermal motion, several lighter nuclei fuse to form a heavier one. Before this process, the involved nuclei j have a total mass ($\sum M_j$) different from that of the product nucleus (M_y). The difference is called the *mass defect*:

$$\Delta M = \sum_j M_j - M_y \, . \tag{18.1}$$

R. Kippenhahn et al., *Stellar Structure and Evolution*, Astronomy and Astrophysics Library, DOI 10.1007/978-3-642-30304-3_18,

It is converted into energy according to Einstein's formula

$$E = \Delta M c^2 \tag{18.2}$$

and is available (at least partly) for the star's energy balance. An example is the series of reactions called "hydrogen burning", where four hydrogen nuclei ^{1}H with a total mass $4 \times 1.0079\, m_u$ (atomic mass units, physical scale) are transformed into one ^{4}He nucleus of $4.0026 m_u$. Atomic masses are given for the neutral atoms, i.e. for the nucleus plus all electrons. However, since the electron mass is only $1/1823\, m_u$, we will assume that masses of nuclei are the same as the atomic masses. Obviously $2.9 \times 10^{-2} m_u$ per produced ^{4}He nucleus have "disappeared" during the fusion of the four protons, which is roughly 0.7 % of the original masses and which corresponds to an energy of about 27.0 MeV according to (18.2). As usual in nuclear physics, as the unit of energy, we take the electron volt eV (1 eV $= 1.6018 \times 10^{-12}$ erg) with the following equivalences:

$$1\,\text{keV} \mathrel{\hat{=}} 1.1606 \times 10^7\,\text{K}\,,$$
$$931.49\,\text{MeV} \mathrel{\hat{=}} 1\, m_u\,. \tag{18.3}$$

The Sun's luminosity corresponds to a mass loss rate of $L_\odot/c^2 = 4.26 \times 10^{12}\,\text{g s}^{-1}$, which appears to be a lot, especially if it is read as "more than four million metric tons per second". If a total of 1 $M_\odot$ of hydrogen were converted into ^{4}He, then the disappearing 0.7 % of this mass would be 1.4×10^{31} g, which could balance the Sun's present mass loss by radiation for about $3 \times 10^{18}\,\text{s} \approx 10^{11}$ years.

The deficiency of mass is just another aspect of the fact that the involved nuclei have different binding energies E_B. This is the energy required to separate the nucleons (protons and neutrons in the nucleus) against their mutual attraction by the strong, but short-range nuclear forces. Or else, E_B is the energy gained if they are brought together from infinity (which starts here at any distance large compared with, say, 10^{-12} cm, the scale of a nuclear size).

Consider a nucleus of mass M_{nuc} and atomic mass number A (the integer "atomic weight"): it may contain Z protons of mass m_p and $(A - Z)$ neutrons of mass m_n. Its binding energy is then related to these masses by (18.2):

$$E_B = [(A - Z)m_n + Z m_p - M_{\text{nuc}}]c^2\,. \tag{18.4}$$

When comparing different nuclei, it is more instructive to consider the *average binding energy per nucleon*,

$$f = \frac{E_B}{A}\,, \tag{18.5}$$

which is also called the *binding fraction*. With the exception of hydrogen, typical values are around 8 MeV, with relatively small differences for nuclei of very different A. This shows that the short-range nuclear forces due to a nucleon mainly affect the nucleons in its immediate neighbourhood only, such that with increasing

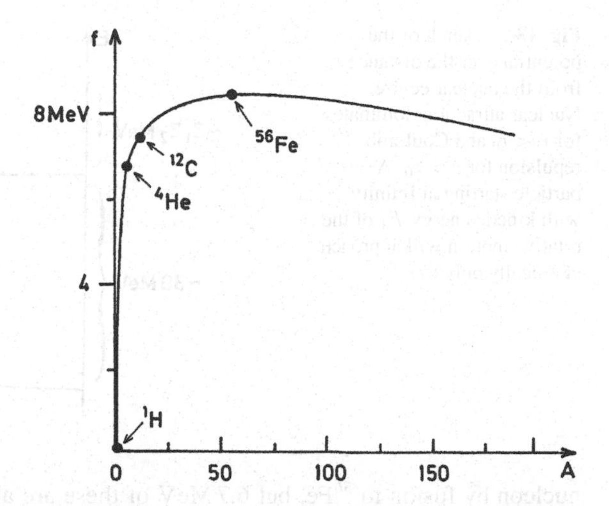

Fig. 18.1 A smoothed run of the fractional binding energy per nucleon, $f = E_B/A$, for stable nuclei, over the atomic mass number A. The curve is smoothed over the wiggles which are due to the nuclear shell structure and pair effects

A, a saturation occurs rather than an increase of f proportional to A. An idealized plot of f against A is shown in Fig. 18.1 (The real curve zigzags around this smoothed curve as a consequence of the shell structure of the nucleus and pair effects.).

With increasing A, $f(A)$ rises steeply from hydrogen, then flattens out and reaches a maximum of 8.5 MeV at $A = 56$ (^{56}Fe), after which it drops slowly with increasing A. The increase for $A < 56$ is a surface effect: particles at the surface of the nucleus experience less attraction by nuclear forces than those in the interior, which are completely surrounded by other particles. And in a densely packed nucleus, the surface area increases with radius slower than the volume (i.e. the number A) such that the fraction of surface particles drops. With increasing A, the number Z of protons also increases (The addition of neutrons only would require higher energy states, because the Pauli principle excludes more than two identical neutrons, and the nuclei would be unstable.). The positively charged protons experience a repulsive force which is far-reaching and therefore does not show the saturation of the nuclear forces. This increasing repulsion by the Coulomb forces brings the curve in Fig. 18.1 down again for $A > 56$.

Around the maximum, at ^{56}Fe, we have the most tightly bound nuclei. In other words, the nucleus of ^{56}Fe has the smallest mass per nucleon, so that any nuclear reaction bringing the nucleus closer to this maximum will be exothermic, i.e. will release energy. There are two ways of doing this:

1. By fission of heavy nuclei, which happens, for example, in radioactivity.
2. By fusion of light nuclei, which is the prime energy source of stars (and possibly ours too in the future).

Clearly, both reach an end when one tries to extend them over the maximum of f, which is therefore a natural finishing point for the stellar nuclear engine. So if a star initially consisted of pure hydrogen, it could gain a maximum of about 8.5 MeV per

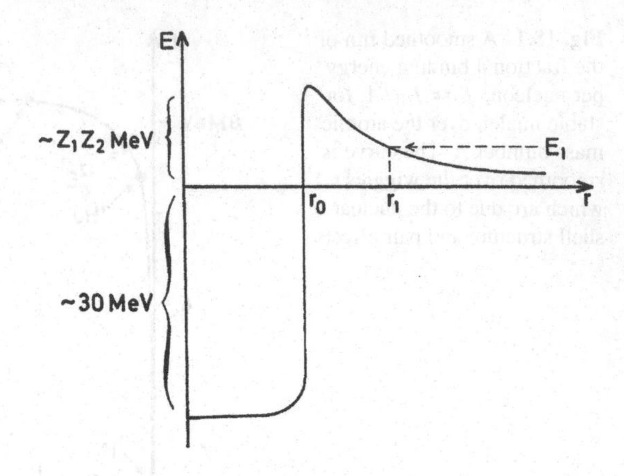

Fig. 18.2 Sketch of the potential over the distance r from the nuclear centre. Nuclear attraction dominates for $r < r_0$ and Coulomb repulsion for $r > r_0$. A particle starting at infinity with kinetic energy E_1 of the relative motion will approach classically only to r_1

nucleon by fusion to ^{56}Fe, but 6.7 MeV of these are already used up when ^{4}He is built up in the first step.

In order to obtain a fusion of charged particles, they have to be brought so close to each other that the strong, but very short-ranged, nuclear forces dominate over the weaker, but far-reaching, Coulomb forces. The counteraction of these two forces leads to a sharp potential jump at the interaction radius (Fig. 18.2):

$$r_0 \approx A^{1/3} 1.44 \times 10^{-13} \mathrm{cm} \tag{18.6}$$

(the "nuclear radius" of the order of femtometer, 1 fm = 10^{-13}cm). For distances less than r_0, the nuclear attraction dominates and provides a potential drop of roughly 30 MeV, while "outside" r_0, the repulsive Coulomb forces for particles with charges Z_1 and Z_2 yield

$$E_{\mathrm{Coul}} = \frac{Z_1 Z_2 e^2}{r} \,. \tag{18.7}$$

The height of the *Coulomb barrier* $E_{\mathrm{Coul}}(r_0)$ is typically of the order

$$E_{\mathrm{Coul}}(r_0) \approx Z_1 Z_2 \,\mathrm{MeV} \,. \tag{18.8}$$

If, in the stationary reference frame of the nucleus, a particle at "infinity" has kinetic energy E_1, it can come classically only to a distance r_1 given by $E_1 = E_{\mathrm{Coul}}(r_1)$ from (18.7), as indicated in Fig. 18.2. Now, the kinetic energy available to particles in stellar interiors is that of their thermal motion, and hence the reactions triggered by this motion are called *thermonuclear*. Since in normal stars we observe a slow energy release rather than a nuclear explosion, we must certainly expect the *average* kinetic energy of the thermal motion, E_{th}, to be considerably smaller than $E_{\mathrm{Coul}}(r_0)$. For the value $T \approx 10^7$ K estimated for the solar centre in Sect. 2.3, according to (18.3), kT is only 10^3 eV, i.e. E_{th} is smaller than the Coulomb barrier (18.8) by a factor of roughly 10^3. This is in fact so low that, with classical effects only, we can scarcely expect any reaction at all. In the high-energy tail of the Maxwell–Boltzmann distribution, the exponential factor drops here to $\exp(-1000) \approx 10^{-434}$, which leaves no chance for the "mere" 10^{57} nucleons in the whole Sun (and even for the $\approx 10^{80}$ nucleons in the whole visible universe)!

The only possibility for thermonuclear reactions in stars comes from a quantum-mechanical effect found by G. Gamow: there is a small but finite probability of penetrating ("tunnelling") through the Coulomb barrier, even for particles with $E < E_{\mathrm{Coul}}(r_0)$. This tunnelling probability varies as

$$P_0 = p_0 E^{-1/2} \mathrm{e}^{-2\pi\eta} \; ; \quad \eta = \left(\frac{m}{2}\right)^{1/2} \frac{Z_1 Z_2 e^2}{\hbar E^{1/2}} \, . \tag{18.9}$$

Here $\hbar$ is $h/2\pi$ and m the reduced mass. The factor p_0 depends only on the properties of the two colliding nuclei. The exponent $2\pi\eta$ is here obtained as the only E-dependent term in an approximate evaluation of the integral over $\hbar^{-1}[2m\,(E_{\mathrm{Coul}} - E)]^{1/2}$, which is extended from r_0 to the distance r_c of closest approach (where $E = E_{\mathrm{Coul}}$). For $Z_1 Z_2 = 1$ and $T = 10^7$ K, P_0 is of the order of 10^{-20} for particles with average kinetic energy E and steeply increases with E and decreases with $Z_1 Z_2$. Therefore, for temperatures as "low" as 10^7 K, only the lightest nuclei (with smallest $Z_1 Z_2$) have a chance to react. For reactions of heavier particles, with larger $Z_1 Z_2$, the energy, i.e. the temperature, has to be correspondingly larger to provide a comparable penetration probability. This will result in well-separated phases of different nuclear "burning" during the star's evolution.

18.2 Nuclear Cross Sections

Consider a reaction of the nucleus X with the particle a by which the nucleus Y and the particle b are formed:

$$a + X \rightarrow Y + b \, , \tag{18.10}$$

represented by the notation *X(a, b)Y*. The reaction probability depends on nuclear details, some of which can be illustrated with the following simplified description. After penetration of the Coulomb barrier, an excited *compound nucleus* C^* may form containing both original particles (The level of excitation is dependent on the kinetic energy and binding energy brought along by the newly added particle.). C^* may decay after a short time, which will still be long enough for the added nucleons to "forget"–owing to interactions within the compound nucleus–their history, a process for which only $\sim 10^{-21}$ s is necessary. The decay then depends only on the energy. C^* can generally decay via one of several "channels" of different probability: $C^* \rightarrow X + a, \rightarrow Y_1 + b_1, \rightarrow Y_2 + b_2, \ldots, \rightarrow C + \gamma$. The first of these would be the reproduction of the original particles, while the last indicates a decay with γ-ray emission; the others are particle decays where the $b_1, b_2, \ldots$ may be, for example, neutrons, protons, and α particles. Compared to these, a decay

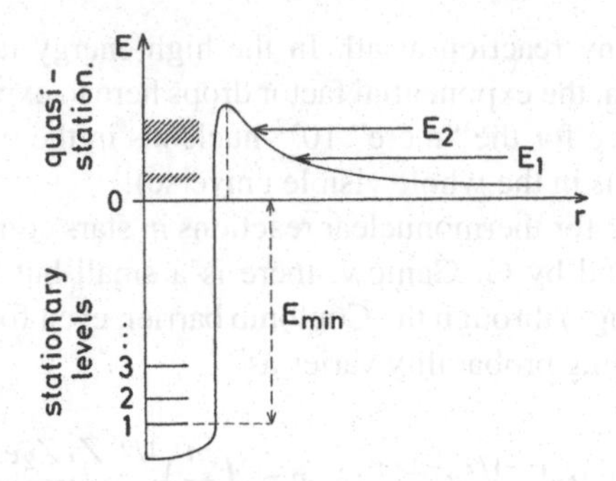

Fig. 18.3 Schematic sketch of energy levels in a compound nucleus C^* formed by particles X and a. The zero of E is here taken as corresponding to zero velocity of X and a at infinity. For initial particle energy E_1, the reaction would be non-resonant, while for E_2, the particles X and a find a resonance in the compound nucleus. $E_{\min}$ is the minimum excitation energy above the ground level for particle emission

with electron emission has negligible probability (β decay times being of order 1 s or larger). Outgoing particles will obtain a certain amount of kinetic energy, which (just as the energy of emitted γ rays) will be shared with the surroundings, though an exception here are the neutrinos, which leave the star without interaction (Sect. 18.7). The possibility that a given energy level of C^* can decay via a certain channel requires fulfilment of the conservation laws (energy, momentum, angular momentum, nuclear symmetries).

It is very important to know the energy levels of the compound nucleus C^*, which can be of different types. Let $E_{\min}$ be the minimum energy required to remove a nucleon from the ground state to infinity with zero velocity (to the level $E = 0$ in Fig. 18.3). This corresponds to the atom ionization energy discussed in Chap. 14. Levels below $E_{\min}$ can obviously only decay by electromagnetic transitions with the emission of γ rays, which are relatively improbable, and hence their lifetime τ is large; these are "stationary" levels of small energy width Γ, since

$$\Gamma = \frac{\hbar}{\tau}, \tag{18.11}$$

as follows from the Heisenberg uncertainty relation. These levels correspond to the discrete, bound atomic states.

The compound nucleus will not, however, immediately expel a particle if its energy is somewhat above $E_{\min}$, since the sharp potential rise holds it back, at least for some time. Eventually it can leave the potential well by the tunnelling effect (which was, in fact, predicted by Gamow for explaining such outward escapes of particles from radioactive nuclei). So there can be "quasi-stationary" levels above $E_{\min}$ that have an appreciably shorter lifetime τ (and are correspondingly wider) than those below $E_{\min}$, since they can also decay via the much more probable particle emission. This probability will clearly increase strongly with increasing energy, which results in corresponding decreases of τ and increases

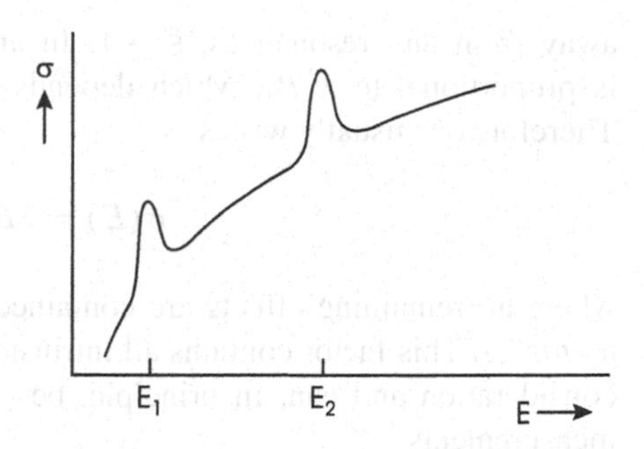

Fig. 18.4 Sketch of the reaction cross section σ over the energy E of the relative motion of the reacting particles, with resonances at E_1 and E_2

of Γ, see (18.11). Above a certain energy $E_{\max}$ the width Γ will become larger than the distance between neighbouring levels, and their complete overlap yields a continuum of energy states, instead of separated, discrete levels.

The possible existence of quasi-stationary levels above $E_{\min}$ requires particular attention. Consider an attempt to produce the compound nucleus C^* by particles $X+a$ with gradually increasing energy E of their relative motion at large distances. The reaction probability will simply increase with the penetration probability (18.9), if E is in a region either without quasi-stationary levels or between two of them. If, however, E coincides with such a level, the colliding particles find a "resonance" and can form the compound nucleus much more easily. At such resonance energies E_{res}, the probability for a reaction (and hence the cross section σ) is abnormally enhanced, as sketched in Fig. 18.4, with resonant peaks rising to several powers of ten above "normal". The energy dependence of the cross section therefore has a factor which has the typical resonance form:

$$\xi(E) = \text{constant}\ \frac{1}{(E-E_{\text{res}})^2 + (\Gamma/2)^2}\ . \tag{18.12}$$

At a resonance, the cross section σ for the reaction of particles X and a can nearly reach its maximum value (geometrical cross section), given by quantum mechanics as $\pi\lambda\!\!\!^{-2}$, where $\lambda\!\!\!^{-}$ is the de Broglie wavelength associated with a particle of relative momentum p:

$$\lambda\!\!\!^{-} = \frac{\hbar}{p} = \frac{\hbar}{(2mE)^{1/2}}\ . \tag{18.13}$$

Here the non-relativistic relation between p and E is used, and m is the reduced mass of the two particles. The meaning of $\pi\lambda\!\!\!^{-2}$ is clear because according to quantum mechanics, the particles moving with momentum p "see" each other not as a precise point but smeared out over a length $\lambda\!\!\!^{-}$. The dependence of σ on E can now be seen from the relation

$$\sigma(E) \sim \pi\lambda\!\!\!^{-2} P_0(E)\xi(E)\ , \tag{18.14}$$

where $\lambda\!\!\!^{-}$ is given by (18.13). For E values well below the Coulomb barrier, P_0 can be taken from (18.9) with a pre-factor $p_0 = E_{\text{Coul}}^{1/2}(r_0)\exp[32mZ_1Z_2e^2r_0/\hbar^2)^{1/2}]$. In the range of a single resonance, $\xi(E)$ is given by (18.12), while far away from any resonances, $\xi \to 1$. In any case, with or without resonances, σ is proportional to $\lambda\!\!\!^{-2} P_0$, which depends on E as shown by (18.9) and (18.13). Therefore one usually writes

$$\sigma(E) = SE^{-1}\mathrm{e}^{-2\pi\eta} , \tag{18.15}$$

where all remaining effects are contained within the here-defined *"astrophysical factor" S*. This factor contains all intrinsic nuclear properties of the reaction under consideration and can, in principle, be calculated, although one rather relies on measurements.

The difficulty with laboratory measurements of $S(E)$–if they are possible at all–is that, because of the small cross sections, they are usually feasible only at rather high energies, say above 0.1 MeV, but this is still roughly a factor 10 larger than those energies which are relevant for astrophysical applications. Therefore one has to extrapolate the measured $S(E)$ downwards over a rather long range of E. This can be done quite reliably for non-resonant reactions, in which case S is nearly constant or a very slowly varying function of E [an advantage of extrapolating $S(E)$ rather than $\sigma(E)$]. The real problems arise from (suspected or unsuspected) resonances in the range over which the extrapolation is to be extended. Then the results can be quite uncertain. Only in underground laboratories, where the experiments are shielded from cosmic rays by hundreds of meters of solid rock, it is sometimes possible to measure the nuclear cross sections of at least a few nuclear reactions at energies as low as 10–30 keV, i.e. at energies relevant for nuclear processes in stellar interiors. The first such measurement was done by Junker et al. (1998) in the *Gran Sasso Laboratory* and concerned the ^{3}He(^{3}He, $2p$)^{4}He reaction (18.62) of the hydrogen burning chains (Sect. 18.5.1). Such experiments sometimes lead to the discovery of resonances, but more importantly reduce the uncertainties of the cross sections considerably, and confirm the near constancy of $S(E)$ at the relevant energies.

18.3 Thermonuclear Reaction Rates

Let us denote the types of reacting particles, X and a, by indices j and k respectively. Suppose there is one particle of type j moving with a velocity v relative to all particles of type k. Its cross section σ for reactions with the k sweeps over a volume σv per second. The number of reactions per second will then be $n_k \sigma v$ if there are n_k particles of type k per unit volume. For n_j particles per unit volume the total number of reactions per units of volume and time is

$$\tilde{r}_{jk} = n_j n_k \sigma v . \tag{18.16}$$

This product may also be interpreted by saying that $n_j n_k$ is the number of pairs of possible reaction partners, and σv gives the reaction probability per pair and second. This indicates what we have to do in the case of reactions between identical particles ($j = k$). Then the number of pairs that are possible reaction partners is $n_j(n_j - 1)/2 \approx n_j^2/2$ for large particle numbers. This has to replace the product $n_j n_k$ in (18.16) so that we can generally write

$$\tilde{r}_{jk} = \frac{1}{1+\delta_{jk}} n_j n_k \sigma v \ , \quad \delta_{jk} = \begin{cases} 0, \ j \neq k \\ 1, \ j = k \end{cases} . \tag{18.17}$$

Now we have to allow for the fact that particles j and k do not move relatively to each other with uniform velocities, which is important since σ depends strongly on v. Excluding extreme densities (as, e.g. in neutron stars) we can assume that both types have a Maxwell–Boltzmann distribution of their velocities. It is then well known that also their *relative velocity* v is Maxwellian. If the corresponding energy is

$$E = \frac{1}{2} m v^2 \tag{18.18}$$

with the reduced mass $m = m_j m_k/(m_j + m_k)$, the fraction of all pairs contained in the interval $[E, E + dE]$ is given by

$$f(E)dE = \frac{2}{\sqrt{\pi}} \frac{E^{1/2}}{(kT)^{3/2}} \mathrm{e}^{-E/kT} dE \ . \tag{18.19}$$

This fraction of all pairs has a uniform velocity and contributes the amount $dr_{jk} = \tilde{r}_{jk} f(E)dE$ to the total rate. The total reaction rate per units of volume and time is then given by the integral $\int dr_{jk}$ over all energies, which formally can be written as

$$r_{jk} = \frac{1}{1+\delta_{jk}} n_j n_k \langle \sigma v \rangle \ , \tag{18.20}$$

where the averaged probability is

$$\langle \sigma v \rangle = \int_0^\infty \sigma(E) v f(E) dE \ . \tag{18.21}$$

Let us replace the particle numbers per unit volume n_i by the mass fraction X_i with

$$X_i \varrho = n_i m_i \ , \tag{18.22}$$

cf. (8.2). If the energy Q is released per reaction, then (18.20) gives the energy generation rate per units of mass (instead of unit volume; obtained by dividing by ϱ) and time:

$$\varepsilon_{jk} = \frac{1}{1+\delta_{jk}} \frac{Q}{m_j m_k} \varrho X_j X_k \langle \sigma v \rangle \ . \tag{18.23}$$

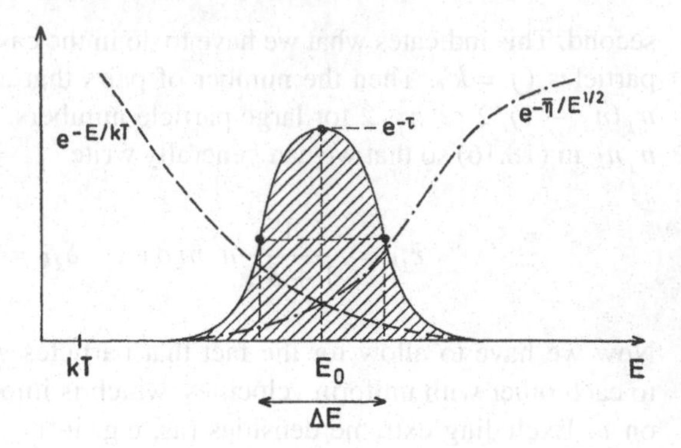

Fig. 18.5 The Gamow peak (*solid curve*) as the product of Maxwell distribution (*dashed*) and penetration factor (*dot-dashed*). The hatched area under the Gamow peak determines the reaction rate. All three curves are on different scales

Using (18.9), (18.15), (18.18) and (18.19) in (18.21), the average cross section $\langle\sigma v\rangle$ can be written as

$$\langle\sigma v\rangle = \frac{2^{3/2}}{(m\pi)^{1/2}}\frac{1}{(kT)^{3/2}}\int_0^\infty S(E)\mathrm{e}^{-E/kT-\bar{\eta}/E^{1/2}}dE\,, \tag{18.24}$$

where

$$\bar{\eta} = 2\pi\eta E^{1/2} = \pi(2m)^{1/2}\frac{Z_j Z_k e^2}{\hbar}\,. \tag{18.25}$$

A further evaluation of $\langle\sigma v\rangle$ requires a specification of $S(E)$. We shall limit ourselves to the simplest but for astrophysical applications very important case of *non-resonant reactions*. Then we can set $S(E) \approx S_0 =$ constant, and take it out of the integral (18.24), since only a small interval of E will turn out to contribute appreciably. The remaining integral may be written as

$$J = \int_0^\infty \mathrm{e}^{f(E)}dE\,, \quad \text{with} \quad f(E) = -\frac{E}{kT} - \frac{\bar{\eta}}{E^{1/2}}\,. \tag{18.26}$$

The integrand is the product of two exponential functions, one of which drops steeply with increasing E, while the other rises. The integrand will therefore have appreciable values only around a well-defined maximum (see Fig. 18.5), the so-called *Gamow peak*. This maximum occurs at E_0, where the exponent has a minimum. From the condition $f' = 0$, where f' is the derivative with respect to E, one finds

$$E_0 = \left(\frac{1}{2}\bar{\eta}kT\right)^{2/3} = \left[\left(\frac{m}{2}\right)^{1/2}\pi\frac{Z_i Z_k e^2 kT}{\hbar}\right]^{2/3}. \tag{18.27}$$

It is usual to introduce now a quantity τ defined by

$$\tau = 3\frac{E_0}{kT} = 3\left[\pi\left(\frac{m}{2kT}\right)^{1/2}\frac{Z_j Z_k e^2}{\hbar}\right]^{2/3} \tag{18.28}$$

and to represent $f(E)$ near the maximum by the series expansion

$$\begin{aligned} f(E) &= f_0 + f_0' \cdot (E - E_0) + \frac{1}{2} f_0'' \cdot (E - E_0)^2 + \cdots \\ &= -\tau - \frac{1}{4}\tau\left(\frac{E}{E_0} - 1\right)^2 + \cdots , \end{aligned} \tag{18.29}$$

from which we retain only these two terms (the linear term vanishes since $f_0' = 0$ at the maximum). Their substitution in (18.26) means to approximate the Gamow peak of the integrand by a Gaussian, as will become particularly clear when we transform J to the new variable of integration $\xi = (E/E_0 - 1)\sqrt{\tau}/2$:

$$J = \int_0^\infty \exp\left[-\tau - \frac{\tau}{4}\left(\frac{E}{E_0} - 1\right)^2\right] dE = \frac{2}{3} kT \tau^{1/2} e^{-\tau} \int_{-\sqrt{\tau/2}}^{\infty} e^{-\xi^2} d\xi . \tag{18.30}$$

The main contribution to J comes from a range close to $E = E_0$, i.e. $\xi = 0$, so that no large errors are introduced when extending the range of integration to $-\infty$, the integral over the Gaussian becoming $\sqrt{\pi}$.

We then have

$$J \approx kT \frac{2}{3} \pi^{1/2} \tau^{1/2} e^{-\tau} , \tag{18.31}$$

and for non-resonant reactions (18.24) becomes

$$\langle \sigma v \rangle = \frac{4}{3}\left(\frac{2}{m}\right)^{1/2} \frac{1}{(kT)^{1/2}} S_0 \tau^{1/2} e^{-\tau} . \tag{18.32}$$

From (18.28) one has $(kT)^{-1/2} \sim \tau^{3/2}$; hence the kT can be substituted in (18.32), which then gives $\langle \sigma v \rangle \sim \tau^2 e^{-\tau}$.

The *properties of the Gamow peak* are so important that we should inspect some of them a bit further. In order to have convenient numerical values, we count the temperature in units of 10^7 K (which is typical for many stellar centres) and denote this dimensionless temperature by $T_7 = T/10^7$ K or generally

$$T_n := \frac{T}{10^n \mathrm{K}} . \tag{18.33}$$

We then have the following relations (some of which will be derived below):

$$\begin{aligned}
W &= Z_j^2 Z_k^2 A = Z_j^2 Z_k^2 \frac{A_j A_k}{A_j + A_k} \,, \\
\tau &= 19.721 W^{1/3} T_7^{-1/3} \,, \\
E_0 &= 5.665\,\mathrm{keV} \cdot W^{1/3} T_7^{2/3} \,, \\
\frac{E_0}{kT} &= \frac{\tau}{3} = 6.574 W^{1/3} T_7^{-1/3} \,, \\
\Delta E &= 4.249\,\mathrm{keV} \cdot W^{1/6} T_7^{5/6} \,, \\
\frac{\Delta E}{E_0} &= 4(\ln 2)^{1/2} \tau^{-1/2} = 0.750 W^{-1/6} T_7^{1/6} \,, \\
\nu &= \partial \ln\langle\sigma v\rangle / \partial \ln T = (\tau - 2)/3 = 6.574 W^{1/3} T_7^{-1/3} - 2/3 \,. \qquad (18.34)
\end{aligned}$$

The value of W is determined by the reaction partners and is at least of order unity. Large W discriminates against the reactions of heavy nuclei so much that only the lighter nuclei can react with appreciable rate. The Gamow peak occurs as a compromise in the counteraction between Maxwell distribution and penetration probability with a maximum at $E = E_0$, which is roughly 5–100 times the average thermal energy kT. This "effective stellar energy range" is, on the other hand, far below the $\gtrsim 100\,\mathrm{keV}$ available to most laboratory experiments. With increasing T, E_0 increases moderately, while the maximum height of the peak $H_0 = \mathrm{e}^{-\tau}$ increases very steeply owing to the decreasing τ.

The width of the effective energy range is described by ΔE, which is the full width of the Gamow peak at half maximum (see Fig. 18.5), i.e. between the points with height $0.5\,\mathrm{e}^{-\tau}$. Equating this to the integrand in the first form of (18.30), we obtain

$$\frac{\Delta E}{E_0} = 4 \frac{(\ln 2)^{1/2}}{\tau^{1/2}} \,. \qquad (18.35)$$

According to (18.34), this is always below unity, and therefore one has a well-defined energy range in which the reactions occur effectively. With ΔE increasing with T only slightly more than E_0, the relative form of the peak remains nearly constant.

The most striking feature of thermonuclear reactions is their strong sensitivity to the temperature. In order to demonstrate this, one represents the T dependence of $\langle\sigma v\rangle$ (and thus of r_{jk} and ε_{jk}) around some value $T = T_0$ by a power law such as

$$\langle\sigma v\rangle = \langle\sigma v\rangle_0 \left(\frac{T}{T_0}\right)^{\nu} \,, \quad \nu = \frac{\partial \ln\langle\sigma v\rangle}{\partial \ln T} \,. \qquad (18.36)$$

From (18.28) we have $\tau \sim T^{-1/3}$, and then from (18.32) $\langle\sigma v\rangle \sim T^{-2/3}\mathrm{e}^{-\tau}$. Therefore

$$\ln\langle\sigma v\rangle = \text{constant} - \frac{2}{3}\ln T - \tau \,, \tag{18.37}$$

and

$$\frac{\partial \ln\langle\sigma v\rangle}{\partial \ln T} = -\frac{2}{3} - \frac{\partial \tau}{\partial \ln T} = -\frac{2}{3} - \tau \frac{\partial \ln \tau}{\partial \ln T} \,. \tag{18.38}$$

Since $\tau \sim T^{-1/3}$, we have $\partial \ln \tau / \partial \ln T = -1/3$, so that finally

$$\nu \equiv \frac{\partial \ln\langle\sigma v\rangle}{\partial \ln T} = \frac{\tau}{3} - \frac{2}{3} \,, \tag{18.39}$$

where for most reactions $\tau/3$ is much larger than 2/3 and $\nu \approx \tau/3$. Then ν decreases with T as $\nu \sim T^{-1/3}$. From (18.34) we see that even for reactions between the lightest nuclei, $\nu \approx 5$, and it can easily attain values around (and even above) $\nu \approx 20$. With such values for the exponent (!) of T, the thermonuclear reaction rate is about the most strongly varying function treated in physics, and this temperature sensitivity has a clear influence on stellar models. Also, since small fluctuations of T (which will certainly be present) must result in drastic changes in the energy production, we have to assume that there exists an effective stabilizing mechanism (a thermostat) in stars (Sect. 25.3.5).

We may easily see how the large ν values are related to the change of the Gamow peak with T: the value $\langle\sigma v\rangle$ is proportional to the integral J in (18.30), and this is given by the area under the Gamow peak, which is roughly $J \approx \Delta E \cdot H_0 (H_0 = \mathrm{e}^{-\tau}$ is the height of the peak). According to (18.34), $\Delta E \sim T^{5/6}$, while H_0 increases strongly with T. In fact it is this height H_0 which provides the exponential $\mathrm{e}^{-\tau}$ in the expressions for $\langle\sigma v\rangle$ and is therefore responsible for the large values of ν.

We should briefly mention a few corrections to the derived formulae for the reaction rates. The first concerns inaccuracies made by evaluating the integral in (18.24) with constant S and with an integrand approximated by a Gaussian. This is usually corrected for by multiplying $\langle\sigma v\rangle$ with a factor

$$g_{jk} = 1 + \frac{5}{12\tau} + \frac{S'}{S}E_0\left(1 + \frac{105}{36\tau}\right) + \frac{1}{2}\frac{S''}{S}E_0^2\left(1 + \frac{267}{36\tau}\right) \,, \tag{18.40}$$

where S and its derivatives with respect to E have to be taken at $E = 0$ (Eq. (17.206b) in Weiss et al. 2004, p. 601).

Another correction factor, f_{jk}, allows for a partial shielding of the Coulomb potential of the nuclei, owing to the negative field of neighbouring electrons. This plays a role only at very high densities; it will be treated separately in Sect. 18.4.

Concerning resonant reactions we shall only remark that the situation depends very much on the location of the resonance. For example, the integral in (18.24) can be dominated by a strong peak at the resonance energy. However, once $S(E)$ is given, (18.24) can in principle always be evaluated.

18.4 Electron Shielding

We have seen that the repulsive Coulomb forces of the nucleus play a decisive role in controlling the rate of thermonuclear reactions. Therefore any modification of its potential by influences from the outside can have an appreciable effect on these rates. An obvious effect to be considered comes from the surrounding free electrons. It is clear that beyond a certain distance an approaching particle will "feel" a neutral conglomerate of the target nucleus plus a surrounding electron cloud rather than the isolated charge of the target nucleus.

The first step is to consider the polarization that the nucleus of charge $+Ze$ produces in its surrounding. The electrons of charge $-e$ are attracted and have a slightly larger density n_e in the neighbourhood of the nucleus; the other ions are repelled and have a slightly decreased density n_i in comparison with their average values $\bar{n}_e$ and $\bar{n}_i$ (without electric fields present). For non-degenerate gases the density of particles with charge q is modified in the presence of an electrostatic potential ϕ according to

$$n = \bar{n}\mathrm{e}^{-q\phi/kT} . \tag{18.41}$$

In most normal cases one will find $|q\phi| \ll kT$ and can then approximate the exponential by $1 - q\phi/kT$. For ions and electrons, (18.41) now yields

$$n_i = \bar{n}_i\left(1 - \frac{Z_i e\phi}{kT}\right), \quad n_e = \bar{n}_e\left(1 + \frac{e\phi}{kT}\right), \tag{18.42}$$

which shows directly the decrease (ions) and increase (electrons) of the two densities.

Considering the n_i for all types of ions present in the gas mixture, one can immediately write down the total charge density σ. For $\phi = 0$ one must have a neutral gas, with $\bar{\sigma} = 0$, i.e.

$$\bar{\sigma} = \sum_i (Z_i e)\bar{n}_i - e\bar{n}_e = 0 , \tag{18.43}$$

whereas for non-vanishing ϕ we have

$$\begin{aligned} \sigma &= \sum_i (Z_i e)n_i - en_e \\ &= \sum_i -\frac{(Z_i e)^2\phi}{kT}\bar{n}_i - \frac{e^2\phi}{kT}\bar{n}_e . \end{aligned} \tag{18.44}$$

Here we have already inserted (18.42) and made use of (18.43) to eliminate the ϕ-independent terms. The second expression (18.44) suggests that we combine the two terms and write

$$\sigma = -\chi \frac{e^2\phi}{kT} n , \tag{18.45}$$

where we have introduced the total particle density $n = n_e + \sum_i n_i$ and the average value χ:

$$\chi := \frac{1}{n}\left(\sum_i Z_i^2 \bar{n}_i + \bar{n}_e\right) . \tag{18.46}$$

If one wishes to use the mass fraction $X_i = A_i \bar{n}_i / n\mu$ (μ = mean molecular weight per free particle, see Sect. 4.2, (4.27)) instead of the particle numbers, the expression follows simply as

$$\chi = \mu\zeta = \mu \sum_i \frac{Z_i(Z_i+1)}{A_i} X_i . \tag{18.47}$$

The charge density σ and the electrostatic potential ϕ are also connected by the Poisson equation

$$\nabla^2 \phi = -4\pi\sigma . \tag{18.48}$$

If we assume spherical symmetry for the charge distribution surrounding the nucleus under consideration, the Laplace operator ∇^2 then reduces to its well-known radial part. Introducing σ from (18.45) on the right-hand side of (18.48), the Poisson equation becomes

$$\frac{r_D^2}{r} \frac{d^2(r\phi)}{dr^2} = \phi , \tag{18.49}$$

where we have scaled the distance r by the so-called Debye–Hückel length

$$r_D = \left(\frac{kT}{4\pi\chi e^2 n}\right)^{1/2} . \tag{18.50}$$

One readily verifies that (18.49) is solved by

$$\phi = \frac{Ze}{r} e^{-r/r_D} , \tag{18.51}$$

and this shows that ϕ tends to the normal (unshielded) potential Ze/r of a point charge Ze for small distances, $r \to 0$, while we have an essential reduction of this "normal" potential at distances $r \gtrsim r_D$. In a certain sense we can call r_D the "radius" of the electron cloud that envelopes the nucleus and shields part of its potential for an outside viewer.

The values of ζ in (18.47) are of order unity. For $T = 10^7$ K and ϱ between 1 and $10^2\,\mathrm{g\,cm^{-3}}$, r_D has typical values of $10^{-8} \cdots 10^{-9}$ cm. In order to judge the influence of the shielding on nuclear reactions between nuclei of types 1 and 2, we should compare r_D with the closest distance r_{c0} to which the particles can classically approach each other if their energy is that of the Gamow peak E_0 [given by (18.27)]. These particles will be the most effective ones for the energy production. According to (18.7) one has $r_{c0} = Z_1 Z_2 e^2/E_0$, and convenient numerical expressions for E_0 are given in (18.34). We then find

$$\frac{r_D}{r_{c0}} \approx 200 \frac{E_0}{Z_1 Z_2} \left(\frac{T_7}{\zeta \varrho} \right)^{1/2} , \tag{18.52}$$

where E_0 is in keV and ϱ in g cm^{-3}. With rough values for the solar centre, $T_7 \approx 1$, $\varrho \approx 10^2$ g cm^{-3}, $\zeta \approx 1$, and for the most important hydrogen reactions, we have $Z_1 Z_2 = 1 \ldots 7$ and $E_0 \approx 5 \ldots 20$ keV; hence (18.52) gives $r_D/r_{c0} \approx 50 \ldots 100$. For all such "normal" stars, $r_D \gg r_{c0}$, which means that the incoming particle even classically (without the tunnelling effect) penetrates nearly the entire electron cloud and the shielding will have little effect at these critical distances.

The decrease of the Coulomb interaction energy E_{Coul} increases the probability P_0 for tunnelling through the Coulomb wall. The decisive exponent η in P_0 [(18.9) and the following] is determined by the function $E_{Coul} - E$. The energy E_{Coul} is now reduced according to (18.51) by the factor $\exp(-r/r_D)$, which is to a first approximation $1 - r/r_D$ for $r/r_D \ll 1$.

This gives

$$E_{Coul} - E \equiv \frac{Z_1 Z_2 e^2}{r} e^{-r/r_D} - E \approx \frac{Z_1 Z_2 e^2}{r} - \frac{Z_1 Z_2 e^2}{r_D} - E , \tag{18.53}$$

which shows that we will obtain the same result as without shielding, but with an enlarged energy:

$$\tilde{E} = E + \frac{Z_1 Z_2 e^2}{r_D} = E + E_D . \tag{18.54}$$

In order to see the influence on simple non-resonant reaction rates, consider the integrand in (18.21) and replace $\sigma(E)$ by $\sigma(\tilde{E})$. With (18.15) and (18.19) and $\tilde{\eta} = \eta (E/\tilde{E})^{1/2}$, we have the proportionality

$$\begin{aligned} \sigma(\tilde{E}) v f(E) &\sim (\tilde{E}^{-1} e^{-2\pi\tilde{\eta}}) E^{1/2} (E^{1/2} e^{-E/kT}) \\ &\sim \left(1 - \frac{E_D}{\tilde{E}}\right) e^{E_D/kT - \tilde{E}/kT - 2\pi\tilde{\eta}} . \end{aligned} \tag{18.55}$$

We assume here that $E_D/kT \ll 1$, which is usually called the case of "weak screening". Considering the fact that only a small range of E at values much larger than kT contributes essentially to $\langle \sigma v \rangle$, we may as well neglect the factor $(1 - E_D/\tilde{E})$ in (18.55) and integrate over $\tilde{E}$ instead of E. The main change is then the additional constant exponent E_D/kT such that $\langle \sigma v \rangle$ is multiplied by a "screening factor"

$$f = e^{E_D/kT} , \tag{18.56}$$

which increases $\langle\sigma v\rangle$, since E_D is positive. For weak screening we have numerically

$$\frac{E_D}{kT} = \frac{Z_1 Z_2 e^2}{r_D kT} = 5.92 \times 10^{-3} Z_1 Z_2 \left(\frac{\zeta \varrho}{T_7^3}\right)^{1/2}, \tag{18.57}$$

with ϱ in g cm^{-3}. For $\zeta \approx 1, \varrho = 1$ g cm^{-3}, and $T_7 = 1$, reactions with $Z_1 Z_2 \lesssim 16$ require correction factors f, which increase the rate by less than 10 %.

Where very large densities are involved, however, one will leave the regime of weak screening. For $E_D/kT \gtrsim 1$, the treatment is much more complicated, and the limiting case of "strong screening" is described approximately by

$$\frac{E_D}{kT} \approx 0.0205[(Z_1 + Z_2)^{5/3} - Z_1^{5/3} - Z_2^{5/3}]\frac{(\varrho/\mu_e)^{1/3}}{T_7}, \tag{18.58}$$

with the molecular weight per free electron $\mu_e = (\sum X_i Z_i / A_i)^{-1}$, see (4.29), and ϱ in g cm^{-3}.

Equations (18.57) and (18.58) show that the screening factor f increases appreciably for increasing ϱ and decreasing T. While f was a minor correction factor to the rate for "normal" stars with weak screening, the situation changes completely in the high-density, low-temperature regime, where screening becomes the dominating factor in the reaction rate.

Consider the shielded reaction rate as represented by

$$f\langle\sigma v\rangle = f_0 \langle\sigma v\rangle_0 \left(\frac{\varrho}{\varrho_0}\right)^{\lambda} \left(\frac{T}{T_0}\right)^{\nu} \tag{18.59}$$

in the neighbourhood of ϱ_0, T_0. In a similar manner to the derivation of ν for the unshielded case in (18.36)–(18.39), we find now that

$$\nu = \frac{\tau}{2} - \frac{2}{3} - \frac{E_D}{kT}; \quad \lambda = 1 + \frac{1}{3}\frac{E_D}{kT}. \tag{18.60}$$

For very high densities and moderate to low temperatures (say $\varrho > 10^6$ g cm^{-3}, $T > 10^7$ K), the temperature sensitivity ν decreases, while the density sensitivity λ becomes larger. This can be seen from Fig. 18.6, where the line of constant ^{12}C–^{12}C burning turns steeply down for large ϱ. Finally, the reaction rates now depend mainly on the density (instead of the temperature) and one speaks of *"pycnonuclear reactions"*. For ^{12}C burning in a pure ^{12}C plasma, (18.60) gives the transition $\lambda = \nu$ at $T_7 = 10$ for $\varrho = 1.60 \times 10^9$ g cm^{-3}.

Pycnonuclear reactions can play a role in very late phases of stellar evolution, where a burning may be triggered by a compression without temperature increase, and they can provide a certain amount of energy release even in cool stars, if only the density is high enough. Of course, other effects, such as the decrease of the mobility of the nuclei because of crystallization, must then also be considered.

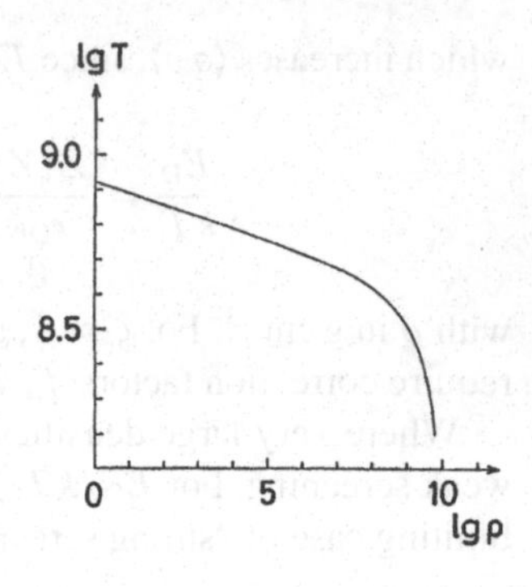

Fig. 18.6 A line of constant energy generation rate $\varepsilon(=10^4 \text{ erg g}^{-1} \text{ s}^{-1})$ for the $^{12}C+^{12}C$ burning in a diagram showing the temperature T (in K) over the density ϱ (in g cm^{-3}). The temperature sensitivity ν and the density sensitivity λ are equal where the slope is -1

18.5 The Major Nuclear Burning Stages

Although no chemical reactions are involved, one usually calls the thermonuclear fusion of a certain element the "burning" of this element. Owing to the properties of thermonuclear reaction rates, different burnings are well separated by appreciable temperature differences. A review of the cross sections for all possible reactions in the major burning stages shows that only very few reactions occur with non-negligible rates during a certain phase. The most important ones will be listed below. Their important properties, such as the astrophysical factors S_0, correction factors to (18.32), or energy release Q, can be found in the literature (e.g. Caughlan and Fowler 1988; Harris and Fowler 1983; Adelberger et al. 2011; Angulo et al. 1999).

The Q values usually contain all of the energy made available to the stellar matter by one such reaction. This includes the energies of the γ rays that are either directly emitted or created by pair annihilation after e^+ emission. Excluded, however, is the energy carried away by neutrinos, since they normally do not interact with the stellar material.

A whole "network" of all simultaneously occurring reactions (8.7) has to be calculated if one is interested in details such as the isotopic abundances produced by the reactions or if the star changes on a timescale comparable with that of one of the reactions. The total ε is then obtained as a sum of (18.23) over all reactions, and one has to ensure the correct bookkeeping of the changing abundances of all nuclei involved. We have encountered nuclear reaction networks also in Sect. 12.3.

If one is interested only in the energy production, often, a much simpler procedure suffices in which only the rate for the slowest of a chain of subsequent reactions is calculated, since it determines essentially the rate of the whole fusion process. An example of such a "bottleneck" is the ^{14}N reaction in the CNO cycle (see below). Then (18.23) has to be used for this reaction, but with Q equal to the sum of all energies released in the single reactions.

In this section, all formulae for ε will be given in units of erg g^{-1}s^{-1}, ϱ in g cm^{-3}, and T in the dimensionless form $T_n = T/10^n$ K. As usual we denote by X_j the mass fraction of nuclei with mass number $A = j$.

18.5.1 Hydrogen Burning

The net result of hydrogen burning is the fusion of four ^{1}H nuclei into one ^{4}He nucleus. The difference in binding energy is almost exactly 27.0 MeV, corresponding to a mass defect of about 0.7 per cent. This is roughly 10 times the energy liberated in any other fusion process, though not all of this energy is available to stellar matter. The fusion requires the transformation of two protons into neutrons, i.e. two β^+ decays, which must be accompanied by two neutrino emissions (conservation of lepton number). The neutrinos carry away 2 ... 30 per cent of the whole energy liberated, the amount depending strongly on the reaction in which they are emitted.

There are different chains of reactions by which a fusion process can be completed and which in general will occur simultaneously in a star. The two main series of reactions are known as the proton–proton chain and the CNO cycle.

The *proton–proton chain* (*pp* chain) is named after its first reaction, between two protons forming a deuterium nucleus ^{2}H, which then reacts with another proton to form ^{3}He:

$$^1\mathrm{H} + {}^1\mathrm{H} \rightarrow {}^2\mathrm{H} + e^+ + \nu \, ,$$
$$^2\mathrm{H} + {}^1\mathrm{H} \rightarrow {}^3\mathrm{He} + \gamma \, . \tag{18.61}$$

The first of these reactions is unusual in comparison with most other fusion processes. In order to form ^{2}H, the protons have to experience a β^+ decay at the time of their closest approach. This is a process governed by the weak interaction and is very unlikely. Therefore the first reaction has a very small cross section.

The completion of a ^{4}He nucleus can proceed via one of three alternative branches (*pp*1, *pp*2, *pp*3) all of which start with ^{3}He. The first alternative requires two ^{3}He nuclei, i.e. the reactions in (18.61) have first to be completed twice. The other alternatives require that ^{4}He already exists (either it is present because of its primordial abundance, or because it was already produced earlier by this burning process). The branching between *pp*2 and *pp*3 exists, since ^{7}Be can react either with e^- or with ^{1}H. All possibilities can be seen from the following scheme:

$$^1\mathrm{H} + {}^1\mathrm{H} \rightarrow {}^2\mathrm{H} + e^+ + \nu$$
$$^2\mathrm{H} + {}^1\mathrm{H} \rightarrow {}^3\mathrm{He} + \gamma$$

(*pp*1): $^3\mathrm{He} + {}^3\mathrm{He} \rightarrow {}^4\mathrm{He} + 2\,{}^1\mathrm{H}$

or: $^3\mathrm{He} + {}^4\mathrm{He} \rightarrow {}^7\mathrm{Be} + \gamma$, followed by either

(*pp*2):
$$^7\mathrm{Be} + e^- \rightarrow {}^7\mathrm{Li} + \nu$$
$$^7\mathrm{Li} + {}^1\mathrm{H} \rightarrow {}^4\mathrm{He} + {}^4\mathrm{He}$$

or (*pp*3):
$$^7\mathrm{Be} + {}^1\mathrm{H} \rightarrow {}^8\mathrm{B} + \gamma$$
$$^8\mathrm{B} \rightarrow {}^8\mathrm{Be} + e^+ + \nu$$
$$^8\mathrm{Be} \rightarrow {}^4\mathrm{He} + {}^4\mathrm{He} \tag{18.62}$$

Owing to the different energies carried away by the neutrinos, the energies released to the stellar matter differ for the three chains. They are $Q = 26.50(pp1)$, $25.97(pp2)$, and $19.59(pp3)$, in MeV per produced ^{4}He nucleus. For each quantity Q released, the first two reactions of (18.61) have to be performed only once in the *pp*2 and *pp*3 branches.

Three reactions in (18.62) release neutrinos, which are given names according to the element being processed in these reactions: *pp*-, ^{7}Be-, and ^{8}B-neutrinos. If they are the only lepton emitted, then their energy is well defined. The ^{7}Be-neutrinos carry away 0.863 MeV in 90 % of the reactions, and 0.386 MeV in the remaining 10 %, depending on the energy state of ^{7}Li produced. If the neutrinos are emitted along with a positron (e^+), the two leptons share the energy, and a spectrum of neutrino energies results. The upper limits are 0.423 MeV for the *pp*-neutrinos and 15 MeV for those of the ^{8}B reaction. The average values are 0.267 and 6.735 MeV respectively.

The relative frequency of the different branches depends on the chemical composition, the temperature, and the density. The ^{3}He–^{4}He reaction has a 14 % larger reduced mass, a 4.6 % larger τ, and thus a slightly larger temperature sensitivity ν than the ^{3}He–^{3}He reaction, cf. (18.34) and (18.39). With increasing T, *pp*2 and *pp*3 will therefore dominate more and more over *pp*1 (say above $T_7 \approx 1$) if ^{4}He is present in appreciable amounts. And with increasing T, the relative importance will gradually shift from the electron capture (*pp*2) to the proton capture (*pp*3) of ^{7}Be.

The energy generation in the *pp* chain should be calculated at small T (say below $T_6 \approx 8$) by calculating all single reactions and their influence on the nuclei involved. For larger T, there will be an equilibrium abundance established for these nuclei (equal rates of consumption and production) and one can simply take the whole ε_{pp} as proportional to that of the *pp*1 branch, which in turn may be calculated from the rate of the first reaction ^{1}H + ^{1}H:

$$\begin{aligned} \varepsilon_{pp} &= 2.57 \times 10^4 \psi f_{11} g_{11} \varrho X_1^2 T_9^{-2/3} \mathrm{e}^{-3.381/T_9^{1/3}} , \\ g_{11} &= (1 + 3.82 T_9 + 1.51 T_9^2 + 0.144 T_9^3 - 0.0114 T_9^4) , \end{aligned} \tag{18.63}$$

where ε_{pp} and ϱ are in cgs and f_{11} is the shielding factor for this reaction. The factor ψ corrects for the additional energy generation in the branches *pp*2 and *pp*3 if there is appreciable ^{4}He present (see Fig. 18.7). For gradually increasing T, ψ starts with the value 1 and can then increase to values close to 2 (at $T_7 \approx 2$), at which point *pp*2 takes over, since then *each* ^{1}H–^{1}H reaction gives one ^{4}He (compared to *every second* such reaction in the branch *pp*1). After this maximum, ψ decreases again to about 1.5 where *pp*3 has taken over owing to its Q being much smaller than those of the other branches.

The formulation of the energy generation ε as given in (18.63) is an analytical fit to measured and tabulated values, based on T-dependences of non-resonant reactions and resonances. They may vary from group to group; the one used here is taken from Angulo et al. (1999).

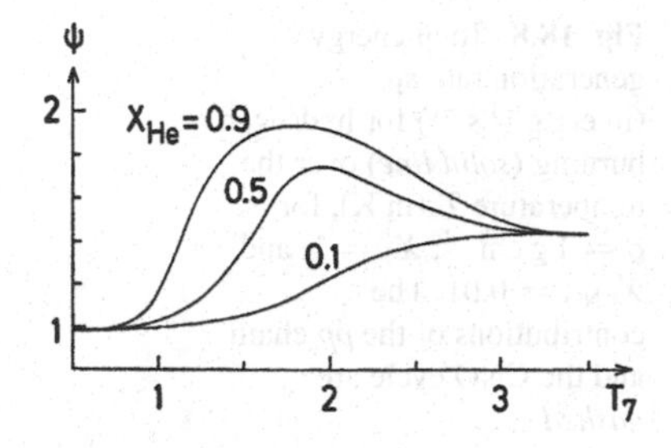

Fig. 18.7 The correction ψ for ε_{pp} as a function of T_7, for three different helium abundances (After Parker et al. 1964)

The temperature sensitivity of the *pp* chain is the smallest of all fusions. At $T_6 = 5$, we have $\nu \approx 6$, which decreases to 3.5 at $T_6 \approx 20$.

The *CNO cycle* is the other main series of reactions in hydrogen burning. It requires the presence of some isotopes of C, N, or O, which are reproduced in a manner similar to catalysts in chemical reactions. The sequence of reactions can be represented as follows:

$$
\begin{array}{rcl}
{}^{12}\mathrm{C} + {}^{1}\mathrm{H} & \rightarrow & {}^{13}\mathrm{N} + \gamma \\
{}^{13}\mathrm{N} & \rightarrow & {}^{13}\mathrm{C} + e^{+} + \nu \\
{}^{13}\mathrm{C} + {}^{1}\mathrm{H} & \rightarrow & {}^{14}\mathrm{N} + \gamma \\
{}^{14}\mathrm{N} + {}^{1}\mathrm{H} & \rightarrow & {}^{15}\mathrm{O} + \gamma \\
{}^{15}\mathrm{O} & \rightarrow & {}^{15}\mathrm{N} + e^{+} + \nu \\
{}^{15}\mathrm{N} + {}^{1}\mathrm{H} & \rightarrow & {}^{12}\mathrm{C} + {}^{4}\mathrm{He} \\
 & \dashrightarrow & {}^{16}\mathrm{O} + \gamma \\
{}^{16}\mathrm{O} + {}^{1}\mathrm{H} & \rightarrow & {}^{17}\mathrm{F} + \gamma \\
{}^{17}\mathrm{F} & \rightarrow & {}^{17}\mathrm{O} + e^{+} + \nu \\
{}^{17}\mathrm{O} + {}^{1}\mathrm{H} & \rightarrow & {}^{14}\mathrm{N} + {}^{4}\mathrm{He}
\end{array}
\qquad (18.64)
$$

The main cycle (CNO-I; upper 6 lines of this scheme) is completed after the initially consumed ^{12}C is reproduced by $^{15}\mathrm{N} + {}^{1}\mathrm{H}$. This reaction shows a branching via ^{16}O into a secondary cycle (CNO-II; connected with the main cycle by dashed arrows), which is, however, roughly 10^3 times less probable. Its main effect is that the ^{16}O nuclei originally present in the stellar matter can also take part in the cycle, since they are finally transformed into ^{14}N by the last three reactions of (18.64). The decay times for the β^{+} decays are of the order of $10^2 \ldots 10^3$ s. As usual, a network of all simultaneous reactions has to be calculated for lower temperatures, rapid changes, or if a detailed knowledge about the abundances of all nuclei involved is desired.

As in the case of the *pp* chains, two protons have to be converted in effect to neutrons in the process, which will release two neutrinos per new helium nucleus.

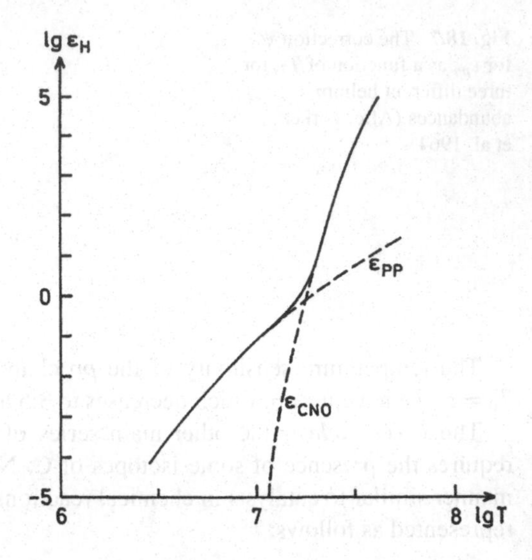

Fig. 18.8 Total energy generation rate ε_{H} (in erg $\mathrm{g}^{-1}\,\mathrm{s}^{-1}$) for hydrogen burning (*solid line*) over the temperature T (in K), for $\varrho = 1\,\mathrm{g\,cm}^{-3}$, $X_1 = 1$, and $X_{\mathrm{CNO}} = 0.01$. The contributions of the *pp* chain and the CNO cycle are *dashed*

The ^{13}N-, ^{15}O-, and ^{17}F-neutrinos of the CNO-cycles have all energy spectra with an upper limit between 1.1 and 1.7 MeV, and an average energy of 0.706, 0.996, and 0.998 MeV.

Most stars change slowly enough that, for sufficiently high temperature (say $T_7 \gtrsim 1.5$), the nuclei involved in the cycle reach their equilibrium abundance (i.e. the rate of production equals that of consumption). Then it suffices to calculate explicitly only the slowest reaction, which is $^{14}\mathrm{N}+^{1}\mathrm{H}$ and which essentially controls the time for completing the cycle. $\varepsilon_{\mathrm{CNO}}$ will then be given by the rate of this reaction and by the energy gain of the whole cycle, which is 24.97 MeV. This slowest reaction acts like a bottleneck where the nuclei involved are congested in their "flow" through the cycle. Nearly all of the initially present C, N, and O nuclei will therefore be found as ^{14}N, waiting to be transformed to ^{15}O. The energy generation rate can be written as (using again the cross section from Angulo et al. 1999 but dropping additional terms important for higher temperatures for simplicity)

$$\begin{aligned} \varepsilon_{\mathrm{CNO}} &= 8.24 \times 10^{25} g_{14,1} X_{\mathrm{CNO}} X_1 \varrho T_9^{-2/3} \mathrm{e}^{(-15.231 T_9^{-1/3} - (T_9/0.8)^2)} , \\ g_{14,1} &= (1 - 2.00 T_9 + 3.41 T_9^2 - 2.43 T_9^3) , \end{aligned} \tag{18.65}$$

where $\varepsilon_{\mathrm{CNO}}$ and ϱ are in cgs. X_{CNO} is the sum of X_{C}, X_{N}, and X_{O}. The temperature sensitivity ν is much higher here than in the *pp* chain. For $T_6 = 10 \ldots 50$, we find $\nu \approx 23 \ldots 13$. This has the consequence that the *pp* chain dominates at low temperatures ($T_6 < 15$), while it can be neglected against $\varepsilon_{\mathrm{CNO}}$ for higher temperatures (see Fig. 18.8). Hydrogen burning normally occurs in the range $T_6 \approx 8 \ldots 50$, since at larger T, the hydrogen is very rapidly exhausted.

18.5.2 Helium Burning

The reactions of helium burning consist of the gradual fusion of several ^{4}He into ^{12}C, ^{16}O, ... This requires temperatures of $T_8 \gtrsim 1$, i.e. appreciably higher than those for hydrogen burning, because of the higher Coulomb barriers.

The first and key reaction is the formation of ^{12}C from three ^{4}He nuclei, which is called the *triple α reaction* (or 3α reaction). A closer look shows that it is performed in two steps, since a triple encounter is too improbable:

$$\begin{aligned} &^4\mathrm{He} + {}^4\mathrm{He} \rightleftarrows {}^8\mathrm{Be}\,, \\ &^8\mathrm{Be} + {}^4\mathrm{He} \rightarrow {}^{12}\mathrm{C} + \gamma\,. \end{aligned} \tag{18.66}$$

In the first step, two α particles temporarily form a ^{8}Be nucleus. Its ground state is nearly 100 keV higher in energy and therefore decays back into the two α's after a few times 10^{-16} s. This seems to be a very short time at a first glance, but it is roughly 10^5 times larger than the duration of a normal scattering encounter. The probability for another reaction occurring during this time is correspondingly enhanced. In fact the lifetime of ^{8}Be is sufficient to build up an average concentration of these nuclei of about 10^{-9} in the stellar matter. The high densities then ensure a sufficient rate of further α captures that form ^{12}C nuclei [the second step in (18.66)]. Both these reactions are complicated owing to the involvement of resonances. The energy release per ^{12}C nucleus formed is 7.274 MeV. This gives an energy release *per unit mass* that is 10.4 times smaller than in the case of the CNO cycle (where only four instead of 12 nucleons are processed): $E_{3\alpha} = 5.8 \times 10^{17}\,\mathrm{erg\,g^{-1}}$. The resulting energy generation rate is

$$\varepsilon_{3\alpha} = 5.09 \times 10^{11}\, f_{3\alpha} \varrho^2 X_4^3 T_8^{-3} \mathrm{e}^{-44.027/T_8} \tag{18.67}$$

(ε and ϱ in cgs), with the screening factor $f_{3\alpha}$. (18.67) is based on an older, simplified analytical fit of $\langle \sigma v \rangle$ by Caughlan and Fowler (1988). A more recent one, taken from the compilation by Angulo et al. (1999), has more terms, reflecting the effect of the several resonances involved:

$$\begin{aligned} \varepsilon_{3\alpha} = {} & 6.272 \varrho^2 X_4^3 \cdot (1 + 0.0158 T_9^{-0.65}) \\ & \times \Big[2.43 \times 10^9 T_9^{-2/3} \exp\left(-13.490 T_9^{-1/3} - (T_9/0.15)^2\right) \cdot (1 + 74.5 T_9) \\ & \quad + 6.09 \times 10^5 T_9^{-3/2} \exp(-1.054/T_9) \Big] \\ & \times \Big[2.76 \times 10^7 T_9^{-2/3} \exp\left(-23.570 T_9^{-1/3} - (T_9/0.4)^2\right) \\ & \quad \times (1 + 5.47 T_9 + 326 T_9^2) + 130.7 T_9^{-3/2} \exp(-3.338/T_9) \\ & \quad + 2.51 \times 10^4 T_9^{-3/2} \exp(-20.307/T_9) \Big]\,. \end{aligned} \tag{18.68}$$

The two terms in square brackets in (18.68) come from the $\alpha + \alpha$ and the $\alpha + {}^8$Be steps of the 3α-process; the first line contains also a conversion factor to go from the mean cross section to the energy production rate according to (18.23). This reaction has an enormous temperature sensitivity. For $T_8 = 1 \ldots 2$, (18.39) gives $\nu \approx 40 \ldots 19$!

Once a sufficient ^{12}C abundance has been built up by the 3α reaction, further α captures can occur simultaneously with (18.66) such that the nuclei ^{16}O, ^{20}Ne, ... are successively formed:

$$\begin{aligned} {}^{12}\mathrm{C} \,+\, {}^{4}\mathrm{He} &\rightarrow {}^{16}\mathrm{O} + \gamma\,, \\ {}^{16}\mathrm{O} \,+\, {}^{4}\mathrm{He} &\rightarrow {}^{20}\mathrm{Ne} + \gamma\,, \\ &\ldots \end{aligned} \tag{18.69}$$

In a typical stellar-interior environment, reactions going beyond ^{20}Ne are rare.

The energy release per ^{12}C$(\alpha,\gamma)^{16}$O reaction is 7.162 MeV, corresponding to $E_{12,\alpha} = 4.320 \times 10^{17}\,\mathrm{erg\,g^{-1}}$ of produced ^{16}O (The whole formation of ^{16}O from the initial four α particles has then yielded $8.71 \times 10^{17}\,\mathrm{erg\,g^{-1}}$.). This is a rather complicated reaction. For moderate temperatures (up to a few 10^8 K), one may use the following simple approximation:

$$\varepsilon_{12,\alpha} = 1.3 \times 10^{27} f_{12,4} X_{12} X_4 \varrho T_8^{-2} \left(\frac{1 + 0.134 T_8^{2/3}}{1 + 0.017 T_8^{2/3}} \right)^2 \mathrm{e}^{-69.20/T_8^{1/3}}\,, \tag{18.70}$$

where ε and ϱ are in cgs. This reaction has been notoriously uncertain by factors of 2 and 3 at stellar temperatures. This has severe consequences for the production of carbon and even for the evolution of stars. The rate has been changed repeatedly within this uncertainty range as a result of new measurements. Kunz et al. (2002) provide the most recent analytical fit.

In each reaction ^{16}O (α,γ) ^{20}Ne, an energy of 4.73 MeV is released. The rate is according to Angulo et al. (1999):

$$\begin{aligned} \varepsilon_{16,\alpha} \approx X_{16} X_4 \varrho f_{16,4} \cdot 1.91 \times 10^{27} T_9^{-2/3} \exp\left(-39.760 T_9^{-1/3} - (T_9/1.6)^2\right) \\ +3.64 \times 10^{18} T_9^{-3/2} \exp(-10.32/T_9) \\ +4.39 \times 10^{19} T_9^{-3/2} \exp(-12.200/T_9) \\ +2.92 \times 10^{16} T_9^{2.966} \exp(-11.900/T_9)\,, \end{aligned} \tag{18.71}$$

where ε and ϱ are in cgs; this rate is also very uncertain.

Summarizing, we can say that during helium-burning reactions, (18.66) and (18.69) occur simultaneously, and the total energy generation rate is given by $\varepsilon_{\mathrm{He}} = \varepsilon_{3\alpha} + \varepsilon_{12,\alpha} + \varepsilon_{16,\alpha}$. If the initial ^{4}He is transformed into equal amounts of ^{12}C and ^{16}O, then the energy yield is $7.28 \times 10^{17}\,\mathrm{erg\,g^{-1}}$.

The general course of helium burning is always according to the following scheme: initially, when burning sets in at temperatures around 10^8 K, the triple-α reaction is dominating, both because of a larger cross section and of the carbon and oxygen abundances, which are low in comparison to that of helium. With progressing conversion of helium to carbon, the $^{12}C(\alpha,\gamma)^{16}O$ becomes more competitive. The increasing temperature is supporting this. When the helium content gets low, the fact that the triple-α reaction is proportional to the third power of the helium abundance disfavours it increasingly, such that the burning of carbon is larger than its creation by triple-α reactions, and its abundance decreases again, while simultaneously that of ^{16}O increases. The final abundances of ^{12}C and ^{16}O will thus depend on the competition between the reactions (18.69) and the exhaustion of ^4He particles. This depends mainly on the $^{12}C(\alpha,\gamma)^{16}O$ rate: if it is higher, the destruction of ^{12}C will set in earlier and a higher O:C ratio will result. Overall, the outcome of helium burning is O:C $\approx$ 1:1–2:1. Neon production is comparably unimportant.

18.5.3 Carbon Burning and Beyond

For a mixture consisting mainly of ^{12}C and ^{16}O (as would be found in the central part of a star after helium burning), *carbon burning* will set in if the temperature or the density rises sufficiently. The typical range of temperature for this burning is $T_8 \approx 5 \ldots 10$.

Here (and in the following types of burning) the situation is already so difficult that one often has to rely on rough approximations and guesses, or on complete nuclear networks. The first complication is that the original $^{12}C+^{12}C$ reaction produces an excited ^{24}Mg nucleus, which can decay via many different channels (the last column gives $Q/1\,\mathrm{MeV}$):

$$\begin{aligned} ^{12}C + ^{12}C &\rightarrow {}^{24}Mg + \gamma \;, && 13.931 \\ &\rightarrow {}^{23}Mg + n \;, && -2.605 \\ &\rightarrow {}^{23}Na + p \;, && 2.238 \\ &\rightarrow {}^{20}Ne + \alpha \;, && 4.616 \\ &\rightarrow {}^{16}O + 2\alpha \;, && -0.114 \end{aligned} \tag{18.72}$$

The relative frequency of the channels is very different, and depends also on the temperature. The γ decay (leaving ^{24}Mg) is rather improbable, and the same is true for the two endothermic decays ($^{23}Mg + n$ and $^{16}O + 2\alpha$). The most probable reactions are those which yield $^{23}Na+p$ and $^{20}Ne+\alpha$. These are believed to occur at about equal rates for temperatures that are not too high (say $T_9 < 3$).

The next problem is that the produced p and α find themselves at temperatures extremely high for hydrogen and helium burning and will immediately react with some of the particles in the mixture (from ^{12}C up to ^{24}Mg). They may even start whole reaction chains, such as $^{12}C(p,\gamma)^{13}N(e^+\nu)^{13}C(\alpha,n)^{16}O$, where the neutron could immediately react further. All these details would have to be evaluated quantitatively in order to find the average energy gain and the final products. For a rough guess one may assume that on average, $Q \approx 13\,\mathrm{MeV}$ are released per ^{12}C–^{12}C reaction (including all follow-up reactions). Then (Caughlan and Fowler 1988),

$$\varepsilon_{CC} \approx 1.86 \times 10^{43} f_{CC} \varrho X_{12}^2 T_9^{-3/2} T_{9a}^{5/6} \cdot \exp[-84.165/T_{9a}^{1/3} - 2.12 \times 10^{-3} T_9^3] \tag{18.73}$$

with ε and ϱ in cgs and with $T_{9a} = T_9/(1+0.0396T_9)$. The screening factor f_{CC} can become important (see Fig. 18.6), since this burning can start in very dense matter. The end products may be mainly ^{16}O, ^{20}Ne, ^{24}Mg, and ^{28}Si.

For *oxygen burning,* $^{16}O+^{16}O$, the Coulomb barrier is already so high that the necessary temperature is $T_9 \gtrsim 1$. As in the case of carbon burning, the reaction can proceed via several channels:

$$\begin{aligned} ^{16}O + ^{16}O &\rightarrow ^{32}S + \gamma\,, \quad 16.541 \\ &\rightarrow ^{31}P + p\,, \quad 7.677 \\ &\rightarrow ^{31}S + n\,, \quad 1.453 \\ &\rightarrow ^{28}Si + \alpha\,, \quad 9.593 \\ &\rightarrow ^{24}Mg + 2\alpha\,, \quad -0.393 \end{aligned} \tag{18.74}$$

Most frequent is the p decay, followed by the α decays. Again, all released p, n, and α are captured immediately, giving rise to a multitude of secondary reactions. Among the end products, one will find a large amount of ^{28}Si. For an average energy $Q \approx 16\,\mathrm{MeV}$ released per $^{16}O+^{16}O$ reaction, the energy generation rate is roughly

$$\varepsilon_{OO} \approx 2.14 \times 10^{53} f_{OO} \varrho X_{16}^2 T_9^{-2/3} \cdot \exp(-135.93/T_9^{1/3} - 0.629 T_9^{2/3} - 0.445 T_9^{4/3} + 0.0103 T_9^2) \tag{18.75}$$

with ε and ϱ in cgs, and the screening factor f_{OO}.

For $T_9 > 1$, one also has to consider the possibility of *photodisintegration* of nuclei that are not too strongly bound. Here the radiation field contains a significant number of photons with energies in the MeV range, which can be absorbed by a nucleus, breaking it up, for example, by α decay. This is a complete analogue of photoionization of atoms, and, in equilibrium, a formula equivalent to the Saha formula [see (14.11)] holds for the number densities n_i and n_j of the final particles (after disintegration), relative to the number n_{ij} of the original (compound) particles:

$$\frac{n_i n_j}{n_{ij}} \sim T^{3/2} \mathrm{e}^{-Q/kT}\,, \tag{18.76}$$

where Q is the difference in binding energies between the original nucleus and its fragments. (Q corresponds to the ionization energy χ; however, it is about $10^2 \dots 10^3$ times larger because of the much stronger nuclear forces.) The proportionality factor contains essentially the partition functions of the three types of particles. Equilibrium is usually not reached, and the details are very complicated and may differ from case to case, which is also true for the amount of energy released or lost.

The photodisintegration itself is, of course, endothermic. But the ejected particles (X_j) will be immediately recaptured. The capture can lead back to the original nucleus X_{ij}, i.e. the reaction would be $X_{ij} \rightleftarrows X_i + X_j$, or it can lead to quite different, even heavier, nuclei X_{jk} that are more strongly bound than the original one $X_j + X_k \rightarrow X_{jk}$. The latter case would be exothermic and can outweigh the endothermic photodisintegration in the total energy balance.

An example is *neon disintegration*, which in stellar evolution occurs even before oxygen burning:

$$^{20}\mathrm{Ne} + \gamma \rightarrow {}^{16}\mathrm{O} + \alpha\,, \quad Q = -4.73\,\mathrm{MeV}\,. \tag{18.77}$$

It dominates over the inverse reaction (known from helium burning) at $T_9 > 1.5$. The ejected α particle reacts mainly with other ^{20}Ne nuclei, yielding $^{24}\mathrm{Mg}+\gamma$. The net result will then be the conversion of Ne into O and Mg:

$$2\,{}^{20}\mathrm{Ne} + \gamma \rightarrow {}^{16}\mathrm{O} + {}^{24}\mathrm{Mg} + \gamma\,, \quad Q = +4.583\,\mathrm{MeV}\,. \tag{18.78}$$

Another example is the photodisintegration of ^{28}Si, which may be the dominant reaction at the end of oxygen burning. Near $T_9 \approx 3$, ^{28}Si can be decomposed by the photons and eject n, p, or α. There follows a large number of reactions in which the thereby created nuclei (e.g. Al, Mg, Ne) will also be subject to photodisintegration, leading to the existence of an appreciable amount of free n, p, and α particles. These react with the remaining ^{28}Si, thus building up gradually heavier nuclei, until ^{56}Fe is reached. Since ^{56}Fe is so strongly bound, it may survive this melting pot as the only (or dominant) species. So, forgetting all intermediate stages, we would ultimately have the conversion of two ^{28}Si into ^{56}Fe, which can be called *silicon burning*.

For $T_9 \gtrsim 5$, photodisintegration breaks up even the ^{56}Fe nuclei into α particles and thus reverses the effect of all prior burnings. Such processes can occur during supernova explosions (see Chap. 36).

18.6 Neutron-Capture Nucleosynthesis

In Fig. 18.9 we show the solar abundances of elements. As on earth, we find all elements from hydrogen to lead and uranium in the Sun. The nuclear burning we discussed so far is able to produce only elements up to iron, since the creation of elements heavier than the "iron peak" is endothermic, and the electrostatic repulsion

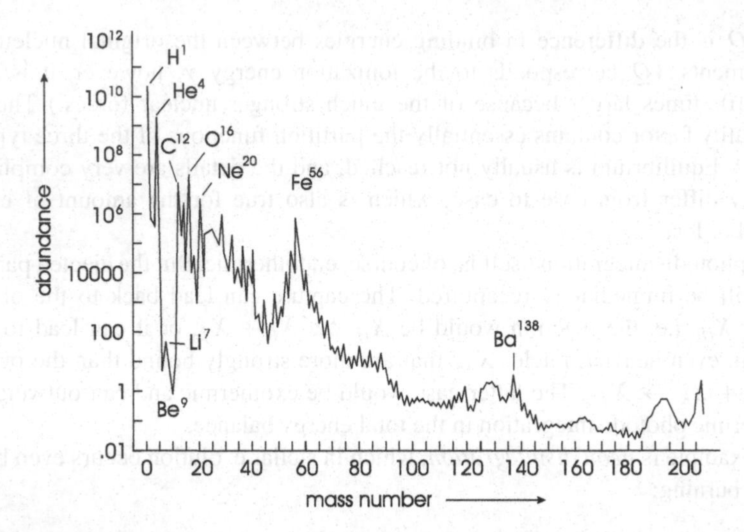

Fig. 18.9 The abundances (particle number fractions) of elements in the Sun, normalized to a value of 10^6 for ^{28}Si

for charged particle reactions is increasing with nuclear charge. The various peaks in this figure reflect the stability of isotopes against further addition of neutrons and protons and are due to the structure of the nuclei, easily explained in the shell model of nuclear physics. In particular isotopes with even and equal numbers of neutrons and protons, such as ^{12}C or ^{40}Ca, are very stable and therefore more abundant than neighbouring ones. If nuclear shells are closed, the stability is even higher, similar to the noble gases in atomic physics. Such isotopes are called "magic nuclei" with "magic" numbers of protons or neutrons. ^{16}O is a "double-magic" nucleus.

During hydrostatic burning phases, the elements beyond the iron peak can be produced only if other reactions with lighter nuclei provide enough energy and, most easily, if the reactions are processing by the capture of neutrons, since they are electrically neutral. Adding neutrons leads initially to heavier isotopes of the same element, which become the more unstable the more neutrons they have. The decay proceeds by emission of an electron, which is temperature-insensitive and therefore is acting as a kind of nuclear clock. β-decay times can reach from minutes to millions of years, and are getting shorter with increasing neutron excess. The decay leads to the creation of a new element of the same mass but with the charge being increased by one.

The general sequence of reactions is therefore

$$
\begin{aligned}
(Z, A) + n &\rightarrow (Z, A+1) + \gamma \\
(Z, A+1) &\rightarrow (Z+1, A+1) + e^- ,
\end{aligned}
\tag{18.79}
$$

where the first reaction can be repeated several times, depending on the number density of available neutrons n and the neutron-capture cross section. If the

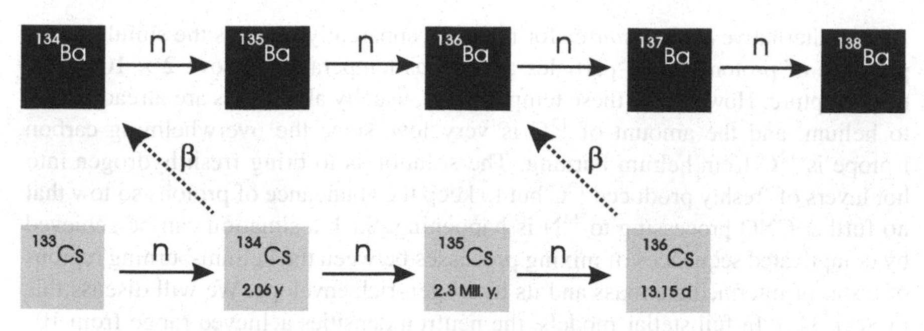

Fig. 18.10 Typical s-process reaction path in the nuclear chart, in the region of Cs and Ba. The laboratory half-life time of the Cs isotopes is given. It can be appreciable shorter at stellar temperatures. ^{134}Cs and ^{136}Cs are so-called branching points, and the relative abundances of isotopes in the various branches allow conclusions about the temperature at the s-process site (after Busso et al. 1999)

neutron-capture time is long compared to the β-decay time, the process is called the *slow neutron-capture process* or simply the *s-process*, and the reaction path remains close to the line of β-stability in the nuclear chart; if it is rapid, such that the first reaction in (18.79) is repeated several times, it is called the *r-process*. Subsequent neutron captures and β-decays will lead to the creation of the heavy elements. The astrophysical site for the r-process is not clearly identified, but is probably to be found in supernova explosions or similar energetic events. The s-process is certainly taking place in stars of intermediate mass ($M \approx 2 \cdots 5 M_\odot$) in an advanced phase of evolution (Sect. 34.3). In the atmospheres of such stars, short-lived isotopes of heavy elements (most importantly ^{99}Tc with a half-life time of only 211,000 years) have been found, which could only have been created in the stars themselves. Although the s-process may drain energy from the star, it is in fact unimportant for the energy budget and the structure of stars, mainly due to the extremely low abundances (see Fig. 18.9) with respect to the elements participating in the burning stages discussed earlier in this chapter. Figure 18.10 shows part of the s-process path in the Cs-Ba region of the nuclear chart. ^{138}Ba has a magic neutron number (82) and therefore is very stable and abundant (Fig. 18.9).

The necessary condition for the s- and r-process is the presence of neutrons. Since free neutrons are both unstable and are easily captured by other nuclei, a constant source of neutron production is needed. Considering the burning phases of Sect. 18.5, we realize that only protons and α-particles were involved. The generation of neutrons is indeed a very rare event in a star's life. However, in Sect. 18.5.3, we already mentioned that during carbon burning, the various reaction channels may lead to whole chains of subsequent reactions, one of them resulted in $^{13}\mathrm{C}(\alpha, n)^{16}\mathrm{O}$. Indeed, this is one of the two neutron sources identified, the other being $^{22}\mathrm{Ne}(\alpha, n)^{25}\mathrm{Mg}$, the end reaction of the sequence $^{14}\mathrm{N}(\alpha, \gamma)^{18}\mathrm{O}(\alpha, \gamma)^{22}\mathrm{Ne}$, which is operating at temperatures of about 4×10^8 K. Such temperatures are encountered during helium burning in massive stars, where the *neon source* may produce s-process elements.

The alternative *carbon source* for neutrons apparently requires the simultaneous presence of protons and α-particles as well as temperatures above 2×10^8 K for the α-capture. However, at these temperatures, usually all protons are already burnt to helium, and the amount of ^{13}C is very low, since the overwhelming carbon isotope is ^{12}C from helium burning. The solution is to bring fresh hydrogen into hot layers of freshly produced ^{12}C but to keep the abundance of protons so low that no further CNO processing to ^{14}N is happening. Such a situation can be achieved by complicated sequences of mixing processes between the helium-burning regions of a star of intermediate mass and its hydrogen-rich envelope. We will discuss this in Sect. 34.3. In full stellar models, the neutron densities achieved range from 10^6 to $10^{10}\,\mathrm{cm}^{-3}$.

The neutron-capture cross section is inversely dependent on velocity (or temperature):

$$\sigma \sim \frac{1}{v}, \tag{18.80}$$

therefore $\langle \sigma v \rangle$ in (18.21) is actually close to a constant (σv) times the integral over $f(E)$ and only slightly dependent on the stellar plasma temperature, except for nuclei close to magic neutron numbers, where σ may be lower by an order of magnitude or more. It generally lies in the range of 100 to 1000 mb ($1\,\mathrm{b} = 10^{-24}\,\mathrm{cm}^2$).

We define

$$\langle \sigma \rangle := \langle \sigma(v) v \rangle / v_T, \tag{18.81}$$

where $v_T = (2kT/\mu_n)^{1/2}$ is the thermal velocity in the system of a nucleus A and a neutron n, μ_n being the reduced mass of it. $\langle \sigma \rangle$ corresponds approximately to the cross section measured at that relative velocity. The rate equation for a nucleus with mass A and density n_A is then

$$\frac{\mathrm{d}n_A}{\mathrm{d}t} = -\langle \sigma(v) v \rangle_A n_n n_A + \langle \sigma(v) v \rangle_{A-1} n_n n_{A-1}. \tag{18.82}$$

With (18.81) this becomes

$$\frac{\mathrm{d}n_A}{\mathrm{d}t} = v_T n_n (-\sigma_A n_A + \sigma_{A-1} n_{A-1}). \tag{18.83}$$

Since the neutron density n_n may vary with time, we define

$$\mathrm{d}\tau = v_T n_n(t) \mathrm{d}t, \tag{18.84}$$

so we obtain

$$\frac{\mathrm{d}n_A}{\mathrm{d}\tau} = -\sigma_A n_A + \sigma_{A-1} n_{A-1}. \tag{18.85}$$

This equation is self-regulating: assume that initially n_A is very small. Then n_A will grow due to the positive second term in (18.85). This will also increase the first term until the right-hand side vanishes and n_A has reached a stationary value. The abundances of A and $A-1$ will therefore reflect the ratio of cross sections σ_A and σ_{A-1}. The integral over $\mathrm{d}\tau$,

$$\tau = v_T \int n_n(t)\mathrm{d}t, \tag{18.86}$$

is called the *neutron exposure* and reflects the integrated flux of neutrons with thermal velocities. Its dimension is that of an inverse area and typically of order mb^{-1}. It is the decisive quantity determining the overall abundances of elements produced by neutron captures, and how far the s- or r-process can proceed. The relative abundances of isotopes produced then reflect the cross section σ.

The synthesis of neutron-capture elements has to be computed with huge nuclear networks consisting of hundreds of isotopes and even more reactions. Simplified models assuming a distribution of neutron exposures on a single initial *seed nucleus*, usually $^{56}\mathrm{Fe}$, can quite successfully reproduce the solar abundance patterns. This distribution $\rho(\tau)$ is

$$\rho(\tau) = \frac{f n_{56}}{\tau_0} \exp(-\tau/\tau_0), \tag{18.87}$$

with f and τ_0 being two free parameters. With (18.87) the rate equation (18.85) can be solved analytically for nucleus A:

$$\sigma_A n_A = \frac{f n_{56}}{\tau_0} \prod_{i=56}^{A} [1 + (\sigma_i \tau_0)^{-1}]^{-1} \tag{18.88}$$

More details about neutron-capture nucleosynthesis for the interested reader can be found in the reviews by Meyer (1994), Arnould and Takahashi (1999), and Busso et al. (1999).

18.7 Neutrinos

Neutrinos require special consideration because their cross section σ_ν for interaction with matter is so extremely small. For scattering of neutrinos with energy E_ν, one has roughly $\sigma_\nu \approx (E_\nu/m_\mathrm{e}c^2)^2 10^{-44}\ \mathrm{cm}^2$. Neutrinos in the MeV range then have $\sigma_\nu \approx 10^{-44}\ \mathrm{cm}^2$, which is a factor 10^{-18} smaller than the cross section for typical photon–matter interactions. The corresponding mean free path in matter of density $\varrho = n\mu m_\mathrm{u}$ and molecular weight $\mu(\approx 1)$ is about

$$\ell_\nu = \frac{1}{n\sigma_\nu} = \frac{\mu m_\mathrm{u}}{\varrho \sigma_\nu} \approx \frac{2\times 10^{20}\mathrm{cm}}{\varrho}, \tag{18.89}$$

with ϱ in cgs. For "normal" stellar matter with $\varrho \approx 1\,\mathrm{g\,cm^{-3}}$, (18.89) would give a mean free path of the neutrinos of $\ell_\nu \approx 100$ parsec, and even for $\varrho = 10^6\,\mathrm{g\,cm^{-3}}$, one has $\ell_\nu \approx 3000 R_\odot$.

Therefore it is safe to say that neutrinos, once created somewhere in the central region, leave a normal star without interactions carrying away their energy. This neutrino energy has then to be excluded from all other forms of energies (e.g. that released by nuclear reactions), which are subject to some diffusive transport of energy according to the temperature gradient.

The situation can be completely different, however, during a collapse in the final evolutionary stage. The density can reach nuclear values, and for $\varrho = 10^{14}\,\mathrm{g\,cm^{-3}}$, (18.89) gives only $\ell_\nu \approx 20\,\mathrm{km}$. Considering the fact that neutrinos can then be rather energetic (which increases σ_ν appreciably) one sees that many of them will be reabsorbed within the star. Then it is necessary to consider a transport equation for neutrino energy and to evaluate the amount of momentum the interacting neutrinos deliver to the overlying layers (see Sect. 36.3.3).

Only electron neutrinos play a role in stellar interiors, and these can be created in quite different processes inside a star. We first recall those processes involving nuclear reactions, which have already been mentioned (Sect. 18.5) in connection with certain nuclear burnings. In this special case one usually allows for the neutrino energy loss by a corresponding reduction of the released energy [This means that in (10.3) ε_n is reduced and no separate ε_ν term is needed.].

We already encountered this situation in the case of hydrogen burning (Sect. 18.5.1), where two neutrinos per fresh helium nucleus are created. The energy loss due to the escaping neutrinos depends on the particular chain or cycle by which the burning proceeds, but on average the energy yield per cycle is 25 MeV or $\approx 4 \times 10^{-5}$ erg. The generation of one solar luminosity ($L_\odot \approx 4 \times 10^{33}\,\mathrm{erg\,s^{-1}}$) by hydrogen burning implies thus a production of about 2×10^{38} neutrinos per second. Those neutrinos coming directly from the central region of the Sun yield a flux of roughly 10^{11} neutrinos per cm^2 each second at the distance of earth. For experiments measuring the *solar neutrinos* see Sect. 29.5.

There are also neutrino-producing nuclear reactions that are not connected with nuclear burnings. For example, at extreme densities, degenerate electrons can be pushed up to energies large enough for *electron capture* by protons in nuclei of charge Z and atomic weight $A : e^- + (Z, A) \rightarrow (Z - 1, A) + \nu$.

Another interesting example is the so-called *Urca process*. For a suitable nucleus (Z, A), an electron capture occurs which is followed by β decay:

$$\begin{aligned}(Z, A) + e^- &\rightarrow (Z - 1, A) + \nu\,,\\ (Z - 1, A) &\rightarrow (Z, A) + e^- + \bar{\nu}\,.\end{aligned} \tag{18.90}$$

The original particles are restored, and two neutrinos are emitted. There are obvious restrictions on the nuclei (Z, A) suitable for this process: they must have an isobaric nucleus $(Z-1, A)$ of slightly higher energy that is unstable to β decay. A possible example would be $^{35}\mathrm{Cl}\,(e^-, \nu)^{35}\mathrm{S}$ (endothermic with $Q = -0.17\,\mathrm{MeV}$), followed

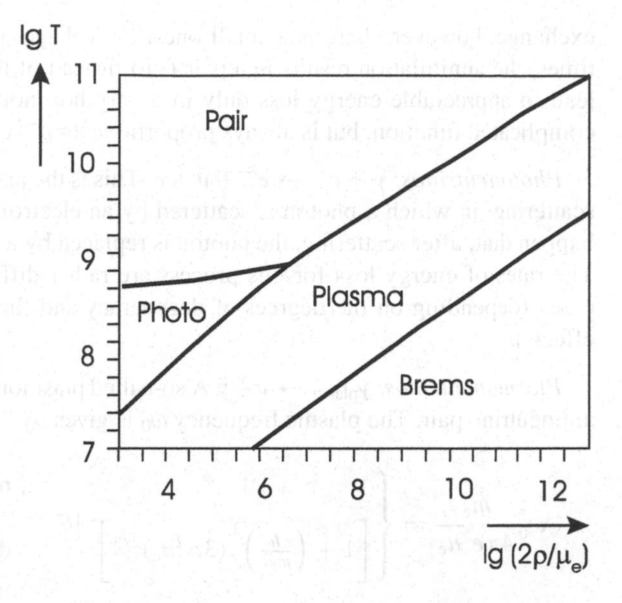

Fig. 18.11 Regions in which different types of neutrino less dominate. The lines indicate where neighbouring processes contribute approximately in equal shares. $2\varrho/\mu_e$ is a suitable quantity proportional to the electron density. It is identical to the mass density if $\mu_e = 2$, for example, in helium cores (After Haft et al. 1994)

by the decay ^{35}S $(e^-\bar{\nu})^{35}$ Cl, the energy for the first reaction being supplied by the captured electron. In this way, thermal energy of the stellar matter is converted into neutrino energy and lost from the star, while the composition remains unchanged (*Urca* is the name of a Rio de Janeiro casino, where Gamow and Schönberg found that, as the only recognizable net effect, similar losses, little by little, occur with visitors' money.). Details depend very much on the stellar material. If appropriate nuclei for this are present, the energy loss will increase with ϱ and T.

The following processes occur *without a nuclear reaction*. These purely leptonic processes were predicted as a consequence of the generalized Fermi theory of weak interaction, which allows a direct electron–neutrino coupling, such that a neutrino pair can be emitted if an electron changes its momentum. It is clear that such processes may be reduced by degeneracy if the electrons do not find enough free cells in phase space.

Several processes of this type can be important for stellar interiors. Figure 18.11 shows the approximate regions of the $\varrho - T$ plane where this is the case. Generally, the energy loss rates are complicated functions of density and temperature. They are calculated from theories of weak interaction and the results obtained as tables, for which approximative analytical fitting formulae are derived, which themselves are too complicated to be reproduced here. A compilation of results is given by Itoh et al. (1996); a somewhat simpler fitting formula for plasma neutrinos was derived by Haft et al. (1994).

Pair Annihilation Neutrinos: $e^- + e^+ \rightarrow \nu + \bar{\nu}$ In very hot environments ($T_9 > 1$), there are enough energetic photons to create large numbers of $(e^- e^+)$ pairs. These will soon be annihilated, usually giving two photons, and a certain equilibrium abundance of e^+ will be reached. In this continuous back and forth exchange, however, there is a small one-way leakage, since roughly once in 10^{19} times, the annihilation results in a pair $(\nu\bar{\nu})$ instead of the usual photons. This can lead to appreciable energy loss only in a very hot, not too dense plasma. ε_ν is a complicated function, but is always proportional to ϱ^{-1}.

Photoneutrinos: $\gamma + e^- \rightarrow e^- + \nu + \bar{\nu}$ This is the analogue of normal Compton scattering, in which a photon is scattered by an electron. In very few cases it may happen that, after scattering, the photon is replaced by a neutrino–antineutrino pair. The rates of energy loss for this process are rather different for different limiting cases (depending on the degrees of degeneracy and the importance of relativistic effects).

Plasmaneutrinos: $\gamma_{\mathrm{plasm}} \rightarrow \nu + \bar{\nu}$ A so-called plasmon decays here to a neutrino–antineutrino pair. The plasma frequency ω_0 is given by

$$\omega_0^2 \frac{m_\mathrm{e}}{4\pi e^2 n_\mathrm{e}} = \begin{cases} 1 & \text{, non-degenerate} \\ \left[1 + \left(\frac{\hbar}{m_\mathrm{e} c}\right)^2 (3\pi^2 n_\mathrm{e})^{2/3}\right]^{-1/2} & \text{, degenerate .} \end{cases} \tag{18.91}$$

This is important for an electromagnetic wave of frequency ω moving through the plasma, since its dispersion relation is

$$\omega^2 = K^2 c^2 + \omega_0^2 , \tag{18.92}$$

where K is the wave number. Here the wave is coupled to the collective motions of the electrons, and a propagating wave can occur only for $\omega > \omega_0$. Multiplication of (18.92) by $\hbar^2$ gives the square of the energy E of a quantum, which therefore behaves as if it were a relativistic particle with a rest mass corresponding to the energy $\hbar\omega_0$. Such a quantum is called a *plasmon.* For the energy rate, one has to add the rates of transversal and longitudinal plasmons: $\varepsilon_\nu^{(\mathrm{plasm})} = \varepsilon_\nu^t + \varepsilon_\nu^l$. The emission rate has an exponential decrease for large ω_0, which is proportional to $\varrho^{1/2}$ at constant T. This comes from the fact that very few plasmons can be excited if kT drops below $\hbar\omega_0$.

Bremsstrahlung Neutrinos Inelastic scattering (deceleration) of an electron in the Coulomb field of a nucleus will usually lead to emission of a "Bremsstrahlung" photon (free–free emission). This photon can be replaced by a neutrino–antineutrino pair. The rate of energy loss for very large ϱ is

$$\varepsilon_\nu^{(\mathrm{brems})} \approx 0.76 \frac{Z^2}{A} T_8^6 , \tag{18.93}$$

(in cgs) where Z and A are the charge and mass number of the nuclei. For smaller densities ε_ν is smaller than this expression, the correction being roughly a factor 10 at $\varrho \approx 10^4\ \mathrm{g\,cm^{-3}}$. This process can dominate, in particular, at low temperature and very high density. The rate $\varepsilon_{\nu}^{(\mathrm{brems})}$ does not decrease with increasing degeneracy (as other processes do), since the lack of free cells in phase space is compensated by an increasing cross section for neutrino emission.

Synchrotron Neutrinos These can only occur in the presence of strong magnetic fields. The normal synchrotron photon emitted by an electron moving in this field is again replaced by a neutrino–antineutrino pair.

Part IV
Simple Stellar Models

While accurate stellar models have to be computed with numerical programmes, for a deeper understanding of stellar properties, general rules and dependencies, and approximative relations, simple stellar models are very useful. They are often based on simplifications of the material functions discussed in Part III or by assuming similarity relations between stars. The polytropes of Chap. 19 were essential for the earliest models of stellar interior but have now gone out of fashion. Nevertheless, we present the definition and the basic properties for those interested in simple models. The homology relations of Chap. 20 are formulated very generally; one usually finds them in more simplified versions, where they are used to derive simple relations like the mass-luminosity-relation for main sequence stars. Simple relations also are useful in clarifying popular misconceptions about stars, such as the assumption that the solar luminosity depends on the nuclear reaction rates (see Sect. 20.2).

In the later sections of this part it will become evident how useful simplified models can be to capture basic principles of stellar structure and evolution. Of course, all these are obtained with high accuracy from numerical solutions, but understanding them is a different issue.

Chapter 19
Polytropic Gaseous Spheres

19.1 Polytropic Relations

As we have seen in Sect. 10.1 the temperature does not appear explicitly in the two mechanical equations (10.1) and (10.2). Under certain circumstances this provides the possibility of separating them from the "thermo-energetic part" of the equations. For the following it is convenient to introduce once again the gravitational potential Φ, as it was defined in Sect. 1.3. We here treat stars in hydrostatic equilibrium, which requires [see (1.11) and (2.3)]

$$\frac{dP}{dr} = -\frac{d\Phi}{dr}\varrho , \tag{19.1}$$

together with Poisson's equation (1.10)

$$\frac{1}{r^2}\frac{d}{dr}\left(r^2\frac{d\Phi}{dr}\right) = 4\pi G\varrho . \tag{19.2}$$

We have replaced the partial derivatives by ordinary ones since only time-independent solutions shall be considered.

In general the temperature appears in the system (19.1) and (19.2) if the density is replaced by an equation of state of the form $\varrho = \varrho(P, T)$. However, we have already encountered examples for simpler cases. If ϱ does not depend on T, i.e. $\varrho = \varrho(P)$ only, then this relation can be introduced into (19.1) and (19.2), which become a system of two equations for P and Φ and can be solved without the other structure equations. An example is the completely degenerate gas of non-relativistic electrons for which $\varrho \sim P^{3/5}$ [see (15.23)].

We shall deal here with similar cases and assume that there exists a simple relation between P and ϱ of the form

$$P = K\varrho^{\gamma} \equiv K\varrho^{1+\frac{1}{n}} , \tag{19.3}$$

R. Kippenhahn et al., *Stellar Structure and Evolution*, Astronomy and Astrophysics Library, DOI 10.1007/978-3-642-30304-3_19, © Springer-Verlag Berlin Heidelberg 2012

where K, γ, and n are constant. A relation of the form (19.3) is called a *polytropic relation*. K is the *polytropic constant* and γ the *polytropic exponent* (which we have to distinguish from the adiabatic exponent γ_{ad}). One often uses, instead of γ, the *polytropic index n,* which is defined by

$$n = \frac{1}{\gamma - 1} \, . \tag{19.4}$$

Obviously for a completely degenerate gas the equation of state in its limiting cases has the polytropic form (19.3). In the non-relativistic limit (15.23) we have $\gamma = 5/3$, $n = 3/2$, while for the relativistic limit (15.26) holds, so that $\gamma = 4/3$, $n = 3$. For such cases, where the equation of state has a polytropic form, the polytropic constant K is fixed and can be calculated from natural constants.

But there are also examples for a relation of the form (19.3) where K is a free parameter which is constant within a particular star but can have different values from one star to another.

Let us consider an isothermal ideal gas of temperature $T = T_0$ and mean molecular weight μ. Its equation of state $\varrho = \mu P/(\Re T)$ can be written in the form (19.3), with $K = \Re T_0/\mu, \gamma = 1$, and $n = \infty$. Here K is not fixed but depends on T_0 and μ, and if we then use (19.3) in the stellar-structure equations, we are free to give K any (positive) value for a certain star.

In a star that is completely convective the temperature gradient (except for that in a region near the surface, which we shall ignore) is given, to a very good approximation, by $\nabla = (d \ln T/d \ln P)_{ad} = \nabla_{ad}$ (see Sect. 7.3). If radiation pressure can be ignored and the gas is completely ionized, we have $\nabla_{ad} = 2/5$ according to (13.12). This means that throughout the star $T \sim P^{2/5}$, and for an ideal gas with μ = constant, $T \sim P/\varrho$, and therefore $P \sim \varrho^{5/3}$. This again is a polytropic relation of the form (19.3) with $\gamma = 5/3, n = 3/2$. But now K is not fixed by natural constants; it is a free parameter in the sense that it can vary from star to star.

The homogeneous gaseous sphere can also be considered a special case of the polytropic relation (19.3). Let us write (19.3) in the form

$$\varrho = K_1 P^{1/\gamma} \, ; \tag{19.5}$$

then $\gamma = \infty$ (or $n = 0$) gives $\varrho = K_1$ = constant.

These examples have shown that we can have two reasons for a polytropic relation in a star. (1) The equation of state is of the simple form $P = K\varrho^\gamma$, with a fixed value of K. (2) The equation of state contains T (as for an ideal gas), but there is an additional relation between T and P (like the adiabatic condition) that together with the equation of state yields a polytropic relation; then K is a free parameter.

On the other hand, if we assume a polytropic relation for an ideal gas, this is equivalent to adopting a certain relation $T = T(P)$. This means that one fixes the temperature stratification instead of determining it by the thermo-energetic equations of stellar structure. For example, a polytrope with $n = 3$ does not necessarily have to consist of relativistic degenerate gases but can also consist of an ideal gas and have $\nabla = 1/(n+1) = 0.25$.

19.2 Polytropic Stellar Models

With the polytropic relation (19.3) (independent of whether K is a free parameter or a constant with a fixed value), (19.1) can be written as

$$\frac{d\Phi}{dr} = -\gamma K \varrho^{\gamma-2} \frac{d\varrho}{dr} . \tag{19.6}$$

If $\gamma \neq 1$ (the case $\gamma = 1, n = \infty$, corresponding to the isothermal model, will be treated in Sect. 19.8), (19.6) can be integrated:

$$\varrho = \left(\frac{-\Phi}{(n+1)K} \right)^n , \tag{19.7}$$

where we have made use of (19.4) and chosen the integration constant to give $\Phi = 0$ at the surface ($\varrho = 0$). Note that in the interior of our model, $\Phi < 0$, giving there $\varrho > 0$. If we introduce (19.7) into the right-hand side of the Poisson equation (19.2), we obtain an ordinary differential equation for Φ:

$$\frac{d^2\Phi}{dr^2} + \frac{2}{r}\frac{d\Phi}{dr} = 4\pi G \left(\frac{-\Phi}{(n+1)K} \right)^n . \tag{19.8}$$

We now define dimensionless variables z, w by

$$z = Ar , \quad A^2 = \frac{4\pi G}{(n+1)^n K^n} (-\Phi_c)^{n-1} = \frac{4\pi G}{(n+1)K} \varrho_c^{\frac{n-1}{n}} ,$$

$$w = \frac{\Phi}{\Phi_c} = \left(\frac{\varrho}{\varrho_c} \right)^{1/n} , \tag{19.9}$$

where the subscript c refers to the centre and where the relation between ϱ and Φ is taken from (19.7). At the centre ($r = 0$) we have $z = 0$, $\Phi = \Phi_c$, $\varrho = \varrho_c$, and therefore $w = 1$. Then (19.8) can be written as

$$\frac{d^2w}{dz^2} + \frac{2}{z}\frac{dw}{dz} + w^n = 0 ,$$

$$\frac{1}{z^2}\frac{d}{dz}\left(z^2 \frac{dw}{dz} \right) + w^n = 0 . \tag{19.10}$$

This is the famous *Lane–Emden equation* (named after J.H. Lane and R. Emden). We are only interested in solutions that are finite at the centre, $z = 0$. Equation (19.10) shows that we then have to require $dw/dz \equiv w' = 0$. Let us assume we have a solution $w(z)$ of (19.10) that fulfils the central boundary conditions $w(0) = 1$ and $w'(0) = 0$; then according to (19.9) the radial distribution of the density is given by

$$\varrho(r) = \varrho_c w^n , \quad \varrho_c = \left[\frac{-\Phi_c}{(n+1)K}\right]^n . \tag{19.11}$$

For the pressure we obtain from (19.3) and (19.4) that $P(r) = P_c w^{n+1}$, where $P_c = K\varrho_c^\gamma$.

Before trying to construct stellar polytropic models we shall discuss some of the mathematical properties of the solutions $w(z)$ of (19.10).

19.3 Properties of the Solutions

The Lane–Emden equation has a regular singularity at $z = 0$. In order to understand the behaviour of the solutions there, we expand into a power series:

$$w(z) = 1 + a_1 z + a_2 z^2 + a_3 z^3 + \dots , \tag{19.12}$$

with $a_1 = w'(0), 2a_2 = w''(0), \dots$. Since the gravitational acceleration $|g| = d\Phi/dr \sim dw/dz$ must vanish in the centre, we have $a_1 = 0$. Inserting (19.12) into the Emden equation (19.10), by comparing coefficients one finds

$$w(z) = 1 - \frac{1}{6}z^2 + \frac{n}{120}z^4 + \dots , \tag{19.13}$$

where again we have excluded the isothermal sphere $n = \infty$. Equation (19.13) shows that $w(z)$ has a maximum at $z = 0$.

Only for three values of n can the solutions be given by analytic expressions. The first case is

$$n = 0 : \qquad w(z) = 1 - \frac{1}{6}z^2 , \tag{19.14}$$

and we have already mentioned that this corresponds to the homogeneous gas sphere. Indeed $\varrho = \varrho_c w^n$ gives constant density for $n = 0$. The two other cases are

$$n = 1 : \qquad w(z) = \frac{\sin z}{z} , \tag{19.15}$$

$$n = 5 : \qquad w(z) = \frac{1}{(1 + z^2/3)^{1/2}} . \tag{19.16}$$

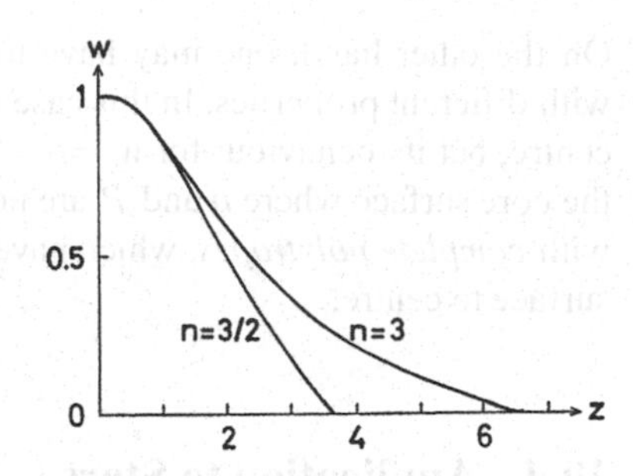

Fig. 19.1 If $n < 5$ the solution of the Lane–Emden equation (19.10) of index n starting with $w(0) = 1$ becomes zero at a finite value of $z = z_n$. Here the solutions for $n = 3/2$ and $n = 3$ are plotted

Table 19.1 Numerical values for polytropic models with index n (after Chandrasekhar 1939)

n	z_n	$\left(-z^2 \frac{dw}{dz}\right)_{z=z_n}$	$\varrho_c/\bar{\varrho}$
0	2.4494	4.8988	1.0000
1	3.14159	3.14159	3.28987
1.5	3.65375	2.71406	5.99071
2	4.35287	2.41105	11.40254
3	6.89685	2.01824	54.1825
4	14.97155	1.79723	622.408
4.5	31.8365	1.73780	6,189.47
5	∞	1.73205	∞

The surface of the polytrope of index n is defined by the value $z = z_n$, for which $\varrho = 0$ and thus $w = 0$. While for $n = 0$ and $n = 1$ the surface is obviously reached for a finite value of z_n, the case $n = 5$ yields a model of infinite radius. It can be shown that for $n < 5$ the radius of polytropic models is finite; for $n \geq 5$ they have infinite radius. This also holds for the limiting case $n = \infty$ (cf. Sect. 19.8).

Apart from the three cases where analytic solutions are known, the Emden equation (19.10) has to be solved numerically, beginning with the expansion (19.13) for the neighbourhood of the centre. Here the solution starts with zero tangent and $w = 1$ and decreases outwards. This can be seen from (19.13) and is illustrated in Fig. 19.1.

For a given value of $n < 5$ the integration comes to a point z where $w(z)$ vanishes, i.e. $\varrho = 0$. This value of z, which corresponds to the surface of the polytrope, will be called z_n. From (19.14)–(19.16) one finds $z_0 = \sqrt{6}, z_1 = \pi, z_5 = \infty$. It is a general property of the solutions that z_n grows monotonically with the polytropic index n. Table 19.1 gives some values of z_n and the values of certain functions at $z = z_n$ which will later turn out to be useful for the construction of models.

So far, we have discussed only solutions that are regular at the centre. But solutions with a singularity at $z = 0$ can also be important if one uses them for stellar regions outside the centre. Let us, for instance, consider a star that is convective in its outer layer, while in the inner part, the energy may be transported by radiation. If the convective envelope is adiabatic, with $\nabla = \nabla_{\mathrm{ad}} = 2/5$, it is polytropic and therefore $\varrho \sim w^{3/2}$ and $P \sim w^{5/2}$. But it is unimportant whether this solution is finite at the centre, since anyway the equations do not hold in the radiative interior.

On the other hand, one may have to fit a polytropic central core to an envelope with different properties. In this case the polytropic solution has to be regular at the centre, but its behaviour for $w = \varrho = 0$ is unimportant, since it is used only up to the core surface where ϱ and P are non-vanishing. In the following we mainly deal with *complete polytropes*, which have a polytropic relation of the form (19.3) from surface to centre.

19.4 Application to Stars

We now construct polytropic models for a given index $n < 5$ and for given values of M and R. This will turn out to be possible as long as K is not fixed by the equation of state. We first derive some more relations for polytropes.

From (10.1) and (19.11) it follows that

$$m(r) = \int_0^r 4\pi \varrho r^2 dr = 4\pi \varrho_c \int_0^r w^n r^2 dr = 4\pi \varrho_c \frac{r^3}{z^3} \int_0^z w^n z^2 dz\,, \qquad (19.17)$$

where we have made use of relations (19.9) and of the fact that r^3/z^3 is constant and can be brought in front of the integral. According to the Lane–Emden equation (19.10) the integrand $w^n z^2$ on the right is a derivative and can immediately be integrated, so that the integral becomes $-z^2 dw/dz$. We obtain

$$m(r) = 4\pi \varrho_c r^3 \left(-\frac{1}{z}\frac{dw}{dz}\right), \qquad (19.18)$$

where the simultaneously appearing z and r are related to each other by $r/z = 1/A = R/z_n$. For the special case of the surface, we have

$$M = 4\pi \varrho_c R^3 \left(-\frac{1}{z}\frac{dw}{dz}\right)_{z=z_n}. \qquad (19.19)$$

The quantity in brackets can be derived from Table 19.1 for several values of n. If we introduce the mean density $\bar{\varrho} := 3M/(4\pi R^3)$, we find

$$\frac{\bar{\varrho}}{\varrho_c} = \left(-\frac{3}{z}\frac{dw}{dz}\right)_{z=z_n}. \qquad (19.20)$$

The right-hand side of this equation depends only on n: for $n = 0$ it is 1–as one can see from (19.11). The higher n, the smaller $\bar{\varrho}/\varrho_c$, which means the higher the density concentration, as can be seen in Table 19.1.

We now have all the means at hand to construct the whole polytropic stellar model for given values of n, M, and R for the case that K is not fixed by the equation of state.

If n is given, a numerical solution of the Lane–Emden equation (19.10) yields the functions $w(z)$, $w'(z)$ and the values of z_n and of $-z_n/(3dw/dz)_n$. If we now use M and R to determine the mean density $\bar{\varrho}$, (19.20) gives ϱ_c. On the other hand, we know the constant $A = z/r = z_n/R$ by which we adjust the dimensionless z scale to the r scale. We therefore know the density distribution in the model $\varrho(r) = \varrho_c w^n(z)$ from (19.11). With ϱ_c and the constant A we can determine K from (19.9) and obtain the pressure distribution $P(r) = K\varrho^{(n+1)/n} = K\varrho_c^{(n+1)/n} w^{n+1}$. The local mass m then follows from (19.18) and the (known) relation between the z scale and the r scale. The whole mechanical structure is now determined. It has to be emphasized that this method of constructing models for given values of n, M, and R is only applicable if K is a free parameter, otherwise the problem would be overdetermined (The case that K has fixed value will be discussed in Sect. 19.6.).

As an example we try to construct a polytropic model of index 3 for the Sun ($M = 1.989 \times 10^{33}$g, $R = 6.96 \times 10^{10}$ cm). For $n = 3$ Table 19.1 gives $z_3 = 6.897$, $\varrho_c/\bar{\varrho} = 54.18$. The mean density becomes $\bar{\varrho} = 1.41\,\mathrm{g\,cm^{-3}}$; consequently the central density $\varrho_c = 76.39\,\mathrm{g\,cm^{-3}}$ and, further, $A = z_3/R = 9.91\times10^{-11}$. From (19.9) we find $K = 3.85 \times 10^{14}$ and consequently $P_c = 1.24 \times 10^{17}$ dyn/cm^2. For the ideal gas equation with $\mu = 0.62$ corresponding to $X \approx 0.7, Y \approx 0.3$ we find for the temperature $T_c = 1.2 \times 10^7$ K. A proper numerical solution of the full set of stellar-structure equations for a chemically homogeneous model of $1M_\odot$ gives $T_c = 1.5\times10^7$ K. We see that a polytropic estimate with $n = 3$ comes considerably closer to the honestly computed value than our crude estimate in Sect. 2.3.

19.5 Radiation Pressure and the Polytrope $n = 3$

We consider here only the case that K is a free parameter. In the example at the end of the previous section we approximated the Sun by a polytrope of $n = 3$. This is formally equivalent to the assumption of an ideal gas ($P \sim \varrho T$) together with a constant temperature gradient $\nabla = 1/4 (T \sim P^{1/4})$. We will now show that this polytropic relation with $n = 3$ can also be obtained by a certain assumption on the radiation pressure. For an ideal gas with radiation pressure

$$P = \frac{\Re}{\mu}\varrho T + \frac{a}{3}T^4 = \frac{\Re}{\mu\beta}\varrho T, \tag{19.21}$$

we assume that the ratio $\beta = P_{\mathrm{gas}}/P$ is constant throughout the star. Now

$$1 - \beta = \frac{P_{\mathrm{rad}}}{P} = \frac{aT^4}{3P} \tag{19.22}$$

shows that $\beta =$ constant means a relation of the form $T^4 \sim P$, which we introduce into (19.21). This gives

$$P = \left(\frac{3\Re^4}{a\mu^4}\right)^{1/3} \left(\frac{1-\beta}{\beta^4}\right)^{1/3} \varrho^{4/3} , \tag{19.23}$$

which indeed is a polytropic relation with $n = 3$ for constant β. Here the polytropic constant K is again a free parameter, since we can choose β in the interval 0, 1.

In Sect. 19.10 we shall apply this to very massive stars. They are fully convective ($\nabla = \nabla_{ad}$) and dominated by radiation pressure.

Relation (19.23) goes back to A.S. Eddington, who obtained it for his famous "standard model". He found that the full set of stellar-structure equations (including the thermo-energetic equations) could be solved very simply by the assumption $\kappa l/m =$ constant throughout the star. One then obtains $\beta =$ constant and therefore the polytropic relation (19.23).

19.6 Polytropic Stellar Models with Fixed *K*

As a typical example we have already mentioned the non-relativistic degenerate electron gas for which the equation of state (15.23) is polytropic with $n = 3/2$ and polytropic constant

$$K = \frac{1}{20}\left(\frac{3}{\pi}\right)^{2/3} \frac{h^2}{m_e} \frac{1}{(\mu_e m_u)^{5/3}} . \tag{19.24}$$

We consider the chemical composition to be given (μ_e fixed). Then in this expression there is no room for the choice of a free parameter as in (19.23). Although $n = 3/2$ is a particularly interesting case, we shall derive our relation for general values of the polytropic index with $n < 5$.

Let us see how to construct a model with index n for a given value of ϱ_c. The functions $w(z)$ and $w'(z)$ can be considered known from an integration of the Emden equation. Then $\varrho = \varrho_c w^n$ is known as a function of z. According to (19.9) the relation between r and z is

$$\left(\frac{r}{z}\right)^2 = \frac{1}{4\pi G}(n+1)K\varrho_c^{\frac{1-n}{n}} . \tag{19.25}$$

This can be used to derive the density also as a function of r, where the radius of the model is $R = z_n/A$ and the value z_n is obtained from the integration. The constant A depends on ϱ_c, as shown by (19.25), and

$$R \sim \varrho_c^{\frac{1-n}{2n}} . \tag{19.26}$$

As long as $n > 1$, the radius R becomes smaller with increasing central density ϱ_c, becoming zero for infinite ϱ_c. On the other hand, the mass M of the model varies with ϱ_c according to (19.19) as $M \sim \varrho_c R^3$ or

$$M = C_1 \varrho_c^{\frac{3-n}{2n}} \; ; \quad C_1 = 4\pi \left(-\frac{w'}{z}\right)_{z_n} z_n^3 \left(\frac{n+1}{4\pi G}\right)^{3/2} K^{3/2} \,. \tag{19.27}$$

Elimination of ϱ_c from (19.26) and (19.27) shows that there is a mass-radius relation of the form

$$R \sim M^{\frac{1-n}{3-n}} \,. \tag{19.28}$$

We see that for given K and n there is a one-dimensional manifold of models only, the parameter being *either M or R* (or ϱ_c), whereas there was a two-dimensional manifold *(M and R* as parameters) when K was a free parameter.

Consider again the case of the non-relativistic degenerateX electron gas, which is not too bad an approximation for white dwarfs of small mass. With $n = 3/2$, (19.28) gives $R \sim M^{-1/3}$ and the surprising result that the larger the mass the smaller the radius (This is made plausible by simple considerations in Sect. 37.1.). The model will shrink with increasing mass and should finally end as a point mass for infinite M. But long before this, our assumed equation of state will not be valid any more, since from (19.27) we see that ϱ_c is proportional to $\sim M^2$. For ever-increasing densities the electrons will become relativistic (see Sect. 16.2), and the equation of state (15.23) has to be replaced by (15.26). This means a transition from a polytrope $n = 3/2$ to one with $n = 3$ (and a different, but also given, polytropic constant K). In this case we shall encounter a new problem, hinted at by the exponent in (19.28).

19.7 Chandrasekhar's Limiting Mass

In Sect. 19.6 we have seen that a polytropic model in which the pressure is provided by a non-relativistic degenerate electron gas reaches higher central and mean densities with growing total mass M. But with increasing density the electrons become gradually more relativistic. This starts in the central region where the density is highest, the outer parts remaining non-relativistic. Although we know that the transition between equations of state (15.23) and (15.26) does not occur abruptly, but smoothly via the more general equation of state (15.13), one can imagine that an idealized stellar model consisting of degenerate matter can be constructed by fitting two regions smoothly together: a (relativistic) polytropic core with $n = 3$ surrounded by a (non-relativistic) polytropic envelope with $n = 3/2$. Indeed Chandrasekhar constructed his first white-dwarf model in this way.

Let us consider how this idealized model changes with growing mass M. At small M the whole model is still non-relativistic. The relativistic core will occur for $\varrho_c \gtrsim 10^6\,\mathrm{g\,cm^{-3}}$ (Fig. 16.1) and gradually encompass larger parts of the model as ϱ_c increases. One would therefore expect the model finally to approach the state where all its mass (except a small surface region) is relativistic, so that a polytrope of index $n = 3$ would describe the whole model properly; however, there is a difficulty. As one can see from (19.27) the mass does not vary with central density in the case of a polytrope of index $n = 3$ if K is fixed. In this case, (19.27) gives $M = C_1$:

$$M = 4\pi \left(-\frac{w'}{z}\right)_{z_3} z_3^3 \left(\frac{K}{\pi G}\right)^{3/2} . \tag{19.29}$$

This is the only possible mass for relativistic degenerate polytropes and is called the *Chandrasekhar mass*, which after insertion of the proper numerical values yields

$$M_{\mathrm{Ch}} = \frac{5.836}{\mu_{\mathrm{e}}^2} M_\odot . \tag{19.30}$$

We therefore can expect that our series of models constructed by fitting an $n = 3/2$ envelope to an $n = 3$ core finds its end at a critical total mass $M = M_{\mathrm{Ch}}$ as given by (19.30). Or in other words our models of increasing central density tend to a finite mass and approach zero radius for $\varrho_{\mathrm{c}} \to \infty$. Of course, this final state is physically unrealistic, since the equation of state is changed by different effects at very high density (see Chaps. 16, 37 and 38).

Although we have discussed the problem only from the standpoint of polytropic models, the result for M_{Ch} remains numerically the same if one uses Chandrasekhar's more general equation of state (15.13) (compare the treatment in Sect. 37.1. The reason is that for extremely high density, (15.13) approaches the polytropic relation (19.3) with $\gamma = 4/3$ or $n = 3$.

It is surprising that the limiting mass not only is finite, but that it is so small that many stars exceed it. But their equation of state is not dominated by degenerate electrons, and therefore Chandrasekhar's limiting mass (19.30) has no meaning for them. White dwarfs seem to be formed of material where all the hydrogen is transformed into helium, carbon, or oxygen, such that we expect $\mu_{\mathrm{e}} = 2$ and therefore $M_{\mathrm{Ch}} = 1.46 M_\odot$. Indeed no white dwarf has been found which exceeds this mass.

In the above considerations we have approached the relativistic degenerate polytrope by way of a sequence with $\varrho_{\mathrm{c}} \to \infty$ (and consequently $R \to 0$). However, this polytrope is a particular case: we have already mentioned that according to (19.27) M and ϱ_{c} are then no longer coupled. In other words, for $M = M_{\mathrm{Ch}}$, the central density can be arbitrary (and therefore also the radius R), i.e. there is a whole series of relativistic degenerate polytropes (having ϱ_{c} or R as parameter) that all have the same mass M_{Ch}. This is a case of neutral equilibrium (see Sect. 25.3.2).

19.8 Isothermal Spheres of an Ideal Gas

We now deal with the case $\gamma = 1$ or $n = \infty$, which we omitted in Sect. 19.2. Here $K = \Re T/\mu$ is a free parameter. If $\gamma = 1$, integration of (19.6) gives

$$-\frac{\Phi}{K} = \ln \varrho - \ln \varrho_{\mathrm{c}} , \tag{19.31}$$

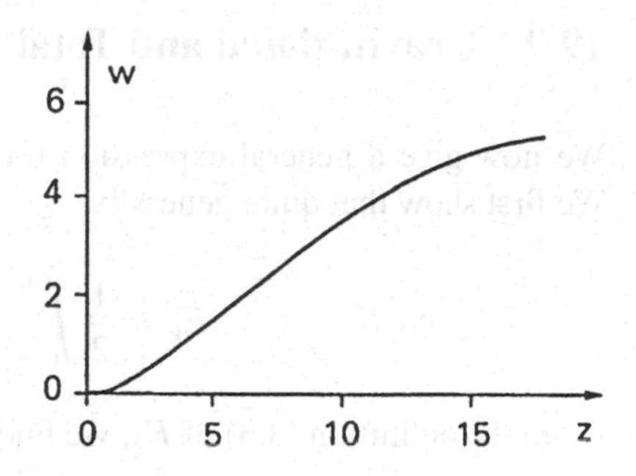

Fig. 19.2 The solution of the Lane–Emden equation (19.35) for the case of an isothermal ideal gas ($n = \infty$)

where we have now chosen the constant of integration in such a way that the gravitational potential is zero at the centre and positive outside it. With

$$\varrho = \varrho_c e^{-\Phi/K} \tag{19.32}$$

and with the Poisson equation (19.2) we find

$$\frac{d^2\Phi}{dr^2} + \frac{2}{r}\frac{d\Phi}{dr} = 4\pi G \varrho_c e^{-\Phi/K} . \tag{19.33}$$

We now introduce dimensionless variables z, w by

$$z = Ar \ , \quad A^2 = \frac{4\pi G \varrho_c}{K} \ , \qquad \Phi = Kw \tag{19.34}$$

and obtain the "isothermal" Lane–Emden equation

$$\frac{d^2w}{dz^2} + \frac{2}{z}\frac{dw}{dz} = e^{-w} , \tag{19.35}$$

which now has to be integrated with the central conditions

$$w(0) = 0 \ , \qquad \left(\frac{dw}{dz}\right)_{z=0} = 0 . \tag{19.36}$$

Again, a power series expansion can be derived and has to be used to describe the behaviour near the centre. The solution is given in Fig. 19.2.

As already mentioned, the isothermal sphere consisting of an ideal gas has an infinite radius, like all polytropes of $n \geq 5$. It also has an infinite mass. Certainly there can be no such stars, but polytropes with $n = \infty$ can be used in order to construct models with non-degenerate isothermal cores. Such models play a role in connection with the so-called Schönberg–Chandrasekhar limit (see Sect. 30.5).

19.9 Gravitational and Total Energy for Polytropes

We now give a general expression for the gravitational energy E_g of polytropes. We first show that quite generally

$$E_g = \frac{1}{2}\int_0^M \Phi \, dm - \frac{1}{2}\frac{GM^2}{R} . \tag{19.37}$$

From the definition (3.3) of E_g, we find

$$E_g = -G\int_0^M \frac{m}{r} dm = -\frac{1}{2}\frac{GM^2}{R} - \frac{1}{2}G\int_0^R \frac{m^2}{r^2} dr , \tag{19.38}$$

where the last expression has been obtained by partial integration and where we have used the fact that m/r vanishes at the centre. But on the other hand

$$\frac{d\Phi}{dr} = \frac{Gm}{r^2} \tag{19.39}$$

and therefore

$$\begin{aligned} E_g &= -\frac{1}{2}\frac{GM^2}{R} - \frac{1}{2}\int_0^R \frac{d\Phi}{dr} m dr \\ &= -\frac{1}{2}\frac{GM^2}{R} + \frac{1}{2}\int_0^M \Phi \, dm , \end{aligned} \tag{19.40}$$

where again we have integrated partially and used the fact that $m\Phi$ vanishes at the centre ($m = 0$) and at the surface [$\Phi = 0$, according to our choice of the integration constant in connection with (19.7)], so we have indeed recovered (19.37). For a polytrope we can use (19.3), (19.7) and write

$$\Phi = -\frac{K\gamma}{\gamma - 1}\varrho^{\gamma-1} = -\frac{\gamma}{\gamma - 1}\frac{P}{\varrho} \tag{19.41}$$

and therefore, with (19.37),

$$E_g = -\frac{1}{2}\frac{GM^2}{R} - \frac{1}{2}\frac{\gamma}{\gamma - 1}\int_0^M \frac{P}{\varrho} dm . \tag{19.42}$$

According to (3.2) and (3.3) the last term on the right can be expressed by E_g. If we replace γ by n, then

$$E_g = -\frac{1}{2}\frac{GM^2}{R} + \frac{1}{6}(n + 1)E_g \tag{19.43}$$

and therefore

$$E_g = -\frac{3}{5-n}\frac{GM^2}{R} . \tag{19.44}$$

We now derive a similar expression for the internal energy E_i. In (3.8) we defined a quantity ζ by

$$\zeta := 3P/(\varrho u) \tag{19.45}$$

(u = internal energy per mass unit).

We saw that for an ideal gas

$$\zeta = 3(\gamma_{ad} - 1) . \tag{19.46}$$

This relation also holds for a more general equation of state as long as ζ is constant.

In order to show this, we take the total differentials from (19.45) and obtain

$$\zeta\, du = 3\frac{dP}{\varrho} - 3\frac{P}{\varrho^2}d\varrho . \tag{19.47}$$

We now assume that the differentials describe adiabatic changes. The first law of thermodynamics gives

$$du = \frac{P}{\varrho^2}d\varrho . \tag{19.48}$$

Then with

$$\gamma_{ad} = \frac{\varrho}{P}\frac{dP}{d\varrho} , \tag{19.49}$$

(19.47) yields

$$\zeta = 3\frac{\varrho}{P}\frac{dP}{d\varrho} - 3 = 3(\gamma_{ad} - 1) . \tag{19.50}$$

For an ideal gas with $\gamma_{ad} = 5/3$ one has $\zeta = 2$, while for an ideal gas with $\gamma_{ad} = 4/3, \zeta = 1$. In the case of a gas dominated by radiation pressure ($P = aT^4/3$ and $u = aT^4$) one finds $\zeta = 1$. Assuming ζ to be constant throughout the star and using (19.44) we find with (3.9)

$$E_i = -\frac{1}{\zeta}E_g = \frac{3}{\zeta(5-n)}\frac{GM^2}{R} . \tag{19.51}$$

The total energy then becomes

$$W = E_i + E_g = \frac{3}{5-n}\left(\frac{1}{\zeta} - 1\right)\frac{GM^2}{R} . \tag{19.52}$$

We can conclude from (19.52) that the total energy for a polytrope of finite radius vanishes when $\zeta = 1$ and in particular for the above cases of an ideal gas with $\gamma_{ad} = 4/3$ and a radiation-dominated gas.

19.10 Supermassive Stars

Let us consider an ideal gas with radiation pressure and assume that $\beta = P_{gas}/P =$ constant throughout the star. We have seen in (19.23) that this yields a polytrope with $n = 3$.

Relation (19.23) defines the polytropic constant K:

$$K = \left(\frac{3\Re^4}{a\mu^4}\right)^{1/3}\left(\frac{1-\beta}{\beta^4}\right)^{1/3} . \tag{19.53}$$

On the other hand, from (19.9) for $n = 3$ we have

$$K = \pi G \varrho_c^{2/3}\frac{R^2}{z_3^2} , \tag{19.54}$$

where we have used $A = z_3/R$. The numerical value of z_3 is 6.897 (Table 19.1). With (19.20) ϱ_c can be expressed by M and R:

$$\varrho_c = 54.18\bar{\varrho} = 54.18\frac{3M}{4\pi R^3} = c_1\frac{M}{R^3} , \tag{19.55}$$

where we have taken the numerical value from Table 19.1. From (19.53) we eliminate K with (19.54) and then ϱ_c with (19.55) and obtain "Eddington's quartic equation":

$$\frac{1-\beta}{\mu^4\beta^4} = \frac{a}{3\Re^4}\frac{(\pi G)^3 c_1^2}{z_3^6}M^2 = 3.02\times10^{-3}\left(\frac{M}{M_\odot}\right)^2 . \tag{19.56}$$

In the interval $0 \le \beta \le 1$ the left-hand side is a monotonically decreasing function of β, which therefore becomes smaller with growing M; this means that radiation pressure becomes the more important the larger the stellar mass.

For a pure hydrogen star of $10^6 M_\odot$ and $\mu = 0.5$, (19.56) gives $(1-\beta)/\beta^4 = 1.9\times10^8$, or $\beta \approx 0.0086$.

Supermassive stars are therefore dominated by radiation pressure. One consequence is that ∇_{ad} is appreciably reduced [$\nabla_{ad} \to 1/4$, for $\beta \to 0$; see (13.12)] and the star becomes convective with $\nabla = \nabla_{ad}$. This can also be seen from an extrapolation of the main-sequence models towards large M (Sect. 22.3). The adiabatic structure requires constant specific entropy s. For a gas dominated by radiation pressure (the density being determined by the gas, the pressure by the photons) the energy u per mass unit and the pressure are given by

$$u = \frac{aT^4}{\varrho} , \quad P = \frac{a}{3}T^4 . \tag{19.57}$$

Then with the first law of thermodynamics we have

$$d_s = \frac{dq}{T} = \frac{1}{T}\left(du - \frac{P}{\varrho^2}d\varrho\right)$$

$$= \frac{4aT^2}{\varrho}dT - \frac{4aT^3}{3\varrho^2}d\varrho = d\left(\frac{4aT^3}{3\varrho}\right) \tag{19.58}$$

and

$$s = \frac{4aT^3}{3\varrho} . \tag{19.59}$$

Constant specific entropy means $\varrho \sim T^3$, which together with the pressure equation $P \sim T^4$ immediately gives $P \sim \varrho^{4/3}$. Indeed supermassive stars are polytropic with $n = 3$ as we assumed initially.

The supermassive star polytropes have a free K, which means that M can be chosen arbitrarily (in contrast to the relativistic degenerate polytrope of the same index, where K and M were fixed). For each mass, $(1 - \beta)/(\mu\beta)^4$ can be obtained from (19.56), and then (19.53) gives the corresponding value of K. But if the mass is given, there still exists an infinite number of models for different R. This is possible in spite of the fact that K is already determined by M: since according to (19.55) $\varrho_c \sim \bar{\varrho} \sim M/R^3$, (19.54) shows K to be independent of R. This is typical for the polytropic index $n = 3$.

Equation (19.59) shows that for an adiabatic change ($ds = 0$) of a given mass element $\varrho \sim T^3$, and therefore with (19.57) $P \sim \varrho^{4/3}$ or $\gamma_{ad} = 4/3$. Then $\zeta = 1$ and (19.52) gives the total energy of the model $W = 0$. The supermassive configuration is in neutral equilibrium. No energy is needed to compress or expand it. In Chap. 25 we will find that $\gamma_{ad} = 4/3$ corresponds to the case of marginal dynamical stability. There a simple interpretation is given for this peculiar behaviour.

19.11 A Collapsing Polytrope

Up to now we have only treated polytropic gaseous spheres in hydrostatic equilibrium. One can also find solutions for polytropes of $n = 3$ for which the inertia term, neglected in (19.1), is important (Goldreich and Weber 1980). Then (19.1) has to be replaced by

$$\frac{\partial v_r}{\partial t} + v_r\frac{\partial v_r}{\partial r} + \frac{1}{\varrho}\frac{\partial P}{\partial r} + \frac{\partial \Phi}{\partial r} = 0 , \tag{19.60}$$

with $v_r = \partial r/\partial t$.

Let us consider a relativistic degenerate polytrope with $n = 3$, or $\gamma = \gamma_{ad} = 4/3$. In a manner similar to that of Sect. 19.2 we define a dimensionless length-scale z by

$$r = a(t)z , \quad v_r = \dot{a}z \tag{19.61}$$

such that z is time independent, the whole time dependence of r being contained in $a(t)$ [Note that a corresponds to $1/A$ in (19.9)]. The form (19.61) describes a homologous change (compare with Sect. 20.3). If we introduce a velocity potential ψ by $v_r = \partial\psi/\partial r$, we can write

$$av_r = a\dot{a}z = a\frac{\partial\psi}{\partial r} = \frac{\partial\psi}{\partial z}\,, \quad \psi = \frac{1}{2}a\dot{a}z^2\,, \tag{19.62}$$

where we have fixed the constant of integration in the velocity potential by $\psi = 0$ at $z = 0$. Note that the time derivative of ψ in the comoving frame is

$$\frac{d\psi}{dt} = \frac{\partial\psi}{\partial t} + v_r\frac{\partial\psi}{\partial r} = \frac{\partial\psi}{\partial t} + (\dot{a}z)^2\,. \tag{19.63}$$

With the new variables, Poisson's equation (19.2) can be written as

$$\frac{1}{z^2}\frac{\partial}{\partial z}\left(z^2\frac{\partial\psi}{\partial z}\right) = 4\pi G\varrho a^2\,, \tag{19.64}$$

while the continuity equation (1.4) becomes with (19.62)

$$\frac{1}{\varrho}\frac{d\varrho}{dt} + \frac{1}{z^2a^2}\frac{\partial}{\partial z}\left(z^2\frac{\partial\psi}{\partial z}\right) \equiv \frac{1}{\varrho}\frac{d\varrho}{dt} + 3\frac{\dot{a}}{a} = 0\,. \tag{19.65}$$

This means that $\varrho \sim a^{-3}$ (in the comoving frame), a result that is obvious from (19.61). As in (19.9) we define $w(z)$ by $\varrho = \varrho_c w^3(z)$. This $w(z)$ will turn out to be related to the Emden function of index 3, as we shall see later. Note that ϱ_c is a function of time. In order to stay as close as possible to the formalism of hydrostatic equilibrium, we fix $a = r/z$ [rather as we did with $1/A$ in (19.9)] by

$$\frac{1}{a^2} = \frac{\pi G}{K}\varrho_c^{2/3} \tag{19.66}$$

such that

$$\varrho = \varrho_c w^3(z) = \left(\frac{K}{\pi G}\right)^{3/2}\frac{1}{a^3}w^3(z)\,. \tag{19.67}$$

We now come to the equation of motion and define

$$h := \int\frac{dP}{\varrho} = 4K\varrho^{1/3}\,, \tag{19.68}$$

where we have made use of (19.3) for $\gamma = 4/3$. Inserting ψ and h from (19.62) and (19.68) into the equation of motion (19.60) gives

$$\frac{\partial^2 \psi}{\partial r \partial t} + \frac{1}{2}\frac{\partial}{\partial r}\left(\frac{\partial \psi}{\partial r}\right)^2 + \frac{\partial \Phi}{\partial r} + \frac{\partial h}{\partial r} = 0\,, \tag{19.69}$$

which can be integrated with respect to r. If we set the integration constant to zero, replace $\partial\psi/\partial r$ by $\dot{a}z$, and consider (19.63), we find that

$$\frac{d\psi}{dt} = \frac{1}{2}\dot{a}^2 z^2 - \Phi - h \tag{19.70}$$

and therefore with (19.62)

$$\frac{1}{2} a\ddot{a} z^2 = -\Phi - h\,. \tag{19.71}$$

From (19.67) and (19.68) follows

$$h = 4K\varrho^{1/3} = 4\frac{K^{3/2}}{(\pi G)^{1/2}}\frac{1}{a}w(z)\,. \tag{19.72}$$

We try a similar dependence of Φ on t and write

$$\Phi = 4\frac{K^{3/2}}{(\pi G)^{1/2}}\frac{1}{a}g(z)\,, \tag{19.73}$$

which defines the dimensionless function $g(z)$. If we insert (19.72) and (19.73) into (19.71) we find

$$\frac{1}{2}a^2\ddot{a} = -\frac{4K^{3/2}}{(\pi G)^{1/2}}(g + w)\frac{1}{z^2}\,. \tag{19.74}$$

Since the left-hand side is a function of t only and the right-hand side is a function of z only, both sides must be constant; therefore

$$\frac{3}{4}\frac{(\pi G)^{1/2}}{K^{3/2}}a^2\ddot{a} = -\lambda\,, \tag{19.75}$$

$$6\frac{g + w}{z^2} = \lambda \tag{19.76}$$

(λ = constant). The first of these equations can be integrated twice. After multiplication with $\dot{a}/a^2$, the first integration gives

$$\dot{a}^2 = \frac{8}{3}\lambda\left(\frac{K^3}{\pi G}\right)^{1/2}\frac{1}{a}\,, \tag{19.77}$$

where the constant of integration is set equal to zero (assuming a zero velocity when the sphere is expanded to infinity). Multiplication of (19.77) with a gives

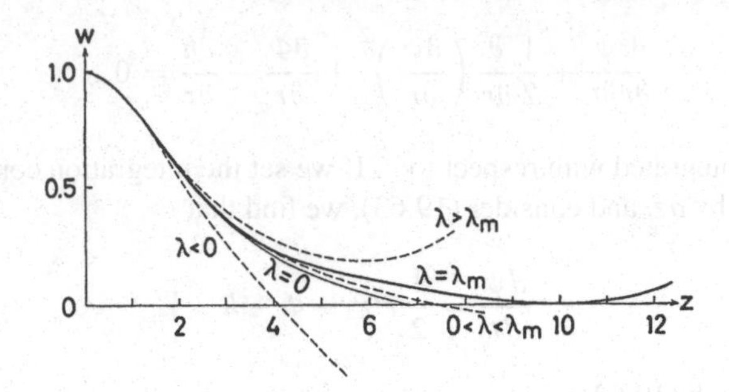

Fig. 19.3 Solutions of (19.81) for different values of λ. In the range $0 < \lambda \leq \lambda_m$, they describe homologously collapsing polytropes of index 3. The solution for $\lambda = \lambda_m$ reaches the abscissa with slope zero. The *broken lines* indicate the behaviour of the solutions for different values of λ

$$a^{1/2}\dot{a} \equiv \frac{2}{3}\frac{d}{dt}(a^{3/2}) = \pm\left[\frac{8\lambda}{3}\left(\frac{K^3}{\pi G}\right)^{1/2}\right]^{1/2} \tag{19.78}$$

(the signs representing exploding or collapsing models, respectively). This can immediately be integrated, yielding for a collapse ($\dot{a} < 0$) that starts at a_0 for $t = 0$:

$$a^{3/2}(t) = a_0^{3/2} - \frac{3}{2}\left[\frac{8\lambda}{3}\left(\frac{K^3}{\pi G}\right)^{1/2}\right]^{1/2} t \ . \tag{19.79}$$

This expression gives the time dependence of the scaling factor $a(t)$ and therefore by way of (19.67), of the density as a function of time.

We now investigate the spatial dependence of our solution. In particular, the function $w(z)$ in (19.67) has to be determined. For this purpose we write Poisson's equation (19.2) in the dimensionless variable z

$$\frac{1}{z^2}\frac{\partial}{\partial z}\left(z^2\frac{\partial \Phi}{\partial z}\right) = 4\pi G \varrho a^2 \ . \tag{19.80}$$

If we here replace Φ by (19.73), $g(z)$ by (19.76), and ϱ by (19.67), we find

$$\frac{1}{z^2}\frac{d}{dz}\left(z^2\frac{dw}{dz}\right) + w^3 = \lambda \ . \tag{19.81}$$

For $\lambda = 0$ this is the classical Emden equation. Solutions for $\lambda \neq 0$ deviate from hydrostatic equilibrium, the value of λ being a measure for this deviation. From numerical integrations it follows that physically relevant solutions $w(z)$ are obtained only for very small values of λ, namely for $\lambda < \lambda_m = 0.0065$. Otherwise the solution $w(z)$ and therefore $\varrho(r)$ do not become zero at a finite radius; they rather increase again to infinity after a minimum has been reached (see Fig. 19.3).

This figure shows also that for $\lambda < \lambda_m$ the solutions deviate appreciably from the "classical" one ($\lambda = 0$) only in the outer layers, where $\lambda \ll w^3$ no longer applies.

The time-dependent solution discussed here has to be understood in the following way. Let us consider a polytrope with $n = 3$ in equilibrium; then the equilibrium is independent of radius. We have already seen that the total energy is $W = 0$ independent of the radius, see (19.52). Therefore the polytrope $n = 3$ is indifferent to radial changes. If we now assume that suddenly the pressure is slightly reduced say, because the constant K is slightly diminished, then the gaseous sphere begins to contract. This contraction can be described by the two equations (19.75) and (19.76). The solution of the first gives the behaviour in time (19.79), while the second is used to derive the modification of the Lane–Emden equation due to the inertia terms. The parameter λ is a measure of the deviation from hydrostatic equilibrium, caused by the assumed reduction of K.

The solutions for collapsing polytropes have been discussed by Goldreich and Weber (1980) with respect to collapsing stellar cores causing supernova outbursts (Chap. 36).

Chapter 20
Homology Relations

In physical problems it often happens that from one solution others can be obtained by simple transformations. When comparing different stellar models that are calculated under similar assumptions (concerning parameters or material functions), one therefore expects to find similarities in the solutions. It would be very helpful if we could find simple analytic expressions that transform one solution into another. It would then only be necessary to produce *one* numerical solution in order to find new ones by a transformation. There is indeed often a kind of "similarity" between different solutions, which is called *homology,* though the conditions for this are so severe that real stars will scarcely match them. There are a few cases, however, for which homology relations offer a rough, but helpful, indication for interpreting or predicting the numerical solutions. We indicate this in two examples, the main-sequence models and the homologous contraction. Except for this classical homology there is another type of homology, which applies to certain red giants (see Sect. 33.2).

20.1 Definitions and Basic Relations

When comparing different models (say of masses M and M', and radii R and R') one considers in particular *homologous points* at which the relative radii are equal: $r/R = r'/R'$. We now speak of *homologous stars* if their homologous mass shells ($m/M = m'/M'$) are situated at homologous points. To be more precise, let us consider all radii as functions of the *relative* mass values ξ, which are the same for homologous masses:

$$\xi := m/M = m'/M'. \tag{20.1}$$

We can then write the homology condition as

$$\frac{r(\xi)}{r'(\xi)} = \frac{R}{R'} \tag{20.2}$$

R. Kippenhahn et al., *Stellar Structure and Evolution*, Astronomy and Astrophysics Library, DOI 10.1007/978-3-642-30304-3_20, © Springer-Verlag Berlin Heidelberg 2012

for all ξ. In homologous stars the ratio of the radii r/r' for homologous mass shells is constant throughout the stars. Going from one homologous star to another, all homologous mass shells are compressed (or expanded) by the same factor R/R' (Note that therefore any two polytropic models of the same index n are homologous to each other.).

Since both models have to fulfil the stellar-structure equations, the transition has, of course, consequences for all other variables. We derive these by comparing two homologous stars of masses M and M' and of two different compositions that are supposed to be homogeneous and represented by the mean molecular weights μ and μ'. The ratio of these basic parameters will be called

$$x = M/M'; \quad y = \mu/\mu'. \tag{20.3}$$

The variables in the two models are always considered functions of the relative mass variable ξ and may be called r, P, T, l (for M, μ), and r', P', T', l' (for M', μ'), respectively. We try the following "ansatz": for homologous mass values ξ (which we omit for clarity in the following equations) the variables are supposed to have the ratios

$$\frac{r}{r'} = z = \frac{R}{R'}; \quad \frac{P}{P'} = p = \frac{P_c}{P'_c}; \quad \frac{T}{T'} = t = \frac{T_c}{T'_c}; \quad \frac{l}{l'} = s = \frac{L}{L'}, \tag{20.4}$$

where z, p, t, s *have the same values for all* ξ and where the subscript c indicates central values.

We start with homologous main-sequence models. Since they evolve within the long nuclear timescale, one can use (10.2), neglecting the inertia term, as well as the time derivatives in the energy equation (10.3). Let us assume that in these two stars in complete equilibrium (hydrostatic and thermal) the energy transport is radiative. The basic equations to be fulfilled are then (10.1), (10.2), (10.4) and (10.16) together with (10.6), where we further set ε for the total energy production rate. We write them for the first star in terms of the relative mass variable ξ as

$$\begin{aligned}
\frac{dr}{d\xi} &= c_1 \frac{M}{r^2 \varrho}, \quad c_1 = \frac{1}{4\pi}, \\
\frac{dP}{d\xi} &= c_2 \frac{\xi M^2}{r^4}, \quad c_2 = -\frac{G}{4\pi}, \\
\frac{dl}{d\xi} &= \varepsilon M, \\
\frac{dT}{d\xi} &= c_4 \frac{\kappa l M}{r^4 T^3}, \quad c_4 = -\frac{3}{64\pi^2 ac}.
\end{aligned} \tag{20.5}$$

Since no time derivatives appear, the differentiations with respect to ξ are written as ordinary derivatives. In these equations we transform the variables r, P, T, l into

r', P', T', l' by use of (20.4). Noting that the *z, p, t, s* are independent of ξ, and that ξ contains the total mass as scaling factor, which has to be transformed by (20.3), one immediately finds the transformed equations:

$$\frac{dr'}{d\xi} = c_1 \frac{M'}{r'^2 \varrho'} \left[\frac{x}{z^3 d} \right],$$
$$\frac{dP'}{d\xi} = c_2 \frac{\xi M'^2}{r'^4} \left[\frac{x^2}{z^4 p} \right],$$
$$\frac{dl'}{d\xi} = \varepsilon' M' \left[\frac{ex}{s} \right], \tag{20.6}$$
$$\frac{dT'}{d\xi} = c_4 \frac{\kappa' l' M'}{r'^4 T'^3} \left[\frac{ksx}{z^4 t^4} \right].$$

$c_1, \ldots, c_4$ are the same constants as before, and we have introduced the additional abbreviations

$$\frac{\varrho}{\varrho'} = d\,; \quad \frac{\varepsilon}{\varepsilon'} = e\,; \quad \frac{\kappa}{\kappa'} = k \tag{20.7}$$

for the ratios of the material functions at homologous points.

Since for the variables r', P', T', l' we could have written the same basic equations (20.5) as for r, P, T, l, a comparison of (20.6) with (20.5) shows immediately that the four factors in brackets in (20.6) must be equal to one:

$$\frac{x}{z^3 d} = 1, \quad \frac{x^2}{z^4 p} = 1, \quad \frac{ex}{s} = 1, \quad \frac{ksx}{z^4 t^4} = 1. \tag{20.8}$$

Without further specification of the material functions, we can obtain two useful relations already from the first and second of equations (20.8). They can be rewritten as

$$\frac{\varrho}{\varrho'} = \frac{M/M'}{(R/R')^3}, \quad \frac{P}{P'} = \frac{(M/M')^2}{(R/R')^4}. \tag{20.9}$$

Therefore, for all homologous points, the density changes simply as the mean density for the whole star, while P varies like $M^2 R^{-4}$.

In order to find solutions for (20.8), we represent the material functions by power laws:

$$\varrho \sim P^\alpha T^{-\delta} \mu^\varphi, \quad \varepsilon \sim \varrho^\lambda T^\nu, \quad \kappa \sim P^a T^b, \tag{20.10}$$

which from (20.7) with (20.4) give

$$d = p^\alpha t^{-\delta} y^\varphi, \quad e = p^{\lambda\alpha} t^{\nu - \lambda\delta} y^{\lambda\varphi}, \quad k = p^a t^b. \tag{20.11}$$

These can be introduced into (20.8), which are then four conditions for the powers of z, p, t, and s. We will try to represent them in terms of x and y, which, according to (20.3), describe the change of the basic parameters M and μ:

$$z = x^{z_1} y^{z_2} \ ; \quad p = x^{p_1} y^{p_2} \ ; \quad t = x^{t_1} y^{t_2} \ ; \quad s = x^{s_1} y^{s_2} . \tag{20.12}$$

Introducing these and (20.11) into (20.8), we obtain four conditions which contain only products of powers of x and y. In each condition, the exponents of x and of y must sum up to zero, since the right-hand sides of (20.8) are independent of x and y. This yields eight linear equations for the exponents $z_1, \ldots, s_2$, which are written in matrix form as

$$\begin{pmatrix} -3 & -\alpha & \delta & 0 \\ -4 & -1 & 0 & 0 \\ 0 & \lambda\alpha & (\nu - \lambda\delta) & -1 \\ -4 & a & (b-4) & 1 \end{pmatrix} \begin{pmatrix} z_1 \\ p_1 \\ t_1 \\ s_1 \end{pmatrix} = \begin{pmatrix} -1 \\ -2 \\ -1 \\ -1 \end{pmatrix} \tag{20.13}$$

and

$$\begin{pmatrix} -3 & -\alpha & \delta & 0 \\ -4 & -1 & 0 & 0 \\ 0 & \lambda\alpha & (\nu - \lambda\delta) & -1 \\ -4 & a & (b-4) & 1 \end{pmatrix} \begin{pmatrix} z_2 \\ p_2 \\ t_2 \\ s_2 \end{pmatrix} = \begin{pmatrix} \varphi \\ 0 \\ -\lambda\varphi \\ 0 \end{pmatrix} . \tag{20.14}$$

The solutions are

$$\begin{aligned} z_1 &= \frac{1}{2}(1 + A), \quad p_1 = -2A, \\ t_1 &= \frac{1}{2\delta}[1 + (3 - 4\alpha)A], \\ s_1 &= 1 + \frac{4-b}{2\delta} + \left[2 + 2a + \frac{3 - 4\alpha}{2\delta}(4 - b)\right] A, \end{aligned} \tag{20.15}$$

and

$$\begin{aligned} z_2 &= \varphi B, \quad p_2 = -4\varphi B, \quad t_2 = \frac{\varphi}{\delta}[1 + (3 - 4\alpha)B], \\ s_2 &= \frac{\varphi}{\delta}(4 - b) + \varphi \left[4 + 4a + \frac{3 - 4\alpha}{\delta}(4 - b)\right] B, \end{aligned} \tag{20.16}$$

$$A = \left[\frac{4\delta(1 + a + \lambda\alpha)}{\nu + b - 4 - \lambda\delta} + 4\alpha - 3\right]^{-1}, \quad B = A\left(1 - \frac{\lambda\delta}{\nu + b - 4}\right)^{-1} . \tag{20.17}$$

20.2 Applications to Simple Material Functions

20.2.1 The Case $\delta = 0$

A special situation arises for the case that the density is independent of T, i.e. $\delta = 0$ in (20.10). The equation of state then is polytropic, the polytropic index being $n = \alpha/(1-\alpha)$, and we must recover the typical properties of polytropic stars (see Sect. 19.3). This can, in fact, be easily verified. To start with, the first two equations of system (20.13) (which represent the mechanical part) can be solved independently of the rest (the thermo-energetic part). For $\delta = 0$ we find from (20.15) and (20.17) that $A = (4\alpha - 3)^{-1}$ and $z_1 = (2\alpha - 1)/(4\alpha - 3)$. The first of (20.12) gives for homologous stars of equal composition ($y = 1$) the mass-radius relation

$$R \sim M^{z_1}. \tag{20.18}$$

For a non-relativistic degenerate electron gas, one has $\alpha = 3/5$, which gives the exponent $z_1 = -1/3$ as already obtained in Sect. 19.6.

20.2.2 The Case $\alpha = \delta = \varphi = 1, a = b = 0$

Further discussion of the above homology solutions will concentrate on the simplest case, an ideal gas ($\alpha = \delta = \varphi = 1$) with constant opacity ($a = b = 0$) [cf. (20.10)]. This extremely rough approximation to reality suffices for outlining some general properties of main-sequence stars (The assumption of homology introduces a much severer limitation on the results.).

From (20.15)–(20.17), one finds

$$\begin{aligned} z_1 &= \frac{\nu + \lambda - 2}{\nu + 3\lambda}, & z_2 &= \frac{\nu - 4}{\nu + 3\lambda}, \\ p_1 &= 2 - 4z_1, & p_2 &= -4z_2, \\ t_1 &= 1 - z_1, & t_2 &= 1 - z_2, \\ s_1 &= 3, & s_2 &= 4. \end{aligned} \tag{20.19}$$

The first surprising result concerns the exponents of the luminosity, s_1 and s_2. In this simple case the square brackets in the equations for s_1 and s_2 in (20.15) and (20.16) vanish, and s_1 and s_2 become simple constant numbers. In particular, *they are independent of* ν and λ, i.e. *of the special mode of energy generation*. In fact the energy equation [giving the third of (20.13)] has no influence on the luminosity, which is determined by hydrostatic equilibrium, the equations of state, and radiative energy *transfer* only. The model has to adjust so that the energy sources (ε) provide

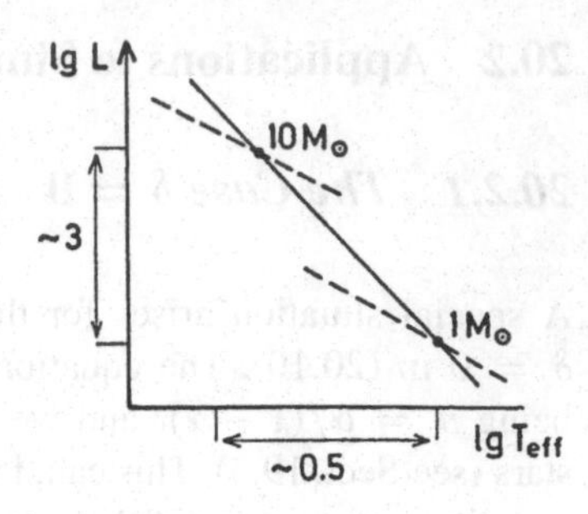

Fig. 20.1 Sketch of the Hertzsprung–Russell diagram with the locus of homologous main-sequence stars (*solid line*) of different masses for a certain constant value of ν. The *dashed lines* indicate lines of R = constant

this luminosity. Introducing the exponents into (20.12), we have from (20.4) that

$$\frac{L}{L'} = \left(\frac{M}{M'}\right)^3 \left(\frac{\mu}{\mu'}\right)^4 . \tag{20.20}$$

There thus exists a mass-luminosity relation that gives a steeply increasing L with increasing M. And L varies even more strongly with the molecular weight μ (The precise values of the exponents vary for other values of a and b roughly in a range from 3 to 6, but the principle result remains.).

All other exponents depend on ν and λ. z_1 and z_2 describe the variation of the radius:

$$\frac{R}{R'} = \left(\frac{M}{M'}\right)^{z_1} \left(\frac{\mu}{\mu'}\right)^{z_2} . \tag{20.21}$$

The exponent z_1 of the $M - R$ relation is positive for all relevant combinations of λ and ν but smaller than one, i.e. R increases slightly with M. Values for typical parameters of hydrogen burning ($\lambda = 1$) via the *pp* chain ($\nu = 4 \ldots 5$) and the CNO cycle ($\nu \approx 15 \ldots 18$) are given in Table 20.1. Over this very large range of ν, z_1 varies relatively little, roughly from 0.4 to 0.8.

The $M - R$ relation together with the $M - L$ relation immediately give the locus of these stars in the Hertzsprung–Russell (HR) diagram, where lg L is plotted over –lg T_{eff} (see Fig. 20.1).

From (20.20) and (20.21) we have $R \sim L^{z_1/3}$ for homologous stars of identical μ. Introducing this into the definition of the effective temperature

$$\sigma T_{\text{eff}}^4 = \frac{L}{4\pi R^2}, \tag{20.22}$$

we obtain the locus as given by

$$\lg L = \frac{12}{3 - 2z_1} \lg T_{\text{eff}} + \text{constant}. \tag{20.23}$$

For an average value $z_1 = 0.6$, the slope is 6.67.

Table 20.1 Exponents in (20.12) for various temperature sensitivities ν of the nuclear reactions, and for $\alpha = \delta = \varphi = 1, a = b = 0, \lambda = 1$, calculated from (20.19)

ν:	4	5	15	18
z_1	0.43	0.5	0.78	0.81
z_2	0	0.13	0.61	0.67
p_1	0.29	0	−1.11	−1.24
p_2	0	−0.5	−2.44	−2.67
t_1	0.57	0.5	0.22	0.19
t_2	1.0	0.88	0.39	0.33
s_1	3	3	3	3
s_2	4	4	4	4

The exponents describe the dependence of R, P, T, L on M and μ ($R \sim M^{z_1}\mu^{z_2}$; $P \sim M^{p_1}\mu^{p_2}$; $T \sim M^{t_1}\mu^{t_2}$; $L \sim M^{s_1}\mu^{s_2}$)

Let us consider how a star of fixed M moves in the HR diagram if μ changes. From (20.20) and (20.21) we have $L \sim \mu^4, R \sim \mu^{z_2}$, which with (20.22) gives $T_{\mathrm{eff}}^8 \sim L^{2-z_2} \approx L^{1.5}$ for $z_2 \approx 0.5$. This defines in the HR diagram a straight line of smaller slope (≈ 5.3) than that of the main sequence. This line for $M =$ constant and μ increasing goes to the upper left with a slope between that of the main sequence and that of the lines $R =$ constant.

The expression for t_1 in (20.19) means that

$$T \sim M/R, \tag{20.24}$$

which simply reflects the virial theorem (thermal energy $\sim$ potential energy). Of special interest are the central values of temperature and density, T_c and ϱ_c, for which one has

$$T_c \sim M^{1-z_1}, \quad \varrho_c \sim M^{1-3z_1}. \tag{20.25}$$

The values in Table 20.1 show that for increasing M, T_c increases relatively slowly, while ϱ_c decreases. This trend is especially pronounced for CNO burning, where T_c scarcely changes at all, typical variations being $T_c \sim M^{0.2}$ and $\varrho_c \sim M^{-1.4}$ (see Fig. 20.2). The predictions of the homology relations are at least qualitatively recovered in the numerical solutions for main-sequence stars (Chap. 22).

20.2.3 The Role of the Equation of State

The procedure by which the homology solutions were obtained shows that their existence rests entirely on the fact that the right-hand sides of (20.5) contain only products of the variables, but no sums. This property is destroyed if the material functions, instead of being products of powers of P and T, contain additive terms as is in general the case with the equation of state. The simplest example is the addition of radiation pressure to an ideal gas such that $P = \Re\varrho T/\mu + aT^4/3$.

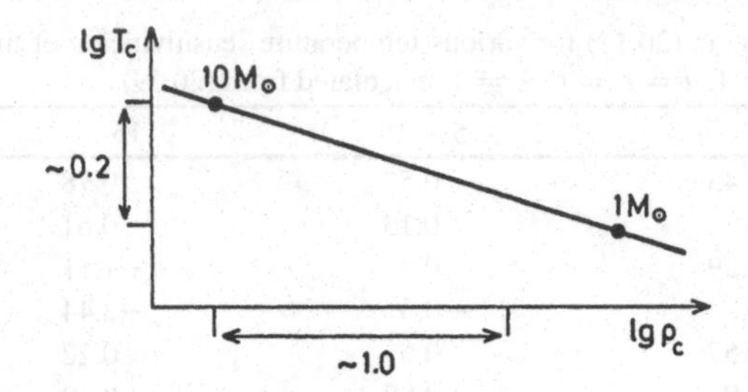

Fig. 20.2 The central values of T and ϱ (both logarithmic) for homologous main-sequence stars of various M. The slope corresponds to a temperature sensitivity ν typical for CNO burning

No strict homology relations are then possible. But one can try to make rough approximations.

One usually writes the corresponding equation of state as

$$\varrho \sim (\mu\beta)\frac{P}{T}, \quad \beta = \frac{P_{\text{gas}}}{P} = \frac{1 - P_{\text{rad}}}{P}. \tag{20.26}$$

The situation would be simple and homology relations would hold if β were constant throughout the model. Then a variation of β obviously has the same effect as that of μ and we would find $R \sim \beta^{z_2}, P \sim \beta^{p_2}, T \sim \beta^{t_2}, L \sim \beta^{s_2}$. In reality β is determined by P and T. For simultaneous variations of M and β, therefore

$$1 - \beta = \frac{P_{\text{rad}}}{P} \sim \frac{T^4}{P} \sim \frac{M^{4t_1}}{M^{p_1}}\frac{\beta^{4t_2}}{\beta^{p_2}}, \tag{20.27}$$

which, if we simply use (20.19), gives

$$\frac{1-\beta}{\beta^4} \sim M^2. \tag{20.28}$$

Now, β is generally *not* constant inside a star [except for the polytrope $n = 3$ as treated in Sect. 19.5; compare with the identical relation (19.56)], but we can consider (20.28) as a relation between M and some kind of mean value of β. One then sees that β decreases strongly with M, i.e. the contribution of the radiation pressure to P increases with mass. Quite similarly we can write

$$L \sim M^{s_1}\beta^{s_2}. \tag{20.29}$$

Since β decreases with increasing M, (20.29) can be written as $L \sim M^{s_1 - c}$ ($c > 0$ for $s_2 > 0$) and the M–L relation becomes less steep. For $\beta \ll 1$ (large P_{rad}), relation (20.28) gives $\beta \sim M^{-1/2}$ such that $L \sim M^{s_1 - s_2/2} = M$. It is generally true that with increasing mass, the pressure in homogeneous stars is increasingly dominated by radiation pressure, and the mass-luminosity relation is less steep than for low-mass stars.

20.3 Homologous Contraction

Now we briefly consider the homologous contraction. This may apply to a chemically homogeneous star of given mass in hydrostatic equilibrium, if its radius is not fixed by an $M-R$ relation but changes in time. Let us assume that consecutive models are homologous to each other. An example in which this assumption is fulfilled is the contraction of a polytrope that does not change its polytropic index n. The solution of the Lane–Emden equation for given n yields the mass value m as a unique function of z only, where z is Emden's dimensionless radius variable, i.e. $z \sim r/R$ (see Sect. 19.2). Therefore the mass elements remain at homologous points, since their values of z do not change in time.

Homologous mass shells (ξ = constant) are here simply those which have the same value of m, since the normalizing factor M remains constant. The radius of any such shell is supposed to change by a rate $\dot{r} = \partial r/\partial t$. In two neighbouring models, separated by a time interval Δt, we have the values r and r' connected by $r' = r + \dot{r}\Delta t$. This gives

$$\frac{r'}{r} = 1 + \frac{\dot{r}}{r}\Delta t. \tag{20.30}$$

For a homologous contraction, we must require that $r'/r = R'/R =$ constant throughout the star. Then also

$$\frac{\dot{r}}{r} = \frac{\dot{R}}{R} \tag{20.31}$$

must be constant, or

$$\frac{\partial}{\partial m}\left(\frac{\partial \ln r}{\partial t}\right) = 0. \tag{20.32}$$

The relative rate of change of the other variables can then be easily expressed in terms of $\dot{r}/r$. From (20.32) we find by exchange of the two derivatives, and by using (10.1),

$$\frac{\partial}{\partial t}\left(\frac{1}{r}\frac{\partial r}{\partial m}\right) = \frac{\partial}{\partial t}\left(\frac{1}{4\pi r^3 \varrho}\right) = \frac{1}{4\pi r^3 \varrho}\left(-3\frac{\dot{r}}{r} - \frac{\dot{\varrho}}{\varrho}\right) = 0, \tag{20.33}$$

which gives

$$\frac{\dot{\varrho}}{\varrho} = -3\frac{\dot{r}}{r}. \tag{20.34}$$

The pressure at a layer of mass value m is given by an integration of the hydrostatic equation as

$$P = \int_m^M \frac{Gm}{4\pi r^4} dm. \tag{20.35}$$

Differentiating this with respect to time and observing that $\dot{r}/r$ is constant throughout the model, we have

$$\dot{P} = \int_m^M \frac{\partial}{\partial t}\left(\frac{1}{r^4}\right)\frac{Gm}{4\pi} dm = -4\frac{\dot{r}}{r}\int_m^M \frac{Gm}{4\pi r^4} dm. \tag{20.36}$$

Equations (20.35) and (20.36) yield

$$\frac{\dot{P}}{P} = -4\frac{\dot{r}}{r}. \tag{20.37}$$

If we have an equation of state with $\varrho \sim P^\alpha T^{-\delta}$, then $\dot{\varrho}/\varrho = \alpha\dot{P}/P - \delta\dot{T}/T$. Solving this for $\dot{T}/T$ and replacing $\dot{\varrho}$ and $\dot{P}$ by (20.34) and (20.37), we have

$$\frac{\dot{T}}{T} = \frac{4\alpha - 3}{\delta}\frac{\dot{r}}{r}. \tag{20.38}$$

The energy generation due to contraction is according to (4.47)

$$\varepsilon_g = c_P T\left(\nabla_{ad}\frac{\dot{P}}{P} - \frac{\dot{T}}{T}\right). \tag{20.39}$$

We introduce (20.37), (20.38) and (20.31), thus obtaining

$$\varepsilon_g = c_P T\left(-4\nabla_{ad} + \frac{4\alpha - 3}{\delta}\right)\frac{\dot{R}}{R}. \tag{20.40}$$

For an ideal monatomic gas ($\nabla_{ad} = 2/5, \alpha = \delta = 1$) this becomes

$$\varepsilon_g = -\frac{3}{5}c_P T\frac{\dot{R}}{R}. \tag{20.41}$$

Therefore $\varepsilon_g > 0$ for contraction ($\dot{R} < 0$). We also see that $|\varepsilon_g| \sim |\dot{R}/R|$; and since ε_g is proportional to T, it represents an energy source that is only rather moderately concentrated towards the centre.

As already mentioned, homology considerations are important for rough interpretations of numerical results, but their strict applicability is very limited. This is ultimately because homology requires a very well concerted action of all mass elements. It can hold approximately only for homogeneous stars. In Sect. 33.2 we will encounter another type of homology which considers only certain parts inside a star, and which applies to some very inhomogeneous stellar configurations.

Chapter 21
Simple Models in the U–V Plane

There are stars in which the nuclear energy generation proceeding close to the centre creates such a high energy flux that the whole central region is convective. These stars can be described by models with a convective core and a radiative envelope. In later stages of stellar evolution the nuclear fuel in the central region of the star is exhausted and nuclear burning takes place only at the surface of a burned-out core. Under certain circumstances these models with shell burning can be described by a core that is isothermal, since no energy has to be transported there, and that is surrounded by a radiative envelope. In both cases a core solution of one type has to be fitted to an envelope solution of another type. In the following we shall deal with a classical fitting procedure which in the past was often used to construct models for such stars (see Schwarzschild 1958; Wrubel 1958) and which gives valuable insight into some of their general properties. Moreover, procedures like this can be helpful in certain special cases where the usual, iterative numerical methods are not practicable.

21.1 The U–V Plane

We define two dimensionless quantities using (1.2) and (2.4):

$$U := \frac{d \ln m}{d \ln r} = \frac{4\pi r^3 \varrho}{m}, \quad V := -\frac{d \ln P}{d \ln r} = \frac{\varrho}{P}\frac{Gm}{r}. \tag{21.1}$$

A solution which is regular in the stellar centre has the central values $U = 3$, $V = 0$, as can easily be seen: a small sphere around the centre has the mass $m = 4\pi r^3 \varrho_c/3$, so that there $U \to 3$ and $V \sim r^2 \to 0$. Near the surface the numerical value of U becomes very small (as ϱ does), as well as P/ϱ ($\sim T$ for the ideal gas or $\sim \varrho^{\gamma-1}$ for polytropes). Therefore V becomes very large.

R. Kippenhahn et al., *Stellar Structure and Evolution*, Astronomy and Astrophysics Library, DOI 10.1007/978-3-642-30304-3_21, © Springer-Verlag Berlin Heidelberg 2012

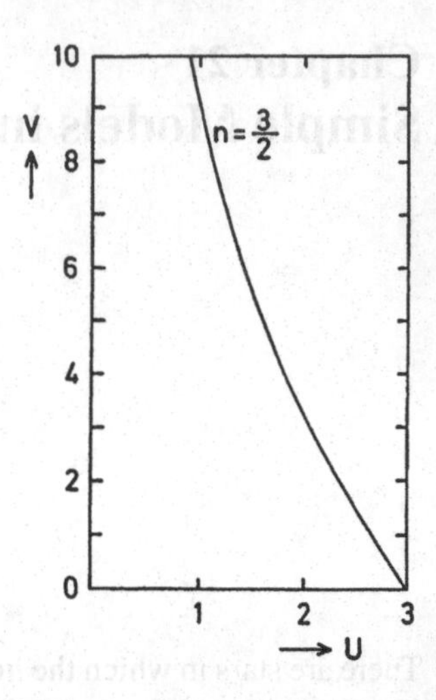

Fig. 21.1 The polytrope $n = 3/2$ in the U–V plane. The stellar centre is in the lower-right corner ($U = 3$, $V = 0$)

Compare two homologous models. Then U as well as V have the same value in homologous mass shells. Indeed with $r/r' = R/R', m/m' = M/M'$, and (20.9) it follows that

$$U = \frac{4\pi r^3 \varrho}{m} = \frac{4\pi r'^3 \varrho'}{m'} = U' \quad \text{and correspondingly } V = V'. \tag{21.2}$$

U and V are therefore also called *homology invariants*.

We now determine the quantities U and V for polytropes. From (19.11) and (19.18), we find

$$U = -w^n \left(\frac{1}{z} \frac{dw}{dz} \right)^{-1}. \tag{21.3}$$

With the expansion (19.12) one can see that indeed $U \to 3$ for $z \to 0$, independent of the value of n. We furthermore find–with $\varrho = \varrho_c w^n$, $P = P_c(\varrho/\varrho_c)^{1+1/n} = P_c w^{n+1}$, and (19.18)–from (21.1) that

$$V = \frac{4\pi G \varrho_c^2 r^2}{P_c} \left(-\frac{1}{z} \frac{dw}{dz} \right) \frac{1}{w}, \tag{21.4}$$

and with (19.3) and (19.9)

$$V = -(n+1) \frac{z}{w} \frac{dw}{dz}, \tag{21.5}$$

which indeed vanishes at the centre and becomes large near the surface where $w \to 0$. Note that the functions $U(z)$ and $V(z)$ depend only on n: they are independent of any other parameter of the model. This is the property which makes a discussion of the U–V plane worthwhile. The function $V = V(U)$ for $n = 3/2$ is plotted in Fig. 21.1.

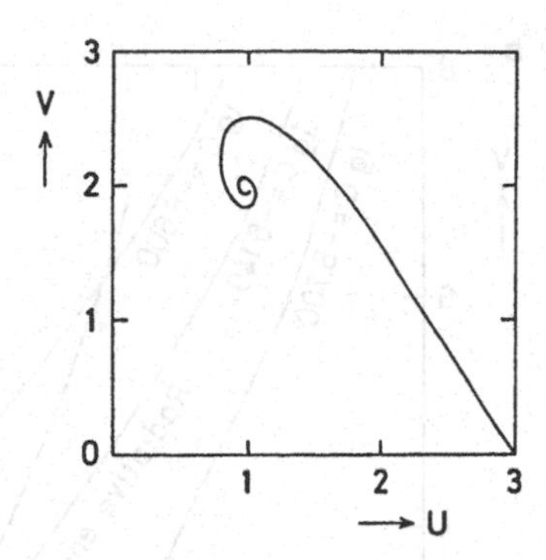

Fig. 21.2 The isothermal sphere for an ideal gas in the U–V plane. The centre ($r = 0$) is in the lower-right comer ($U = 3$, $V = 0$), while for the surface ($r \to R = \infty$) the curve spirals into the point $U = 1, V = 2$

The above polytropic relations hold for finite n only. The isothermal polytrope for an ideal gas ($n = \infty$) again is an exceptional case. Instead of (21.3) and (21.5) one finds from (21.1) and the relations of Sect. 19.8

$$U = e^{-w} \left(\frac{1}{z} \frac{dw}{dz} \right)^{-1} , \quad V = z \frac{dw}{dz}, \tag{21.6}$$

where w now is the solution of (19.35). This case is shown in Fig. 21.2: although the corresponding polytropic model has an infinite radius, its image curve in the U–V plane spirals into the point $U = 1$, $V = 2$, which represents the surface ($z = \infty$). The spiral of the isothermal gaseous sphere unwinds and reaches higher and higher values of V if degeneracy becomes important. In the limit case of complete non-relativistic degeneracy, the image curve approaches that of the polytrope $n = 3/2$ of Fig. 21.1.

The U–V plane has often been used to construct simple stellar models by fitting core and envelope solutions. Clearly this is most profitable when the core is polytropic with given index n and therefore all possible cores are represented by a single, known curve in the plane. This is the case for stars with convective cores (polytropic with $n = 3/2$) or with non-degenerate isothermal cores ($n = \infty$).

The fitting requires continuity of r, P, T, l at the interface. If μ is continuous, then also ϱ–and according to (21.1)–U and V have to be continuous at the fitting point: core and envelope curves intersect (compare Figs. 21.3 and 21.4). If μ is discontinuous at the interface having there the values μ_1, μ_2, then the continuity of P and T for an ideal gas requires $\varrho_1/\varrho_2 = \mu_1/\mu_2$, and (21.1) shows that

$$\frac{U_1}{U_2} = \frac{V_1}{V_2} = \frac{\varrho_1}{\varrho_2} = \frac{\mu_1}{\mu_2}, \tag{21.7}$$

where subscripts 1 and 2 refer to core and envelope solutions at the interface respectively. This means that the points (U_1, V_1) and (U_2, V_2) lie on a straight line through the origin.

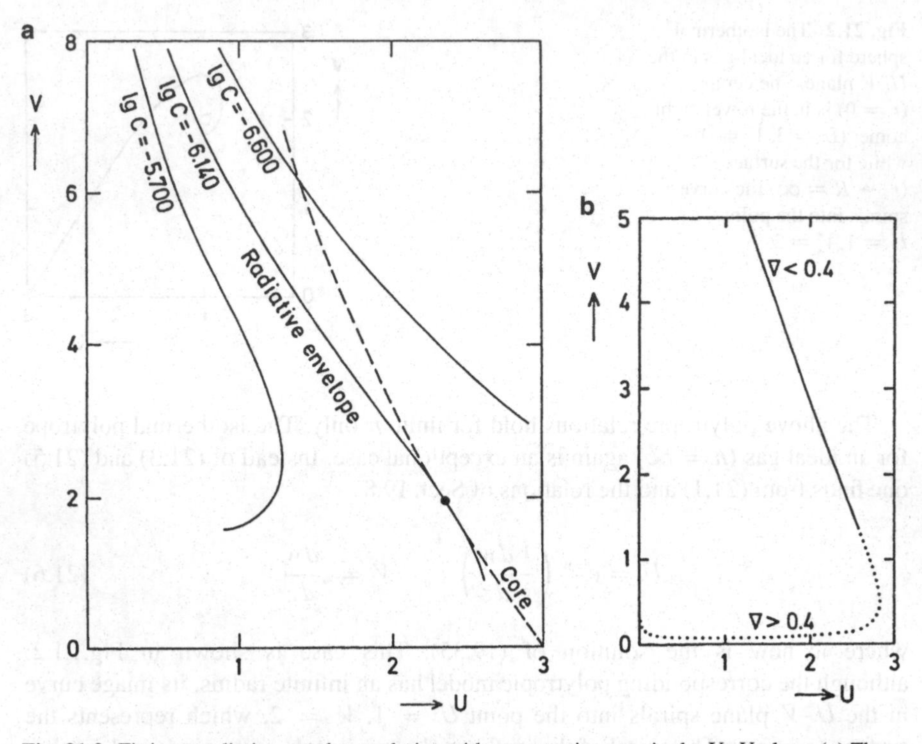

Fig. 21.3 Fitting a radiative-envelope solution with a convective core in the U–V plane. (**a**) Three envelope solutions with different values of the parameter C come from the upper left downwards (*solid lines*). One of them fits to the convective-core solution (*dashed line*), which is given by the polytrope of $n = 3/2$ and starts in the centre at $U = 3$, $V = 0$. At the fitting point, both curves have the same gradient $\nabla = \nabla_{ad} = 0.4$ and the same tangent. (**b**) A radiative-envelope solution in the U–V plane. The solution is shown by a *solid line* as far as $\nabla < 0.4$, and by a *dotted line* where $\nabla > 0.4$ such that the assumption of radiative transport breaks down (After Schwarzschild 1958)

21.2 Radiative Envelope Solutions

We first consider solutions for the envelope where $\varepsilon = 0$ and therefore $l = \text{constant} = L$. The gas is supposed to be ideal, and the opacity is approximated by a power law

$$\kappa = \kappa_0 \varrho^a T^{-b}, \tag{21.8}$$

where $\kappa_0 = \text{constant}$ (Note that here a representation in ϱ and T is used which gives a different exponent b than a representation in P and T.).

We want to obtain many different solutions from a given one by simple scaling. For this aim we replace P, T, m, r by the dimensionless Schwarzschild variables y, t, q, x (Schwarzschild 1946):

$$P = \frac{GM^2}{4\pi R^4} y\,, \quad T = \frac{\mu}{\Re}\frac{GM}{R} t\,, \quad m = qM, \quad r = xR. \tag{21.9}$$

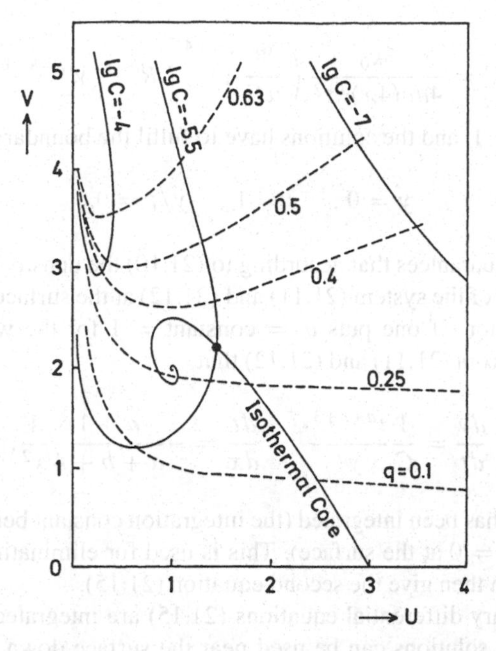

Fig. 21.4 Three envelope solutions with different parameters C and the curve of the non-degenerate isothermal core in the U–V plane. *The dashed lines* combine those points of the envelope solutions where $q = m/M$ reaches certain values. Since, in the case of a homogeneous model, envelope and core solution must be fitted continuously in the U–V plane, one can see that no complete models are possible for isothermal cores with more than about 0.38 M (This limit is even lower if the core has a higher molecular weight than the envelope.). A possible fit for $q \approx 0.3$ between the envelope curve for $\lg C = -5.5$ and the isothermal-core curve is indicated by a *heavy dot*

The equation of state gives the density as

$$\varrho = \frac{M}{4\pi R^3} \frac{y}{t}. \tag{21.10}$$

One can easily see that then the homology variables become $U = x^3 y/(qt)$ and $V = q/(tx)$. The stellar-structure equations (10.1) and (10.2) give

$$\frac{dx}{dq} = \frac{t}{x^2 y}, \quad \frac{dy}{dq} = -\frac{q}{x^4}, \tag{21.11}$$

while the equation for energy transport (10.4) with expression (10.6) gives

$$\frac{dt}{dq} = -C \frac{y^a}{t^{a+b+3} x^4}, \tag{21.12}$$

with

$$C = \frac{3\kappa_0}{4ac(4\pi)^{a+2}} \left(\frac{\Re}{\mu G}\right)^{b+4} L R^{b-3a} M^{a-b-3}. \tag{21.13}$$

At the surface $q = 1$, and the solutions have to fulfil the boundary conditions

$$y = 0\,, \quad x = 1\,, \quad y/t = 0, \tag{21.14}$$

the last of which guarantees that according to (21.10) the density vanishes there.

The singularity of the system (21.11) and (21.12) at the surface can be overcome by an approximation. If one puts q = constant = 1 for the whole near-surface region, one finds from (21.11) and (21.12) that

$$\frac{dy}{dt} = \frac{1}{C}\frac{t^{a+b+3}}{y^a}\,, \quad \frac{dt}{dx} = -\frac{a+1}{a+b+4}\frac{1}{x^2}. \tag{21.15}$$

The first equation has been integrated (the integration constant being chosen in such a way that $y = t = 0$ at the surface). This is used for eliminating y from (21.11) and (21.12), which then give the second equation (21.15).

The two ordinary differential equations (21.15) are integrated by separation of the variables. The solutions can be used near the surface down to a safe distance from the singularity. From there on the normal equations (21.11) and (21.12) can be numerically integrated inwards.

Obviously one obtains a one-parameter set of solutions, the parameter being C. Three such envelope solutions in the U–V plane are shown in Fig. 21.3a. All of them come from the upper left and miss the central boundary condition ($U = 3$, $V = 0$), since they have a singularity there. This does not matter, since anyway we have to fit them to a core solution (compare also with Sect. 12.1). From (21.11) and (21.12) it results that

$$\nabla \equiv \frac{d\ln T}{d\ln P} = \frac{y}{t}\frac{dt}{dy} = C\frac{y^{a+1}}{t^{a+b+4}q}, \tag{21.16}$$

from which one can see that owing to the factor q^{-1} the value of ∇ tends to infinity near the centre. In fact ∇ is small near the surface and increases inwards until it reaches the critical value ∇_{ad} (see Fig. 21.3b). Further inwards the Schwarzschild criterion (6.13) requires convection and the radiative-envelope solutions are no longer valid.

21.3 Fitting of a Convective Core

In order to obtain a model with a convective core inside a radiative envelope we have to fit the solutions of Sect. 21.2 with a polytropic solution of $n = 3/2$ starting at the centre ($U = 3$, $V = 0$). The fit has to be done at the point where the envelope solution reaches $\nabla = \nabla_{\text{ad}}$. Joining all these points on the different envelope solutions (different C) gives a line $\nabla = \nabla_{\text{ad}}$ in the U–V plane, which intersects the core polytrope at the fitting point U^*, V^*. The envelope solution through this point has the value $C = C^*$. Because of the condition that the gradient ∇ is also continuous there, the solutions for core and envelope are tangential to each other, as can be seen in Fig. 21.3a. At the fitting point the variables of the envelope solution may be q^*, y^*, x^*, t^*, while the core polytrope has the variables z^*, w^*.

Let us assume a certain value for the mean molecular weight μ in the envelope. The fit has fixed $C = C^*$, which according to (21.13) gives a relation between L, R, and M. But L is determined by the energy generation in the core, for which we assume a rate of

$$\varepsilon = \varepsilon_0 \varrho T^\nu . \tag{21.17}$$

In the convective core we can connect the Emden variable z with r by $r = zr^*/z^*$, where $r^* = x^* R$ from the outer solution. Then $r^* dl/dr = z^* dl/dz$, and with $\varrho = \varrho_c w^{3/2}$, $T = T_c w$, we have the energy equation with $\lambda = l/L$

$$\frac{d\lambda}{dz} = Bz^2 w^{\nu+3}, \quad B = \frac{4\pi\varepsilon_0}{L}\left(\frac{x^* R}{z^*}\right)^3 \varrho_c^2 T_c^\nu . \tag{21.18}$$

Continuity of ϱ and T in core and envelope solutions requires

$$\varrho^* = \varrho_c w^{*3/2} = \frac{M}{4\pi R^3}\frac{y^*}{t^*}, \tag{21.19}$$

$$T^* = T_c w^* = \frac{\mu}{\Re}\frac{GM}{R} t^* . \tag{21.20}$$

With these two equations we can express ϱ_c, T_c as functions of w^*, y^*, t^* (all known from the integrations) and of M and R. The expressions inserted into (21.18) give

$$B = B_0 \varepsilon_0 \left(\frac{\mu G}{\Re}\right)^\nu \frac{M^{\nu+2}}{LR^{\nu+3}}, \tag{21.21}$$

where B_0 is known from the numerical integrations to the fitting point. Since L is to be generated in the core, $\lambda = l/L = 1$ at the fitting point. Therefore integration of (21.18) gives

$$1 = \int_0^{z^*} \frac{d\lambda}{dz} dz = B \int_0^{z^*} z^2 w^{\nu+3} dz. \tag{21.22}$$

This fixes the value $B = B^*$, since z^* is known, and the integral follows from a simple quadrature.

The fitting procedure now has yielded two numerical values C^*, B^*. Therefore for a given value of M one obtains L and R from (21.13) and (21.21). Of course, one has to check afterwards that (21.17) only gives negligible contributions to L in the envelope solution (where l = constant was assumed).

Models of this type were first constructed by Cowling (1935). They have the advantage that l appears in the structure equations only for the envelope where it is constant ($= L$).

21.4 Fitting of an Isothermal Core

In stellar evolution we shall have to discuss models with an isothermal helium core surrounded by a hydrogen-rich envelope. The luminosity is generated in a thin shell at the interface. This will be idealized by assuming a discontinuity of l (from 0 to L) at the interface.

Let us discuss here a model in which μ is continuous at the interface so that the image curve in the U–V plane is continuous at the fit.

In Fig. 21.4 we have plotted envelope solutions together with the isothermal-core solution for an ideal gas. Along each envelope curve the value of q decreases inwards. We have also plotted some lines q = constant. As one can see from the figure there are no fits possible with $q > q_{max} \approx 0.38$, i.e. when more than 38 % of the total mass lies within the isothermal core. For given $q < q_{max}$ a fit is possible. An example for a fit at $q \approx 0.3$ is shown in Fig. 21.4. One can show that such a fit determines a model completely for given M. Physically more realistic is a model in which μ is higher in the core than in the envelope, which we idealize by a jump of μ at the interface. Then the curve in the $U - V$ plane is discontinuous, fulfilling the conditions (21.7) at the interface ($\mu_1 > \mu_2$). If one tries to fit core and envelope with this condition, and say $\mu_1/\mu_2 = 1.333/0.62$, one finds that q_{max} is considerably smaller: no fits are possible at $q > q_{max} \approx 0.1$. This gives the Schönberg–Chandrasekhar limit for isothermal cores consisting of an ideal gas (see Sect. 30.5) enclosed by the stellar envelope.

Chapter 22
The Zero-Age Main Sequence

We consider here a sequence of chemically homogeneous models in complete (mechanical and thermal) equilibrium with central hydrogen burning. All of them are composed of the same hydrogen-rich mixture, while the stellar mass M varies from model to model along the sequence.

These models can represent very young stars which have just formed from the interstellar medium, and in which the foregoing contraction (see Chap. 28) has raised the central temperature so far that hydrogen burning has started. This provides a long-lasting energy source, and consequently the stars change only on the very long nuclear timescale τ_n. Within the much shorter Kelvin–Helmholtz timescale (see Sect. 3.3), the stars will "forget" the details of their thermal history long before the nuclear reactions have noticeably modified the composition. The only nuclear changes that have taken place during the previous phase are the burning of the light elements deuterium, lithium, beryllium and boron in the largest part of the star, and the conversion of carbon to nitrogen in the centre. The latter reactions consume approximately 1 % of the protons in the stellar core; the former ones are orders of magnitude less important due to the very low abundances of the mentioned elements. This is why one can reasonably treat them as homogeneous models in thermal equilibrium. The now-beginning evolution, in which hydrogen is slowly consumed in the stellar core, has such a long duration that most visible stars are presently found in this phase. Our homogeneous models define its very beginning, and their sequence is therefore more precisely called the *zero-age main sequence* (ZAMS), since one usually counts the age of a star from this point on.

22.1 Surface Values

Homogeneous, hydrogen-burning equilibrium models can be very easily calculated and are available for many different chemical compositions. We limit ourselves to discussing a set of calculations with $X_H = 0.70$, $X_{He} = 0.28$, such that all heavier

R. Kippenhahn et al., *Stellar Structure and Evolution*, Astronomy and Astrophysics Library, DOI 10.1007/978-3-642-30304-3_22, © Springer-Verlag Berlin Heidelberg 2012

Fig. 22.1
Hertzsprung–Russell diagram with the zero-age main sequence computed for a composition with $X_H = 0.70$, $X_{He} = 0.28$. The locations of models for several masses between 0.1 and $55 M_\odot$ are indicated

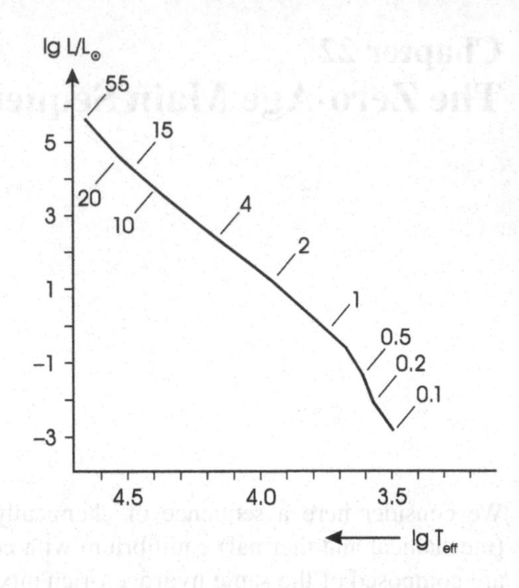

elements amount only to 0.02 of the mass.[1] This is a chemical composition typical for the younger population of stars found in the spiral arms of the Milky Way. The metallicity Z is slightly higher than that of the Sun.

Figure 22.1 shows the Hertzsprung–Russell diagram for the models in the wide range of stellar masses from $0.1 M_\odot$ to more than $50 M_\odot$. L and T_{eff} increase with increasing M, thus forming the ZAMS, which coincides more or less with the lower border of the observed main-sequence band. The slope of the ZAMS below $\approx 0.6 M_\odot$ depends sensitively on the equation of state, the opacities, and the atmospheric boundary conditions.

The important mass-radius and mass-luminosity relations for these models are shown in Figs. 22.2 and 22.3 by the solid lines. In both cases they should constitute a lower envelope to the distribution of stars, since radius as well as luminosity are increasing during the main-sequence evolution and mass remains constant or decreases slightly. Those objects in Fig. 22.2 clearly detached from the bulk of objects are stars that have already developed off the main sequence and therefore have considerably larger radii. Note the very good agreement with theory, although the stars shown do not have identical composition and, in particular, not exactly that of the models. Points below the theoretical sequence may also be due to measurement errors. As predicted already by the simple homology relations for main sequence models [see (20.20) and (20.21)], R increases slowly, and L

[1]Note that we will also use the notation X, Y, and Z for the mass fractions of hydrogen, helium, and the sum of all remaining elements, commonly labelled "metals", as is the case in the astrophysical literature.

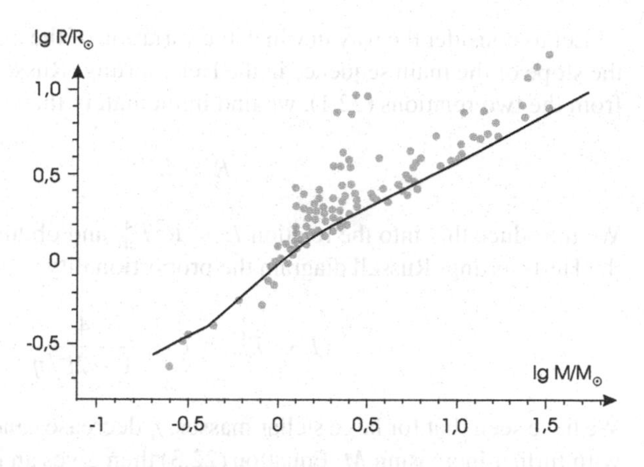

Fig. 22.2 The *line* shows the mass-radius relation for the models of the zero-age main sequence plotted in Fig. 22.1. For comparison, the best measurements (as collected by Malkov et al. 2006, containing the very important catalogue of Andersen 1991) of main sequence primary components of detached and visual binary systems are shown as *grey dots*

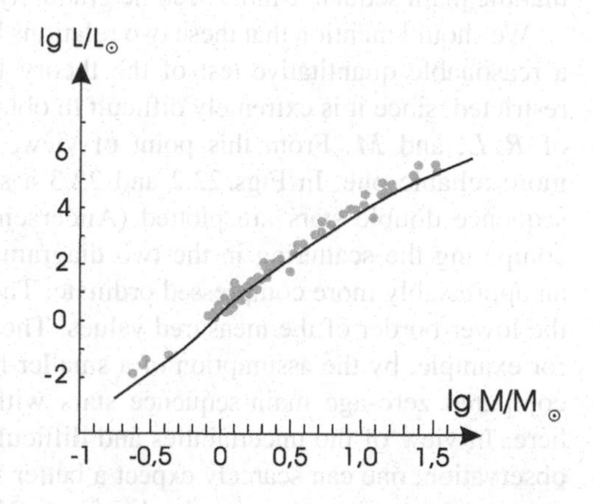

Fig. 22.3 The *line* gives the mass-luminosity relation for the models of the main sequence shown in Fig. 22.1. Measurements of binary systems are plotted for comparison as in Fig. 22.2

increases strongly with increasing M. For an interpolation over a certain range of M we may again write

$$R \sim M^{\xi}, \quad L \sim M^{\eta} . \tag{22.1}$$

From the slopes of the curve in Fig. 22.2 we find roughly $\xi = 0.56$ and 0.79 in the upper and lower mass ranges, respectively. In the range of small values of M, there is a pronounced maximum of the slope around $M = 1M_{\odot}$, indicating a remarkable deviation from homologous behaviour in this range. With decreasing effective temperature these models have outer convective zones of strongly increasing extension (cf. Sects. 11.3.2, 11.3.3 and Fig. 22.7). This tends to decrease R, in addition to other effects.

Also the slope of the $M - L$ relation in Fig. 22.3 varies with M. Over the whole mass range plotted, the average of η is about 3.37. For $M = 1 \ldots 10M_{\odot}$ the average exponent is 3.89, while in the larger range $M = 1 \ldots 50M_{\odot}$ it is 3.35. The decreasing slope towards larger M is an effect of the increasing radiation pressure (see below and Sect. 20.2.3).

Let us consider the way in which the variation of the exponents ξ and η influences the slope of the main sequence in the Hertzsprung–Russell diagram. Eliminating M from the two relations (22.1), we find immediately that

$$R \sim L^{\xi/n} . \tag{22.2}$$

We introduce this into the relation $L \sim R^2 T_{\text{eff}}^4$ and obtain for the main sequence in the Hertzsprung–Russell diagram the proportionality

$$L \sim T_{\text{eff}}^{\zeta}, \quad \zeta = \frac{4}{1 - 2\xi/\eta} . \tag{22.3}$$

We have seen that for large stellar masses, η decreases and ξ remains about constant with further increasing M. Equation (22.3) then gives an increase of ζ, which means that the main sequence must become gradually steeper towards high luminosities.

We should mention that these two relations belong to the rare instances for which a reasonable quantitative test of the theory is possible. Even here one is rather restricted, since it is extremely difficult to obtain sufficiently precise measurements of R, L, and M. From this point of view, the $M - R$ relation should be the more reliable one. In Figs. 22.2 and 22.3 a selection of the best observed main-sequence double stars are plotted (Andersen 1991; Malkov et al. 2006). When comparing the scattering in the two diagrams one should note that Fig. 22.3 has an appreciably more compressed ordinate. The theoretical curves map out roughly the lower border of the measured values. They would be shifted slightly upwards, for example, by the assumption of a smaller hydrogen content. However, we have compared zero-age main-sequence stars with real stars of varying composition here. In view of the uncertainties and difficulties involved in theory as well as in observation, one can scarcely expect a better fit, particularly when considering the enormous range of values involved (a factor 250 in M, nearly 8 powers of 10 in L).

22.2 Interior Solutions

The behaviour of the interior may be illustrated by characteristic variables as functions of m/M. They are plotted in Fig. 22.4 for two stellar masses in order to demonstrate typical dependencies of the solutions on M.

The density ϱ (Fig. 22.4a) increases appreciably towards the centre where we have $\varrho_c \approx 10^2\,\text{g}\,\text{cm}^{-3}$ for 1 $M_\odot$, i.e. roughly a factor 10^9 larger than in the photosphere. For $10 M_\odot$, the central density is smaller by more than a factor 10. The inward increase of ϱ indicates a very strong concentration of the mass elements towards the centre, illustrated in Fig. 22.4b. For 1 $M_\odot$, the inner 30 % of the radius (i.e. only 3 % of the total volume) contains 50 % of the mass; and in the outer 50 % of R (i.e. 88 % of the volume) only about 15 % of M can be found.

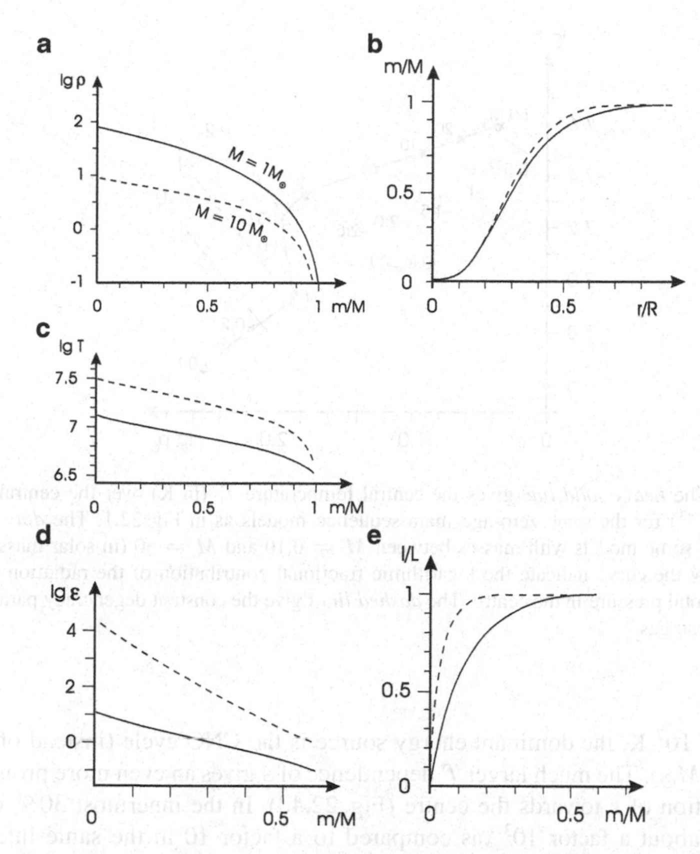

Fig. 22.4 The run of some functions inside zero-age main-sequence models for $M = 1 M_\odot$ (*solid lines*) and $M = 10 M_\odot$ (*dashed lines*) with the same composition as in Fig. 22.1 ($X_\mathrm{H} = 0.70$, $X_\mathrm{He} = 0.28$); (**a**) density ϱ (in g cm^{-3}), (**b**) radial mass distribution $m(r)/M$, (**c**) temperature T (in K), (**d**) nuclear energy production (in erg g^{-1} s^{-1}), (**e**) local luminosity l

The temperature (Fig. 22.4c) also increases towards the centre. For 1 $M_\odot$, the central value of 1.36×10^7 K is a factor 2,400 larger than the photospheric value. Values of $T > 3 \times 10^6$ K extend to $m \approx 0.95M$, so that the average T value (averaged over the mass elements) is roughly 7.7×10^6 K. In a $10 M_\odot$ star, T has slightly more than twice the values of corresponding mass elements for 1 $M_\odot$.

The behaviour of T is necessarily reflected by that of the rate of energy generation due to hydrogen burning (Fig. 22.4d). The dependence of ε on T (cf. Sect. 18.5.1), together with the T gradient, yields a strong decrease of ε from the centre outwards. In the $1 M_\odot$ star, ε has dropped by a factor 10^2 from the centre to $m = 0.6M$, and still further outward it is quite negligible. This is particularly well seen in Fig. 22.4e: 90 % of L is generated in the inner 30 % of M; and l reaches about 99 % of L at $m/M = 0.53$. In the central part of the $10 M_\odot$ star, where

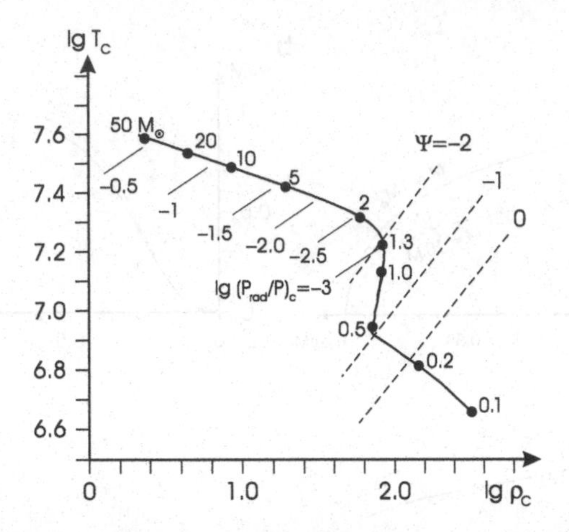

Fig. 22.5 The *heavy solid line* gives the central temperature T_c (in K) over the central density ϱ_c (in $g\,cm^{-3}$) for the same zero-age main-sequence models as in Fig. 22.1. The *dots* give the positions of some models with masses between $M = 0.10$ and $M = 50$ (in solar masses). The *labels* below the curve indicate the logarithmic fractional contribution of the radiation pressure P_{rad} to the total pressure in the centre. The *dashed lines* give the constant degeneracy parameter ψ of the electron gas

$T_c = 3 \times 10^7$ K, the dominant energy source is the CNO cycle (instead of the *pp* chain in $1M_\odot$). The much larger T dependence of ε gives an even more pronounced concentration of ε towards the centre (Fig. 22.4d). In the innermost 30 % of M, ε drops by about a factor 10^3 (as compared to a factor 10 in the same interval of $1M_\odot$). This corresponds to an ε with an exponent of T roughly three times larger. Further outwards, where T is low enough for the *pp* chain to dominate, the slope of ε becomes the same in both stars. In the $10M_\odot$ star, 90 % of the total luminosity is generated within the innermost 10 % of the mass (Fig. 22.4e).

We have seen that in spite of all similarities there are characteristic differences between the interior solutions for different values of M. Some of these can be found in the plot of the central values of temperature and density (Fig. 22.5). This diagram exhibits at least qualitatively another prediction of the homology considerations in Chap. 20: with increasing M, there is a slight increase of T_c together with a substantial decrease of ϱ_c. Between $M = 2M_\odot$ and $50M_\odot$ the differences are $\Delta \lg T_c = +0.28$ and $\Delta \lg \varrho_c = -1.43$. The striking change of the curve around $1.3M_\odot$ is a direct consequence of the transition from the CNO cycle to the *pp*-chains as the dominating energy source. At $M = 1.4M_\odot$, the CNO cylce dominates at the centre, which reaches the critical temperature of $\lg T_c = 7.25$ (Fig. 18.8), while at $M = 1.9M_\odot$ the *pp*-contribution to the total energy production has fallen below 50 % (see also Fig. 22.6). From the homology relations (20.25) and Table 20.1 the slope of the curve in Fig. 22.5 can be predicted: it has a small negative value at

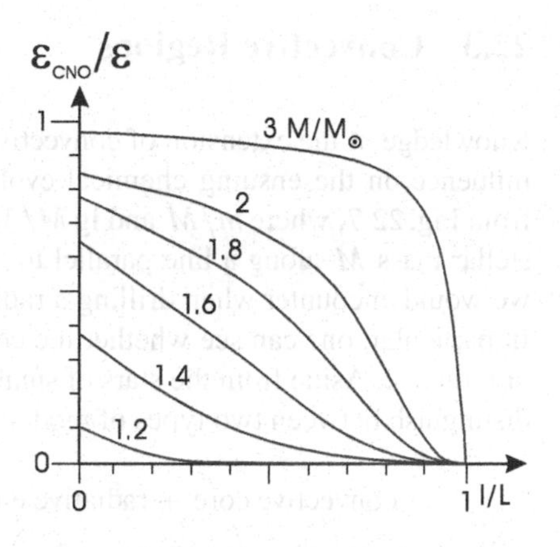

Fig. 22.6 For six zero-age main-sequence models of the same composition as in Fig. 22.1 (mass in solar units indicated along each curve), the fraction that the CNO cycle contributes to the total energy generation rate at different places inside the model (characterized by the corresponding local luminosity l at the abscissa) is shown

the high mass end (−0.16 compared to −0.20 from the numbers given above) and a large negative value ($\lesssim -2$) for $M < 1.3M_{\odot}$. In addition, there are deviations from homology, partly due to the appearance of the outer convective zone (the homology relations were derived under the assumption of radiative energy transport), which is deepening with decreasing mass. The extension of convective regions should certainly influence the centre, since they have a less pronounced mass concentration than radiative regions. Note that both flat parts of the $T_c - \varrho_c$ curve in Fig. 22.5 belong to models in which the central part is convective (cf. Fig. 22.7). At the lowest masses the stars are fully convective and follow the relations for a polytrope of index $n = 3/2$ (Chap. 19 and Sect. 24.2).

In the upper range of masses degeneracy is negligible, while it becomes increasingly important towards smaller M owing to the increasing density. Below $0.5M_{\odot}$, say, other deviations from the ideal gas approximation also become important in the equation of state, for example, electrostatic interaction between the ions.

On the other hand, the radiation pressure P_{rad} must increase towards larger M owing to the increasing T, since $P_{rad} \sim T^4$. At $M = 1M_{\odot}$, radiation contributes only the negligible fraction of a few 10^{-4} to the total central pressure. This fraction becomes about 1 % at $4M_{\odot}$, while in the centre of the $50M_{\odot}$ star, P_{rad} contributes no less than 1/3 to the total pressure (see Fig. 22.5).

Another effect of the growing T_c, which also occurs around 1 $M_{\odot}$, is the transition from the *pp* chain to the CNO cycle as the dominant energy source (compare also Fig. 18.8). For models in the transition region from $M = 1\ M_{\odot}$ to 3 $M_{\odot}$, Fig. 22.6 shows the contribution of ε_{CNO} to the local energy generation rate as a function of l/L. The integral over such a curve gives the fraction of L due to burning in the CNO cycle. This amounts only to a few percent for $M = 1.2\ M_{\odot}$. In the 1.6 $M_{\odot}$ star, the CNO cycle already contributes 65 % at the centre, and nearly one half of the total luminosity. It clearly dominates the whole energy generation for $1.8M_{\odot}$ and more massive stars.

22.3 Convective Regions

Knowledge of the extension of convective regions is very important in view of their influence on the ensuing chemical evolution. A rough overview can be obtained from Fig. 22.7, where m/M and $\lg M/M_\odot$ are ordinate and abscissa. For any given stellar mass M along a line parallel to the ordinate it is indicated what conditions we would encounter when drilling a radial borehole from the surface to the centre. In particular, one can see whether the corresponding mass elements are convective or radiative. Aside from the stars of smallest mass ($M < 0.25M_\odot$), we can roughly distinguish between two types of model:

convective core + radiative envelope (upper main sequence);

radiative core + convective envelope (lower main sequence).

The transition from one type to the other again occurs near $M = 1M_\odot$.

The distinction between convective and radiative regions is made here by using the Schwarzschild criterion (see Sect. 6.1), which predicts convection if the radiative gradient of temperature $\nabla_{\rm rad}$ exceeds the adiabatic gradient $\nabla_{\rm ad}$ (The gradient ∇_μ of the molecular weight appearing in the Ledoux criterion is zero in these homogeneous models. Possible effects of overshooting will be discussed in Chap. 30.). The variation of these gradients (together with that of the actual gradient ∇) throughout the star is plotted in Fig. 22.8 for $M = 1M_\odot$ and $10M_\odot$. For the abscissa, $\lg T$ is chosen, since this conveniently stretches the scale in the complicated outer layers.

Let us start with the simpler situation concerning the convective core. When comparing Fig. 22.8a, b, we see that the convective core in the more massive models is caused by a steep increase of $\nabla_{\rm rad}$ towards the centre. The reason for this is that the dominating CNO cycle , with its extreme temperature sensitivity, concentrates the energy production very much towards the centre (cf. the curve $l/L = 0.5$ in Fig. 22.7, and Fig. 22.4e). Therefore we find in these stars very high fluxes of energy $(l/4\pi r^2)$ at small r, which produce large values of $\nabla_{\rm rad}$. Figure 22.7 shows a remarkable increase in the extent of the convective core for increasing M. The core covers as much as 65 % of the stellar mass in a star of $50M_\odot$, an increase caused by the increasing radiation pressure (cf. Sect. 22.2 and Fig. 22.5), which depresses the value of $\nabla_{\rm ad}$ well below its standard value of 0.4 for an ideal monatomic gas [see (13.12)]. In the centre of the $50M_\odot$ model, roughly 1/3 of P is radiation pressure, and $\nabla_{\rm ad} \approx 0.27$. From Fig. 22.8b it is clear that a depression of $\nabla_{\rm ad}$ in the central region will shift the intersection with $\nabla_{\rm rad}$ (i.e. the border of the convective core) outwards to smaller T. When we increase M to much larger values still, the top of the convective core will finally approach the surface such that we should obtain fully convective stars. We then approach models of the so-called supermassive stars (see Sect. 19.10).

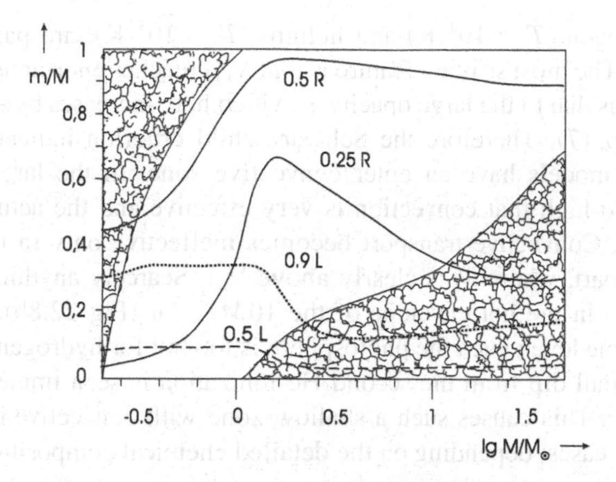

Fig. 22.7 The mass values m from centre to surface are plotted against the stellar mass M for the same zero-age main-sequence models as in Fig. 22.1. "Cloudy" areas indicate the extension of convective zones inside the models. Two *solid lines* give the m values at which r is 1/4 and 1/2 the total radius R. The *dashed and dotted lines* show the mass elements inside which 50 % and 90 % of the total luminosity L are produced

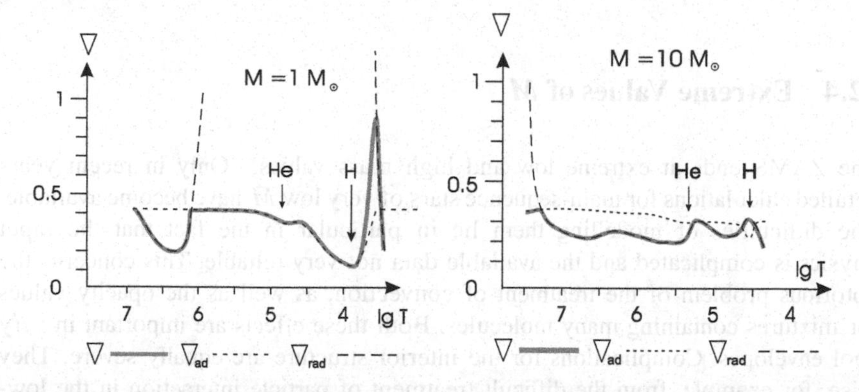

Fig. 22.8 The *grey solid lines* show the actual temperature gradient $\nabla = d\ln T/d\ln P$ over the temperature T (in K) inside two zero-age main-sequence models of $1M_\odot$ (*left panel*) and $10M_\odot$ (*right panel*). The corresponding adiabatic gradients ∇_{ad} (*dotted lines*) and radiative gradients ∇_{rad} (*dashed lines*) are also plotted, and the location of the ionization zones of hydrogen and helium are indicated (*arrows*). The chemical composition of the models is the same as for those of Fig. 22.1

In less massive stars, the *pp* chain with its smaller temperature sensitivity dominates. This distributes the energy production over a much larger area, so that the flux and ∇_{rad} are much smaller in the central region, which thus remains radiative.

Outer convective envelopes can generally be expected to occur in stars of low effective temperature, as the discussion of the boundary conditions in Sect. 11.3.2 has already shown. When studying the different gradients in the outer layers of cool stars (Fig. 22.8a), one finds a variety of complicated details. The variation of ∇_{ad} clearly shows depressions in those regions where the most abundant elements, hydrogen ($T \gtrsim 10^4$ K) and helium ($T \approx 10^5$ K), are partially ionized (see Chap. 14). The most striking feature is that $\nabla_{\rm rad}$ reaches enormous values (more than 10^5). This is due to the large opacity κ, which here increases by several powers of 10 (cf. Chap. 17). Therefore the Schwarzschild criterion indicates convective instability: the models have an outer convective zone. In the largest part of it, the density is so high that convection is very effective and the actual gradient ∇ is close to $\nabla_{\rm ad}$. Convective transport becomes ineffective only in the outermost, superadiabatic part, where ∇ is clearly above $\nabla_{\rm ad}$. Scarcely anything of all these features appears in the hot envelope of the $10 M_\odot$ star (Fig. 22.8b). $\nabla_{\rm rad}$ remains nearly at the same level; even the photosphere is too hot for hydrogen to be neutral, and only the small dip from the second He ionization is seen immediately below the photosphere. This causes such a shallow zone with convective instability that only for special cases, depending on the detailed chemical composition, convective motions set it.

The outer convection zone gradually penetrates deeper into the star with decreasing $T_{\rm eff}$. Its lower border finally reaches the centre at $M \lesssim 0.25 M_\odot$ (left end of Fig. 22.7), such that the main-sequence stars of even smaller masses are fully convective.

22.4 Extreme Values of *M*

The ZAMS ends at extreme low and high mass values. Only in recent years detailed calculations for main-sequence stars of very low M have become available. The difficulties of modelling them lie in particular in the fact that the input physics is complicated and the available data not very reliable. This concerns the notorious problem of the treatment of convection, as well as the opacity values for mixtures containing many molecules. Both these effects are important in very cool envelopes. Complications for the interior structure are equally severe. They arise, for example, from the difficult treatment of particle interaction in the low-temperature high-density regime and influence the equation of state and the electron screening of nuclear reactions. Progress has been made in the equation of state under such conditions (Sect. 16.6), in the treatment of the opacities (Sect. 17.8) and the calculation of the atmospheric structure. The latter is very important since stars below $\approx 0.2 M_\odot$ are fully convective (Fig. 22.7) and their interior structure therefore depends very much on the outer boundary conditions (Fig. 11.2).

Quite another problem concerns the relevance of the calculated equilibrium models for real, evolving stars. At the low central temperatures in models of extremely small masses, for example, the time for reaching equilibrium burning can become exceedingly long. A preceding phase in which the original ^{3}He is burned may be at least equally important, but this ^{3}He content is very uncertain. And below about $M = 0.1 M_\odot$, even the original contraction leads so far into electron degeneracy that hydrogen burning is no longer ignited (refer to Chap. 28).

In this sense one may speak of the "lower end of the main sequence" at this mass value. Disregarding this evolutionary argument, however, one can ask whether solutions for main-sequence models (homogeneous, hydrogen burning, complete equilibrium) exist down to arbitrary small values of M. It turns out that such models end to exist at $M \approx 0.08 M_\odot$. Real stars simply fail to provide all the luminosity from nuclear burning alone and need thermal energies to supply the rest of the energy. Such objects are called *brown dwarfs* and are no longer considered as "real stars". Details about very low-mass stars and brown dwarfs, their physical properties and how they are modelled, can be found in the review by Chabrier and Baraffe (2000). Although they are extremely faint, they are now routinely found with large telescopes. A decisive test to confirm that a "star" is indeed a brown dwarf is the *lithium test*: going down in mass along the main sequence, stars become fully convective. Any change in element abundances due to hydrogen burning is therefore reflected in the surface abundances. This includes lithium, which, as part of the *pp2* chain (18.62), is destroyed due to proton captures at temperatures above $\approx 2.5 \times 10^6$ K. Its surface abundance is therefore very low on the lower main sequence, as it is almost completely destroyed throughout the star. If the mass is however low enough such that the critical temperature is not reached even at the centre, lithium can survive and "reappears" for the very faintest main-sequence stars. The mass at which such low central temperatures are reached is $\approx 0.06 M_\odot$, which is lower than the 0.08 $M_\odot$, which denotes the transition from stars to brown dwarfs. The lithium test has lead to the first definite detection of brown dwarfs.

In the direction towards large M, on the other hand, the sequence of equilibrium models can principally be continued up to the "supermassive" stars (see Sect. 19.10). Long before they are reached, however, an instability occurs which sets in between $M \approx 60$ and $100 M_\odot$ (depending on the composition). It is a vibrational instability caused by the so-called ε mechanism (see Sect. 41.5) and supported by the large amount of radiation pressure. Such stars, instead of sitting quietly at their proper place on the main sequence, will start to oscillate with growing amplitude. This may go so far as to throw off matter from the surface, until the mass is reduced below the critical value for the instability.

22.5 The Eddington Luminosity

For massive, hot stars there exists another physical limit for hydrostatic stability, which results from the increasingly important radiation pressure. According to (13.1)

$$P_{\text{rad}} = \frac{1}{3} U = \frac{a}{3} T^4 .$$

Therefore there exists a gradient of the radiation pressure

$$\frac{dP_{\text{rad}}}{dr} = \frac{a}{3} T^3 \frac{dT}{dr} , \tag{22.4}$$

which exerts, just like the gas pressure gradient, an outward acceleration ($dP_{\mathrm{rad}}/dr < 0$)

$$g_{\mathrm{rad}} = -\frac{1}{\rho}\frac{dP_{\mathrm{rad}}}{dr} . \tag{22.5}$$

(This outward force is already included in the hydrostatic equation, if the total pressure is considered according to (13.2). Here we consider it separately only for clarifying the effect.)

Using (5.8) we see that we can rewrite (22.5) as

$$g_{\mathrm{rad}} = \frac{\kappa F_{\mathrm{rad}}}{c} = \frac{\kappa L_r}{4\pi r^2 c} . \tag{22.6}$$

In case that radiation pressure completely dominates over gas pressure, a star can no longer be in hydrostatic equilibrium if $g_{\mathrm{rad}} > -g$. The sum of both accelerations can be written as

$$g + g_{\mathrm{rad}} = -\frac{Gm}{r^2}\left[1 - \frac{\kappa L_r}{4\pi c G m}\right] = -\frac{Gm}{r^2}\left[1 - \Gamma_r\right] , \tag{22.7}$$

where Γ_r can be understood as the ratio of the luminosity relative to the critical luminosity at which the bracket changes sign, and thus the star becomes unbound. For $m = M$ this critical luminosity is called the *Eddington luminosity* and is

$$L_{\mathrm{E}} = \frac{4\pi c G M}{\kappa} . \tag{22.8}$$

Expressed in solar units it is

$$\frac{L_{\mathrm{E}}}{L_\odot} = 1.3 \times 10^4 \frac{1}{\kappa}\frac{M}{M_\odot} \tag{22.9}$$

and grows linearly with stellar mass. Since $L \sim M^3$, stars obviously reach a limit, where radiation pressure is able to drive a strong stellar wind, and which depends on the opacity.

For hot, massive stars electron scattering is the dominating opacity source, which can be approximated by (17.1), and is $\kappa_{\mathrm{sc}} = 0.20(1 + X)$. For a mass fraction of hydrogen of 0.70 (22.9) simplifies to

$$\frac{L_{\mathrm{E}}}{L_\odot} = 3.824 \times 10^4 \frac{M}{M_\odot} . \tag{22.10}$$

For $M \approx 200\, M_\odot$ the luminosity of massive main-sequence stars reach the Eddington limit and disperse. This is a rough estimate for an upper limit. In reality the instability of the ε mechanism occurs at lower mass. However, the Eddington limit can become quite important in other situations.

Chapter 23
Other Main Sequences

The simplicity and the importance of the results obtained for the main sequence suggest the extension of this concept to stars of quite different composition. We can then describe a main sequence as any sequence of homogeneous models with various masses M in complete equilibrium, consisting (mainly) of a certain element which burns in the central region. In this sense, *the* (normal) main sequence as treated before is a special case and is more precisely called the *hydrogen main sequence* (H-MS). In a further step of generalization, we will even drop the assumption of chemical homogeneity, thus arriving at the so-called generalized main sequences (GMS) (Sect. 23.3). Of course, compared with the H-MS, the other sequences are far less important for real, observed stars. But their properties yield valuable information for understanding certain types of evolved stars, for example.

The numerical models shown in this chapter have been calculated with an older equation of state and simpler opacities. Also, the chemical composition for the hydrogen-rich models differs from that used in the last chapter and is for a slightly higher (= 0.021) metallicity. But since we will discuss fundamental properties of stars in this section, these details are of no relevance.

23.1 The Helium Main Sequence

The helium main sequence (He-MS) contains chemically homogeneous equilibrium models that consist almost completely of He (with the usual few per cent of heavier elements) and have central helium burning. In principle, one could imagine them to be the descendants of perfectly mixed hydrogen-burning stars (however, perfect mixing during evolution is very improbable). Or they may represent the remnants of originally more massive stars that have developed a central helium core and then lost their hydrogen-rich envelope.

In the Hertzsprung–Russell diagram (Fig. 23.1) the He-MS is situated far to the left of the (normal) H-MS at fairly high luminosities. If we compare the same stellar

R. Kippenhahn et al., *Stellar Structure and Evolution*, Astronomy and Astrophysics Library, DOI 10.1007/978-3-642-30304-3_23, © Springer-Verlag Berlin Heidelberg 2012

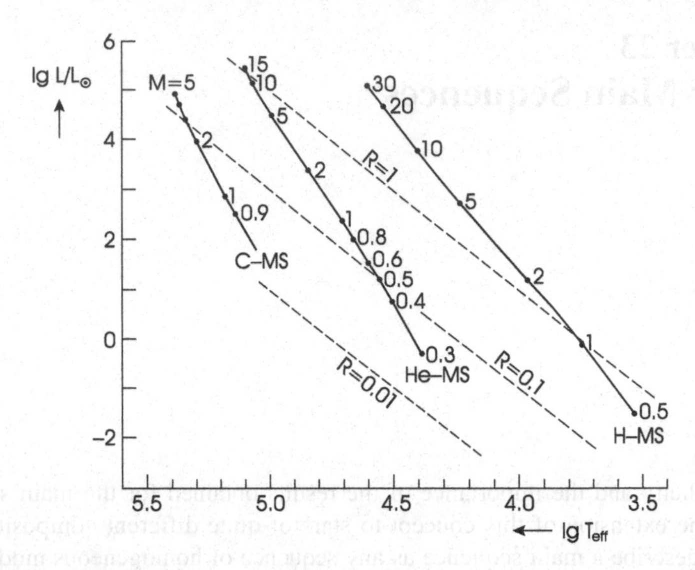

Fig. 23.1 In the Hertzsprung–Russell diagram the *solid lines* show the normal hydrogen main sequence (H-MS; $X_{\mathrm{H}} = 0.685$, $X_{\mathrm{He}} = 0.294$), the helium main sequence (He-MS; $X_{\mathrm{H}} = 0$, $X_{\mathrm{He}} = 0.979$), and the carbon main sequence (C-MS; $X_{\mathrm{H}} = X_{\mathrm{He}} = 0$, $X_{\mathrm{C}} = X_{\mathrm{O}} = 0.497$). The *labels* along the sequences give stellar masses M (in units of $M_\odot$). Three lines of constant stellar radius (R in units of $R_\odot$) are plotted (*dashed*)

mass M on each sequence, we see that the helium stars have smaller radii and much higher luminosities. The remarkable difference in L for given M is particularly well illustrated by the $M - L$ relations in Fig. 23.2. The main cause is certainly the difference in the mean molecular weight μ, which is 0.624 for the mixture used for the stars on the H-MS and 1.343 for the helium stars. If everything else were the same and the models were homologous, then we would expect from (20.20) for stars with the same M a difference in luminosity given by $\Delta \lg L = 4\Delta \lg \mu = 1.33$. This is in fact very nearly the shift between the two $M - L$ relations in Fig. 23.2 at $M = 10M_\odot$, while for $M = 1M_\odot$, we even have $\Delta \lg L \approx 2.5$.

The interior structure resembles roughly that of models on the upper H-MS. The extreme temperature sensitivity of helium burning concentrates the energy production into a small central sphere where the large energy flux produces a convective core. This contains about $0.27M$ in the $1M_\odot$ star, and nearly $0.7M$ for $10M_\odot$. The increase of the convective core is again a consequence of the increasing radiation pressure: it contributes 1.5 % to the total pressure in the centre of the $1M_\odot$ star, 18 % for $5M_\odot$, and 32 % for $10M_\odot$, which is very much more than for the corresponding stars on the H-MS (6×10^{-4}, 0.018, and 0.063, respectively). The difference is due to the fact that helium burning requires temperatures roughly six times higher, as can be seen in Fig. 23.3, which shows the central values of T and ϱ. The high radiation pressure provides relatively large amplitudes of pulsation in the central region. This again produces a vibrational instability due to the ε mechanism,

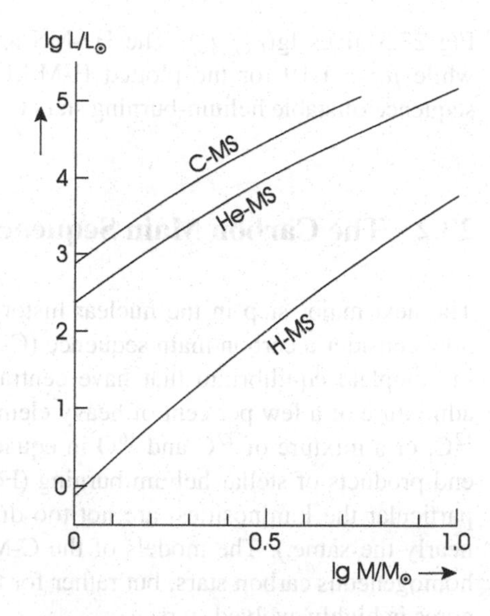

Fig. 23.2 Mass–luminosity relations for the models of the hydrogen, helium, and carbon main sequences of Fig. 23.1

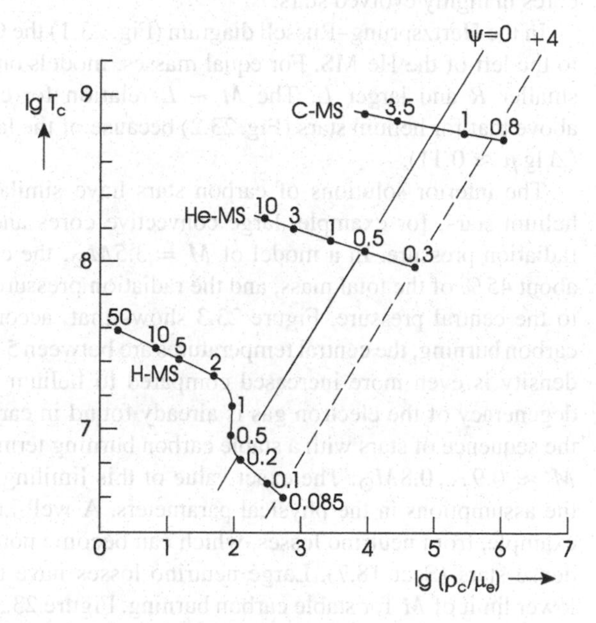

Fig. 23.3 Central temperature T_c (in K) and central density ϱ_c/μ_e (ϱ_c in $\mathrm{g\,cm^{-3}}$, μ_e = molecular weight per electron) of the models on the hydrogen, helium, and carbon main sequences of Fig. 23.1. The *labels* along the lines give the stellar mass M (in $M_\odot$). The *dashed lines* indicate constant degeneracy parameters ψ of the electron gas

the onset of which occurs around $M = 15M_\odot$, depending somewhat on the content of heavier elements.

Another property of the helium stars to be seen in Fig. 23.3 is their much larger central density: for $M = 0.3M_\odot$, ϱ_c reaches $10^5\,\mathrm{g\,cm^{-3}}$, and, in spite of the larger T, the electron gas has about the same degree of degeneracy as at the lower end of the H-MS [In order to plot a unique degeneracy parameter ψ (see Chap. 15) for compositions with different molecular weight per electron μ_e, the abscissa of

Fig. 23.3 gives $\lg(\varrho_c/\mu_e)$. The He-MS and the C-MS (see below) have $\mu_e = 2$, while $\mu_e = 1.19$ for the plotted H-MS.]. The increasing degeneracy causes the sequence of stable helium-burning stars to terminate at about $M \approx 0.3 M_\odot$.

23.2 The Carbon Main Sequence

The next major step in the nuclear history of a star is carbon burning. Thus, we now consider a carbon main sequence (C-MS) consisting of homogeneous models in complete equilibrium that have central carbon burning. Except for the usual admixture of a few per cent of heavy elements, the composition can be either pure ^{12}C, or a mixture of ^{12}C and ^{16}O in equal amounts, which represents roughly the end products of stellar helium burning (For both assumptions the basic results, in particular the luminosities, are not too different, since the molecular weights are nearly the same.). The models of the C-MS are not so much used for describing homogeneous carbon stars, but rather for the purpose of surveying carbon-burning cores in highly evolved stars.

In the Hertzsprung–Russell diagram (Fig. 23.1) the C-MS is at $T_{\text{eff}} > 10^5$ K even to the left of the He-MS. For equal masses, models on the C-MS have remarkably smaller R and larger L. The $M - L$ relation for carbon stars is $\Delta \lg L \approx 0.5$ above that for helium stars (Fig. 23.2) because of the larger mean molecular weight ($\Delta \lg \mu \approx 0.11$).

The interior solutions of carbon stars have similar properties to those of the helium stars, for example, large convective cores and an appreciable amount of radiation pressure. In a model of $M = 3.5 M_\odot$, the convective core encompasses about 45 % of the total mass, and the radiation pressure contributes more than 20 % to the central pressure. Figure 23.3 shows that, according to the requirements of carbon burning, the central temperatures are between 5 and 8×10^8 K. But the central density is even more increased compared to helium stars. Therefore appreciable degeneracy of the electron gas is already found in carbon stars around $1 M_\odot$. And the sequence of stars with a stable carbon burning terminates at masses in the range $M \approx 0.9 \ldots 0.8 M_\odot$. The exact value of this limiting mass depends somewhat on the assumptions in the physical parameters. A well-known uncertainty comes, for example, from neutrino losses, which can become noticeable in these very hot and dense stars (Sect. 18.7). Large neutrino losses have the tendency to increase the lower limit of M for stable carbon burning. Figure 23.3 shows that in all three main sequences the limiting mass occurs at roughly the same degree of degeneracy of the electron gas ($\psi \approx 4.5$). The C-MS and the He-MS have a much simpler structure than the H-MS, which is affected by the complications occurring near $1 M_\odot$, namely the transition from convective to radiative cores and the growth of outer convection zones with decreasing T_{eff}.

23.3 Generalized Main Sequences

The logical next step in extending the concept of main sequences is to drop the condition of chemical homogeneity. This is suggested by the chemical evolution we encounter in all stars: the conversion of hydrogen to helium by nuclear reactions (which are concentrated towards the centre) produces a central helium core, while the outer envelope retains its original hydrogen-rich mixture. If the temperatures are high enough, helium burning will occur around the centre, and hydrogen burning continues in a so-called shell source, i.e. a concentric shell starting at the bottom of the hydrogen-rich envelope. Based on this picture, different types of significant sequences may be defined. We will limit ourselves in the following to the simplest case, which nevertheless finds useful applications.

For these *generalized main sequences* (GMS), we consider models in complete equilibrium, with a chemical profile as shown in Fig. 23.4: a central helium core of mass $M_{\rm He}$, i.e. of the mass fraction $q_0 = M_{\rm He}/M$, is surrounded by an envelope of mass $(1 - q_0)M$ with the usual hydrogen-rich mixture of unevolved stars. At the interface of the two regions, the hydrogen content $X_{\rm H}$ changes discontinuously ("step profile"), while the hydrogen content in the envelope as well as the small admixture of heavier elements in both regions is assumed to be fixed at some reasonable values. The energy is supplied by central helium burning and (possibly) by an additional hydrogen burning in a shell source at q_0.

Each of these models is characterized by two parameters, the stellar mass M, and the relative core mass q_0. We then obtain a generalized main sequence by keeping q_0 constant and varying M as a parameter. For each value of q_0 there is one GMS. In the evolution the value of q_0 is not constant: q_0 can slowly increase because of the shell source burning, and it can increase by mass loss from the surface. We will therefore consider GMS of various values of q_0.

The upper limit is obviously $q_0 = 1$, implying that the "core" encompasses the whole star, which is then a homogeneous helium star. The GMS for $q_0 = 1$ is therefore identical with the well-known He-MS discussed in Sect. 23.1.

For values of q_0 slightly below 1, the GMS are shifted appreciably to the right in the Hertzsprung–Russell diagram (Fig. 23.5). They have already passed the H-MS for $q_0 \approx 0.9 \ldots 0.85$, depending on the value of M. In other words, the addition of a relatively small hydrogen-rich layer on top of a helium star will remarkably increase its radius and decrease $T_{\rm eff}$.

This behaviour changes completely if q_0 drops below a certain value, which is about $0.8 \ldots 0.7$, depending on M. Figure 23.5 shows that the GMS are then compressed towards a limiting line far to the right-hand side of the Hertzsprung–Russell diagram. This will turn out to be the Hayashi line, a limit for all stars in hydrostatic equilibrium (Chap. 24). The closest approach to it is found roughly for the GMS with $q_0 = 0.5$. For even smaller q_0, the GMS move slowly back to the left in the Hertzsprung–Russell diagram. We conclude that the upper part of this diagram can be covered at least once by these GMS, i.e. by very simple equilibrium models depending on two parameters (M, q_0) only.

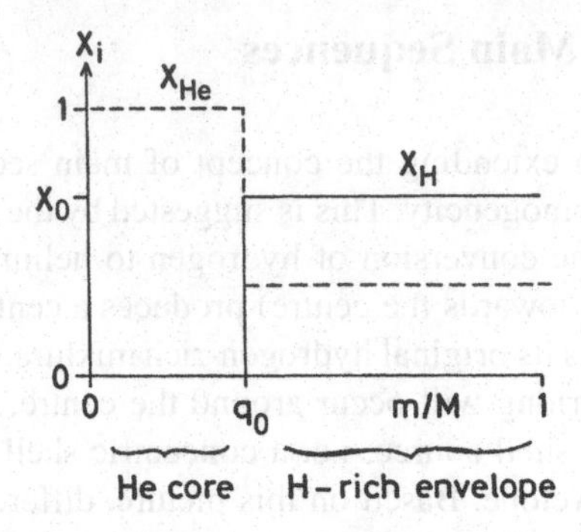

Fig. 23.4 Chemical composition inside the models on the generalized main sequences. The mass concentrations of hydrogen X_H (*solid line*) and helium X_{He} (*dashed line*) are plotted over the mass variable m/M from centre to surface. X_0 is the hydrogen content in the envelope. The relative core mass is $M_{He}/M = q_0$

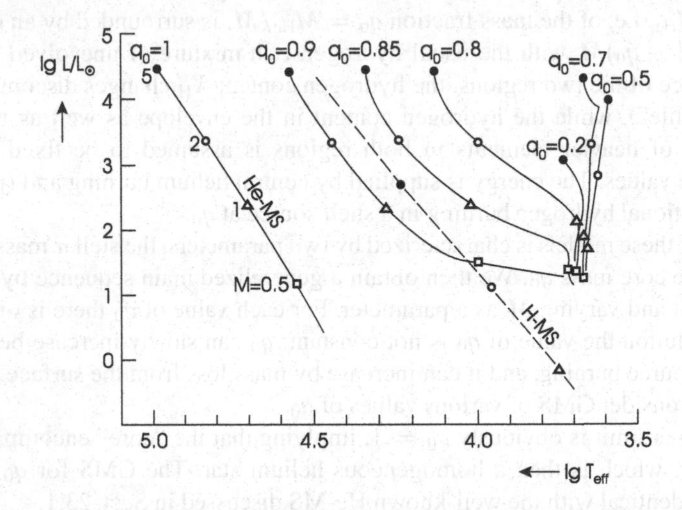

Fig. 23.5 Hertzsprung–Russell diagram with generalized main sequences for models with helium cores of relative mass q_0 and hydrogen-rich envelopes of relative mass $1 - q_0$ (cf. Fig. 23.4). The sequences plotted here cover only the range from $q_0 = 1$ (helium main sequence) to $q_0 = 0.2$. For comparison, the limiting case of the hydrogen main sequence ($q_0 = 0$, *dashed*) is shown. Models with a stellar mass $M = 5$ (in $M_\odot$) are indicated by *solid dots*, $M = 2$ by *open circles*, $M = 1$ by *triangles*, and $M = 0.5$ by *squares* (After Giannone et al. 1968)

Let us compare models with the same M on different GMS. If we connect their points in Fig. 23.5, we obtain curves such as those plotted in Fig. 23.6 for two values of M. This shows that the luminosity remains roughly constant in the range $q_0 = 1 \ldots 0.7$. This is caused by two opposite effects nearly cancelling each other: when we decrease q_0 at M = constant, M_{He} decreases, which reduces the luminosity of the core, L_{He}, approximately as given by the $M - L$ relation for the He-MS (Fig. 23.2, if here we take M_{He} for M). At the same rate, the mass of the

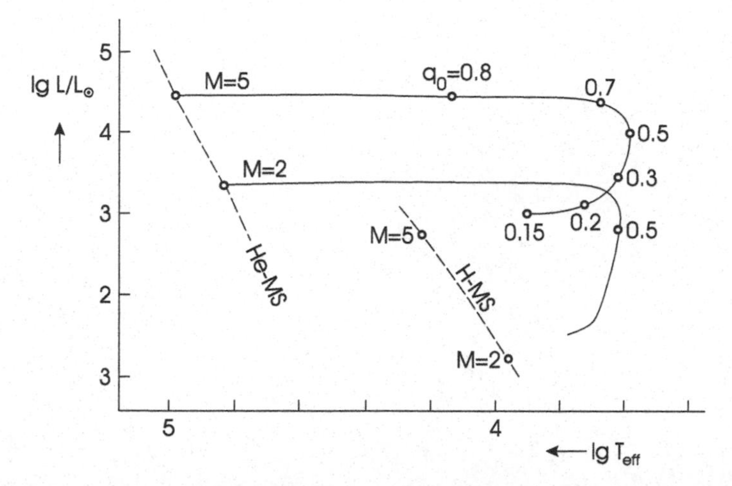

Fig. 23.6 The *solid lines* connect models of the same stellar mass M (in $M_{\odot}$) on the different generalized main sequences of Fig. 23.5. *Labels* along the lines give the q_0 values of the generalized main sequences (After Lauterborn, Refsdal, Weigert, 1971a)

envelope $M(1 - q_0)$ increases, which gives an increasing energy production L_H of the hydrogen shell source, such that the total luminosity $L = L_{He} + L_H$ can remain almost constant. The situation changes when q_0 drops below, say, 0.7. The "helium luminosity" L_{He} then decreases so strongly that it is compensated no longer by the increase of L_H, which eventually dominates L completely.

Not only the cases $q_0 = 0$ and $q_0 = 1$ which give the ZAMS and the helium main sequence, but also the cases in between sometimes give insight how stars behave, for instance, in the case where in a close binary system mass flows from one star to its companion. If a primary of, say, one solar mass evolves it forms a helium core, so it resembles a star on a generalized main sequence with a certain value of q_0. While the evolution goes on q_0 grows while simultaneously the star becomes a red giant. If before the onset of helium burning the surface of it comes close to the companion (to be more precise: when it fills the *Roche lobe*), mass flows from the red giant onto the surface of the still unevolved secondary until only the helium core is left and the original primary after thermal adjustment has become a star of the helium main sequence. In the HR diagram the star has moved from the Red Giant branch to the helium main sequence while its value of q_0 has grown during the mass loss.

Chapter 24
The Hayashi Line

We have seen that convection can occur in quite different regions of a star. In this section we consider the limiting case of *fully convective stars,* i.e. stars which are convective in the whole interior from centre to photosphere, while only the atmosphere remains radiative.

The Hayashi line (HL) is defined as *the locus in the Hertzsprung–Russell diagram of fully convective stars of given parameters* (mass M and chemical composition). Note that for each set of the parameters, such as mass or chemical composition, there is a separate Hayashi line. These lines are located far to the right in the Hertzsprung–Russell diagram, typically at $T_{\mathrm{eff}} \approx 3,000 \ldots 5,000\,\mathrm{K}$, and they are very steep, in large parts almost vertical.

From the foregoing definition one may not immediately realize the importance of this line. However, the HL also represents *a borderline between an "allowed" region* (on its left) *and a "forbidden" region* (on its right) in the Hertzsprung–Russell diagram for all stars with these parameters, provided that they are in hydrostatic equilibrium and have a fully adjusted convection. The latter means that, at any time, the convective elements have the properties (for instance the average velocity) required by the mixing-length theory. Changes in time of the large-scale quantities of the stars are supposed to be slow enough for the convection to have time to adjust to the new situation; otherwise one would have to use a theory of time-dependent convection. Since hydrostatic and convective adjustment are very rapid, stars could survive on the right-hand side of the HL only for a very short time.

In addition, parts of the early evolutionary tracks of certain stars may come close to, or even coincide with, the HL. It is certainly significant for the later evolution of stars, which is clearly reflected by observed features (e.g. the ascending branches of the Hertzsprung–Russell diagrams of globular clusters). One may even say that the importance of the HL is only surpassed by that of the main sequence. It is all the more surprising that its role was not recognized until the early 1960s when the work of Hayashi (1961) appeared. The late recognition of the HL may partly be because its properties are derived from involved numerical calculations. In the following we will use extreme simplifications in order to make some basic characteristics of the HL plausible.

R. Kippenhahn et al., *Stellar Structure and Evolution*, Astronomy and Astrophysics Library, DOI 10.1007/978-3-642-30304-3_24, © Springer-Verlag Berlin Heidelberg 2012

24.1 Luminosity of Fully Convective Models

Let us consider the different ways in which the luminosity is coupled to the pressure-temperature stratification of radiative and convective stars.

For regions with radiative transport of energy, we can write the "radiative luminosity" $l_{\rm rad} = 4\pi r^2 F_{\rm rad}$ according to (7.2) as

$$l_{\rm rad} = k'_{\rm rad} \nabla, \tag{24.1}$$

with the usual notation $\nabla = d \ln T / d \ln P$ and the "radiative coefficient of conductivity"

$$k'_{\rm rad} = \frac{16\pi acG}{3} \frac{T^4 m}{\kappa P}. \tag{24.2}$$

If a stratification of P and T is given, then the luminosity l_{rad} is obviously determined and can be easily calculated from (24.1).

For convective transport of energy by adiabatically rising elements we can write accordingly from (7.7) the convective luminosity as

$$l_{\rm con} = k'_{\rm con} (\nabla - \nabla_{\rm ad})^{3/2} \tag{24.3}$$

with the coefficient

$$k'_{\rm con} = \frac{\pi}{\sqrt{2}} \left(\frac{\ell_m}{H_P} \right)^2 r^2 c_P T (\varrho P \delta)^{1/2}. \tag{24.4}$$

Here we have made use of the hydrostatic equation and the definition (6.8) of the pressure scale height. The mixing length $\ell_{\rm m}$ was defined in Sect. 7.1.

In principle, we can again assume the luminosity to be determined using (24.3) for a given P–T stratification. In practice, however, we would never be able to calculate $l_{\rm con}$ from this equation for the stellar interior, since it would require the knowledge of the value of ∇ with inaccessible accuracy. The point is that $l_{\rm con}$ is not proportional to the gradient ∇ itself but rather to a power of the excess over the adiabatic gradient, $\nabla - \nabla_{\rm ad}$, which may be as small as 10^{-7} for very effective convection (see Sect. 7.3). Therefore the convective conductivity $k'_{\rm con}$ must be very high, since large luminosities $l_{\rm con}$ are carried. This may be looked at in another way: by solving (24.3) for ∇ and writing

$$\nabla = \nabla_{\rm ad}(1 + \varphi), \tag{24.5}$$

we see that the luminosity influences the T gradient only through the tiny correction $\varphi (\approx 10^{-7})$:

$$\varphi = \left[\frac{l_{\rm con}}{\nabla_{\rm ad}^{3/2} k'_{\rm con}} \right]^{2/3}. \tag{24.6}$$

Therefore one usually neglects this correction in the case of effective convection and takes simply

$$\nabla = \nabla_{ad}, \tag{24.7}$$

which is equivalent to assuming an infinite conductivity k'_{con}. Then de facto the luminosity is decoupled from the $T-P$ structure.

In order to fix the luminosity of a fully convective star, we have to appeal to the only region where the gradient is sufficiently non-adiabatic. This is the radiative atmosphere and a layer immediately below where the convection is ineffective, i.e. strongly superadiabatic. We have seen that then the transport of energy is essentially radiative (in spite of violent convective motions), and we can again use (24.1). By this argumentation one arrives at the statement that the structure of the outermost layers determines the luminosity of a fully convective star. This means, on the other hand, that such stars are very sensitive to all influences and uncertainties near their outer boundary.

Of course, if the energy production is prescribed, one would rather say that the outer layers have to adjust to this value of L (for this point of view, see Sect. 24.5).

24.2 A Simple Description of the Hayashi Line

In order to derive some typical properties of the HL analytically, we shall use an extremely crude model for fully convective stars (Further refinements of the picture, though possible, would not be worth the large additional complications involved.).

We have seen that nearly all of the interior part of convective stars has an adiabatic stratification, such that $d \ln T / d \ln P = \nabla_{ad}$. We shall assume that this simple relation between P and T holds for the whole interior up to the photosphere, i.e. we neglect the superadiabaticity in the range immediately below the photosphere. We also neglect the depression of ∇_{ad} in those regions near the surface where H and He are partially ionized (see Figs. 11.2 and 14.1). We thus simply assume ∇_{ad} to be constant throughout the star's interior, say $\nabla_{ad} = 0.4$, which is the value for a fully ionized ideal gas. With these simplifications we certainly introduce errors in the $P-T$ stratification. However, they will be nearly the same for neighbouring models, and we can hope to obtain at least the correct differential behaviour.

We then have for the whole interior the simple $P-T$ relation

$$P = CT^{1+n}, \tag{24.8}$$

i.e. the star is polytropic with an index $n = 1/\nabla_{ad} - 1 = 3/2$, and we can use the earlier results for such stars (see Chap. 19). The constant C is related to the polytropic constant K defined in (19.3). With $P = \Re \varrho T/\mu$, one finds $C = K^{-n}(\Re/\mu)^{1+n}$. K and C are constant only within one model, but vary from star to star, which means that we do not have a mass-radius relation. From (19.9) and (19.19) it follows that

$$K \sim \varrho_c^{1/3} A^{-2} \sim \varrho_c^{1/3} R^2 \sim M^{1/3} R, \tag{24.9}$$

so that

$$C = C' R^{-3/2} M^{-1/2}, \tag{24.10}$$

where the constant C' is known for given n and μ.

Relation (24.8) is now assumed to hold as far as the photosphere, where the optical depth $\tau = 2/3$, $P = P_0$, $T = T_{\rm eff}$, $r = R$, and $m = M$. Above this point we suppose to have a radiative atmosphere with a simple absorption law of the form

$$\kappa = \kappa_0 P^a T^b. \tag{24.11}$$

Integration of the hydrostatic equation through the atmosphere yields the photospheric pressure [cf. (11.13), where $\bar{\kappa}$ is replaced by (24.11)] as

$$P_0 = \text{constant} \left(\frac{M}{R^2} T_{\rm eff}^{-b} \right)^{\frac{1}{a+1}}. \tag{24.12}$$

We now fit this to the interior solution by setting $P = P_0, T = T_{\rm eff}$ in (24.8) and then eliminating P_0 with (24.12). For given values of M and μ this yields a relation between R and $T_{\rm eff}$, or between R and L, since $L \sim R^2 T_{\rm eff}^4$. Thus, any value of R corresponds to a certain point in the Hertzsprung–Russell diagram. The interior solutions form a one-dimensional manifold, since the constant C contains the free parameter R for given M [and given μ, see (24.10)]. In the Hertzsprung–Russell diagram this is reflected by a one-dimensional manifold of points defining the Hayashi line.

The fitting procedure is illustrated in Fig. 24.1. Each interior solution of the form (24.8) with $n = 3/2$ is represented in this diagram by a straight line:

$$\lg T = 0.4 \lg P + 0.4 \left(\frac{3}{2} \lg R + \frac{1}{2} \lg M - \lg C' \right). \tag{24.13}$$

For fixed values of M and μ, each of these lines is characterized by a value of R. The atmospheric solutions (24.12) are another set of straight lines in Fig. 24.1:

$$(a+1) \lg P_0 = \lg M - 2 \lg R - b \lg T_{\rm eff} + \text{constant}. \tag{24.14}$$

The intersection of a line of the first set with a line of the second set, both with the same value of R, fixes the corresponding value of $T_{\rm eff}$ (and of P_0). From R and $T_{\rm eff}$ we have L, i.e. a point in the Hertzsprung–Russell diagram. We then obtain the Hayashi line by a continuous variation of R.

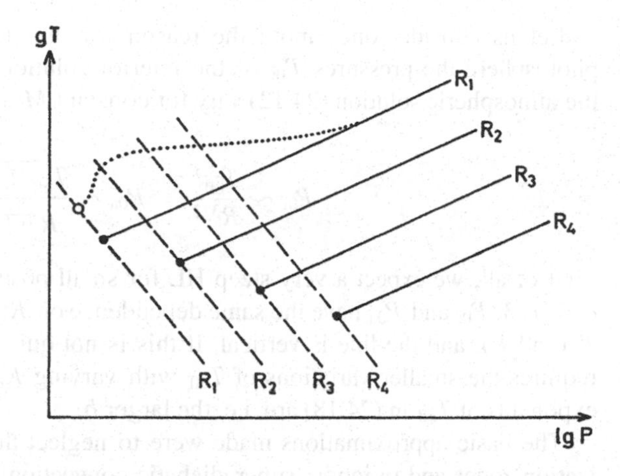

Fig. 24.1 Fit of a polytropic ($n = 3/2$) interior solution (*solid line*) with an atmospheric condition (*dashed line*) for different values of R ($R_1 > R_2 > R_3 > R_4$). The photospheric points obtained by this fit are marked by *dots*. The *dotted line* illustrates schematically the effects of superadiabatic convection and depression of ∇_{ad} in an ionization zone for $R = R_1$

The formalism for this procedure, as described, yields immediately an equation for the Hayashi line in the Hertzsprung–Russell diagram:

$$\lg T_{\text{eff}} = A \lg L + B \lg M + \text{constant} \tag{24.15}$$

with the coefficients

$$A = \frac{0.75a - 0.25}{b + 5.5a + 1.5}, \quad B = \frac{0.5a + 1.5}{b + 5.5a + 1.5}. \tag{24.16}$$

We now need typical values for the exponents a and b in the atmospheric absorption law (24.11). An important property of fully convective stars can immediately be concluded from the discussion in Sect. 11.3: such stars must have very low values of T_{eff}, i.e. *the Hayashi line must be far to the right in the Hertzsprung–Russell diagram.* For atmospheres this means that in most parts $T \lesssim 5 \times 10^3$ K, and H^- absorption will provide the dominant contribution to κ. If hydrogen is essentially neutral, the free electrons necessary for the formation of H^- ions are provided by the heavier elements (see Sect. 17.5). A very rough interpolation gives $a \simeq 1, b \simeq 3$. With these values (24.16) yields the coefficients

$$A = 0.05, \quad B = 0.2. \tag{24.17}$$

According to (24.15), the slope of the Hayashi line in the Hertzsprung–Russell diagram is $\partial \lg L / \partial \lg T_{\text{eff}} = 1/A$. Since $A \ll 1$, we conclude that *the Hayashi line must be very steep.* The value of $B \equiv \partial \lg T_{\text{eff}} / \partial \lg M$ means that *the Hayashi line shifts slightly to the left in the Hertzsprung–Russell diagram for increasing M.* These qualitative predictions, although derived from very crude assumptions, are fully supported by the numerical results (see Fig. 24.3).

Let us consider once more the reason for the steepness of the HL. At the photosphere the pressures P_{0i} of the interior solution (24.8), (24.10) and P_{0a} of the atmospheric solution (24.12) vary for constant M as

$$P_{0i} \sim \frac{T_{eff}^{2.5}}{R^{3/2}}, \qquad P_{0a} \sim \frac{T_{eff}^{-\frac{b}{a+1}}}{R^{\frac{2}{a+1}}}. \tag{24.18}$$

First of all, we expect a very steep HL for small positive values of a. In fact, for $a = 1/3$, P_{0i} and P_{0a} have the same dependence on R; then T_{eff} does not vary with R (and L), and the line is vertical. If this is not quite fulfilled, the fit $P_{0i} = P_{0a}$ requires the smaller variations of T_{eff} with varying R, the more different the two exponents of T_{eff} in (24.18) are, i.e. the larger b.

The basic approximations made were to neglect the depression of ∇_{ad} in ionization zones and to ignore superadiabatic convection. The dotted line in Fig. 24.1 indicates how these effects change the $P-T$ structure relative to a simple polytrope. One sees that they tend to increase the effective temperature. The precise value of T_{eff} obviously depends on the detailed structure of the outermost envelope. The extension and the depth of the ionization zones and the superadiabatic layers change systematically with L. This has the consequence that, in better approximations, the coefficient A in (24.15) changes sign at $L \simeq L_\odot$. It is positive for smaller L, and negative for larger L, so that the HL is convex relative to the main sequence.

Another important conclusion is that the whole uncertainty which remained in the mixing-length theory of ineffective convection must occur as a corresponding uncertainty in the precise value of T_{eff} for the HL.

Finally, we note that the chemical composition enters into the position of the HL in two ways. The interior is affected, since the polytropic constant C depends on μ via C' [see (24.10)], and the outer layers are particularly affected via the opacity κ.

24.3 The Neighbourhood of the Hayashi Line and the Forbidden Region

We now consider stars in hydrostatic equilibrium that are close to, but not exactly on, their HL. Certainly the stars cannot be fully convective with an adiabatic interior (otherwise they would be on the HL). Their interior is then no longer a simple polytrope. They do not even have to be chemically homogeneous, since they are not fully mixed by the turbulent motions. We must therefore expect that an analytical treatment will be much more complicated. We will nevertheless try to give some simple arguments which may help to make the numerical results plausible. In the following, we treat models with a fixed value of M and the same chemical composition (at least in their outer layers).

An important indication can be obtained from the discussion of the envelope integrations in Sect. 11.3. When integrating inwards into models with different T_{eff} (but with the same parameters M and μ and, say, the same L), we will reach a radiative region the earlier, the larger T_{eff}. In other words, in models left of the HL we will encounter a radiative region before reaching the centre. In these regions, the gradient $\nabla < \nabla_{\mathrm{ad}}$. Let us consider some average $\bar{\nabla}$ obtained by averaging over the whole interior (where we again neglect the complications in the outermost parts of the envelope). On the HL we have $\bar{\nabla} = \nabla_{\mathrm{ad}}$. In a model to the left of the HL the radiative part decreases the average value such that $\bar{\nabla} < \nabla_{\mathrm{ad}}$. This suggests that we would have to allow $\bar{\nabla} < \nabla_{\mathrm{ad}}$ in models to the right of the HL.

In order to prove this we treat models with a constant gradient $\nabla = \bar{\nabla}$ in the interior and vary $\bar{\nabla}$ slightly around ∇_{ad}. We then have again polytropic stars with slightly different n (around 3/2). The interior solution is written as

$$P = C_n T^{1+n}, \tag{24.19}$$

where $\bar{\nabla} = (1+n)^{-1}$ and, similarly to (24.10),

$$C_n = C_n' \mu^{-n-1} M^{1-n} R^{n-3}. \tag{24.20}$$

From now on we measure R and M in solar units. Then

$$C_n' = \frac{\Re^{n+1}}{4\pi G^n}(n+1)^n \left[-\left(\frac{dw}{dz}\right)_{z=z_n} \right]^{n-1} z_n^{n+1} R_\odot^{n-3} M_\odot^{1-n}. \tag{24.21}$$

We extend relation (24.19) to the photosphere ($P = P_0, T = T_{\mathrm{eff}}$), where we again eliminate P_0 by (24.12) and R by the relation $R = c_2 L^{1/2} T_{\mathrm{eff}}^{-2}$. This gives the locus in the Hertzsprung–Russell diagram. The factor of proportionality in (24.12) may be called c_1. Choosing for simplicity $a = 1, b = 3$ in the opacity law, we obtain

$$\lg T_{\mathrm{eff}} = \alpha_1 \lg L + \alpha_2 \lg M + \alpha_3 \lg \mu + \alpha_4 \lg C_n' + \alpha_5 \lg c_1 + \alpha_6 \lg c_2, \tag{24.22}$$

where the coefficients depend on n:

$$\alpha_1 = \frac{2-n}{13-2n}, \quad \alpha_2 = \frac{2n-1}{13-2n}, \quad \alpha_3 = \frac{2(1+n)}{13-2n},$$
$$\alpha_4 = \frac{-2}{13-2n}, \quad \alpha_5 = -\alpha_4, \quad \alpha_6 = 2\alpha_1. \tag{24.23}$$

The α_i do not vary too much with small deviations of n from 3/2. This means, for example, since α_1 determines the slope, lines of neighbouring values of n are nearly parallel to the HL. Without loss of generality, we may consider particular models on and close to the HL with $L = M = \mu = 1$. The variation of $\lg T_{\mathrm{eff}}$ with n is then only due to the variation of the last three terms in (24.22). One finds that

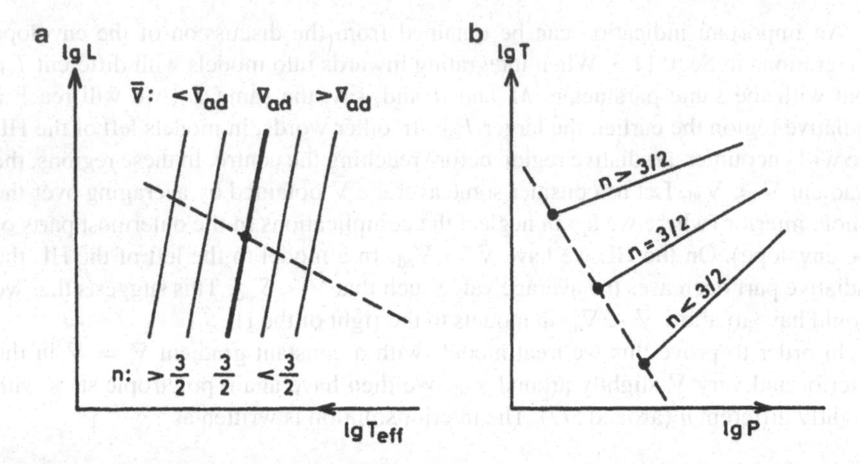

Fig. 24.2 (**a**) In the Hertzsprung–Russell diagram, the Hayashi line ($n = 3/2$, *heavy line*) is indicated, together with some neighbouring lines for interior polytropes with $n > 3/2$ and $< 3/2$. (**b**) The same as Fig. 24.1 but with three different polytropic interior solutions for the same value of R

$\partial \lg T_{\mathrm{eff}}/\partial n > 0$: the stars move to the right in the Hertzsprung–Russell diagram with decreasing n (i.e. increasing $\bar{\nabla}$).

Thus, we have to expect the following situation (see Fig. 24.2): left of the HL we have $\bar{\nabla} < \nabla_{\mathrm{ad}}$ and some part of the model is radiative. On the HL, the model is fully convective with $\bar{\nabla} = \nabla_{\mathrm{ad}}$. Models to the right of the HL should have $\bar{\nabla} > \nabla_{\mathrm{ad}}$, which means that they should have a superadiabatic stratification in their very interior (aside from the outermost zone of ineffective convection).

The mixing-length theory has shown that a negligibly small excess of ∇ over ∇_{ad} suffices in order to transport any reasonable luminosity in the deep interior of stars. Then, what happens with a star that by some arbitrary means (e.g. initial conditions) has been brought to a place to the right of the HL, such that some region in its deep interior has remarkably large values of $\nabla - \nabla_{\mathrm{ad}} > 0$? The results are large convective velocities $v_{\mathrm{conv}} \sim (\nabla - \nabla_{\mathrm{ad}})^{1/2}$ and corresponding convective fluxes [cf. (24.3)]. These cool the interior and heat the upper layers rapidly until the gradient is lowered to $\nabla \approx \nabla_{\mathrm{ad}}$ and the star has moved to the HL. This will happen within the short timescale for the adjustment of convection.

Another possibility for a star being situated to the right of its HL is, of course, that it is not in hydrostatic equilibrium (which is assumed for the interior solution). But a deviation from this equilibrium will be removed in the timescale for hydrostatic adjustment, which is even shorter.

Therefore the HL is in fact a borderline between an "allowed" region (left) and a "forbidden" region (right) for stars of given M and composition that are in hydrostatic equilibrium and have a fully adjusted convection.

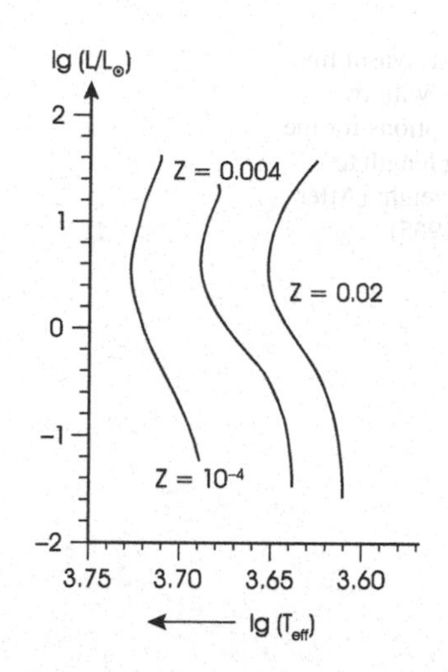

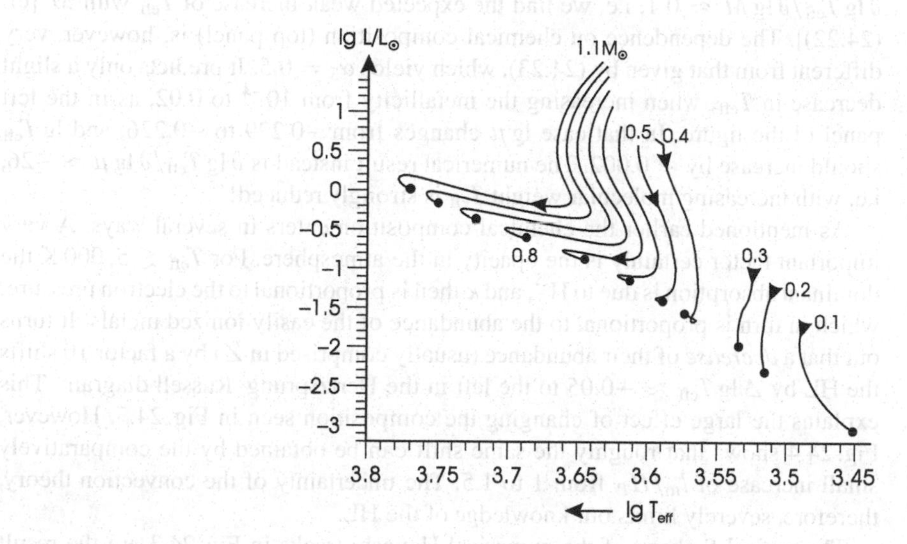

Fig. 24.3 *Top*: The position of Hayashi lines for stars of $M = 0.8M_\odot$ but different composition. The helium content is always 0.245, while Z varies from 10^{-4} to 0.02. *Bottom*: Pre-main-sequence evolution along the Hayashi line to the zero-age main sequence for stars between 0.1 and 1.1 $M_\odot$ and a solar-like composition (Data courtesy S. Cassisi)

24.4 Numerical Results

There are many results available giving the position of Hayashi lines for stars of widely ranging mass and chemical composition and for different assumptions in the convection theory. The latter concerns in particular the ratio of mixing length to pressure scale height used for calculating the superadiabatic envelope.

Figure 24.3 shows typical results of calculations for stellar masses of up to 1.1 $M_\odot$. One sees that indeed the HLs plotted here are very steep, the exact slope depending mainly on L. The dependence on M (lower panel) is roughly given by

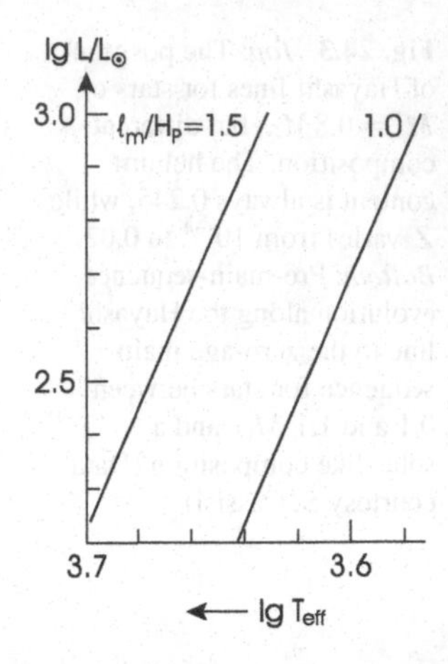

Fig. 24.4 The Hayashi line for $M = 5M_\odot$ with two different assumptions for the ratio of mixing length to pressure scale height (After Henyey et al. 1965)

$\partial \lg T_{\rm eff}/\partial \lg M \approx 0.1$, i.e. we find the expected weak increase of $T_{\rm eff}$ with M [cf. (24.22)]. The dependence on chemical composition (top panel) is, however, very different from that given by (24.23), which yields $\alpha_3 = 0.5$. It predicts only a slight decrease in $T_{\rm eff}$, when increasing the metallicity from 10^{-4} to 0.02, as in the left panel of the figure. In that case $\lg \mu$ changes from -0.229 to -0.226, and $\lg T_{\rm eff}$ should increase by ≈ 0.002. The numerical result instead is $\partial \lg T_{\rm eff}/\partial \lg \mu \approx -26$, i.e. with increasing molecular weight $T_{\rm eff}$ is strongly reduced!

As mentioned earlier the chemical composition enters in several ways. A very important factor certainly is the opacity in the atmosphere. For $T_{\rm eff} \lesssim 5{,}000\,{\rm K}$ the dominant absorption is due to H^-, and κ then is proportional to the electron pressure, which in turn is proportional to the abundance of the easily ionized metals. It turns out that a *decrease* of their abundance (usually comprised in Z) by a factor 10 shifts the HL by $\Delta \lg T_{\rm eff} \approx +0.05$ to the left in the Hertzsprung–Russell diagram. This explains the large effect of changing the composition seen in Fig. 24.3. However, Fig. 24.4 shows that roughly the same shift can be obtained by the comparatively small increase of $l_{\rm m}/H_P$ from 1 to 1.5. The uncertainty of the convection theory, therefore, severely limits our knowledge of the HL.

The typical S-shape of the numerical Hayashi tracks in Fig. 24.3 are the result of the sign change of coefficient A in (24.15), which was mentioned at the end of Sect. 24.2. At the lowest end of the Hayashi tracks the models develop a radiative core and begin to bend back to the main sequence, where they end once nuclear burning has started at the centre, supplying the energy radiated from the surface. This is the situation discussed in Sect. 24.3.

Thus, the HLs are far away from the main sequence in the upper part of the diagram, and approach it in the lower part. This fact will turn out to influence the evolutionary tracks of stars of different M. Recall that the main-sequence stars were found to be fully convective for $M \lesssim 0.25M_\odot$ (see Sect. 22.3). This obviously means that the corresponding Hayashi lines cross the main sequence there.

24.5 Limitations for Fully Convective Models

In order to describe the HL, we have considered models for which the convection was postulated to reach from centre to surface. This provided a polytropic interior structure with typical decoupling from the luminosity. We have not yet asked whether the physical situation will in fact allow the onset of convection throughout the star. This depends on the distribution of the energy sources.

According to the Schwarzschild criterion (6.13), a chemically homogeneous layer will be convective if

$$\nabla_{rad} \geq \nabla_{ad}, \tag{24.24}$$

where the radiative gradient [see (5.28)] is

$$\nabla_{rad} \sim \frac{\kappa l P}{T^4 m}. \tag{24.25}$$

If the energy sources were completely arbitrary, we could choose their distribution so that (24.24) is violated at some point and the model could not be *fully* convective. A trivial example would be a central core without any sources, with the result that there $l = 0$, i.e $\nabla_{rad} = 0$. Then the core must be radiative. On the other hand, we have the best chance of finding convection throughout a star of given L if the sources are highly concentrated towards the centre (in the extreme: a point source), which gives almost $l = L$ everywhere.

We consider a contracting polytrope (see Sect. 20.3) without nuclear energy sources, which is of interest for early stellar evolution. According to (20.41) the energy generation rate is then only proportional to T, which means a rather weak central concentration. For the sake of simplicity we even go a step further and assume constant energy sources with

$$\frac{l}{m} = \frac{L}{M} = \text{constant}. \tag{24.26}$$

We again use the opacity law (24.11) and the polytropic relation (24.8) with $n = 1.5$ (corresponding to $\nabla = \nabla_{ad} = 0.4$). Equation (24.25) then gives

$$\nabla_{rad} \sim \frac{L}{M} C^{1+a} T^{b-4+2.5(1+a)}. \tag{24.27}$$

For a typical Kramers opacity with $a = 1, b = -4.5$ this becomes $\nabla_{rad} \sim T^{-3.5}$. Indeed, for all reasonable interior opacities, ∇_{rad} has a minimum at the centre and increases outwards. Therefore the centre is the first point in a fully convective star where ∇_{rad} drops below ∇_{ad} (and a radiative region starts to develop) if L decreases below a minimum value L_{min}.

The constant C depends on M and R as given by (24.10), and $T \sim T_c \sim M/R$ after (20.24). Introducing this into (24.27) we obtain

$$\nabla_{\rm rad} \sim LM^{b-5+2(1+a)}R^{-b+4-4(1+a)}. \tag{24.28}$$

Let us again set $a = 1, b = -4.5$, which gives

$$\nabla_{\rm rad} \sim LM^{-5.5}R^{0.5}. \tag{24.29}$$

For models on the HL, the effective temperatures vary only a very little and we simply take $R \sim L^{1/2}$. Then,

$$\nabla_{\rm rad} \sim L^{1.25}M^{-5.5}. \tag{24.30}$$

For any given value of M the luminosity reaches $L_{\rm min}$ if the central value of $\nabla_{\rm rad}$ has dropped to 0.4. According to (24.30), $L_{\rm min}$ depends on M as

$$L_{\rm min} \sim M^{4.4}. \tag{24.31}$$

This minimum luminosity (down to which models of the specified type on the HL remain fully convective) decreases strongly with M. The decrease is in fact steeper than that given by the $M - L$ relation of the main sequence. This provides the possibility that the HL for very small M can cross the main sequence without reaching $L_{\rm min}$.

Note, however, that strictly speaking a "minimum luminosity" always refers to a fixed distribution of the energy sources.

Chapter 25
Stability Considerations

Even the most beautiful stellar model is not worth anything if one does not know whether it is stable or not. Stability is discussed again and again throughout this book. Here we review the different types of stability considerations necessary for stars. We intend to make the basic mechanisms and concepts plausible rather than present the full formalism; the reader will find this, for example, in the review article by Ledoux (1958).

25.1 General Remarks

It is not easy to give a very general concept of stability that is applicable to all possible cases. Different definitions are discussed in La Salle and Lefschetz (1961). We may use, for example, the following: let the solution of a system of (time-dependent) differential equations be a set of functions $y_1(t)$, $y_2(t)$, ... which we comprise in the symbol $y(t)$. We define a "distance" between two such solutions $y^a(t)$, $y^b(t)$ by

$$||y^a(t) - y^b(t)|| := \sum_i \left[\left(y_i^a(t) - y_i^b(t) \right)^2 \right]. \tag{25.1}$$

We then call the solution $y^a(t)$ stable at $t = t_0$ if for any $t_1 > t_0$ and for any small positive number δ there exists a small positive number δ such that any other solution $y^b(t)$ having the distance $||y^a(t_0) - y^b(t_0)|| < \delta$ at $t = t_0$ will keep a distance $||y^a(t_1) - y^b(t_1)|| < \varepsilon$.

This definition in plain words says that a solution is stable at a given point t_0 if all solutions that at $t = t_0$ are in its neighbourhood remain neighbouring solutions. The problems we are interested in can be reduced to first-order systems in time. Therefore the above definition of neighbouring solutions also guarantees neighbouring derivatives.

R. Kippenhahn et al., *Stellar Structure and Evolution*, Astronomy and Astrophysics Library, DOI 10.1007/978-3-642-30304-3_25, © Springer-Verlag Berlin Heidelberg 2012

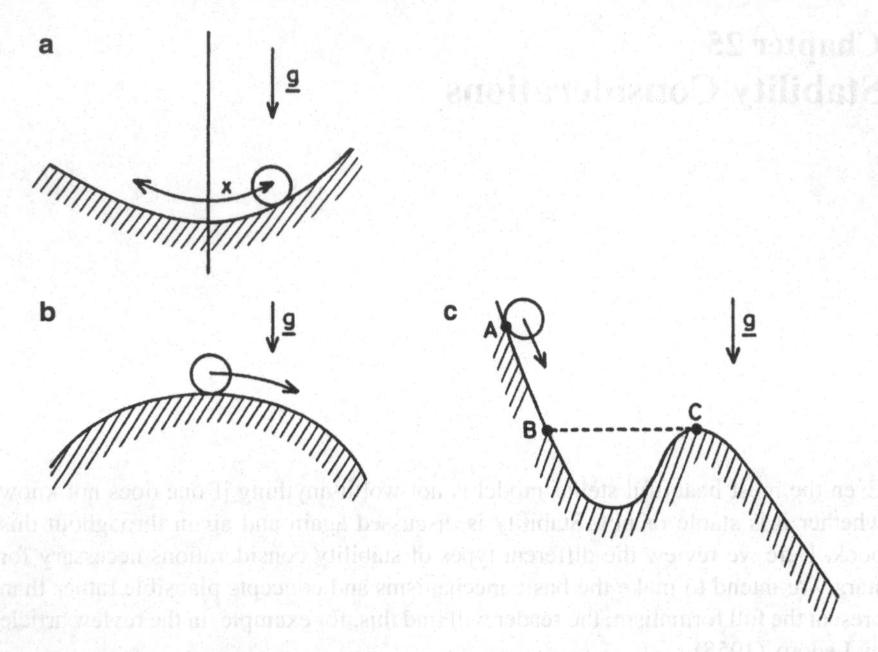

Fig. 25.1 An example of stability in mechanics. A *ball* on a surface under the influence of gravity (**a**) in stable and (**b**) in unstable equilibrium. In (**c**) the motion starting at point A is stable, but, starting with zero velocity at point B, the motion is unstable

One normally is familiar with stability problems in mechanics. We recall a few simple examples, the first being the freely rolling ball on a curved surface which is concave in the direction opposite to gravity (see Fig. 25.1a). One solution is that of equilibrium, where the ball rests in the lowest position. The initially neighbouring position is obtained by a small perturbation, say, by a slight horizontal displacement. The ball will then move about the equilibrium position, but it will never increase its distance above its initial value: the equilibrium position is stable and friction would merely restore the ball to its equilibrium position. In the case of a convex surface (see Fig. 25.1b) the equilibrium is unstable, since after a small displacement the ball will move further and further from the equilibrium position. While these examples deal with the stability of an equilibrium in which the solution is time independent, our general definition also concerns time-dependent solutions. The motion of a ball rolling on the surface in Fig. 25.1c can be stable or unstable. The motion is stable if it starts with zero velocity at a point A *above* B (non-periodic motion), or *below* B (periodic motion). But a motion starting exactly at B with zero velocity and ending at rest at C is unstable: a slight perturbation of the initial conditions can either produce a periodic motion (the ball never overcomes the summit C) or cause the ball to roll beyond C and never come back.

When considering the influence of friction, one may naïvely expect that it stabilizes an otherwise unstable motion, since it uses up energy. But the following example will show that friction can also produce instability.

We again consider the ball in the spherical bowl (Fig. 25.1a). But now we assume that the bowl is rotating with an angular velocity ω around a vertical axis through the minimum. Without friction no angular momentum can be transferred to the ball which therefore does not know anything of rotation and behaves as in the non-rotating case: the lowest position is stable. If there is friction, however, and the ball is "kicked" out of its lowest (equilibrium) position, it will take up angular momentum from the rotating bowl. For sufficiently large ω the ball goes to a new equilibrium position outside the axis around which it rotates with ω and where the tangential components of centrifugal and gravity forces balance each other. The lowest position has obviously become unstable by the inclusion of friction.

25.2 Stability of the Piston Model

Closer to stars than the above mechanical examples is the piston model introduced in Sect. 2.7, since it also incorporates thermal effects. We consider the stability of an equilibrium solution with a certain constant height h. Will a solution originating from a small displacement of the piston remain in its neighbourhood? This stability problem has already been discussed in Sect. 6.6, where we made approximations appropriate for the illustration of the stability of convective blobs. We now improve the model by adding some complications typical of stars.

25.2.1 *Dynamical Stability*

In this case one assumes that there is no heat leakage, no nuclear energy generation, and no absorption, i.e. $\varepsilon = \kappa = \chi = 0$ in (5.39). Therefore the entropy of the gas remains constant during the displacement of the piston. In Sect. 6.6, we investigated the resulting (adiabatic) oscillations of the model around the equilibrium position, though with constant weight G^* only. We now allow G^* to vary with height $[G^* = G^*(h)]$ as we did in Sect. 3.2. This can be achieved, for instance, by putting the piston model into an inhomogeneous gravitational field. Then the equation of motion (2.34)

$$M^* \frac{d^2 h}{dt^2} = -G^* + PA \tag{25.2}$$

with the perturbations (6.30) gives after linearization, instead of (6.32),

$$M^* h_0 \omega^2 x + P_0 A_p - G_h^* G_0^* x = 0. \tag{25.3}$$

Here $G_h^* := d \ln G^* / d \ln h (< 0)$, while $G_0^* = P_0 A = g_0 M^*$ is the equilibrium value of G^* and g_0 is that of g. With the perturbed perfect gas equation (6.31) we find

$$\left[\frac{\omega^2 h_0}{g_0} - G_h^* - 1\right] x + \vartheta = 0. \tag{25.4}$$

This together with the adiabatic equation (6.36),

$$(\gamma_{\text{ad}} - 1)x + \vartheta = 0, \tag{25.5}$$

gives for the eigenvalues of adiabatic oscillations $\omega = +\omega_{\text{ad}}$ and $\omega = -\omega_{\text{ad}}$ with

$$\omega_{\text{ad}} = \left[(\gamma_{\text{ad}} + G_h^*)\frac{g_0}{h_0}\right]^{1/2}, \tag{25.6}$$

which replaces (6.37). Recall that the perturbation changes with time as $e^{i\omega t}$. We see that ω_{ad} is a real number only as long as $\gamma_{\text{ad}} > -G_h^*$. In this case the small perturbation is followed by a periodic oscillation which remains small for all times. It is therefore stable in the sense of our definition of stability at the beginning of this paragraph. But if $\gamma_{\text{ad}} < -G_h^*$, then ω_{ad} is imaginary and one of the eigenvalues ω gives an amplitude growing exponentially in time: the equilibrium solution is unstable (We will see in Sect. 25.3.2 that for stars the analogue of $\gamma_{\text{ad}} > -G_h^*$ is $\gamma_{\text{ad}} > 4/3$.).

25.2.2 *Inclusion of Non-adiabatic Effects*

We now drop the assumption of strict adiabaticity. Non-adiabatic changes were previously included in Sect. 5.4 (refer also to the last part of Sect. 6.6). The energy equation of the piston model (5.39) includes the non-adiabatic terms for nuclear generation ε, absorption κ, and heat leakage χ. We consider ε and κ as functions of P and T, while χ shall be constant. Let F be the radiative flux through the gas. In the case of thermal equilibrium (vanishing time derivatives) we have [see (5.37)]

$$\varepsilon_0 m^* + \kappa_0 m^* F = \chi(T_0 - T_s), \tag{25.7}$$

where subscript 0 indicates the equilibrium and subscript s the surroundings. If we perturb this equilibrium according to (6.30), we find for the perturbations after linearization

$$\begin{aligned} &i\omega(c_v m^* T_0 \vartheta + P_0 A h_0 x) \\ &\quad = \varepsilon_0 m^*(p\,\varepsilon_P + \vartheta \varepsilon_T) + \kappa_0 m^* F(p\kappa_P + \vartheta\kappa_T) - \chi T_0 \vartheta, \end{aligned} \tag{25.8}$$

where the derivatives

$$\varepsilon_P = \left(\frac{\partial \ln \varepsilon}{\partial \ln P}\right)_T, \quad \varepsilon_T = \left(\frac{\partial \ln \varepsilon}{\partial \ln T}\right)_P,$$

$$\kappa_P = \left(\frac{\partial \ln \kappa}{\partial \ln P}\right)_T, \quad \kappa_T = \left(\frac{\partial \ln \kappa}{\partial \ln T}\right)_P \tag{25.9}$$

are taken at the values P_0, T_0.

The equation of motion (25.2) yielded (25.4) for which we now assume constant weight of the piston ($G_h^* = 0$, giving dynamical stability):

$$\left(\frac{\omega^2 h_0}{g_0} - 1\right) x + \vartheta = 0. \tag{25.10}$$

Since $\varrho \sim h^{-1}$, the equation of state for an ideal (or "perfect") gas gives (6.31)

$$p = \vartheta - x. \tag{25.11}$$

System (25.8), (25.10) and (25.11) comprises three linear homogeneous algebraic equations for the perturbations p, ϑ, x. To find a solution it is necessary that the determinant of the coefficients vanishes:

$$\frac{h_0}{g_0} \mathrm{i} u_0 \omega^3 - \frac{h_0}{g_0}(e_P + e_T)\omega^2 - \frac{5}{3} u_0 \mathrm{i}\omega + e_T = 0 \tag{25.12}$$

with

$$e_P = \varepsilon_0 \varepsilon_P + \kappa_0 F \kappa_P, \quad e_T = \varepsilon_0 \varepsilon_T + \kappa_0 F \kappa_T - \frac{\chi T_0}{m^*}, \quad u_0 = c_v T_0, \tag{25.13}$$

where for the last relation we have assumed the gas to be ideal and monatomic (Note that $P_0 A h_0 / m^* = P_0/\varrho_0 = 2u_0/3$.). Equation (25.12) becomes one with *real* coefficients if instead of ω we use the eigenvalue $\sigma := \mathrm{i}\omega$,

$$\frac{h_0}{g_0} u_0 \sigma^3 - \frac{h_0}{g_0}(e_P + e_T)\sigma^2 + \frac{5}{3} u_0 \sigma - e_T = 0. \tag{25.14}$$

This is a third-order equation for the eigenvalue σ (or ω). While in the adiabatic case ($e_P = e_T = 0$) we obtained two solutions $\sigma = \pm\sigma_{\mathrm{ad}} = \pm i\omega_{\mathrm{ad}}$ (where ω_{ad} was real), we now have *three* eigenvalues. If the non-adiabatic terms e_P, e_T are small, we can expect that two (conjugate complex) eigenvalues lie near the adiabatic ones:

$$\sigma = \sigma_{\mathrm{r}} \pm \mathrm{i}\omega_{\mathrm{ad}}, \quad \omega_{\mathrm{ad}} = \left(\gamma_{\mathrm{ad}} \frac{g_0}{h_0}\right)^{1/2}, \tag{25.15}$$

where σ_{r} is real and $|\sigma_{\mathrm{r}}| \ll \omega_{\mathrm{ad}}$. While in the adiabatic case the oscillation was strictly periodic, the real part σ_{r} causes the amplitude of the oscillation to grow or decrease in time, depending on the sign of σ_r. Because of $|\sigma_r| \ll \omega_{ad}$ these changes take place over a time much longer than the oscillation period, actually on a scale corresponding to τ_{adj} in (5.41). This type of stability behaviour is called the *vibrational stability* (compare Sect. 6.6). If the oscillation grows in time, the solution leaves the neighbourhood of equilibrium, which therefore is unstable.

We now turn to the third root of (25.12) or (25.14), which occurs necessarily with the dissipative terms e_P, e_T. Instead of solving the third-order equation (25.14), we will follow some heuristic arguments. The addition of non-adiabatic terms has changed the rapid oscillations only to the extent that their amplitude varies on long timescales (of the order of σ_r^{-1}). We now look for the existence of a third solution changing with this long timescale only. Then the inertia terms can be neglected and, consequently, the terms with σ^3 and σ^2 disappear in (25.14). The solution of (25.14) for this so-called *secular stability* problem is

$$\sigma = \sigma_{sec} = i\omega_{sec} = \frac{3}{5}\frac{e_T}{u_0}. \tag{25.16}$$

For sufficiently small non-adiabaticity e_T, we can achieve $|\sigma_{sec}| \ll \omega_{ad}$, and neglecting the σ^2 and σ^3 terms in (25.14) was justified. If $\sigma_{sec} < 0$, any perturbation will decay within a kind of thermal adjustment time $\tau_{adj} \approx \sigma_{sec}^{-1}$ and the equilibrium is secularly stable. But if $\sigma_{sec} > 0$, then it will grow on that timescale (independently of vibrational stability): The equilibrium is secularly unstable.

We have now found the three well-known types of stability behaviour: dynamical, vibrational, and secular stability . This classification is possible since $|\omega_{ad}| \gg |\omega_{sec}|$, which is equivalent to saying that $\tau_{hydr} \ll \tau_{adj}$. From one type of stability one cannot draw any conclusions about the behaviour of another type, for example, a dynamically stable model can still be vibrationally or secularly unstable. If the model were dynamically unstable, the other instabilities would be of no interest since the model would move out of equilibrium long before any other instability can develop.

We will find more or less the same behaviour in stars where also $\tau_{hydr} \ll \tau_{adj} \approx \tau_{KH}$. However, there we cannot solve the eigenvalue problem analytically any more. This is the reason why we dwelt in such length on the stability of the piston model.

25.3 Stellar Stability

For the problem of *stellar* stability a very general definition, like that given at the beginning of Sect. 25.1, has to be taken with care. For example, a star may be stable in one phase (e.g. on the main sequence) and later on become unstable (e.g. in the Cepheid phase). At any stage of evolution the solution (the stellar model) is obtained for certain parameters, for instance, a certain chemical composition or a certain distribution of entropy. It is reasonable to ask whether this solution is stable in the following sense: Does a small perturbation decay rapidly compared to the change of the parameters of the model (e.g. its chemical composition)? Then we would call the model stable. Therefore, the question of the Cepheid stability is irrelevant for the stability of its main-sequence progenitor since the chemical composition is different. The solution for a certain phase of evolution, in general, is obtained by solving approximate equations. For example, complete equilibrium may be assumed in the case of the main sequence, while only the inertia terms are dropped for the evolution through the Cepheid phase. If such approximate models approach an instability in the run of their evolution, the neglected time derivatives become important and have to be taken into account. In general, then, the solution obtained from better approximations tells us in which direction the evolution really goes.

25.3.1 Perturbation Equations

We want to investigate the stability of a stellar model in complete equilibrium for given input parameters M and chemical composition. Let the model be described by $r_0(m)$, $P_0(m)$, $T_0(m)$, $l_0(m)$, which solve the time-independent stellar structure equations. We test its stability by investigating how a neighbouring (perturbed) solution evolves in time. We here restrict ourselves to spherically symmetric perturbations which depend on m and t in such a way that the perturbed variables become

$$\begin{aligned} r(m,t) &= r_0(m)\left[1 + x(m)e^{i\omega t}\right], \\ P(m,t) &= P_0(m)\left[1 + p(m)e^{i\omega t}\right], \\ T(m,t) &= T_0(m)\left[1 + \vartheta(m)e^{i\omega t}\right], \\ l(m,t) &= l_0(m)\left[1 + \lambda(m)e^{i\omega t}\right], \end{aligned} \tag{25.17}$$

where the absolute values of x, p, ϑ, and λ are $\ll 1$. These variables have to fulfill the time-dependent equations (10.1)–(10.4). As an example let us introduce (25.17) into the equation of motion (10.2). If we linearize with respect to p and x, this becomes

$$\begin{aligned} &P_0'\left(1 + pe^{i\omega t}\right) + P_0 p' e^{i\omega t} \\ &= -\frac{Gm}{4\pi r_0^4}\left(1 - 4xe^{i\omega t}\right) + \frac{\omega^2}{4\pi r_0} xe^{i\omega t}, \end{aligned} \tag{25.18}$$

where primes indicate derivatives with respect to m. Since P_0, r_0 obey (10.2), we have $P_0' = -Gm/(4\pi r_0^4)$: The time-independent terms in (25.18) cancel each other, the exponentials drop out, and we are left with (25.19). By a similar procedure, we find for the case of a radiative layer and an equation of state of the form $\varrho \sim P^\alpha T^{-\delta}$

from (10.1), (10.3), (10.4) the equations (25.20)–(25.22):

$$p' = -\frac{P_0'}{P_0}\left[p + \left(4 + \frac{r_0^3}{Gm}\omega^2\right)x\right], \tag{25.19}$$

$$x' = -\frac{1}{4\pi r_0^3 \varrho_0}(3x + \alpha p - \delta\vartheta). \tag{25.20}$$

$$\lambda' = -\frac{\varepsilon_0}{l_0}(\lambda - \varepsilon_P p - \varepsilon_T \vartheta) - \mathrm{i}\omega \frac{P_0 \delta}{l_0 \varrho_0}\left(\frac{\vartheta}{\nabla_{\mathrm{ad}}} - p\right), \tag{25.21}$$

$$\vartheta' = \frac{P_0'}{P_0}\nabla_{\mathrm{rad}}[\kappa_P p + (\kappa_T - 4)\vartheta + \lambda - 4x]. \tag{25.22}$$

Equations (25.19)–(25.22) are four linear homogeneous differential equations of first order for the variables p, ϑ, x, λ which have to obey certain boundary conditions corresponding to those of the unperturbed solutions. They have to be regular in the centre and to be fitted to an atmosphere. We will deal with the boundary conditions in Chaps. 40 and 41, where they are shown to be equivalent to four linear homogeneous equations. Therefore, solutions exist only for certain *eigenvalues* of ω, which have to be found numerically. There exists an infinite number of eigenvalues for which the system can be solved. For each eigenvalue ω^* one obtains a set of *eigenfunctions* $p^*(m)$, $\vartheta^*(m)$, $x^*(m)$, $\lambda^*(m)$.

The term with $\omega^2(\sim \ddot{r})$ in (25.19) comes from the inertial terms in the equation of motion, while in (25.21), the term with $\mathrm{i}\omega(\sim \dot{P}, \dot{T})$ is due to the time derivatives in the energy equation. The two corresponding timescales are τ_{hydr} and $\tau_{\mathrm{adj}} = \tau_{\mathrm{KH}}$. Since $\tau_{\mathrm{hydr}} \ll \tau_{\mathrm{KH}}$, we have a situation similar to that described for the piston model in Sect. 25.2. Correspondingly, in general, we can speak of dynamical, vibrational, and secular stability.

There are, however, more complicated cases where this classification of stability behaviour is not possible. For example, the relevant thermal timescale may not be that of the whole star but a much shorter one for a small subregion. If the characteristic wavelength of a thermal perturbation is short enough, the corresponding adjustment time can become comparable or shorter than τ_{hydr} (of the whole star). Another example is the case of a dynamically stable model which evolves in such a way that it approaches marginal stability ($\omega_{\mathrm{ad}} \to 0$). Then the oscillations become so slow that they certainly will not be adiabatic anymore: $1/\omega_{\mathrm{ad}} \gg \tau_{\mathrm{KH}}$ (although $\tau_{\mathrm{hydr}} \ll \tau_{\mathrm{KH}}$ still).

25.3.2 Dynamical Stability

Since in Chap. 40 we will treat this problem thoroughly, we merely present some general results here. Instead of solving all four equations (25.19)–(25.22), one can consider oscillations taking place on the timescale τ_{hydr}. Since $\tau_{\mathrm{hydr}} \ll \tau_{\mathrm{adj}}$, the temperature of the matter changes almost adiabatically. Instead of solving (25.21) and (25.22) one just replaces ϑ by $p\nabla_{\rm ad}$ in (25.20). Therefore (25.19) and (25.20) present two equations for p and x with the eigenvalue ω^2. As we will see in Chap. 40 the eigenvalue problem is self-adjoint. Then there exists an infinite series of eigenvalues ω_n^2 which are real. (ω_n is either real or purely imaginary). Therefore, they either correspond to periodic oscillations ($\omega_n^2 > 0$) or exponentially decreasing/increasing solutions ($\omega_n^2 < 0$). The same behaviour was found for the adiabatic case of the piston model. But now, with an infinite number of eigenvalues, stability demands that for *all* eigenvalues $\omega_n^2 > 0$, while even a *single* eigenvalue with $\omega_n^2 < 0$ is sufficient for instability.

How a star behaves after it is adiabatically compressed or expanded depends on the numerical value of $\gamma_{\rm ad}$. This can be most easily seen in the case of homologous changes. Let us consider a concentric sphere $r = r(m)$ in a star of hydrostatic equilibrium.

The pressure there is equal to the weight of the layers above a unit area of the sphere, as shown by integrating the hydrostatic equation:

$$P = \int_m^M \frac{Gm}{4\pi r^4} dm. \tag{25.23}$$

We now compress the star artificially and assume the compression to be adiabatic and homologous. In general, after this procedure, the star will no longer be in hydrostatic equilibrium.

If a prime indicates values after the compression, then homology demands that the right-hand side of (25.23) varies like $(R'/R)^{-4}$ [cf. (20.37)] where R is the stellar radius, while adiabaticity *and* homology demand that the left-hand side varies as

$$(\varrho'/\varrho)^{\gamma_{\rm ad}} = (R'/R)^{-3\gamma_{\rm ad}} \tag{25.24}$$

according to (20.9). Therefore, if $\gamma_{\rm ad} = 4/3$, the pressure on the left-hand side of (25.23) increases stronger with the contraction than the weight on the right: The resulting force is directed outwards, and the star will move back towards equilibrium: it is dynamically stable.

For $\gamma_{\rm ad} < 4/3$ the weight increases stronger than the pressure and the star would collapse after the initial compression (dynamical instability). For $\gamma_{\rm ad} = 4/3$, the compression leads again to hydrostatic equilibrium: One has neutral equilibrium. The condition $\gamma_{\rm ad} > 4/3$ corresponds to the dynamical stability condition $\gamma_{\rm ad} > -G_h^*$ for the piston model (Sect. 25.2.1).

In Chap. 40 we will see that $\gamma_{\rm ad} = 4/3$ is also a critical value for non-homologous perturbations. If $\gamma_{\rm ad}$ is not constant within a star, for instance, because of ionization, then marginal stability occurs if a certain mean value of $\gamma_{\rm ad}$ over the star reaches the critical value 4/3.

It should be noted that radiation pressure can bring $\gamma_{\rm ad}$ near the critical value 4/3 (see Sect. 13.2). This is the reason why supermassive stars are in indifferent equilibrium, i.e. they are marginally stable (see Sect. 19.10).

The critical value 4/3 depends strongly on spherical symmetry and Newtonian gravitation. The 4 in the numerator comes from the fact that the weight of the envelope in Newtonian mechanics varies as $\sim r^{-2}$ and has to be distributed over the surface of our sphere, giving another r^{-2}. The 3 in the denominator comes from the r^3 in the formula for the volume of a sphere. Therefore, effects of general relativity change the critical value (see Sect. 38.2) of γ_{ad} and make the models less stable. Since we have assumed spherical symmetry in deriving the critical value of γ_{ad}, rotation changes it, too. It can decrease the critical value of γ_{ad} and make the models more stable.

25.3.3 Non-adiabatic Effects

The inclusion of non-adiabatic effects in a dynamically stable model brings us to the question of its vibrational and secular stability (A dynamical instability makes a perturbation grow so rapidly that any other possible instability of vibrational or secular type is irrelevant because of their much longer timescales.). Vibrational stability means an oscillation with nearly adiabatic frequency but with slowly decreasing (stability) or increasing amplitude (instability). Such oscillations describe the behaviour of pulsating stars and therefore are treated in detail in Chap. 41.

Secular (or thermal) stability is governed by thermal relaxation processes. In general these proceed on timescales long compared to τ_{hydr} and, therefore, the inertia terms in the equation of motion can be dropped. This means that the term $\sim \omega^2$ in (25.19) can be omitted. Equations (25.19)–(25.22) together with proper boundary conditions can then be solved, yielding an infinite number of secular eigenvalues ω_{sec}. Normally they are purely imaginary (as in the case of the piston model). This is what one expects from a thermal relaxation process, such as in the problem of diffusion of heat. It is therefore all the more surprising that in certain cases a few complex eigenvalues occur (Aizenman and Perdang 1971). The oscillatory behaviour here comes from heat flowing back and forth between different regions in the star (Obviously this could not occur in the single layer of the piston model.). If instead of ω we again use $\sigma := \mathrm{i}\omega$, the system (25.19)–(25.22) has real coefficients. Therefore the eigenvalues σ, if complex, appear in conjugate complex pairs. Again, the sign of the real part of σ (the imaginary part of ω) distinguishes between secular stability or instability.

The most important application of the secular problem to stellar evolution concerns the question whether a nuclear burning is stable or not. Secular instability in degenerate regions leads to the flash phenomenon, while in thin (nondegenerate) shell sources, it results in quasiperiodic thermal pulses.

In order to make the secular stability of a central burning plausible, we treat a simple model of the central region, assuming homologous changes of the rest of the star. Other secular instabilities which occur in burning shells or which are due to nonspherical perturbations will be discussed later (Sects. 33.5 and 34.2).

25.3.4 The Gravothermal Specific Heat

Let us consider a small sphere of radius r_s and mass m_s around the centre of a star in hydrostatic equilibrium. If the sphere is sufficiently small, then P at r_s and the mean density in the sphere are good approximations for the central values P_c, ϱ_c. Suppose that, as a reaction to the addition of a small amount of heat to the central sphere, the whole star is slightly expanding and let the expansion be homologous. Then any mass shell of radius r after expansion has the radius $r + dr = r(1 + x)$, where x is constant for all mass shells. If after the expansion the pressure in the sphere is $P_c + dP_c$, then, similarly to (20.34) and (20.37), the resulting changes of ϱ_c and P_c are

$$\frac{d\varrho_c}{\varrho_c} = -3x, \quad p_c := \frac{dP_c}{P_c} = -4x. \tag{25.25}$$

We now write the equation of state in differential form,

$$\frac{d\varrho_c}{\varrho_c} = \alpha p_c - \delta\vartheta_c, \tag{25.26}$$

($\vartheta c := dT_c/T_c$) as in (6.5) but here with constant chemical composition. Elimination of $d\varrho_c/\varrho_c$ and of x from (25.25) and (25.26) gives

$$p_c = \frac{4\delta}{4\alpha - 3}\vartheta_c. \tag{25.27}$$

According to the first law of thermodynamics the heat dq per mass unit added to the central sphere is

$$dq = du + P dv = c_P T_c(\vartheta_c - \nabla_{ad} p_c) := c^* T_c \vartheta_c, \tag{25.28}$$

where we have used (4.18), (4.21) and where according to (25.27)

$$c^* = c_P \left(1 - \nabla_{ad} \frac{4\delta}{4\alpha - 3}\right). \tag{25.29}$$

This quantity has the dimension of a specific heat per mass unit. Indeed, $dT = dq/c^*$ gives the temperature variation in the central sphere if the heat dq is added. In thermodynamics we are used to defining specific heats with some mechanical boundary conditions, for example, c_P and c_v. For c^* the mechanical condition is that the gas pressure is kept in equilibrium with the weight of all the layers with $r > r_s$. This c^* is called the *gravothermal specific heat.*

For an ideal monatomic gas ($a = \delta = 1$, $\nabla_{ad} = 2/5$), as we have approximately in the central region of the Sun, one finds from (25.29) that $c^* < 0$. This is fortunate, since if in the Sun the nuclear energy generation is accidentally enhanced for a moment ($dq > 0$), then $dT < 0$, the region cools, thereby reducing the overproduction of energy immediately. Therefore the negative specific heat acts as a stabilizer. At first glance it seems as if the decrease of temperature after an injection of heat contradicts energy conservation. But one has also to take into account the Pdv work done by the central sphere. Indeed, while the centre cools ($\vartheta_c < 0$), the whole star expands, since elimination of p_c and $d\varrho_c/\varrho_c$ from (25.25) and (25.26) gives $x = -\delta\vartheta_c/(4\alpha - 3)$, which in the case $\alpha = \delta = 1$ yields $x > 0$. It turns out that, if heat is added to the central sphere, more energy is used up by the expansion, and therefore some must be taken from the internal energy. This behaviour is essentially connected with the virial theorem (see Sect. 3.1). A corresponding property can be found for the piston model by assuming a variable weight G^* of the piston as in Sect. 3.2.

For a nonrelativistic degenerate gas ($\delta \to 0, \alpha \to 3/5$) equation (25.29) gives $c^* > 0$: the addition of energy to the central sphere heats up the matter, which can lead to thermal runaway.

25.3.5 Secular Stability Behaviour of Nuclear Burning

Having derived a handy expression for dq, we shall now use it in the energy balance of the central sphere considered in Sect. 25.3.4. Energy is released in the sphere by nuclear reactions and transported out of it by radiation (we assume here that the central region is not convective). In the steady state gains and losses compensate each other. Let ε be the mean energy generation rate, and l_s the energy per unit time which leaves the sphere; then $\varepsilon m_s - l_s = 0$. Now the equilibrium is supposed to be perturbed on a timescale τ, such that τ is much larger than τ_{hydr} but short compared to the thermal adjustment time of the sphere. Then, while hydrostatic equilibrium is maintained, the thermal balance is perturbed.

For the perturbed state the energy balance is

$$m_s d\varepsilon - dl_s = m_s \frac{dq}{dt} \equiv m_s c^* \frac{dT_c}{dt}. \tag{25.30}$$

Here, dq is the heat gained per mass unit, which is expressed by $c^* dT_c$ according to (25.28).

If we now perturb the equation for radiative heat transfer (5.12),

$$l \sim \frac{T^3 r^4}{\kappa} \frac{dT}{dm}, \tag{25.31}$$

we obtain for l_s

$$\frac{dl_s}{l_s} = 4\vartheta_c + 4x - \kappa_P p_c - \kappa_T \vartheta_c. \tag{25.32}$$

For the perturbation of dT/dm we have made use of the fact that for homology $\vartheta = dT/T =$ constant and therefore $d(dT/dm) = d(T\vartheta)/dm = \vartheta dT/dm$.

From (25.25), (25.27) and (25.32) it follows that

$$\frac{dl_s}{l_s} = \left[4 - \kappa_T - \frac{4\delta}{4\alpha - 3}(1 + \kappa_P)\right]\vartheta_c. \tag{25.33}$$

This, introduced into (25.30), gives

$$\frac{m_s}{l_s}\frac{dq}{dt} = (m_s d\varepsilon - dl_s)l_s = \varepsilon_T\vartheta_c + \varepsilon_P p_c - \frac{dl_s}{l_s}$$

$$= \left[(\varepsilon_T + \kappa_T - 4) + \frac{4\delta}{4\alpha - 3}(\varepsilon_P + \kappa_P + 1)\right]\vartheta_c, \tag{25.34}$$

where we have made use of $l_s = \varepsilon m_s$ and of (25.27). Then with (25.30) we find

$$\frac{m_s c^* T_c}{l_s}\frac{d\vartheta_c}{dt} = \left[(\varepsilon_T + \kappa_T - 4) + \frac{4\delta}{4\alpha - 3}(1 + \varepsilon_P + \kappa_P)\right]\vartheta_c. \tag{25.35}$$

The sign of the bracket tells us whether for $dT_c > 0$ the additional energy production exceeds the additional energy loss of the sphere ($[\ldots] > 0$). The sign of c^* tells us whether in this case the sphere heats up ($c^* > 0$) or cools ($c^* < 0$). Normally ε_T is the leading term in the bracket, so that indeed $[\ldots] > 0$. We first assume an ideal gas ($\alpha = \delta = 1, c^* < 0$) and obtain

$$\frac{m_s c^* T_c}{l_s}\frac{d\vartheta_c}{dt} = [\varepsilon_T + \kappa_T + 4(\varepsilon_P + \kappa_P)]\vartheta_c. \tag{25.36}$$

Since $c^* < 0$, one finds from (25.36) that $(d\vartheta_c/dt)/\vartheta_c < 0$, meaning that the perturbation dT_c decays and the equilibrium is stable if

$$\varepsilon_T + \kappa_T + 4(\varepsilon_P + \kappa_P) > 0. \tag{25.37}$$

This criterion is normally fulfilled. The only "dangerous" term is κ_T, which can be as low as to -4.5 for Kramers opacity. But then, even $\varepsilon_T = 5$ for the *pp* chain suffices to fulfill (25.37), since the other terms are positive.

Any temperature increase $dT_c > 0$ would cause a large additional energy overproduction $\varepsilon_0\varepsilon_T dT_c/T_c$. But since the gravothermal heat capacity $c^* < 0$, the sphere reacts with $dT_c < 0$, and this cooling brings energy production back to normal. We then can say that the burning in a sphere of ideal gas proceeds in a stable manner, the negative gravothermal specific heat acts like a thermostat. This, for example, is the case in the Sun.

We go back to (25.35) for the general equation of state. Since normally ε_T dominates the other terms in the square bracket (in some case $\varepsilon_T > 20$), we neglect them for simplicity. Then (25.35) can be written

$$\frac{d\vartheta_c}{dt} = \frac{l_s \varepsilon_T}{m_s T_c c^*} \vartheta_c := \frac{1}{D} \vartheta_c. \tag{25.38}$$

Obviously $D < 0$ indicates stability, $D > 0$ instability. Since $\varepsilon_T > 0$ and, for an ideal gas, $c^* < 0$, the quantity D is negative: The nuclear burning is stable.

For a nonrelativistic degenerate gas we have $\delta = 0$, $\alpha = 3/5$. Therefore, $c^* > 0$ and $D > 0$: Any nuclear burning with a sufficiently strong temperature dependence will then be unstable. This is the reason, for instance, why in the central regions of a white dwarf there can be no strong nuclear energy source [as first shown by Mestel (1952)]; the star would be destroyed by thermal runaway, or at least heat up until it was not degenerate and then expand. Of course, then it would no longer be a white dwarf. The same instability is also responsible for the phenomenon of the so-called flash (compare Sect. 33.4) which occurs if a new nuclear burning starts in a degenerate region. Note that the appearance of $4\alpha - 3$ in the denominator in several equations, including (25.29) for c^* and (25.35), does not become serious even if $\alpha \to 3/4$ for partial nonrelativistic degeneracy, since the singularity can be removed from the equation which one obtains if c^* is inserted in (19.35) by multiplication with $4\alpha - 3$.

From (25.38) one can draw another conclusion. Let us assume that in the central region of a star there is no nuclear burning but that energy losses by neutrinos (Sect. 18.7) are important. The nuclear energy production in the star may take place in a concentric shell of finite radius. Part of this energy flows outwards, providing the star's luminosity, while part of it flows from the shell inwards towards the centre where it goes into neutrinos. The maximum temperature then is in the shell and not in the stellar centre. In Sect. 33.5 we shall see that this really can be the case in models of evolved stars. If we now again look at (25.38), we have to be aware that $l_s < 0$. If $\varepsilon_T > 0$, as it is for neutrino losses (see Sect. 18.7), all the above conclusions are contradicted because of the different sign of l_s: The equilibrium is stable if $c^* > 0$, i.e. for degeneracy, but unstable if $c^* < 0$, which is the case for an ideal gas.

All our discussions here were based on the assumption of homologous changes in the stellar model. Although stars clearly never change precisely in such a simple way, it turns out that the above conclusions describe qualitatively correctly the secular stability behaviour of stars. Deviations from homology only influence the factors [e.g. in the bracket in (25.36)], thus modifying the exact position of the border between secular stability and instability.

Part V
Early Stellar Evolution

Chapter 26
The Onset of Star Formation

Stars form out of interstellar matter. With modern telescopes and instruments this can nowadays be observed directly and in many phases of the formation process. Indeed a homogeneous cloud of compressible gas can become gravitationally unstable and collapse. In this section we shall deal with gravitational instability and then discuss some of its consequences. But before we do so it may be worth comparing this instability with those discussed in Chap. 25. For gravitational instability the inertia terms are important as well as heat exchange of the collapsing mass with its surroundings. But it is not a vibrational instability, since the classification scheme of Chap. 25 holds only if the free-fall time is much shorter than the timescale of thermal adjustment. As we will see later, just the opposite is the case here, during the earliest phases of star formation.

26.1 The Jeans Criterion

26.1.1 An Infinite Homogeneous Medium

We start with an infinite homogeneous gas at rest. Then density and temperature are constant everywhere. However, we must be aware that this state is not a well-defined equilibrium. For symmetry reasons the gravitational potential Φ must also be constant. But then Poisson's equation $\nabla^2\Phi = 4\pi G\varrho$ demands $\varrho = 0$. Indeed the gravitational stability behaviour should be discussed starting from a better equilibrium state, as we will do later. Nevertheless we first assume a medium of constant non-vanishing density. If we here apply periodic perturbations of sufficiently small wavelength, the single perturbation will behave approximately like one with the same wavelength in an isothermal sphere in hydrostatic equilibrium (which is a well-defined initial state).

R. Kippenhahn et al., *Stellar Structure and Evolution*, Astronomy and Astrophysics Library, DOI 10.1007/978-3-642-30304-3_26, © Springer-Verlag Berlin Heidelberg 2012

The gas has to obey the equation of motion of hydrodynamics

$$\frac{\partial \boldsymbol{v}}{\partial t} + (\boldsymbol{v} \cdot \nabla)\boldsymbol{v} = -\frac{1}{\varrho}\nabla P - \nabla \Phi \tag{26.1}$$

(Euler equation), together with the continuity equation

$$\frac{\partial \varrho}{\partial t} + \boldsymbol{v}\nabla\varrho + \varrho\nabla \cdot \boldsymbol{v} = 0\,. \tag{26.2}$$

In addition we have Poisson's equation

$$\nabla^2 \Phi = 4\pi G \varrho \tag{26.3}$$

and the equation of state for an ideal gas

$$P = \frac{\Re}{\mu}\varrho T = v_s^2 \varrho\,, \tag{26.4}$$

where v_s is the (isothermal) speed of sound. For equilibrium we assume $\varrho = \varrho_0 =$ constant, $T = T_0 =$ constant, and $\boldsymbol{v}_0 = \boldsymbol{0}$. Φ_0 may be determined by $\nabla^2\Phi_0 = 4\pi G\varrho_0$ and by boundary conditions at infinity.

We now perturb the equilibrium

$$\varrho = \varrho_0 + \varrho_1\,, \quad P = P_0 + P_1\,, \quad \Phi = \Phi_0 + \Phi_1\,, \quad \boldsymbol{v} = \boldsymbol{v}_1\,, \tag{26.5}$$

where the functions with subscript 1 depend on space and time. In (26.5) we have already used that $\boldsymbol{v}_0 = \boldsymbol{0}$. If we substitute (26.5) in (26.1) and (26.4), assuming that the perturbations are isothermal (v_s is not perturbed), and if we ignore non-linear terms in these quantities, we find

$$\frac{\partial \boldsymbol{v}_1}{\partial t} = -\nabla\left(\Phi_1 + v_s^2\frac{\varrho_1}{\varrho_0}\right), \tag{26.6}$$

$$\frac{\partial \varrho_1}{\partial t} + \varrho_0 \nabla \cdot \boldsymbol{v}_1 = 0\,, \tag{26.7}$$

$$\nabla^2 \Phi_1 = 4\pi G \varrho_1\,. \tag{26.8}$$

The terms with index 0, describing the equilibrium part, have vanished, as usual. This is a linear homogeneous system of differential equations with constant coefficients. We therefore can assume that solutions exist with the space and time dependence proportional to exp $[\mathrm{i}(kx + \omega t)]$ such that

$$\frac{\partial}{\partial x} = \mathrm{i}k\,, \qquad \frac{\partial}{\partial y} = \frac{\partial}{\partial z} = 0\,, \qquad \frac{\partial}{\partial t} = \mathrm{i}\omega\,. \tag{26.9}$$

With $v_{1x} = v_1, v_{1y} = v_{1z} = 0$ we find from (26.6)–(26.8) that

$$\omega v_1 + \frac{k v_s^2}{\varrho_0} \varrho_1 + k\Phi_1 = 0, \tag{26.10}$$

$$k\varrho_0 v_1 + \omega \varrho_1 = 0, \tag{26.11}$$

$$4\pi G \varrho_1 + k^2 \Phi_1 = 0. \tag{26.12}$$

This homogeneous linear set of three equations for v_1, ϱ_1, Φ_1 can only have nontrivial solutions if the determinant

$$\begin{vmatrix} \omega & \frac{kvs^2}{\varrho 0} & k \\ k\varrho_0 & \omega & 0 \\ 0 & 4\pi G & k^2 \end{vmatrix}$$

is zero. Assuming a non-vanishing wave number k we obtain

$$\omega^2 = k^2 v_s^2 - 4\pi G \varrho_0. \tag{26.13}$$

For sufficiently large wave numbers the right-hand side is positive, i.e. ω is real. The perturbations vary periodically in time. Since the amplitude does not increase, the equilibrium is stable with respect to perturbations of such short wavelengths.

In the limit $k \to \infty$, (26.13) gives $\omega^2 = k^2 v_s^2$, which corresponds to isothermal sound waves. Indeed for very short waves gravity is not important, any compression is restored by the increased pressure, and the perturbations travel with the speed of sound through space.

If $k^2 < 4\pi G \varrho_0 / v_s^2$, the eigenvalue ω is of the form $\pm i\xi$, where ξ is real. Therefore there exist perturbations $\sim \exp(\pm \xi t)$ which grow exponentially with time, so that the equilibrium is unstable. If we define a characteristic wave number k_J by

$$k_J^2 := \frac{4\pi G \varrho_0}{v_s^2}, \tag{26.14}$$

or a corresponding characteristic wavelength

$$\lambda_J := \frac{2\pi}{k_J}, \tag{26.15}$$

then perturbations with a wave number $k < k_J$ (or a wavelength $\lambda > \lambda_J$) are unstable; otherwise, they are stable with respect to the perturbations applied here. The condition for instability $\lambda > \lambda_J$, where

$$\lambda_J = \left(\frac{\pi}{G \varrho_0} \right)^{1/2} v_s, \tag{26.16}$$

is called the *Jeans criterion* after James Jeans, who derived it in 1902. Depending on the detailed geometrical properties of equilibrium and perturbation, the factors on the right-hand side of (26.16) can differ.

For our special choice of perturbations the case of instability can be described as follows: after a slight compression of a set of plane-parallel slabs, gravity overcomes pressure and the slabs collapse to thin sheets. If we estimate ω for the collapsing sheets only from the gravitational term in (26.13) (which indeed is larger than the pressure term), we have $i\omega \approx (G\varrho_0)^{1/2}$ and the corresponding timescale is $\tau \approx (G\varrho_0)^{-1/2}$, which corresponds to the free-fall time, as defined in Sect. 2.4.

26.1.2 A Plane-Parallel Layer in Hydrostatic Equilibrium

We have already mentioned the contradictions connected with the assumption of an infinite homogeneous gas as initial condition. One way out of this difficulty is to investigate the equilibrium of an isothermal plane-parallel layer stratified according to hydrostatic equilibrium in the z direction. Perpendicular to the z direction all functions are constant, the layer extending to infinity. This defines a one-dimensional problem: ϱ_0, P_0, T_0 depend only on one coordinate, say z. Poisson's equation then is

$$\frac{d^2\Phi_0}{dz^2} = 4\pi G\varrho_0, \tag{26.17}$$

while hydrostatic equilibrium, $dP_0/dz = -\varrho_0 d\Phi_0/dz$, can be written with (26.4) as

$$v_s^2 \frac{d\ \ln \varrho_0}{dz} = -\frac{d\Phi_0}{dz} . \tag{26.18}$$

After differentiation of (26.18) one obtains from (26.17)

$$\frac{d^2 \ln \varrho_0}{dz^2} = -\frac{4\pi G}{v_s^2}\varrho_0 . \tag{26.19}$$

With the boundary condition $\varrho_0 = 0$ for $z = \pm\infty$, (26.19) has the solution

$$\varrho_0(z) = \frac{\varrho_0(0)}{\cosh^2(z/H)} , \tag{26.20}$$

with

$$H = \left(\frac{\Re T}{2\pi\mu G\varrho_0(0)}\right)^{1/2} = \frac{v_s}{[2\pi G\varrho_0(0)]^{1/2}} , \tag{26.21}$$

which can be seen if (26.20) and (26.21) are inserted into (26.19). The (stratified) disc does not cause problems similar to those encountered in the case of the infinite homogeneous gas.

In order to investigate the stability of this disc one defines cartesian coordinates x, y in the plane perpendicular to the z-axis and considers perturbations of the form $\varrho_1 \sim f(z) \exp[\mathrm{i}(kx + \omega t)]$. Since the perturbations do not depend on y the layer collapses in the x-direction to a set of plane-parallel "sticks" in y-direction in the case of instability. We shall not go into the details of the stability analysis, which has been described by Spitzer (1968). The result is that again there is a critical wave number

$$k_{\mathrm{J}} = \frac{1}{H} = \frac{[2\pi G\varrho_0(0)]^{1/2}}{v_{\mathrm{s}}} \tag{26.22}$$

and that instability occurs for wave numbers $k < k_{\mathrm{J}}$, while perturbations with $k > k_{\mathrm{J}}$ remain finite. This is very similar to what we have obtained in the homogeneous case, as can be seen by comparing (26.22) and (26.14). The difference in the numerical factors is due to the different geometry.

The two cases discussed above have in common that for smaller wave numbers (larger wavelengths and therefore larger amounts of mass involved in the resulting collapse) the equilibrium is unstable, while for larger wave numbers, it is stable. In hydrostatic equilibrium the force due to the pressure gradient and the gravitational force cancel each other. In general this balance is disturbed after a slight compression. If only a small amount of mass is compressed, the pressure increases more than the force due to gravity, and the gas is pushed back towards the equilibrium state. This is the case if a toy balloon is slightly compressed. Only the increase of pressure counts, since the gravity of the trapped gas is negligible. The same is true for the compressions which occur in sound waves where gravity plays no role. But if a sufficient amount of gas is compressed simultaneously, the increase of gravity overcomes that of pressure and makes the compressed gas contract even more.

26.2 Instability in the Spherical Case

In order to investigate the Jeans instability for interstellar gas in a configuration more realistic than the two examples of Sect. 26.1, we now consider an isothermal sphere of finite radius imbedded in a medium of pressure $P^* > 0$. The sphere is supposed to consist of an ideal gas. The structure of the sphere can be obtained from a solution of the Lane–Emden equation (19.35) for an isothermal polytrope. The solution is cut off at a certain radius where P has dropped to the surface pressure $P = P^*$. The stratification outside the sphere is not relevant as long as it is spherically symmetric with respect to the centre, since then there is no gravitational influence of the outside on the inside. Its only influence will be via the surface pressure, which we assume to be constant during the perturbation.

The essential points of this problem can be easily seen if one discusses the virial theorem for the sphere, as described in Sect. 3.4. Since our sphere of mass M and radius R is isothermal, its internal energy is $E_i = c_v MT$. For the gravitational energy we write $E_{\mathrm{g}} = -\Theta GM^2/R$, where Θ is a factor of order one. It can be obtained by numerical integration of the Lane–Emden equation and is related to the polytropic index n by $\Theta = 3/(5-n)$. For a fully convective sphere ($n = 3/2$), for example, its value would be 6/7; for the homogenous sphere with $n = 0$, $\Theta = 3/5$, and for an $n = 3$ polytrope (see Sect. 19.4) it is 1.5. Here, however, we use it as a general factor that depends on the actual density distribution within the sphere. With these expressions and with $\zeta = 2$ [ideal monatomic gas; (3.8)] the virial theorem (3.21) can be solved for the surface pressure P_0 giving

$$P_0 = \frac{c_v M T}{2\pi R^3} - \frac{\Theta G M^2}{4\pi R^4} . \tag{26.23}$$

The first term on the right is due to the internal gas pressure, which tries to expand the sphere. It is proportional to the mean density. The second term is due to the self-gravity of the sphere, which tries to bring all matter to the centre.

At this point we introduce two scaling factors for radius and pressure which allow us to write (26.23) in dimensionless form

$$\widetilde{R} = \frac{\Theta G M}{2 c_v T}, \qquad \widetilde{P} = \frac{c_v M T}{2\pi \widetilde{R}^3}, \tag{26.24}$$

and write

$$R = x\widetilde{R}, \qquad P_0 = y\widetilde{P} . \tag{26.25}$$

We then obtain instead of (26.23)

$$y = \frac{1}{x^3}\left(1 - \frac{1}{x}\right) . \tag{26.26}$$

We now discuss how P_0 varies with R for fixed values of M, T, and Θ (Fig. 26.1). For small x the value of y is negative. It changes sign with increasing x at $x = 1$ (or $R = \widetilde{R}$), and approaches zero from positive values for x (or R) $\to \infty$. x has a (positive) maximum at 4/3 (or P_0 at $R = R_\mathrm{m}$), a value which can be obtained by differentiation of (26.26) or (26.23). After replacing c_v by $3\Re/(2\mu)$ we find that dP_0/dR vanishes at

$$R_\mathrm{m} = \frac{4\Theta}{9} \frac{G\mu M}{\Re T} = \frac{4}{3}\widetilde{R} . \tag{26.27}$$

Suppose the sphere to be in equilibrium with the surroundings: $P_0 = P^*$. For $R < R_\mathrm{m}$, the surface pressure P_0 decreases with decreasing R. Therefore, after a slight compression, $P_0 < P^*$ and the sphere will be compressed even more; it is unstable. For $R > R_\mathrm{m}$, the pressure P_0 increases during a slight compression and the sphere will expand back to equilibrium; it is stable (These simple plausibility arguments are supported by the results of a decent stability analysis.). We have obviously recovered the Jeans instability discussed in Sect. 26.1. This can be seen if in (26.27) M is replaced by $4\pi R_\mathrm{m}^3 \bar{\varrho}/3$, where $\bar{\varrho}$ is the mean density of the sphere. We then obtain

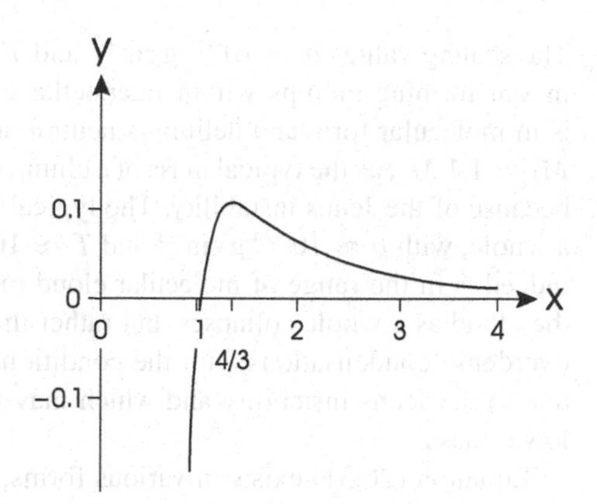

Fig. 26.1 The function given in (26.23) in dimensionless form (26.26). The variable y and therefore also P_0 change sign at $x = 1$ (or at $R = \tilde{R}$). It has a positive maximum at $x = 4/3$ (or at $R = R_m = 4\tilde{R}/3$)

$$R_\mathrm{m}^2 = \frac{27}{16\pi\Theta}\frac{\Re T}{G\mu\bar{\varrho}}. \tag{26.28}$$

Here R_m is the critical radius of a gaseous mass of mean density $\bar{\varrho}$ and temperature T which is marginally stable. We compare it with the critical Jeans wavelength obtained in (26.16), which with $v_\mathrm{s}^2 = \Re T/\mu$ becomes

$$\lambda_\mathrm{J}^2 = \frac{\Re T\pi}{G\mu\bar{\varrho}}. \tag{26.29}$$

Clearly λ_J and R_m are of the same order of magnitude.

Obviously for a given equilibrium state, defined by a radius R and a surface pressure P_0, there exists a critical mass M_J, the so-called *Jeans mass*, where $R = R_\mathrm{m}$. Masses larger than M_J are gravitationally unstable because R would be smaller than the corresponding R_m, which grows linearly with M according to (26.27). If slightly compressed they fall together. According to (26.28)

$$M_\mathrm{J} = \frac{4\pi}{3}\bar{\varrho}R_\mathrm{m}^3 = \frac{27}{16}\left(\frac{3}{\pi}\right)^{1/2}\left(\frac{\Re}{\Theta G}\right)^{3/2}\left(\frac{T}{\mu}\right)^{3/2}\left(\frac{1}{\bar{\varrho}}\right)^{1/2}. \tag{26.30}$$

Depending on the treatment of the perturbation problem and its geometry, one finds slightly differing pre-factors in the expression for M_J, but they all give the same order of magnitude.

We can rewrite (26.30) into a more convenient form (setting $\Theta = 1$):

$$\begin{aligned} M_\mathrm{J} &= \frac{27}{16}\left(\frac{3}{\pi}\right)^{1/2}\left(\frac{\Re}{G\mu}\right)^{3/2} T^{3/2}\varrho^{-1/2} \\ &= 1.1\,M_\odot\left(\frac{T}{10\mathrm{K}}\right)^{3/2}\left(\frac{\varrho}{10^{-19}\,\mathrm{g\,cm^{-3}}}\right)^{-1/2}\left(\frac{\mu}{2.3}\right)^{-3/2}. \end{aligned} \tag{26.31}$$

The scaling values $\varrho = 10^{-19}\,\mathrm{g\,cm^{-3}}$ and $T = 10\,\mathrm{K}$ are typical for the conditions in star-forming clumps within interstellar clouds. We assumed that all hydrogen is in molecular form and helium is neutral, and therefore $\mu \approx 2.3$. We thus obtain $M_\mathrm{J} \approx 1.1\,M_\odot$ as the typical mass of a clump of molecular gas from which stars form because of the Jeans instability. The typical Jeans mass for the molecular cloud as a whole, with $\varrho \approx 10^{-24}\,\mathrm{g\,cm^{-3}}$ and $T \approx 100\,\mathrm{K}$, would be around $10^5\,M_\odot$, which indeed is in the range of molecular cloud masses. However, it is believed that not the cloud as a whole collapses, but rather that turbulence within the cloud leads to overdense condensations with the conditions outlined above, which then collapse due to the Jeans instability and which may fragment further to form stars of even lower mass.

Equation (26.31) exists in various forms, which differ in the numerical factors. This can be the result of different assumptions about Θ, or ζ not being equal to 2. For example, for a bimolecular gas, it would be 6/5. Sometimes also half of the Jeans wavelength λ_J is used instead of R_m. All this can amount to a variation of the typical Jeans mass by a factor of a few.

We have already shown, following (26.16), that the timescale for the growth of the instability is $\tau \approx (G\varrho)^{-1/2}$, the free-fall time. This is of course also valid for the present spherical case. For a density of $\varrho \approx 10^{-19}\,\mathrm{g\,cm^{-3}}$, the collapse takes place on a timescale of some 10^5 years. During collapse, τ becomes shorter, since the density increases.

This timescale τ is long compared to that for thermal adjustment τ_adj. Since the cloud is optically thin, τ_adj is the internal energy per unit mass divided by the rate of energy losses owing to radiation. For typical neutral hydrogen clouds, Spitzer (1968) and Low and Lynden-Bell (1976) estimate a loss Λ of the order $1\,\mathrm{erg\,g^{-1}\,s^{-1}}$. With $T = 10\,\mathrm{K}$ we find $\tau_\mathrm{adj} \approx c_v T/\Lambda \approx 10$ years. Comparison with the free-fall time of some 10^5 years shows that the collapse proceeds in thermal adjustment (which turns out to mean that it is almost isothermal). In Sect. 26.3 we will show where this breaks down. As a rough estimate a molecular cloud is optically thin for particle densities below $10^{-10}\mathrm{cm^{-3}}$, or mass densities below $10^{-14}\,\mathrm{g\,cm^{-3}}$, and optically thick, if the density is higher.

So far, the external pressure P^* has not entered our discussion, because we have asked for the maximum pressure for given mass M at the cut-off radius R. If $P(R)$ is given by the external pressure P^*, one can turn around the question and ask for the maximum mass an isothermal sphere of given T and M can have before it has to collapse. Such a sphere is called *Bonnor-Ebert sphere*, and the critical mass, the Bonnor-Ebert mass M_BE, is given here without derivation (Ebert 1955; Bonner 1956):

$$M_\mathrm{BE} = 1.18 \frac{\Re^2}{\mu^2 G^{3/2}} T^2 (P^*)^{-1/2}\, M_\odot. \tag{26.32}$$

As expected, M_BE is increasing with its temperature because the thermal pressure of the sphere can better balance the outer pressure and is decreasing with increasing external pressure.

26.3 Fragmentation

For a long time it was believed that large molecular clouds of $10^4 \cdots 10^5\, M_\odot$ were collapsing because they exceeded their Jeans mass. To actually form stars of much lower mass from such clouds, fragmentation into smaller clumps, which are collapsing faster than the cloud as a whole, is required. Due to progress of theories, numerical simulations, and observations of molecular clouds the picture has changed. Molecular clouds are highly turbulent, with supersonic motions of gas streams depositing kinetic energy into the cloud, stabilizing it against gravitational collapse. The same shock waves, on smaller scales, result in a local compression of gas. This process is called gravoturbulent cloud fragmentation (Mac Low and Klessen 2004) and leads to overdense gas filaments and clumps. Some of them remain gravitationally bound and may collapse if they exceed their Jeans mass.

Even then, the question remains whether out of clumps of several solar masses many stars of lower mass can form, or how stars with masses below $1\, M_\odot$ are formed. Under what circumstances can fragments of a collapsing cloud become unstable and collapse faster than the cloud?

At first glance it seems to be a natural mechanism for producing collapsing objects with masses smaller than the initial M_J. Indeed, if a clump collapses isothermally, then M_J decreases as $\varrho^{-1/2}$. If, however, the gas were to change adiabatically, then for a monatomic ideal gas, $\nabla_{ad} = (d \ln T/d \ln P)_{ad} = 2/5$ or $T \sim P^{2/5}$, and from $P \sim \varrho T$, the temperature would change as $T \sim \varrho^{2/3}$, and therefore $M_J \sim T^{3/2}\varrho^{-1/2} \sim \varrho^{1/2}$. So the Jeans mass would *grow* during an adiabatic collapse. But already in Sect. 26.2 we have seen that under interstellar conditions the thermal adjustment timescale is much shorter than the free-fall time, which is of the order $(G\varrho)^{-1/2}$, and this also holds when the density increases during collapse. One can therefore assume the collapse to be isothermal rather than adiabatic. Then the Jeans mass becomes smaller than the mass of the originally collapsing cloud. If it has dropped, say, to one half its original value, the clump can split into two independently collapsing parts. This kind of fragmentation can go on as long as the collapse remains roughly isothermal. It will stop as soon as matter becomes opaque and the heat gained by gravothermal contraction can no longer be radiated away (Note that in principle it is not justified to apply the concept of the Jeans mass to an already collapsing medium, since it has been derived for an equilibrium state. But we may do it for order-of-magnitude estimates.).

What are the final products of this fragmentation process? Will the collapsing clump finally fall apart into a swarm of clumplets of planetary masses or even smaller? Even detailed multidimensional simulations of the hydrodynamics and thermodynamics of this complicated process cannot follow it in all details. But we may just estimate when the thermal adjustment time of the fragments becomes comparable with the free-fall time. Then the collapse can certainly not be isothermal anymore and must approach an adiabatic one. As we have seen, then the Jeans mass no longer decreases with increasing ϱ. This means that subregions of the fragments do not fall together on their own and fragmentation stops.

For a detailed estimate, one has to know the radiation processes that cool the gas during collapse. One can then find how long the gained work $-Pdv$ can be radiated away, as is done in modern radiation-hydrodynamical simulations of star formation. Instead, we give a rough estimate of the mass limit of fragmentation based on simple physical arguments, following Rees (1976), without specifying the detailed radiation processes.

The characteristic time of the free-fall of a fragment is $(G\varrho)^{-1/2}$, and the total energy to be radiated away during collapse is of the order of the gravitational energy $E_g \approx GM^2/R$ (see Sect. 3.1), where M and R are the mass and radius of the fragment. Therefore the rate A of energy to be radiated away in order to keep the fragment always at the same temperature is of the order

$$A \approx \frac{GM^2}{R}(G\varrho)^{1/2} = \left(\frac{3}{4\pi}\right)^{1/2} \frac{G^{3/2}M^{5/2}}{R^{5/2}}. \tag{26.33}$$

But the fragment at temperature T cannot radiate more than a black body of that temperature (This implies approximate thermal equilibrium, which is not too bad an assumption for the final stage of fragmentation, where matter starts to become opaque.). Therefore the rate of radiation loss of the fragment is

$$B = 4\pi f \sigma T^4 R^2, \tag{26.34}$$

where $\sigma = 2\pi^5 k^4/(15c^2h^3)$ is the Stefan–Boltzmann constant, while f is a factor less than 1 taking into account that the fragment radiates less than the corresponding black body. For isothermal collapse it is necessary that $B \gg A$. The transition to adiabatic collapse will occur if $A \approx B$. From (26.33) and (26.34) we find that this is the case when

$$M^5 = \frac{64\pi^3}{3} \frac{\sigma^2 f^2 T^8 R^9}{G^3}. \tag{26.35}$$

We assume that fragmentation has reached its limit when M_J is equal to this M. We therefore replace M in (26.35) by M_J, R by

$$R = \left(\frac{3}{4\pi}\right)^{1/3} \frac{M_J^{1/3}}{\varrho^{1/3}}, \tag{26.36}$$

and eliminate ϱ with the help of (26.31). The Jeans mass at the end of fragmentation is then obtained as

$$\begin{aligned} M_J &= \frac{81}{64}\left(\frac{3}{\pi}\right)^{3/4} \frac{1}{(\sigma G^3)^{1/2}} \left(\frac{\Re}{\mu}\right)^{9/4} f^{-1/2}T^{1/4} \\ &= 6.2 \times 10^{30}\,\mathrm{g}\, f^{-1/2}T^{1/4} = 0.003 M_\odot \frac{T^{1/4}}{f^{1/2}}, \end{aligned} \tag{26.37}$$

where T is in K and where we have set $\mu = 1$.

Let us assume that the temperature T of the smallest elements is 10 K and, further, that appreciable deviations from isothermal collapse occur when the radiation losses have to exceed 10 % of the maximal possible (black-body) radiation losses ($f = 0.1$). We then find from (26.37) that $M \approx 0.001\, M_{\odot}$. This rough estimate is surprisingly close to the mass of the smallest optically thick, pressure-supported protostellar cores that were found in numerical simulations. These objects in fact grow in mass by accretion from the surrounding clump.

It should be noted that our result is dependent on the chemical composition because the efficiency of cooling is higher the more heavy elements with rich spectral line systems are present. In particular for stars of the first generation, which are formed shortly after the Big Bang (also called Population III stars), cooling is very inefficient and proceeds mainly via hydrogen molecules, which are even dissociated easily at temperatures around 2,000 K. As a consequence, the smallest condensations in a collapsing cloud of primordial material is of the order of 100 $M_{\odot}$ (see Bromm and Larson 2004 for a review on first stars).

In the above considerations a number of complicating effects have been ignored. The role of magnetic fields is manifold. They may stabilize clouds against collapse, as long as there are ions in the gas, but are usually found to be too weak to do so. However, they may help to mediate the *angular momentum problem*, which is due to the fact that the initially present angular momentum in the cloud works against gravitational collapse. Magnetic fields may allow the transport of angular momentum away from collapsing clumps. Nevertheless, matter does not accrete spherically onto the smallest condensation objects, but accumulates in an accretion disc around it. Disc, protostellar object, and surrounding matter interact in a complicated way through matter in- and outflow, where magnetic fields and angular momentum influence the geometrical shape and efficiency. Magnetic fields also appear to help in keeping clumps together, after the effects of turbulence, which has created them in the first place, have faded. Zinnecker and Yorke (2007) as well as McKee and Ostriker (2007) give comprehensive reviews of star formation theory.

Chapter 27
The Formation of Protostars

The Jeans criterion derived in the foregoing section follows from a first-order perturbation theory and gives conditions under which perturbations of an equilibrium stage will grow exponentially. But the linear theory does not give information, for instance, about the fully developed collapse, to say nothing about the final product. For this, one has to follow the perturbation into the non-linear regime. We first begin with some very simple cases, assuming always spherical symmetry for the collapsing cloud.

27.1 Free-Fall Collapse of a Homogeneous Sphere

If, according to the Jeans criterion, a gaseous mass has become unstable and the collapse has started, gravity increases relatively more than the pressure gradient. The collapse is more and more governed by gravity alone, which is easily seen from the following arguments. For spherical symmetry, the gravitational acceleration is of the order GM/R^2, where M and R are the mass and radius of the cloud. On the other hand, an estimate of the acceleration due to the pressure gradient is

$$\left|\frac{1}{\varrho}\frac{\partial P}{\partial r}\right| \approx \frac{P}{\varrho R} \approx \frac{\Re}{\mu}\frac{T}{R}. \tag{27.1}$$

The ratio of gravitational force to pressure gradient is therefore $\sim M/(RT)$, which during isothermal collapse increases as $1/R$. Consequently we here neglect the gas pressure.

The free collapse of a homogeneous sphere can be treated analytically. At a distance r from the centre the gravitational acceleration is Gm/r^2, where m is the mass within the sphere of radius r. If the pressure can be neglected, the sphere collapses in free fall, according to the equation of motion

$$\ddot{r} = -\frac{Gm}{r^2}, \tag{27.2}$$

R. Kippenhahn et al., *Stellar Structure and Evolution*, Astronomy and Astrophysics Library, DOI 10.1007/978-3-642-30304-3_27, © Springer-Verlag Berlin Heidelberg 2012

where the dots indicate the time derivatives of the radius $r(m,t)$. We now replace m by $4\pi\,\varrho_0 r_0^3/3$, where the subscript zero indicates the values at the beginning of the collapse, by assumption ϱ_0 = constant. Multiplication of (27.2) by $\dot r$ and integration gives

$$\frac{1}{2}\dot r^2 = \frac{4\pi r_0^3}{3r} G\varrho_0 + \text{constant}. \tag{27.3}$$

Choosing the integration constant so that $\dot r = 0$ at the beginning, when $r = r_0$, we get

$$\frac{\dot r}{r_0} = \pm\left[\frac{8\pi G}{3}\varrho_0\left(\frac{r_0}{r} - 1\right)\right]^{1/2}. \tag{27.4}$$

In order to obtain only real values of r, it must always be less than r_0, which means that only the minus sign on the right of (27.4) gives relevant solutions.

For the solution of (27.4) we introduce a new variable ζ, defined by

$$\cos^2\zeta = \frac{r}{r_0}. \tag{27.5}$$

Therefore

$$\frac{\dot r}{r_0} = -2\,\dot\zeta\,\cos\zeta\,\sin\zeta, \quad \frac{r_0}{r} - 1 = \frac{\sin^2\zeta}{\cos^2\zeta}, \tag{27.6}$$

and (27.4) gives

$$2\,\dot\zeta\,\cos^2\zeta = \left(\frac{8\pi G\varrho_0}{3}\right)^{1/2}. \tag{27.7}$$

With the identity

$$2\,\dot\zeta\,\cos^2\zeta = \frac{d}{dt}\left(\zeta + \frac{1}{2}\sin 2\zeta\right), \tag{27.8}$$

which is easily verified, we can write instead of (27.7) that

$$\zeta + \frac{1}{2}\sin 2\zeta = \left(\frac{8\pi G\varrho_0}{3}\right)^{1/2} t, \tag{27.9}$$

where the integration constant is chosen such that the beginning of the collapse (when $r = r_0$ or $\zeta = 0$) coincides with $t = 0$. It should be noted that r_0 no longer explicitly appears in the solution (27.9) and that ϱ_0 = constant. Therefore the solution $\zeta(t)$ is the same for all mass shells. Then, according to (27.6), r/r_0 and also $\dot r/r_0$ at a given time t are the same for all mass shells. This means that the sphere undergoes a *homologous contraction*. Since $\dot r/r_0$ is independent of r_0, the relative density variation is independent of r_0, and the sphere, which was homogeneous at $t = 0$, remains homogeneous. The time it takes to reach the centre ($r = 0$ or $\zeta = \pi/2$) is the free-fall time

$$t_{\rm ff} = \left(\frac{3\pi}{32G\varrho_0}\right)^{1/2}, \tag{27.10}$$

which follows from (27.9) and is the same for all mass shells. With $\varrho_0 = 4 \times 10^{-23}$ g/cm^3, corresponding to a slightly enhanced interstellar density, one obtains $t_{ff} \approx 10^7$years. For a typical protostellar clump as in Sect. 26.2, with $\varrho_0 = 4 \times 10^{-19}$ g/cm^3, (27.10) results in $t_{ff} \approx 2 \times 10^5$ years. It should be noted that expression (27.10) is very similar to the free-fall time τ_{ff} for a star we estimated in (2.17), if there g is replaced by $GM/R^2 = 4\pi G\varrho_0 R/3$.

Of course, before the centre is reached the pressure will become relevant as the gas becomes opaque and T increases. Then the free-fall approximation has to be abandoned, and finally the collapse will be stopped.

27.2 Collapse onto a Condensed Object

As the collapsing cloud becomes opaque the heating will first start in the central parts, since radiation can escape more easily from gas near the surface. Therefore the collapse will be stopped first in the central region. In order to see what then happens we consider a core which has already reached hydrostatic equilibrium, surrounded by a still-free-falling cloud. We emphasize that usually matter carries angular momentum, which is conserved, with the result that matter is first accumulated in an accretion disc around the central object, from where it finally flows onto the accreting body. This fact is ignored here.

Now let M be the mass of the core. For the sake of simplicity we neglect the self-gravity of the free-falling matter. The simplest case is that for the steady state. This would mean that the core is surrounded by an infinite reservoir of matter from which a steady flow rains down. Then the mass flow with absolute radial velocity v,

$$\dot{M} = 4\pi r^2 \varrho v, \tag{27.11}$$

must be constant in space and time. Differentiation of (27.11) with respect to r gives the continuity equation

$$\frac{2}{r} + \frac{1}{\varrho}\frac{d\varrho}{dr} + \frac{1}{v}\frac{dv}{dr} = 0. \tag{27.12}$$

If for v we take the free-fall velocity $v = v_{ff} = [GM/(2r)]^{1/2}$ and assume $M \approx$ constant, we find

$$\frac{1}{\varrho}\frac{d\varrho}{dr} = -\frac{3}{2r}, \tag{27.13}$$

or

$$\varrho(r) = \frac{\text{constant}}{r^{3/2}} \tag{27.14}$$

If R is the radius of the core, then at impact the free-falling matter has the velocity $v_{ff}(R) = [GM/(2R)]^{1/2}$.

The matter falling onto the core is stopped at its surface. The kinetic energy is then transformed into heat, part of which is used to heat up the core, the rest being radiated away. If we ignore the heating of the core, the radiation losses are

$$L_{\text{accr}} = \frac{1}{2} v_{\text{ff}}^2(R) \dot{M} = \frac{1}{4} \frac{GM}{R} \dot{M}. \tag{27.15}$$

L_{accr} is called the *accretion luminosity.* Since for the steady-state solution we have assumed constant M in the expression for v_{ff}, (27.15) is only valid if the accretion timescale

$$\tau_{\text{accr}} := M/\dot{M} \tag{27.16}$$

is long compared to the free-fall time t_{ff}.

27.3 A Collapse Calculation

The collapse of an unstable interstellar cloud can in principle be followed numerically. We will describe the first, meanwhile classical collapse calculations of a spherical, homogeneous cloud of one solar mass by Larson (1969). Although in the meantime three-dimensional hydrodynamical calculations have become possible, Larson's work is nicely illustrating basic effects and remains conceptionally very instructive. Modern one-dimensional calculations (e.g. Ogino et al. 1999) of collapsing Bonnor-Ebert spheres (see Sect. 26.2) give in fact results very similar to Larson's original models. The mass fractions of hydrogen, helium, and heavier elements were taken to be $X = 0.651$, $Y = 0.324$, and $Z = 0.025$, respectively. The boundary conditions assumed that the surface of the sphere remained fixed. The equations to be solved are the continuity equation

$$\frac{\partial m}{\partial t} + 4\pi r^2 v \varrho = 0 \tag{27.17}$$

(with the radial velocity v having positive values in outward direction), the equation of motion

$$\frac{\partial v}{\partial t} + v\frac{\partial v}{\partial r} + \frac{Gm}{r^2} + \frac{1}{\varrho}\frac{\partial P}{\partial r} = 0, \tag{27.18}$$

and the energy equation

$$\frac{\partial u}{\partial t} + P\frac{\partial}{\partial t}\left(\frac{1}{\varrho}\right) + v\left[\frac{\partial u}{\partial r} + P\frac{\partial}{\partial r}\left(\frac{1}{\varrho}\right)\right] + \frac{1}{4\pi\varrho r^2}\frac{\partial l}{\partial r} = 0, \tag{27.19}$$

where u is the internal energy per unit mass. Here the terms on the left (except for the last one) give the substantial derivative $du/dt + Pd(1/\varrho)/dt$ according to $d/dt = \partial/\partial t + v\partial/\partial r$. In addition we have the relation

$$\frac{\partial m}{\partial r} = 4\pi r^2 \varrho. \tag{27.20}$$

Finally we need an equation which describes the energy transport by radiation. Although the diffusion approximation is certainly not good in those parts of the cloud which are optically thin (see Chap. 5), the equation

$$l = -\frac{16\pi acr^2}{3\kappa\varrho} T^3 \frac{\partial T}{\partial r} \tag{27.21}$$

was used, which is identical with our equation (5.11). The errors introduced do not change the qualitative (and maybe even the quantitative) results too much.

For the absorption properties of a gas at extremely low temperatures, other effects than those due to atomic absorption and scattering discussed in Chap. 17 have to be considered. As long as they exist, dust grains are the dominant source of opacity. With increasing temperature (above 1,000 K) the dust particles evaporate. Then the collapsing material becomes more transparent, the opacity being dominated by molecules (Sect. 17.8).

With (27.17)–(27.21), one has five equations for the five unknown variables $m(r,t)$, $v(r,t)$, $P(r,t)$, $T(r,t)$, and $l(r,t)$, while ϱ, κ, and u are given material functions of, say, P and T. The equation of state is assumed to be that of an ideal gas (including effects of dissociation and ionization). The numerical solution now has to be determined with one of the methods described in Sect. 12.3. The outer boundary condition at $r = R$ in these calculations is $v(R,t) = 0$. Since the equations show a singularity at the centre, one has to demand as inner boundary condition that the solutions remain regular there. The initial conditions are $v(r,0) = 0$, while $P(r,0)$ and $T(r,0)$ are constant, and therefore $l(r,0) = 0$. The initial values were $T(r,0) = 10\,\mathrm{K}$, $\varrho(r,0) \approx 10^{-19}\,\mathrm{g/cm^3}$. It should be noted that then almost all hydrogen is in molecular form. These are exactly the conditions we used for the derivation of the typical Jeans mass for a realistic collapsing clump in Sect. 26.2.

In order to have instability at the beginning, the cloud of one solar mass must be sufficiently dense and therefore small. Instability was found numerically by Larson (1969) for $R < 0.46 GM\mu/(\Re T)$. The close resemblance to the critical radius (26.27) for *homologous* collapse should be noted. The calculations began with a slightly compressed cloud with $R = 1.63 \times 10^{17}$ cm. With the density $10^{-19}\,\mathrm{g\,cm^{-3}}$ the free-fall time is $6.6 \times 10^{12}\,\mathrm{s} \approx 210{,}000$ years, according to (27.10), where we already estimated such a value.

In the following we describe the different phases of the collapse.

27.4 The Optically Thin Phase and the Formation of a Hydrostatic Core

In the very first phase the whole collapsing cloud remains optically thin and therefore nearly isothermal with $T \approx 10\,\mathrm{K}$.

When the instability evolves into the non-linear regime the collapse becomes non-homologous, which is not surprising in view of the outer boundary condition. It holds the outer layers of the sphere at a fixed radius while the inner part is free to collapse. Indeed during collapse the density increases rapidly in the central part, while it remains practically constant in the outer regions. A small central concentration, once formed, will necessarily enhance itself. The free-fall time of a certain mass shell at distance r from the centre is of the order $[G\bar{\varrho}(r)]^{-1/2}$, where $\bar{\varrho}(r)$ is the mean density inside the sphere of radius r. If $\bar{\varrho}$ increases towards the centre, then the (local) free-fall time decreases in this direction. Therefore the inner shells fall faster than the outer ones, and the central density concentration becomes even more pronounced.

The calculations show that the density distribution–starting from $\varrho =$ constant–approaches the form $\varrho \sim r^{-2}$ over gradually increasing parts of the cloud (see Fig. 27.1). It is not surprising that it does not follow (27.14), since there we have made assumptions (steady state, a free fall determined only by the gravity of a central object, ignoring gas pressure) which are not fulfilled here.

The density profiles in Fig. 27.1 can be described as follows. A smaller and smaller homogeneous mass collapses more and more rapidly, continuously leaving behind more matter in the inhomogeneously contracting envelope. There the timescale of collapse remains much larger because (1) the density is smaller and (2) pressure gradients brake the free fall.

The collapse of the homogeneous central part resembles a free fall as long as the matter can get rid of the released gravitational energy via radiation. The central region becomes opaque once a central density of 10^{-13} g cm^{-3} is reached. Now the further increase of density in the centre causes an adiabatic increase of temperature. As a consequence the pressure there increases until the free fall is stopped.

This leads to the formation of a central core in hydrostatic equilibrium surrounded by a still-falling envelope. Immediately after the core has reached hydrostatic equilibrium, its mass and radius are 10^{31} g and 6×10^{13} cm, similar to the values estimated in Sect. 26.3 for the Jeans mass at the end of fragmentation, and the central values are $\varrho_c = 2 \times 10^{-10}$ g cm^{-3}, $T_c = 170$ K. The free-fall velocity at the surface of the core is 75 km/s. With increasing core mass and decreasing core radius, the velocity of the falling material exceeds the velocity of sound in the core surface regions. Therefore a spherical shock front is formed which separates the supersonic "rain" from the hydrostatic interior. In this shock front the falling matter comes to rest, releasing its kinetic energy. If all the energy released is radiated away (which is approximately the case), the luminosity of the accreting core is given by (27.15).

In certain respects the hydrostatic core resembles a star. But while the surface pressure is virtually zero for a star, here it has to balance the pressure exerted by the infalling material. If v_e and ϱ_e are the velocity relative to the shock front and the density of the falling gas just above it, respectively, and if P_i is the surface pressure, then conservation of momentum demands that

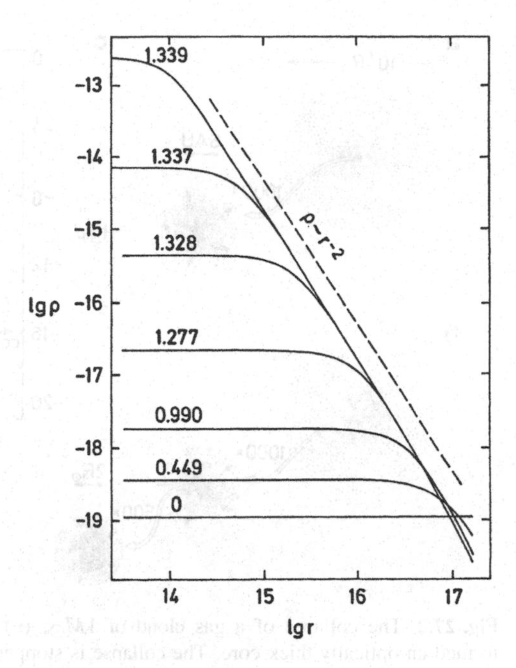

Fig. 27.1 The density ϱ (in $\mathrm{g\,cm^{-3}}$) against the distance from the centre r (in cm) in a collapsing cloud. The density distribution is shown by *solid lines* for different times (labels in 10^{13} s after the onset of the collapse). Regions with homologous changes remain homogeneous ($\partial\varrho/\partial r = 0$); regions in free fall approach a distribution with $\varrho \sim r^{-2}$ (i.e. a slope indicated by the *dashed line*) (After Larson 1969)

$$P_{\mathrm{i}} = \varrho_{\mathrm{e}} v_{\mathrm{e}}^2 = \varrho_{\mathrm{e}} \frac{GM}{2R}, \tag{27.22}$$

where M and R are the mass and radius of the core. This equation is a special case of the more general condition for shock fronts (see Landau and Lifshitz 1987, Vol. 6, p. 320) according to which the quantity $P + \varrho v^2$ must have the same values on both sides of the front. In (27.22) P is neglected outside the front and v inside.

Another difference between an accreting core and a real star is that the accretion energy is released in a thin surface layer, while in a star, the energy source is in the deep interior.

At first glance one would expect the whole core to be isothermal. But while matter is raining down on its surface the core is contracting. This has the consequence that L_{accr} as given by (27.15) increases for $\dot{M} \approx$ constant (since M grows and R decreases). Since during contraction gravitational energy is released in the deep interior of the core, there must be a finite temperature gradient in order to transport this energy outwards. The diameter of the accreting core in hydrostatic equilibrium is already comparable to the dimensions of the solar system (see Fig. 27.2).

27.5 Core Collapse

The accreting hydrostatic core heats up in its interior. We have to keep in mind that the gas consists mainly of hydrogen that at low temperatures is in molecular form as H_2. When the central temperature reaches about 2,000 K, the hydrogen

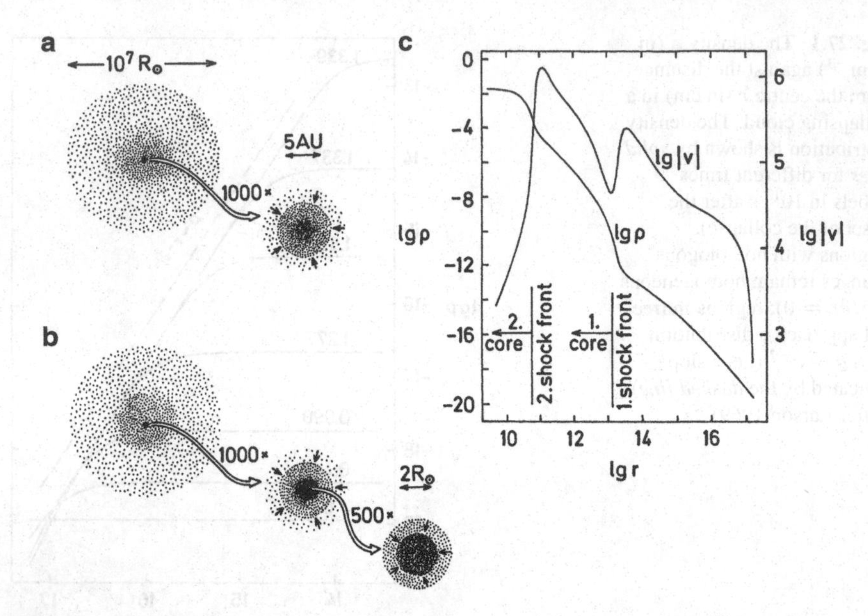

Fig. 27.2 The collapse of a gas cloud of $1M_\odot$. (**a**) After about 1.3×10^{13} s, the cloud has formed an optically thick core. The collapse is stopped there, and a shock front develops at the interface between the core, which is in hydrostatic equilibrium, and the still freely falling envelope. (**b**) When the core has become dynamically unstable owing to dissociation of H_2, a second collapse occurs within the core, forming a second shock front at much smaller r. (**c**) Schematic plot of the absolute value of the velocity v (in cm s^{-1}) and the density ϱ (in g cm^{-3}) against r (in cm), for a time shortly after the formation of a second core within the first one. The regions of the shock fronts are characterized by steep (positive) slopes in the velocity curve

molecules dissociate. The equilibrium between molecular and atomic hydrogen is governed by an equation similar to the Saha equation (see Sect. 14.1). Like ionization, dissociation influences the specific heat, since not all the energy injected into a gas goes into kinetic energy, a fraction being used to break up the molecules into atoms. This decreases γ_{ad}. For hydrogen molecules there are $f = 5$ degrees of freedom, three belonging to translation and two to rotation around two possible axes. Consequently $\gamma_{ad} = (f + 2)/f = 7/5 = 1.40$. This is much closer to the critical value $4/3 = 1.33$ (see Sect. 25.3.2) than in the case of a monatomic gas ($\gamma_{ad} = 5/3 = 1.667$). Only a slight reduction of γ_{ad} owing to dissociation therefore brings it below the critical value 4/3. Then the hydrostatic equilibrium becomes dynamically unstable, and the core starts to collapse again.

In Larson's calculations this happened when the core has, compared to the initial values, twice the mass and half the radius. It collapses as long as the gas is partially dissociated. When almost all hydrogen in the central region is in atomic form, γ_{ad} increases above 4/3 (approaching the value 5/3 for a monatomic gas) and the collapsing core forms a dynamically stable subcore in its interior. This core, which is generally called *protostar* has an initial mass of $1.5 \times 10^{-3} M_\odot$ and an initial radius of $1.3R_\odot$. Its central density is 2×10^{-2} g cm^{-3} and the central temperature is

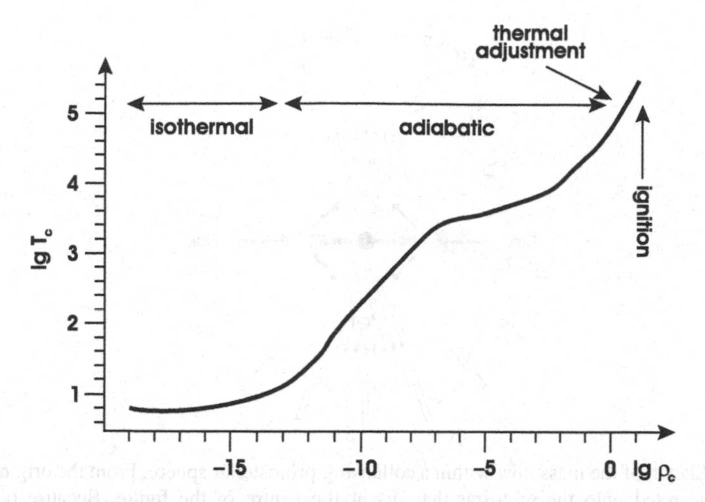

Fig. 27.3. The central evolution of a $1 M_\odot$ cloud from the isothermal collapse to the ignition of nuclear burning. The central temperature T_c (in K) is plotted over the central density ϱ_c(in g cm^{-3}). (After Masunaga and Inutsuka 2000)

2×10^4 K. At the surface of this protostar there is another shock front. The situation is illustrated in Fig. 27.2b, c. As a consequence of the second collapse the density below the outer shock front decreases, and the outer shock finally disappears. More recent calculations by Masunaga et al. (1998) and Masunaga and Inutsuka (2000), which follow the collapse of a $1 M_\odot$ clump through the whole sequence outlined above, confirm it to high degree. Their calculations include a better treatment of the radiative transport and can follow the collapse for a longer time due to a higher spatial resolution. The main difference with respect to Fig. 27.2c is that the velocities at the second shock front reach final values a factor of 10 higher than shown here, while the first shock front has already disappeared. The density profile is, however, very similar to that of Larson's original calculation.

The evolution of the centre of the $1 M_\odot$ cloud, as it results from the radiation-hydrodynamical simulation by Masunaga and Inutsuka (2000), starting from the original Jeans instability, is given in Fig. 27.3. The curve starts on the left during the isothermal collapse. After the matter has become opaque, T rises adiabatically. The slope is at first close to 0.4 (corresponding to $\gamma_{ad} = 1.4$ for H_2), but then becomes considerably less owing to partial dissociation ($\gamma = 1.1$), and finally approaches $2/3$ (corresponding to $\gamma_{ad} = 5/3$ for a monatomic gas).

The central compression is adiabatic as long as the accretion timescale τ_{accr} of the core (or of the innermost core, if there are two) is short compared to its Kelvin–Helmholtz timescale τ_{KH}. But the more the envelope is depleted the more the accretion rate will diminish and consequently τ_{accr} will grow. When it exceeds τ_{KH} the core can adjust thermally and the evolution of the central region ceases to be adiabatic. Since then $\dot{M}$ has become very small, the protostar has practically

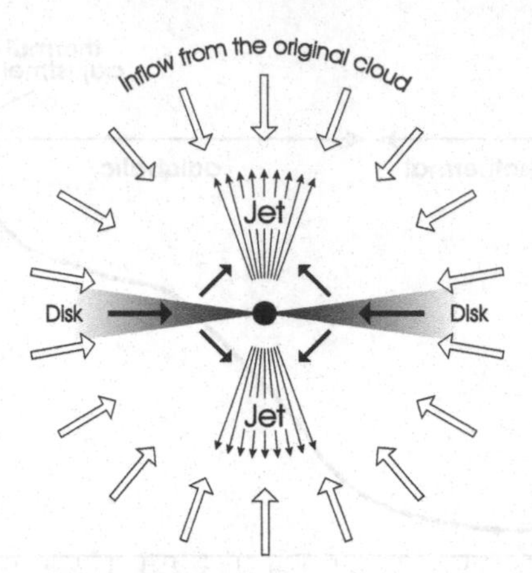

Fig. 27.4. Sketch of the mass flow within a collapsing protostellar sphere. From the original cloud, matter is accreted onto the protostar that sits at the centre of the figure. Because of angular momentum conservation, most of it accumulates in an accretion disc. Part of the matter finally falling onto the star is however ejected in a bipolar jet along the axis of rotation. The jet also may gain additional material directly from the disc due to heating of the inner disc (According to Zinnecker and Yorke 2007)

constant mass. We shall discuss its further evolution with constant M in the next section.

We repeat once more that these calculations were made without considering the fact that the angular momentum of the prestellar cloud leads to the formation of an accretion disc. Most of the matter falling onto the central protostar has first circled the star in this disc.

At the same time the protostar may start to lose mass due to stellar winds and bipolar outflows and jets. All this has already been revealed by observations. The interaction between cloud, protostar, and disc is complicated and also depends on the presence of magnetic fields. This phase has to be investigated by three-dimensional magnetohydrodynamical simulations (Banerjee and Pudritz 2007). The situation is illustrated in Fig. 27.4.

27.6 Evolution in the Hertzsprung–Russell Diagram

A plot of the evolution of a collapsing cloud in the Hertzsprung–Russell (HR) diagram has to be made with care. The radiation emitted by the core is absorbed in the falling envelope, particularly by dust grains, which heat up and reradiate in the infrared. One can assign an effective temperature to the protostellar models. Defining an effective radius R at the optical depth 2/3 one can derive an effective temperature $T_{\rm eff}$ from $L = 4\pi R^2 \sigma T_{\rm eff}^4$. Evolutionary tracks for initial masses of $1 M_\odot$ and $60 M_\odot$ are given in Fig. 27.5. Although the numerical

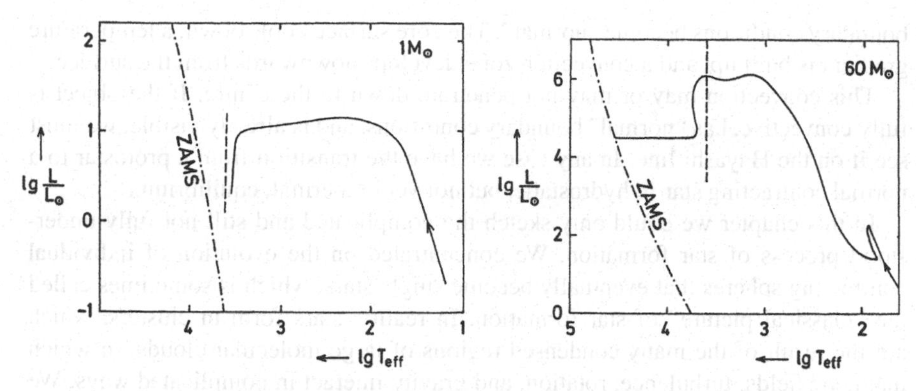

Fig. 27.5. Hertzsprung–Russell diagrams with evolutionary tracks for protostars of $1M_\odot$ and $60M_\odot$. The tracks start at the lower right, where the thermal radiation of the clouds is in the infrared, and they finally approach the zero-age main sequence (ZAMS, *dot-dashed*). In the case of $60M_\odot$, part of the mass of the envelope is blown away so that a star of only $17M_\odot$ settles down on the main sequence. The corresponding Hayashi lines are indicated by *broken lines* (After Appenzeller and Tscharnuter 1974, 1975a,b)

results shown in this figure are quite old, the newer calculations by Wuchterl and Tscharnuter (2003) have confirmed the overall picture very well. To an outside observer the collapsing cloud remains an infrared object as long as the envelope is opaque to visible radiation. The evolutionary track, therefore, starts extremely far to the right in this diagram. This, of course, is no contradiction to the statements about a forbidden region to the right of the Hayashi line (Chap. 24) since the falling envelope (including the "photosphere") is far from being in hydrostatic equilibrium. Even if we could see the already hydrostatic core, we would not observe a normal star, since its boundary conditions are still perturbed by infalling matter.

The thinning out of the envelope has several effects: the first is that it becomes more transparent, and the photosphere ($\tau = 2/3$) moves downwards until it has reached the surface of the hydrostatic core. With decreasing radius of the photosphere, $T_{\rm eff}$ must increase in order to radiate away the energy. In the whole first phase (through the maximum of L in the evolutionary tracks of Fig. 27.5) the luminosity is produced by accretion: $L = L_{\rm accr} \sim \dot{M}$. With decreasing $\dot{M}$, the luminosity L decreases until it is finally provided by contraction of the core.

It is generally found that for low-mass stars accretion onto the protostar stops well before central temperatures for hydrogen ignition is reached. For massive stars, however, accretion continues while central hydrogen burning has already set in. Therefore, when the newborn star finally separates from its surrounding cloud and becomes visible it has already consumed part of its hydrogen fuel and has evolved on the main sequence. Massive stars are therefore unlikely to be found on the ZAMS.

Another effect is the influence of accretion on the boundary conditions of the core. Strong accretion heats up the surface of the core so much that the core is nearly isothermal and the ram pressure $\varrho_{\rm e} v_{\rm e}^2$ is appreciable. With decreasing $\dot{M}$ the boundary conditions become "normal". The core surface cools down, a temperature gradient is built up, and a convection zone develops downwards from the surface.

This convection may or may not penetrate down to the centre. If the object is fully convective, has "normal" boundary conditions, and is already visible, we must see it on the Hayashi line. In any case we have the transition from a protostar to a normal contracting star in hydrostatic, but not yet in thermal, equilibrium.

In this chapter we could only sketch the complicated and still not fully understood process of star formation. We concentrated on the evolution of individual contracting spheres that eventually become single stars, which is sometimes called the "classical picture" of star formation. In reality, stars form in clusters, which are the result of the many condensed regions of large molecular clouds, in which magnetic fields, turbulence, rotation, and gravity interact in complicated ways. We refer the reader to the reviews by Mac Low and Klessen (2004) and Zinnecker and Yorke (2007) for more details about this field.

Chapter 28
Pre-Main-Sequence Contraction

In the last section we left the newly born star while it was still contracting in hydrostatic, but not yet thermal, equilibrium. Essential features of this contraction can already be understood by assuming simple homologous changes. It will turn out that the fate of such a sphere is mainly determined by the equation of state.

28.1 Homologous Contraction of a Gaseous Sphere

A star which has not yet reached the temperature for nuclear burning has to supply its energy loss by contraction. This is a consequence of the virial theorem and of energy conservation as discussed in Sect. 3.1. We have seen, in particular, that part of the released gravitational energy goes into internal energy, while the rest supplies the luminosity [see (3.12)]. The characteristic timescale is τ_{KH}, as shown in Sect. 3.3.

In the following we will be concerned with the centre of the star. For this we can use the relations of Sect. 20.3, which hold for any mass shell of a homologously contracting star. The equation of state (for fixed chemical composition) was written there as $d\varrho/\varrho = \alpha dP/P - \delta dT/T$. According to (20.34) and (20.38), the variation of the central temperature, dT_c, is related to the variation of the central density, $d\varrho_c$, by

$$\frac{dT_c}{T_c} = \frac{4\alpha - 3}{3\delta}\frac{d\varrho_c}{\varrho_c}. \tag{28.1}$$

This defines a field of directions in the $\lg \varrho_c$–$\lg T_c$ plane as displayed in Fig. 28.1. Each arrow there indicates how T_c changes during contraction ($d\varrho_c > 0$). According to (28.1) the slope depends on the equation of state via α and δ. For an ideal gas $\alpha = \delta = 1$ and (28.1) becomes

$$\frac{dT_c}{T_c} = \frac{1}{3}\frac{d\varrho_c}{\varrho_c}. \tag{28.2}$$

R. Kippenhahn et al., *Stellar Structure and Evolution*, Astronomy and Astrophysics Library, DOI 10.1007/978-3-642-30304-3_28, © Springer-Verlag Berlin Heidelberg 2012

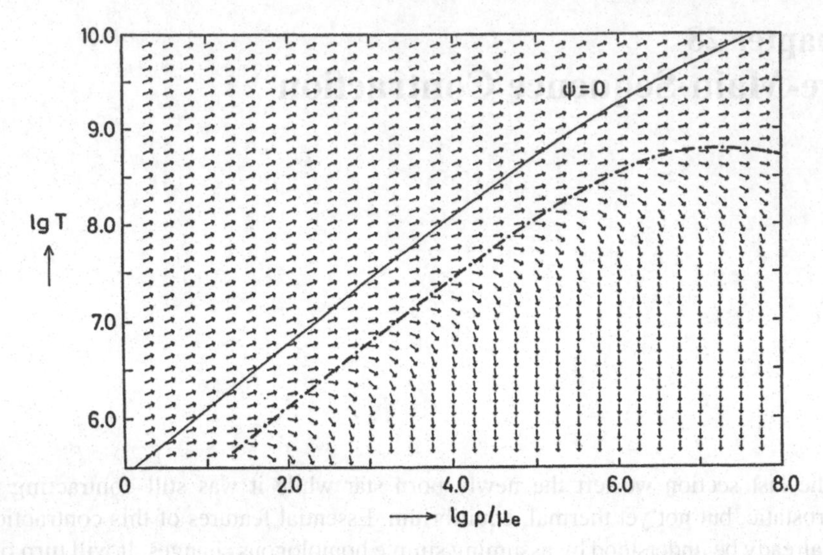

Fig. 28.1 The vector field given by (28.1) in a diagram showing the temperature T (in K) over the density ϱ/μ_e (in g cm^{-3}). The *arrows* indicate the direction in which the centre of a homologously contracting star would evolve. In the upper-left part the equation of state is that of an ideal gas, and therefore the arrows have a slope of 1/3. The *thin solid line* at which the degeneracy parameter $\psi = 0$ indicates roughly the transition from the ideal gas to degeneracy of the electrons. The critical line along which $\alpha = 3/4$ is *dot-dashed*. On this curve the *arrows* point horizontally while below it the *arrows* point downwards

Here the slope is 1/3, a contracting ideal gas heats up (the latter conforms with the conclusions drawn from the virial theorem in Sect. 3.1). The same slope also holds for non-negligible radiation pressure ($\beta < 1$) as can be seen if (13.7) is introduced into (28.1). In Fig. 28.1 the evolutionary track of a (homologously) contracting ideal gaseous sphere is a straight line with slope 1/3. This necessarily leads closer to the regime of degeneracy, which is separated from that of ideal gas by a line of slope 2/3 [see (16.6) and Fig. 16.1]. The onset of degeneracy changes α and δ and decreases the slope of the arrows in Fig. 28.1. In the limit of complete non-relativistic degeneracy one has $\alpha \to 3/5$ and $\delta \to 0$. What happens to a sphere which is contracting and becomes more and more degenerate? Then α will pass the value 3/4 when δ is still finite and the slope given by (28.1) will change sign. Further contraction leads to cooling: the stronger the degeneracy the steeper will be the then negative slope, until finally the stellar centre tends to cool off at almost constant density. In the case of complete relativistic degeneracy, with $\alpha = 3/4$ and $\delta = 0$, the factor on the right of (28.1) becomes indeterminate. Then the ion gas - although its pressure is negligible compared to that of the degenerate electrons - will

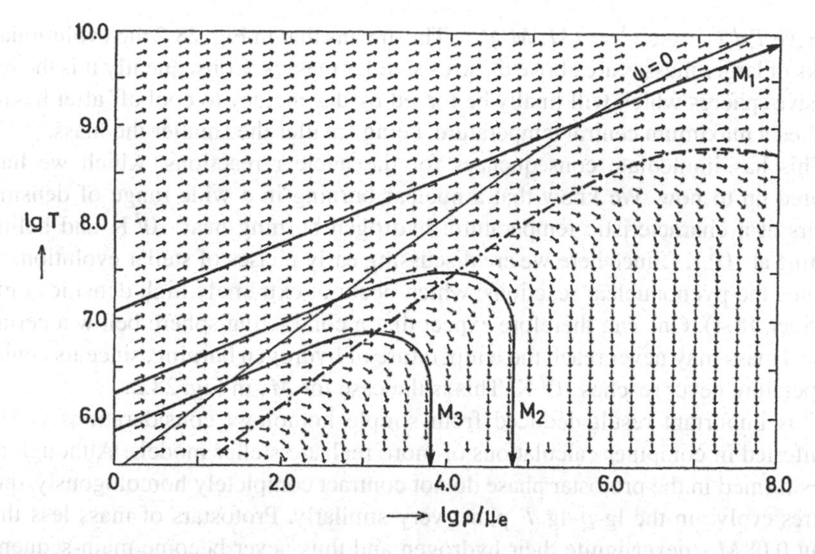

Fig. 28.2 Temperature T (in K) over density ϱ/μ_e(in g cm^{-3}) with the vector field and the lines $\psi = 0$ and $\alpha = 3/4$ as in Fig. 28.1. The *heavy lines* give the "evolutionary tracks" of the centres of three homologously contracting stars of different masses. Mass M_1 is so large that the evolution is not remarkably influenced by degeneracy, and the centre continuously heats up during contraction. For mass $M_2(< M_1)$ degeneracy becomes important in the centre, and consequently a homologous contraction cannot bring the central temperature above a few 10^7 K (which is not sufficient to start helium burning). Mass $M_3(< M_2)$ while contracting will start to cool off even before the temperature of hydrogen burning is reached

determine the slope. A dash-dotted line in Fig. 28.1 connects the points of vanishing slope ($\alpha = 3/4$).[1]

For the sake of simplicity let us first ignore the fact that nuclear reactions set in at certain temperatures. Obviously, the evolutionary track of a contracting gaseous sphere in the lg ϱ_c–lg T_c diagram depends very much on the starting point at the left-hand border, as can be seen from Fig. 28.2. If a stellar centre starts there sufficiently low it will reach a maximum temperature and begin to cool again after entering the domain of degeneracy. But if it started on the left at a sufficiently high temperature, it will never be caught by degeneracy and thus will continue to heat up.

Which types of spheres do reach a maximum temperature, and which types have the privilege of heating up forever? This depends on the mass of the sphere. In order to show this we consider two homologous spheres of an ideal gas with masses M and $M' = M/x$ and radii R and $R' = R/z$. Then, according to (20.9), $\varrho_c/\varrho_c' = xz^{-3}$, $P_c/P_c' = x^2z^{-4}$, and therefore, for an ideal gas, $T_c/T_c' = x/z$. If we now compare states in which the two spheres have the same central density ($xz^{-3} = 1$),

[1]Since the dash-dotted line in Fig. 28.1 gives the impression of delineating a hill, this kind of figure is sometimes called *Thomas-mountain* after H.-C. Thomas who first used it to illustrate the evolution of homologously contracting stellar cores.

we have $T_c/T'_c = x^{2/3} = (M/M')^{2/3}$. This means that in Fig. 28.2 the evolutionary tracks of larger masses are above those of smaller masses. Consequently it is the less massive spheres which will finally be forced by degeneracy to cool off after having reached a maximum central temperature, being smaller the smaller the mass.

This has immediate consequences for the nuclear reactions, which we have ignored up to now. We know that a nuclear burning in a wide range of densities occurs at a characteristic temperature: hydrogen burning near 10^7 K and helium burning at 10^8 K (Since here we are discussing early phases of stellar evolution, we exclude the pycnonuclear reactions, which occur at extremely high densities only; see Sect. 18.4). One can therefore expect that a contracting sphere below a certain critical mass may never reach the temperature of hydrogen burning, since its central temperature never reaches 10^7 K. This is the case for M_3 in Fig. 28.2.

This important result deduced from simple homology considerations is also manifested in computer calculations of more realistic stellar models. Although the cores formed in the protostar phase do not contract completely homologously, their centres evolve in the lg ϱ–lg T plane very similarly. Protostars of mass less than about $0.08 M_\odot$ never ignite their hydrogen and thus never become main-sequence stars. These are the *brown dwarfs* we already introduced in Sect. 22.4. Here we have encountered an evolutionary aspect of the lower end of the main sequence: protostars born with too little mass never reach the state of complete equilibrium by which the main-sequence models are defined. Even if some nuclear reactions have started, they are so slow at these low temperatures that equilibrium abundances (rate of destruction = rate of production) of the involved nuclei are not reached even in the lifetime of the galaxy.

We shall see later that analogous considerations can be used to explain critical masses for the ignition of each higher nuclear burning in contracting cores of evolved stars. Helium burning is not reached by stars of an initial mass below approximately $0.5 M_\odot$; for carbon burning, it has to be above $6 M_\odot$. And masses above $\approx 8 M_\odot$ will never be caught by degeneracy in this way (see Sect. 35.2).

28.2 Approach to the Zero-Age Main Sequence

We have seen that a contracting star of more than $0.08 M_\odot$ ignites hydrogen in its centre and becomes a star on the zero-age main sequence (ZAMS). While the luminosity of the star was originally due to contraction, it now originates from nuclear energy. These two energy sources are quite differently distributed in the star. According to (20.41), $\varepsilon_g \sim T$ is not so much concentrated towards the centre, while hydrogen burning with $\varepsilon_{pp} \sim T^5$ and $\varepsilon_{CNO} \sim T^{18}$ has strong central concentration. Clearly the transition from contraction to hydrogen burning requires a rearrangement of the internal structure. The protostar becomes a zero-age main-sequence star with properties very close to those described in Chap. 22. The difference arises from the fact that some nuclear reactions, for example, the proton captures on ^{2}H, ^{7}Li, or ^{12}C, start at temperatures lower than those of core hydrogen burning.

The way in which nuclear reactions take over the energy production can now be followed by detailed numerical models following the approach to the main sequence of contracting protostars. We first discuss the results for one solar mass. Some reactions of the CNO cycle as given in (18.64) become important before the central temperature has reached that of equilibrium hydrogen burning (where the participating nuclei have equilibrium abundances). At a central temperature of about 10^6 K, all the ^{12}C that had been in the interstellar cloud will burn into ^{14}N via the reactions of the first three lines in (18.64). However, the following $^{14}N(p,\gamma)^{15}O$ reaction is much slower–and therefore often called the bottleneck reaction–such that the full CNO cycle cannot be completed. Once switched on, this process will take over the energy generation and stop the contraction. Because of the high temperature sensitivity of ε, the energy is released close to the centre. Consequently the energy flux $l/4\pi r^2$ is large, and a convective core that contains almost 10 % of the total mass develops. At the same time, the first reactions of the pp chain become relevant, transforming H into ^3He [see the first two lines of (18.62)]. With decreasing ^{12}C the pp reactions become more important, and ^3He can be destroyed by $^3He+^3He$ and $^3He+^4He$ [the two reactions in the third line of (18.62)]. As a consequence the concentration of ^3He reaches a maximum at $m = 0.6M$. Outside, the temperature is too low to form ^3He, while inside, ^3He is used up to form ^4He. This characteristic shape of the ^3He abundance curve remains throughout the main-sequence evolution (see Fig. 29.3). With the depletion of ^{12}C in the central region the convective core disappears and the pp chain becomes the dominant energy source.

The situation is similar for more massive stars. But then instead of the pp chain, the CNO cycle finally takes over and the abundance of ^{12}C becomes that of equilibrium. For stars of $M > 1.5M_\odot$ the effect of pre-main-sequence ^{12}C burning can even be seen in the computed evolutionary tracks in the Hertzsprung–Russell diagram: there seems to be another, relatively short-lived main sequence to the right of the ordinary (hydrogen) main-sequence. Contracting protostars stay there until their ^{12}C fuel is used up before they move on to the main sequence. This somewhat prolongs the time a protostar needs to reach the ZAMS.

The numbers quoted here are from pre-main-sequence evolution calculations that ignore the detailed results of Chap. 27. They start out with a cool protostar on the Hayashi line and follow the ensuing quasi-hydrostatic contraction until the model reaches the hydrogen main sequence. The errors introduced by this simplification are not too large and certainly become negligible towards the end of pre-main-sequence contraction when the thermal history of accretion is forgotten by the star.

This has to do with the fact that, whatever the thermal history of the protostar, its structure has adjusted to thermal equilibrium after a Kelvin–Helmholtz time. Since the main-sequence timescale (which is relevant for the ensuing evolution) is much longer, the stars settle on the ZAMS quite independently of their past. Whatever their detailed history, tracks of protostars of the same mass and chemical composition lead to the same point on the ZAMS.

We now turn to the question of how rapidly stars of different M approach the ZAMS. Decisive for this is the Kelvin–Helmholtz timescale $\tau_{KH} \approx c_v \bar{T} M/L$. The mean temperature $\bar{T}$ does not vary too much with M, since T_c is anyway just below

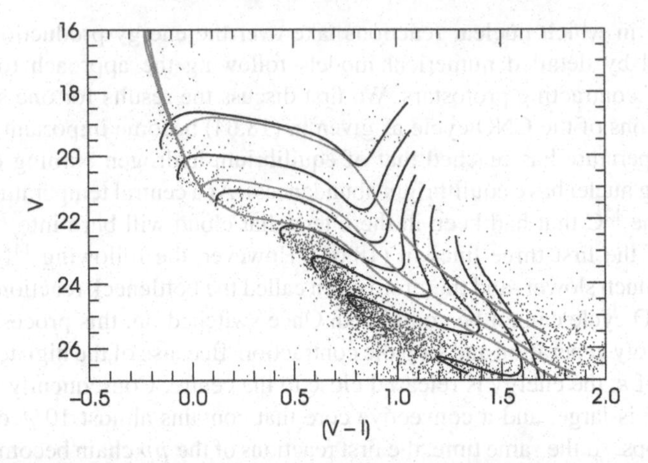

Fig. 28.3 Colour-magnitude diagram of the young open cluster NGC 602. The *dots* are the cluster stars. Overlaid are pre-main-sequence evolutionary tracks for masses between $3.0M_\odot$ and $0.5M_\odot$ (*black lines*; *top to bottom*) as well as isochrones obtained from these tracks with ages of 1, 5, and 10 Myr (*grey lines*; *top to bottom*) (After Cignoni et al. 2009)

the ignition temperature of hydrogen. As a rough estimate for L, we may take the corresponding ZAMS luminosity, since the evolutionary tracks in their final parts are at about that luminosity (see Fig. 27.5). Then $L \sim M^{3.5}$ and $\tau_{KH} \sim M^{-2.5}$. This means that massive protostars reach the ZAMS much faster than their low-mass colleagues.

In the Hertzsprung–Russell diagrams of very young stellar clusters one finds that only massive stars are on the main sequence, while the low-mass stars lie to the right of it. As an example we show in Fig. 28.3 the case of NGC 602 (Cignoni et al. 2009), a very young star cluster with a population of stars born only a few million years (Myrs). From comparison with low-mass pre-main-sequence evolutionary tracks and isochrones[2] it is obvious that many stars have not yet reached the main sequence. Similar cases in the Milky Way are the Pleiades (80 Myrs) and NGC 2264 (5 Myrs). It seems that, because of their longer τ_{KH}, these stars are still in the contraction phase and have not yet begun with nuclear burning. Among them are flare stars (UV Ceti stars), Herbig AE/BE, and T Tauri variables. The cause of their (irregular) variability is not yet fully understood, but is ascribed to circumstellar or chromospheric activity in connection with rotation as well as internal pulsations.

---

[2] An *isochrone* is the locus of stellar models of identical age, but different mass in the Hertzsprung–Russell diagram.

Chapter 29
From the Initial to the Present Sun

There is evidence on Earth that the Sun has shone for more than 3,000 million years with about the same luminosity. From radioactive decay in different materials of the solar system, one nowadays assumes that it was formed 4.57 Gigayears (Gyrs; 10^9 years) ago. Since then, the Sun has lived on hydrogen burning, predominantly according to the *pp* chain, and its interior has been appreciably enriched in ^{4}He. In the following we show how a model of the present Sun can be constructed.

29.1 Known Solar Data

Although the Sun is a very ordinary star of average mass and in a quiet state of main-sequence hydrogen burning, it is a unique object for stellar evolution theorists. For no other star so many quantities are known with comparable accuracy obtained by so many different and independent methods. From Kepler's laws and known distances within the solar system we can derive its mass and radius as well as the total luminosity. This yields the effective temperature by application of the Stefan-Boltzmann law. Neutrino experiments on Earth (see Sect. 29.5) allow the determination of conditions in the innermost energy producing core. And the art of (helio-)seismology has returned with high accuracy the run of the sound speed throughout most of the solar interior, the helium content of the outer convective envelope, and its depth. These quantities restrict the modelling of the present Sun and allow a comparison with stellar evolution theory at a degree of precision which is almost unique in astrophysics. Table 29.1 summarizes the fundamental solar parameters and the method to derive them. Note that the rather large uncertainty in the solar mass is the result of the uncertainty in Newton's constant of gravity G. Kepler's third law returns their combination, $GM_{\odot}$, with a precision of 10^{-7}!

There is still one significant uncertainty in the solar quantities, and this is the present surface (or convective envelope) composition. The determination by Grevesse and Noels (1993) was considered to be very close to the real composition, as it also agreed very well with meteoritic values in those elements

R. Kippenhahn et al., *Stellar Structure and Evolution*, Astronomy and Astrophysics Library, DOI 10.1007/978-3-642-30304-3_29, © Springer-Verlag Berlin Heidelberg 2012

Table 29.1 Solar quantities and how they are derived

Quantity	Value	Method
Mass	$(1.9891 \pm 0.0004) \times 10^{33}$ g	Kepler's third law
Radius	$695,508 \pm 26$ km	Angular diameter plus distance
Luminosity	$(3.846 \pm 0.01) \times 10^{33}$ erg s^{-1}	Solar constant
Effective temp.	$5,779 \pm 2$ K	Stefan-Boltzmann law
Z/X	0.0245 ± 0.001	Meteorites and solar spectrum
	0.0165	(new determination)
Age	4.57 ± 0.02 Gyr	Radioactive decay in meteorites
Depth of conv. env.	$0.713 \pm 0.001\, R_\odot$	Helioseismology
Env. helium content	0.246 ± 0.002	Helioseismology

(Z/X) is given twice: the more traditional value by Grevesse and Noels (1993) and the more recent one by Asplund et al. (2005)

that can be compared. However, new analyses of the solar spectrum (Asplund et al. 2005, 2009), done with sophisticated three-dimensional, non-LTE radiation-hydrodynamics methods, returned (see Table 29.2) much lower values in particular for the volatile elements C, N, and O, which cannot be measured accurately in meteorites. The difference is a reduction of the total amount of metals relative to hydrogen, (Z/X), by 30 %! The latest revision by Asplund et al. (2009) for the solar element composition resulted in somewhat higher abundances than the 2005 values but still distinctively lower than the Grevesse and Noels numbers. Consequently, the structure of solar models using the 2009 abundances lies between the two other cases, which we will present in the following. This issue is not yet settled, but since the solar composition is the yardstick for all abundance determinations in astrophysics, the outcome will certainly be of great importance.

As we will see, the older abundance data yielded solar models in very good agreement with helioseismology. The lower abundances deteriorate this. In the following we will present a standard solar model based on the older Grevesse and Noels (1993) abundances, since such a model appears to be closer to the real solar structure, even if this could be due to coincidence.

A *standard solar model* is derived under the assumptions of spherical symmetry and hydrostatic equilibrium, ignoring effects of rotation and the influence of magnetic fields. Convection is usually treated in mixing-length theory, and no overshooting is assumed. The only effect beyond these most basic assumptions is the inclusion of atomic diffusion, since it turned out that models which disregarded this disagree more with seismic results. This is true for both sets of solar abundances mentioned above.

A solar model has to match the solar radius, luminosity, and surface abundance of metals at the solar age. The evolution is started from the pre-main-sequence hydrostatic contraction until the solar age. The mass can be kept fixed because mass loss is known to be unimportant.

Table 29.2 Solar atmospheric and meteoritic abundances of the most important elements, as determined by Grevesse and Noels (1993; "GN93") and Asplund et al. (2005, "AGS05")

Element	GN93	AGS05	Meteorites
H	12.00	12.00	8.25
C	8.55	8.39	7.40
N	7.97	7.78	6.25
O	8.87	8.66	8.39
Ne	8.08	7.84	−1.06
Na	6.33	6.17	6.27
Mg	7.58	7.53	7.53
Al	6.47	6.37	6.43
Si	7.55	7.51	7.51
S	7.21	7.14	7.16
Cl	5.50	5.50	5.23
Ar	6.52	6.18	−0.45
Ca	6.36	6.31	6.29
Ti	5.02	4.90	4.89
Cr	5.67	5.64	5.63
Mn	5.39	5.39	5.47
Fe	7.50	7.45	7.45
Ni	6.25	6.23	6.19
Z/X	0.0245	0.0165	

Abundances are given in logarithms of particle abundance on a scale on which hydrogen has the abundance of 10^{12}

The abundance of helium cannot be determined from the spectrum and is therefore missing

Errors are for most elements in the range of 0.02–0.06 dex

Neon and argon can be determined only indirectly from coronal abundance ratios with respect to oxygen, and are basically absent in meteorites

The bottom line gives the total metallicity in mass fractions relative to hydrogen

29.2 Choosing the Initial Model

While the observations yield information about the mass abundance Z of heavier elements, it is difficult to determine spectroscopically the helium content Y of the solar surface. One therefore uses Y as a free parameter. This is actually the initial helium content Y_i, which will change during the evolution due to the effects of nuclear burning and diffusion. Its value cannot be compared directly with an observed value. Sedimentation–the main effect of diffusion–will also lead to a reduction of Z/X with time. Therefore also the initial metallicity Z_i has to be chosen such that after 4.57 Gyr the present Z/X is obtained. Furthermore, there is no information about the mixing length ℓ_m to be used in the convection theory (see Chap. 7). One normally expresses ℓ_m in units of the local pressure scale height H_P and treats the dimensionless quantity ℓ_m/H_P as another free parameter.

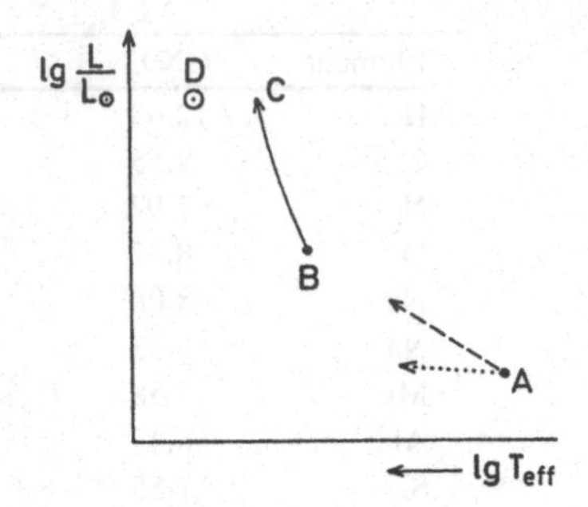

Fig. 29.1 Finding a model that for given values of $Z = 1 - X - Y$ describes the present Sun. For arbitrary values of Y, ℓ_m one obtains a ZAMS model at A, from where it shifts along the *broken* and *dotted arrow* as a result of independent changes of Y and ℓ_m, respectively. Based on this, one guesses the values of Y, ℓ_m that yield the model at B. Its evolution is calculated from age zero (B) to $t = 4.57 \times 10^9$ years (C). The guessed values Y, ℓ_m are modified until C coincides with D (present Sun)

We now sketch the way to obtain a solar model using some simplifications. These are not done in numerical calculations, but they allow us to use properties of simplified models. We first ignore diffusion. Then $(Z/X)_i$ is known from the present photospheric abundances. We now start the construction of an initial solar model with trial values of Y_i and ℓ_m/H_P. Since the model changes only on the (long) nuclear timescale, it can well be approximated by assuming complete equilibrium. This means that in addition to the inertia term in (10.2) the time derivatives in the energy equation (10.3) can be neglected. The evolution can then be followed from the ZAMS until a time of 4.57×10^9 years after the onset of hydrogen burning has elapsed. During this time interval the molecular weight in the central regions increases owing to the enrichment of helium. Consequently, the luminosity increases slightly, as can be expected from the homology relation (20.20) according to which the luminosity should increase like μ^4. (The fact that the solar evolution is not homologous changes the result only quantitatively.) At the same time, the point in the Hertzsprung–Russell (HR) diagram moves slightly to the left. If our choice of the free parameters were correct, the model after 4.57×10^9 years should resemble the present Sun. But, in general, this will not be the case, and the evolutionary track will miss the image point of the present Sun. One therefore has to adjust the two free parameters in order to end up with the present Sun.

A variation of the mixing length changes the radius slightly, but turns out to have almost no influence on the luminosity. Therefore, while varying ℓ_m, the initial model will move almost horizontally (Fig. 29.1). If, on the other hand, Y_i is changed, the mean molecular weight μ varies. With increasing helium content, μ also increases, and since the computed models roughly behave as the homologous models of Sect. 20.2.2, the image point of the model moves to the upper left on a line below the main sequence [see the arguments after (20.23)].

Since small changes in the two parameters do not modify the form of the evolutionary track very much, the whole track makes an approximately parallel

Table 29.3 Dependence of solar model quantities on model parameters

	ℓ_m/H_P	Y_i	Z_i
$L/L_\odot$	0.038	*8.515*	−38.60
$R/R_\odot$	*−0.129*	2.019	−7.05
$(Z/X)/(Z/X)_\odot$	0.043	0.523	*56.0*

The table entries are to be read as the partial derivative of the column quantity with respect to the row quantity
The strongest influence by each parameter is in italics

shift. Therefore one can find values for Y_i and ℓ_m/H_P for which the end point of the evolutionary track coincides with the point of the (observed) present Sun. The procedure is illustrated in Fig. 29.1. A model constructed in this way, and by using the standard assumptions for the input physics, is often called a "standard solar model".

Table 29.3 gives an overview of the partial derivatives $\partial y/\partial x$ in the vicinity of the final, calibrated solar model, where y corresponds to the solar observable (rows) and x to the model parameter (columns). The values were obtained from the solar model calculations presented in the next section. While the absolute numbers depend a lot on the individual calculation, the relative ratios are very similar for all solar model calibrations. Clearly, the initial helium content affects mostly the luminosity and the mixing length the radius of the solar model. The initial metallicity Z_i has not only an obviously direct effect on Z/X but also, due to the change of μ, on luminosity.

The values of the initial Y and ℓ_m/H_P, which after 4.57×10^9 years lead to the present Sun, depend sensitively on the details of the computations, for instance, on the opacities used and the equation of state applied.

29.3 A Standard Solar Model

After the procedure to compute a standard solar model has been outlined, we now show the results of such a detailed computation. The final model agrees with the present solar luminosity and effective radius (Table 29.1) to 1 part in 10^4 or better and has $Z/X = 0.0245$ according to the analysis of Grevesse and Noels (1993), from which also the chemical composition (Table 29.2) was taken. The effect of diffusion was included. Up-to-date tables for the equation of state (Sect. 16.6) and the opacities (Sect. 17.8) were used, the latter for the same metal composition of Grevesse and Noels (1993). The calculation starts with a homogenous pre-main-sequence model, since the assumption of a homogenous ZAMS model in complete equilibrium would already be too inaccurate (see Sect. 28.2).

All modern stellar evolution codes using the same physical input data are able to produce a standard solar model that reproduces known properties of the Sun at a similar accuracy and agree very well with each other.

The evolution in the Hertzsprung-Russell diagram is shown in Fig. 29.2. It begins with a fully convective, contracting pre-main-sequence model. At an age of 1.7 Myr

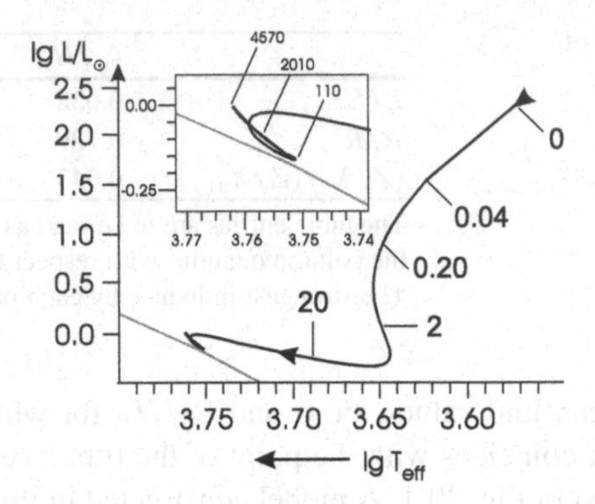

Fig. 29.2 Evolution of the standard solar model from the initial pre-main-sequence contraction to the present age. Ages in million years are indicated along the track. Notice the drastic slowdown of the evolution as soon as the nuclear timescale has become the dominant one and the rapid pre-main-sequence evolution on thermal timescales during the hydrostatic contraction along the Hayashi line. The main-sequence evolution and the phase immediately preceding it are magnified in the *inset* for clarity. The zero-age main sequence of Fig. 22.1 is shown as well. Since the solar composition is not exactly the same, the evolutionary path of the Sun is slightly offset from this ZAMS

the centre begins to become radiative; at that time $\log L/L_\odot$ has already dropped to 0.163. The evolution slows down considerably in the following. At 28.3 Myr and a luminosity of $\log L/L_\odot = -0.038$ and $\log T_{\text{eff}} = 3.748$ a transient convective core begins to develop due to the strongly peaked energy release of the CN conversion. It lasts for about 120 Myr at which time the luminosity minimum has been reached. From there on the evolution proceeds on the very long nuclear timescale as the very short linear part of the track that ends at the solar position.

The initial homogeneous composition of this standard solar model is $X_\text{i} = 0.7058$, $Y_\text{i} = 0.2743$, and $Z_\text{i} = 0.0199$. Z/X therefore was initially 0.0282 and has dropped at the photosphere to 0.0245 due to the settling of all heavier elements and the corresponding increase of hydrogen in the convective envelope. This effect is visible in Fig. 29.3, where the hydrogen and metal content X and Z as functions of m/M are plotted; the final surface hydrogen abundance is 0.7377 and that of metals 0.0181. They are higher, respectively lower than the initial ones because of the sedimentation of all elements heavier than hydrogen below the thin convective envelope. This leads to the sudden increase of hydrogen abundance to the higher and constant value in the convectively mixed outermost layers. Accordingly, the abundance of metals (dashed curve in Fig. 29.3) decreases at the beginning of the convective envelope.

In the central region of the present Sun, quite an appreciable percentage of the original hydrogen has already been converted into helium. The central value of X has dropped to 0.338. The abundance of ^{3}He, also shown in Fig. 29.4, displays the characteristic shape discussed in Sect. 28.2 due to its evolution towards an equilibrium value within the *pp* chain.

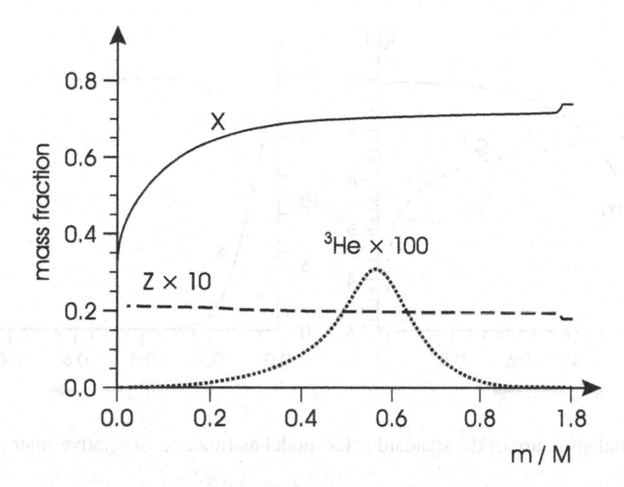

Fig. 29.3 Element abundances in a model for the present Sun (age 4.57×10^9 years) as a function of m/M. Shown are the mass fractions of hydrogen, (X, *solid line*), metals (Z, multiplied by a factor 10, *dashed line*), and of ^{3}He (multiplied by 100, *dotted line*). The initial values for X and Z in the homogeneous model were 0.7058, respectively 0.0199. The increase in X and the decrease in Z in the outermost regions is due to the effect of diffusion

Some details of the solar structure are shown in Fig. 29.4. The left panel shows the concentration of mass. More than 80 % are contained within 40 % of the solar radius or just 6.4 % of the volume. Temperature rises over two orders of magnitude within the outermost 20 % of the radius, but then only by another factor of ten until the centre. Pressure and density profiles have a similar shape like that of temperature. In the right panel the strongly peaked energy generation is shown, which results in the fact that over 90 % of the total luminosity are reached already at $r/R_\odot = 0.2$, corresponding to a mass coordinate of only 0.3. Note the similarity of the solar structure to that of the 1 $M_\odot$ ZAMS model of Fig. 22.4. Although the Sun has burnt hydrogen for almost 5 billion years, it still has the shape of a young star.

Had we used the abundances of Asplund et al. (2005; middle column in Table 29.2) with the corresponding present value for Z/X of 0.0165, the structure of that solar model would hardly be distinguishable from the one shown in Fig. 29.4. However, initial and present composition would be different: the initial abundances in that case are $X_\mathrm{i} = 0.7261$, $Y_\mathrm{i} = 0.2599$, and $Z_\mathrm{i} = 0.0140$, and the present solar surface values are $X = 0.7578$, $Y = 0.2297$, and $Z = 0.0125$.

The outer convective zone of our standard solar model reaches down to a temperature of 2.2×10^6 K. The radius of its inner boundary is $r = 0.713\ R_\odot$, and the corresponding mass is $0.9761\ M_\odot$.

The temperature gradients ∇, ∇_ad, and ∇_rad as defined in Chaps. 5–7 are plotted in Fig. 29.5. In the near-surface regions where $\lg P < 5.0$, one finds $\nabla_\mathrm{rad} < \nabla_\mathrm{ad}$ and the layer is stable (Fig. 29.5a). Then convection sets in where ∇_rad exceeds ∇_ad. In the outermost part of the convective zone the convection is very ineffective and ∇ is close to ∇_rad, according to the considerations in Sect. 7.3. But ∇ does not follow

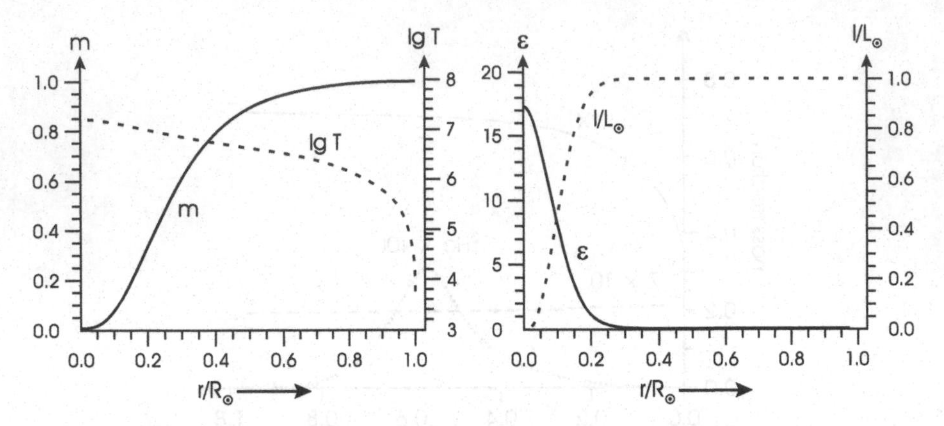

Fig. 29.4 Internal structure of the standard solar model as function of relative solar radius $r/R_\odot$

∇_{rad} to the extreme values (which at $\lg P = 9$ reach a maximum of 4.1×10^5). It never exceeds 0.9. Owing to partial ionization of the most abundant elements, ∇_{ad} is not constant in the outer region of the solar model, as we have already shown in Fig. 14.1b. The deeper inside, the more the actual gradient approaches the adiabatic one, following it up and down (Fig. 29.5a, b). In Fig. 29.5c the convective velocity obtained from U, ∇_{rad}, and ∇ according to (7.6) and (7.15) is given in units of the (isothermal) velocity of sound $v_{\mathrm{s}} = (\Re T/\mu)^{1/2}$. At the top of the convection zone, v/v_{s} reaches its maximum of about 0.4.

29.4 Results of Helioseismology

It is not surprising that one can produce models for the present Sun which have the correct position in the HR diagram, since three free parameters, Y_{i}, ℓ_{m} and Z_{i}, can be varied to adjust the quantities L, R, and Z/X. Therefore obtaining a solar model with the right age at the right position in the HR diagram and the right surface composition is not much of a test of stellar evolution theory.

At present there are two observational tests to compare the solar interior with model calculations. One are solar neutrino experiments, which will be discussed in the next section. They test the conditions at the solar centre, where nuclear reactions take place. The other one allows an almost complete "view" of the solar interior and is based on the investigation of non-radial solar oscillations, commonly called helioseismology. We shall deal with such oscillations later (see Sect. 42.4). For the moment it is sufficient to state that the frequencies of thousands of non-radial solar oscillation modes, measured with extremely high precision, depend in particular on the sound speed profile throughout the Sun. In the following we discuss the most important results for the solar interior.

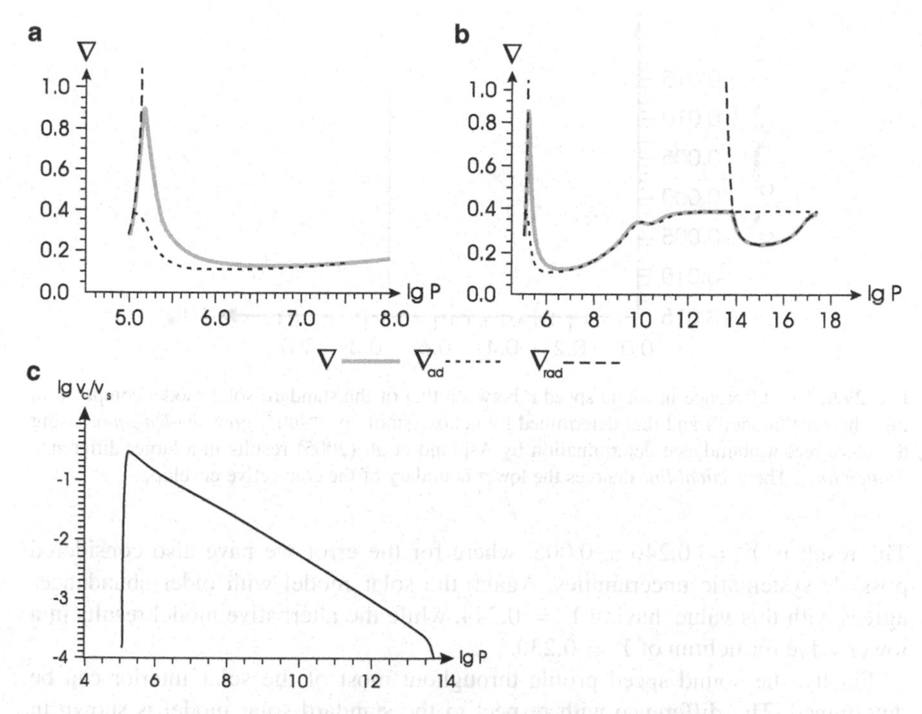

Fig. 29.5 Some properties of the model for the present Sun described in the text. (**a**) The temperature gradients in the outer layers, against the pressure P (in dyn cm^{-2}). In the outermost layers the actual gradient ∇ (*gray-shaded line*) coincides with $\nabla_{\rm rad}$ (*dashed line*), which then, however, goes up to values above the range of the ordinate. The strong depression of $\nabla_{\rm ad}$ (*lower short-dashed line*) for $\lg P > 5$ is due to hydrogen ionization. (**b**) The same curves as in (**a**) but with compressed scales, such that the whole interior of the model is covered. $\nabla_{\rm rad}$ is still out of the range for almost all of the outer convective zone. The depression of $\nabla_{\rm ad}$ is caused by the ionization of H, He, and He$^+$ (at values of lg P around 6, 8, and 10). Note that the centre of the Sun is close to convective instability. (**c**) The convective velocity v in units of the local velocity of sound, $v_{\rm s}$, in the outer convective zone of the Sun

The first one is that the transition from the nearly adiabatic temperature gradient to the radiative one at the bottom of the convective zone leaves a significant change in the slope of the square of the sound speed divided by the gravitational acceleration (Gough 1986). This allows a very accurate determination of the bottom of the convective envelope, which is at $0.713 \pm 0.001\, R_\odot$. The solar model of Sect. 29.3 has exactly the same depth of the convective zone. The solar model with the newer abundance determination, in contrast, is convective to $0.731\, R_\odot$. This would favour the older abundances, provided that the physical input (equation of state, opacities, diffusion theory) is correct.

The second envelope quantity that can be determined by seismology is the envelope helium content and is based on the fact that the quantity Γ_1, defined in (13.18), depends on the chemical composition and therefore allows the determination of the helium abundance as the ionization of helium modifies Γ_1.

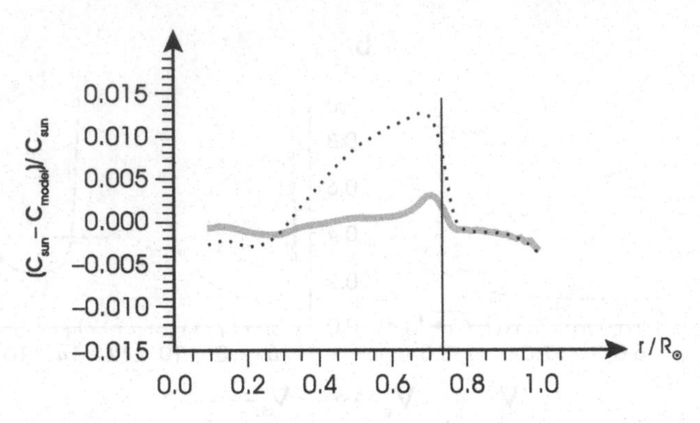

Fig. 29.6. The difference in sound speed c between that of the standard solar model computed in this chapter ("model") and that determined by helioseismology ("sun"; *grey shaded line*). Using the more recent abundance determination by Asplund et al. (2005) results in a larger difference (*dotted line*). The *vertical line* denotes the lower boundary of the convective envelope

The result is $Y = 0.246 \pm 0.005$, where for the error we have also considered possible systematic uncertainties. Again, the solar model with older abundances agrees with this value, having $Y = 0.244$, while the alternative model results in a lower value for helium of $Y = 0.230$.

Finally, the sound speed profile throughout most of the solar interior can be determined. The difference with respect to the standard solar model is shown in Fig. 29.6 for both determinations of the solar abundances. As before, the older one results in a model closer to the seismic results, even if the reasons for this good agreement are not clear. The uncertainty of the seismic sound speed is below 0.002 for $r/R_\odot$ between 0.2 and 0.7. Towards the centre it is increasing due to the small number of modes extending into the core, and in the outermost layers it is larger because of the uncertainties concerning the damping of oscillations in the atmosphere. Therefore the only significant deviation of the solar model sound speed profile from the seismically determined one is the maximum of the grey line just below the convective envelope.

Although the discrepancy between model and seismic data appears to be large for the alternative model, one should keep in mind that the agreement is still within one per cent everywhere. Ignoring the effect of diffusion in the solar model calculation, which, when looking at its effect in Fig. 29.3, appears to be rather small, would result in a very similar discrepancy between model and seismic result. Overall, helioseismology confirms that stellar evolution theory can reproduce the structure of the Sun with an accuracy that is much higher than usually found in astrophysical situations.

29.5 Solar Neutrinos

Some of the nuclear reactions of the *pp* chain, as well as of the CNO cycle, produce neutrinos (Sect. 18.5.1). In addition, there are also neutrinos due to the very rare *pep* and *hep* reactions

$$
\begin{aligned}
{}^{1}\mathrm{H} + {}^{1}\mathrm{H} + e^{-} \rightarrow &\ {}^{2}\mathrm{H} + \nu\ (pep) \\
{}^{3}\mathrm{He} + {}^{1}\mathrm{H} \rightarrow &\ {}^{4}\mathrm{He} + e^{+} + \nu\ (hep),
\end{aligned}
\tag{29.1}
$$

the latter one being the trivial way to produce ^{4}He after the reactions of (18.61), but it is occurring in only 10^{-8} of all cases. However, the energy of the emitted neutrino is close to 10 MeV, and it is therefore necessary to consider this reaction. As already discussed in Sect. 18.7, the neutrinos leave the star practically without interacting with the stellar matter. The energy spectrum of neutrinos from β decay is continuous, since the electrons can take part of the energy away, while neutrinos released after an inverse β decay are essentially monochromatic. Therefore most reactions of the *pp* chain have a continuous spectrum, while the *pep*-reaction (29.1) and the electron capture on ^{7}Be (18.62) have a line spectrum. Since ^{7}Be can decay into ^{7}Li either in the ground state or in an excited state, this reaction gives two spectral lines. The neutrino spectrum of the Sun as predicted from the reactions of the *pp* chain, computed from our standard solar model, is given in Fig. 29.7. In order to obtain the neutrino spectrum of the present Sun one cannot use the simple (equilibrium) formulae (18.63) and (18.65), but must compute the rates of all the single reactions given in (18.62), (18.64) and in addition the reactions of (29.1) in a nuclear network.

Since the solar neutrinos can leave the Sun almost unimpeded they can in principle be measured in terrestrial laboratories and thus be used to learn directly about conditions in the innermost solar core. This difficult task indeed has been undertaken since 1964, when John Bahcall and Raymond Davies began to plan for an underground neutrino detector in a mine in Homestead, North Dakota. Forty years later the experiments finally have confirmed the standard solar model, and R. Davies received the Nobel Prize for his work. The time in between, however, was characterized by the "solar neutrino problem". The history of solar neutrino physics and the resolution of the problem is summarized in detail in Chap. 18 of the textbook by Weiss et al. (2004) and in Bahcall and Davies (2000).[1]

The solar neutrino problem consisted in the fact that since the first results from the so-called *chlorine* experiment by Davies there was a lack of neutrinos compared to solar model predictions. The chlorine experiment is sensitive to neutrinos with energies above 0.814 MeV and therefore, as can be seen in Fig. 29.7 mainly to the ^{8}B neutrinos, with some contribution from *pep*, *hep*, and ^{7}Be neutrinos. The experiment is based on the reaction $^{37}\mathrm{Cl} + \nu \rightarrow {}^{37}\mathrm{Ar}$, where the decays of radioactive argon nuclei are counted. The rate of neutrino captures is commonly measured in *solar neutrino units* (SNU). One SNU corresponds to 10^{-36} captures per second and per target nucleus. The predicted counts amount to 7.5 SNU for the chlorine experiment, the measurements averaged over several decades to only 2.5 ± 0.2 SNU. The deficit could indicate that the solar centre is cooler than in the models

[1]There is also a review by Bahcall (*Solving the Mystery of the Missing Neutrinos*) at the electronic library of the Nobel prize committee (URL: `nobelprize.org/nobel_prizes/physics/articles/bahcall`).

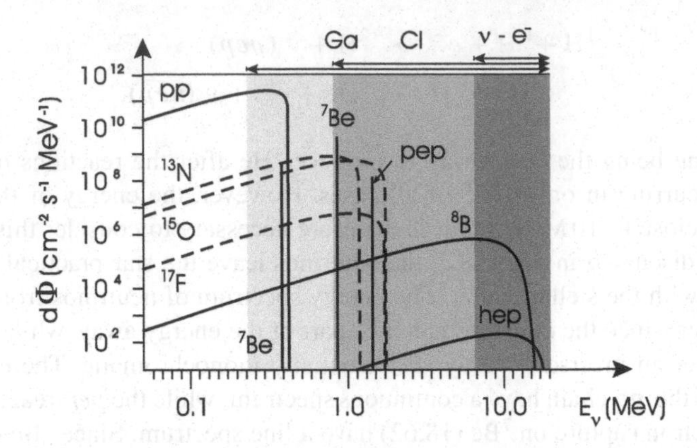

Fig. 29.7. The neutrino spectrum of the Sun as predicted from the theoretical standard solar model. The *solid lines* belong to reactions of the *pp* chain while the *broken lines* are due to reactions of the CNO cycle. The neutrinos from most of the reactions have continuous spectra, while monoenergetic neutrinos come from ^{7}Be and from the *pep*-reaction (29.1). The flux ϕ for the continuum sources is given in $\mathrm{cm^{-2}\,s^{-1}\,MeV^{-1}}$ and for the line sources in $\mathrm{cm^{-2}\,s^{-1}}$. The sensitivity of the three types of neutrino experiments is indicated above the figure and by the *shaded* regions

To improve the experimental evidence, additional experiments were started. First, another kind of radiochemical detector using gallium in the detector fluid measured, due to a much lower energy threshold, the majority of neutrinos, including those from the *pp*-reaction. Later, electron-scattering detectors were developed, which are sensitive to the highest energies only, but which provide directional information about the neutrino source (For these detectors the *hep*-neutrinos of (29.1) have to be taken into account.). All experiments confirmed that the solar neutrino flux was of the right order of magnitude, and therefore that indeed the Sun shines by the nuclear fusion of hydrogen, but they also consistently measured a deficit of neutrinos. This deficit, however, varied between different kinds of detectors.

The various ideas on how to solve the solar neutrino problem are discussed in Chap. 18 of Weiss et al. (2004). With more and more experimental data it became evident that even hypothetical changes to the solar centre cannot solve the problem and that the solution is most likely to be found in the properties of neutrinos. All nuclear reactions emit *electron neutrinos*, and these are the only ones that were measured in terrestrial experiment, with the exception of the electron-scattering *Sudbury Neutrino Observatory* experiment in Canada, where heavy water (with a high percentage of deuterium isotopes) was used as the detector. Here also reactions with the two other types (*flavours*) of neutrinos, *muon* and *tau neutrinos* can be detected. Summing these and the *electron neutrinos* up, the total number of detections is completely consistent with the solar model prediction, within a few per cent. What created the apparent solar neutrino deficit is the fact that neutrinos can change their flavour, both while travelling through vacuum and more efficiently in the presence of electrons in the solar interior. A similar effect was also confirmed for muon neutrinos arising in the Earth's upper atmosphere from high-energy cosmic radiation, when measured before or after they have travelled through the Earth's interior. The modelling of the solar interior, together with sophisticated experiments, has therefore resulted in new knowledge about fundamental properties of neutrinos. In particular, these so-called *neutrino oscillations* are possible only if neutrinos have mass.

Chapter 30
Evolution on the Main Sequence

30.1 Change in the Hydrogen Content

In the main-sequence phase, the large energy losses from a star's surface are compensated by the energy production of hydrogen burning (see Sect. 18.5.1). These reactions release nuclear binding energy by converting hydrogen into helium. This chemical evolution of the star concerns primarily its central region, since the energy sources are strongly concentrated towards the centre (Sect. 22.2).

Somewhat larger volumes are affected simultaneously if there is a convective core in which the turbulent motions provide a very effective mixing. If the extent of convective regions and the rate of energy production ε_H for all mass elements are known, the rate of change of the hydrogen content X_H can be calculated according to Sect. 8.2.3.

The situation is particularly simple for stars of rather small mass (say $0.1 M_\odot < M \lesssim 1 M_\odot$) that have a radiative core. In the absence of mixing, the change of X_H at any given mass element is proportional to the local value of ε_H. After a small time step Δt, the change of hydrogen concentration is $\Delta X_H \sim \varepsilon_H \Delta t$ everywhere (with a well-known factor of proportionality). Following the chemical evolution in this way over many consecutive time steps, one obtains "hydrogen profiles" [i.e. functions $X_H(m)$] as shown in Fig. 30.1. At the end of the main-sequence phase, $X_H \to 0$ in the centre.

With the change in the hydrogen profile also a change in the energy generation rate ε_H takes place (Fig. 30.2). Initially, it has a maximum at the centre, since there temperature is highest and the abundance of hydrogen almost the same everywhere in the core. However, in the course of evolution, though temperature rises in the centre, the hydrogen abundance drops, and after some time, the maximum ε_H, which depends on both of these quantities, is larger outside the centre. This can be seen first for the model of 8.2×10^9 years (dotted line in Fig. 30.2). When the centre is completely depleted of hydrogen, $\varepsilon_H = 0$ there and the energy generation profile looks like the strongly peaked (solid) line, corresponding to the final model of Fig. 30.1. Energy is now being produced effectively in a shell around the exhausted

R. Kippenhahn et al., *Stellar Structure and Evolution*, Astronomy and Astrophysics Library, DOI 10.1007/978-3-642-30304-3_30, © Springer-Verlag Berlin Heidelberg 2012

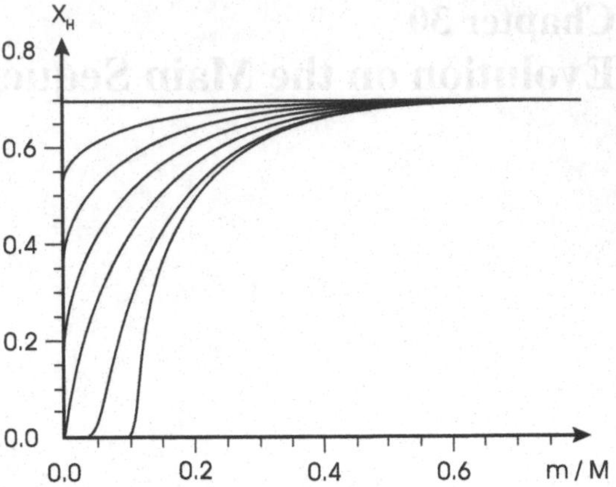

Fig. 30.1 Hydrogen profiles showing the gradual exhaustion of hydrogen in a star of $1 M_\odot$. The homogeneous initial model consists of a mixture with $X_{\mathrm{H}} = 0.700$ and $X_{\mathrm{He}} = 0.280$. The hydrogen content X_{H} over m/M is plotted for seven models which correspond to an age of 0.0, 2.2, 4.2, 6.2, 8.2, 10.2, and 11.2×10^9 years after the onset of hydrogen burning

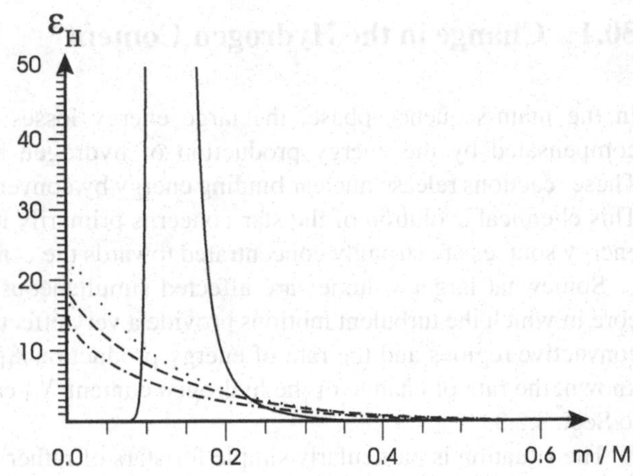

Fig. 30.2 Profiles of nuclear energy production ε_{H} from hydrogen burning (in erg/gs) for some of the models for which hydrogen profiles are shown in Fig. 30.1. These are the ones at the very beginning of hydrogen burning (*dash-dotted line*) and at ages of 6.2 (*dashed*), 8.2 (*dotted*), and 11.2×10^9 years (*solid line*). The maximum energy generation rate in this latter model is 170 erg/gs

core. This "hydrogen-shell burning" is taking place within the much larger region, in which core hydrogen burning has reduced the hydrogen content. It leads to a steepening of that profile and to a narrowing of the burning shell. Hydrogen burning is now even for this $1 M_\odot$ star proceeding via the CNO cycle.

In more massive stars, the helium production is even more concentrated towards the centre because of the large sensitivity to temperature of the CNO cycle. But the mixing inside the central convective core is so rapid compared to the local production of new nuclei that the core is virtually homogeneous at any time. Inside the core, $\Delta X_{\mathrm{H}} \sim \bar{\varepsilon}_{\mathrm{H}} \Delta t$ with an energy production rate $\bar{\varepsilon}_{\mathrm{H}}$ averaged over the whole core. The only difficulty comes from the fact that the border of the convective core may change during the time step Δt. The numerical calculations show that for stars below $10 M_\odot$ the mass M_{c} of the convective core decreases with progressive hydrogen consumption, which leads to a hydrogen profile $X_{\mathrm{H}}(m)$, as shown in Fig. 30.3 for a $5 M_\odot$ star. At the end of central hydrogen burning, one has a helium core with $M_{\mathrm{He}} \approx 0.1 M$, and the envelope in which X_{H} still

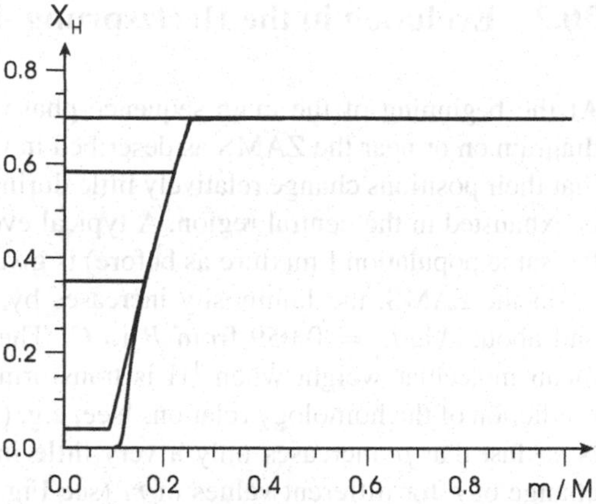

Fig. 30.3 The hydrogen profile $X_H(m)$ that is established in a 5 $M_\odot$ star of the same composition as in Fig. 30.1 during and at the end of hydrogen burning in a shrinking convective core. With decreasing central hydrogen content the age of the models is 0.7, 23, 55, 78, and 82 × 10⁶ years

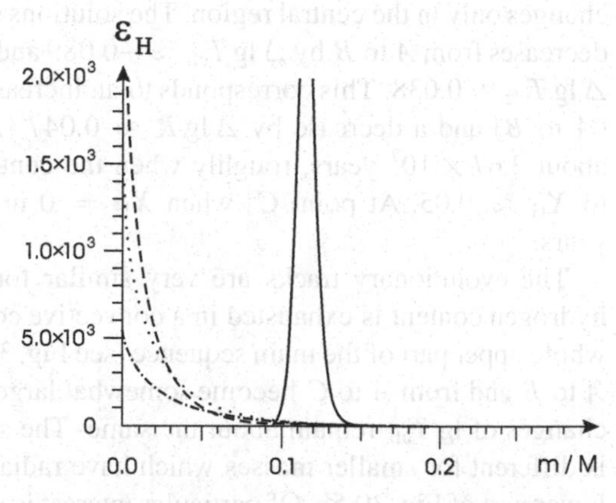

Fig. 30.4 Energy production profiles for the models with an age of 0.7 (*dash-dotted*), 55 (*dotted*), 78 (*dashed*), and 82 × 10⁶ years (*solid line*) of Fig. 30.3. ε_H, the energy generation rate by hydrogen burning (in units of erg/gs), is about 10^4 times larger than in the 1 $M_\odot$ star (Fig. 30.2) and has a maximum of 2.6×10^4 erg/gs in the last model

has almost its original value. The corresponding energy production is shown in Fig. 30.4. Notice that it is more and more concentrated towards the centre with progressing main-sequence evolution and that the energy producing shell is located just outside the helium core left after the end of central hydrogen burning. Similar hydrogen and energy production profiles are established in stars with other values of M. The main difference is that with increasing M the hydrogen profile is gradually shifted to larger values of m/M, i.e. the relative mass of the produced helium core increases with M. The corresponding increase of the convective core with increasing M for zero-age main-sequence (ZAMS) models has already been shown in Fig. 22.7.

This simple scenario is seriously complicated, particularly for rather massive stars, by two uncertainties in the theory of convection (convective overshoot and semiconvection). These effects will be dealt with separately in Sect. 30.4.

30.2 Evolution in the Hertzsprung–Russell Diagram

At the beginning of the main-sequence phase the models are located in the HR diagram on or near the ZAMS as described in Chap. 22. Numerical solutions show that their positions change relatively little during the long phase in which hydrogen is exhausted in the central region. A typical evolutionary track (for a $7M_\odot$ star of the same population I mixture as before) is given in Fig. 30.5a. Starting from point A on the ZAMS, the luminosity increases by about $\Delta \lg L = 0.240$ to point B and about $\Delta \lg L = 0.059$ from B to C. The rise of L is due to the increasing mean molecular weight when ^{1}H is transformed to ^{4}He, in accordance with the prediction of the homology relations [see, e.g. (20.20)]. The evolution from B to C is so fast that μ increases only a very little in this short time interval. From the change of r for different values of m (see Fig. 31.3) one clearly sees that the star evolves non-homologously, which ultimately is because the chemical composition changes only in the central region. The solutions show that the effective temperature decreases from A to B by $\Delta \lg T_{\text{eff}} \approx -0.089$ and then increases again to point C by $\Delta \lg T_{\text{eff}} \approx 0.038$. This corresponds to an increase of the radius by $\Delta \lg R \approx 0.299$ (A to B) and a decrease by $\Delta \lg R \approx 0.047$ (B to C). Point B is reached after about 3.67×10^7 years, roughly when the central hydrogen content has dropped to $X_{\text{H}} \approx 0.05$. At point C, when $X_{\text{H}} = 0$ in the centre, the age is 3.74×10^7 years.

The evolutionary tracks are very similar for all stellar masses for which the hydrogen content is exhausted in a convective core of appreciable mass, i.e. on the whole upper part of the main sequence (see Fig. 30.5b). The increments of $\lg L$ from A to B and from A to C become somewhat larger for larger values of M, while the changes of $\lg T_{\text{eff}}$ remain about the same. The structure of the evolutionary tracks is different for smaller masses which have radiative cores. This can be seen in the lower part of Fig. 30.5b. Of particular interest is the star with $M = 1.2\,M_\odot$, since it barely develops a convective core of only $0.05\,M_\odot$. This is also visible in the shape of its track in the Hertzsprung–Russell diagram, which appears to be a transition between those for lower and higher masses.

A common feature of all evolutionary tracks described here is that they point in some direction *above* the ZAMS. This is the case only for an evolution producing chemically inhomogeneous models (composed of a helium core and a hydrogen-rich envelope). In an evolution assuming complete mixing of the whole model, μ would have a constant spatial distribution and would increase in time. Then the star would evolve below the ZAMS, in accordance with the discussion after (20.23). Aside from all details, the observations (e.g. cluster diagrams) show that evolved stars are in fact above and to the right of the ZAMS, i.e. the stars obviously develop chemical inhomogeneities in their interior. This conclusion is very important, in particular, for the theory of stellar rotation. It excludes, for example, a complete mixing by the large-scale currents of rotationally driven meridional circulations (Chap. 44).

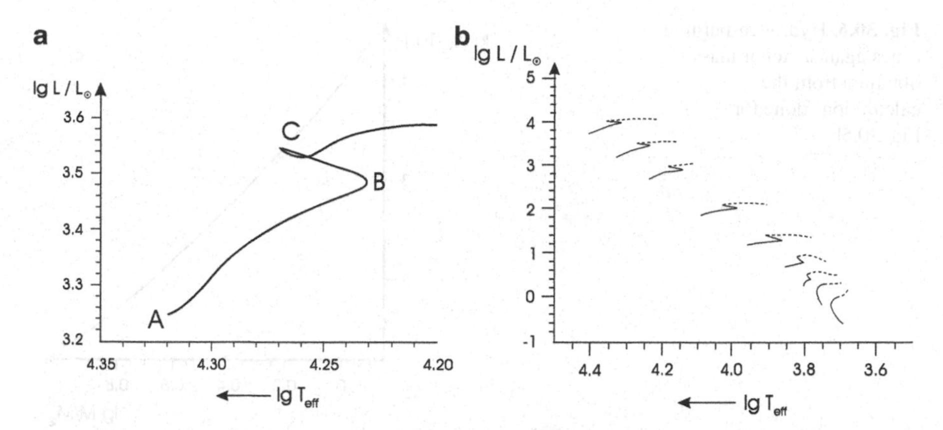

Fig. 30.5 Hertzsprung–Russell diagrams with evolutionary tracks for population I stars during central hydrogen burning (main-sequence phase). The tracks start on the zero-age main-sequence and extend into the post-main-sequence phase. (**a**) For stellar mass $M = 7M_\odot$. Some characteristic models are labelled by A (age zero), B (minimum of T_{eff}), and C (exhaustion of central hydrogen). (**b**) For stellar masses $M = 0.8, 1.0, 1.2, 1.5, 2.0, 3.0, 5.0, 7.0$, and $10\,M_\odot$. The *dotted parts* of the tracks indicate their continuation into the ensuing phase after central hydrogen exhaustion

30.3 Timescales for Central Hydrogen Burning

The time τ_{H} a star spends on the main sequence while burning its central hydrogen depends on M. This is because its luminosity L increases so strongly with M. Let us consider this timescale:

$$\tau_{\text{H}} = \frac{E_{\text{H}}}{L}, \tag{30.1}$$

where E_{H} is the nuclear energy content that can be released by central hydrogen burning. As a rough estimate, we assume that the same fraction of the total mass of hydrogen M_{H} in the star is consumed in all stars. Then we have $E_{\text{H}} \sim M_{\text{H}} \sim M$. Since L does not vary very much in this phase, we take the M–L relation of the ZAMS, $L \sim M^{\eta}$ [cf. (22.1)]. Introducing these proportionalities into (30.1), we have for the dependence of τ_{H} on M

$$\tau_{\text{H}}(M) \sim \frac{M}{L} \sim M^{1-\eta}. \tag{30.2}$$

For an average exponent in the M–L relation of, say, $\eta = 3.5$, one has $\tau_{\text{H}} \sim M^{-2.5}$, i.e. a strong decrease of τ_{H} towards larger values of M.

Of course, the numerical results are influenced and modified by a variety of details. The sequence of calculations made for Fig. 30.5b yields $\tau_{\text{H}}/(10^6\,\text{years}) =$ 23,283.89, 2,420.24, 303.32, 37.42, and 18.91 for $M/M_\odot = 0.8, 1.5, 3.0, 7.0$, and 10, respectively. In all the cases with a convective core, by far the largest part of τ_{H}

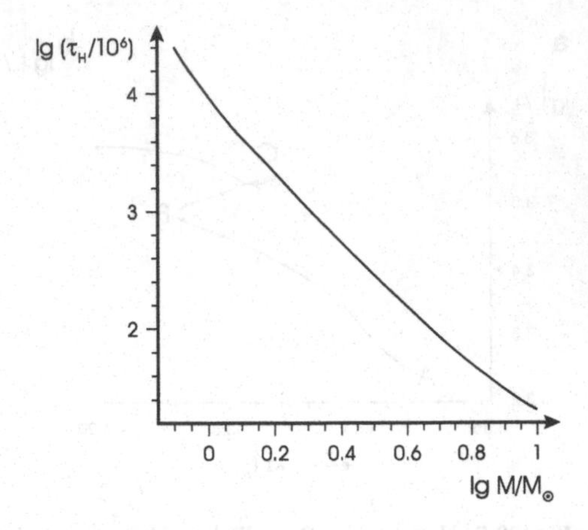

Fig. 30.6. Hydrogen-burning times against stellar mass obtained from the calculations done for Fig. 30.5b

is spent in the first phase between points A and B, while the last part (B–C) covers only about 3 % ... 5 %. Figure 30.6 shows the main-sequence lifetime as function of mass in a double-logarithmic plot. The mean slope is ≈ -2.8, corresponding to $\eta \approx 3.8$.

Although the absolute values are very uncertain (Sect. 30.4), the general trend is clear and has remarkable consequences for the observed HR diagrams of star clusters, by which it is confirmed. Assume that all stars of such a cluster were formed at the same time, i.e. that they now have the same age τ_{cluster}. We must then conclude that all stars with masses larger than a limiting mass M_0 have already left the main-sequence region, while stars with $M < M_0$ are still on the main sequence. M_0 is given by the condition $\tau_{\text{cluster}} = \tau_{\text{H}}(M_0)$. This is the basis for the age determination of such clusters.

30.4 Complications Connected with Convection

The seemingly nice and clear picture of the main-sequence phase as described above is unfortunately blurred by the notorious problem of convection. Questionable points include the precise determination of those regions in the deep interior in which convective motions occur and therefore the extent to which the chemical elements are mixed. The mixing influences the later evolution, since the chemical profile, which is established and left behind, is a long-lasting memory. We briefly mention two problems, the first of which concerns all main-sequence stars having convective cores, while the second occurs only in the more massive of these stars.

30.4.1 Convective Overshooting

We consider the situation in the surroundings of the outer boundary of a convective core of mass M_{bc}, as calculated without allowance for overshooting. This means that here we have defined the boundary to be at the position of neutral stability, i.e. where

$$\nabla_{rad} = \nabla_{ad} \tag{30.3}$$

according to the classical criterion (6.13). (Without much loss of generality, we may here treat a chemically homogeneous layer, for example, in the model for a ZAMS star.) Complete mixing and a nearly adiabatic stratification with $\nabla = \nabla_{ad} + \varepsilon (0 < \varepsilon \ll 1)$ is assumed in the convective region below M_{bc}, while no mixing and $\nabla = \nabla_{rad}$ is assumed for the radiative region above M_{bc} (cf. Chaps. 6 and 7, in particular Sect. 7.3).

This model implies an obvious problem: the boundary between the regimes in which convective motions are present ($v > 0$) and absent ($v = 0$) is determined by the criterion (30.3), which essentially relies on buoyancy forces, and therefore describes the *acceleration* $\dot{v}$ rather than the velocity v (cf. Sect. 6.1). Rising elements of convection are accelerated until they have reached M_{bc}; the braking starts only beyond this border, which is passed by elements owing to their inertia. The situation is the same as if we were to hope that a car would come to a full stop at the very point where one switches from acceleration to braking. The only way to substantiate this would be to try it (once) right in front of a hard and solid enough wall.

Simple estimates (e.g. Saslaw and Schwarzschild 1965) indeed give the impression that there is such a hard wall for elements passing the border M_{bc}. We have seen in Sect. 7.4 that in the deep interior of the star the elements rise adiabatically such that $\nabla_e = \nabla_{ad}$. From (7.5) we then see that the buoyancy force k_r acting on an element is

$$k_r \sim \nabla - \nabla_{ad} \, , \tag{30.4}$$

with a positive factor of proportionality. Below the border, k_r is small and positive (small acceleration) since $\nabla - \nabla_{ad}$ is extremely small and positive ($\approx 10^{-6}$). In contrast to this, the braking *above* the border is by orders of magnitude more efficient. We have assumed that there ∇ is equal to ∇_{rad}, which drops rapidly below ∇_{ad} (in Fig. 22.8b by about 0.1 within a scale height). So the force k_r due to $\nabla - \nabla_{ad}$ soon reaches rather large and negative values: therefore an overshooting element can be stopped within a negligible fraction of the pressure scale height.

A significant overshoot, therefore, could result only if the braking were substantially reduced (the "wall" softened). A possibility for this was outlined by Shaviv and Salpeter (1973), who pointed to the recoupling of the overshoot on the thermal structure of the layer. Consider the temperature excess DT of a moving element ($\nabla_e = \nabla_{ad}$) over the surroundings (gradient ∇). According to (7.4), we have $DT \sim \nabla - \nabla_{ad}$, and DT becomes negative above the border, i.e. the overshooting elements become cooler than the surroundings, which results in a cooling of the upper layers and an increase of the gradient ∇. We may describe it in terms of the convective flux (positive, if it points outwards), which according to (7.3) is

$$F_{con} \sim v \cdot DT \tag{30.5}$$

(with positive factors of proportionality). Above the border, the upward motion ($v > 0$) of cooler elements ($DT < 0$) represents a negative F_{con}. In order to maintain a constant total flux

$$F = F_{con} + F_{rad} = \frac{l}{4\pi r^2}, \tag{30.6}$$

with $F_{con} < 0$, the radiative flux F_{rad} must become larger than the total flux F. From (7.1) and (7.2) we immediately have

$$\frac{F_{rad}}{F} = \frac{\nabla}{\nabla_{rad}}, \tag{30.7}$$

which shows that $\nabla > \nabla_{rad}$ for $F_{rad} > F$. The increase of ∇, however, reduces the absolute values of $\nabla - \nabla_{ad}$ and of the braking force k_r compared with the situation without overshooting; the elements can penetrate farther into the region of stability than originally estimated, etc.

To find out whether or not this provides an appreciable amount of overshooting is a difficult problem and one that is still far from being solved. In order to find the point where the velocity v vanishes, one needs a self-consistent and detailed solution (including velocities, fluxes, gradients) for the whole convective core. This can only be obtained by using a theory of convection, the uncertainties of which now enter directly into the interior solution of the star. Even if we want to apply the mixing-length theory, the procedure is not clear. Instead of the usual local version of the theory, one needs a non-local treatment. At a given point, for example, the velocity of an element or its temperature excess depends not only on quantities at that point, but on the precise amount of acceleration (and braking) which the element has experienced along its whole previous path. All prescriptions for evaluating this and for averaging quantities like v or DT are as arbitrary as the choice of the mixing length. In fact any detailed modelling of the convective core by a mixing-length theory is necessarily ambiguous. For example, it encounters the difficulty that a core extends over less than a pressure scale height [the local expression of which, $H_P = -dr/d \ln P$, becomes ∞ at the centre according to (11.7)]. Different authors using different prescriptions have arrived at answers ranging from virtually no overshoot to rather extensive overshoot; and all of them have been questioned (see Renzini 1987). In the following we present a physically motivated treatment by Maeder (1975). Figure 30.7 shows the typical run of some characteristic functions as obtained from such calculations for $M = 2M_\odot$ and $\alpha = \ell_m/H_P = 1$. Below the "classical" border of stability ($\nabla_{rad} = \nabla_{ad}$), one has typically $\nabla - \nabla_{ad} \approx +10^{-4}$ which is enough to accelerate the convective elements

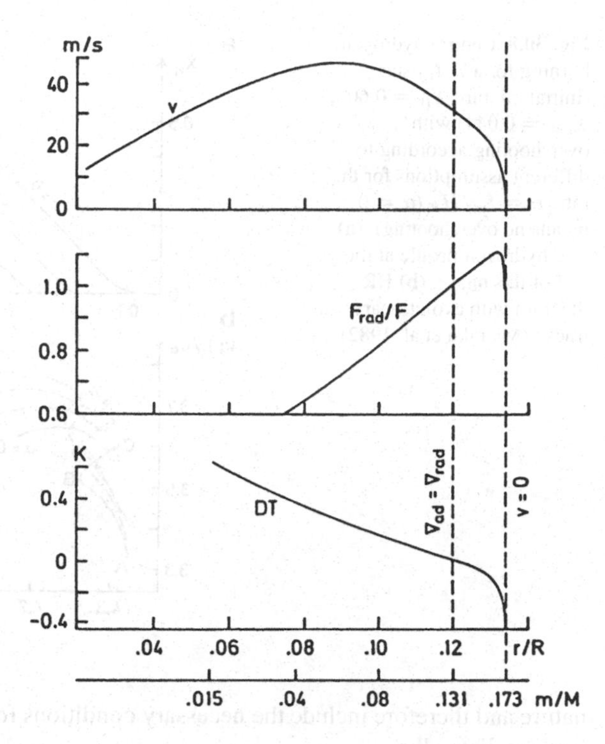

Fig. 30.7. Velocity v and temperature excess DT of rising convective elements and the ratio of the radiative flux F_{rad} relative to the total flux F around the border of stability ($\nabla_{\mathrm{rad}} = \nabla_{\mathrm{ad}}$) in a star of $2M_\odot$. Overshooting calculated with $\alpha = \ell_{\mathrm{m}}/H_P = 1$ extends to the point where $v = 0$ (after Maeder 1975)

to $30 \cdots 40\,\mathrm{m\,s^{-1}}$. Above the border, where still $v > 0$, but $DT < 0$, F_{rad} exceeds the total flux F by about 10 %, while $\nabla - \nabla_{\mathrm{ad}}$ ranges from -10^{-4} to -10^{-2}. The overshooting reaches to the point with $v = 0$, which occurs at about 14 % of the local scale height H_P above the border, corresponding to an increase of the mass of the convective core M_{c} of more than 30 %. This amount depends on the assumed value of α, because the velocity of the convective elements depends on the mixing length ℓ_{m} according to (7.6). Figure 30.8a shows the hydrogen profile established during hydrogen burning in a $7M_\odot$ star calculated with such overshooting for different α (The limit case $\alpha = 0$ is the model calculated without overshooting.). The influence of overshooting on the evolutionary tracks is shown in Fig. 30.8b. The consequences of an increased helium core at the end of this phase are an increased luminosity, an increased age (by about 25 % for $\alpha = 1$) due to the enlarged reservoir of nuclear fuel, and lower effective temperatures reached during the main-sequence evolution. This leads to a broadening of the upper main-sequence compared to calculations without overshooting. Indeed, the observed width of the upper main-sequence is one test to estimate the amount of overshooting from convective cores in massive stars (Maeder and Meynet 1991). However, if such overshooting occurs, its main effect will show up only later, during the phase of helium burning (see Sect. 31.4).

As mentioned at the beginning of Chap. 7 efforts to develop more realistic convection models, based on either the Reynolds stress approach or on multi-dimensional simulations, have been made. Such models would be non-local by

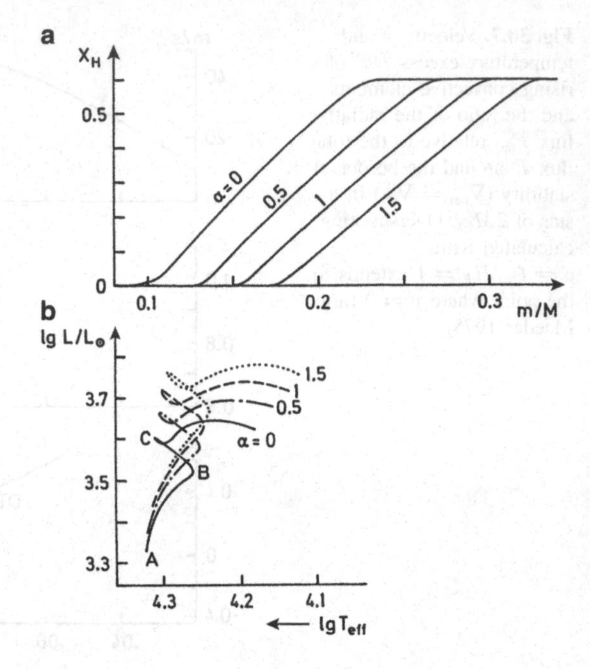

Fig. 30.8. Central hydrogen burning for a $7M_\odot$ star (initial mixture $X_H = 0.602$, $X_{rest} = 0.044$) with overshooting according to different assumptions for the ratio $\alpha = \ell_m/H_P$ ($\alpha = 0$ means no overshooting). (**a**) The hydrogen profile at the end of this phase. (**b**) HR diagram with evolutionary tracks (Matraka et al. 1982)

nature and therefore include the necessary conditions for treating also overshooting more realistically.

Up to the present time two standard methods for modelling overshooting are being used in numerical calculations. The first one is based on a simple extension of the convectively mixed region above the boundary defined by the Schwarzschild criterion. This extension l_{ov} is parametrized in terms of the local pressure scale height at the boundary

$$l_{ov} = \alpha_{ov} H_P. \tag{30.8}$$

The parameter α_{ov} is typically of order $0.1 \cdots 0.2$ for modern stellar models. It has no relation to the mixing-length parameter α_{MLT}, and is most often determined by fitting models to observed colour-magnitude diagrams (e.g. Stothers and Chin 1992). For the overshooting region the assumption $\nabla = \nabla_{ad}$ is usually made. One sometimes speaks of "convective penetration" instead of overshooting. Strictly speaking, the temperature gradient should be at least slightly subadiabatic, otherwise convective elements would not be decelerated.

In an alternative approach, convective overshooting is considered to be a diffusive process with a diffusion constant

$$D(z) = D_0 \exp \frac{-2z}{f_{ov} H_P}, \tag{30.9}$$

where z is the radial distance from the formal Schwarzschild border and f_{ov} the free parameter of this description. D_0 sets the scale of diffusive speed and is

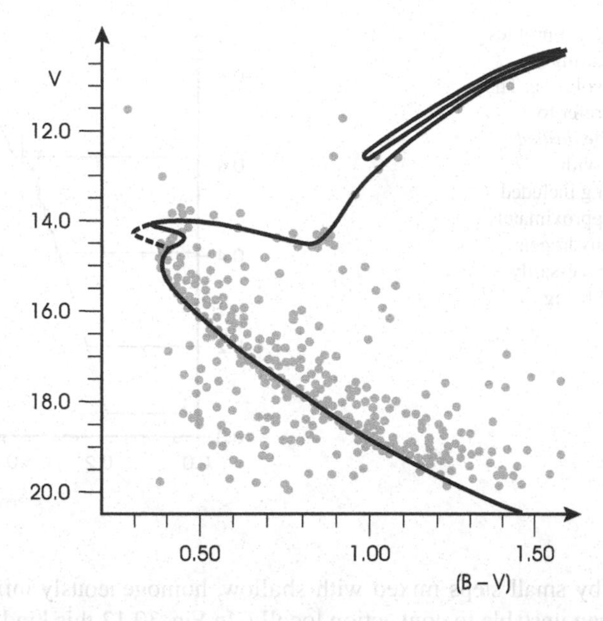

Fig. 30.9. Fit of isochrones to the colour-magnitude diagram of the open cluster NGC 2420 (adopted from Pietrinferni et al. 2004). The *dashed line* is obtained from stellar models without overshooting, and the *solid line* for models with overshooting, treated in the approach of (30.8), with a parameter α_{ov} of about 0.1. The "hook" at the end of the main-sequence can be reproduced correctly only with overshooting taken into account. Note that the isochrone age is 3.2 Gyr for this case, while it is lower (2 Gyr) for the case without overshooting to balance the fact that in this case main sequence luminosities are lower. Except for the turn-off region the two isochrones are almost identical

derived from the convective velocity obtained from mixing-length theory and taken below the Schwarzschild boundary. This approach is based on two-dimensional hydrodynamical simulations of thin convective envelopes in A-type stars and cool white dwarfs. Although its theoretical foundation is therefore limited, it has been used in a variety of situations and been shown that it also can be used to reproduce the width of the upper main sequence, and the colour-magnitude diagrams of open clusters (Fig. 30.9), with a numerical value of f_{ov} in the range of 0.02. The hydrodynamical models also indicate that the temperature gradient in the overshooting layers is close to the radiative one. A further advantage of this approach is that it can easily be added to a stellar evolution code that already has implemented diffusion (Sect. 8.2.2). In both cases the extent of the overshooting region has to be limited for small convective cores because of the divergence of H_P near the centre, as one otherwise gets unrealistically large mixed cores.

The "diffusive" approach (30.9) leads to smoother chemical profiles than those resulting from (30.8). This is quite obvious from the example we show in Fig. 30.10 for a star of $15 M_\odot$. The solid black lines are for the calculation without overshooting taken into account. The receding convective core leaves behind a profile

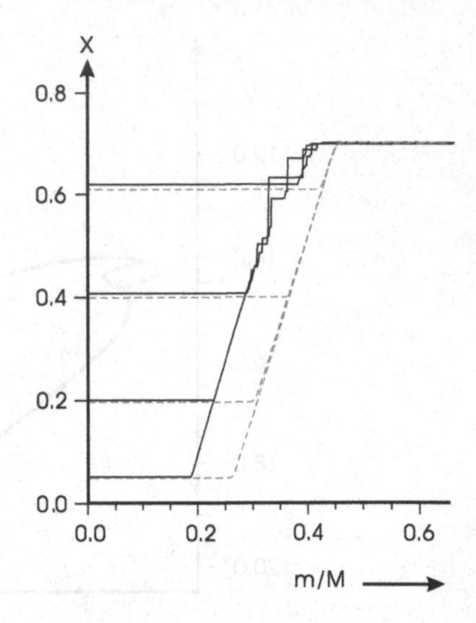

Fig. 30.10. Hydrogen profiles in a $15M_\odot$ star during the main-sequence evolution. The *solid black lines* refer to a model without, the *dashed grey lines* to one with overshooting being included. Four models at approximately the same central hydrogen content, but not necessarily the same age, are being compared

characterized by small steps mixed with shallow, homogeneously mixed regions, which have been unstable to convection locally. In Fig. 30.12 this kind of structure shows up as "convective tongues" in the upper panel. The dashed grey lines are the resulting chemical profiles if overshooting according to (30.9) is included. Due to the exponentially declining mixing speed the profiles are very smooth. It is also clearly seen how overshooting extends the homogeneously mixed core by about $0.05M$ in this case. For a large part of the main-sequence evolution the Schwarzschild boundary of the convective core remains at about the same mass coordinate as in the case without overshooting. Only in later main-sequence phases it changes (see Fig. 30.12, top and middle panel).

But as mentioned before the question of overshooting is quite open and can be settled only by use of a better theory of convection. This also concerns the question how the amount of overshooting varies with stellar parameters, such as mass and composition, and whether it occurs at all convective boundaries. So far, both issues can be addressed only tentatively by comparison with observations.

30.4.2 Semiconvection

Another phenomenon related to convection introduces a large amount of uncertainty in the evolution of rather massive stars, say, for $M > 10M_\odot$ (This limit depends on the chemical composition; it can even be around $7M_\odot$ for hydrogen-rich mixtures of extreme population I stars.).

In these stars during central hydrogen burning the convective core retreats, leaving a certain hydrogen profile behind; the radiative gradient ∇_{rad} outside the core starts to rise and soon exceeds the adiabatic gradient ∇_{ad}. This happens in a

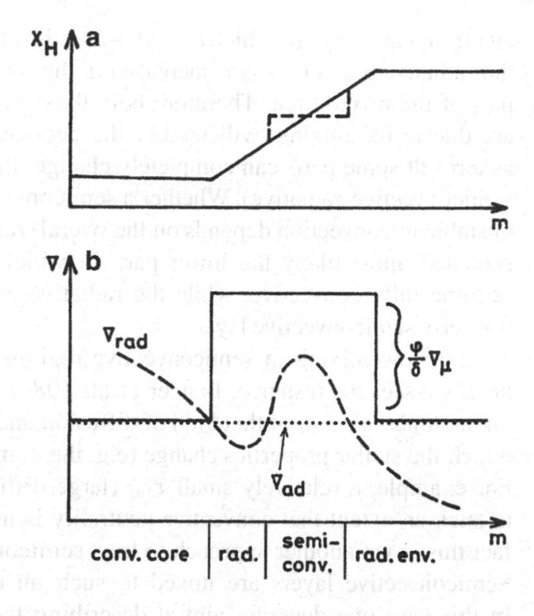

Fig. 30.11. Schematic illustration of the example for semiconvection discussed in the text. The *solid line* in (**a**) shows a hydrogen profile in which semiconvection occurs. Complete mixing in this layer would lead to the *dashed* "plateau". The gradients in the same range of m are sketched in (**b**), indicating the radiative-semiconvective-convective properties of the different layers

region with outwardly increasing hydrogen content (decreasing molecular weight μ); therefore $\nabla_\mu \equiv d \ln \mu / d \ln P > 0$, which makes the layer dynamically stable (Sect. 6.1). Considering the classical criteria for convective stability according to Schwarzschild and Ledoux we find

$$\nabla_{\text{ad}} < \nabla_{\text{rad}} < \nabla_{\text{ad}} + \frac{\varphi}{\delta}\nabla_\mu . \tag{30.10}$$

As described in Sect. 6.3 a layer in which (30.10) is fulfilled is vibrationally unstable ("overstable"). A slightly displaced mass element starts to oscillate with slowly growing amplitude and penetrates more and more into regions of different chemical composition. This results in a rather slow mixing which is called *semiconvection*. The treatment of this process is complicated, one difficulty being that any degree of mixing must have a noticeable reaction on the stratification in the mixed layer.

Suppose that semiconvection occurs in some region of an originally very smooth hydrogen profile (solid line in Fig. 30.11a). The corresponding gradients are schematically sketched in Fig. 30.11b. The solid line is the decisive gradient of the Ledoux criterion. The region is fully convective in the innermost part, because $\nabla_{\text{ad}} < \nabla_{\text{rad}}$ and $\nabla_\mu = 0$. Next follows a radiative zone because of the drop of ∇_{rad}, above which a semiconvective layer exists, which would be convective according to the Schwarzschild criterion, but is stabilized due to the positive ∇_μ-term. If the mixing in the semiconvective region were very efficient, we would obtain a "plateau" in the profile (dashed line in Fig. 30.11a). There are obviously two main effects of such a mixing on the gradients. Firstly, any change of profile changes the value of ∇_μ, which goes to zero in the plateau. Secondly, the mixing increases the hydrogen content X_{H} in the lower part and decreases X_{H} in the upper part of the mixed region. In massive main-sequence stars the opacity is largely dominated by electron scattering, for which $\kappa \sim (1+X_H)$, [cf. (17.2)]. Since $\nabla_{rad} \sim \kappa$, [cf. (5.28)], the radiative gradient ∇_{rad} is increased in the lower part and decreased in the upper part of the mixed area. Therefore both these changes (of ∇_μ and of ∇_{rad}), which are due to the mixing, will modify the decisive terms entering into (30.10), and as a result some parts can completely change their stability properties (convective-semiconvective-radiative). Whether a semiconvective layer becomes more stable or unstable to convection depends on the overall result of both effects. In the situation sketched, most likely the lower part, in which hydrogen content increases, will become fully convective, while the radiative envelope will grow deeper into the formerly semiconvective layer.

The slow mixing in semiconvective regions can be considered as a diffusion process (see, for instance, Langer et al. 1985). The resulting profile will depend on the timescale τ_{diff} of that kind of diffusion and its ratio to the typical timescale in which the stellar properties change (e.g. the composition due to nuclear reactions). For example, a relatively small τ_{diff} (large diffusion coefficient) will tend to mix to such an extent that convective neutrality is nearly reached with $\nabla_{rad} \approx \nabla_{ad}$. In fact this is yet another approach to treat semiconvection in numerical calculations: Semiconvective layers are mixed to such an extent that neutrality is achieved. In this case one does not aim at describing the physical properties in detail but rather aims at a likely final situation. In general one should expect a continuous change of the profile and radiative, semiconvective, and fully convective regions moving slowly through the star. Unfortunately the coefficient of diffusion cannot yet be determined satisfactorily, which is rather serious, since, as in the case of overshooting, the details of the established profile are very decisive for the later evolution of these stars. In Fig. 30.12 we show an example for the different convective and semiconvective layers establishing in a 15 $M_\odot$ star during the main-sequence evolution, when different approaches to overshooting and semiconvection are employed. The semiconvective layers outside the fully convective core in the Schwarzschild case (top panel in the figure) change their character–convective or radiative–with time, depending on changes in the thermal structure and on mixing. The result is a typical tongue-like extent of convective layers, separated by radiative "tongues", and a H/He profile that shows many irregular steps. This kind of structure can already be seen in the early works Langer (1989, 1991) in stars of 30 and 20 $M_\odot$.

Additional complications can arise from the interaction of semiconvection and overshooting. Note that semiconvection can also play a role in later phases, for example, if a convective core increases during helium burning and expands into a region of different chemical composition.

30.5 The Schönberg–Chandrasekhar Limit

Since the nuclear timescale for central hydrogen burning is large compared to the Kelvin–Helmholtz timescale, stars can be well represented by models in complete equilibrium throughout this phase. The question is now whether this continues to

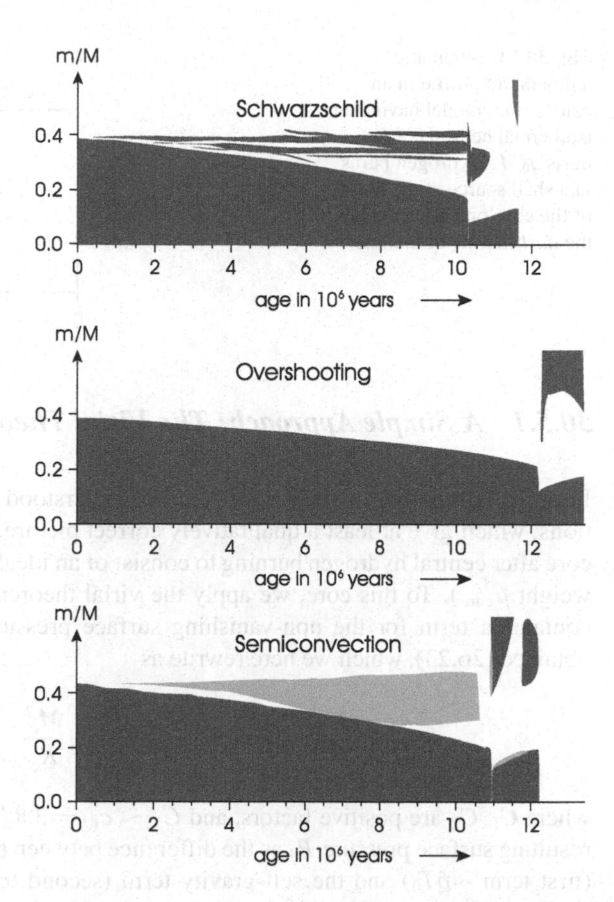

Fig. 30.12. Convective and semiconvective regions in a star of 15 $M_\odot$ for three different treatments of convection. The *top figure* shows fully convective regions (*dark grey*) if the Schwarzschild criterion is applied. The *bottom panel* shows the result if the Ledoux criterion is used, and slow semiconvective mixing is done according to the diffusive approach by Langer et al. (1985; with the free parameter in this description set to 0.1). Semiconvective regions are indicated by *light grey*. The *central panel* finally shows the case with overshooting considered as a diffusive process according to (30.9), with $f_{\rm ov} = 0.02$. Note that only the convective core ($\nabla_{\rm rad} \geq \nabla_{\rm ad}$) is shown; the region of overshooting, which extends over more than 5 % of the mass, is not visible in this figure (but see Fig. 30.10)

be valid also for the subsequent evolution. At the end of central hydrogen burning, the star is left with a helium core without nuclear energy release surrounded by a hydrogen-rich envelope. At the bottom of this envelope, the temperature is just large enough for further hydrogen burning, which continues at this place in a *shell source* (see Figs. 30.2 and 30.4). The problematic part is the possible structure and change of the helium core. A core almost in thermal equilibrium without nuclear energy sources cannot have a considerable luminosity, and hence must be nearly isothermal, since $dT/dr \sim l$.

Therefore we consider here equilibrium models consisting of an isothermal helium core of mass $M_{\rm c} = q_0 M$ and a hydrogen-rich envelope of mass $(1 - q_0)M$ (see Fig. 30.13). For simplicity the chemical composition is taken to change discontinuously at the border of the two regions. The luminosity is supplied by hydrogen-shell burning at the bottom of the envelope. In the following, solutions for the core (subscript 0 at its surface $q = q_0$) and solutions for the envelope (subscript e at the lower boundary $q = q_0$) are first discussed separately and then fitted to each other. In view of their importance we will look at the surprising results from different points of view.

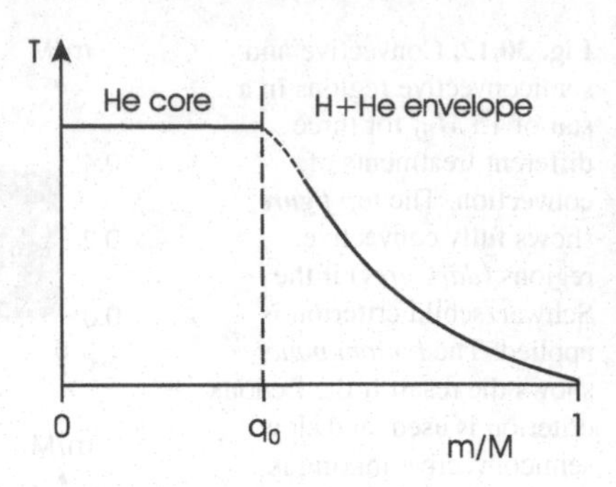

Fig. 30.13. Schematic temperature profile in an equilibrium model having an isothermal helium core of mass $q_0 M$. Hydrogen burns in a shell source at the bottom of the envelope, indicated by the *dashed part* of the line

30.5.1 A Simple Approach: The Virial Theorem and Homology

Important properties of such models can be understood by rather simple considerations, which give at least a qualitatively correct picture. We assume the isothermal core after central hydrogen burning to consist of an ideal monatomic gas (molecular weight μ_{core}). To this core, we apply the virial theorem in the form (3.21) which contains a term for the non-vanishing surface pressure P_0. Solving for P_0, we obtained (26.23), which we here rewrite as

$$P_0 = C_1 \frac{M_c T_0}{R_c^3} - C_2 \frac{M_c^2}{R_c^4}, \tag{30.11}$$

where C_1, C_2 are positive factors, and $C_1 \sim c_v = 3\Re/(2\mu_{\text{core}})$. This describes the resulting surface pressure P_0 as the difference between the average interior pressure (first term $\sim\bar{\varrho}T_0$) and the self-gravity term (second term $\sim R_c\bar{g}\bar{\varrho}$), when we use $\bar{\varrho} \sim M_c/R_c^3$ and $\bar{g} \sim M_c/R_c^2$.

For simplicity we assume T_e to be kept at a constant value by the thermostatic action of hydrogen burning. The fitting condition at q_0 then requires

$$T_0 = T_e = \text{constant}, \tag{30.12}$$

and P_0 depends only on M_c and R_c. As explained in Sect. 26.2 the counteraction of the two terms in (30.11), which depend on different powers of R_c, has the result that, for $M_c =$ constant, P_0 has a *maximum value* $P_{0\,\max}$ at $R_c = R_{c\max}$ [see (26.27)],

$$R_{c\max} = C_3 \frac{M_c}{T_0}, \quad P_{0\max} = C_4 \frac{T_0^4}{M_c^2}, \tag{30.13}$$

with some positive constants C_3, C_4. This can be obtained by solving $\partial P_0/\partial R_c = 0$ (for constant T_0) from (30.11). The function $P_0(R_c)$ for given M_c and T_0 is sketched in Fig. 30.14. From (30.13) we see that $P_{0\max} \sim M_c^{-2}$, i.e. the *maximum surface pressure of the core decreases strongly with the mass* M_c *of the core.*

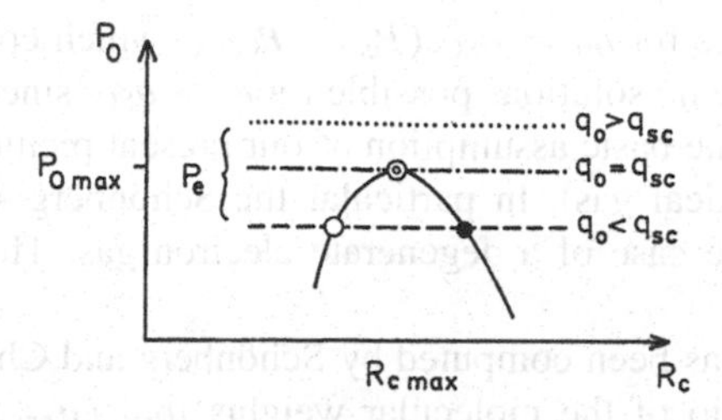

Fig. 30.14. The *solid line* shows schematically the pressure P_0 at the surface of the isothermal core as a function of the core radius R_c. *Horizontal lines* indicate the pressure P_c at the bottom of the envelope for three different relative core masses q_0. The stable solution is marked by a *dot* and the unstable solution by an *open circle*; the solution at $P_{0\max}$, is marginally stable

For the functions at the bottom of the envelope we simply assume that all possible envelopes are homologous to each other. Then from (20.9) and (20.24) follow $P_e \sim M^2/R^4$ and $T_e \sim M/R$. The latter relation together with (30.12) means that M/R = constant, such that the relation for P_e becomes

$$P_e = C_5 \frac{T_0^4}{M^2}. \tag{30.14}$$

We see that P_e is independent of R_c and has the same dependence on T_0 as $P_{0\max}$, but decreases with M instead of M_c. This can lead to difficultics! In Fig. 30.14 the envelope pressure P_e according to (30.14) is given by a horizontal straight line, the height of which depends on M.

The remaining fitting conditions for a complete solution of the star require $R_c = r_e$ and $P_0 = P_e$, i.e. we look for an intersection of the two types of curves in Fig. 30.14. Obviously this can be obtained only if $P_e \leq P_{0\max}$, which together with (30.13) and (30.14) gives the condition

$$q_0 \equiv \frac{M_c}{M} \leq q_{SC}, \tag{30.15}$$

i.e. the *relative* core mass q_0 must not exceed a certain limiting value, which is the *Schönberg–Chandrasekhar limit* q_{SC}. This limit was already derived in Sect. 21.4 from fitting solutions for isothermal cores and for envelopes in the U–V plane.

For $q_0 < q_{SC}$ we have $P_e < P_{0\max}$, and there are two intersections in Fig. 30.14. The solution for the smaller value of R_c is thermally unstable, the other one is stable. This can be made plausible by a simple argument. Figure 30.14 shows that, if we slightly increase the core radius of the stable solution, P_0 drops below P_e and the envelope tends to compress the core, thus restoring the equilibrium state. The opposite behaviour (further increase of an initial expansion, since P_0 exceeds P_e) can be seen to result from the perturbation of the unstable equilibrium state, and this rough argument is confirmed by a strict eigenvalue analysis.

The solutions merge for $q_0 = q_{SC}$ ($P_e = P_{0max}$) which corresponds to neutral stability. And there are no solutions possible for $q_0 > q_{SC}$, since P_e always exceeds P_0. In such a case some basic assumption of our present picture has to be dropped (e.g. equilibrium or ideal gas). In particular the Schönberg–Chandrasekhar limit does not apply for the case of a degenerate electron gas. This will be discussed later.

The value of q_{SC} has been computed by Schönberg and Chandrasekhar (1942). It depends on the ratio of the molecular weights μ_{core}/μ_{env}, since the envelope pressure depends on μ_{env}, while P_0 depends on μ_{core} via C_1. One can write roughly

$$q_{SC} = 0.37 \left(\frac{\mu_{env}}{\mu_{core}} \right)^2 , \tag{30.16}$$

which means for a pure helium core $\mu_{core} = 4/3$ and for a hydrogen-rich envelope $q_{SC} \approx 0.09$. This value is certainly exceeded by the helium cores that are left after central hydrogen burning in stars of the upper main sequence. Stars of somewhat smaller mass may encounter the same difficulty later, when the shell source burns outwards, thus increasing the mass of the helium core above the critical value. The Schönberg–Chandrasekhar limit is therefore quite relevant for the evolution in any phases in which at a first glance one would expect isothermal cores of ideal gas to appear.

30.5.2 Integrations for Core and Envelope

More reliable curves in the $P - R_c$ diagram (Fig. 30.14) can be easily obtained by numerical integrations for core and envelope (Roth 1973).

An envelope solution can be calculated for given M and M_c by requiring the lower boundary conditions $l = 0, r = R_0$ to hold at $M = M_c$. The solution gives P_e and T_e at $m = M_c$. By varying R_c, one obtains a set of solutions which gives $P_e(R_c), T_e(R_c)$. Two typical envelope curves $P_e(R_c)$ are shown in Fig. 30.15a. It turns out that these curves, in their important parts, are nearly independent of M_c but are raised essentially by a decrease of M [This is qualitatively the same as in the approximation (30.14).]. The temperature T_e varies, in fact, very little along such an envelope curve. For later applications (Sect. 31.1) we briefly mention the surface values of these envelope solutions. Those with large values of R_c are located near the main-sequence. With decreasing R_c they move to the right in the HR diagram, and envelopes with the smallest values of R_c are close to the Hayashi line.

The solution for an isothermal core with temperature T_0 can be obtained by a straightforward integration starting at the centre with an assumed value of $P = P_c$ and continued until $m = M_c$ is reached. At this point one finds a pair of values $P = P_0$ and $r = R_c$. Many such integrations for different values of the parameter P_c then give the curve $P_0(R_c)$ for the core. The solid line in Fig. 30.15b gives such a curve for cores of mass $M_c = 0.18 M_\odot$ and $T_0 = 2.24 \times 10^7$ K. The lower-right part (small P_0, large R_c) corresponds to small central pressures P_c. With increasing

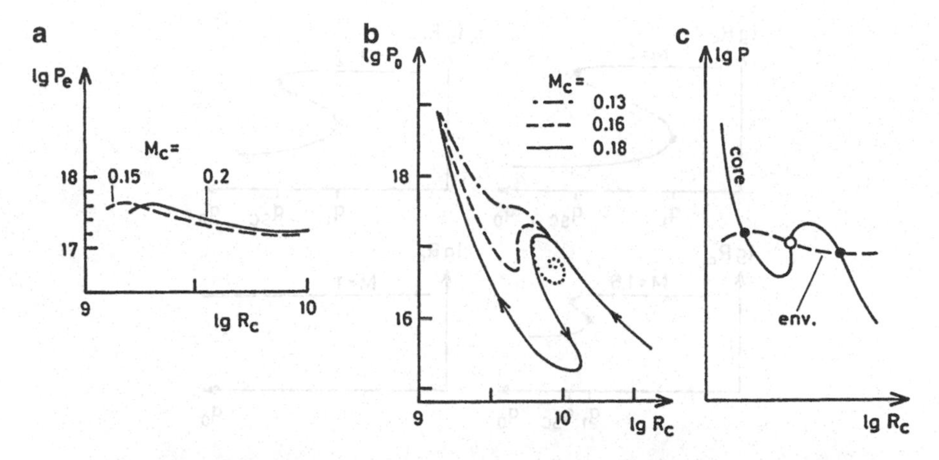

Fig. 30.15. Some typical curves of the pressure P (in dyn cm^{-2}) against the core radius R_c (in cm). (**a**) The pressure P_e at the lower boundary of the envelope for a stellar mass $M = 2M_\odot$ and two values of the core mass M_c (in $M_\odot$). (**b**) The pressure P_0 at the surface of isothermal cores of different mass M_c (in $M_\odot$). The *arrows* along the *solid curve* indicate the direction of increasing central pressure. The *dotted spiral* is with neglect of degeneracy. (**c**) Sketch of core and envelope curves for the case of three intersections giving three complete solutions (*filled circles* stable, *open circle* unstable) (After Roth 1973)

P_c the curve leads up to the maximum and decreases again (This corresponds to the maximum of the core curve in Fig. 30.14, while the horizontal envelope curves there are now replaced by envelope curves like those in Fig. 30.15a.). Then it would follow the dotted spiral, if we artificially suppress the deviation from the ideal-gas approximation in the equation of state. This may be compared with the spiral in the U–V plane obtained for an isothermal core in Fig. 21.2. An increasing P_c, however, implies an increasing degeneracy of the electron gas. This "unwinds" the spiral and P_0 drops, while a gradually increasing fraction of the core becomes degenerate. When degeneracy encompasses practically the whole core, P_0 rises again strongly with decreasing R_c (upper-left end of the solid curve in Fig. 30.15b). The dashed and dot-dashed lines demonstrate how the curve changes when M_c is decreased. As predicted by (30.13) the maximum shifts to smaller R_c and larger P_0. The main effect, however, is that the minimum is less and less pronounced. This goes so far that finally the maximum, which is decisive for the existence of a Schönberg–Chandrasekhar limit, has disappeared. A similar change of the structure of the curve is obtained if, instead of decreasing M_c, we increase the temperature T_0.

30.5.3 Complete Solutions for Stars with Isothermal Cores

As mentioned, each sequence of envelope solutions yields a relation $T_e = T_e(R_c)$. Assume now that along a corresponding sequence of isothermal-core solutions

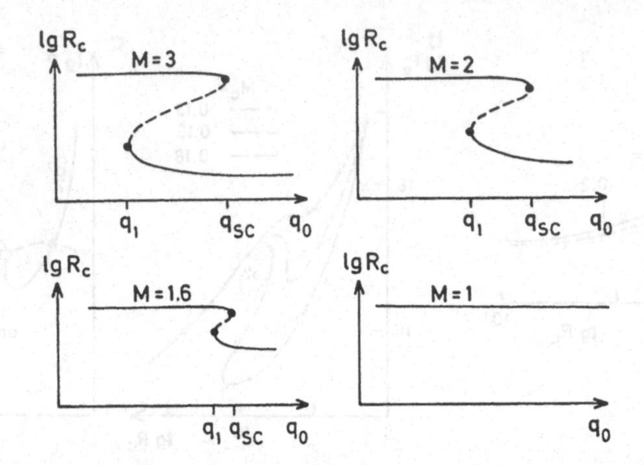

Fig. 30.16. Complete equilibrium solutions for four different stellar masses M (in $M_\odot$) having an isothermal core of mass $M_c = q_0 M$. Each solution here is characterized by its core radius R_c and its relative core mass q_0. Branches with thermally stable solutions are shown by *solid lines* and branches with unstable solutions by *dashed lines*. The turning point at $q_0 = q_{SC}$ defines the Schönberg–Chandrasekhar limit (After Roth 1973)

T_0 is varied such that $T_0(R_c) = T_e(R_c)$ for all R_c. This deforms a core curve in Fig. 30.15b only slightly. Any intersection of this new core curve with a corresponding envelope curve gives a complete solution, since we then have at $m = M_c$

$$r_e = R_c, \quad P_e = P_0, \quad T_e = T_0, \quad l_e = l_0 = 0, \tag{30.17}$$

i.e. continuity of all variables.

Suppose that the core curve has a pronounced maximum. We can then obviously expect to have up to three solutions (see Fig. 30.15c), one with an ideal gas (largest R_c), the second with partial degeneracy (intermediate R_c), and the third with large degeneracy (smallest R_c) in the core. If the envelope curve passes below the minimum or above the maximum of the core curve, there will be only one solution. And there can also be only one solution with a monotonic core curve.

The resulting solutions for different values of M and M_c can best be reviewed by representing them as models in which $q_0 = M_c/M$ varies as a parameter while M is fixed (Fig. 30.16). Each model is represented here by its core radius R_c in order to give an easy connection with the foregoing fitting procedure.

Figure 30.16 shows that for larger M the sequence of equilibrium solutions consists of three branches. Two of them contain thermally stable models (solid lines), the other unstable models (dashed). On the upper and lower stable branches, the isothermal cores have no or strong degeneracy respectively. The branches are connected by two turning points (at q_1, and q_{SC}) where the models have marginal stability. A real star would first evolve along the upper stable branch, increasing its core at nearly constant radius. When the mass of the core reaches the turning point, the core will contract on a thermal timescale (q_0 staying constant because of the much longer nuclear timescale) to the lower branch. The turning point with the larger q_0 defines the Schönberg–Chandrasekhar limit. Its value q_{SC} turns out to be nearly independent of M. For $q_1 < q_0 < q_{SC}$, there are three solutions, otherwise one solution. When going to gradually smaller M, we see that q_1 approaches q_{SC}, until both turning points merge and finally disappear for $M < 1.4 M_{\odot}$. For such small M, therefore, one has only one (stable) branch and no Schönberg–Chandrasekhar limit. This agrees with what one expects from the core curves given in Fig. 30.15b. It shows, for example, that the curves are already monotonic for $M = 1.3 M_{\odot}$ and $q_0 \leq 0.1$ (i.e. $M_c \leq 0.13 M_{\odot}$) (The exact mass values depend not only on the chemical composition, but also on the detailed physical input of the stellar models. Those given here are from the model calculations by Roth 1973, but are representative.). Instead of R_c, we might have plotted the stellar radius R over the parameter q_0. As mentioned above, small R_c corresponds to large R and vice versa. The sequences for large enough M would then exhibit a stable dwarf branch for $q_0 < q_{SC}$, a stable giant branch for $q_0 > q_{SC}$ and an unstable intermediate branch.

In evolutionary models one will encounter a smooth profile rather than a discontinuity of the chemical composition. In such a case various definitions of the core mass are possible: it can be the point at which $X_H > 0$, or where the maximum of shell source burning is located. Since the shell is comparably thin, the various definitions do not differ too much from each other, anyhow. The Schönberg–Chandrasekhar limit can be identified by the departure from thermal stability, i.e. by a higher fraction of thermal to nuclear energy. In any case, one finds again that $q_{SC} \approx 0.1$.

Part VI
Post-Main-Sequence Evolution

Chapter 31
Evolution Through Helium Burning: Intermediate-Mass Stars

31.1 Crossing the Hertzsprung Gap

After central hydrogen burning, the star has a helium core, which in the absence of energy sources tends to become isothermal. Indeed thermal equilibrium would require that the models consist of an isothermal helium core (of mass $M_c = q_0 M$, radius R_c), surrounded by a hydrogen-rich envelope [of mass $(1 - q_0)M$] with hydrogen burning in a shell source at its bottom. Such models were discussed in detail in Sect. 30.5. We now once more consider the case of $M = 3M_\odot$, which is typical for stars on the upper part of the main sequence (say $M > 2.5M_\odot$). The possible solutions were comprised in a series of equilibrium models consisting of three branches. This is shown in the first graph of Fig. 30.16, and again in Fig. 31.1, which also gives schematically the position in the HR diagram.

Suppose that the relative mass of the core q_0 has not yet reached the Schönberg–Chandrasekhar limit $q_{SC}(\approx 0.1)$ at the end of central hydrogen burning. The model then can easily settle into a state contained in the uppermost branch of Fig. 31.1a, which consists of stars close to the main sequence (Fig. 31.1b). Let us imagine a "quasi-evolution" of this simple model by assuming that M_c grows because of shell burning while complete equilibrium is maintained. The result is that the model is shifted towards the right in Fig. 31.1a. This proceeds continuously until the model reaches the Schönberg–Chandrasekhar limit, represented by the turning point which terminates the uppermost branch. Further increase of M_c would require the model to jump discontinuously onto the lower branch in Fig. 31.1a. This decrease of R_c (i.e. compression of the core) would be accompanied by a large jump in the HR diagram, from the main sequence to the region of the Hayashi line (Fig. 31.1b). This means that such equilibrium stars have to become giants because the main-sequence solutions (which the stars had selected owing to their history) cease to exist, while the red giant solutions (which have coexisted for a long time) are still available. In Fig. 31.1a, b the quasi-evolution of increasing M_c is indicated by solid lines, while those parts which can obviously not be reached are broken. We will see that basic features of this jump in the simple

R. Kippenhahn et al., *Stellar Structure and Evolution*, Astronomy and Astrophysics Library, DOI 10.1007/978-3-642-30304-3_31, © Springer-Verlag Berlin Heidelberg 2012

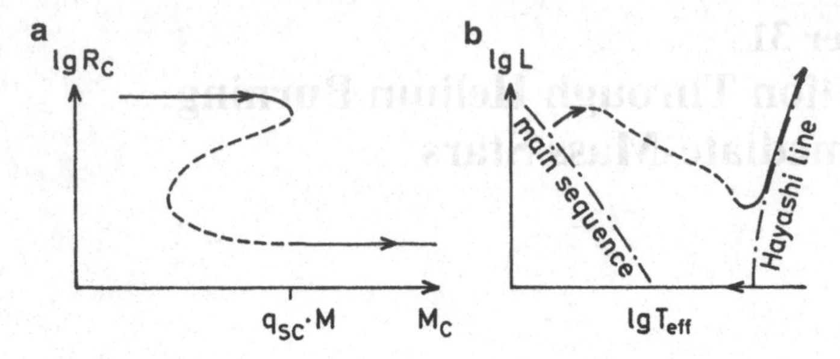

Fig. 31.1 (**a**) The same series of equilibrium models for $M = 3M_{\odot}$ as in Fig. 30.16. The core radius R_c is plotted against the core mass M_c. In a quasi-evolution with increasing mass M_c of the isothermal helium core, the model shifts along the *solid lines*, as indicated by the *arrows*. (**b**) The corresponding position in the HR diagram

quasi-evolution (particularly the compression of the core together with an expansion of the envelope to a red giant stage) are recovered in the real evolution which, of course, leads through non-equilibrium models. In any case, a phase of thermal non-equilibrium must follow after central hydrogen burning since a continuation via suitable equilibrium models would involve a discontinuity.

As an example for the real evolution we take numerical solutions obtained for upper main-sequence stars with our standard Pop. I initial mixture ($X_H = 0.70, X_{He} = 0.28, X_{rest} = 0.02$). This model was calculated neglecting overshooting from the convective core completely and is a continuation of the main-sequence evolution shown, for example, in Fig. 30.3. The transition from central to shell burning can be seen from Fig. 31.2a. Any line parallel to the ordinate indicates what one would encounter in different layers when moving along the radius of the star at that moment of the evolution. Figure 31.2b gives the corresponding evolutionary track in the HR diagram. The first part of Fig. 31.2a (from A to C) shows the phase of central hydrogen burning which exhausts ^{1}H in the core within about 7.9×10^7 years for $5M_{\odot}$. With hydrogen being depleted there, the burning together with the convection ceases rather abruptly in the central region. At the same time, hydrogen burning intensifies in an initially rather broad shell around the core, i.e. in the mass range of the outwards-increasing hydrogen content left by the shrinking convective core (cf. Fig. 30.3a). Later this shell source narrows remarkably in mass scale, particularly when it has consumed the lower tail of the hydrogen profile. After phase C the evolution is so much accelerated that the abscissa had to be expanded. The models are no longer in thermal equilibrium, i.e. the time derivatives ($\varepsilon_g = -T\partial s/\partial t$) in the energy equation are not negligible [cf. (4.47) and (4.48)]. The star has now encountered the situation outlined earlier in this section.

The radial motion of different mass elements in this phase is shown in Fig. 31.3 for the same star. After a short resettling at the end of central hydrogen burning (point C) we see that core and envelope change in opposite directions: an expansion of the layers above the shell source (at $m \approx 0.14M$) is accompanied by a contraction of the layers below. The fact that $\dot{r}$ changes sign at the maximum of

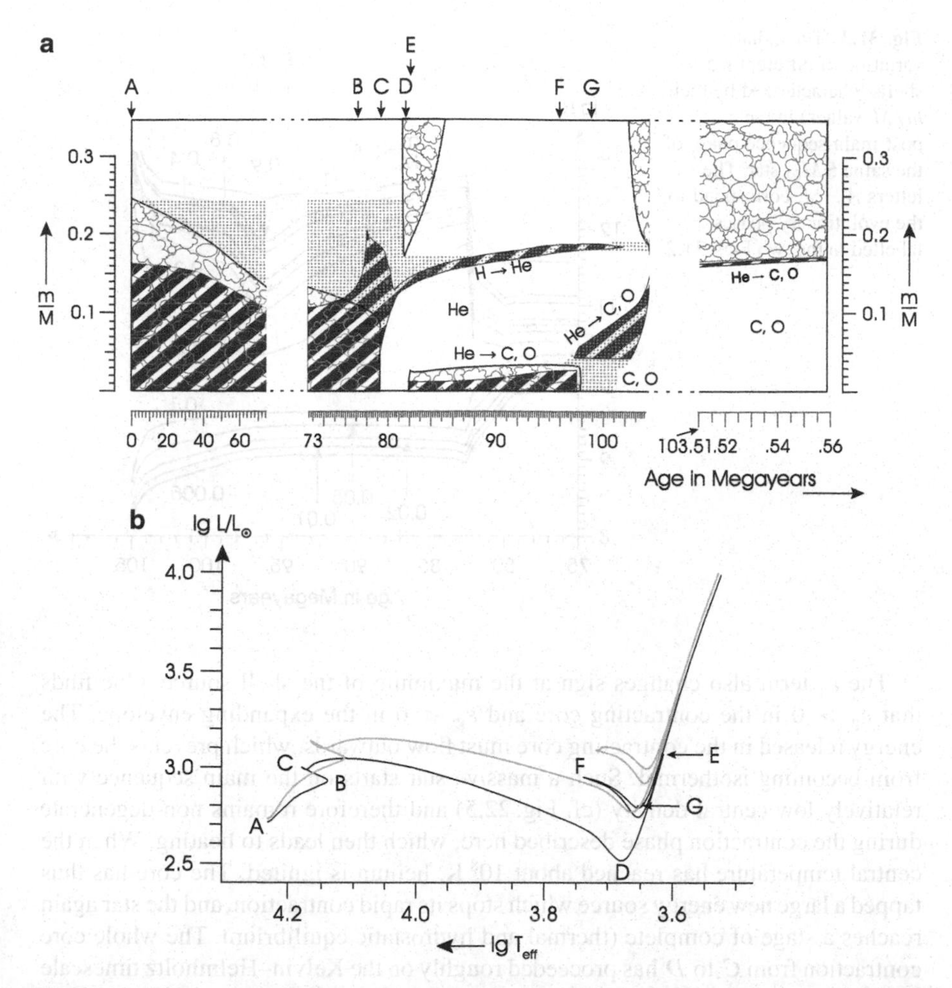

Fig. 31.2 **(a)** The evolution of the internal structure of a star of $5M_\odot$ of Population I. The abscissa gives the age (in units of 10^6 years) since the beginning of the evolution on the main sequence. Each *vertical line* corresponds to a model at a given time. The different layers are characterized by their values of m/M. "Cloudy" regions indicate convective areas. *Heavily hatched* regions indicate where the nuclear energy generation (ε_H or ε_{He}) exceeds 10^2 erg g^{-1} s^{-1}. Regions of variable chemical composition are *dotted*. The letters *A* ... *G* above the upper abscissa indicate the corresponding points in the evolutionary track, which is plotted in Fig. 31.2 **(b)** with a *solid line*. The *grey line* in the Hertzsprung-Russell diagram shows the evolution of a star of the same mass and composition, but with convective overshooting included

a shell source is a pattern very characteristic for models with strongly burning shell sources; it can occur in quite different phases of evolution, for contracting or expanding cores, for one or two shell sources. Such shell sources seem to represent a kind of mirror in the pattern of contraction and expansion inside a star ("mirror principle" of radial motion).

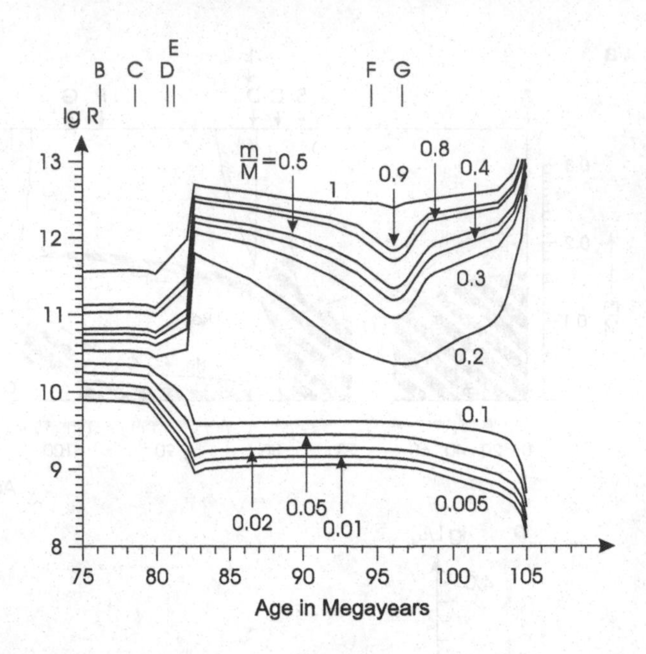

Fig. 31.3 The radial variation of different mass shells (characterized by their m/M values) in the post-main-sequence phase of the same $5\,M_\odot$ star. The letters $A \ldots E$ correspond to the evolutionary phases labelled in the two Figs. 31.2

The ε_g term also changes sign at the maximum of the shell source. One finds that $\varepsilon_g > 0$ in the contracting core and $\varepsilon_g < 0$ in the expanding envelope. The energy released in the contracting core must flow outwards, which prevents the core from becoming isothermal. Such a massive star starts on the main sequence with relatively low central density (cf. Fig. 22.5) and therefore remains non-degenerate during the contraction phase described here, which then leads to heating. When the central temperature has reached about 10^8 K, helium is ignited. The core has thus tapped a large new energy source which stops its rapid contraction, and the star again reaches a stage of complete (thermal and hydrostatic equilibrium. The whole core contraction from C to D has proceeded roughly on the Kelvin–Helmholtz timescale of the core (in 32.3×10^6 years for $5M_\odot$). In the same time, the outer layers have rapidly expanded, and the stellar radius is increased appreciably (roughly by a factor 15 in Fig. 31.3).

The evolutionary path in the HR diagram for the $5M_\odot$ star is shown in Fig. 31.2b. The expansion transforms the star into a red giant so rapidly that there is little chance of observing it during this short phase of evolution. This explains the existence of the well-known *Hertzsprung gap,* an area between main sequence and red giants with a striking deficiency of observed stars. It is a direct consequence of stars with cores reaching the Schönberg–Chandrasekhar limit after central hydrogen burning.

The evolution is qualitatively similar for all stars in which helium burning is ignited before the core becomes degenerate and in which possible complications due to semiconvection cannot prevent the star from moving close to the Hayashi line. This includes stellar masses of, say, $2.5M_\odot < M < 10M_\odot$. A set of evolutionary tracks in this phase for different M is shown in Fig. 31.4.

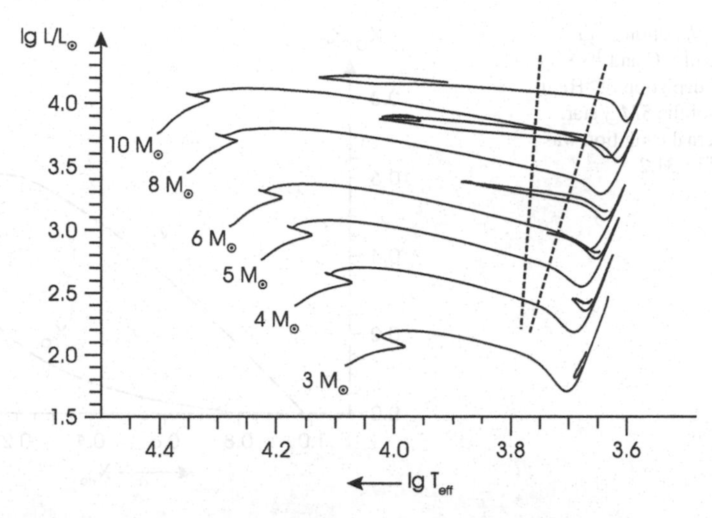

Fig. 31.4 The HR diagram with evolutionary tracks from the zero-age main sequence to the end of central helium burning for stars with different masses M (from $3M_\odot$ to $10M_\odot$) and for an initial composition with $X_H = 0.70$, $X_{He} = 0.28$. The instability strip of Cepheids is indicated by the *broken line*

In Fig. 31.2b the second track (grey line) is that of the same $5\,M_\odot$ star, but here overshooting from the convective cores during hydrogen and helium burning is taken into account. The extent of the overshooting corresponds to about 0.2 pressure scale heights at the Schwarzschild boundary, equivalent to an increase in mass by more than 20 %. The region in which material is mixed to the stellar centre is thus extended, the main-sequence evolution lasts longer, and the extension of the main-sequence phase in the HR diagram is larger. Point B along the track is 470 K cooler than for the evolution without overshooting and 26 % more luminous. Point C, the end of core hydrogen burning, which now lasts 96.3 instead of 79.9 Myr, moves in a similar way. The wider main sequence is observationally confirmed, and indeed the width of the upper main sequence required the inclusion of convective overshooting in stellar models for massive stars. All the subsequent evolution takes place at higher luminosity, which can be explained according to (20.20) by the larger increase of the mean molecular weight for the whole star during core hydrogen burning.

31.2 Central Helium Burning

As a consequence of the rapid contraction and heating of the core, central helium burning sets in (at the age of 8.3×10^7 years for our $5\,M_\odot$ star). The star is then in the red giant region of the HR diagram, close to the Hayashi line (D–E in Fig. 31.2b). Correspondingly it has a very deep outer convection zone, which can be seen in Fig. 31.2a to reach down to $m/M \approx 0.17$. The larger M, the deeper

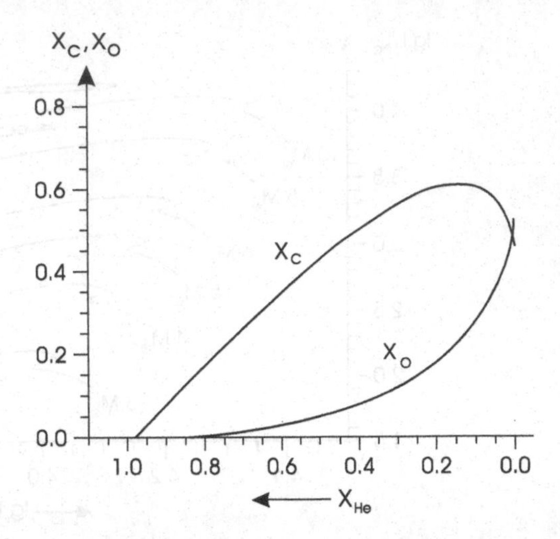

Fig. 31.5 Variation of the abundance of ^{12}C and ^{16}O during the depletion of ^{4}He in the centre of the $5M_\odot$ star, whose internal evolution was shown in Fig. 31.2

the convection zone penetrates, and it reaches into layers in which the composition was already slightly modified by the earlier hydrogen burning. Therefore some products of this burning are now dredged up by the convection and distributed all over the envelope. This is called the *first dredge-up* event. We here encounter one of the mechanisms by which nuclear species produced in the very deep interior can be lifted to the stellar surface. For example, the $^{12}\mathrm{C}/^{13}\mathrm{C}$ isotope ratio drops from its initial value of approximately 90 to values close to 20. This is the result of mixing material that has undergone CNO burning, where the equilibrium value of $^{12}\mathrm{C}/^{13}\mathrm{C} \approx 5$ was established, with pristine matter in the envelope having the initial value. Furthermore, the nitrogen abundance increases at the expense of that of carbon, and even the surface helium abundance increases slightly by a few per mille. The carbon isotope ratio can be determined rather accurately from stellar spectra of giants and confirms the presence of the first dredge-up and thus also the CNO burning in the deep stellar interior.

The high temperature sensitivity of helium burning causes a strong concentration of the energy release towards the centre and therefore the existence of a convective core. The core contains roughly 3–4 % of M, i.e. much less mass than during hydrogen burning.

At first the dominant reaction is $3\alpha \rightarrow {}^{12}\mathrm{C}$ (cf. Sect. 18.5.2). With increasing abundance of ^{12}C the reaction $^{12}\mathrm{C} + \alpha \rightarrow {}^{16}\mathrm{O}$ gradually takes over. When ^{4}He has already become rather rare the depletion of ^{12}C on account of ^{16}O is larger than the production of ^{12}C by the 3α reaction, and ^{12}C decreases again after having reached a maximum abundance. This is explained by the fact that the production of ^{12}C is proportional to X_α^3, while its depletion is proportional to $X_\alpha X_{12}$. The change of the abundances can be seen from Fig. 31.5, which shows the final composition for such stars to be ^{12}C and ^{16}O in roughly equal amounts with only a very small admixture of ^{20}Ne. In the example shown, the ratio $^{16}\mathrm{O}/^{12}\mathrm{C}$ is 51:47; if overshooting is included in the models, it changes slightly to 55:42 due to the different history of the temperature stratification of the core. Note, however, that the final ratio of $^{16}\mathrm{O}/^{12}\mathrm{C}$ depends strongly on the rather uncertain reaction rate for $^{12}\mathrm{C}(\alpha,\gamma)^{16}\mathrm{O}$.

The experimentally determined values have been varying by factors of 2–3 over the years. One of the most recent and widely used rate is that by Kunz et al. (2002), which–at 100 million K–is about 40 % higher than the classical one by Caughlan and Fowler (1988), but 40 % lower than the one recommended in the NACRE compilation (Angulo et al. 1999). Generally, the $^{16}O/^{12}C$ ratio as well as $^{20}Ne/^{16}O$ increases with increasing stellar mass, since T increases.

The phase of central helium burning lasts roughly 1.6×10^7 years, which is about 20 % of the duration of the main-sequence phase. This fraction seems to be surprisingly large in view of the facts that now L is somewhat higher, the exhausted core is much smaller, and the specific gain of energy (per unit of mass of the fuel) is only 1/10, as compared with hydrogen burning. The simple reason is that most of the total energy output in this phase comes from hydrogen-shell-source burning. For a star of $5M_\odot$ helium burning contributes only about 7 %, 26 %, and 42 % at points E, F, and G, respectively: a rather small release of nuclear energy inside the core is obviously sufficient to prevent it from contraction and to bring the whole star nearly into thermal equilibrium. The luminosity L_{He} produced between points E and F by helium burning in a helium core of mass M_{He} is roughly equal to the luminosity a pure helium star of $M = M_{He}$ would have on the helium main sequence (cf. Sect. 23.1). In fact the helium-burning core resembles in several respects a star on the helium main sequence with $M = M_{He}$. For later applications we note that the radius R_{He} of the core changes rather little during most of this phase. It increases very slowly until the central helium content has dropped to $X_{He} \approx 0.3$. It is only in the final phase of central helium burning ($X_{He} < 0.1$) that the core contracts and R_{He} drops appreciably (cf. Fig. 31.3). It should be mentioned that the evolution will be affected by convective overshooting, which enlarges the convective core also during central helium burning, but does not extend its duration appreciably. The larger supply of nuclear fuel is compensated by the higher luminosity. In case of the star shown in Fig. 31.2b it lasts 1.59×10^7 years.

Let us now look at the HR diagram in Fig. 31.2b. After point E the star (the one calculated without overshooting, but this discussion applies also to the more general case, as can be seen from Fig. 31.4) goes at first down along the Hayashi line, then leaves this line and moves back to the left. The "bluest" point F, for $5M_\odot$, is reached after 1.4×10^7 years (88 % of the helium-burning phase) when the central helium content is down to about $X_{He} \approx 0.15$. The track then leads back towards point G in the vicinity of the Hayashi line. The further evolution in which another loop may occur will be discussed in Sect. 31.5.

The extension of the loops, i.e. the distance of their bluest points from the Hayashi line, depends on the stellar mass M. We limit the discussion to a range of not too large masses, say $M < 10M_\odot$, where the situation is relatively simple and clear. Large loops are obtained for stars with large M. With decreasing M the loops become gradually smaller and finally degenerate to a mere down and up along the Hayashi line. This can be seen in Fig. 31.4, which gives the evolutionary tracks for a comparable set of computations. The loops for different stellar masses cover a roughly wedge-shaped area which is bordered by the Hayashi line and the connection of the bluest points of the loops (i.e. points F where T_{eff}

Table 31.1 Characteristic points and the time elapsed after the zero-age main-sequence stage in the evolutionary tracks of the models shown in Fig. 31.4

	t (in 10^6a)	lg $L/L_\odot$	lg $T_{\rm eff}$ (K)
$3M_\odot$	E 319.95	2.459	3.633
	E' 327.24	1.809	3.692
	F 337.69	1.810	3.692
	G' 409.83	2.012	3.672
$4M_\odot$	E 145.71	2.787	3.626
	E' 158.74	2.343	3.674
	F 163.48	2.439	3.695
	G' 176.65	2.437	3.661
$5M_\odot$	E 82.62	3.082	3.615
	E' 91.92	2.758	3.654
	F 96.05	2.963	3.739
	G' 98.32	2.816	3.648
$6M_\odot$	E 52.97	3.344	3.605
	E' 58.82	3.069	3.640
	F 61.51	3.369	3.886
	G' 63.21	3.138	3.635
$8M_\odot$	E 28.82	3.787	3.585
	E' 31.50	3.532	3.617
	F 32.54	3.869	4.024
	G' 34.40	3.739	3.676
$10M_\odot$	E 19.09	4.126	3.569
	E' 20.31	3.854	3.603
	F 20.85	4.198	4.125
	G' 22.00	4.165	3.909

The meaning of the points E, E', F, G' is explained in the text

has a maximum). The duration of characteristic phases as obtained from these calculations can been seen from Table 31.1. Point E' corresponds to the minimum of L after E, where the leftwards motion starts, while G' indicates the end of the central helium burning [As with most numerical values obtained up to now from evolutionary calculations, these data should be taken as an indication of typical relative properties, rather than as absolutely reliable. For other data see the original literature.]. The situation is much more complicated for still larger masses, where the loops do not continue to grow with M and the tracks remain well separated from the Hayashi line. Unfortunately this depends on the uncertain details of the mixing during the earlier main-sequence phase (compare Sects. 30.4 and 31.4), and of the poorly known mass loss rates.

The importance of the loops comes from the fact that they occur during a nuclear, slow phase of evolution in which the star has a sufficient chance of being observed (contrary to the foregoing phase of core contraction). We therefore expect to find

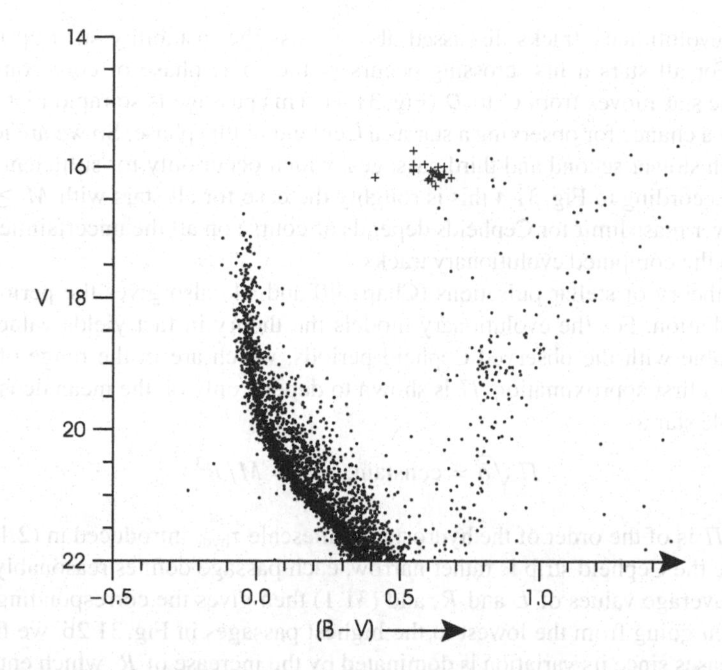

Fig. 31.6 Equivalent of an HR diagram (magnitude V against colour index $B - V$) for the cluster NGC 1866 (after Musella et al. 2006; data courtesy E. Brocato). *Crosses* indicate identified Cepheid variables with well-observed light curves and known period

helium-burning stars as red giants in the area of the HR diagram covered by the loops. This is in fact the case, as can be seen from HR diagrams of open clusters (see, e.g. Fig. 31.6). They often show a more or less extended giant branch, which is clearly separated from the main sequence by the Hertzsprung gap, and which sets out nicely the range of loops for the corresponding values of M.

31.3 The Cepheid Phase

It is of particular significance that the loops are necessary for explaining the observed δ Cephei variables. The observations show that these stars are giants, located in the HR diagram in a narrow strip roughly parallel to the Hayashi line and a few 10^2 K wide (cf. Fig. 31.4). Indeed the theory of stellar pulsations which will be described in Chap. 40 predicts that a star is vibrationally unstable if it is located in the "instability strip" of the HR diagram, where the observed Cepheids are found (Fig. 31.6). This is a consequence of the way in which the outer stellar envelope (particularly the helium ionization zone) reacts on small perturbations. When a stellar model has evolved into the instability strip, the oscillation will grow to finite, observable amplitudes. This phenomenon does not show up in the normal evolutionary calculations which are carried out by neglecting the inertia terms in the equation of motion, since these terms are necessary to obtain an oscillation at all. The calculated evolution therefore gives only the unperturbed solution.

The evolutionary tracks discussed above cross the instability strip up to three times. For all stars a first crossing occurs in the short phase of core contraction when the star moves from C to D (Fig. 31.4). This passage is so rapid that there is scarcely a chance for observing a star as a Cepheid in this phase. So we are left with the much slower second and third passages, which occur only for sufficiently large loops. According to Fig. 31.4 this is roughly the case for all stars with $M \geq 5M_\odot$. This lower mass limit for Cepheids depends of course on all the uncertainties of the loops in the computed evolutionary tracks.

The theory of stellar pulsations (Chaps. 40 and 41) also gives the period Π of the oscillation. For the evolutionary models the theory in fact yields values of Π comparable with the observed Cepheid periods, which are in the range of 1–100 days. In a first approximation, Π is shown to depend only on the mean density $\bar{\varrho}$ of the whole star as

$$\Pi\sqrt{\bar{\varrho}} = \text{constant}, \quad \bar{\varrho} \sim M/R^3. \tag{31.1}$$

Indeed Π is of the order of the hydrostatic timescale $\tau_{\rm hydr}$ introduced in (2.19).

Since the Cepheid strip is rather narrow, each passage defines reasonably well a pair of average values of L and R, and (31.1) then gives the corresponding period Π. When going from the lowest to the highest passages in Fig. 31.2b, we find that Π increases since its variation is dominated by the increase of R, which enters into $\bar{\varrho}$ with the third power. In fact, this, together with the properties of the instability strip discussed in Chap. 41, will be shown to lead to the famous Π–L relation of Cepheids, which is the basic standard for the determination of extragalactic distances.

During a passage through the Cepheid strip from right to left ($E \rightarrow F$), the radius R decreases, which means that Π must also decrease according to (31.1). During a passage in the opposite direction ($F \rightarrow G$), the period Π will increase. From (31.1) and the Stefan-Boltzmann law (11.14) one derives for the period change

$$\frac{d\log\Pi}{dt} = \frac{3}{4}\frac{d\log L}{dt} - 3\frac{d\log T_{\rm eff}}{dt}. \tag{31.2}$$

If we take as an example the 5 $M_\odot$ model of Table 31.1, we can calculate the average $\frac{d\Pi}{dt}$ (in the conventional units of s/year) for both passages. The pulsation periods at points E', F, and G' are approximately 4.4, 2.7, and 5.8 days, and the period changes are predicted to be -9.5×10^{-3} s/year for the first crossing and 2.8×10^{-2} s/year for the return passage. For more massive stars these rates can be higher by a factor of 10^4, as can easily be inferred from Table 31.1. These predicted changes of the period can in fact be measured by high-precision photometric observations covering many periods. An analysis of over 200 Milky Way Cepheids by Turner et al. (2006) confirmed the generally good agreement between observations and theoretical predictions. Two Cepheids, α UMi (Polaris) and DX Gem, show such high period changes that they are believed to belong to that rare group of stars which are currently crossing the Hertzsprung gap.

The duration of a passage $\tau_{\rm cep}$ increases strongly towards lower values of L (i.e. of Π). For an assumed width of $\Delta \lg T_{\rm eff} = 0.05$ for the strip, the crossing on the

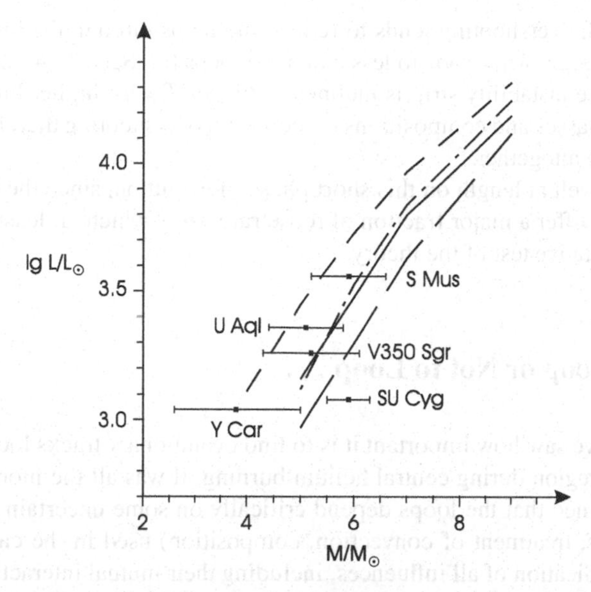

Fig. 31.7 Well-determined masses of Cepheids in comparison with the luminosity of the tip of blue loops as function of mass from different theoretical stellar evolution calculations. The *long-dashed line* was obtained without the inclusion of overshooting; the *solid* and *dot-dashed lines* with moderate overshooting (from two different calculations), and finally the *short-dashed* one from models with strong overshooting. The empirical Cepheid masses seem to indicate slightly more overshooting than used in the two "moderate" cases (after Evans et al. 1998)

way from E to F takes $\tau_{cep} = 2.4 \times 10^6$ years for $5M_\odot$. For $8M_\odot$ the strip is wider ($\Delta \lg T_{eff} \approx 0.1$) and $\tau_{cep} = 0.25 \times 10^6$ years. From τ_{cep} one can draw conclusions on the number of Cepheids to be expected. It turns out that this number should increase substantially towards smaller values of Π, reach a maximum (at a period of a few days), and then drop steeply, since the loops no longer reach the Cepheid strip. This is at least qualitatively in agreement with the observations.

A less favourable result concerns the masses of the Cepheids. One value, called the "evolutionary mass" M_{ev}, can be obtained with the help of evolutionary calculations essentially by comparing the luminosities. On the other hand, non-linear pulsation calculations show that the form of the light curves should depend on M, and a comparison with observed light curves gives a "pulsational mass" M_{pul}. Now one finds that M_{ev} notoriously exceeds M_{pul} by 15–20 %. This result is confirmed by a handful of Cepheids for which the mass can be determined from the dynamics of binary systems. This problem has been amply discussed in the literature (Cox 1980; Keller 2008). Two solutions are considered: either the stars have lost the "excessive" mass by stellar winds prior to becoming Cepheids, or convective overshooting increases the mass of the helium core and the luminosity of Cepheids (see Fig. 31.2b), such that for given Π, hence L, a lower mass M_{ev} is deduced. The latter solution appears to be the more realistic one (Keller 2008), as it appears to solve the problem for Cepheids with dynamic masses (Fig. 31.7). On the other hand, overshooting tends to reduce the loops through the instability strip both because larger M_{He} leads to less extended loops (cf. Sect. 31.4) and because of the fact that the instability strip is inclined to higher T_{eff} for higher luminosity. For some stellar masses and compositions models with overshooting therefore avoid the Cepheid phase altogether.

We have dwelt at length on this short phase of evolution, since the Cepheids are important and offer a major fraction of those rare cases which, at least in principle, allow a quantitative test of the theory.

31.4 To Loop or Not to Loop . . .

In Sect. 31.3 we saw how important it is to find evolutionary tracks looping through the red giant region during central helium burning. It was all the more noteworthy when one learned that the loops depend critically on some uncertain input parameters (e.g. κ, ε, treatment of convection, composition) used in the calculations. A detailed classification of all influences, including their mutual interaction, is far too involved. Rather we point out a few characteristic properties of the models which allow a phenomenological prediction on the looping (We here follow the discussion of Lauterborn et al. 1971a,b; for other descriptions see Robertson 1971; Fricke and Strittmatter 1972).

For not too large masses (say, $M \lesssim 7M_{\odot}$), the evolution through the loops is so slow that the ε_g terms scarcely play a role. So we can reproduce the loops sufficiently well by models in complete equilibrium. Let us again consider solutions for the helium core (mass M_c, radius R_c, luminosity l_0) and for the hydrogen-rich envelope separately before fitting them to a full solution. The core luminosity l_0 is supplied by central helium burning; hydrogen-shell burning at the bottom of the envelope gives the additional luminosity L–l_0.

For given chemical composition a solution for the envelope can be obtained after specifying a pair of values R_c, l_0 as inner boundary conditions at $m = M_c$ (This is quite analogous to the usual central conditions $r = l = 0$ at $m = 0$.). Any solution gives a point in the HR diagram as well as pressure and temperature a M_c, i.e. values for $L, T_{\text{eff}}, P_0, T_0$. For the first part of the loop, helium burning contributes relatively little to L. Consequently we may approximate the envelope by setting $l_0 = 0$ (This can be done, of course, only for the calculation of the envelope which is dominated by hydrogen burning; in the core, l_0 cannot be neglected since it represents the whole local luminosity of this region.). The envelope solutions there form a two-parameter set in which we treat M_c, R_c as free parameters.

Next we look for a simple description of the chemical composition in the envelope. Figure 31.8a shows a typical hydrogen profile. A rather moderate increase of X_H is the relic of hydrogen consumption in the convective core during the main-sequence phase. The very narrow shell source has already eaten away the lower part of this profile (dashed) and produced a steep increase of X_H above the momentary

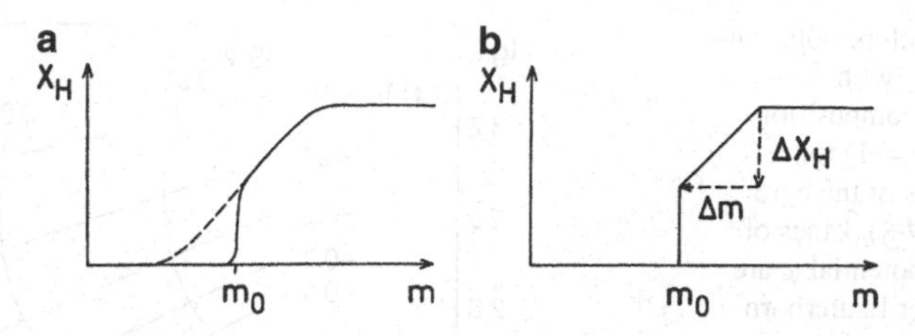

Fig. 31.8 The hydrogen abundance in an evolved star. (**a**) The convective core has left a fairly smooth profile (*dashed line*) which afterwards is steepened by shell burning. The shell is centred at m_0. Consequently $X_H = 0$ for $m < m_0$. For $m > m_0$, there is still a region in which X_H is not constant. (**b**) Schematic description of the chemical profile given by the *solid line* in (**a**)

helium core. We idealize this by a profile described by the parameters Δm and ΔX, as shown in Fig. 31.8b. The further shell burning will obviously increase M_c and decrease Δm and ΔX.

Now the envelope solution and its position in the HR diagram depend on the four parameters M_c, R_c Δm, ΔX. We would like to have a simple function of these parameters which can serve as a measure for the separation from the Hayashi line. The back-and-forth motion in the loop would then correspond to a non-monotonic variation of this function. A hint for a suitable procedure can be found in Fig. 31.1. The envelopes there shift monotonically to the right in the HR diagram, while the cores move through all three branches of the series of equilibrium models with increasing ratio M_c/R_c. This is essentially the surface potential of the core and plays a decisive role in many descriptions of radial expansion and contraction during the evolution. So we consider an "effective core potential":

$$\varphi := h\frac{M_c}{R_c}, \tag{31.3}$$

where we count M_c, R_c in solar units. The function $h = h(\Delta m, \Delta X)$ takes account of the influence of the chemical profile. We normalize it by setting $h = 1$ for a simple step profile ($\Delta m = \Delta X = 0$) and specify it later for other profiles. For a step profile and for $M = 5M_\odot$ five sequences of envelope solutions with constant M_c are shown in Fig. 31.9. The plotted lines $\varphi =$ constant illustrate that φ may indeed serve as an indicator of the distance from the Hayashi line. In particular we can find a critical value φ_{cr} such that all envelopes with

$$\varphi > \varphi_{cr} \tag{31.4}$$

are close to, and move upwards along, the Hayashi line with increasing φ. The line $\varphi = \varphi_{cr}$ may therefore roughly connect the minima of the envelope curves, and from Fig. 31.9 we see that $\lg \varphi_{cr} = 0.93$ for $5M_\odot$. For $M = 3M_\odot$ and $7M_\odot$, it is 0.83 and 0.99, respectively.

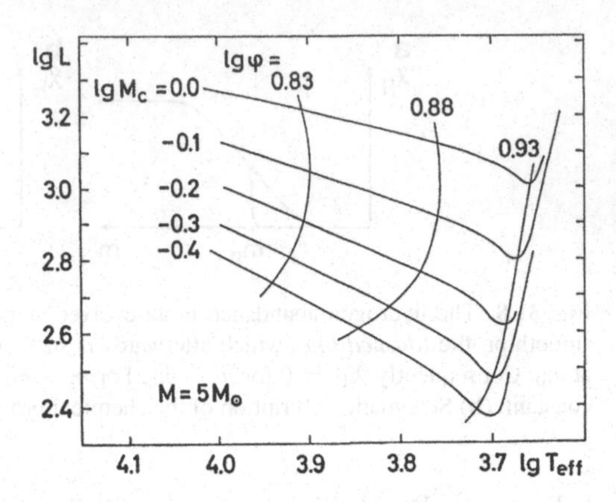

Fig. 31.9 Envelope solutions for $M = 5M_\odot$ with homogeneous composition down to $M_c(h = 1)$ for different values of the core mass M_c (in $M_\odot$). Lines of constant core potential φ are indicated (After Lauterborn et al. 1971a)

The function h is defined so that models with different profiles but equal distances from the Hayashi line have the same φ. Numerical experiments with different profiles have shown that the simple approximation

$$h = e^{\text{constant}\cdot\Delta m\cdot\Delta X} \tag{31.5}$$

is sufficient. Here h depends only on the product $\Delta m \cdot \Delta X$, i.e. to say on the amount of excess helium in and just above the shell source. The profile influences the envelope mainly through a hydrostatic effect.

Finally, relations between M_c and R_c have to be derived from solutions for the core. Each solution of an envelope of given M_c, R_c yields a pair of values P_0, T_0. For each M_c we vary R_c and get the functions $P_0(R_c)$ and $T_0(R_c)$, which are taken as outer boundary conditions for the core. For a specified composition and different M_c the core solutions then give the required relation $R_c(M_c)$, which is quite different for φ larger or smaller than φ_{cr}, namely $M_c/R_c \sim M_c^{0.4}$ $(\varphi < \varphi_{cr})$ and $\sim M_c^{0.25}(\varphi > \varphi_{cr})$. Therefore this factor tends to increase φ when the shell source burns outwards. We then have, in addition, the influence of the chemical evolution of the core on R_c. As mentioned earlier, an appreciable effect occurs only after the central helium content has dropped below, say, 0.1. The following rapid decrease of R_c tends to increase φ. Both these effects (the increase of M_c and the decrease of R_c) tend to shift the model to the right in the HR diagram and may therefore finish a loop, but they can never start it.

Obviously the responsibility for the onset of a loop rests with the function h. In fact, when the shell source burns farther into the profile, Δm and ΔX (cf. Fig. 31.8) become smaller and h decreases according to (31.5). This outweighs the increase of M_c/R_c in the first phase after E, and φ becomes smaller (Fig. 31.10). Sooner or later, however, the factor M_c/R_c takes over, since it continues to grow steadily, while h will level off near its maximum $h = 1$ when the shell source has "crunched up" almost the entire profile. Therefore φ reaches a minimum φ_{min} and then increases again. The turning of φ at φ_{min} can be caused either by the growth of M_c or by

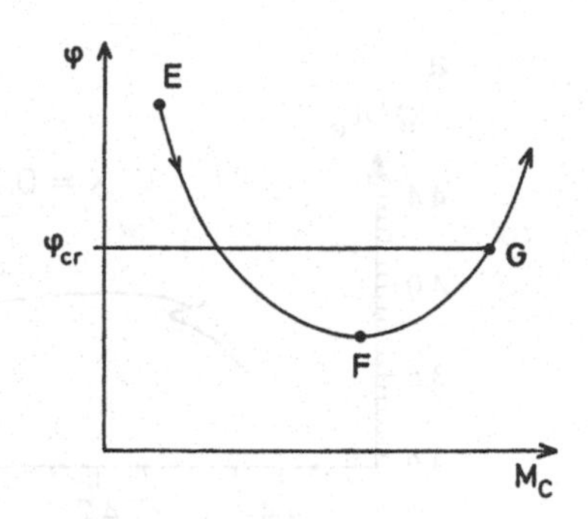

Fig. 31.10 Sketch of the effective potential φ as a function of the core mass M_c for an evolution through a loop. The points E, F, G refer to those in Fig. 31.2

the drop of R_c due to helium depletion. Which of these effects occurs earlier will depend on the ratio of the timescales for shell source and central burning.

So we have found a non-monotonic variation of φ. Whether this results in a loop, and if so the length of the loop, will depend on φ_{cr} and the starting value $\varphi(E)$ by which we denote the value of φ at point E. For small $M, \varphi(E)$ exceeds φ_{cr} by so much that even φ_{min} remains above φ_{cr}, and no loop occurs. The variation of φ then is reflected only in a motion down and up near the Hayashi line (Fig. 31.4, for $M \lesssim 4M_\odot$). High values of M bring $\varphi(E)$ close to φ_{cr}, and therefore in the further evolution φ goes below φ_{cr}. A case with $\varphi_{min} < \varphi_{cr}$ is illustrated in Fig. 31.10. When φ drops below φ_{cr} the model detaches from the Hayashi line and starts looping to the left. The turn to the right begins at point F when $\varphi = \varphi_{min}$.

Now it is obvious that many factors can modify the loops. For example, all properties changing the ratio of the timescales for central helium burning and shell burning can shift φ_{min} and thus the bluest point of the loop. In particular, we have to mention all the uncertainties concerning convection. Appreciable overshooting on the main sequence shifts the whole profile outwards. This can increase M_c and consequently $\varphi(E) - \varphi_{cr}$ such that the loop becomes smaller if it is not completely suppressed. Other factors affect the decisive upper part of the hydrogen profile. Aside from careless integrations during the main-sequence phase there are also physical uncertainties which can leave faulty profiles in the models. An example is the mixing by the outer convection zone during its deepest penetration, which in turn depends on the chosen mixing length in the superadiabatic layer. A similar problem causes the semiconvective region in main-sequence stars of large M (cf. Sect. 30.4.2). The assumption that this region is fully mixed leads to a plateau in the calculated profile with a discontinuous drop of X_H at its bottom. The presence or absence of this plateau must strongly influence the function h. Correspondingly the literature presents quite different evolutionary tracks for massive stars during helium burning (some with loops near the Hayashi line, others more to the left and completely detached from this line) for different assumptions on the semiconvective mixing.

In the following we present some examples for these effects. In Fig. 31.11 the evolution of a $9\,M_\odot$ star is shown for two different chemical compositions and different treatment of convection as well as a numerical aspect. In the top panel

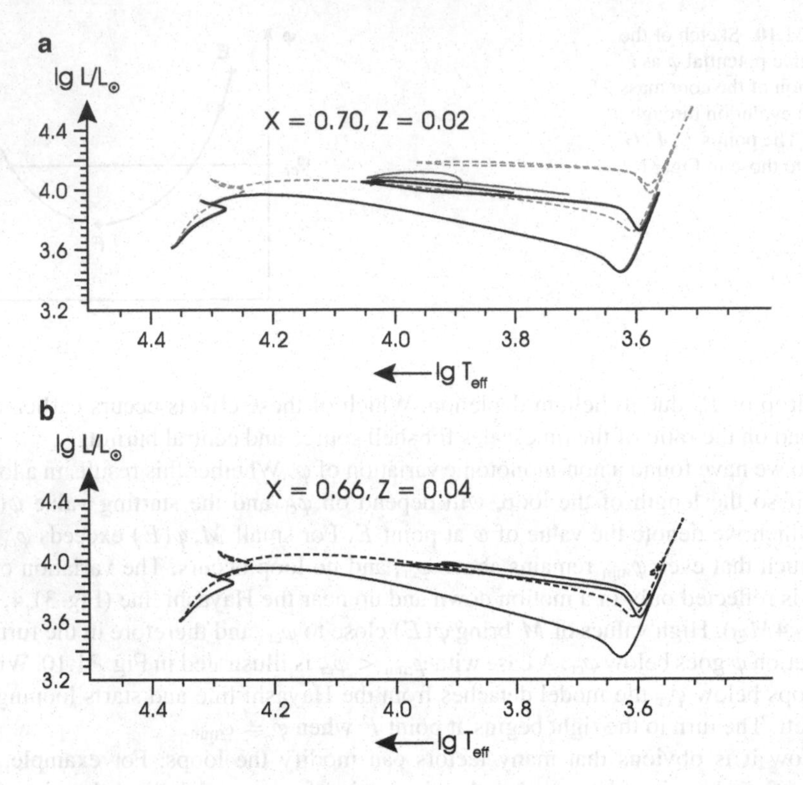

Fig. 31.11 Sensitivity of blue loops during core helium burning for a star of $9\,M_\odot$ and two compositions. (**a**) (*top panel*): for $X = 0.70$, $Z = 0.02$. The *black solid line* is a calculation with Schwarzschild criterion and without overshooting, which is included in the model shown by the *dashed line*. The *solid grey line* is using the same physical assumptions as in the first case, but with a coarse spatial resolution during the core hydrogen burning phase. (**b**) (*bottom panel*): for $X = 0.66$, $Z = 0.04$. The two lines correspond to the cases without (*solid*) and with (*dashed*) overshooting

(for a case with $Z = 0.02$) the inclusion of overshooting (dashed line) simply shifts the track to higher luminosity, as expected, with respect to the "standard" case without overshooting and using the Schwarzschild criterion for convection. However, when reducing the spatial resolution of the model (solid grey line), the chemical profile at the end of core hydrogen burning differs from the standard case, such that near the end of core helium burning convection brings fresh helium from outside the convective helium core, leading to a rejuvenation of the helium burning and resulting is a second loop before the star finally returns to the Hayashi line. Note that all calculations were terminated once helium at the centre was exhausted.

In the lower panel of Fig. 31.11, for $Z = 0.04$, the standard case is shown again (solid line), which is similar to the standard case for the previous mixture. However, in this case, overshooting (dashed line) enlarges the convective core sufficiently such that the loop is reduced to a very small "looplet" close to the Hayashi line. These numerical results agree well with the analysis done above.

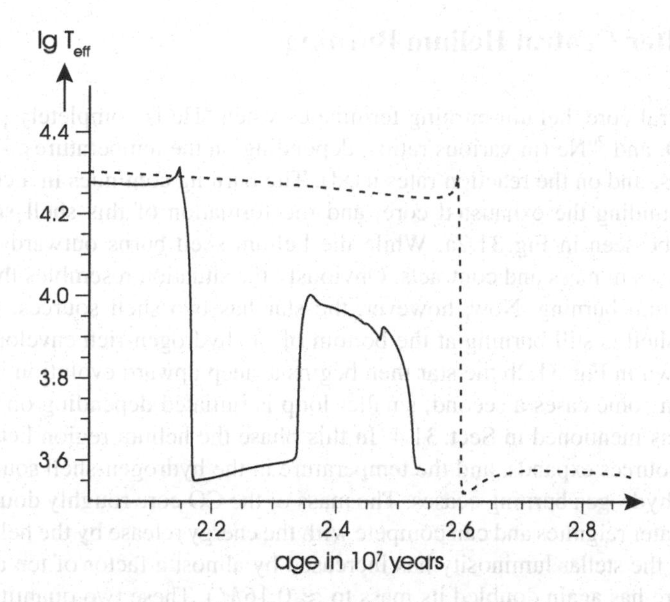

Fig. 31.12 T_{eff} as function of age (years) for the 9 $M_{\odot}$ models of Fig. 31.11b

A further example how different assumptions concerning convection affect the loops can be found in Fig. 32.1 for a star of 15 $M_{\odot}$. In this particular case, the use of the Ledoux criterion and semiconvective mixing lead to a quite extended loop, while the application of the Schwarzschild criterion as well as the inclusion of overshooting suppress it.

We see that details, which have originated from different regions and from earlier phases when the effects were scarcely recognizable, can now pop up and modify the evolution appreciably. The present phase is a sort of magnifying glass, also revealing relentlessly the faults of calculations of earlier phases.

Finally, we demonstrate in Fig. 31.12 that the crossing of the Hertzsprung gap and the onset of the blue loop happen on a thermal timescale as discussed in the preceding paragraphs and that the "looplet" for the case with overshooting of Fig. 31.11b (dashed line) is taking place on a much longer, nuclear timescale. We show the effective temperature as function of age. The nearly vertical drop after the main sequence (hottest points reached during the evolution) corresponds to the Hertzsprung gap crossing on a thermal timescale of about 1.1×10^5 years (taken as the time between $\log T_{\mathrm{eff}} = 4.2$ and 3.7); the leftward evolution on the blue loop (from $\log T_{\mathrm{eff}} = 3.65$ to 4.0) lasts 9×10^4 years. The looplet, in contrast, needs almost 3×10^6 years.

31.5 After Central Helium Burning

In the central core, helium burning terminates when ^{4}He is completely processed to ^{12}C, ^{16}O, and ^{20}Ne (in various ratios, depending on the temperatures, i.e. on the stellar mass, and on the reaction rates used). The burning continues in a concentric shell surrounding the exhausted core, and the formation of this shell source for $5M_\odot$ can be seen in Fig. 31.2a. While the helium shell burns outwards, the CO core increases in mass and contracts. Obviously the situation resembles that before central helium burning. Now, however, the star has two shell sources, since the hydrogen shell is still burning at the bottom of the hydrogen-rich envelope. In the model shown in Fig. 31.2b the star then begins a steep upward evolution in the HR diagram. In some cases a second, smaller loop is initiated depending on the same quantities as mentioned in Sect. 31.4. In this phase the helium region between the two shell sources expands, and the temperature in the hydrogen-shell source drops so far that hydrogen burning ceases. The mass of the CO core roughly doubles. The hydrogen later reignites and can compete with the energy release by the helium shell only when the stellar luminosity has increased by almost a factor of ten and when the CO-core has again doubled its mass to $\approx 0.16M$). These two quantities are in fact correlated, as we shall see in Chaps. 33 and 35.

From Fig. 31.2a we see that the outer convective envelope gradually reaches further down until it contains more than 80 % of the stellar mass. Its lower boundary clearly penetrates into a range of mass through which the hydrogen-shell source has burned during the preceding $\sim 10^7$ years, processing all ^{1}H to ^{4}He, and nearly all ^{12}C and ^{16}O to ^{14}N. These nuclei are now dredged up by the outer convection zone and can appear at the surface. This is usually called the phase of the *second dredge-up,* during which also the surface helium content, which has risen from an initial value of 0.28 to 0.29 during the first dredge-up, increases again to 0.31 in this example. It happens only in a mass range above 3–5 $M_\odot$, but not for massive stars.

With the inward motion of the lower border of convection, the H–He discontinuity has come rather close to the helium-shell source where $T \approx 2 \times 10^8$ K. This hot helium shell moves outwards until it is close enough to the hydrogen-rich layers so that they heat up and hydrogen is ignited–the hydrogen-shell source is reactivated. This mixing of hydrogen at the same time reduces the mass of the hydrogen-exhausted core, which will later become the white-dwarf remnant of such stars. The second dredge-up therefore prevents the formation of more massive white dwarfs, too.

Before we continue discussing the stars in this mass range in Chap. 35, we have to describe the evolution of massive and low-mass stars through central helium burning in the next chapters.

Chapter 32
Evolution Through Helium Burning: Massive Stars

The evolution of massive stars (stars with $M \gtrsim 8 \cdots 10$) through the phases of central hydrogen and helium burning would be quite similar to that of intermediate-mass stars (Chap. 31) if it were not for a few effects that influence it appreciably and which are specific for this mass range. The fact that the size of the convective core is encompassing large fractions of the star (Fig. 22.7) makes uncertainties connected with the treatment of convection even more important. These are twofold: semiconvection and overshooting, which we introduced already in Sect. 30.4. Furthermore, massive stars are known to have intensive mass loss, which in some cases are able to uncover the cores such that layers with a composition modified by nuclear fusion processes become visible. These are the so-called Wolf-Rayet stars. Finally, massive stars can rotate with surface rotation speeds of up to a few hundred km/s, or to an appreciable fraction of the break-up speed. This leads, as we will see in Chap. 44, to additional mixing processes beyond convective mixing, and further effects. The modelling of massive star evolution therefore becomes quite complicated and uncertain because of these physical effects which are not well understood. We will discuss their general influence in the following.

32.1 Semiconvection

The problem of semiconvection was already introduced and illustrated in Sect. 30.4.2 and Fig. 30.12. It is of particular importance for the evolution of stars above, say, 10 $M_{\odot}$, and results from the fact that the convective core contains less and less mass during the main-sequence evolution. This "shrinking" of the core is due to the increasing concentration of nuclear energy production, taking place via the CNO cycles, towards the centre with increasing core temperature. It leaves behind a region of varying chemical composition around the convective core, with material that experienced only some amount of nuclear fusion surrounding inner layers, where the conversion of hydrogen to helium proceeded further. In these layers, both the stabilizing molecular gradient ∇_{μ} and the radiative temperature

R. Kippenhahn et al., *Stellar Structure and Evolution*, Astronomy and Astrophysics Library, DOI 10.1007/978-3-642-30304-3_32, © Springer-Verlag Berlin Heidelberg 2012

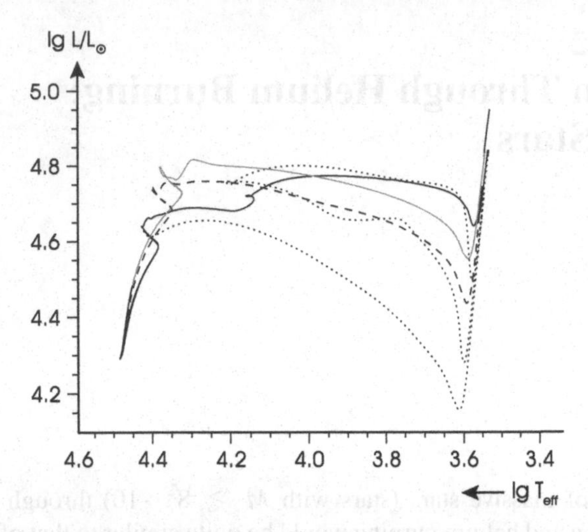

Fig. 32.1 Evolution of a $15\,M_{\odot}$ star of initial composition $X = 0.70$, $Y = 0.28$, $Z = 0.02$ with different treatments of core convection. The *solid black line* is the resulting evolution if the Schwarzschild criterion is applied, the *dotted* one in case of the Ledoux criterion, with slow semiconvective mixing, and the *dashed* one for the case with the inclusion of overshooting and the Schwarzschild criterion. The *grey solid line*, finally, has been computed with more overshooting and additional mass loss

gradient $\nabla_{\rm rad}$ are strongly oscillating functions of depth, depending on the exact chemical profile left behind by the shrinking core. In numerical models this profile depends on the spatial and temporal resolution of the models and their evolution, but also, for example, on the detailed interpolation in sets of opacity tables. In particular, it is quite important how accurately the varying chemical composition at each position inside the star is represented by these tables. As a consequence of these fluctuating terms in (30.10), the stability condition may be fulfilled in some parts of these critical regions, but not in others. The result is a region above the core with fluctuating radiative and convective layers, the exact structure of which is rather uncertain to compute.

If the Schwarzschild criterion for convection is used, the stabilizing molecular weight gradient in (30.10) is omitted and the layers become convective more easily and earlier in the evolution. This is the situation displayed in the top panel of Fig. 30.12 for a sample calculation of a $15\,M_{\odot}$ star. Since the separation between the convective core and the semiconvective layers outside of it may be rather small, a connection of both may occur, which "rejuvenates" the core by mixing fresh hydrogen into the burning region. The main-sequence evolution is thus extended and happens at higher luminosities. In Fig. 32.1 we show the resulting evolutionary track in the HR diagram (solid line) of a calculation, in which this effect was avoided. In this case the core helium burning is starting already during the evolution towards the red region, at a temperature of $\log T_{\rm eff} = 4.17$ (we have set, rather arbitrarily, this phase to the point, when the helium luminosity has reached 20 % of the hydrogen luminosity). This is at an age of 9.35 Myr, and the core helium burning lasts for another 1.43 Myr. At that time the star is beyond the luminosity minimum close to the red (super-)giant region. The onset of core helium burning is connected with the small loop at $\log T_{\rm eff} \approx 4.15$.

The lower panel in Fig. 30.12 and the dotted line in Fig. 32.1 correspond to the case of the Ledoux criterion. In this case the semiconvective layers above the shrinking core mix only slowly, and a more gradual chemical gradient between the hydrogen-exhausted core and the envelope is maintained. This is illustrated in Fig. 32.2, which compares the hydrogen profile at the end of the main sequence for the two criteria for convective stability. In the Ledoux case, the profile in the outer parts of the initially convective core (out to $m/M \approx 0.35$) is very smooth, whereas in the case of the Schwarzschild criterion it shows steps due to the sporadic appearance of localized convective regions.

In the Ledoux case, the main sequence lasts longer for 1.4 Myr, and the helium burning starts only 40,000 years later, at $\log L/L_\odot = 4.41$ and $\log T_{\rm eff} = 3.78$, i.e. after the star has crossed the Hertzsprung gap on a thermal timescale and is approaching the luminosity minimum close to the Hayashi line, along which it quickly ascends within a few 10^4 years. Due to the deep convective envelope the chemical composition is homogeneous down to $m/M \approx 0.25$ (Fig. 32.2, lines showing the helium profile), while in the Schwarzschild case, steps still exist, because at this stage, when the central helium content is 0.48, the star has not yet reached the red giant region.

The most striking difference is the blue loop that the Ledoux model performs during core helium burning. At its hottest extension, the central helium content is reduced to 19 %; it is exhausted when the star is about halfway back to the giant region. This phase lasts for 1.23 Myr, comparable to the duration of central core helium burning in the Schwarzschild case. The fact that stars in that mass range, calculated using the Ledoux criterion for convection and under the assumption of slow semiconvective mixing, first become red giants and then perform blue loops was very crucial in explaining the pre-explosion evolution of the progenitor of supernova SN1987A, a star known as Sanduleak $-69°202$ (see, e.g. Woosley et al. 1988; Langer 1989).

32.2 Overshooting

The effect of overshooting on the interior evolution of the same star is visible in the middle panel of Fig. 30.12. The evolutionary path in the HR diagram (Fig. 32.1) is shifted in a similar way as was shown in Fig. 31.2b, i.e. to higher luminosities. Due to the enlarged convectively mixed core, central hydrogen burning lasts now for 12.12 Myr, i.e. about 2.8 Myr (30 %) longer than for the case without overshooting. This corresponds roughly to the increased amount of fuel for the nuclear fusion, which can also be recognized from the hydrogen profile shown in Fig. 32.2. Note also that overshooting has the additional effect of creating a

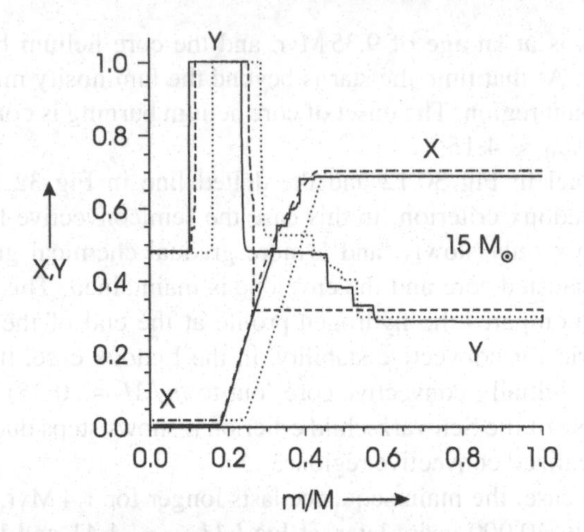

Fig. 32.2 Chemical composition profiles for selected models along the evolution shown in Fig. 32.1. Shown are the hydrogen and helium mass fractions (indicated by the usual symbols X and Y) as function of relative mass. The *solid lines* refer to the models calculated with the Schwarzschild criterion, the *dashed lines* to those with the Ledoux criterion for convection, and finally the *dotted* ones to the case with convective overshooting. The models were taken at the end of the main sequence when the central hydrogen abundance had been reduced to 0.01, and during core helium burning, when the central helium abundance is at 0.48

smooth chemical profile. The increase in luminosity of $\Delta \log L/L_{\odot} \approx 0.11$ at the end of the main sequence agrees well with a simple estimate using (20.20), which predicts $L \sim \mu^4$, where μ is the mean molecular weight obtained from that of the hydrogen-rich envelope and of the helium-rich core. μ increases from 0.83 to 0.90 when overshooting is enlarging the core, and therefore $\log L$ by approximately 0.14. Core helium burning starts again halfway through the crossing of the HR diagram, but without a visible feature in the track, and lasts for another 1.2 Myr. Although overshooting enlarges the convective helium-burning core, too, the increased luminosity leads to an overall reduced duration of this nuclear phase. This star does not perform any loop. This agrees with the similarity of the helium profile with that of the Schwarzschild case (Fig. 32.2).

Overshooting is even more important for more massive stars. Figure 32.3 shows evolutionary tracks for stars of 40 and 50 $M_{\odot}$. For reference, the solid track of the 40 $M_{\odot}$ star was calculated without any overshooting, while the dotted grey line does include it. The broadening of the main-sequence phase and the increase in luminosity are obvious. The core hydrogen burning phase is extended from 4.47 to 4.69 Myrs, too. The models for the 50 $M_{\odot}$ star all include overshooting, but differ in the criterion for convection. One realizes that until the end of core hydrogen burning (after 4.14 respectively 4.12 Myrs), both tracks are almost identical, but differ afterwards, when the newly established hydrogen shell encounters the hydrogen profile, which, due to the use of the Schwarzschild (dotted line) or Ledoux (solid line) criterion, is different. As we discussed in the previous section, the treatment of convection influences strongly the post-main-sequence evolution!

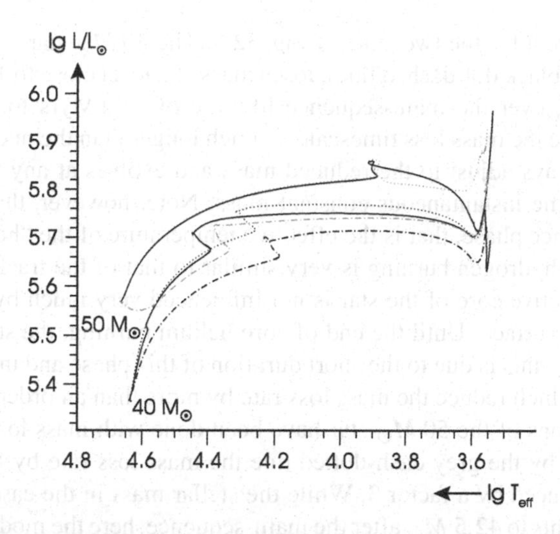

Fig. 32.3 Evolution of 40 and 50 $M_\odot$ stellar models calculated with various assumptions concerning semiconvection, overshooting, and mass loss. For the 40 $M_\odot$ star three cases are shown: one with neither overshooting nor mass loss using the Schwarzschild criterion for convection (*black solid track*), one with strong overshooting (*grey dotted*), and one with additional mass loss (*black dot dashed*). In total, this last model loses about 5 $M_\odot$. The *black solid line* for the 50 $M_\odot$ star refers to a case with overshooting and mass loss; the Ledoux criterion for convection was used here. For comparison, using the Schwarzschild criterion results in the *grey dotted line*. Finally, the *grey dash-dotted line* corresponds to a case with significantly enhanced mass loss. The final mass of this model is 28 $M_\odot$ compared to 37.5 and 42.4 $M_\odot$ in the former cases

32.3 Mass Loss

In addition to the complications of the interior evolution due to convection, the evolution of massive stars is also much stronger influenced by mass loss due to stellar winds than that of stars of low and intermediate mass. These strong stellar winds are driven by the radiation field and therefore increase with luminosity and effective temperature (the energy density of radiation scales with T_{eff}^4). For a review of winds from hot stars, see Kudritzki and Puls (2000). In the following we used the empirical mass loss formula by Vink et al. (2001) in our models.

In Fig. 32.1 (grey solid line) we show the evolution of the 15 $M_\odot$ model when mass loss is added and the amount of overshooting is increased even further, to now about 0.3 H_P. Accordingly, luminosity increases even further, and the main-sequence phase extends over 12.8 Myr. The mass loss rate on the main sequence is of the order of $1 - 2 \times 10^{-8}\ M_\odot/\text{year}$ and drops by a factor of a few when the star gets cooler. At the end of helium burning 1.15 $M_\odot$ is lost due to stellar winds. Since the amount of mass loss is below 10 % of the initial mass, the influence on the track, when compared to the case with overshooting, but no mass loss (dashed), is very small.

This is different for the two stars of Fig. 32.3. The 40 $M_\odot$ star, when calculated with mass loss (black dot-dashed line), loses mass at a level close to 10^{-6} $M_\odot$/year, which amounts, over the main-sequence lifetime of 4.54 Myrs to a reduction to 35.52 $M_\odot$. Since the mass loss timescale is much longer than the nuclear timescale, the star can always adjust to the reduced mass and evolves at any time similar to a star of the same instantaneous constant mass. Note, however, that the width of the main-sequence phase, that is the effective temperature of the "hook" indicating the end of core hydrogen burning is very similar to that of the track without mass loss. The convective core of the star is not influenced very much by the mass loss from the stellar surface. Until the end of core helium burning, the star loses only a further 0.03 $M_\odot$; this is due to the short duration of this phase and the cooler stellar temperatures, which reduce the mass loss rate by more than an order of magnitude.

All calculations of the 50 $M_\odot$ star have been done with mass loss. However, in the case shown by the grey dash-dotted line the mass loss rate by Vink et al. was artificially enhanced by a factor 3. While the stellar mass in the cases with normal mass loss amounts to 42.5 $M_\odot$ after the main-sequence, here the model loses 12 $M_\odot$ over 4.47 Myrs. Mass loss after the main sequence is negligible in all cases. Since this enhanced mass loss is so strong, the star can no longer evolve unperturbed. One can see this in the early part of the main sequence: instead of increasing in luminosity, the track bends down trying to follow a sequence of unevolved stars of decreasing mass. Only during the second half of core hydrogen burning, when the nuclear timescale is further reduced, the usual evolution proceeds, but at lower luminosity and also with a smaller extension of the main sequence. Indeed, the convective core is smaller than in the cases with normal mass loss (17 instead of 27 $M_\odot$ in this phase; compare this to 22 $M_\odot$ of the 40 $M_\odot$ star with Schwarzschild criterion, the evolution of which resembles most closely this one).

The maximum mass loss rate is 5×10^{-6} $M_\odot$/year for the last case presented. With even more extreme mass loss, up to 10^{-4} $M_\odot$/year and above, the track would even turn around and the star evolve to temperatures higher than the main sequence (compare this to the generalized main sequences of Sect. 23.3 for large values of q_0). During this evolution, the wind would uncover nuclear-processed layers of the star: first hydrogen-rich layers with high nitrogen abundance (from CNO-burning), later helium-rich layers, and even later possibly carbon-rich, hydrogen-free layers that experienced helium burning. These different surface compositions define the sequence of different types of Wolf-Rayet stars (WN, WC). Such models have been computed and presented by, for example, Maeder and Meynet (1987). However, stars with such strong winds can no longer be considered as having an optically thin atmosphere on top of the opaque interior. Instead, interior, atmosphere, circumstellar envelope and hot, fast stellar wind should be treated together and consistently (see, e.g. Schaerer 1996).

The evolution of massive stars is further influenced significantly by rotation and the mixing of the interior induced by rotation (see Chap. 43). Massive stars are known to rotate with surface velocities of several hundred km/s, sometimes close to break-up velocities. Modelling rotating stars is an active field of research going beyond the scope of this book. Therefore we refer the reader to the monograph by Maeder (2009)

Chapter 33
Evolution Through Helium Burning: Low-Mass Stars

33.1 Post-Main-Sequence Evolution

Compared to more massive stars, those of lower masses (typically $M < 2.3M_\odot$) evolve in a qualitatively different way after the exhaustion of hydrogen in their central regions. There are several reasons for this difference. Low-mass main-sequence stars have small, or no, convective cores, and degeneracy is important, if not on the main sequence, then shortly afterwards. In addition they start at a point on the main sequence much closer to the Hayashi line than the starting points of massive stars.

For example, if hydrogen is consumed in a well-mixed convective core, there will be a helium core of appreciable mass at the very end of central hydrogen burning. However, stars of around $1M_\odot$ have no convective cores; they consume hydrogen as illustrated in Fig. 30.1. Consequently they produce a growing helium core starting at zero mass. Therefore there is a smooth transition from central to shell burning. These stars start with such large central densities ($\gtrsim 10^2\,\mathrm{g\,cm^{-3}}$) that the electron gas is at the border of degeneracy, which has several consequences. The Schönberg–Chandrasekhar limit (Sect. 30.5) is not important: initially, the core mass M_c is below $0.1M$. When, however, with outward burning shell source $M_c > 0.1M$, the core contraction has produced sufficient degeneracy, making this limit irrelevant. The stars can then well exist in thermal equilibrium with a degenerate, isothermal helium core. This means that there is no "need" for a *rapid* core contraction as described in Sect. 31.1 and no equivalent of the Hertzsprung gap. Another consequence of degeneracy is that core contraction is not connected with heating. This is in contrast to the pre-main-sequence contraction (Sect. 28.1) and to post-main-sequence core contraction, which leads to helium ignition in massive stars.

At least in the first phases to be discussed here, the growth of the core mass is slow (since the productivity of the shell source is low), and the whole core settles at the temperature of the surrounding hydrogen-burning shell. This means that the core temperature is far from that of the ignition of helium ($\approx 10^8$ K). In low-mass stars, helium burning will be seen to start much later owing to secondary effects, after

R. Kippenhahn et al., *Stellar Structure and Evolution*, Astronomy and Astrophysics Library, DOI 10.1007/978-3-642-30304-3_33, © Springer-Verlag Berlin Heidelberg 2012

the core mass has grown up to a certain limit. Therefore the shell-burning phase between the central hydrogen and helium burning is a nuclear, slow phase, and one can expect to find many such stars in the sky.

The contraction of the core is (as in the case of larger M) accompanied by an expansion of the hydrogen-rich envelope outside the shell source. However, as long as the luminosity does not change drastically, the expansion cannot carry the star far away from its starting point on the main sequence. The reason is that this point is already close to the Hayashi line, which cannot be crossed (Chap. 24).

Any further expansion of the envelope is only possible if the luminosity increases. In fact the calculations show that L now increases by more than a factor 10^2 while M_c grows.

Surprisingly enough it turns out that L soon depends on the properties of the core only and is practically independent of the mass of the envelope (and therefore of M). In this phase the models can be well described analytically by a generalized form of homology.

33.2 Shell-Source Homology

Consider a model in complete equilibrium consisting of a degenerate helium core (mass M_c, radius R_c) surrounded by an extended envelope of hydrogen with abundance X_H and mass $M_{env} = M - M_c$. The core mass M_c grows owing to hydrogen-shell burning, which provides the luminosity L:

$$\dot{M}_c = \frac{L}{X_H E_H} \tag{33.1}$$

(where E_H is the energy gain per unit mass of hydrogen). This equation could easily be integrated if L were constant. However, while evolution proceeds, L grows too since there is a relation between L and M_c. The properties of the shell (and therefore L) are mainly determined by M_c and R_c, while they are almost independent of the properties of the envelope. This can be understood from the fact that the core is highly concentrated and the gravity at its surface is very large. Then, according to hydrostatic equilibrium, $|dP/dm| \sim m/r^4$ is very large, and P drops by powers of 10 within a thin mass shell just above the core surface. The typical situation is illustrated in Fig. 33.1. In other words, the extended envelope above this layer is nearly weightless and has no influence on the burning shell.

We now present an analytic approach of Refsdal and Weigert (1970) giving relations between the properties of the core and the physical variables in the hydrogen-burning shell. For this purpose we will generalize the homology considerations of Chap. 20 and use again the power approximations for κ and ε:

$$\kappa = \kappa_0 P^a T^b, \quad \varepsilon = \varepsilon_0 \varrho^{n-1} T^{\nu}. \tag{33.2}$$

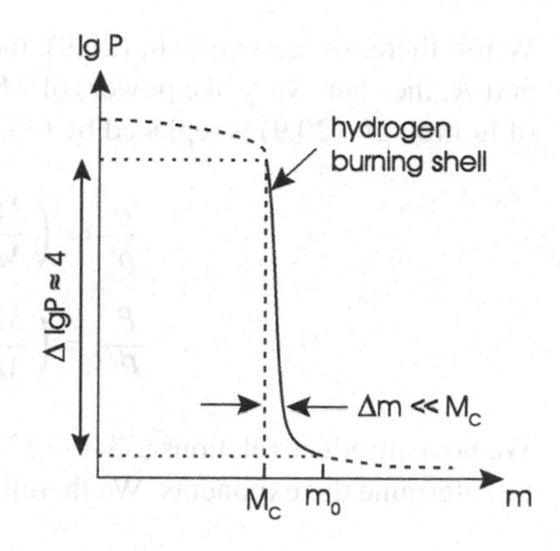

Fig. 33.1 A schematic sketch of the run of pressure in the vicinity of a thin shell. M_c is the mass of the core; the shell extends to $M_0(r_0)$. Its thickness is Δr, i.e. $M_0 = M_r(R_c + \Delta r)$ (After H. Ritter, priv. communication)

Here we have replaced the exponent λ used in Chap. 20 by $n - 1$.

For the gas pressure we will use the ideal-gas equation

$$P = \frac{\Re}{\mu} \varrho T \; , \tag{33.3}$$

since we only want to apply it to regions outside the core, where the gas is not degenerate. We also neglect radiation pressure since it is not important for low-mass stars. In Chap. 35 we shall apply the relations derived here to more massive stars and then take radiation pressure into account.

We now assume for the density, temperature, pressure, and local luminosity in the region of the hydrogen-burning shell (i.e. for $R_c \le r \le R_c + \Delta r$) that there exists a simple dependency on M_c and R_c:

$$\varrho(r/R_c) \sim M_c^{\varphi_1} R_c^{\varphi_2} \; , \tag{33.4}$$

$$T(r/R_c) \sim M_c^{\psi_1} R_c^{\psi_2} \; , \tag{33.5}$$

$$P(r/R_c) \sim M_c^{\tau_1} R_c^{\tau_2} \; , \tag{33.6}$$

$$l(r/R_c) \sim M_c^{\sigma_1} R_c^{\sigma_2} . \tag{33.7}$$

These homology-type relations have the following meaning: we compare two stellar models of different core masses M_c and M_c' and core radii R_c and R_c'. We define homologous points, r and r', in the two models by

$$\frac{r}{R_c} = \frac{r'}{R_c'} \; ; \tag{33.8}$$

the physical quantities at homologous points in the two models shall then be connected by relations (33.4)–(33.7). This indeed is very similar to the considerations of Sect. 20.1, though there the homologous points were defined with respect to the total radius R, whereas we here define them with respect to the core radius R_c.

While there, for example, in (20.9), the physical quantities vary like powers of M and R, they here vary like powers of M_c and R_c. For example, with our new concept of homology, (20.9) is replaced by (33.4) and (33.6), which are written explicitly as

$$\frac{\varrho}{\varrho'} = \left(\frac{M_c}{M'_c}\right)^{\varphi_1} \left(\frac{R_c}{R'_c}\right)^{\varphi_2} , \tag{33.9}$$

$$\frac{P}{P'} = \left(\frac{M_c}{M'_c}\right)^{\tau_1} \left(\frac{R_c}{R'_c}\right)^{\tau_2} . \tag{33.10}$$

We now introduce relations (33.4)–(33.7) into the stellar-structure equations in order to determine the exponents. We therefore write (2.4), (5.11), and (4.42) in the form

$$dP \sim M_c \varrho d(1/r) , \tag{33.11}$$

$$d(T^4) \sim \kappa \varrho l \, d(1/r) = \kappa_0 \varrho P^a T^b l \, d(1/r) , \tag{33.12}$$

$$dl \sim \varepsilon \varrho d(r^3) = \varepsilon_0 \varrho^n T^\nu d(r^3) , \tag{33.13}$$

with positive factors of proportionality. In (33.11) we have assumed that $m \approx M_c$ = constant, which is a sufficient approximation in the region in which P drops to negligible values. This assumption yields decisive differences from the relations discussed in Chap. 20. Introducing (33.4)–(33.6) into (33.3) we easily obtain for the exponents

$$\tau_1 = \varphi_1 + \psi_1 , \quad \tau_2 = \varphi_2 + \psi_2 . \tag{33.14}$$

We now integrate (33.11)–(33.13) over the shell, starting with (33.11): we choose a radius r_0 sufficiently larger than R_c that $P(r_0/R_c) \ll P(r/R_c)$, and find from (33.11) that

$$P(r/R_c) = P(r_0/R_c) + \int_{1/r_0}^{1/r} GM_c \varrho d(1/r) \approx \frac{GM_c}{R_c} \int_{x_0}^{x} \varrho \, dx , \tag{33.15}$$

with $x = R_c/r$. If we do the same for another model with M'_c, R'_c, we find for the pressure at the homologous radius r'

$$P'(r'/R'_c) \approx \frac{GM'_c}{R'_c} \int_{x_0}^{x} \varrho' \, dx = \frac{GM'_c}{R'_c} \left(\frac{M'_c}{M_c}\right)^{\varphi_1} \left(\frac{R'_c}{R_c}\right)^{\varphi_2} \int_{x_0}^{x} \varrho \, dx , \tag{33.16}$$

where (33.9) has been introduced into the integral. Comparing (33.16) with (33.15) yields

$$P(r/R_c) \sim M_c^{\varphi_1+1} R_c^{\varphi_2-1} , \tag{33.17}$$

and if we compare this with (33.6) we find

$$\tau_1 = \varphi_1 + 1 , \quad \tau_2 = \varphi_2 - 1 . \tag{33.18}$$

The same procedure can be carried out using (33.12) and (33.13). For the integration in the first case we again choose r_0 sufficiently far outside, where the temperature is small compared to its values in the shell; for the integration of (33.13) we take $r_0 = R_c$, where the local luminosity vanishes. We then obtain

$$(4-b)\psi_1 = \varphi_1 + a\tau_1 + \sigma_1 , \tag{33.19}$$

$$(4-b)\psi_2 = \varphi_2 + a\tau_2 + \sigma_2 - 1 , \tag{33.20}$$

$$\sigma_1 = n\varphi_1 + \nu\psi_1, \quad \sigma_2 = n\varphi_2 + \nu\psi_2 + 3 . \tag{33.21}$$

Equations (33.14) and (33.18)–(33.21) are eight linear inhomogeneous algebraic equations for the eight exponents in (33.4)–(33.7). The solutions are

$$\varphi_1 = -\frac{\nu - 4 + a + b}{N}, \quad \varphi_2 = \frac{\nu - 6 + a + b}{N}, \quad \psi_1 = 1, \quad \psi_2 = -1 ,$$
$$\tau_1 = 1 + \varphi_1, \quad \tau_2 = \varphi_2 - 1, \quad \sigma_1 = \nu + n\varphi_1, \quad \sigma_2 = 3 - \nu + n\varphi_2 , \tag{33.22}$$

with

$$N = 1 + n + a . \tag{33.23}$$

Equations (33.22) allow us to determine the variations of the physical quantities from one model (characterized by M_c, R_c) to another (characterized by M'_c, R'_c). The temperature and the local luminosity at homologous points vary as

$$T \sim M_c^{\psi_1} R_c^{\psi_2} = M_c/R_c , \tag{33.24}$$

$$l \sim M_c^{\nu + n\varphi_1} R_c^{3-\nu+n\varphi_2} . \tag{33.25}$$

This holds for all homologous points, also for those at the upper border of the range of integration where $l = L$. Therefore the luminosity of these shell-source models depends on M_c (rather than on M) and on the mode of energy generation (in striking contrast to main-sequence type models, cf. Chap. 20). As an illustration we assume $a = b = 0$ (electron scattering, see Sect. 17.1) and $\nu = 13, n = 2$ (CNO cycle, see Sect. 18.5.1). Then $\varphi_1 = -3, \varphi_2 = 7/3$, and we find

$$L \sim M_c^7 R_c^{-16/3} . \tag{33.26}$$

We have obtained relations $T(M_c, R_c)$ and $L(M_c, R_c)$ independent of M. In order to see how T and L vary along an evolutionary sequence of models with increasing M_c, one has to know how R_c varies with M_c. Since the cores in the evolution under consideration are degenerate, they resemble white dwarfs whose radii decrease with increasing mass (see Sect. 19.6, Chap. 37). We therefore can expect from (33.24) that the temperature in the shell source increases with M_c, and according to (33.26), the luminosity increases strongly with M_c even with R_c = constant (this increase being much steeper than the $L(M)$ relation for main-sequence stars).

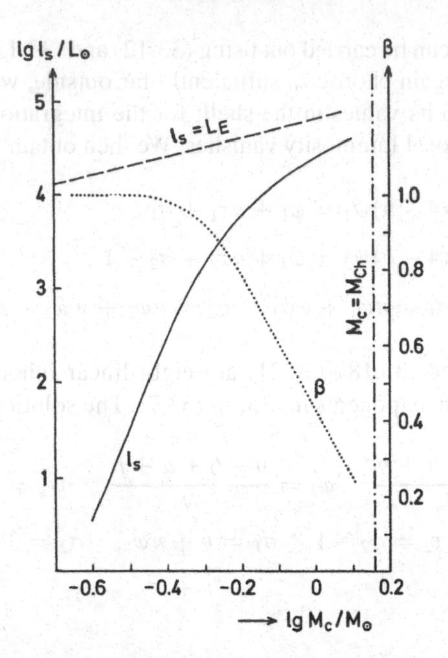

Fig. 33.2 The luminosity l_s (*solid curve, left ordinate*) at the top of the hydrogen-burning shell around a degenerate helium core of mass M_c. The *dotted line* indicates the importance of the radiation pressure, the value of $\beta(= P_{gas}/P_{total})$ being given by the ordinate at the *right*. When M_c approaches the Chandrasekhar mass M_{Ch} (Sect. 19.7; *dot-dashed vertical line*), the luminosity curve has the tendency to approach the Eddington luminosity L_E (*dashed line*) for which gravity equals the radiation-pressure gradient (for an opacity dominated by electron scattering; see Sect. 22.5)

We now need a relation $R_c(M_c)$. The classical mass-radius relation for white dwarfs (Chap. 37) is, of course, not directly applicable to these cores. Below the shell there must be a transition from complete through partial to no degeneracy. Compared to the outer layers of white dwarfs, this transition region is very hot (like the shell source) and may occupy an appreciable fraction of the core volume (For a discussion of this problem, see Refsdal and Weigert 1970.). Nevertheless, as a simple example for $R_c(M_c)$ we here take the relation for the cold white dwarfs of Table 37.1, yielding $d\ln R_c/d\ln M_c$ for different values of M_c. This can be used in

$$\frac{d\ln L}{d\ln M_c} = \sigma_1 + \sigma_2 \frac{d\ln R_c}{d\ln M_c}, \tag{33.27}$$

which follows from (33.7). The coefficients σ_1 and σ_2 are determined by (33.22). For $a = b = 0, n = 2, \nu = 14$, one finds $d\ln L/d\ln M_c \approx 8\cdots10$. We can also integrate (33.27) numerically when starting from a correctly computed model, which gives an initial value L for a given M_c. The results of such an integration, l_s, are shown in Fig. 33.2 by the left part of the solid curve where radiation pressure can be neglected ($\beta \approx 1$).

For the temperature at homologous points, say at the bottom of the hydrogen-burning shell, instead of (33.27), we obtain from (33.24)

$$\frac{d\ln T}{d\ln M_c} = 1 - \frac{d\ln R_c}{d\ln M_c} , \tag{33.28}$$

and we get $d\ln T/d\ln M_c$ somewhat larger than 1. Since the cores are assumed to be isothermal, this also gives the increase of the central temperature T_c. We see that in this way T_c can be raised to helium ignition even by models in complete equilibrium.

33.3 Evolution Along the Red Giant Branch

In the following we describe the evolution of a star of $1.3\,M_\odot$ as calculated by Thomas (1967) in a pioneering paper. The chemical composition of the initial model on the ZAMS is $X_H = 0.9, X_{He} = 0.099, Z = 0.001$, which at that time seemed to be the appropriate mixture for a star of population II. The essential results, however, do not depend too much on the chosen chemical composition, as we will show later and in Fig. 33.5. The initial model has $L = 1.91L_\odot, T_{eff} = 6,760\,\mathrm{K}$. Nuclear energy is released in the central region at $T_c = 1.48 \times 10^7\,\mathrm{K}$. There is a small convective core containing 4.3 % of the total mass, which disappears long before the exhaustion of hydrogen in the centre. There is also an outer convective zone, which reaches inwards from the photosphere to about $r \approx 0.95R$.

The evolutionary track in the HR diagram is shown in Fig. 33.3, while the internal evolution is illustrated by Fig. 33.4. In the HR diagram the image point of the model first moves upwards and then to the right. At the same time, the model switches from central nuclear burning to shell burning, as can be seen in Fig. 33.4. We have already learned from the shell-source homology of Sect. 33.2 that the luminosity must grow with increasing core mass. The calculated evolution confirms these predictions once the core is sufficiently compressed. The track is very close to the Hayashi line, leading up along the giant branch to higher luminosities and correspondingly larger radii. The neighbourhood of the line of fully convective stars can also be seen from the internal structure of the models. Figure 33.4 shows that the outer convective zone penetrates deeply inwards until more than 70 % of the total mass is convective. It then reaches into layers which are already contaminated by products of nuclear reactions (see dotted area in Fig. 33.4). The processed material is distributed over the whole convective region and therefore also brought to the surface. This type of partial mixing, the *first dredge-up*, we have already encountered for more massive stars in Chap. 31.

The monotonic increase of the luminosity is interrupted when the hydrogen-burning shell reaches the layer down to which the outer convective zone has mixed at the moment of deepest penetration. At this point the mixing has produced a discontinuity in molecular weight between the homogeneous hydrogen-rich outer layer and the helium-enriched layers below. When the shell source reaches the

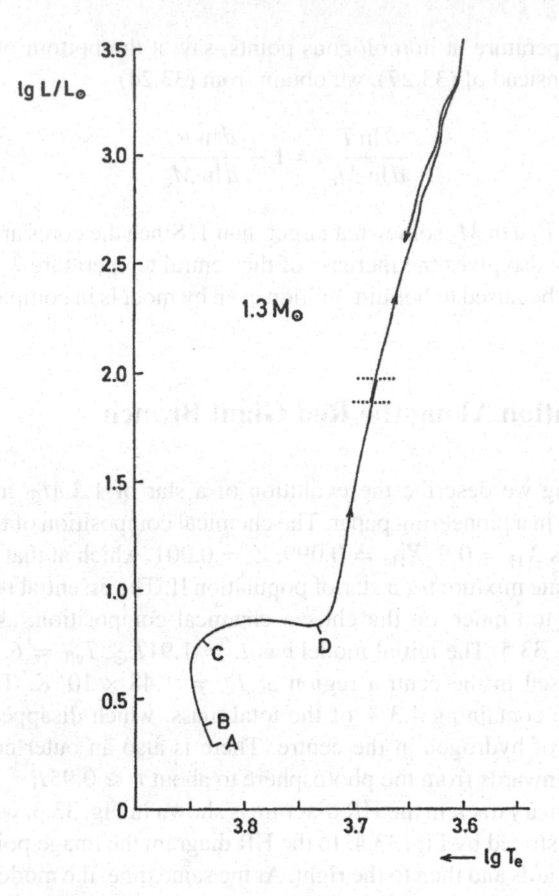

Fig. 33.3 The evolutionary track of a star of $1.3M_\odot$ with the initial composition $X_H = 0.9, X_{He} = 0.099, Z = 1 - X_H - X_{He} = 0.001$ as computed by Thomas (1967). The letters $A - D$ refer to the corresponding evolutionary states in Fig. 33.4. The *arrows* indicate the direction of the evolution. This direction is reversed for a short period between the *dotted horizontal lines*. This transient drop in luminosity at about lg $L/L_\odot = 2$ occurs when the hydrogen-burning shell crosses the chemical discontinuity left behind when the bottom of the outer convective zone moves outwards again in the mass scale after it has reached its deepest extension (see Fig. 33.4)

discontinuity, the molecular weight of the shell material becomes smaller. This causes the drop of luminosity at $L \approx 100L_\odot$ (see Fig. 33.3) as can easily be understood.

For this purpose we follow the considerations of Sect. 33.2, but this time, we vary the molecular weight μ at homologous points while keeping M_c, R_c, and all other parameters constant. Analogously to (33.4)–(33.7) we write

$$\varrho(r/R_c) \sim \mu^{\varphi_3} , \tag{33.29}$$

$$T(r/R_c) \sim \mu^{\psi_3} , \tag{33.30}$$

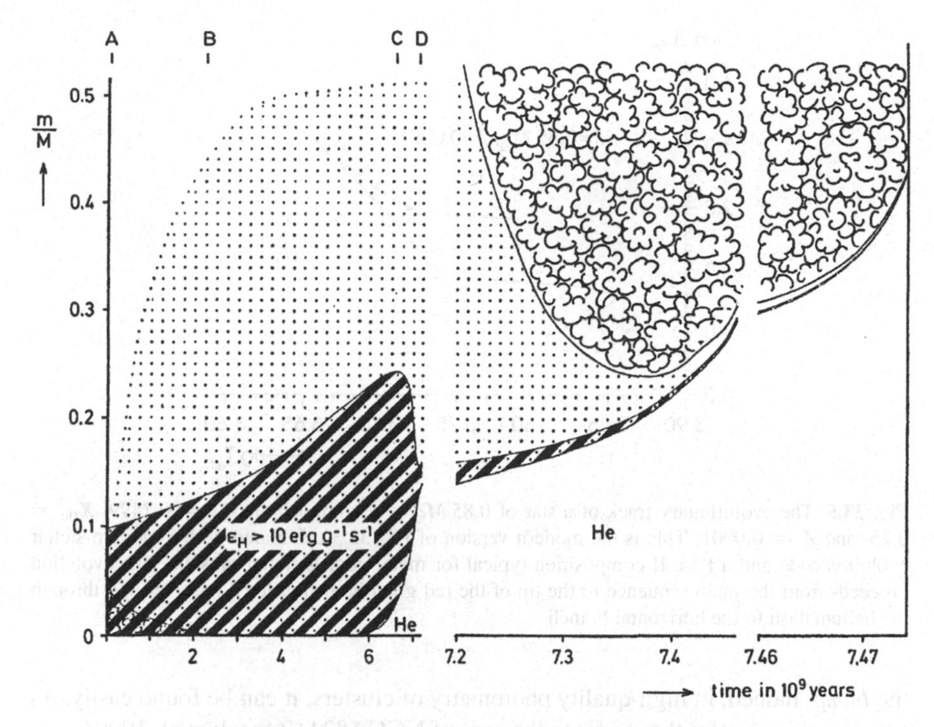

Fig. 33.4 The evolution of the internal structure of a star of $1.3M_\odot$ plotted in the same manner as in Fig. 31.2a. The main region of hydrogen burning is hatched; "cloudy" areas indicate convection. Regions of variable hydrogen content are *dotted* (After Thomas 1967)

$$P(r/R_c) \sim \mu^{\tau_3}, \tag{33.31}$$

$$l(r/R_c) \sim \mu^{\sigma_3}, \tag{33.32}$$

and with the same procedure as in Sect. 33.2 we find

$$\varphi_3 = \frac{4-b-\nu}{N}, \quad \psi_3 = 1, \quad \tau_3 = \varphi_3, \quad \sigma_3 = \nu + n\varphi_3, \tag{33.33}$$

with $N = 1+n+a$. For example, using again the values $\nu = 13, n = 2, a = b = 0$ as in Sect. 33.2, we see that (33.32) becomes $l \sim \mu^7$. Therefore the luminosity decreases with decreasing μ, which explains the transient reduction of L. After the shell source has passed the discontinuity, μ remains at its reduced value and the luminosity grows again with increasing core mass. As the star passes three times through the region between the two dotted horizontal lines in Fig. 33.3, observations have a higher probability of finding stars in this luminosity range than in the neighbouring ones. In luminosity functions of globular clusters, which give the number of stars per brightness bin, this event in the evolution of low-mass stars shows up as a localized peak, which is either called *Thomas peak*, or simply

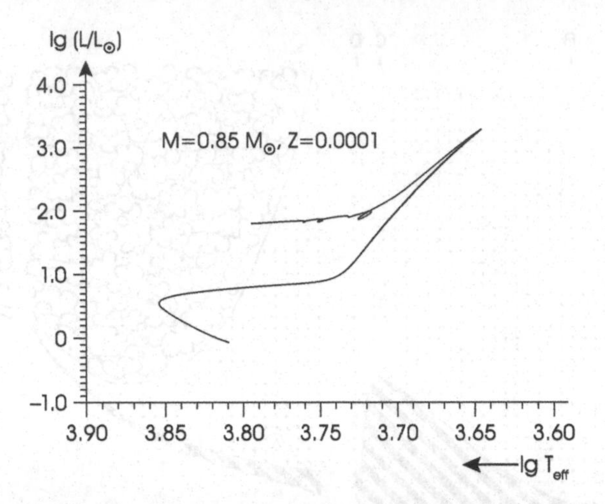

Fig. 33.5 The evolutionary track of a star of 0.85 $M_\odot$ and composition $X_{\rm H} = 0.7499$, $X_{\rm He} = 0.25$, and $Z = 0.0001$. This is the modern version of Fig. 33.3 calculated with a modern stellar evolution code and a Pop. II composition typical for metal-poor globular clusters. The evolution proceeds from the main sequence to the tip of the red giant branch at $\log L/L_\odot \approx 3.2$, through the helium flash to the horizontal branch

the *bump*. Indeed, in high-quality photometry of clusters, it can be found easily. As an example, we refer the reader to the case of NGC 5824 (Zoccali et al. 1999).

Evolutionary calculations for somewhat different total masses M yield similar results. Near the main sequence the tracks are shifted relative to each other according to their different starting points on the ZAMS. When approaching the Hayashi line the tracks merge (This is not exactly true, since different total masses have slightly different Hayashi lines.). After the cores are sufficiently condensed they are virtually independent of the envelope (and therefore of the total mass M). However, they determine the total luminosity according to the $L(M_{\rm c})$ relation. Consequently stars of different M but the same $M_{\rm c}$ have the same L and are practically at the same point in the HR diagram.

The same convergence of the evolution for different M must occur for all properties of the shell source and the core. For example, the central values of density and temperature converge to the same evolutionary track in the $\varrho_{\rm c}$–$T_{\rm c}$ plane.

Numerical calculations show that with growing core mass the temperature in the core rises. This is due to two effects which are of approximately the same order. The first is the increase of the temperature in the surrounding shell source where $T \sim M_{\rm c}/R_{\rm c}$ after (33.24). While this effect already occurs in models of complete equilibrium, there is an additional effect due to non-stationary terms. With growing $M_{\rm c}$ the core contracts, releasing energy. If this occurs rapidly enough, it heats up the transition layer below the shell, and therefore the whole core. An inward-directed temperature gradient is built up in the transition region, such that the energy released by $\varepsilon_{\rm g}$ terms is carried away. However, this is not the whole story, since at the same time conditions in the core are such that cooling by plasma neutrinos, which were discussed in Sect. 18.7 (see also Fig. 18.11), becomes important and modifies the temperature gradient, as will be seen in Sect. 33.5. The core evolution is enhanced by increasing L: the rate $\dot{M}_c$ is proportional to L, which in turn increases by a high power of M_c, and the process speeds up more and more. Both these effects, controlled by the growth of M_c, finally increase the core temperature to $\approx 10^8$ K at which helium is ignited. This happens when $M_c \approx 0.48 M_\odot$, almost independently of M, but slightly decreasing with increasing metallicity. The matter in the core is highly degenerate, and the nuclear burning is unstable. The resulting thermal runaway terminates the slow and quiet evolution along the giant branch.

33.4 The Helium Flash

We start with some analytic considerations and assume that helium is ignited in the centre, where the electron gas is assumed to be non-relativistic and degenerate. In Sect. 25.3.5 we have discussed the secular stability of nuclear burning in a small central sphere of mass m_s, "luminosity" $l_s = \varepsilon m_s$, and gravothermal specific heat c^*. Assuming a homologous reaction of the layers above, a small relative temperature perturbation $\vartheta_c (= dT_c/T_c)$ was shown in (25.35) to evolve according to

$$\dot{\vartheta}_c = \frac{l_s}{c_P m_s T_c}(\varepsilon_T + \kappa_T - 4)\vartheta_c \,, \tag{33.34}$$

where we have set $\delta = 0$ and therefore $c^* = c_P$ according to (25.29). For helium burning we have $\varepsilon_T > 19$ (see Sect. 18.5.2), which certainly dominates the other terms in the parenthesis which thus is positive: the onset of helium burning in the degenerate core is unstable and results in a thermal runaway. The timescale of the thermal runaway is of the order $c_P m_s T_c / l_s = c_P T_c / \varepsilon$, i.e. of the order of the thermal timescale of the helium-burning region.

The homologous linear approximation which yielded (33.34) can only give a very rough picture of the events after helium ignition. Nevertheless we can try to discuss the consequences which follow from our simple formalism. From (25.25) and (25.26) one obtains

$$\frac{d\varrho_c}{\varrho_c} = \frac{3\delta}{4\alpha - 3}\vartheta_c \,, \tag{33.35}$$

and for the completely degenerate non-relativistic gas, where $\alpha = 3/5, \delta = 0$, we find $d\varrho_c = 0$. Therefore, while during the thermal runaway the central temperature is rising, the matter neither expands nor contracts. The central density remains constant, and in the $\lg \varrho_c$–$\lg T_c$ diagram, the centre evolves vertically upwards as indicated in Fig. 33.6. The reason is that in the (fully) degenerate gas the pressure does not depend on temperature and therefore remains constant during the thermal runaway. But only an increase of pressure could lift the weight of the mass above and cause an expansion. Since the $P dv$ work is zero, all nuclear power goes into internal energy. During the thermal runaway there is an enormous overproduction of nuclear energy. The local luminosity l at maximum comes to $10^{11} L_\odot$, about that of

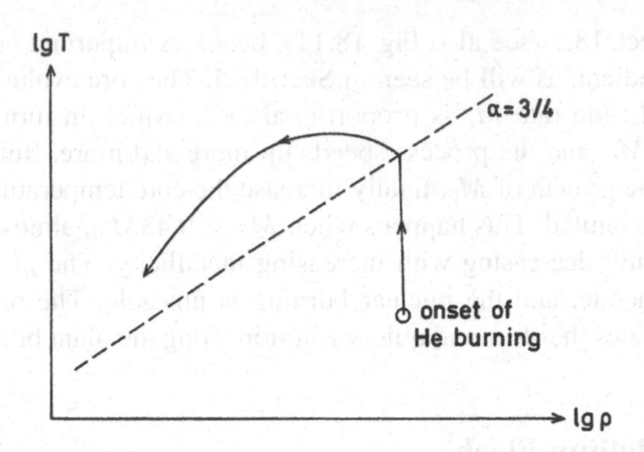

Fig. 33.6 Schematic sketch of the changes of temperature and density during the helium flash. After the ignition temperature is reached in the regime of degeneracy the temperature rises almost without a change of density until degeneracy is removed near the *broken line*. Then a phase of almost isothermal expansion ensuesfollowed by a phase of stable helium burning in the non-degenerate regime

a whole galaxy, but only for a few seconds (The expression "helium flash" is quite appropriate indeed!). However, almost nothing of it reaches the surface, since it is absorbed by expansion of the non-degenerate layers above.

With increasing temperature at constant density, the degeneracy is finally removed. This happens roughly when in Fig. 33.6 the border ($\alpha = 3/4$) between degeneracy and ideal gas is crossed. Then with further increase of T the core expands. With the removal of degeneracy the gravothermal specific heat becomes negative again and central helium burning becomes stable; the expansion stops the increase of temperature. The overproduction then is gradually removed by cooling until the temperature has dropped to "normal" values for quiet (stable) helium burning. In the lg ϱ_c–lg T_c plane the core settles near the image point of a homogeneous helium star of mass M_c, which is of the order of $0.48 M_\odot$.

There is another prediction we can make for the changes in the HR diagram. Until the onset of helium burning the total luminosity of the star (which is just the power produced in the shell) increases with increasing core mass as expected from (33.26). After degeneracy is removed in the central region, the core expands and R_c increases. During the short phase of the flash, M_c remains practically unchanged. From (33.26) we therefore expect the luminosity to be appreciably reduced after the flash phase, and this indeed can be seen from Figs. 33.3 and 33.5.

33.5 Numerical Results for the Helium Flash

In Sect. 33.4 we have tacitly assumed that the maximum temperature is in the centre. This, however, is not the case if neutrinos–as we discussed them in Sect. 18.7–are created in the very interior of the core and provide an energy sink there, since they

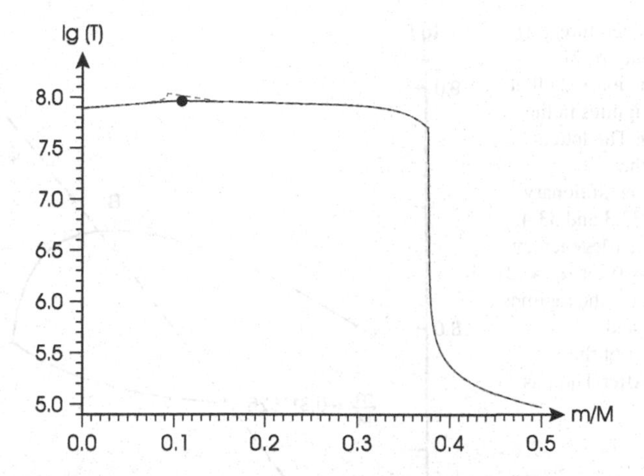

Fig. 33.7 The temperature T (in K) as a function of the mass variable m in a $1.2M_{\odot}$ model with solar metallicity, shortly before the onset of (unstable) helium burning (*solid line*). Owing to neutrino losses the maximum temperature does not occur in the centre but near $m/M = 0.1$ (indicated by the *dot*). The *dashed line* (deviating from the *solid* one only near the *dot*) shows how the ignition of helium burning has raised the temperature at this position inside the star only 10^4 years later

leave the star without noticeable interaction. Then the maximum of temperature is not in the centre but at a finite value of m (see Fig. 33.7). From there, energy flows outwards ($l > 0$) and inwards ($l < 0$). This energy is released by core contraction in the transition zone below the burning shell as mentioned in Sect. 33.3. The transport mechanisms are radiation and conduction. The inward-going energy is carried away by neutrinos. Then the ignition of helium and the flash will not take place in the centre but in the concentric shell of maximum temperature. This is near $m/M = 0.1$ according to Fig. 33.7 [Note that in the calculations shown in Fig. 33.4 an unusually low value of μ in the envelope was assumed. Therefore, according to (33.30) and (33.32), T in the shell source and L are smaller for the same M_c and R_c, and helium ignites at correspondingly larger M_c, in this case at $m/M \approx 0.3$, cf. Fig. 33.10.].

In Fig. 33.8, the evolution is shown in a lg ϱ–lg T diagram for the shell in which helium is ignited. We see that the shell behaves roughly as predicted in Fig. 33.6 for the centre. When the temperature of helium burning is reached at point A, the core matter heats up. After degeneracy is removed near point B, the core expands and a non-degenerate phase follows with stable helium burning, roughly at the same temperature at which the flash phase had started but at much lower densities. The internal structure of the model after the ignition of helium is indicated in Fig. 33.10.

The calculations by Thomas (1967) were carried out with neutrino rates which turned out to be too high. In calculations for $1.3M_{\odot}$, with more realistic neutrino rates (and composition), the igniting shell was at $m/M = 0.11$, similar to the value in Fig. 33.7, which is based on calculations with the most recent neutrino loss rates. Sweigart and Gross (1978), and more recently Salaris and Cassisi (2005), investigated in detail the dependence of the core mass and the location of the temperature maximum as function of mass, helium content, and metallicity. For

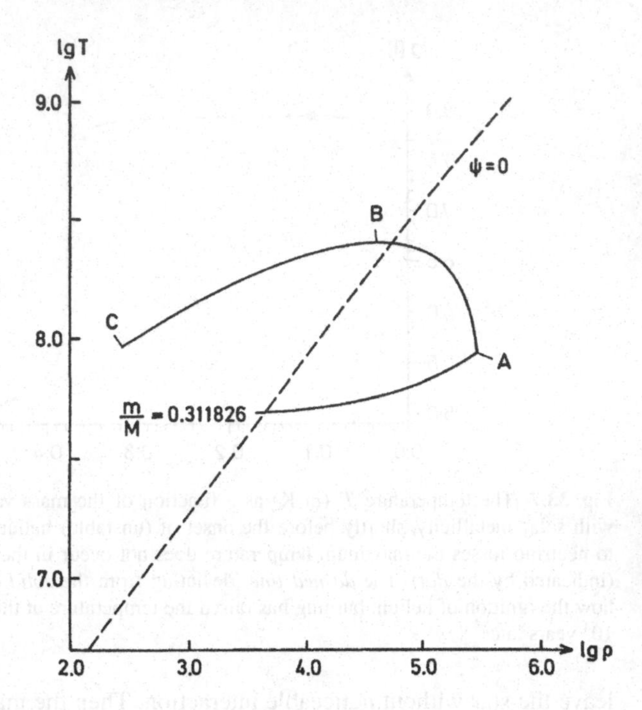

Fig. 33.8 Temperature T (in K) versus density ϱ (in $\mathrm{g\,cm^{-3}}$) for the mass shell at which helium ignites in the $1.3M_\odot$ model. The letters *A–C* refer to the corresponding evolutionary slates in Figs. 33.3 and 33.4. The *dashed line* (degeneracy parameter $\psi = 0$ for $\mu_e = 2$) roughly separates the regimes of degeneracy and non-degeneracy of the electron gas (After Thomas 1967)

stellar masses in the range $0.7 \le M/M_\odot \le 2.2$ helium ignites at $m/M \approx 0.17$ for $M = 0.7M_\odot$, while with increasing total mass the shell of ignition moves closer to the centre.

The changing luminosity provided by the hydrogen shell and helium burning is displayed in Fig. 33.9. The timescale in this figure changes several times due to the different phases of the flash, which starts very slowly over several 10^5 years (not shown), but accelerates dramatically once $\log L_{\mathrm{He}}/L_\odot > 3$. Within a few years $\log L_{\mathrm{He}}/L_\odot > 10$, but drops equally fast after the peak. Around this time, when the layers above the flash location begin to expand, the hydrogen shell basically extinguishes due to the drop in temperature, while the total luminosity remains almost constant. In the following 10^5 years the hydrogen shell reignites and L drops as predicted in Sect. 33.4. The flash and the resulting convection heat up the core such that the lower boundary of helium fusion is moving inwards within approximately one million years. In all hydrostatic calculations this progression is connected with small, secondary helium flashes, as can be seen in the figure. When the star finally settles on the horizontal branch, the core, which is now burning helium under non-degenerate conditions, is already enriched in carbon by about 5 %.

Although the properties of the regions in which the flash occurs can change drastically within a few seconds, it seems as if inertia terms can be neglected even in the most violent phases of the flash. Another open question is, how convection behaves during the rapid evolution of the helium flash and whether the ignition of helium and the flash in a shell proceeds in strict spherical symmetry. Such question can only be answered with 2- and 3-dimensional hydrodynamical calculations.

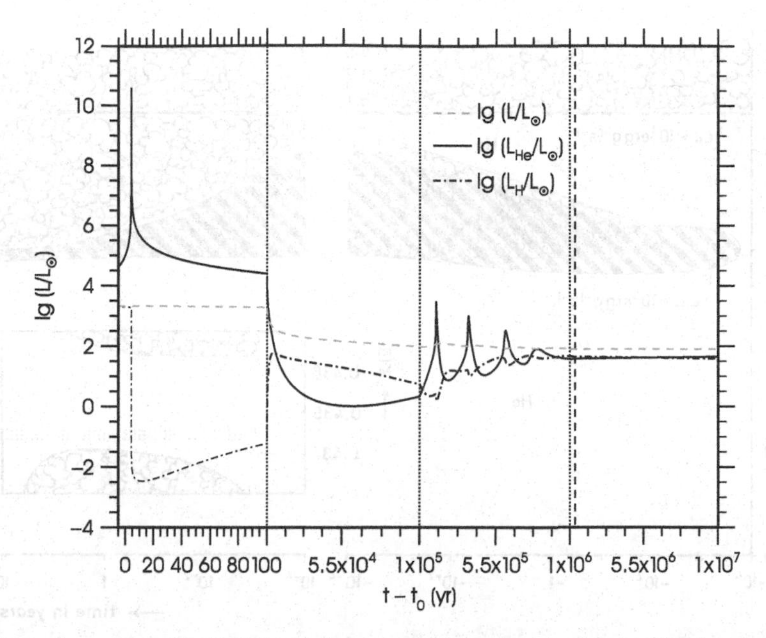

Fig. 33.9 Changes in total (L), hydrogen ($L_{\rm H}$), and helium ($L_{\rm He}$) luminosity with time during the helium flash in the 0.85 $M_{\odot}$ star of Fig. 33.5. $t = 0$ is defined for the moment when $L_{\rm He}/L_{\odot} = 5$, and the zero-age horizontal branch is reached 1.325×10^6 years later. This is defined as the point of minimal total thermal energy and indicated by the *vertical dashed line*. *Vertical dotted lines* delimit ranges of different scale for the time axis

These, however, need such enormous computational resources that so far only parts of stars can be modelled and their evolution be followed for only a short period of a few hours to days. This restricts such simulations to the peak of the helium flash and the inner core of the star. Mocák et al. (2008, 2009) have done such simulations and, apart from some details, confirmed the overall applicability of the 1-dimensional, spherical, and hydrostatic stellar models. However, they seem to predict a much faster heating of the core and the absence of the secondary flashes. This is one of the several unanswered questions connected with the helium flash. Another one is, whether during the flash matter is expelled from the surface.

If helium is ignited off centre, then the burning forms a shell enriched in carbon and oxygen which surrounds a helium sphere. But if the molecular weight decreases in the direction of gravity, the layer is secularly unstable: a mass element pushed down so slowly that it could adjust its pressure and temperature to that of the new surroundings ($DP = 0, DT = 0$, in the terms of Chap. 6) would have a higher density ($D\varrho > 0$, because $D\mu > 0$) and would sink deeper. This corresponds to the "salt finger instability" discussed in Sect. 6.5. In the case discussed here it will cause mixing between the shell in which carbon and oxygen are produced, and the helium region below. The linear stability analysis is rather easy, though it is difficult to follow the instability into the non-linear regime and, for instance, to determine

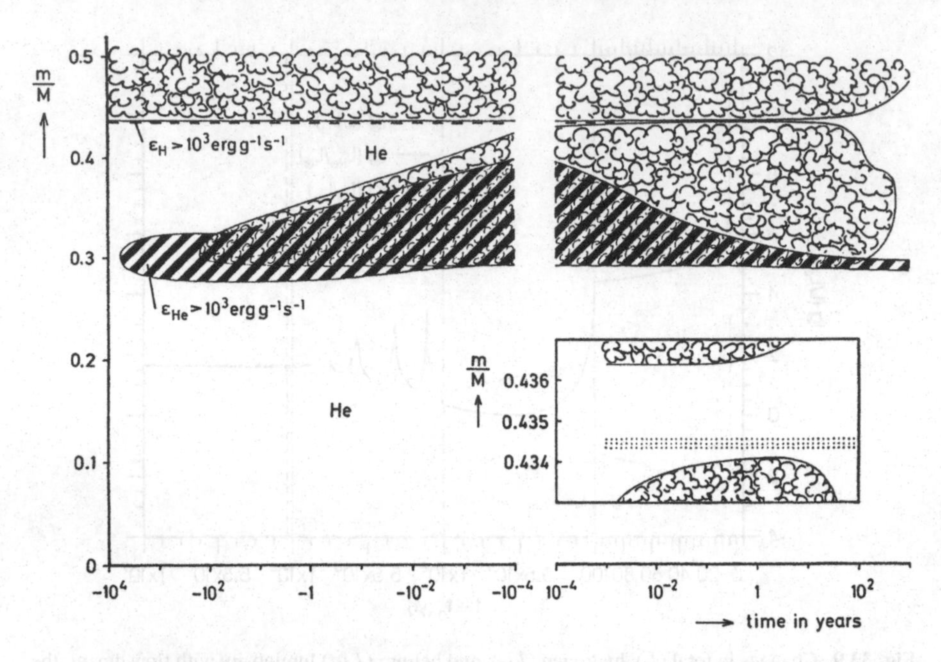

Fig. 33.10 The evolution of the internal structure of a star of $1.3 M_{\odot}$ during the helium flash. The zero point of the abscissa corresponds to the age 7.474×10^9 years of the abscissa of Fig. 33.4. The main regions of nuclear energy release are hatched; the hydrogen-burning shell is, in the mass scale of the ordinate, so narrow that it appears as a *broken line*. It extinguishes at $t \approx 10^{-3}$ years. "Cloudy" areas indicate convection. The close approach of the outer convective envelope and the convective region above the helium-burning shell is shown with a strongly enlarged ordinate in a window at the *lower right*. There the *dotted area* indicates the transition region of the chemical composition left by the (then extinguished) hydrogen-burning shell

the characteristic time for this mixing process. Simple assumptions about the flow pattern suggest that mixing due to the inwardly decreasing molecular weight is slow compared to the nuclear timescale and can therefore be neglected (Kippenhahn et al. 1980a,b). The multidimensional hydrodynamical models by Mocak and co-workers mentioned above indeed show the occurrence of such fingers, which, however, the authors ascribe to Rayleigh-Taylor instabilities. They could be followed for less than 2 days only.

More spectacular mixing than in the case just discussed can occur if the convective shell, forming above the helium-burning shell during the flash, merges with the outer convective layer. Then hydrogen-rich matter will be mixed down to regions with high temperatures where simultaneous helium and hydrogen burning give rise to quite unusual nuclear reactions and chemical compositions. Although the boundaries between the two convective zones come very close to each other, they do not merge usually. This can be seen in the detailed picture on the lower right of Fig. 33.10. But there are situations where such "flash-induced mixing" indeed happens. The first example is stars with zero initial metallicity, so-called Population III stars, and with $M \lesssim 1.0\, M_{\odot}$. In such stars only the *pp* chains can produce helium, and this leads to a different temperature stratification, which allows the penetration of the hydrogen/helium discontinuity by the convective layers above the helium ignition shell (Fujimoto et al. 1990). The result of the flash-induced mixing are surface abundances drastically enhanced in carbon produced by triple-alpha reaction and in nitrogen resulting from proton captures on some of this carbon (Schlattl et al. 2001). The second case where this was encountered is Pop. II stars with extremely thin hydrogen envelopes (of order $10^{-4} M_\odot$), which could be the result of enhanced mass loss on the red giant branch (RGB). Due to the low envelope mass the hydrogen shell is extinguishing and the star leaves the RGB, returning first to hotter temperatures and then entering the white dwarf cooling phase. If on its way across the Hertzsprung-Russell diagram the helium flash sets in (such stars are also called "hot flashers"), convection can penetrate into the envelope to engulf protons into the hot helium-burning regions, which leads to a "CNO flash". As a consequence the surface is enriched both in helium and carbon, and the star resembles, both in composition and its location, stars at the very hot end of the horizontal branch (see Sect. 33.6). For more details we refer the reader to Cassisi et al. (2003).

33.6 Evolution After the Helium Flash

After the violent phase of the helium flash there follows a phase of quiet burning in non-degenerate matter. The transition to this is not particularly well covered by calculations; one of the few exceptions is shown in Figs. 33.5 and 33.9. Most authors prefer to start with models that belong to a later state in which the models already resemble the horizontal-branch stars of globular clusters. These methods and how accurately they reproduce the full calculations carried through the complete flash event can be found in Serenelli and Weiss (2005). One should keep in mind, though, that the comparison is done with spherical symmetric, hydrostatic models. Once multidimensional hydrodynamical models are available for all phases of the helium flash, one will see how accurate the hydrostatic models are themselves.

Although during the flash helium is ignited in a shell, it will also burn in the central region after some time, and the stars can be approximated by models on generalized main sequences (cf. Sect. 23.3). For example, a $0.9M_\odot$ star, having a helium core of $0.45\, M_\odot$ after the flash, corresponds to the generalized main sequence for $q_0 = 0.5$. Then from Fig. 23.5 we expect that the model should lie in the HR diagram near the Hayashi line at a luminosity of about $L \approx 100L_\odot$, appreciably lower than just before the flash. This is also what we had expected from the analytic discussion at the end of Sect. 33.4, and the historical evolutionary track in Fig. 33.3 in fact already pointed downwards in the right direction. The modern calculation of Fig. 33.5 covers the whole post-flash part. When in the subsequent phase q_0 increases with growing M_c, the model should cross over to generalized main sequences of larger q_0, i.e. move to the left with slightly increasing luminosity. This also applies when comparing models of the same core, but different total mass, and therefore different values of q_0. Thus, the analytic discussion and the generalized main sequences already sketch basic properties of the phase following the helium flash, which is identified as the (zero-age) horizontal branch.

Detailed calculations, first carried out by Faulkner (1966) in order to reproduce the horizontal branch of globular clusters, show that the models after the helium flash depend not only on q_0 but also on the chemical composition. He compared models of different mass M in complete equilibrium at the onset of quiet helium burning in a core of $M_c = 0.5 M_\odot$ with a hydrogen-burning shell at the bottom of the envelope. For $M > 0.75\, M_\odot$ (at about solar metallicity) they were close to the Hayashi line, but for a smaller mass, they were located considerably to the left. In order to cover the whole observed horizontal branch with such models for a fixed metallicity, one has to assume that the models differ in mass. In a globular cluster, where all stars have the same age and all stars at the tip of the RGB and on the zero-age horizontal branch (ZAHB) had nearly the same initial mass, the horizontal extent of the horizontal branch provides stringent evidence for different mass loss during the previous phase, either before or during the helium flash. This question still awaits a final answer, but the most likely scenario is the following:

During the slow evolution before the helium flash the stars lose an appreciable, but from star to star different, amount of mass from their surfaces. Then the stars start their evolution after the flash with the same core masses but different envelope masses: those which have lost more mass lie on the left, while those which have lost only little mass lie in the red region (Fig. 33.12).

To some degree, however, the observed horizontal branches reflect the evolution of stars after their appearance on the zero-age branch. When their cores grow owing to shell hydrogen burning, and the helium is consumed in their central part, their evolutionary tracks loop back and forth, populating the horizontal branch. The observed branches are not simply the locus of zero-age models. We will come to the further evolution in Sect. 33.7. Since the horizontal branch crosses the instability strip (see Chap. 41) we can expect pulsating horizontal-branch stars. Indeed there one finds the RR Lyrae variables.

Faulkner's results revealed another important property of zero-age models. If one keeps the total mass constant but decreases their metal content, then the models move to the left of the HR diagram. This helped to understand an observed correlation between horizontal-branch characteristics of different globular clusters and their composition: the concentration of stars on the horizontal branch shifts from left to right with increasing contents of heavier elements. This is usually called the *first parameter* effect for horizontal-branch morphology. Observations point to a *second parameter*, which so far has not been identified undisputedly. There exist pairs of globular clusters of (almost) identical age and the same metallicity, but different numbers of stars on the red and blue part of the horizontal branch. Examples for such famous "twins" are M13 and M3, and NGC 362 and NGC 288. Among the candidates for the second parameter are age, helium content, and the density of stars inside the cluster.

Detailed, full evolutionary calculations confirm all these dependencies. We show in Fig. 33.11 the location of ZAHB models for three different metallicities, ranging

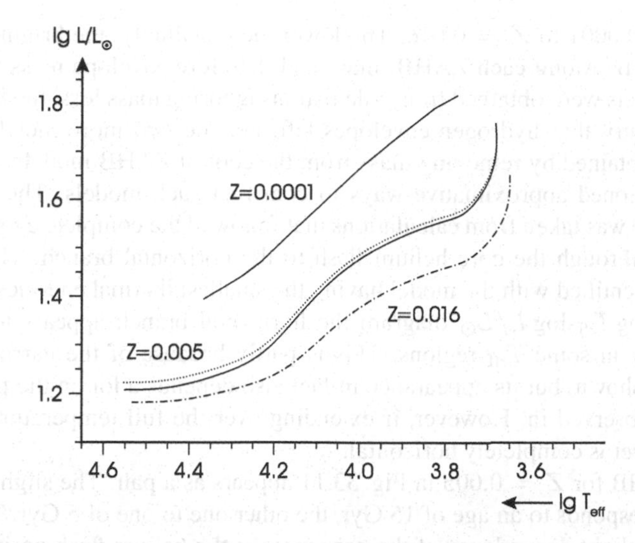

Fig. 33.11 Zero-age horizontal-branch (ZAHB) models for different compositions. The ZAHB brightness increases with decreasing metallicity, which is in this version $Z = 0.016$, 0.005, and 0.0001. The two lines for $Z = 0.005$ correspond to ages of 15 Gyr (*dotted*) and 5 Gyr (*solid*)

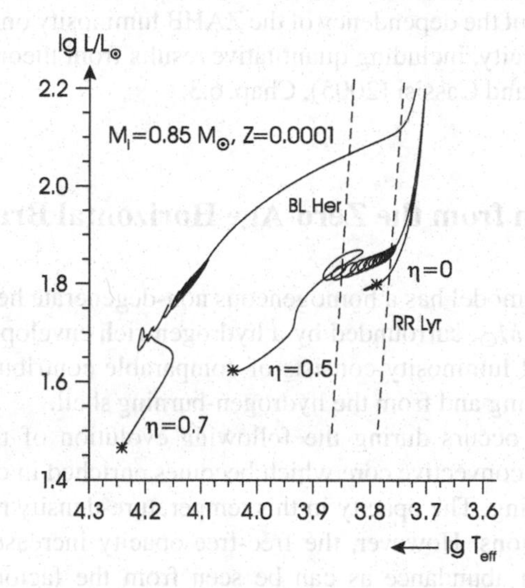

Fig. 33.12 Evolution on the horizontal branch starting at the zero-age position for models with different mass loss rates during the preceding red-giant phase, indicated by the η-parameter in the mass loss formula (9.1). The main-sequence mass was $M_\mathrm{i} = 0.85 M_\odot$ in all cases; at the beginning of the horizontal-branch evolution, the models have 0.85, 0.66, and $0.56 M_\odot$ (*right to left*). The *loops* correspond to the so-called "mini pulses". The *dashed lines* indicate the location of the instability strip, continuing that for classical Cepheids (Fig. 31.4). In this strip, the RR Lyr and BL Her variable stars are found; they are obviously stars either in or after the HB phase

from $Z = 0.0001$ to $Z = 0.016$. The lower the metallicity, the brighter the stars on the ZAHB. Along each ZAHB, mass and therefore envelope mass varies. The coolest models were obtained from calculations ignoring mass loss, the hottest ones have only very thin hydrogen envelopes left. For the two more metal-rich cases they were obtained by removing mass from the coolest ZAHB models. This is one of the mentioned approximative ways to construct such models. The ZAHB for $Z = 0.0001$ was taken from calculations that followed the complete evolution from the ZAMS through the core helium flash to the horizontal branch. The zero-age stage was identified with the model having the smallest thermal energies.

In this $\log T_{\text{eff}}$-$\log L/L_\odot$ diagram the horizontal branch appears to be "horizontal" only in some T_{eff} regions. This is partly because of the narrow range in luminosity shown, but its appearance in fact also depends a lot on the photometric band it is observed in. However, if extending over the full temperature or colour range, it never is completely horizontal.

The ZAHB for $Z = 0.005$ in Fig. 33.11 appears as a pair. The slightly brighter branch corresponds to an age of 15 Gyr, the other one to one of 5 Gyr. This in turn reflects the slight dependence of the core mass at the helium flash on initial mass: the brighter branch originates from stars below $\approx 1.0\,M_\odot$, the dimmer one from $M \approx 1.3\,M_\odot$.

More details about the dependency of the ZAHB luminosity on core mass, helium content, and metallicity, including quantitative results from theoretical models, can be found in Salaris and Cassisi (2005), Chap. 6.3.

33.7 Evolution from the Zero-Age Horizontal Branch

A so-called ZAHB model has a homogeneous non-degenerate helium core of mass $M_c \approx 0.45 - 0.50\,M_\odot$, surrounded by a hydrogen-rich envelope of mass $M_{\text{env}} = M - M_c$. The total luminosity consists of comparable contributions from (quiet) central helium burning and from the hydrogen-burning shell.

A complication occurs during the following evolution of these models. The stars have a central convective core which becomes enriched in carbon and oxygen during helium burning. The opacity in this temperature-density range is dominated by free-free transitions. However, the free-free opacity increases with increasing carbon and oxygen abundance as can be seen from the factor B in (17.5) and (17.6), which depends on the square of the nuclear charge. As a consequence the radiative gradient inside the Schwarzschild boundary grows during core helium burning, and a discontinuity in ∇_{rad} at the edge of the convective core develops. The situation is similar to that in massive stars on the main sequence (see Sect. 30.4.2) where the opacity is governed by electron scattering and decreases with increasing helium abundance. In this case the core is therefore shrinking during the main-sequence evolution. The radiative layers of increasing hydrogen content above the core can locally become convective if some mixing increases the hydrogen content (Fig. 30.3).

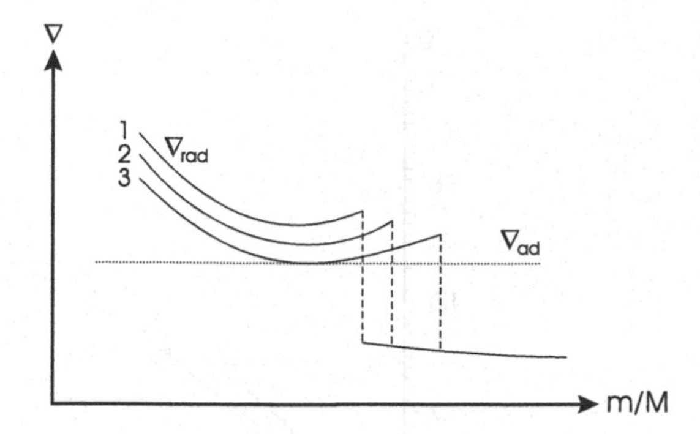

Fig. 33.13 Change in the run of the radiative temperature gradient with time during the evolution on the horizontal branch (see text). Three situations 1–3 for increasing time are sketched (after Salaris and Cassisi 2005)

A similar semiconvective situation is given outside the convective helium-burning core on the horizontal branch. If some mixing, for example, due to overshooting, mixes C/O-enriched material outside the formal Schwarzschild border, the radiative gradient will increase there and convection sets in. Such cores therefore have the tendency to grow, and a jump in $\nabla_{\rm rad}$ at the convective core boundary cannot develop in the early phase of HB evolution. Detailed models show that the radiative gradient then tends to increase and to develop a minimum inside the growing convective core, which, due to the continuing mixing of helium-rich layers and the combined effect of changing physical quantities, at some point begins to become smaller, until it drops to the value of the adiabatic value. This change in the radiative gradient with time is sketched in Fig. 33.13 for three consecutive times. Further mixing at the core border then would lead to a stabilization of an intermediate region, and therefore to the development of a radiative zone inside the core. In real stars one expects therefore mixing up to a composition that leads to a marginally unstable layer with $\nabla_{\rm rad} = \nabla_{\rm ad}$. This partially mixed layer constitutes another case of semiconvection, and was discussed first by Castellani et al. (1971). The development of this situation is indicated by the line labelled "3" in Fig. 33.13, and the real situation in a stellar model calculated with semiconvection is shown in Figs. 33.15 and 33.16.

If a continuous, slow growth of the convective core is inhibited, a strong discontinuity at its edge is developing. As in the case of massive stars, during the further evolution, a sudden mixing between the core and the overlying, helium-rich layers may occur, which leads to a sudden increase in the core's helium content and a loop in the HR diagram. The occurrence of these so-called "breathing pulses" depends a lot on the details of the treatment of convection and of the border of the convective core. The most favourable situation for their occurrence is when the Schwarzschild criterion for convection is used, but they are probably an artefact of the calculations (see the discussion in Salaris and Cassisi 2005). In most of the models we show in Figs. 33.12 and 33.14 they are not present.

Figure 33.14 shows evolutionary tracks for the horizontal-branch evolution of the same initial pre-ZAHB model with $M = 0.6856\,M_\odot$ ($Z = 0.0001$), calculated with either the Schwarzschild criterion for convection (solid line), the Ledoux criterion and semiconvection (dotted line), or with overshooting (dashed line). The

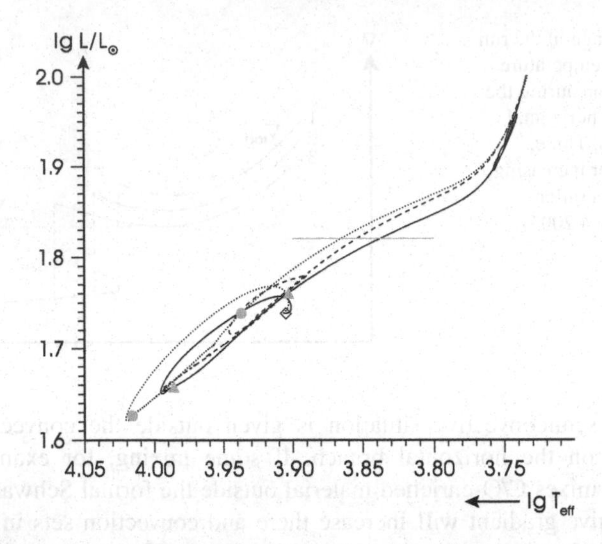

Fig. 33.14 An example for horizontal-branch evolution using different treatments for convection. The evolution begins at a pre-ZAHB position (*diamond symbol*) for a mass of $0.6856\,M_\odot$ ($Z = 0.0001$). The *grey triangles* and *circles* refer to models used for Fig. 33.16. *Solid line*: Schwarzschild criterion; *dashed*: Schwarzschild and overshooting; *dotted*: Ledoux criterion and semiconvection. The *horizontal line* indicates where the interior hydrogen profile of the models of Fig. 33.15 was taken

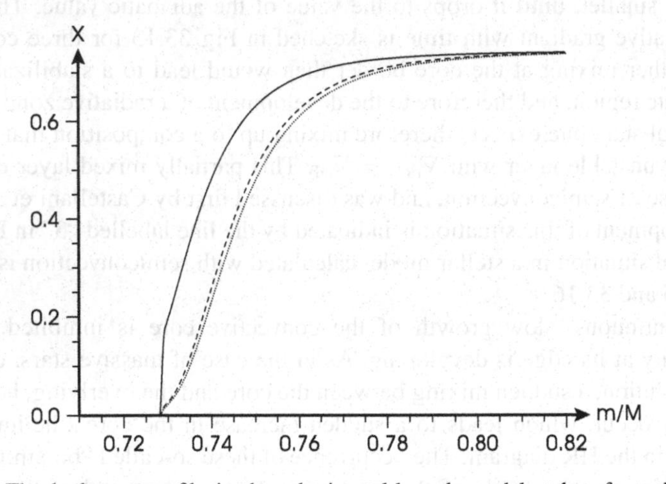

Fig. 33.15 The hydrogen profile in three horizontal-branch models taken from the tracks of Fig. 33.14, taken at approximately the same luminosity. The linestyles refer to the same cases as in the previous figures

overshooting model initially follows the track of the Schwarzschild case, until after the hottest point on the evolution the core is expanding; generally higher luminosities are reached, but also a second loop close to the initial ZAHB position takes place. After this the track with overshooting is slowly approaching the one

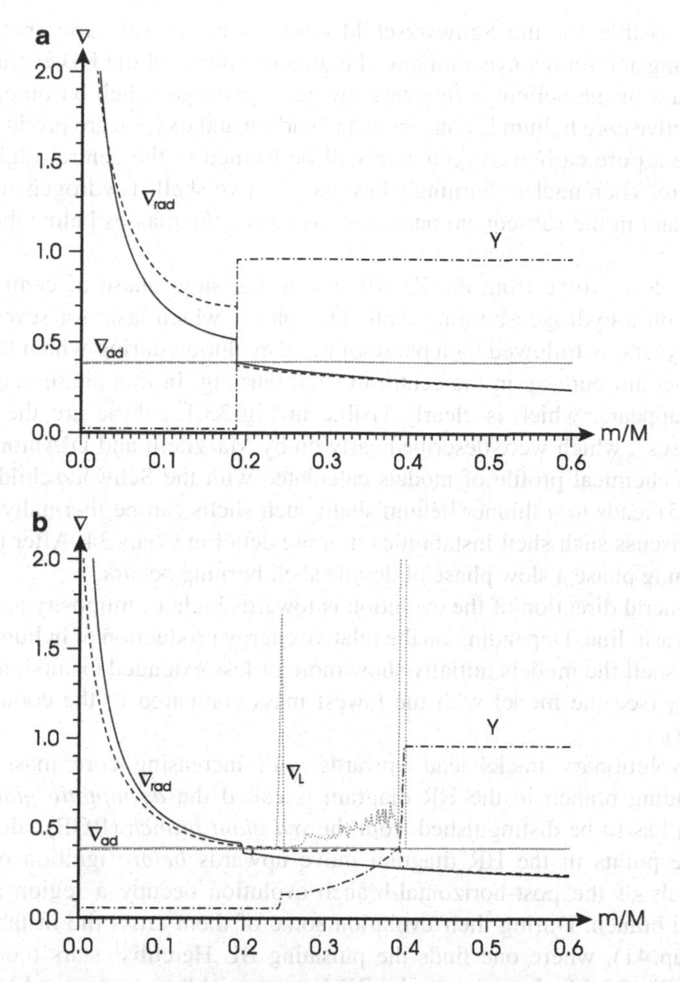

Fig. 33.16 **(a)** The radiative (*black lines*) and adiabatic (*grey lines*) temperature gradients in two models of Fig. 33.14, indicated there by *triangles*, calculated with the Schwarzschild criterion. The *solid line* refers to a state at the beginning and the *dashed line* to the end of horizontal-branch evolution. The *dash-dotted line* shows the helium mass fraction in the latter model. **(b)** The same for the case using the Ledoux criterion and semiconvection. The two models are indicated by *circles* in Fig. 33.14. Additionally, the Ledoux-gradient ∇_L (6.12) is shown as the *grey dotted line*. Notice the fact that ∇_{rad} is almost identical to ∇_{ad} as the result of semiconvective mixing

calculated with semiconvection. In this latter case, only one extended loop takes place in the final phase of approaching the ZAHB (which here can be identified with the hottest point on the track), and then the evolution is proceeding smoothly to higher luminosities and cooler temperatures. In Fig. 33.15 we show the run of hydrogen abundance inside models taken at approximately the same luminosity of $\log L/L_{\odot} = 1.82$. The composition jump at the edge of the convective core is clearly visible for the Schwarzschild case, as is the similarity between the overshooting and semiconvection model in this late phase of the HB evolution.

The mass of the helium core grows owing to hydrogen-shell burning, while in the convective core helium is consumed and carbon and oxygen are produced. After some time a pure carbon-oxygen core will be formed in the central region of the helium core. Then nuclear burning takes place in two shells (hydrogen and helium burning), and in the subsequent phases of evolution, the masses below these shells will grow.

The models evolve from the ZAHB first in the slow phase of central helium burning with a hydrogen-burning shell. This phase, which lasts for several 10^7 to up to 10^8 years, is followed by a phase of rapid evolution during which the models go from helium burning in the centre to shell burning. In this phase another kind of loops appears, which is clearly visible in Fig. 33.12; these are the so-called "mini-pulses", which were described early on by Mazzitelli and D'Antona (1986). The steep chemical profile of models calculated with the Schwarzschild criterion (Fig. 33.15) leads to a thinner helium shell; such shells can be thermally unstable. We will discuss such shell instabilities in more detail in Chap. 34. After this initial shell-burning phase a slow phase of double shell burning occurs.

The general direction of the evolution is towards higher luminosity and a return to the Hayashi line. Depending on the relative energy production of helium core and hydrogen shell the models initially show more or less extended excursions towards higher T_{eff} (see the model with the lowest mass compared to the coolest one in Fig. 33.12).

The evolutionary tracks lead upwards with increasing core mass, and the corresponding branch in the HR diagram is called the *asymptotic giant branch* (AGB). It has to be distinguished from the *red giant branch* (RGB), along which the image points in the HR diagram move upwards *before* ignition of helium. The models of the post-horizontal-branch evolution occupy a region above the horizontal branch. During their evolution some of them cross the instability strip (see Chap. 41), where one finds the pulsating BL Herculis[1] stars (compare the sketch in Fig. 33.12). In contrast, the RR Lyrae variables are stars, which are still on the horizontal branch, and which are located in the region, where it crosses the instability strip.

[1]BL Her stars belong to the class of *type II Cepheids*, which used to be called collectively W Virginis stars. Nowadays the latter term is used for even brighter stars in the same instability strip, which cross it during an excursion from the AGB.

Part VII
Late Phases of Stellar Evolution

The more advanced the evolution of stars is, the less it is possible to treat it with simple models or even analytic descriptions. Instead, numerical calculations are the only way to follow the evolution. Therefore, in the following chapters, we will rely almost solely, with a few exceptions, on results from models produced in the computer. At the same time, the complications from poorly known physics are becoming more and more important. This is in particular true for the treatment of convection and of mass loss, not to speak of the influence of rotation and additional mixing mechanisms. Unfortunately, the late evolution of stars depend a lot on exactly these physical effects. The following chapters therefore give an overview of our current understanding of stellar evolution after the core helium-burning phase. It is very likely that this will change–hopefully improve–in the future.

Chapter 34
Evolution on the Asymptotic Giant Branch

34.1 Nuclear Shells on the Asymptotic Giant Branch

In stars of low and intermediate mass, i.e. in stars of initial mass $\lesssim 8\,M_\odot$, the phase following the end of core helium burning is of special interest. It is characterized by the presence of two nuclear burning shells around a carbon-oxygen core, of which one–the helium shell–is thermally unstable. Stars in this mass range and phase of the evolution populate the so-called *asymptotic giant branch* (AGB), previously also known as "second-ascent branch". In this chapter we give an overview over the important physical effects which are characteristic for AGB stars. As we will see, the evolution is highly complicated and the numerical models far from being perfect. For more details and a much more thorough discussion, we refer the reader to the review by Herwig (2005) and to the textbook by Habing and Olofsson (2003). The classical review by Iben and Renzini (1983) is still worth being studied, too.

After the end of core helium burning and after the hydrogen shell has burned outwards for some time, the temperature in this shell drops, and hydrogen-shell burning extinguishes. This phase of the evolution is often called the *early AGB* (E-AGB). The layer of transition between the hydrogen-rich envelope and the region of helium stays now at a fixed value of m. In stars above $M \approx 4\,M_\odot$ convection may reach below the H–He discontinuity and mix more ashes of hydrogen shell burning to the surface. This is the *second dredge-up* we already encountered in Chap. 31. But there is still the active helium-burning shell moving to higher values of m and therefore approaching the bottom of the hydrogen-rich envelope. Since helium burning proceeds at a temperature of $\gtrsim 10^8$ K, which is about ten times the temperature of hydrogen ignition, hydrogen burning starts again, and once more there are two shell sources. In this phase, shell burning becomes secularly unstable, resulting in a thermal runaway. This leads to a cyclic phenomenon (reoccurring here within some 10^4–10^5 years) known as *thermal pulses (TP)*. Their general properties will be discussed in Sect. 34.3 in connection with their appearance in intermediate-mass stars where the unstable shells initially are in the deep interior, and the response of the surface is moderate. In the case of low-mass stars, the

R. Kippenhahn et al., *Stellar Structure and Evolution*, Astronomy and Astrophysics Library, DOI 10.1007/978-3-642-30304-3_34, © Springer-Verlag Berlin Heidelberg 2012

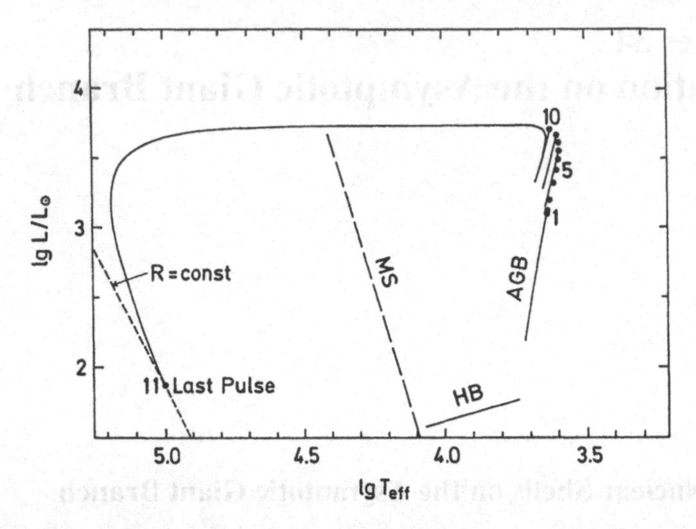

Fig. 34.1 Schematic evolutionary track of a star of $0.6M_\odot$ ($X_\mathrm{H} = 0.749, X_\mathrm{He} = 0.25$) for the phases after central helium burning. The model moves upwards along the asymptotic giant branch (AGB) until thermal pulses occur (indicated by *full circles*). The changes during a pulse are shown only for pulse 9 and pulse 10. Before the last pulse (11), for which only the onset is shown, the track has reached the white-dwarf area of the HR diagram. The main sequence (MS), the horizontal branch (HB), and a line of constant radius in the white-dwarf region are indicated (after Iben and Renzini 1983)

luminosity and the surface temperature can vary appreciably with each pulse. This is the more pronounced the less mass is left above the unstable shells, as we will see in Sect. 34.7. If a thermal pulse occurs in certain critical phases (with neither too much nor too little mass above the shells) the models can even move rapidly through large regions of the HR diagram (Kippenhahn et al. 1968; Schönberner 1979). The evolution displayed in Fig. 34.1, shown as an illustrative example and taken from the review article "Asymptotic Giant Branch Evolution and Beyond" by Iben and Renzini (1983) goes through 11 pulses, the onsets of which are indicated by heavy dots. The variation of the surface values is not very pronounced, since there is enough mass above the nuclear shells to damp the changes caused by the instability. This phase is also called the *thermally pulsing AGB* (TP-AGB) to discriminate it from the E-AGB.

The pulses are more or less an envelope phenomenon and are of no influence on the core. The inner part of the CO-core resembles more and more a white dwarf. Only the hydrogen-rich envelope, small in mass but thick in radius, at first gives the star the appearance of a red giant. After the envelope mass has dropped below, say, one per cent, the envelope starts to shrink. With decreasing envelope mass the star moves typically within a few thousand to 10^4 years to the left of the main sequence (see Fig. 34.1). This is the *post-AGB* phase. Then shell burning extinguishes and the star becomes a white dwarf. In the case shown, the star experiences a final thermal pulse (11), which will lead to large excursion in the HRD. This will be discussed further in Sect. 34.9.

It is clear that the mass in the envelope is diminished by two effects: the hydrogen burning at the bottom and mass loss from the surface. Therefore the stage at which the star leaves the asymptotic branch, turning to the left, is sensitive also to the amount of mass loss in the red giant phase. This influences the mass of the final white dwarf (cf. Sect. 35.2) and limits the number of thermal pulses (see Sect. 34.6).

34.2 Shell Sources and Their Stability

Stars on the AGB are the first to have more than one nuclear shell. Their productivity may change considerably and even go to zero for some time. Neighbouring shell sources can influence each other, since each type of burning requires a separate range of temperature. For example, if a helium shell source operating at roughly 2×10^8 K approaches a hydrogen-rich layer, we can expect an enormous increase of hydrogen burning, which usually proceeds at $T \lesssim 3 \times 10^7$ K. It is also clear that different shell sources will generally move with different "velocities" $\dot{m}_i$ through the mass, unless their contributions L_i to the total luminosity are in certain ratios. If X_i denotes the mass concentration of the reacting element ahead of the shell source, and q_i the energy released by the fusion of one unit of mass, then $\dot{m}_i = L_i/(q_i X_i)$. For example, on the AGB, the relative motion of the hydrogen and helium shell sources through the mass is given by the ratio

$$\frac{\dot{m}_{\rm H}}{\dot{m}_{\rm He}} = \frac{L_{\rm H}}{L_{\rm He}} \frac{q_{\rm He}}{q_{\rm H}} \frac{X_{\rm He}}{X_{\rm H}} . \tag{34.1}$$

This gives a stationary situation with roughly equal velocities only if $L_{\rm H} \approx 7 L_{\rm He}$, since typically $X_{\rm H} \approx 0.7$, $X_{\rm He} \approx 1$, and $q_{\rm H}/q_{\rm He} \approx 10$. Otherwise the two shell sources approach each other or the inner one falls behind.

Shell-source models for several evolutionary phases can be approximated well by solutions obtained by assuming complete equilibrium. While burning outwards, a shell source has the tendency to concentrate the reactions over steadily decreasing mass ranges. One then has to deal with rather short *local* nuclear time scales, defined as those time intervals in which the burning shifts the very steep chemical profile over a range comparable to its own extension. This would require computations of tens of thousands models with very short time steps, if it were not for the influence of mass loss (see Sect. 34.6).

All changes become much more rapid, and the assumption of complete equilibrium certainly has to be dropped if the shell source is thermally unstable. The reasons for such instabilities will be made plausible by considering a very simple model for the shell source and its perturbation. The procedure is completely analogous to that used in Sect. 25.3.5 for the stability of a central nuclear burning. The only difference between the two cases is that the burning regions are geometrically different and the density reacts differently to an expansion.

Let us compare the two cases of a central burning and a shell-source burning in Fig. 34.2. In the central case, the mass of the burning region is $m \sim \varrho r^3$, and an

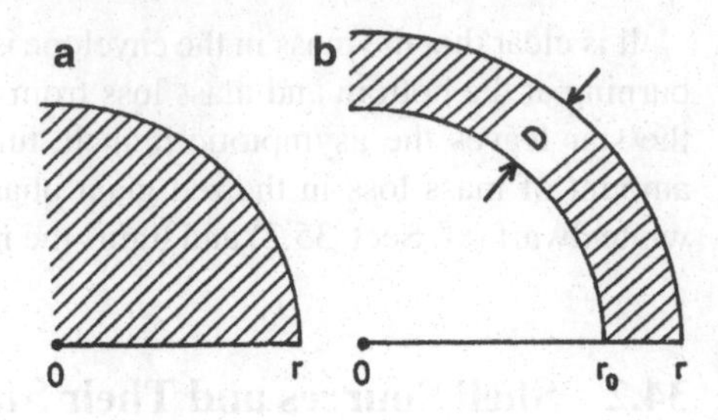

Fig. 34.2 The main region of nuclear energy production (*hatched*) in the cases of (**a**) central burning and (**b**) shell source burning

expansion $dr > 0$ with $dm = 0$ requires a relative change of the density [compare with (25.25)]

$$\frac{d\varrho}{\varrho} = -3\frac{dr}{r}. \tag{34.2}$$

In the case of a shell source of thickness D, we write the upper boundary of the burning region as $r = r_0 + D$ (cf. Fig. 34.2b). For relatively small D the mass in the burning shell is $m \sim \varrho r_0^2 D$. If the burning region expands with roughly $r_0 =$ constant as a reaction to an energy perturbation, we have $dr = dD$, and the condition $dm = 0$ now leads to

$$\frac{d\varrho}{\varrho} = -\frac{dD}{D} = -\frac{r}{D}\frac{dr}{r}. \tag{34.3}$$

We now assume that the mass outside $r_0 + D$ expands or contracts homologously. Then for the pressure in the shell we can use the relation $dP/P = -4dr/r$ as in (25.25). When comparing (34.3) with (34.2) we see that we only have to replace the factor 3 by the factor r/D when going from the central case to that of a shell source. This can be done directly in expression (25.29) for the gravothermal heat capacity c^*. For simplicity we neglect the perturbation of the flux dl_s and have from (25.30)

$$c^*\frac{dT}{dt} = d\varepsilon\,; \quad c^* = c_P\left(1 - \nabla_{\rm ad}\frac{4\delta}{4\alpha - r/D}\right). \tag{34.4}$$

(Note that the time derivative dT/dt represents a differential perturbation; it could be replaced by $d(dT/dt)$ since $T = T_0 + dt$ with time-independent T_0.) If c^* is positive, then the shell source is unstable, since an additional energy input ($d\varepsilon > 0$) leads to higher T and further increased burning.

We first recover the well-known flash instability in the case of strong degeneracy of the electron gas with $\delta \to 0$. Indeed we have seen in Chap. 33 that the helium flash occurs in a shell rather than in the centre if the central part is cooled by neutrino emission.

In addition, (34.4) shows that there is a new instability which can occur even for an ideal monatomic gas ($\alpha = \delta = 1$, $\nabla_{\rm ad} = 2/5$) and which has no counterpart in the case of central burning. It depends only on the geometrical thickness D of the shell source. If D/r is small enough (in our simple representation smaller than 1/4),

c^* is positive and the shell source is secularly unstable. This instability of a shell source is called *pulse instability* for reasons which will become obvious very soon.

It is amazing that such a simple geometrical property can cause a thermal instability, though it becomes more plausible if we consider the change of the pressure in the shell source as a hydrostatic reaction to the lifting of the layers above (for which we simply assume homology). Suppose that the shell tries to get rid of the perturbation energy by expansion. A substantial relative increase of the thickness $dD/D > 0$ gives the same absolute value for the relative decrease of the density $d\varrho/\varrho < 0$, but only a very small relative increase dr/r, if $D/r \ll 1$ [cf. (34.3) and Fig. 34.2b]. This means that the layers above are scarcely lifted, so that their weight remains about constant and hydrostatic equilibrium requires $dP/P \approx 0$. In fact with the homology relation $dP/P = -4dr/r$ and (34.3) we find the connection between dP and $d\varrho$ to be

$$\frac{dP}{P} = 4\frac{D}{r}\frac{d\varrho}{\varrho}. \tag{34.5}$$

Considering the equation of state

$$\frac{d\varrho}{\varrho} = \alpha\frac{dP}{P} - \delta\frac{dT}{T}, \tag{34.6}$$

we see that expansion ($d\varrho/\varrho < 0$) necessarily leads to an *increase* of the temperature ($dT/T > 0$), since $dP/P \rightarrow 0$ for $D/r \rightarrow 0$:

$$\frac{d\varrho}{\varrho} = -\delta\frac{dT}{T}. \tag{34.7}$$

Therefore the expansion of a thin shell source does not stabilize it, but rather enforces the liberation of energy by heating. This means that the shell source reacts just as if the equation of state were $\varrho \sim 1/T$, which, of course, gives instability [cf. (34.4) with $\alpha = 0$ and $\delta = 1$].

While the foregoing discussion provides the main points correctly, it can easily be completed by also considering the perturbation of the local luminosity. Then some of the surplus energy can flow away, and instability requires, in addition, that the temperature sensitivity of the burning exceeds a certain limit, which is usually fulfilled. The eigenvalue analysis of such stellar models has shown that they are indeed thermally unstable and that the unstable modes are complex (Härm and Schwarzschild 1972).

The pulse instability was first found (Schwarzschild and Härm 1965) for a helium shell source in calculations for a $1M_\odot$ star. The same type of instability was encountered independently in a model for $5M_\odot$ during the two-shell phase, and here it turned out that the instability leads to nearly periodic relaxation oscillations, which were called *thermal pulses,* as described below (Weigert 1966). They are now known to be a genuine property of those low- and intermediate-mass stars, which are massive enough to ignite helium and evolve into the double-shell burning phase of the AGB.

A unified scheme for the stability of shell sources has been developed by Yoon et al. (2004). It includes the present case of geometrically thin shells as well as the flash instability of Sect. 33.4 and demonstrates that shells are more stable, if they are geometrically thick, non-degenerate, or hotter.

34.3 Thermal Pulses of a Shell Source

Thermal pulses occur in models containing one or more shell sources, and in stars of different masses and compositions. We start by describing their properties according to the calculation of the first six pulses in a 5 $M_\odot$ model, found for the first time in a star in this mass range by Weigert (1966). Although the physical details and numerical treatment of the models have changed a lot since then, the basic picture of thermal pulses is still the same. The instability occurs in the helium shell source after it has reached $m/M \approx 0.1597$. It then contributes only a little to the surface luminosity L, which is almost completely supplied by the nearby hydrogen shell source located at $m/M \approx 0.1603$.

The instability results immediately in a thermal runaway: the shell source reacts to the surplus energy with an increase in T, which enhances the release of nuclear energy, etc. The increase of T is connected with an expansion according to (34.7). This can be seen from Fig. 34.3a, b which give T and ϱ at maximum $\varepsilon_{\mathrm{He}}$ in the unstable shell source as functions of time (Note that the thermal runaway in a *flash* instability would proceed with ϱ = constant.). Since helium burning has an extreme temperature sensitivity, the increase of T strongly enhances the productivity L_{He} of the shell source, in later pulses even to many times the surface value L. But most of this energy is used up by expansion of the layers above, and this expansion reduces considerably the temperature in the hydrogen shell source, such that L_{H} decreases significantly. After starting rather slowly the thermal runaway accelerates more and more until reaching a sharp peak within a few years. The helium shell source is now widely expanded and is therefore no longer unstable. The whole region then starts to contract again, which heats up the hydrogen shell source so that it regains its large productivity. Within a time of a few 10^3 years the whole region has asymptotically recovered its original overall structure, the helium shell source becomes unstable again and the next pulse starts. Figure 34.3 shows that the amplitude of the pulses and the time between consecutive pulses grows (in these calculations from 3,200 to 4,300 years). The reason for these changes is that the chemical composition around the shells changes considerably from pulse to pulse. Later calculations (for an early review, see Iben and Renzini 1983; for more recent results Wagenhuber 1996) showed that a nearly periodic behaviour is usually reached after roughly 20 pulses. The amplitude of a pulse has then become so large that during the maximum L_{He} exceeds L by orders of magnitude. The changes of the chemical composition still provide a small deviation from periodicity. Otherwise we would expect strictly periodic relaxation oscillations, i.e. the solution would have reached a limit cycle.

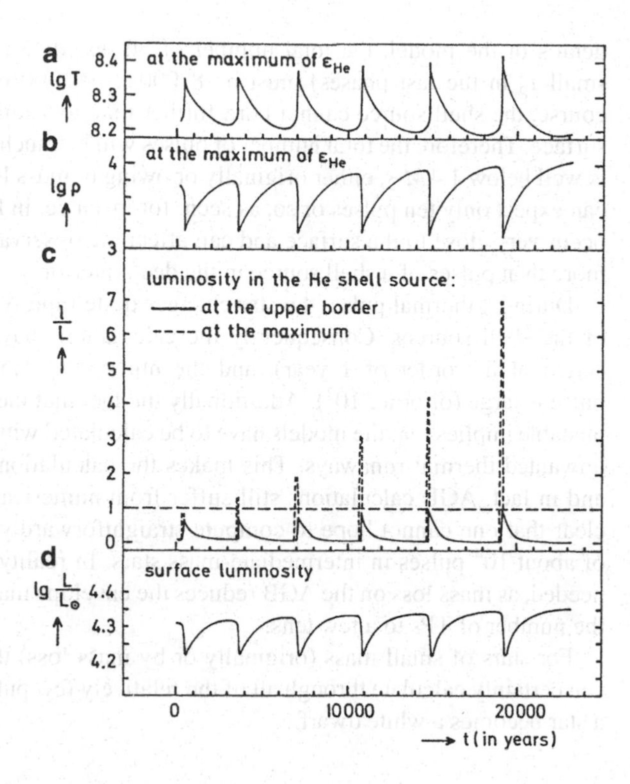

Fig. 34.3 Thermal pulses of the helium shell source in a $5M_\odot$ star after central helium burning. For the first six pulses, some characteristic functions are plotted against time from the onset of the first pulse. T is in K, ϱ in g cm^{-3} (After Weigert 1966)

The surface luminosity (Fig. 34.3d) drops in each pulse by typically $\Delta \lg L \approx 0.1 \ldots 0.2$ for models with rather massive outer envelopes. The visible reaction of the surface is much more pronounced if the pulses occur in a shell source close to the surface. Such models can move quite spectacularly through the HR diagram (compare with Sect. 34.9).

The properties of the thermal pulses depend on the type of star in which they occur. The cycle time τ_p (between the peaks of two consecutive pulses) becomes smaller with increasing mass M_c of the degenerate CO-core inside the helium shell source. From a large sequence of calculations Paczyński (1975) derived the following rough relation:

$$\lg\left(\frac{\tau_p}{1\text{year}}\right) \approx 3.05 + 4.50\left(1 - \frac{M_c}{M_\odot}\right). \tag{34.8}$$

For $M_c \approx 0.5M_\odot$ the cycle time is of the order of 10^5 years, while near the limit mass $M_c \approx 1.4M_\odot$ it would be of the order of 10 years only. We now consider the number of pulses that can occur until M_c has reached 1.4 $M_\odot$. Suppose that the hydrogen shell source moves outwards by Δm per cycle time and produces most of the energy $L\tau_p$. Although $L \sim M_c$ (cf. Sect. 34.4), Δm decreases strongly with growing M_c owing to the decrease of τ_p. One can estimate that, depending on the details of the model, the total number of pulses (determined mainly by the very small τ_p in the last phases) must be $8,000\ldots 10,000$ before $M_c \approx 1.4M_\odot$. Of course, the shell source cannot burn further than to within a few $10^{-3}M$ from the surface. Therefore the total number of pulses will be much smaller if the stellar mass is well below $1.4M_\odot$, either originally or owing to mass loss. In low-mass stars one can expect only ten pulses or so, as seen, for instance, in Fig. 34.1. These, however, occur very close to the surface and can affect the observable values certainly much more than pulses of a shell source in the deep interior.

During a thermal pulse, the star changes quite rapidly, particularly in the layers of the shell sources. Consequently the calculations have to use short time steps (often of the order of 1 year), and the number of models to be computed per pulse is large (or order 10^3). Additionally the fact that the helium shell is thermally unstable implies that the models have to be calculated with high precision to prevent unwanted thermal runaways. This makes the calculations even more challenging, and in fact, AGB calculations still suffer from numerical problems. It is therefore clear that one cannot hope to compute straightforwardly through the whole phase of about 10^4 pulses in intermediate-mass stars. In reality this is–fortunately–never needed, as mass loss on the AGB reduces the envelope mass quickly enough to limit the number of TPs to a few tens.

For stars of small mass (originally or by mass loss) the situation is better. One can certainly calculate through all of the relatively few pulses that occur before such a star becomes a white dwarf.

34.4 The Core-Mass-Luminosity Relation for Large Core Masses

Since the direct computation of TP-AGB models is so difficult, one may try to suppress the pulses artificially by neglecting the time-dependent terms (ε_g) in the energy equation and computing models in complete equilibrium. This gives (hopefully) an average evolution which might suffice in order to describe the evolution of the central core, and therefore of the final fate of the star.

An alternative approach are the so-called *synthetic AGB models* (see, e.g., Renzini 1981, or Marigo et al. 1996), where the global properties of AGB stars are followed using fitting functions such as (34.8) for the pulse durations. Extensive analytical fitting functions were derived by Wagenhuber and Groenewegen (1998) for quantities such as the luminosity, the pulse duration and interpulse time, the core mass at the first pulse, and many more. These functions were derived by fitting them to numerical models and are valid for various masses and chemical compositions, but simpler versions can be derived analytically. This is particularly true for the important core-mass-luminosity relation on the AGB.

We have seen that medium-mass stars, after central helium burning, develop a degenerate CO-core which is separated from the hydrogen-rich envelope by a thin helium layer. At its bottom there is helium-shell burning, which contributes only, say, 10 % to L. Most of the luminosity is produced in a hydrogen shell source at the bottom of the envelope. It is not too bad an approximation if we simply assume $L \approx L_H$, the hydrogen luminosity generated above a condensed core of mass M_c and radius R_c. We also have seen that L increases with increasing M_c (giving the upwards motion along the asymptotic branch) and here face the same situation as for low-mass stars on the ascending giant branch. One can again derive the dependence of the properties of the shell on M_c and R_c by homology relations as in Sect. 33.2, assuming the simple power laws (33.2) for κ and ε. But since we are dealing with rather massive cores and high temperatures here, the radiation pressure cannot be neglected. We therefore have to replace (33.3) by

$$P = \frac{\Re}{\mu}\varrho T + \frac{a}{3}T^4 = \frac{1}{\beta}\frac{\Re}{\mu}\varrho T. \tag{34.9}$$

If again we write in the neighbourhood of given P and T the equation of state as a power law, $\varrho \sim P^\alpha T^{-\delta}$, we know from (13.7) that $\alpha = 1/\beta, \delta = (4-3\beta)/\beta$. Therefore we have as equation of state

$$P \sim \varrho^\beta T^{4-3\beta}. \tag{34.10}$$

As in (33.4)–(33.7), we write the quantities ϱ, T, P, and l in the shell as powers of M_c and R_c. By the same procedure as in Sect. 33.2 we can derive equations for the exponents. For the sake of simplicity we restrict ourselves to the case $a = b = 0$ and obtain, instead of (33.22),

$$\begin{aligned}
\varphi_1 &= \frac{4-\nu}{N}, \quad \varphi_2 = \frac{\nu - 12 + 6\beta}{N},\\
\psi_1 &= \frac{1+n}{N}, \quad \psi_2 = \frac{2\beta - n - 3}{N},\\
\tau_1 &= \beta\varphi_1 + (4-3\beta)\varphi_1, \quad \tau_2 = \beta\varphi_2 + (4-3\beta)\varphi_2,\\
\sigma_1 &= \frac{4n+\nu}{N}, \quad \sigma_2 = \frac{3-\nu-3n}{N}\beta
\end{aligned} \tag{34.11}$$

with

$$N = (4-3\beta)(1+n) + (1-\beta)(\nu - 4). \tag{34.12}$$

For $\beta = 1$ the relations (34.11) and (34.12) agree with (33.22) and (33.23) for $a = b = 0$.

With increasing core mass, β in the shell must decrease strongly, as can be seen from the following considerations. From (34.9) and (34.10) we have

$$\beta \sim \frac{\varrho T}{P} \sim \varrho^{1-\beta} T^{-3(1-\beta)}. \tag{34.13}$$

If we here replace ϱ, T by (33.4) and (33.5), then the dependence of β on M_c, R_c is given by

$$\frac{d\ln\beta}{d\ln M_c} = (1-\beta)\left[(\varphi_1 - 3\psi_1) + (\varphi_2 - 3\psi_2)\frac{d\ln R_c}{d\ln M_c}\right]. \tag{34.14}$$

One may start from an initial model that has been computed by solving the stellar structure equations numerically. This gives initial values for M_c, R_c, L, and β. Starting from these initial values we want to integrate (34.14). For simplicity, let us take for the derivative on the right-hand side of (34.14) Chandrasekhar's mass-radius relation of white dwarfs, and for the exponents in the energy generation $n=2,\ \nu=14$. The result of such an integration is shown by a dotted line in Fig. 33.2. In the same way, (33.27) can be integrated with σ_1, σ_2 from (34.11) and $\beta(M_c)$ as derived from the solution of (34.14). This gives the solid curve in Fig. 33.2. In spite of all approximations used, the integrated curves illustrate clearly the essential points.

For small core masses, $\beta \approx 1$ and the relation (33.25) holds, giving a steep increase of L with M_c $[L \sim M_c^7$ after (33.26)]. For larger M_c, radiation pressure becomes more and more important and β decreases. This gives a much smaller slope of the $L(M_c)$ curve. Indeed in the limit $\beta = 0$ (34.11) gives $\sigma_1 = 1, \sigma_2 = 0$, independent of n and ν:

$$L \sim M_c. \tag{34.15}$$

The L-M_c relation has become extremely simple, and we do not have to worry about the correct R_c-M_c relation. Indeed from numerical models Paczyński (1970) derived

$$\frac{L}{L_\odot} = 5.92 \times 10^4 \left(\frac{M_c}{M_\odot} - 0.52\right) \tag{34.16}$$

as an interpolation formula for sufficiently large M_c. The corresponding formula by Wagenhuber and Groenewegen (1998) contains a linear term as well, but has additional correction terms which improve the fit also for low core masses and which take into account different metallicities of the models. It is therefore much more complicated than (34.16).

34.5 Nucleosynthesis on the AGB

We now turn to the change of the chemical composition by a combination of burning and convection. Figure 34.4 shows (with expanded scales) m against t during the peak of two pulses of the models by Weigert (1966). The high fluxes near the maximum of helium burning create a short-lived, pulse-driven intershell convection zone (ISCZ), which, in the later pulses of this calculation, comes very close to the H–He discontinuity. For a short time, almost the entire matter between the two shells

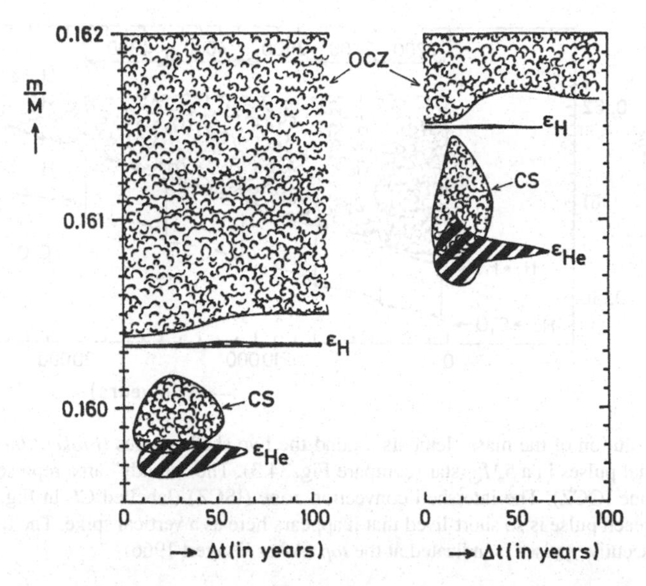

Fig. 34.4 Evolution of the mass shells around the two shell sources in a $5M_\odot$ star near the maxima of the first and sixth thermal pulses of the helium shell source (compare Fig. 34.3). The mass variable m is plotted against time, starting from an arbitrary zero point. Note the strongly expanded scales on both axes. *Cloudy* areas indicate the intershell convection zone (labelled here as CS) and the outer convective zone (OCZ); *hatched* areas show the regions of strongest nuclear energy production ($\varepsilon > 3 \times 10^7$ erg g^{-1} s^{-1}) (After Weigert 1966)

is mixed into the helium-burning shell, the products of which are spread over the intershell region. The outer convection zone (OCZ), which extends to the surface, can be seen to reach down nearly to the hydrogen shell source. The lower boundary of the OCZ moves during each pulse at first somewhat outwards, and then back again (compare also with Fig. 34.5, where the t axis is more compressed). Depending on mass, composition, and in particular on the assumption of mixing by convection or rotation, the ISCZ may even reach beyond the H–He discontinuity and dredge hydrogen into the intershell region but also enrich the outer layers with carbon. Similarly, the lower border of the OCZ can descend beyond the former location of the H–He discontinuity into the intershell region. Also in this case, hydrogen-rich material is transported downwards, while intershell material is dredged up by the OCZ and distributed over the whole outer envelope. This event is called the *third dredge-up*, and its reality is witnessed by the existence of carbon stars, stars on the AGB, in which the ratio of carbon-to-oxygen abundance is C/O > 1. In models with no mixing processes apart from convection according to the Schwarzschild criterion, the third dredge-up occurs only in stars of low mass and very low metallicity. This is not in agreement with observations, which found modifications of the surface composition also in more metal-rich and more massive stars. They can only be explained by the third dredge-up, and this requires additional mixing processes in the models, which could be overshooting or mixing induced by rotation. In Fig. 34.6

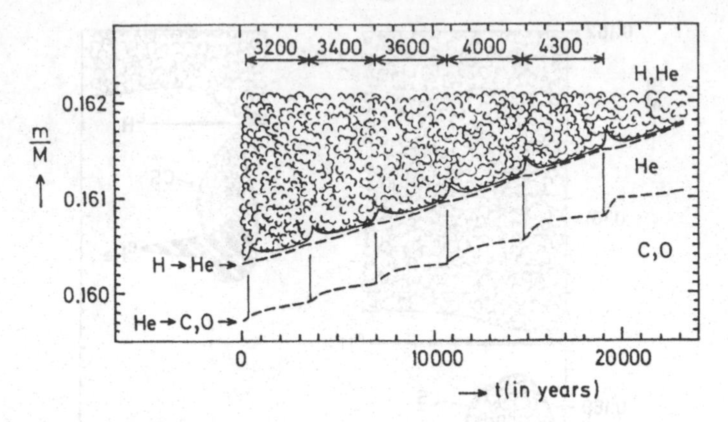

Fig. 34.5 Evolution of the mass elements around the two shell sources (*broken lines*) during the first six thermal pulses in a $5M_{\odot}$ star (compare Fig. 34.3). The "cloudy" area represents the outer convective zone (OCZ). The intershell convection zone (ISCZ) (labelled CS in Fig. 34.4) at the maximum of each pulse is so short-lived that it appears here as a vertical spike. The time (in years) between consecutive pulses is indicated at the *top* (After Weigert 1966)

we show a sketch of the sequence of mixing episodes leading to dredge-up and the formation of a so-called ^{13}C-pocket. In some models this additional mixing is added ad hoc with an efficiency tuned to reproduce the observations. The third dredge-up is one of the major problems of stellar evolution theory.

The mixing between layers containing protons and those burning helium at high temperatures is the beginning of interesting nucleosynthesis in stars on the AGB. Helium burning transforms ^{4}He into ^{12}C and ^{16}O, and the hydrogen shell source converts ^{16}O and ^{12}C into ^{14}N, which is left behind when the shell burns outwards between two pulses (between, e.g. t_3 and t_5 in Fig. 34.6). The ISCZ of the next pulse sweeps these ^{14}N nuclei down into the helium shell source where they are burned in the chain ^{14}N (α,γ) ^{18}F $(\beta^+\nu)^{18}$O $(\alpha,\gamma)^{22}$Ne. During a pulse in fairly massive stars, and therefore within the pulse-driven convective zone, the helium shell source attains a temperature so high that ^{22}Ne is also burned in the reaction ^{22}Ne $(\alpha,n)^{25}$Mg. This can provide a neutron source sufficiently strong to build up elements beyond the iron peak in the s-process (i.e. with neutron captures being *slow* compared with beta decay; see Sect. 18.7; Iben 1975; Truran and Iben 1977).

In other cases a corresponding neutron source may be provided by ^{13}C nuclei, which are burned via the chain ^{12}C (p,γ) ^{13}N $(\beta^+\nu)$ ^{13}C (α,n) ^{16}O in the helium shell. This happens, in contrast to the neon neutron source, between pulses and in a radiative environment (beginning at t_3 and t_6 in Fig. 34.6). For this neutron source to operate it is necessary to bring a sufficient amount of ^{13}C into the helium shell, which is achieved by mixing hydrogen-rich material from the envelope into the ^{12}C-rich region during the pulse phase in which the hydrogen-burning shell is extinguished (Fig. 34.6; $t_2 - t_3$). The protons are then captured by the ^{12}C nuclei to form ^{13}C. The region of high ^{13}C abundance is known as the *^{13}C-pocket*, and will provide the neutron source later in the pulse cycle. According to theoretical models

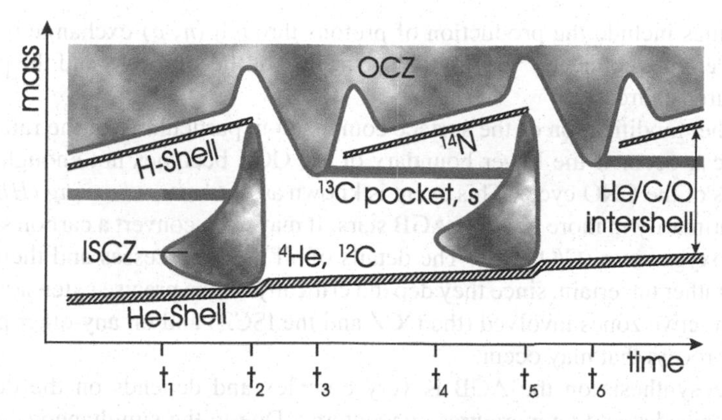

Fig. 34.6 Schematic sketch of the mixing episodes during thermal pulses, following similar representations in the reviews by Busso et al. (1999; Fig. 5) and Herwig (2005; Fig. 3). In contrast to Fig. 34.5, dredge-up is occurring here due to suitable assumptions about mixing processes, for example, due to the inclusion of overshooting. Shown is the region between the two shell sources and the bottom of the outer convection zone (OCZ). The mass scale is of order a few hundredths of a solar mass. At time t_1 the thermal pulse of the He-shell starts and triggers the intershell convection zone (ISCZ), which grows in mass and may reach the H-shell location at time t_2. The H-shell, however, has extinguished at this time due to the radial expansion of the intershell. After the pulse terminates, the OCZ can extend deeper than before (at t_3) and mixes both protons into the intershell as well as carbon, produced in the He-shell and transported upwards by the ISCZ, into the envelope. Upon contraction of the intershell, the OCZ recedes, the H-shell reignites, and the proton-enriched intershell layers heat to sufficiently high temperatures to allow $^{12}\mathrm{C}(p,\gamma)^{13}\mathrm{C}$ reactions in a radiative environment, forming the ^{13}C-*pocket*, which marks the start of s-process nucleosynthesis (see text; "s" in the figure). At t_4 the next pulse cycle starts, eventually leading to dredge-up of s-process elements and carbon to the surface of the AGB star. The interpulse time $t_4 - t_1$ is of order 10^4 years, the timescale $t_3 - t_1$ (or $t_6 - t_4$) is a few hundred years

this is the preferred neutron source in most AGB stars. Only in stars of higher mass the neon source may act as well. The main source for s-process elements appear to be AGB stars of lower mass. This is in agreement with theoretical models, which predict the third dredge-up to happen more easily in the lower mass range of AGB stars.

Such mixing and nuclear burning episodes may lead to modifications of the surface composition of AGB stars. We already mentioned that they may become enriched in carbon, initially produced in the helium shell, but subsequently mixed by the ISCZ and the OCZ to the surface. If a neutron source is operating and s-process elements are created, they, too, may appear at the photosphere. The detection of ^{99}Tc in the atmosphere of Mira variables (pulsating AGB stars) by Merrill (1952) proved the in situ production of this rare-earth element, as this isotope is unstable with a half-life time of only 211,000 years. Another signature of mixing between the convective envelope and nuclear burning regions is the presence of ^{19}F in AGB stars. It is created by the reaction $^{15}\mathrm{N}(\alpha,\gamma)^{19}\mathrm{F}$, which is taking place in the helium shell. The necessary production of ^{15}N can happen through different paths, one of them being $^{18}\mathrm{O}(p,\alpha)^{15}\mathrm{N}$ (see the neon neutron source, p. 428). Other possibilities include the production of protons through (n, p)-exchange reactions, which therefore connects ^{19}F with the occurrence of the s-process and the presence of a neutron source.

Another modification of the surface composition, particularly of the ratio of ^{12}C to ^{14}N, can occur if the lower boundary of the OCZ becomes hot enough to start reactions of the CNO cycle. This event is known as *hot bottom burning (HBB)*, and occurs primarily in more massive AGB stars. It may even convert a carbon star back into an oxygen star (C/O $<$ 1). The details of all these processes and their results are still rather uncertain, since they depend critically on the precise extensions of the two convective zones involved (the OCZ and the ISCZ) and on any other potential mixing process that may occur.

Nucleosynthesis on the AGB is very complex and depends on the details of mixing episodes and temperatures encountered. Due to the simultaneous presence of protons, α-particles, and possibly neutrons at temperatures of several 10^7 K up to $\approx 2 \times 10^8$ K, elements from C to Al are both produced and destroyed by proton and α captures. The primary production site is the helium shell, which creates C and O. If these elements encounter protons, ^{14}N will be the result. AGB stars can therefore be the source of primary nitrogen (*Primary* elements are produced directly from the basic building blocks, hydrogen and helium. *Secondary* elements, in contrast, are the result of nucleosynthesis of pre-existing heavier elements. An example would be the nitrogen resulting from CNO processes on the main sequence.). If ^{14}N is further exposed to α-particles in the helium shell, ^{18}O and then ^{22}Ne will result. In the more massive AGB stars, temperatures can be high enough for further α-captures, resulting in various Mg isotopes. Proton captures on Ne, Na, and Mg may change the isotope ratios and eventually lead to Al, including ^{26}Al, which has a half-life time of almost a million years. It decays under emission of 1.81 MeV γ-photons, contributing partially to the galactic γ-rays. Isotope ratios may further be modified by neutron capture reactions. ^{14}N, for example, acts as a so-called *neutron poison*, as it very effectively captures neutrons, thereby reducing the neutron flux needed for the s-process.

The detailed analysis of the abundances of these and the s-process elements is a very important way to learn about the internal evolution of AGB stars. An extensive discussion of nucleosynthesis in AGB stars can be found in the review by Lattanzio and Wood (in Habing and Olofsson 2003, p. 24). Chemical yields from AGB stars have been published by Karakas (2010), and Busso et al. (1999) reviewed in particular the s-process in AGB stars.

34.6 Mass Loss on the AGB

There is plenty of observational evidence that AGB stars suffer significant mass loss through stellar winds such that mass loss is, next to the thermal pulses, the second major factor determining the evolution on the AGB. Direct evidence comes from observations of circumstellar envelopes which enshroud luminous AGB stars, and may make them visible only in the infrared. The analysis of winds and circumstellar shells shows that mass loss rates range from $10^{-8}\,M_{\odot}$/year to $10^{-5}\,M_{\odot}$/year and are strongly correlated with increasing luminosity and decreasing effective temperature. The highest rates effectively terminate the AGB evolution by removing the envelope within several thousands of years to a level, where the star leaves the giant region (as in Fig. 34.1). In this phase, the stellar wind is often called a *superwind*, a term coined by Renzini in 1981, to indicate that it is orders of magnitudes higher than the standard Reimers wind (9.1). We will not go into the details of the observations and the physics of AGB winds, but give, as an illustrative example, a fit formula by van Loon et al. (2005) that describes the mass loss rate (in $M_{\odot}$/year) for oxygen stars (C/O < 1) as a function of the star's position in the HRD, and demonstrates the high sensitivity to effective temperature:

$$\log \dot{M}_{\rm AGB} = -5.65 + 1.05 \cdot \log\left(10^{-4}\frac{L}{L_{\odot}}\right) - 6.3 \cdot \log\left(\frac{T_{\rm eff}}{3500K}\right). \tag{34.17}$$

Generally, it is believed that the winds of AGB stars are due to the coupling of the radiation field to dust forming in the outer atmospheres. The formation of dust is favoured by very low temperatures, which are achieved during and due to large-amplitude stellar pulsations (see Chap. 40). Indeed, many AGB stars are known to be pulsating stars of type Mira or semi-regular and long-period pulsators (periods are between 100 and 1,000 days). In addition, temperature variations during TPs also modulate the mass loss. The complex interplay between pulsating envelopes, the chemistry of dust formation, and the interaction with radiation poses an extremely difficult chemistry-radiation-hydrodynamics problem. So far, such models have been successful for carbon-rich atmospheres, and theoretical dust-driven wind models and mass loss rates are available for carbon stars. Again, as an example that is sufficient to make order-of-magnitude estimates for the mass loss rate, we give the fitting formula to theoretical carbon-dust models by Wachter et al. (2002):

$$\log \dot{M}_{\rm AGB} = -4.52 + 2.47 \cdot \log\left(10^{-4}\frac{L}{L_{\odot}}\right) \\ -6.81 \cdot \log\left(\frac{T_{\rm eff}}{2600K}\right) - 1.95 \cdot \log\left(\frac{M}{M_{\odot}}\right). \tag{34.18}$$

Such fitting formulae may be accurate within one or two orders of magnitude. Within this accuracy, the similarity of the dependencies on L and $T_{\rm eff}$ in (34.17) and (34.18) is interesting. More about AGB mass loss can be found in the textbook by Habing and Olofsson (2003).

A more indirect but very convincing fact that demonstrates the importance of mass loss from AGB stars comes from the *initial-final-mass relation*, pioneered by Weidemann (1977; revised 2000). We describe briefly the general idea: spectrosopy of white dwarfs allows to determine surface gravity $g = GM/R$ and $T_{\rm eff}$. The

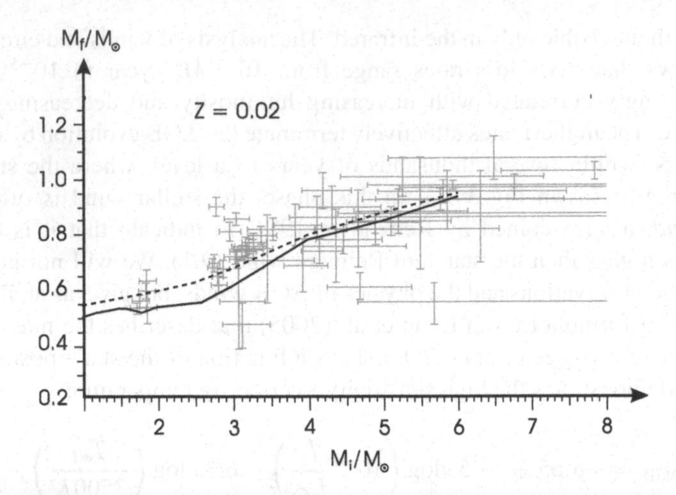

Fig. 34.7 The initial-final mass relation. The data points and their error bars are taken from Salaris et al. (2009). The *dashed line* is the empirical relation derived by Weidemann (2000). The *solid line* is the relation predicted from theoretical AGB evolution models for $Z = 0.02$ (After Weiss and Ferguson 2009)

former quantity, which can in principle also be determined from gravitational photon redshift, together with known mass-radius relations (Chap. 37) yields the white dwarf's mass. Theoretical cooling curves for the so determined mass, together with T_{eff}, give the cooling age, t_{cool}. If the white dwarfs are in a stellar cluster, the age t of this cluster can be determined from comparison with isochrones of appropriate composition. Since the white dwarfs are the descendants of former main-sequence stars in the cluster, the difference $t - t_{\mathrm{cool}}$ is the age the progenitor spent in the pre-white dwarf stages. This time is dominated by the main-sequence lifetime, which is depending on the initial mass, as we have estimated in (30.2) and plotted in Fig. 30.6. In this way, the initial mass can be determined. Since the more massive AGB stars and their mass loss are more interesting, open clusters of several 100 Myrs are mainly investigated. However, also binary systems, in which one component is a white dwarf, are suitable. A famous example is Sirius B; popular clusters are the Hyades, the Pleiades, and Praesepe.

Figure 34.7 shows the empirical initial-final-mass relation (Salaris et al. 2009; data points), the previous analytical fit by Weidemann (2000), and the prediction from theoretical AGB evolution (Weiss and Ferguson 2009), which included overshooting and mass loss according to (34.17) and (34.18). Within the errors the theoretical models lie well within the empirical data. This is an indication that the total mass loss is described well by the models. Notice that a 6 $M_\odot$ star ends as a white dwarf of only 1 $M_\odot$. It has lost a total of 5 $M_\odot$, and this happens mostly on the TP-AGB. The second remarkable result is that even for the highest masses ($\approx$ 8 $M_\odot$) the final white-dwarf mass is well below the critical Chandrasekhar mass of $\approx 1.4\,M_\odot$. This will turn out to be an important fact in relation to the progenitors of supernovae (Chap. 36).

34.7 A Sample AGB Evolution

In this section, we will present the complete evolution of a star of $2\,M_\odot$ and a standard composition ($X = 0.695$, $Y = 0.285$, $Z = 0.02$) from the main-sequence until the final white-dwarf state. The calculations for this and many other values for mass and composition were done by A. Kitisikis (PhD thesis, Munich University, 2008) and published by Weiss and Ferguson (2009). They are the first attempt to include as many crucial physical aspects of AGB evolution as possible. Overshooting is treated according to (30.9), and applied at all convective boundaries. The opacity tables used include variations of C and O abundance and therefore are sensitive to dredge-up processes and in general lead to lower T_{eff} in case of carbon enhancement of the envelope due to the third dredge-up. Mass loss is included in parametrized form following (34.17) and (34.18), depending on the C/O ratio. Figure 34.8a gives an overview of the full evolution, and Fig. 34.8b shows the TP-AGB phase. Figure 34.9 summarizes details during the TPs.

The evolution starts on the main sequence, which lasts for 1.075 Gyr. A further 58 Myr are spent on the RGB, before helium ignites in a moderately energetic core helium flash, with a peak helium luminosity of "only" $\log L_{\text{He}}/L_\odot = 7.2$. The surface luminosity of $\log L/L_\odot = 2.87$ is also lower than that of the RGB tip of low-mass stars indicating that at this mass, we are already in the transition region to intermediate-mass stars. Core helium burning, which lasts for 177 Myr, takes place in a barely visible loop around $\log L/L_\odot = 1.8$. Then the star starts to climb the E-AGB, and the first thermal pulses set in around $\log L/L_\odot = 3.0$. Figure 34.8b shows this part of the HRD in more detail. In the course of the TPs the peak luminosity increases and the effective temperature drops. This leads to stronger mass loss. The last TP, which sets in after C/O > 1 is reached, leads to a strong excursion to temperatures as low as 2,000 K. After this final pulse, due to the fast shedding of the envelope, the star contracts and crosses the HRD within 4,100 yrs. Its mass, and therefore the final white-dwarf mass, is $0.543\,M_\odot$. The E- and TP-AGB last for 15.4 and 2.6 Myr.

Details of the TP-AGB phase are given in Fig. 34.9. As mentioned in Sect. 34.3, an asymptotic pulse behaviour is slowly approached after about 10 TPs, but not completely reached, even after all 15 pulses. The first pulse is, as is very often found in such calculations, different from the subsequent ones. The surface luminosity drops more than for the $5\,M_\odot$ star of Fig. 34.3, due to the less massive envelope. However, it also shows a very short-lived peak of 500 'years' duration, which is not visible in Fig. 34.3. This is found in many models with massive envelopes and a deep penetration of the OCZ down to the hydrogen-burning shell.

Figure 34.9 also shows the interaction of dredge-up, effective temperature and mass loss. With increasing pulse number, the star gets increasingly cooler. At the

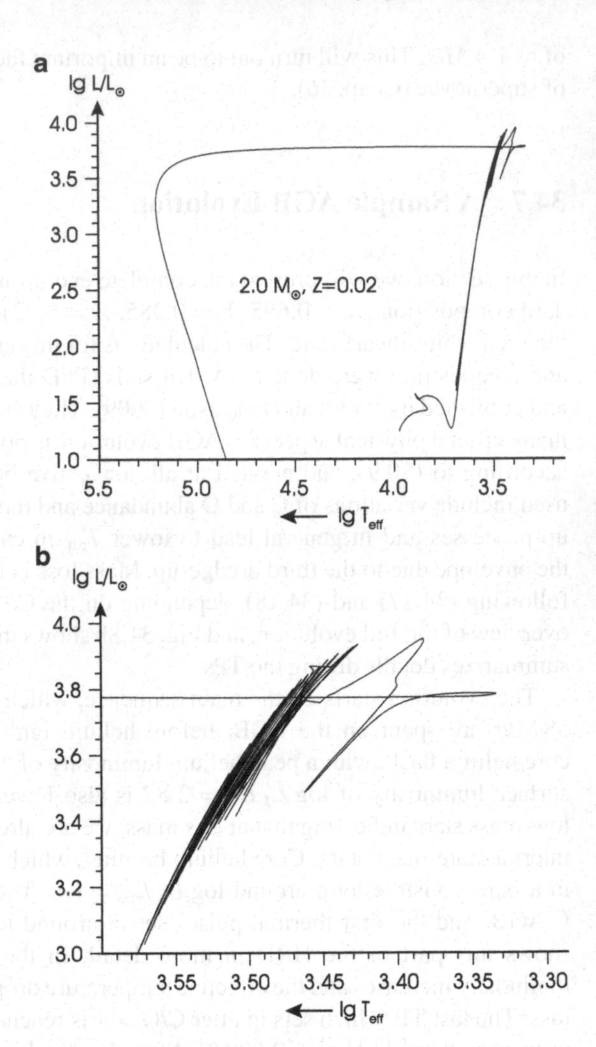

Fig. 34.8 (**a**) Evolution of a star of $2\,M_\odot$ and $Z = 0.02$ from the ZAMS to the white-dwarf cooling stage (Weiss and Ferguson 2009). (**b**) Detail view of the TP-AGB phase, with 15 TPs, the last one leading to an excursion to very low T_{eff}, the final expulsion of the stellar envelope, and the beginning of the post-AGB transition to hot T_{eff} at nearly constant L. For details about the calculations, see text

same time, the carbon abundance in the envelope increases in each pulse due to dredge-up, ensured by the application of convective overshooting from the ISCZ.

The dredge-up in fact starts already with the first thermal pulse, but is interrupted until TP 9. Then a significant increase up to $C/O = 1$ takes place. Note that also some oxygen–the second result of helium burning–is dredged up. The increase in nitrogen is due to protons ingested from the envelope. The abundances in Fig. 34.9 are given in mass fractions, while the C/O ratio is in number fractions. This is the reason why the carbon abundance remains below that of oxygen, but $C/O > 1$ after pulse 14. At this moment, a strong increase in opacity due to the carbon-rich atmosphere leads to a drop in T_{eff}, and thus, due to the high sensitivity of the mass loss rate to temperature, to a strong increase in $\dot{M}$. In the corresponding panel $\dot{M}$ increases to levels above $10^{-5}\,M_\odot/\text{year}$, such that the total mass (middle panel) is

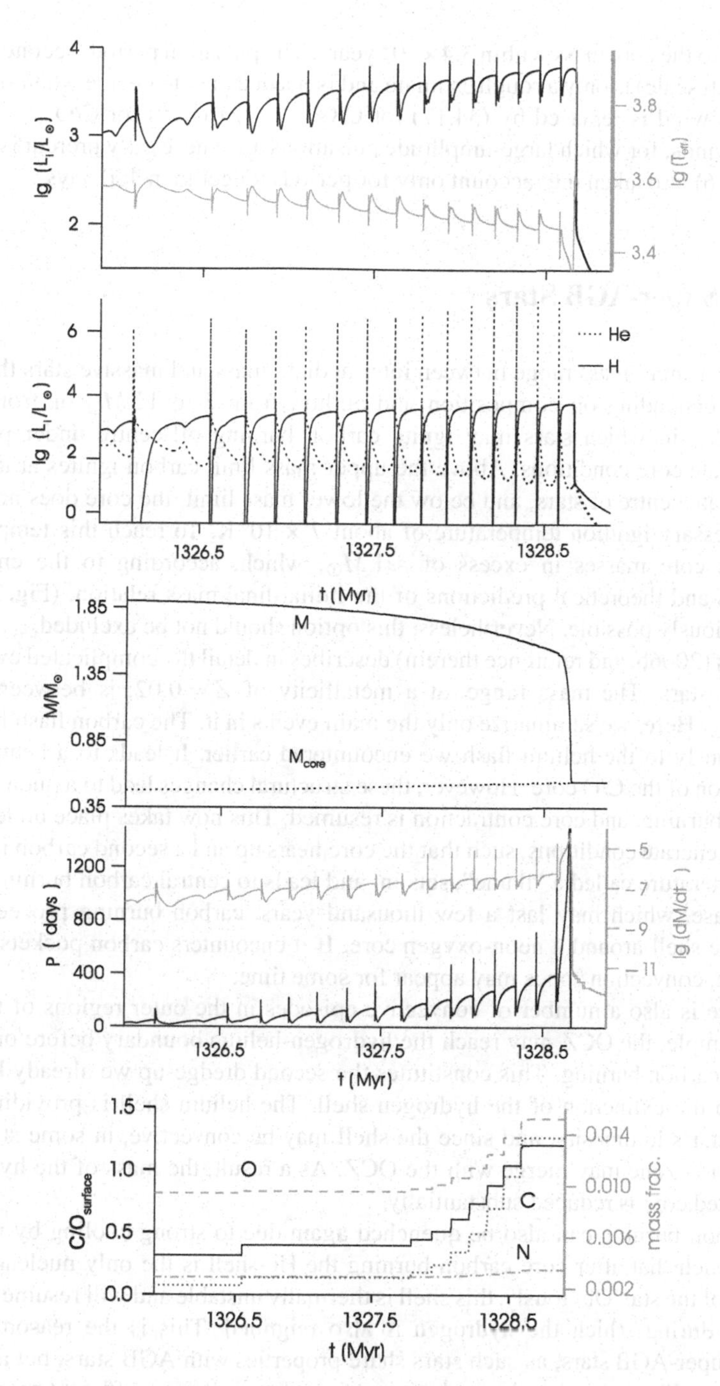

Fig. 34.9 Physical quantities during the TPs of the same model as in Fig. 34.8. The *panels* show, from *top* to *bottom*: total luminosity L (*left scale*) and T_{eff} (*right scale*); helium and hydrogen shell luminosity; total and C/O core mass; pulsation period (*left scale*) and mass loss rate (*right scale*); and finally C/O ratio at the surface (*left scale*) and C, N, and O mass fractions (*right scale*)

reduced to the core mass within 3.4×10^4 years. The pulsation period (second-to-last panel, left scale) is only a rough estimate and is needed only to decide when standard Reimers wind is replaced by (34.17) for C/O < 1 or (34.18) for C/O > 1. Dust-driven winds, for which large-amplitude pulsations are a necessary prerequisite (see Sect. 34.6), are taken into account only for periods longer than 400 days.

34.8 Super-AGB Stars

There is a small mass range between intermediate-mass and massive stars that may extend, depending on composition and author, from 7 to 12 $M_\odot$ or from $\approx$ 8 to 10 $M_\odot$, in which stars may ignite carbon burning off-centre under partially degenerate core conditions. Above the upper mass limit carbon ignites at the non-degenerate centre of stars, and below the lower mass limit, the core does not reach the necessary ignition temperature of about 7×10^8 K. To reach this temperature requires core masses in excess of $\approx 1\,M_\odot$, which, according to the empirical findings and theoretical predictions of the initial-final mass relation, (Fig. 34.7) is not obviously possible. Nevertheless, this option should not be excluded.

Siess (2006b, and reference therein) describes in detail the complicated evolution of such stars. The mass range, at a metallicity of $Z = 0.02$, is between 9 and 11.3 $M_\odot$. Here, we summarize only the main events in it. The carbon flash happens analogously to the helium flash we encountered earlier. It leads to a heating and expansion of the C/O core. However, these structural changes lead to a quenching of carbon burning, and core contraction is resumed. This now takes place under much less degenerate conditions, such that the core heats up and a second carbon ignition, in the literature called a "flame", sets in, and leads to central carbon burning. After this phase, which may last a few thousand years, carbon burning proceeds in a radiative shell around a neon-oxygen core. If it encounters carbon pockets around the core, convection zones may appear for some time.

There is also a number of convective episodes in the outer regions of the star. For example, the OCZ may reach the hydrogen-helium boundary before or during central carbon burning. This constitutes the second dredge-up we already know. It leads to the extinction of the hydrogen shell. The helium shell is providing most of the star's luminosity, and since the shell may be convective, in some stars, this convection zone may merge with the OCZ. As a result, the mass of the hydrogen-exhausted core is reduced substantially.

Carbon burning can also be quenched again due to strong cooling by neutrino losses such that after core carbon burning the He-shell is the only nuclear energy source of the star. Obviously, this shell is thermally unstable and will resume thermal pulses, during which the hydrogen is also reignited. This is the reason for the name super-AGB stars, as such stars share properties with AGB stars, but are more luminous. Since mass loss has to be minor in order to allow a sufficient core growth during the AGB phase, the envelope mass is still substantial and super-AGB stars may indeed suffer up to thousand TPs.

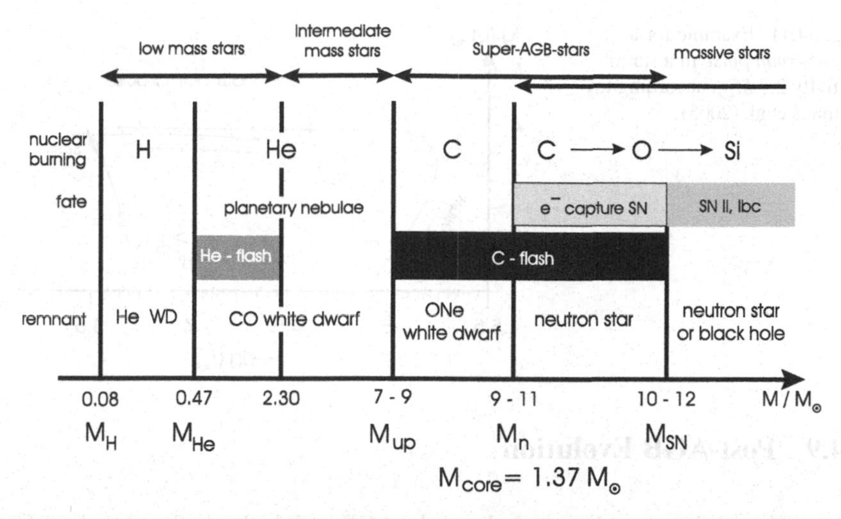

Fig. 34.10 Transition masses between different evolutionary paths for stars, and their final fate. M_{H}, M_{He}, M_{up}, M_{n}, and M_{SN} correspond to the minimum initial stellar mass for hydrogen, helium, and carbon ignition, the formation of a neutron star, and for stars undergoing a type II supernova explosion. The final fate of the star and its remnant are indicated for each mass range. Note that these mass limits depend crucially on the initial composition and on the detailed computations (After Siess 2006b)

The final fate of super-AGB stars depends on the mass of the NeO core. If it exceeds $\approx 1.37\,M_{\odot}$ after carbon burning, the nuclear evolution will go through all phases just as for massive stars. If it is slightly lower than this value initially, Ne ignition can be avoided and the core can grow due to shell burning and develop a highly degenerate NeO core of $1.37\,M_{\odot}$. It will not undergo the thermal pulses, but will end as a so-called *electron-capture supernova* (see Sect. 36.3.4), which is initiated by electron-capture reactions on ^{24}Mg, ^{24}Na, and other isotopes. This leads, among other effects, to the reduction of electron pressure, and a subsequent collapse of the core. A low-mass neutron star will be the remnant after the supernova explosion. If the NeO core mass always stays below this critical core mass, the star will end–after envelope expulsion–as a NeO white dwarf. In Fig. 34.10 an overview of the various mass limits separating the different evolutionary paths is given (adapted from Siess 2006b). These limits are very uncertain and depend a lot not only on the details of the computations, but also on the initial composition. In fact some limits may not even exist at all metallicities, because of effects of mass loss or overshooting.

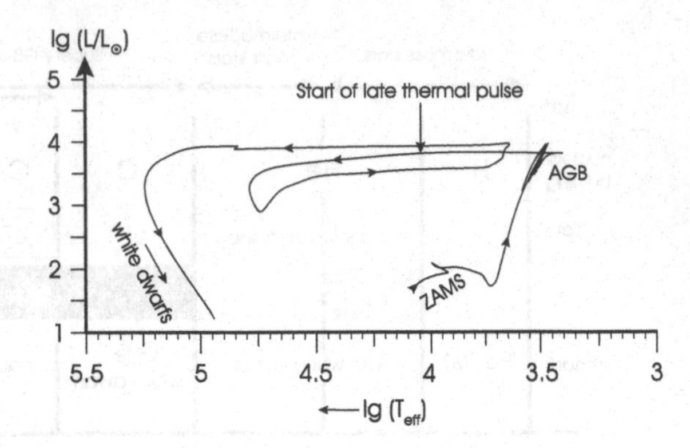

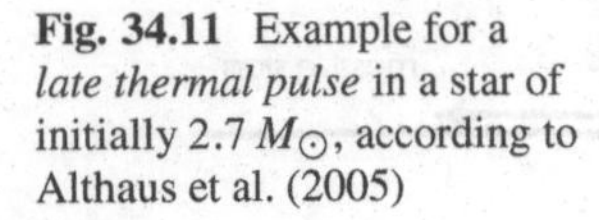
Fig. 34.11 Example for a *late thermal pulse* in a star of initially 2.7 $M_\odot$, according to Althaus et al. (2005)

34.9 Post-AGB Evolution

Stars of low and intermediate mass leave the AGB mainly due to the very short-lived superwind phase. Depending on the phase of the TP cycle this happens, the star may have a hydrogen- or helium-shell providing the majority of the total luminosity. The post-AGB stars are therefore divided into *hydrogen and helium burners*. The latter group generally has longer HRD-crossing timescales.

With increasing $T_{\rm eff}$, mass loss is quickly dying out. Depending on the density and expansion velocity of the circumstellar shell which was lost during the TP-AGB phase, the increasingly higher number of UV photons emerging from the star, which is crossing the HRD to very high temperatures, may ionize the circumstellar matter and lead to the creation of a planetary nebula. The critical values for $T_{\rm eff}$ are 30,000 K and 60,000 K for hydrogen and helium ionization. In addition, a hot wind with velocities of the order of 1,000 km/s ploughs into the slowly expanding shell (of a few tens of km/s), compressing it and creating shocks.

At the bluest point in the evolution (see Figs. 34.1 and 34.8), the shell is finally fading away and the star begins its final cooling phase and becomes a white dwarf.

However, in some cases, the star may still suffer a last thermal pulse. Such late pulses are found in numerical calculations both during the HRD crossing and during the earliest cooling phases. In the course of such a late TP the star returns to the AGB and resumes a second HRD crossing. We show such an excursion in the HRD in Fig. 34.11. Due to the mixing and burning episodes connected with the pulse the thin envelope undergoes drastic changes in its composition. In fact, there is a small number of stars which evolved drastically over a few decades including changes of the surface composition. These are generally connected with late TP events. Famous examples are Sakurai's object (V4334 Sgr), FG Sge, and V650 Aql. There is still a discrepancy between the timescales for the changes between the models and the objects, but this may be due to insufficient theories for time-dependent convection. This kind of objects and their relation to post-AGB evolution have been reviewed by Schönberner (2008).

Chapter 35
Later Phases of Core Evolution

35.1 Nuclear Cycles

The stellar evolution described above may seem to be rather complicated with regard to the nuclear shell instabilities, but also where the changes of the surface layers are concerned, for example, in the case of evolutionary tracks in the HR diagram. The processes appear much simpler and even become qualitatively predictable if we concentrate only on the central evolution. Extrapolating from central hydrogen and helium burning of sufficiently massive stars, we can imagine that the central region continues to pass through cycles of nuclear evolution which are represented by the following simple scheme:

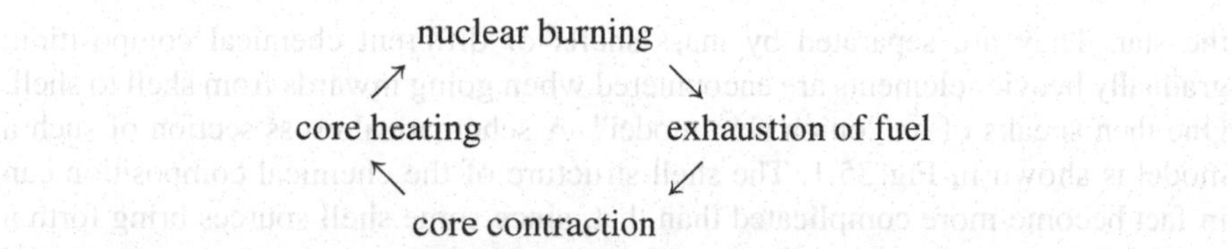

The momentary burning will gradually consume all nuclei inside the convective core that serve as "fuel". The exhausted core then contracts. This raises the central temperature until the next higher burning is ignited etc.

As long as this scheme works, gradually heavier elements are built up near the centre from cycle to cycle. The new elements are evenly distributed in convective cores which usually become smaller with each step. For example, in the first cycle (hydrogen burning), the star develops a massive helium core, inside which a much smaller CO core is produced in the next cycle (helium burning), and so on.

We have also seen that after the core is exhausted the burning usually continues in a concentric shell at the hottest place where the fuel is still present. A shell source can survive several of the succeeding nuclear cycles, each of which generates a new shell source, such that several of them can simultaneously burn outwards through

R. Kippenhahn et al., *Stellar Structure and Evolution*, Astronomy and Astrophysics Library, DOI 10.1007/978-3-642-30304-3_35, © Springer-Verlag Berlin Heidelberg 2012

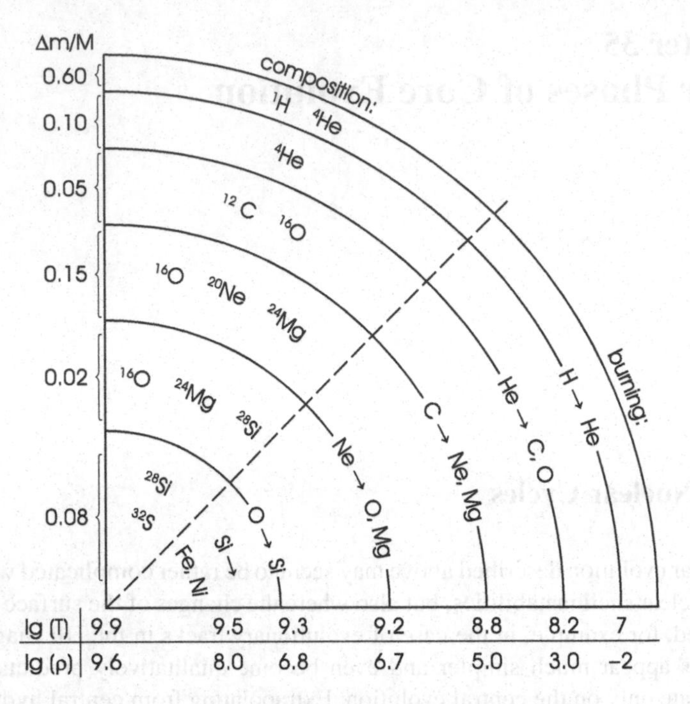

Fig. 35.1 Schematic illustration (not to scale) of the "onion skin structure" in the interior of a highly evolved massive star. Along the vertical radius and below the horizontal radius some typical values of the mass, the temperature (in K), and the density (in g cm^{-3}) are indicated

the star. They are separated by mass shells of different chemical composition; gradually heavier elements are encountered when going inwards from shell to shell. One then speaks of an "onion skin model". A schematical cross section of such a model is shown in Fig. 35.1. The shell structure of the chemical composition can in fact become more complicated than that, since some shell sources bring forth a convective (or semiconvective) subshell, inside which the newly processed material is completely (or partially) mixed. This can be recognized in Fig. 36.4, which shows the interior composition of a model for a $25 M_\odot$ star in a very advanced stage (just before core collapse, see Chap. 36). We have also seen that, depending on the change of T in certain regions, a shell source may stop burning for some time and be reignited later.

The simple evolution through nuclear cycles as sketched above can obviously be interrupted, either temporarily or for good. From the discussion of the nuclear reactions in Chap. 18 we know that the cycles must come to a termination, at the latest, when the innermost core consists of ^{56}Fe (or neighbouring nuclei) and no further exothermic fusions are possible. However, it is easily seen that the sequence of cycles can be interrupted much earlier by another effect. Each contraction between consecutive burnings increases the central density ϱ_c. Assuming homology for the contracting core (cf. Sect. 28.1) and ignoring the influence of the rest of the star, we obtain from (28.1) the change of the central temperature T_c

$$\frac{dT_c}{T_c} = \left(\frac{4\alpha - 3}{3\delta}\right)\frac{d\varrho_c}{\varrho_c}. \tag{35.1}$$

The decisive factor, in parenthesis on the right-hand side, depends critically on the equation of state which is written as $\varrho \sim P^{\alpha}T^{-\delta}$. For an ideal gas with $\alpha = \delta = 1$, we have $dT_c/T_c = (1/3)(d\varrho_c/\varrho_c)$. This means that each contraction of the central region increases the temperature, as well as the degeneracy parameter ψ of the electron gas [ψ = constant for $dT/T = (2/3)(d\varrho/\varrho)$ (cf. Sects. 15.4 and 16.2)]. With increasing degeneracy the exponents α and δ become smaller. When the critical value $\alpha = 3/4$ is reached (δ is then still > 0), the contraction ($d\varrho_c > 0$) no longer leads to a further increase of T_c according to (35.1). The degeneracy in the central region has obviously decoupled the thermal from the mechanical evolution, and the cycle of consecutive nuclear burnings is interrupted. In this case the next burning can be ignited only via more complicated secondary effects, which originate, for example, in the evolution of the surrounding shell source (cf. Sect. 33.2).

Other complications may arise if the central region of a star suffers an appreciable loss of energy by strong neutrino emission (cf. Sect. 18.7). We have already seen (Sect. 33.5) that this can decrease the central temperature and, therefore, influence the onset of a burning.

In any case, the nuclear cycles tend to develop central regions with increasing density and with heavier elements. We should note, however, that the later nuclear burnings are not capable of stabilizing the star long enough for us to observe many stars in such phases (as is the case with central hydrogen burning and helium burning). The main reason for this is the strongly decreasing difference in binding energy per nucleon (Fig. 18.1). Table 35.1 on page 447 gives typical durations for the various hydrostatic burning phases. From carbon burning on, these are comparable, respectively much shorter than the thermal timescale of the star. This means that any change in the core is no longer reflected by a change of surface properties, and therefore the star remains at its position in the Hertzsprung-Russell diagram. From the outside, one cannot see whether the star is 10,000 years or 10 h before the final core collapse!

35.2 Evolution of the Central Region

The description of the nuclear cycles in Sect. 35.1 has already given a rough outline of the central evolution of a star. We recognize it easily in Fig. 35.2, where the evolution of the centre is plotted in the lg ϱ_c–lg T_c plane according to evolutionary calculations for different stellar masses M, covering the full range from brown dwarfs to the most massive stars. We see that T_c indeed rises roughly $\sim\varrho_c^{1/3}$ [cf. (35.1)] as long as the central region remains non-degenerate. Of course, the

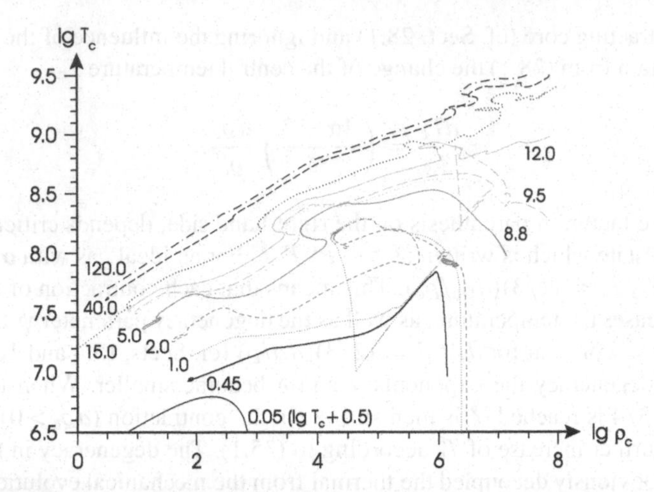

Fig. 35.2 Evolution of the central values of temperature T_c (in K) and density ϱ_c (in g cm^{-3}) for stars of all masses (from $0.05 M_\odot$ to $120 M_\odot$). The tracks are labelled with the stellar mass M (in $M_\odot$). The tracks have been collected from different sources but are all for a metallicity of approximately $Z = 0.02$. The brown dwarf track ($M = 0.05\,M_\odot$) includes the pre-main-sequence phase and is from Baraffe et al. (2003). Stars with M from 0.45 to $5\,M_\odot$ are from the authors, those with $M = 8.8$, 9.5, and $12.0\,M_\odot$ (the super-AGB range) from Siess (2006a), and the massive stars ($M = 15$, 40, and $120\,M_\odot$) from Limongi and Chieffi (2006) (Data courtesy of I. Baraffe, M. Limongi, L. Siess)

details of the central evolution are much more complicated than predicted by the simple vector field in Fig. 28.1. During the burnings the curves bulge out to the upper left. This is not surprising, since then the changes are far from homologous [which is assumed in (35.1) and for Fig. 28.1], for example, owing to the restratification from a radiative to a convective core. After these interludes of burning, the evolution returns more or less to the normal slope. A parallel shift of the track from one to the next contraction is to be expected, since the contracting region (the core) will in general have a larger molecular weight, but a smaller mass.

We have already mentioned in Sect. 28.1 and in Sect. 35.1 the important fact that each contraction with $T_c \sim \varrho_c^{1/3}$ brings the centre closer to the regime of electron degeneracy. The degree of non-relativistic degeneracy is constant on the steeper lines $T \sim \varrho^{2/3}$. Once the central region has reached a certain degree of degeneracy (where $\alpha = 3/4$ in the simple model of Sect. 28.1), T_c no longer increases, and the next burning is not reached in this way (if at all), as we have already seen in Fig. 28.2. This happens the earlier in nuclear history, the closer to degeneracy a star has been at the beginning, i.e. the smaller M is (cf. Fig. 35.2). Recall (Sect. 22.2) that with increasing mass, also T_c increases, but ϱ_c decreases (Fig. 22.5). Therefore which nuclear cycle is completed before the star develops a degenerate core depends on the stellar mass M.

If the evolution were to proceed with complete mixing, we would only have to consider homogeneous stars of various M and different compositions, and to see whether their contraction leads to ignition ($M > \widetilde{M}_0$) of a certain burning or to a degenerate core ($M < \widetilde{M}_0$). These limits for reaching the burning of H, He, and C are $\widetilde{M}_0 \approx 0.08, 0.3$, and $0.8 M_\odot$, respectively.

We know that the evolution lies far from the case of complete mixing, and only the innermost core of a star is processed by nuclear burning. But for sufficiently concentrated cores, the central contraction proceeds independently of the conditions at its boundary, i.e. independently of the non-contracting envelope. Therefore the above values $\widetilde{M}_0$ give roughly the limits for the masses of the corresponding cores.

Standard evolutionary calculations (assuming a typical initial composition, no convective overshooting, and no mass loss) give the following characteristic ranges of M, which we already mentioned earlier. After central hydrogen burning, *low-mass stars* with $M < M_1(\mathrm{He}) \approx 2.3 M_\odot$ develop degenerate He cores. After central helium burning, *intermediate-mass stars* with $M < M_1(\mathrm{CO}) \approx 9 M_\odot$ develop a degenerate CO core. And in *massive stars* with $M > M_1(\mathrm{CO})$ even the CO core remains non-degenerate while contracting for the ignition of the next burning. The precise values of the limiting masses M_1 depend, for example, not only on the assumed initial composition but also on details of the physical effects considered. Another important influence is the downwards penetration of the outer convection zone after central helium burning (in the second dredge-up phase). This lowers the mass of the core and therefore encourages the evolution into stronger degeneracy, i.e. it lowers M_1 (cf. Sect. 34.8). The depth of the second dredge-up depends on the choice of the mixing length parameter and the inclusion of convective overshooting. In Fig. 35.2 we see that the models with $M = 0.05, 0.45$, and $8.8\ M_\odot$ just miss the ignition of H, He, and C, respectively.

After a star has developed a strongly degenerate core it has not necessarily reached the very end of its nuclear history. This is only the case if the shell-source burning cannot sufficiently increase the mass of the degenerate core. However, the next burning is only delayed, and it will be ignited later in a "flash" if the shell source is able to increase the mass of the core to a certain limit M_c'. We have seen in Sect. 33.3 that the critical mass for ignition of helium in a degenerate core is $M_c'(\mathrm{He}) \approx 0.48 M_\odot$, which agrees with the case shown in Fig. 35.2. The corresponding critical mass of a degenerate CO core is $M_c'(\mathrm{CO}) \approx 1.4 M_\odot$ as we shall see immediately. Note that these limits are appreciably larger than the corresponding lower limits ($\widetilde{M}_0$) for reaching a burning by non-degenerate contraction, as described above. This indicates the possibility that the evolution depends discontinuously on M around the limits $M_1(\mathrm{He})$ and $M_1(\mathrm{CO})$. For example, stars with $M = M_1(\mathrm{He}) - \Delta M$ ignite helium via a flash in a degenerate core of mass $0.48\ M_\odot$, while stars with $M = M_1(\mathrm{He}) + \Delta M$ can ignite helium burning via core contraction in (nearly) non-degenerate cores of about $0.3 M_\odot$ (cf. the idealized scheme in Fig. 35.3). Here one could imagine a bifurcation at $M = M_1$, where fluctuations would decide into which of the two regimes the star turns. In reality (by which we mean numerical models) the limit is "softened up" (a little bit of degeneracy leading to a baby flash, etc.), as can be seen in Fig. 5.19 of Salaris and Cassisi (2005). Nevertheless, the transition range is narrower than $\approx 0.5\ M_\odot$ between the two regimes.

The ignition of He and C under degenerate conditions in the 1.0 and $8.8\ M_\odot$ stars of Fig. 35.2 first leads to a strong cooling and expansion of the core, followed by a

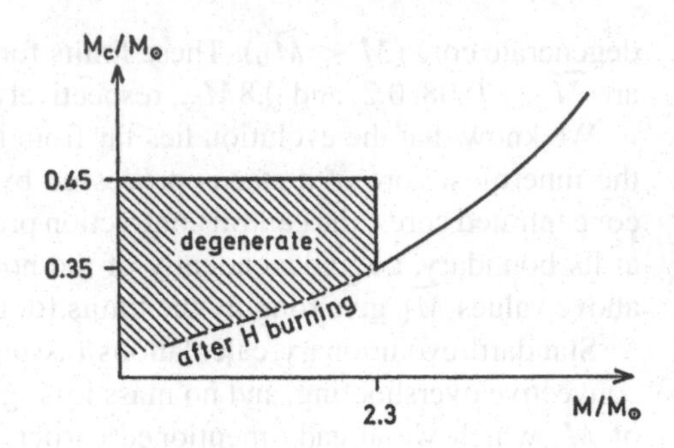

Fig. 35.3 The *solid line* shows schematically the mass M_c of the helium core at the onset of helium burning as a function of the stellar mass M. The *broken line* shows the core mass at the end of hydrogen burning in low-mass stars, before the electron gas in the core becomes degenerate

temperature increase and further expansion until stable core He burning is reached. This realistic evolution is more complicated than the simple picture illustrated in Fig. 33.6.

The evolution of degenerate CO cores is similar to that of degenerate helium cores in low-mass stars (Sects. 33.3 and 33.4). The structure of the core is more or less independent of the details of the envelope. Therefore the evolution of the central values converges for stars of different M as long as the core mass is the same (cf. Fig. 35.2, $M = 0.45$ and $1.0\,M_\odot$). While the mechanical structure of such a core is determined by its mass M_c, its thermal properties depend on the surrounding shell source and on the neutrino losses. If the shell source were extinguished, the core would simply cool down with $\varrho_c =$ constant (on a vertical line in Fig. 35.2) to the white-dwarf state, as can be seen in this figure for the lower and intermediate-mass values. The brown dwarf will end as a hydrogen white dwarf, that with $M = 0.45\,M_\odot$ as a helium white dwarf, those with higher masses up to $8.8\,M_\odot$ as carbon-oxygen white dwarfs, and the one with $9.5\,M_\odot$ possibly as a oxygen-neon white dwarf.

The continuous burning of the shell source increases M_c, which in turn increases the temperature in the shell source (cf. Sect. 33.2). It also increases the central density, as we know from the discussion of the structure of degenerate configurations (Sect. 19.6), i.e. the evolution goes to the right in Fig. 35.2. The contraction due to this effect releases a large amount of gravitational energy, which, in the absence of energy losses (by conduction or neutrinos), would heat the core adiabatically.

However, there are strong neutrino losses ε_ν in this part of the T-ϱ diagram (cf. Fig. 18.11), which modify the whole situation. Since ε_ν increases appreciably with T, we should first make sure that there is no thermal runaway in the degenerate core (a "neutrino flash"), in analogy to a flash at the onset of a burning. This can be easily shown by the stability consideration presented in Chap. 25, where we analysed the reaction of the central region on an assumed increase $d\varepsilon$ of the energy *release*. This led to (25.30) with gravothermal heat capacity c^* (25.29). Now we replace $d\varepsilon$ by the small energy *loss* $-d\varepsilon_\nu$. If we neglect the perturbation of the flux ($dl_s = 0$) for simplicity, (25.29) and (25.30) become

$$c^* \frac{dT}{dt} = -d\varepsilon_\nu, \quad c^* = c_P \left(1 - \nabla_{\rm ad} \frac{4\delta}{4\alpha - 3}\right). \tag{35.2}$$

Obviously the reversal of the sign of the right-hand side in the first equation (35.2) has reversed the conditions for stability. An ideal gas with $\alpha = \delta = 1$ has the gravothermal heat capacity $c^* < 0$, and neutrino losses are unstable since $\dot{T} > 0$ (a thermal runaway with ever increasing neutrino losses). Degenerate cores with $\alpha \to 3/5, \delta \to 0$ have $c^* > 0$, i.e. $\dot{T} < 0$, and these cores are stable: a small additional energy loss reduces T and ε_ν such that the core returns to a stable balance. In the following scheme we summarize the different properties of thermal stability we have encountered:

	Burning ($\varepsilon > 0$)	Neutrinos ($-\varepsilon_\nu < 0$)
Ideal gas	Stable	Unstable
Degeneracy	Unstable	Stable

According to Sect. 34.3 the scheme also holds for burning in shell sources, where we have in addition the pulse instability for thin shells. We recall that a general treatment of the shell source stability is possible (Yoon et al. 2004).

Numerical calculations approve the above conclusions: instead of leading to a thermal runaway, the neutrino losses cool the central region of a degenerate core such that ε_ν remains moderate. Typical "neutrino luminosities" L_ν (= total neutrino energy loss of the star per second) remain only a fraction of the normal "photon luminosity" L. In Fig. 35.4 we show a very instructive example from an early model by Paczyński (1971). Although a star of only 3 $M_\odot$ is almost certainly never able to develop a CO core of more than 0.8 $M_\odot$, the figure still shows all principle effects: The temperature profiles inside the cores of two different M_c are shown in Fig. 35.4 by the broken S-shaped curves. They follow roughly lines of ε_ν = constant. With increasing M_c the point for the centre moves along the solid line to the right, and extremely high values of ϱ_c would necessarily occur if M_c could go to the Chandrasekhar limit of $1.44M_\odot$. Shortly before this limit, at $M_c \approx 1.4M_\odot$, the central values reach the dotted line $\varepsilon_\nu = \varepsilon_C$ to the right of which pycnonuclear carbon burning dominates over the neutrino losses, $\varepsilon_C > \varepsilon_\nu$. Now carbon burning starts with a thermal runaway. If this happens in the centre, then explosive carbon burning will finally disrupt the whole star, such that one should expect a supernova outburst that does not leave a remnant (a neutron star); compare this also with Chap. 36. We have already seen (Sect. 34.8) that in more massive stars carbon ignition starts in a shell, such that the star survives this event, but the principal story remains that the degenerate CO core is ignited when its mass $M_c \approx 1.4M_\odot$, although it depends on the initial mass and varies, according to Siess (2006b), between 1.1 and 1.5 $M_\odot$ for ZAMS masses between 9.0 and 11.5 $M_\odot$. With increasing mass and decreasing core degeneracy the location of carbon ignition is moving towards the centre. At the end of the rather complicated carbon burning phase an ONeMg core is left over. For the initial mass range for which this core is degenerate, its mass is between 1.05 and 1.12 $M_\odot$, again far from the Chandrasekhar mass.

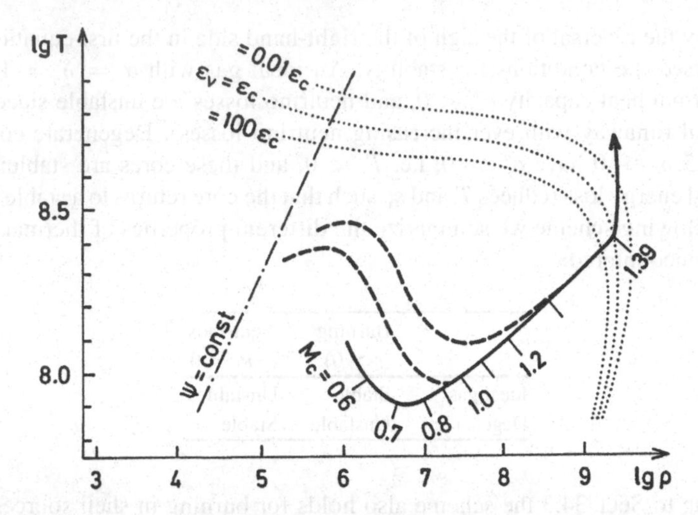

Fig. 35.4 Temperature T (in K) and density ϱ (in g cm^{-3}) in the CO core of a $3M_\odot$ star after central helium burning. The *solid line* gives the evolution of the centre with increasing core mass M_c (in $M_\odot$). The carbon flash starts at about $M_c = 1.39M_\odot$ when the energy production by carbon burning (ε_C) exceeds the neutrino losses (ε_ν). Some lines of constant ratio $\varepsilon_\nu/\varepsilon_C$ are *dotted*. The *broken lines* show the T stratification in the core for two consecutive stages; neutrino losses have produced a maximum of the temperature outside the centre (After Paczyński 1971)

The just-described central evolution is the same for all stars that are able to develop a degenerate CO core of $M_c \approx 1.4M_\odot$. The obvious condition for this is that the stellar mass M is larger than that limit. For $\dot{M} = 0$ this would include all stars in the range $1.4M_\odot < M < 9M_\odot$, i.e. the intermediate-mass stars ($M \approx 2.3 \ldots 9M_\odot$) and the low-mass stars with $M > 1.4M_\odot$. More precisely the stellar mass M must be larger than $1.4M_\odot$ *at the moment of ignition* (which does not occur before $M_c \approx 1.4M_\odot$). This can require that the initial stellar mass M_i (on the main sequence) was much larger than $1.4M_\odot$ if M has been reduced in the meantime by a strong mass loss.

Obviously there are two competing effects, the increase of M_c due to shell-source burning and the simultaneous decrease of the stellar mass M due to mass loss. Their changes in time are schematically shown in Fig. 35.5, and the outcome of this race decides the final stage of the star. The two values (M and M_c) reach their goal at $1.4M_\odot$ simultaneously if the initial mass has the critical value M_i(min). Stars with $M_i > M_i$(min) will ignite the CO core, since M_c can reach $1.4M_\odot$. For stars with initial masses $M_i < M_i$(min), the mass loss will win and M_c never reaches $1.4M_\odot$. Such stars will finally cool down to the white-dwarf state after the shell source has died out near the surface (cf. Sect. 33.7). Unfortunately the total loss of mass during the evolution is not well known. The various mass loss formulae (Sect. 32.3) and the initial-final mass relation (Fig. 34.7) predict a total mass loss of up to $\Delta M \approx 6 \ldots 7\,M_\odot$, which would mean a critical initial mass above $M_i(\text{min}) \approx 7\,M_\odot$ at least. Of course, if the mass loss were so large that even stars

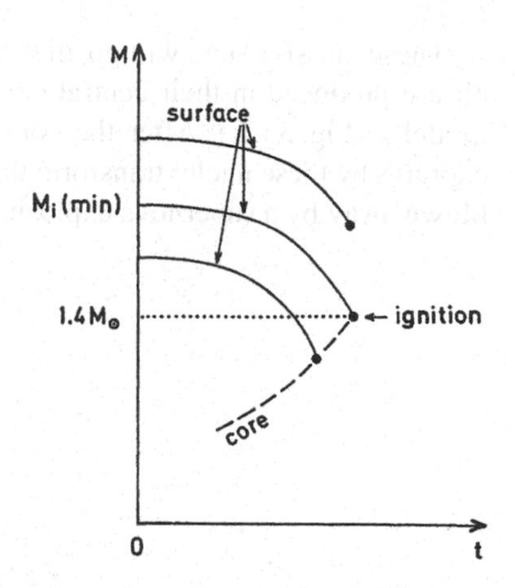

Fig. 35.5 For three different initial masses M_i the *solid lines* show schematically the decrease of the stellar mass M due to mass loss, while the mass of their degenerate CO cores (*dashed line*) increases owing to helium-shell burning. Carbon burning is ignited when the core mass reaches about $1.4 M_\odot$. This never occurs for $M_i < M_i(\text{min})$, since then the surface reaches the core before it can grow to $1.4\ M_\odot$

Table 35.1 The duration of burning stages (in years) in three models of different mass, taken from Limongi and Chieffi (2006)

Burning:	$M = 15\ M_\odot$:	$M = 40\ M_\odot$:	$M = 120\ M_\odot$:
H	1.31×10^7	4.88×10^6	2.80×10^6
He	9.27×10^5	3.82×10^5	2.96×10^5
C	3.25×10^3	1.86×10^2	3.62×10^1
Ne	6.67×10^{-1}	1.34×10^{-1}	6.56×10^{-2}
O	3.59×10^0	1.59×10^{-1}	2.57×10^{-2}
Si	6.65×10^{-2}	1.47×10^{-3}	3.63×10^{-4}

The beginning and end of each burning stage is defined as the times when 1 % of the fuel has been burnt, respectively when its abundance has dropped to below 10^{-3} (Data courtesy M. Limongi)

with $M_i \approx 10\ M_\odot$ were reduced to $M < 1.4\ M_\odot$ before carbon ignition, then all intermediate stars (developing a degenerate CO core) would become white dwarfs. In any case, there are drastic differences between the final stages (white dwarfs or explosions) to be expected for stars in a narrow range of M_i near $M_i(\text{min})$. Current models (Sect. 34.8 and Fig. 35.2) put this mass range between ≈9 and 11 $M_\odot$. These numbers are all depending on the composition.

It is clear that we have the same competition between $\dot{M}_c > 0$ and $\dot{M} < 0$ in the analogous problem of determining initial masses for which the degenerate helium cores are ignited (at $M_c \approx 0.48\ M_\odot$). In this case the bifurcation of the evolution concerns mainly the composition of the final white dwarfs (He or CO).

Finally, we have to consider the massive stars with $M > 9 \ldots 11\ M_\odot$, in which the CO core does not become degenerate during the contraction after central helium burning. Therefore T_c rises sufficiently during this contraction to start the (non-explosive) carbon burning. Here the neutrino losses can become very large, carrying away most of the energy released by carbon burning. In the later burnings, massive stars can have neutrino luminosities up to 10^6 times larger than L; but these stages are very short-lived: for example, silicon burning lasts just a few days (see Table 35.1).

These massive stars will go all the way through the nuclear burnings until Fe and Ni are produced in their central core (Such a case is illustrated in the onion skin model in Fig. 35.1.). After the core has become unstable and collapses, electron captures by these nuclei transform the core into a neutron star, while the envelope is blown away by a supernova explosion (see Chap. 36).

Chapter 36
Final Explosions and Collapse

We have seen that stars can evolve to the white dwarf stage through a sequence of consecutive hydrostatic states if they develop a degenerate core and have final masses less than the Chandrasekhar limit M_{Ch}. It is not well known, however, how much mass the stars can have initially (on the main sequence) in order to end this way. From what was discussed in Chap. 34, it seems that except for a very narrow mass range at the upper end, all stars that develop degenerate cores end as white dwarfs. The main uncertainty here is the total amount of mass lost by stellar winds.

Other stars certainly undergo explosions, ejecting a large part of their mass, if not disrupting completely. In the case where a neutron star is left as a remnant the core must have undergone a collapse, since it cannot reach the neutron-star stage by a hydrostatic sequence. Collapse and explosions are connected with supernova events, and although the theory and the numerical models are well developed and far advanced, not all questions concerning the different mechanisms have been answered, and not all different observed phenomena can be explained so far. The singular event of SN 1987A and the ongoing large-scale supernovae searches, which have returned hundreds of such objects throughout the universe, have led to a much better understanding of stellar explosions, but have also raised new questions. In this section we only discuss some basic effects which certainly play an important role in late phases of more massive stars, and that will probably remain to be an important part of full theories of supernovae.

Since we will not go into the details of the physics of collapse and explosion, and neither into the interesting question of explosive nucleosynthesis in supernovae, we refer the reader to the respective reviews on the subject, such as Hillebrandt and Niemeyer (2000; on supernovae of type I), Smartt (2009; on core collapse supernovae progenitors), Janka et al. (2007; on the theory of core collapse supernovae), Heger et al. (2003; on the fate of massive stars), and others.

R. Kippenhahn et al., *Stellar Structure and Evolution*, Astronomy and Astrophysics Library, DOI 10.1007/978-3-642-30304-3_36,

36.1 The Evolution of the CO-Core

After central helium burning, the further evolution depends critically on the question whether or not the CO-core becomes degenerate in the ensuing contraction phase. Clearly this will depend on the mass of the core. Since its contraction is practically independent of the envelope, the core can be considered as if it were a contracting gaseous sphere with zero surface pressure, as discussed in Chap. 28.

We first estimate the critical core mass that separates the case where the contraction leads to increasing temperatures from the case where degeneracy prevents further heating. For this purpose we replace the equation of state by an interpolation formula between different asymptotic behaviours. In the cores of evolved stars the molecular weight per electron is $\mu_e \approx 2$, while that per ion is $\mu_0 \geq 12$, and therefore the pressure of non-degenerate electrons ($\sim 1/\mu_e$) dominates the ion pressure ($\sim 1/\mu_0$). This holds even more so if the electrons are degenerate. For simplicity we here neglect radiation pressure, as well as the creation of electron–positron pairs, which can also lead to partial degeneracy at very high temperatures and low densities (see Sect. 36.3.5). We then approximate the equation of state by the simple form

$$P \approx P_e = \frac{\Re}{\mu_e}\varrho T + K_\gamma \left(\frac{\varrho}{\mu_e}\right)^\gamma . \tag{36.1}$$

In the second term the exponent γ is not a constant, allowing for non-relativistic and relativistic degeneracy. It varies from $\gamma = 5/3$ for $\varrho \ll 10^6\,\mathrm{g\,cm^{-3}}$ to $\gamma = 4/3$ for $\varrho \gg 10^6\,\mathrm{g\,cm^{-3}}$, while K_γ varies from the constant in (15.23) to that in (15.26).

The equation of hydrostatic equilibrium (2.4) yields as a rough estimate for the central values (which we denote by subscript 0):

$$P_0 \approx \frac{GM_c\bar{\varrho}}{R_c} = fGM_c^{2/3}\varrho_0^{4/3} . \tag{36.2}$$

Here we have used the fact that P_0 is almost given by the weight of the core material alone and $\bar{\varrho} = 3M_c/(4\pi R_c^3)$ is assumed to be proportional to ϱ_0. The dimensionless factor f, containing, for example, the ratio $\bar{\varrho}/\varrho_0$, is kept constant in this consideration. Using (36.1) for the centre and eliminating P_0 from (36.2) yields

$$\frac{\Re}{\mu_e}T_0 = fGM_c^{2/3}\varrho_0^{1/3} - K_\gamma \varrho_0^{\gamma-1}\mu_e^{-\gamma} . \tag{36.3}$$

On the right-hand side, the first term dominates in the non-degenerate case, while the two terms are about equal for high degeneracy.

For a given mass M_c, (36.3) gives an evolutionary track in the $\lg\varrho_0$–$\lg T_0$ plane in Fig. 36.1, similar to the tracks shown in Fig. 28.2. Starting with rather small ϱ_0 and $\gamma = 5/3$, the central temperature T_0 grows with ϱ_0 and has a maximum at $\varrho_{0\,\max}$, after which T_0 decreases again until $T_0 = 0$ is reached at a density of $8\varrho_{0\,\max}$.

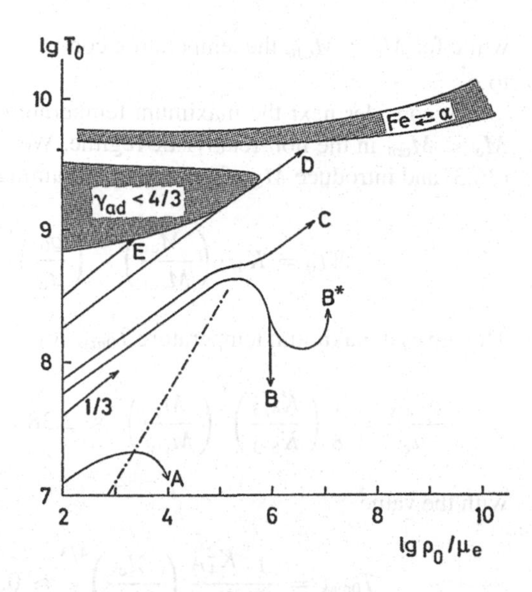

Fig. 36.1 Schematic evolution of the central values T_0 (in K) and ϱ_0 (in g cm^{-3}) for different core masses. The *dot-dashed line* corresponds to the left-hand part of the *dot-dashed line* in Figs. 28.1 and 28.2. Five evolutionary tracks are plotted which illustrate the different cases discussed in the text: A and B correspond to case 1. B^* illustrates case 2, where the core gains mass after it has become degenerate and undergoes a carbon flash. The curves C, D correspond to case 3, while curve E corresponds to case 4

The behaviour of these evolutionary tracks is the same as that discussed in Chap. 28, if there M is replaced by M_c (The way we have made our estimate here, keeping f constant during contraction, is equivalent to the assumption of homology there.). For example, in the non-degenerate case [first term on the right of (36.3) dominant], the slope of the tracks is 1/3 as indicated on the left-hand side of Fig. 36.1, and the tracks for different M_c are shifted at the same values of ϱ_0 like $T_0 \sim M_c^{2/3}$, in analogy to Sect. 28.1.

With sufficiently growing central density, relativistic degeneracy becomes important, and $\gamma \to 4/3$, $K_\gamma \to K_{4/3}$. If we now write $\gamma = 4/3 + \chi$ (where $\chi \to 0$ for $\varrho/\mu_e > 10^7$ g cm^{-3}), we can replace (36.3) by

$$\frac{\Re}{\mu_e} T_0 = \varrho_0^{1/3} \left(f G M_c^{2/3} - K_{(4/3+\chi)} \mu_e^{-(4/3+\chi)} \varrho_0^{\chi} \right) . \tag{36.4}$$

This shows that with increasing ϱ_0 the temperature T_0 does not become zero, but rises again $\sim \varrho^{1/3}$ if

$$M_c > M_{\text{crit}} = \left(\frac{K_{4/3}}{f G} \right)^{3/2} \mu_e^{-2} . \tag{36.5}$$

Obviously the critical value of M_c obtained in (36.5) is of the order of the Chandrasekhar mass M_{Ch} as in (19.29) and (19.30) [Note that a comparison of (36.1) with (19.3) shows that $K_{4/3} = K \mu_e^{4/3}$.]. In fact if $M_c = M_{\text{crit}}$ as defined here, then the core at zero temperature is fully relativistic, degenerate, and in hydrostatic equilibrium, which requires $M_c = M_{\text{Ch}}$.

We can therefore say that during contraction of a core with $M_c \lesssim M_{\text{Ch}}$ the central temperature reaches a maximum and afterwards decreases because of degeneracy, while for $M_c \gtrsim M_{Ch}$, the temperature continues to increase, roughly proportionally to $\varrho_0^{1/3}$.

We consider next the maximum temperature an evolutionary track reaches for $M_c < M_{crit}$ in the non-relativistic regime. We simply set $\gamma = 5/3$, $K_\gamma = K_{5/3}$ in (36.3) and introduce M_{crit} from (36.5), obtaining

$$\Re T_0 = K_{4/3}\left(\frac{M_c}{M_{crit}}\right)^{2/3}\left(\frac{\varrho_0}{\mu_e}\right)^{1/3} - K_{5/3}\left(\frac{\varrho_0}{\mu_e}\right)^{2/3}. \tag{36.6}$$

This gives a maximum temperature T_{0max} for

$$\frac{\varrho_{0max}}{\mu_e} = \frac{1}{8}\left(\frac{K_{4/3}}{K_{5/3}}\right)^3\left(\frac{M_c}{M_{crit}}\right)^2 \approx 2.38 \times 10^5 \mathrm{g\,cm^{-3}}\left(\frac{M_c}{M_{crit}}\right)^2, \tag{36.7}$$

with the value

$$T_{0max} = \frac{1}{4\Re}\frac{K_{4/3}^2}{K_{5/3}}\left(\frac{M_c}{M_{crit}}\right)^{4/3} \approx 0.5 \times 10^9 \mathrm{K}\left(\frac{M_c}{M_{crit}}\right)^{4/3}. \tag{36.8}$$

(Note that $K_{4/3}$ and $K_{5/3}$ have different dimensions.) For cores with $M_c \lesssim M_{crit}$, therefore, T_0 cannot exceed $\approx 0.5 \times 10^9$ K. This is in rough agreement with the "summit" of the dotted line in Fig. 28.1.

The events in the following stages depend sensitively on details of the material functions, the initial models, and the numerical calculations. These factors can decide, for example, whether core collapse is followed by an explosion, whether a remnant is left, etc. In view of the uncertainties involved and the many complications which can occur, it is not surprising that the present picture is not too clear (see Heger et al. 2003 for an overview of possibilities). Nevertheless we will tentatively classify the different evolutionary scenarios according to the core mass M_c after helium burning. As can be seen, for example, from (36.3), the tracks for lower mass are below those for higher mass. We distinguish four cases, each of which is represented by one or more schematic evolutionary tracks in Fig. 36.1.

Case 1. If $M_c < M_{crit} \approx M_{Ch}$, and if the envelope is not massive enough (due either to the original mass or to mass loss), so that M_c cannot approach M_{Ch} during the shell burning phase, T_0 first grows in the non-degenerate regime until a maximum is reached. Then the core becomes degenerate, starts to cool, and the star must become a white dwarf. This is most likely the fate for most intermediate mass stars, which evolve as single stars (Chap. 34; Fig. 34.7). Only if a star is a member of a binary system and accretes sufficient mass at certain rates carbon can finally be ignited in a flash. A very popular scenario is the *double-degenerate* one, in which a CO-WD and a He-WD in a binary system merge due to the loss of angular momentum by gravitational radiation. It would explain how the CO core would reach M_{Ch} and why the spectrum would be devoid of hydrogen lines (the definition of type I supernovae). From the shell in which the flash occurs, a helium detonation wave (see Sect. 36.2.4) starts, moving both out- and inwards (This is in fact a simplified one-dimensional picture; in reality the flash occurs at a certain location, and the front travels in all directions and even around the star.). When it arrives near the centre, carbon will be ignited and a second (carbon) detonation front moves outwards, too. In this double-detonation model the star will finally be disrupted (for a summary see, for instance, Hillebrandt and Niemeyer 2000). Alternatively, in the *single degenerate scenario*, the CO-WD accretes matter from a non-degenerate companion. This could be a main-sequence star or a giant, for example, on the RGB. If the mass transfer rate is favourable, the accreted matter is burnt hydrostatically and the core grows in mass up to M_{Ch}. Again, the envelope is hydrogen free and the spectrum would classify the supernova as of type I. In this scenario the previous loss of matter in the pre-WD evolution is effectively reversed. A third possibility is that explosions in the accreted helium layers shock the CO-core sufficiently to trigger a nuclear runaway, although the core has not reached the Chandrasekhar mass. These are the *Sub-Chandrasekhar models*. In all three cases the explosion of the star is due to a thermonuclear runaway resulting from the carbon flash. These are the type Ia supernovae. We will discuss basic facts of the carbon flash briefly in Sect. 36.2.

Case 2. If initially $M_\mathrm{c} < M_{\mathrm{crit}}$, but if the remaining envelope is sufficiently massive, so that because of shell burning, M_c can grow to M_{Ch}, the core becomes degenerate and cools after having reached a maximum temperature. But ϱ_0 increases with M_c, and finally carbon burning begins (e.g. by pycnonuclear reactions; compare with Sect. 35.2). It starts in a highly degenerate state and is therefore explosive. This carbon flash can occur in stars that have started on the main sequence in the range $4 \lesssim M/M_\odot < 8$, if their mass loss has not been too strong. However, as we have mentioned before, this seems to be unlikely, although it cannot be excluded completely, for example, in the case of extremely metal-poor or metal-free stars (so-called Population III or *First Stars*), where mass loss may be significantly lower than in stars of solar metallicity (Chap. 9). Since the spectrum in such an event would contain hydrogen lines, as is the definition for type II supernovae, but the explosion mechanism is that of a thermonuclear runaway, typical for type Ia supernovae, such events are called supernovae of type 1.5. Whether they exist remains unclear.

Case 3. If $M_{\mathrm{crit}} < M_\mathrm{c} \lesssim 40 M_\odot$, the evolutionary track misses the non-relativistic region of degeneracy. The core heats up, reaching successively higher nuclear fusion phases. In a small mass range above the minimum mass to start carbon burning (this critical mass is usually refered to as M_{up}; see Fig. 34.10), electron captures by Mg, Na, and Ne reduce the pressure and a central collapse ensues. This is the fate of some of the super-AGB stars of Sect. 34.8, if the mass of the resulting NeO-core is initially below $1.37 M_\odot$ to avoid Ne ignition, but reaches this critical value due to shell burning. This will be discussed further in Sect. 36.3.4. The corresponding CO core mass limit is of order 2–$4 M_\odot$. For $M_\mathrm{c} \gtrsim 4 M_\odot$, photodisintegration of Ne and Mg nuclei brings γ_{ad} below $4/3$ and triggers a collapse. Both types of collapse may lead to neutron-star formation and to the ejection of the envelope, the latter mechanism also to black holes as the stellar remnants. It is assumed to cause the standard type II supernovae, and will be introduced in Sect. 36.3.

Case 4. : If $M_c \gtrsim 40M_\odot$, the cores also reach the carbon burning in a non-degenerate state as in Case 3. This mass limit is, as always, metallicity dependent, and corresponds to helium-core mass of $\approx 65M_\odot$ and an initial mass of $\approx 140M_\odot$. After carbon burning the evolutionary tracks in Fig. 36.1 cross the region of pair creation, which also reduces γ_{ad}. If $\gamma_{ad} < 4/3$ in an appreciable fraction of the core, say, within 40 % of its mass, then the core collapses adiabatically until the temperature of oxygen burning is reached. This may stop the collapse and make the star explode; if not, the collapse would lead into the region of instability because of photodisintegration, and the events would be as in Case 3. We will discuss this in Sect. 36.3.5. The remnants of stars in this mass range will be neutron stars (for lower masses), black holes by fallback on the proto-neutron star, or black holes by direct formation. Pair-instability supernovae, a subclass of type II core collapse supernovae, leave no remnant at all.

36.2 Carbon Ignition in Degenerate Cores

Consider stars starting with masses in the range $4 \lesssim M/M_\odot \lesssim 8$ and assume that they have almost no mass loss. After helium burning, they will form a CO-core that is degenerate, and in the subsequent evolution, M_c grows owing to shell burning until it comes close to M_{Ch}. During this phase the central density increases with increasing M_c (similar to a sequence of white dwarfs with increasing mass). The energy released in the core during this contraction is transported by electron conduction in the direction of the centre, where the temperature is smaller and neutrino losses (see Sect. 18.7) carry away the energy. The increase of the central density or of the temperature at the place of its maximum finally ignites carbon burning.

36.2.1 The Carbon Flash

The ignition of carbon in degenerate CO-cores of mass $M_c \approx M_{Ch}$ has already been discussed in Sect. 35.2. As described there, the ignition of carbon may occur in the centre or in the shell of maximum temperature. The general properties of the flash are the same in both cases. We discuss here the central ignition in the case of strong degeneracy, but recall that most likely only stars above $8M_\odot$ will reach carbon ignition and this will happen off-centre at very modest degeneracy. The carbon flash under such circumstances is described in the literature about super-AGB stars, for example, by García-Berro and Iben (1994).

In Fig. 36.2 the lg ϱ_0-lg T_0 plane is shown again with an evolutionary path of the centre. The stability behaviour of the degenerate core depends critically on the question whether the energy balance is dominated by neutrino losses ($\varepsilon_{CC} - \varepsilon_\nu < 0$: stable) or by carbon burning ($\varepsilon_{CC} - \varepsilon_\nu > 0$: unstable).

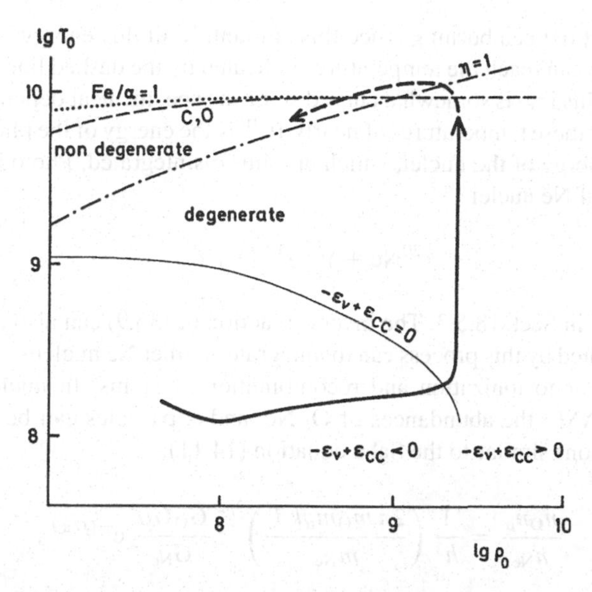

Fig. 36.2 Schematic evolution of the central region during and after the carbon flash (*heavy*). It corresponds to the evolution of type B^* in Fig. 36.1. The flash starts when the central density ϱ_0 (in g cm^{-3}) or the central temperature T_0 (in K) is so high that the neutrino losses do not overcome the energy generation by carbon burning. The temperature then rises almost at constant density until degeneracy is removed. The *dot-dashed line* labelled $\eta = 1$ indicates where the gas pressure is twice the (degenerate) pressure at temperature zero; it roughly separates the regions of degeneracy and non-degeneracy. The *broken line* labelled C, O gives the temperature reached if all the energy released by carbon burning is used to increase the internal energy. The *dotted line* labelled Fe/α = 1 shows the points for which statistical equilibrium gives equal abundances of iron and helium

The borderline $\varepsilon_{CC} - \varepsilon_\nu = 0$ bends down at a few $10^9\,\mathrm{g\,cm^{-3}}$, since ε_{cc} here increases mainly with increasing density (pycnonuclear reactions, see Sect. 18.4). Numerical calculations indicate that CO-cores reach the critical border $\varepsilon_{CC} - \varepsilon_\nu = 0$ between stability and instability at a density of $2 \times 10^9\,\mathrm{g\,cm^{-3}}$.

The slightest increase in temperature now makes $\varepsilon_{CC} - \varepsilon_\nu > 0$. Because of degeneracy the pressure does not increase and there is no consumption of energy through expansion. Therefore the temperature rises even more: a violent flash occurs. As in the case of the helium flash (see Sect. 33.4) the involved matter heats up at constant density until degeneracy is removed. Then it expands.

36.2.2 Nuclear Statistical Equilibrium

How violent the carbon flash can become is seen from a simple estimate. In a mixture of equal parts of C and O the carbon burning can release 2.5×10^{17} erg/g and the subsequent oxygen burning twice this amount. If all this energy is used to heat the material, it can reach the temperatures indicated by the dashed line labelled C, O in Fig. 36.2. This line is somewhat curved since the specific heat depends slightly on the density. At these temperatures of nearly 10^{10} K the energy of the photons exceeds the binding energy of the nuclei, which are thus disintegrated. Photodisintegration, for example, of Ne nuclei

$$^{20}\mathrm{Ne} + \gamma \rightarrow ^{16}\mathrm{O} + \alpha, \tag{36.9}$$

was discussed in Sect. 18.5.3. The inverse reaction of (36.9) can also occur, and the photon generated by this process can disintegrate another Ne nucleus. The processes are very similar to ionization and recombination of atoms. In nuclear statistical equilibrium (*NSE*) the abundances of O, Ne, and α particles can be derived from a set of equations similar to the Saha equation (14.11):

$$\frac{n_\mathrm{O} n_\alpha}{n_\mathrm{Ne}} = \frac{1}{h^3}\left(\frac{2\pi m_\mathrm{O} m_\alpha kT}{m_\mathrm{Ne}}\right)^{3/2} \frac{G_\mathrm{O} G\alpha}{G_\mathrm{Ne}}\, \mathrm{e}^{-Q/kT}, \tag{36.10}$$

where G_O, G_α, and G_Ne are the statistical weights, while Q is the difference of binding energies

$$Q = (m_\mathrm{O} + m_\alpha - m_\mathrm{Ne})c^2. \tag{36.11}$$

In addition to (36.10) there are two other conditions, one of which relates the particle numbers to the density, the other one describing the initial composition, since (36.9) and its inverse cannot change $n_\mathrm{O} - n_\alpha$. Of course, one cannot consider a single reaction only, but has to take into account all reactions that can take place simultaneously. For example, α particles generated by (36.9) can also be captured by ^{12}C or ^{20}Ne (The problem is similar when ionization of different elements takes place simultaneously. They are not independent of each other, since all of them produce electrons which influence all recombination rates.).

If the temperatures are sufficiently high, many nuclei are disintegrated by photons and their fragments react again. The abundances of the different elements are then determined by a set of "Saha formulae" of the type (36.10). The nucleus $^{56}_{26}$Fe as the most stable one plays a crucial role in this statistical equilibrium. It can be disintegrated by photons into α particles and neutrons:

$$\gamma + ^{56}_{26}\mathrm{Fe} \rightleftarrows 13\alpha + 4\mathrm{n}. \tag{36.12}$$

In order to determine the ratio n_Fe/n_a we consider quite general reactions of the type

$$\gamma + (Z, A) \rightleftarrows (Z-2, A-4) + \alpha, \tag{36.13}$$

$$\gamma + (Z, A) \rightleftarrows (Z, A-1) + \mathrm{n}. \tag{36.14}$$

We start with the nucleus $(26, 56) = {}^{56}\mathrm{Fe}$ and consider 13 reactions of type (36.13) and four of type (36.14). Then the abundance ratios are all given by equations like (36.10), and they can be combined to

$$\frac{n_\alpha^{13} n_\mathrm{n}^4}{n_\mathrm{Fe}} = \frac{G_\alpha^{13} G_\mathrm{n}^4}{G_\mathrm{Fe}} \left(\frac{2\pi kT}{h^2} \right)^{24} \left(\frac{m_\alpha^{13} m_\mathrm{n}^4}{m_\mathrm{Fe}} \right)^{3/2} \mathrm{e}^{-Q/kT}, \qquad (36.15)$$

with

$$Q = (13 m_\alpha + 4 m_\mathrm{n} - m_\mathrm{Fe}) c^2. \qquad (36.16)$$

If one assumes that the numbers of protons to neutrons (independently of whether they are free or in nuclei) have a ratio $n_\mathrm{p}/n_\mathrm{n} = 13/15$, as it is in the nucleus ${}^{56}\mathrm{Fe}$, then

$$n_\mathrm{n} = \frac{4}{13} n_\alpha . \qquad (36.17)$$

This, for instance, would be approximately the case in a mixture in which ${}^{56}\mathrm{Fe}$ is by far the most abundant heavy nucleus and its disintegration yields almost all neutrons and α particles. Then the left-hand side of (36.15) can be replaced by

$$\left(\frac{4}{13} \right)^4 \frac{n_\alpha^{17}}{n_\mathrm{Fe}} . \qquad (36.18)$$

Ignoring the binding energies, we can write the density as

$$\varrho = (56 n_\mathrm{Fe} + 4 n_\alpha + n_\mathrm{n}) m_\mathrm{u}, \qquad (36.19)$$

where m_u is the atomic mass unit. For given values of ϱ, T, and the ratio n_n/n_α [corresponding to (36.17)] with (36.15), (36.18) and (36.19) we have two equations for n_Fe and n_α.

Suppose again that the ratio of protons to neutrons per unit volume, normally called $\bar{Z}/\bar{N}$, is 13/15. Then equilibrium demands that all matter goes into ${}^{56}\mathrm{Fe}$ (the nucleus of the highest binding energy per nucleon) for temperatures that are not too high, and into ${}^{4}\mathrm{He}$ for high temperatures (see Fig. 36.3a). However, if we assume $\bar{Z}/\bar{N} = 1$, then for the former temperatures ${}^{56}_{28}\mathrm{Ni}$ is the dominant nucleus, since it has the highest binding energy per nucleon of all nuclei with $Z = N$. With increasing temperature the equilibrium shifts from ${}^{56}\mathrm{Ni}$ to ^{54} Fe+2p and finally to 14 ${}^{4}\mathrm{He}$. For very high temperatures it may even shift to the basic constituents, protons and neutrons (see Fig. 36.3b).

The value $\bar{Z}/\bar{N}$ at the occurrence of photodisintegration depends on the weak interaction processes (β decays) during the nuclear history of the stellar matter. In any case, in equilibrium at moderate temperatures, one expects nuclei of the iron group, which with increasing temperature disintegrate to α particles and at temperatures around 10^{10} K, which can also be reached in exploding cores, even to protons and neutrons. In this case, (36.12)–(36.19) would have to be written for ${}^{56}\mathrm{Fe}$, n, and p.

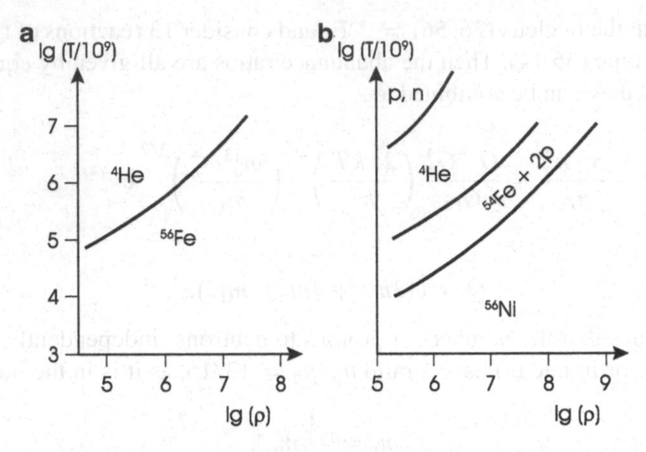

Fig. 36.3 (**a**) In the temperature-density diagram (T in 10^9 K, ϱ in g cm^{-3}) the curve separates the regions in which equilibrium demands matter to be in the form of ^{4}He and ^{56}Fe, respectively, for the case of $\bar{Z}/\bar{N} = 13/15$. (**b**) The corresponding equilibrium regions for $\bar{Z}/\bar{N} = 1$

36.2.3 Hydrostatic and Convective Adjustment

Even during the rapid helium flash the star remains very nearly in hydrostatic equilibrium, and convection can carry away all the released nuclear energy without becoming appreciably superadiabatic. The situation is completely different if unstable carbon burning proceeds in a degenerate core on a time-scale of milliseconds.

Consider the events after the onset of the carbon flash in the centre. The rapid rise of the central temperature is sufficient for immediately starting higher nuclear reactions, such as oxygen burning, which release additional energy. In one single runaway the central temperature rises so much that statistical equilibrium between Fe and He is reached, and eventually degeneracy is removed (see Fig. 36.2). Then the pressure increases and the central region starts to expand. This will occur roughly on a timescale τ_ε, in which the central temperature and the internal energy u rise. Since $\dot{T}/T \approx \varepsilon_{\mathrm{CC}}/u$, we have

$$\tau_\varepsilon = \frac{c_P T}{\varepsilon_{\mathrm{CC}}}. \tag{36.20}$$

The other regions of the core react on the central expansion on the hydrostatic timescale $\tau_{\mathrm{hydr}} \approx (G\bar{\varrho})^{-1/2}$ [compare with (2.19)], where $\bar{\varrho}$ is the mean density of the core. As long as $\zeta := \tau_\varepsilon/\tau_{\mathrm{hydr}} \gg 1$ the core follows the central expansion quasi-hydrostatically. If, however, $\zeta \ll 1$, then the layers above cannot react rapidly enough, and a compression wave will move outwards with the speed of sound. If the push by the suddenly expanding burning region is sufficiently strong, an outwards travelling shock wave may develop.

Owing to the energy release in the flash, a central convective core will form, which has two effects. Part of the surplus energy is carried away (reducing the intensity of the flash), and new nuclear fuel is brought to the region of carbon burning (enhancing the flash). A characteristic timescale for convection is $\tau_{conv} \approx \ell_m / v_s$, where ℓ_m is the mixing length and v_s the local velocity of sound. Indeed turbulent elements will scarcely move faster than v_s, since otherwise shock waves would strongly damp the motion. If $\xi := \tau_\varepsilon / \tau_{conv} \gg 1$, convection is able to carry away all the nuclear energy released. If, however, $\xi \ll 1$, then convection cannot carry away the released energy.

The timescales τ_{hydr} and τ_{conv} are very short indeed. For the central parts of the core with $\varrho > 10^8\,\mathrm{g\,cm^{-3}}$, one finds typically $\tau_{hydr} \approx 0.1\,\mathrm{s}$, and τ_{conv} is of the same order. However, for $T = 2 \ldots 3 \times 10^9\,\mathrm{K}$, the local timescale τ_ε for the flash is of the order of $10^{-6}\,\mathrm{s}$. Therefore ζ and ξ are both $\ll 1$. This means that, instead of hydrostatic adjustment, a compression wave will start outwards and that "convective blocking" prevents a rapid spread of released energy in the core. The changes caused by the flash in one mass element propagate comparatively slowly to other parts.

These estimates clearly show that the carbon flash and the following explosion can be treated accurately only when the full hydrodynamical equations are solved. The important role of convection and the propagation of burning fronts necessitate three-dimensional models. Such models are the current state of the art, but will not be the subject of this book.

36.2.4 Combustion Fronts

The local nuclear timescale τ_ε at the onset of the flash is rather short. If a flash is started somewhere in a degenerate CO-core, the burning proceeds at such high rates that the fuel in this mass element is used up almost instantaneously. To be more precise, the consumption is completed locally before the layers above can adjust. Only then is the unburnt material ahead heated to ignition (either by compression or by energy transport, which may be by convection), and the flash proceeds outwards. But the burning is always confined to a layer of (practically) zero thickness. We have an outward-moving *combustion front*, which can be of two different types.

We have seen that a shock wave develops. Matter in front penetrates the discontinuity with supersonic velocity and is compressed and heated. If this suffices to ignite the fuel, then the combustion front coincides with the shock front moving outwards supersonically. This is called a *detonation front.* It releases enough energy to lead to a complete stellar explosion, but matter ahead of the blast wave cannot expand and ignites under typical white-dwarf conditions. Temperatures and densities are so high that NSE is reached and is peaking around Ni. This is in conflict with the presence of intermediate-mass elements seen in the spectra of some type Ia supernovae.

If the compression in the shock does not ignite the fuel, then the ignition temperature is reached owing to energy transport (convection or conduction). This gives a slower, subsonic motion for the burning front and contains a discontinuity in which density and pressure drop. This is a *deflagration front.* Since it allows the nucleosynthesis to occur at lower density and pressure, NSE peaks at lower mass numbers and intermediate-mass elements can be created. Whether such deflagration models can unbind completely the white dwarf is unclear. Higher-dimensional simulations seem to be more promising to achieve this.

Obviously the speed of a deflagration front is controlled by that of energy transport. This in turn depends on the conductivity (thermal or convective) and on the temperature difference between the deflagration front and the material ahead. Numerical modelling thus needs hydrodynamical simulations of convection. Simple mixing length theory (Chap. 7), which has been useful for hydrostatic stellar evolution phases, certainly will not suffice to compute accurate models.

In both cases the deviations from hydrostatic equilibrium are mainly confined to a thin shell across which the pressure is discontinuous and all nuclear energy is released. The momentum of the matter approaching a detonation front supersonically is balanced by the higher pressure behind the front; the momentum of the matter approaching a deflagration front subsonically is balanced by the recoil of the matter moving away from it behind the front. The front in both cases is unstable to spatial perturbations, the scale of which is well below any numerical resolution in the simulations, and must be represented by a physical subscale model.

For an account of the theory of the two types of combustion fronts, see Courant and Friedrichs (1976), Landau and Lifshitz (1987), and Hillebrandt and Niemeyer (2000). As with normal shock waves, the theoretical results follow from the conservation of mass, momentum, and energy of the matter going through the discontinuity. For energy conservation, however, it also has to be taken into account that energy is released at the discontinuity. This makes the *two* types of solutions (detonation and deflagration waves) possible, while the theory of normal shock waves allows only that solution in which the density of matter going through the discontinuity increases.

In principle, detonation fronts as well as deflagration fronts can occur in stars. Which of the two will develop depends on the details of the transport mechanism, which determines the motion of a deflagration front and of the preceding shock.

In some cases the explosion may start out with a slow deflagration front, which allows some expansion of the layers ahead of the front, but which switches at some point into a detonation, when the front begins to progress supersonically. These are the *delayed detonation models*, which combine the advantage of allowing intermediate-mass elements to be created and deliver typical supernova energies, which are of order 10^{51} erg.

The details of type Ia supernova explosions can be uncovered only by very complicated and challenging numerical simulations that take into account both hydrodynamics and nuclear processes, and should resolve scales ranging from that of white dwarfs ($\sim 1,000$ km) down to that of the burning flame of a few cm. Early one-dimensional hydrodynamical calculations for detonation models were done by Arnett (1969) and Ivanova et al. (1977). A classical deflagration model, still in one dimension, is by Nomoto et al. (1976) and Nomoto (1984). State-of-the-art models are mostly 3-dimensional; a summary and discussion can be found in the reviews by Hillebrandt and Niemeyer (2000) and Roepke (2008).

36.2.5 *Carbon Burning in Accreting White Dwarfs*

Rather similar phenomena to those described above for CO-cores of single stars can occur in CO white dwarfs which are members of binary systems. They can receive appreciable amounts of matter from their companions. The accreted matter is compressed and heated, and its ignition can give rise to various phenomena.

For example, if helium is accreted with relatively low rates (about $10^{-8} M_\odot$/year), a helium flash will be ignited in a shell of high density. The result can be a double detonation wave: a helium detonation front running outwards and a carbon detonation front going to the centre. As a result the white dwarf will be disrupted.

For higher accretion rates the new material can burn quietly near the surface, thus simply increasing the mass of the CO white dwarf. When it approaches M_{Ch}, the density in the inner parts becomes so large that carbon burning starts either in the centre, or in the shell of maximum temperature. This results in a flash, and a deflagration (or detonation) front starts, as discussed above for single stars. The white dwarf will also be disrupted. Both possibilities correspond to Case 1 of Sect. 36.1. It is this mechanism which is generally believed to cause the Type Ia supernovae. Note that it has to be invoked, since the spectra of these supernovae show no hydrogen, and because evolving single stars of $M < 10 M_\odot$ may lose so much mass that their CO-core can never come close to M_{Ch}.

36.3 Collapse of Cores of Massive Stars

According to Fig. 36.1 one can expect that the cores of massive stars will not cool, because of non-relativistic degeneracy, but will heat up during core contraction until the next type of nuclear fuel is ignited. The core then is either non-degenerate (larger core mass M_{c}) or degenerate but to the upper right of the "summit" of the line $\alpha = 3/4$ in Fig. 28.1. In both cases the gravothermal heat capacity is negative, and the burning is self-controlled. In the following we discuss stars with core masses in the range $M_{\mathrm{ch}} < M_{\mathrm{c}} < 40 M_\odot$. The evolutionary paths of these stars will avoid the region of $\alpha < 3/4$, where in Fig. 28.1 the arrows point downwards.

After going through several cycles of nuclear burning and contraction, the core will heat up to silicon burning. Nuclear burning in several shell sources has produced layers of different chemical composition, as shown in Fig. 36.4. Finally the central region of the core reaches a temperature at which the abundances are determined by nuclear statistical equilibrium. In this stage the core is in a peculiar state in several respects. Since the electron gas dominates the pressure, and since at temperatures of $T_9 \approx 10$ the electrons are relativistic ($kT \approx 1.7 m_{\mathrm{e}} c^2$), the adiabatic exponent γ_{ad} is close to 4/3. In the more massive stars photodisintegration of heavy nuclei reduces γ_{ad} even more (like partial ionization). In addition general relativistic effects increase the critical value of γ_{ad} above 4/3, and the core becomes dynamically unstable. As a consequence core collapse sets in. For less massive stars the relativistic electrons are degenerate with high Fermi energies. Then electron

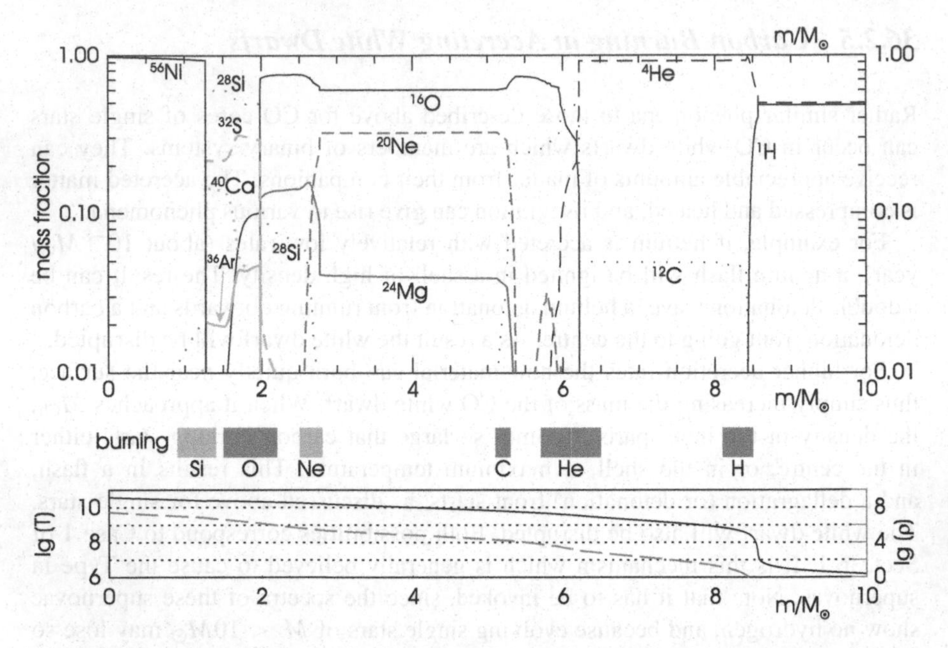

Fig. 36.4 The chemical composition in the interior of a highly evolved model of a population I star with an initial mass of $25M_{\odot}$, close to the end of hydrostatic nuclear burning. The mass at this time is reduced to $16M_{\odot}$ due to mass loss. In the *upper panel* the mass concentrations of important elements are plotted against the mass variable m. *Below* the abscissa, in the *middle* of the figure, the approximate location of shell sources in different nuclear burning phases is indicated by the *grey rectangles*. In the *lower panel* the run of temperature ($\lg T$: scale at *left axis*) and density ($\lg \varrho$: *right axis*) is given to identify typical burning conditions for these nuclear shells (data courtesy R. Hirschi, published in Hirschi et al. 2004)

captures by heavy nuclei reduce the pressure and start the collapse. For this stage we now discuss a simple solution.

36.3.1 Simple Collapse Solutions

Suppose we have a core at the onset of collapse, say, with central values $\varrho_0 = 10^{10}\,\mathrm{g\,cm^{-3}}$, $T_0 \approx 10^{10}\,\mathrm{K}$. The electrons are relativistically degenerate. Then the equation of state is polytropic and can be written as

$$P = K'\varrho^{4/3}, \qquad (36.21)$$

where $K' = K_{4/3}/\mu_e^{4/3}$ [compare with (15.26)]. Therefore the core can be described by a polytrope of index 3. We have already discussed the collapse of such a polytrope in Sect. 19.11. As we have seen there, the parameter λ appearing in the modified Emden equation (19.81) is a measure of the deviation from hydrostatic equilibrium, which corresponds to the value $\lambda = 0$. Solutions with finite radius are possible only for values $0 < \lambda < \lambda_m = 6.544 \times 10^{-3}$, where $\lambda = \lambda_m$ corresponds to the strongest deviation from equilibrium. For $\lambda = \lambda_m$ no homologous collapse of a polytrope of $n = 3$ is possible.

We now adapt the formalism of Sect. 19.11 for application to the collapse of stellar cores. The solution of the spatial structure is given by the function $w(z)$, which obeys (19.81). We denote the value of z at the surface of the collapsing core by z_3, so that $w(z_3) = 0$; for $\lambda = 0$ one has $z_3 = 6.897$. It increases with λ and reaches the maximum value 9.889 for $\lambda = \lambda_m$. The limit $\lambda = \lambda_m$ is reached when the surface of the core collapses with the acceleration of free fall.

If we apply (19.75) to the surface we have

$$z_3 \ddot{a} = -\frac{4}{3}\lambda \frac{(K')^{3/2}}{\sqrt{\pi G}} \frac{z_3}{a^2}. \tag{36.22}$$

If this is equal to the free-fall acceleration $-GM_c/(az_3)^2$, then

$$\lambda = \lambda_m = \frac{3}{4}\sqrt{\frac{\pi G}{K'^3}} \frac{GM_c}{z_3^3}. \tag{36.23}$$

On the other hand, (19.67) and (19.81) give

$$\frac{\varrho}{\varrho_0} = w^3 = \lambda - \frac{1}{z^2}\frac{d}{dz}\left(z^2 \frac{dw}{dz}\right), \tag{36.24}$$

and therefore with $r = az$, $R_c = az_3$, and

$$\bar{\varrho} = \frac{3}{R_c^3}\int_0^{R_c} \varrho r^2 dr, \tag{36.25}$$

after some manipulation we find

$$\frac{\bar{\varrho}}{\varrho_0} = \lambda - \left[\frac{3}{z}\left(\frac{dw}{dz}\right)\right]_{z=z_3}. \tag{36.26}$$

If we apply this to the limit case $\lambda = \lambda_m$ in which dw/dz vanishes at the surface (compare with Fig. 19.3), we find $\bar{\varrho}/\varrho_0 = \lambda_m$.

The core may start out from the (marginally stable) equilibrium for which $\lambda = 0$. Here the actual acceleration at the surface is zero, since gravity and pressure gradient cancel each other. But if the pressure is slightly decreased, the core will start to collapse ($\lambda > 0$). The numerical integration of (19.81) for different values of λ in the range $0 \leq \lambda \leq \lambda_m$ gives values for z_3 and $\bar{\varrho}/\varrho_0$ in the ranges $6.897 \leq z_3 \leq 9.889$ and $0.01846 \leq \bar{\varrho}/\varrho_c \leq 0.0654$ (Goldreich and Weber 1980). If we determine the masses for different collapsing polytropes, we can use the expression

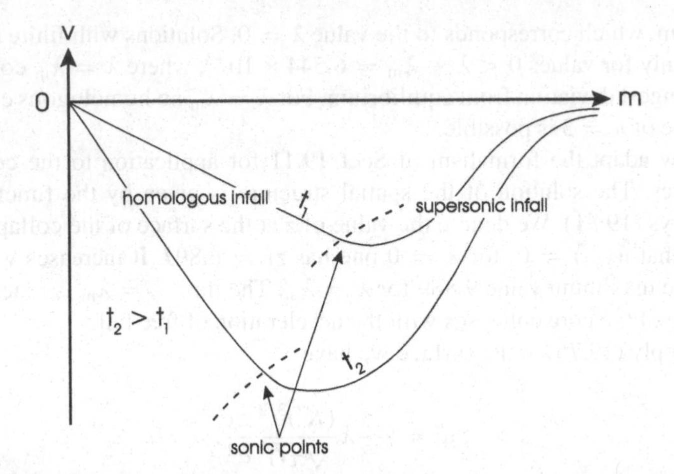

Fig. 36.5 Schematic picture of the velocity distribution in a collapsing stellar core obtained in numerical calculations, and taken at two subsequent times t_1 and t_2. Note the two regimes: on the *left* $|v_r|$ increases in the outward direction. It corresponds to a (roughly) homologously collapsing part, while on the *right* $|v_r|$ decreases with m. This corresponds to the free-fall regime, where velocities are supersonic. The run of the (negative) sound speed is indicated by the *dashed lines*, and the location, where the infall becomes supersonic, by *arrows*. With time the mass of the "inner core" (defined as the part left of the maximum infall velocity) is decreasing. Velocities are of the order of 10^9 cm s^{-1} (after Müller 1997)

$$M_c \equiv \frac{4\pi a^3 z_3^3 \varrho_0}{3} \frac{\bar{\varrho}}{\varrho_0} = \frac{4\pi z_3^3}{3} \left(\frac{K'}{\pi G}\right)^{3/2} \frac{\bar{\varrho}}{\varrho_0}, \tag{36.27}$$

which has been derived with the help of (19.67). Equation (36.27) for $\lambda = 0$ gives the Chandrasekhar mass M_{Ch}, as can be seen from (19.29), (19.30) and (36.26). In fact all masses obtained for different values of λ in the narrow interval $0 \leq \lambda \leq \lambda_m$ are close to the Chandrasekhar mass, namely $M_{Ch} \leq M_c \leq 1.0499 M_{Ch}$.

Only core masses in this small interval can collapse homologously. Now we know that $M_{Ch} \sim \mu_e^{-2}$. Electron captures during the collapse increase μ_e and reduce M_{Ch}. Therefore the upper bound for M_c for homologous collapse decreases. If initially $\mu_e = \mu_{e0}$ and $M_{Ch} = M_{Ch0}$, then after some time not more than the mass

$$M_c = 1.0499 \left(\frac{\mu_{e0}}{\mu_e}\right)^2 M_{Ch0} \sim \mu_e^{-2} \tag{36.28}$$

can collapse homologously (Note that, strictly speaking, the whole formalism should be repeated for a time-dependent K'.). Numerical integrations in fact indicate that during collapse the mass of the homologously collapsing part of the core decreases with increasing μ_e as given by (36.28).

This simple collapse model has been generalized by Yahil and Lattimer (1982) for values of the polytropic index in (36.21) between $6/5 < \gamma_{ad} < 4/3$. Figure 36.5 shows the infall velocity as a function of m as obtained from numerical computations, in agreement with the models by Goldreich and Weber, and Yahil and Lattimer. The maximum separates the homologously collapsing inner core (left) from the nearly free-falling outer part of the core (right). The outer core collapses supersonically; the sound speed is exceeded at a location somewhat interior to the maximum collapse velocity. During collapse the boundary between the two regimes is not fixed but moves to smaller m values: mass from the inner core is released into the free-fall regime. This corresponds to the decrease of M_{Ch} with increasing μ_{e} as discussed above.

The collapse is extremely short-lived; it takes a time which is of the order of the free-fall time. If the core starts with an initial density of $10^{10}\,\mathrm{g\,cm^{-3}}$ one obtains $\tau_{\mathrm{ff}} \approx (G\bar{\varrho})^{-1/2} \approx 40$ ms at the onset of collapse, while it is 0.4 ms for $\bar{\varrho} = 10^{14}\,\mathrm{g\,cm^{-3}}$.

36.3.2 The Reflection of the Infall

Because of the collapse, the density finally approaches that of neutron stars (nuclear densities of the order $10^{14}\,\mathrm{g\,cm^{-3}}$). Then the equation of state becomes "stiff," i.e. the matter becomes almost incompressible. This terminates the collapse.

If the whole process were completely elastic, then the kinetic energy of the collapsing matter would be sufficient to bring it back after reflection to the state just before the collapse began. This energy can be estimated roughly from

$$E \approx GM_{\mathrm{c}}^2 \left(\frac{1}{R_{\mathrm{n}}} - \frac{1}{R_{\mathrm{wd}}} \right) \approx \frac{GM_{\mathrm{c}}^2}{R_{\mathrm{n}}} \approx 3 \times 10^{53}\,\mathrm{erg}, \tag{36.29}$$

where M_{c} is the mass of the collapsing core, while R_{n} and R_{wd} are the typical radii of a neutron star and of a white dwarf. We compare this with the energy E_{e} necessary to expel the envelope, which had no time to follow the core collapse,

$$E_{\mathrm{e}} = \int_{M_{\mathrm{wd}}}^{M} \frac{Gm\,dm}{r} \ll \frac{GM^2}{R_{\mathrm{wd}}} \approx 3 \times 10^{52}\,\mathrm{erg} \tag{36.30}$$

for $M = 10 M_{\odot}$. Realistic estimates bring E_{e} down to 10^{50} erg, and therefore only a small fraction of the energy involved in the collapse of the core is sufficient to blow away the envelope. In predicting what happens after the bounce, one has to find out what (small) fraction of the energy of the collapse can be transformed into kinetic energy of outward motion. Remember that the energy estimated in (36.29) would suffice only to bring back the whole collapse to its original position–and no energy would be left for expelling the envelope. But if a remnant (neutron star) of mass M_{n} remains in the condensed state, the energy of its collapse is available. The question is how this can be used for accelerating the rest of the material outwards.

A possible mechanism would be a shock wave moving outwards. The remnant is somewhat compressed by inertia beyond its equilibrium state and afterwards, acting like a spring, it expands, pushing back the infalling matter above. This creates a pressure wave, steepening when it travels into regions of lower density. The kinetic energy stored in such a wave may be sufficient to lift the envelope into space. However, the following problem arises. One can imagine that the neutron star formed has a mass of the order of the final Chandrasekhar mass M_{ChF}. The rest of the collapsing matter still consists mainly of iron. When, after rebounce, this region is passed by the shock wave, almost all of its energy is used up to disintegrate the iron into free nucleons. Therefore only a small fraction of the initial kinetic energy remains in the shock wave and is available for lifting the envelope.

In fact the major part of the energy estimated in (36.29) of order 10^{53} erg is lost in the form of neutrinos (Sect. 36.3.3). Only 1% of it–10^{51} erg–is actually converted into kinetic energy, and only a few per cent of this is escaping from the supernova in the form of light. Nevertheless, this tiny part of the collapse energy makes supernovae the brightest stellar objects in the universe.

36.3.3 Effects of Neutrinos

Before collapse, neutrinos were created by the processes described in Sect. 18.7, and their energy is of the order of the thermal energy of the electrons. During collapse, neutrino production by neutronization becomes dominant. As soon as the density approaches values of $10^{12}\,\mathrm{g\,cm^{-3}}$, inverse β decay becomes more pronounced, and the equilibrium shifts to increasingly neutron-rich nuclei. During this neutronization neutrinos are released. In connection with supernova SN 1987A, neutrinos have been observed in underground neutrino detectors–manifest evidence that core collapse is indeed connected with the supernova phenomenon. The typical energy of the neutrinos released during collapse is of the order of the Fermi energy of the (relativistic) electrons. Therefore when using the relation $\varrho = \mu_e n_e m_u$ and (15.11) and (15.15) one finds

$$\frac{E_\nu}{m_e c^2} \approx \frac{E_F}{m_e c^2} = \frac{p_F}{m_e c} = \left(\frac{3}{8\pi m_u}\right)^{1/3} \frac{h}{m_e c}\left(\frac{\varrho}{\mu_e}\right)^{1/3} \approx 10^{-2}\left(\frac{\varrho}{\mu_e}\right)^{1/3}. \tag{36.31}$$

If heavy nuclei are present, the neutrinos interact predominantly through the so-called "coherent" scattering (rather than scattering by free nucleons):

$$\nu + (Z,A) \rightarrow \nu + (Z,A). \tag{36.32}$$

The cross section is of the order of

$$\sigma_\nu \approx \left(\frac{E_\nu}{m_e c^2}\right)^2 A^2 10^{-45}\mathrm{cm}^2, \tag{36.33}$$

which with (36.31) gives

$$\sigma_\nu \approx A^2 \left(\frac{\varrho}{\mu_e}\right)^{2/3} 10^{-49}\mathrm{cm}^2. \tag{36.34}$$

This allows an estimate of the mean-free-path ℓ_ν of neutrinos in the collapsing core. If $n = \varrho/(Am_u)$ is the number density of nuclei, then with (36.34)

$$\ell_\nu \approx \frac{1}{n\sigma_\nu} = \frac{1}{\mu_e A}\left(\frac{\varrho}{\mu_e}\right)^{-5/3} 1.7 \times 10^{25}\mathrm{cm}. \tag{36.35}$$

Can ℓ_ν become comparable with the dimension of the collapsing core, say, 10^7 cm? With $\mu_e = 2, A \approx 100$, we obtain from (36.35) $\ell_\nu = 10^7$ cm for $(\varrho/\mu_e) = 3.6 \times 10^9\,\mathrm{g\,cm^{-3}}$. We may bring (36.35) into a more convenient form by putting in such typical values for μ_e, A, and ϱ such that it yields a typical length

$$\ell_\nu \approx \frac{1}{n\sigma_\nu} = \frac{200}{\mu_e A}\left(\frac{2\varrho}{10^{10}\mu_e}\right)^{-5/3} 5.8 \times 10^6\,\mathrm{cm}. \tag{36.36}$$

Obviously we cannot simply assume that the neutrinos escape without interaction. The more the density rises, the smaller ℓ_ν, and the collapsing core becomes opaque for neutrinos. Then they can only diffuse through the matter via many scattering processes. For sufficiently high density the diffusion velocity becomes even smaller than the velocity of the collapse. Calculations show that the neutrinos cannot escape by diffusion within the free-fall time τ_{ff} of the core if $\varrho \gtrsim 3 \times 10^{11}\,\mathrm{g\,cm^{-3}}$: the neutrinos are then trapped.

In the schematic picture of the core structure (Fig. 36.6), the place where the infall velocity of matter equals the velocity of outward neutrino diffusion is indicated as the "neutrino trapping surface". Below it the neutrinos are trapped; above it they diffuse outwards until reaching the so-called "neutrinosphere". This provides the boundary of the opaque part of the core and is located one mean free path ℓ_ν beneath the surface. From here the neutrinos leave the core almost without further interaction.

Detailed calculations have to deal with a radiation-hydro-problem, where one has to solve the neutrino transport problem in a six-dimensional phase space, defined by three spatial and three momentum coordinates, one of the latter being the neutrino energy, for example. In particular one has to consider and calculate the detailed distribution function of the neutrinos (rather than their average energy). This is obvious since the cross section as given in (36.33) depends on the energy of the neutrinos: those with low energy can escape more easily than those of high energy. This problem, which is essentially that of solving the Boltzmann equation, requires very extensive and challenging computations and has not been solved so far in all aspects for realistic physical conditions. Another important aspect is the detailed consideration of all neutrino interactions with matter, including neutrino oscillations and cross sections for the different neutrino families. The neutrino transport is essential for modelling type II supernova explosions as the neutrinos deposit part of

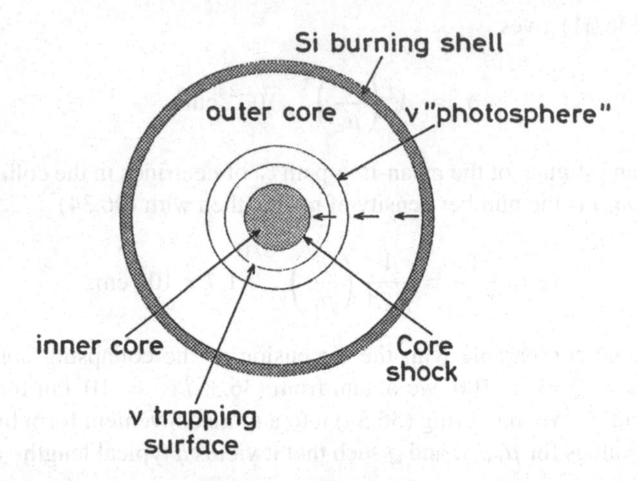

Fig. 36.6 Schematic picture of a collapsing stellar core at bounce. The *short arrows* correspond to the velocity field. At the sphere labelled *core shock*, the shock is formed. Inside this sphere the *matter* is almost at rest. Above the shock there is a still collapsing shell in which neutrinos are trapped. But on top there is a shell from which neutrinos can escape. One can define a neutrino photosphere, a *neutrinosphere*, analogous to the photosphere in a stellar atmosphere

their energy below the neutrinosphere, possibly raising the energy to levels sufficient to expel the envelope.

The congestion of the neutrinos, resulting from the opaqueness of the core, influences the further neutronization. With increasing density the neutrinos become degenerate with a high Fermi energy. Electron capture becomes less probable, since the new neutrinos have to be raised to the top of the Fermi sea. Around a density of $3 \times 10^{12}\,\mathrm{g\,cm^{-3}}$ the so-called β-equilibrium is reached, where the reaction $p + e^- \leftrightarrow n + \nu_e$ proceeds in both directions. However, the neutrino capture reaction is also subject to the requirement that the resulting electron has to have an energy above the Fermi energy of the degenerate electron gas. In total, with increasing density, β-equilibrium shifts to the right-hand side. Since the neutrinos can no longer escape, the number of leptons (electrons and neutrinos) stays constant. γ_{ad} has increased to a value close to $4/3$, which corresponds to relativistic degeneracy. The collapse continues until $\varrho > 10^{14}\,\mathrm{g\,cm^{-3}}$, the nuclear density. At such densities the equation of state is very stiff, and $\gamma_{\mathrm{ad}} \gtrsim 2$ due to the repulsive nuclear forces of the strong interaction. Therefore the collapse is stopped. Further neutronization can proceed only as far as the neutrinos diffuse outwards. This enables further electron captures on protons, lowering the proton-to-neutron ratio. Most of this takes place in the neutronization shell between trapping surface and neutrino photosphere (Fig. 36.6) where the density is several $10^{11}\,\mathrm{g\,cm^{-3}}$. During this phase, which can last a few to 10 s, the proto-neutron star evolves into a neutron star.

As in the case of thermonuclear supernova explosions caused by the carbon flash detailed models are possible only by two- and three-dimensional hydrodynamical simulations, taking into account nucleosynthesis and the problem of neutrino transport in full detail. Another major ingredient is the equation of state at nuclear matter density, finite temperature, and for extremely neutron-rich matter. Such computations are extremely demanding, require always the latest generation of supercomputers, but are still from giving final answers concerning the details of core collapse supernova explosions. Some of these simulations have resulted in successfully exploding pre-SN models, other have failed or remained inconclusive. We therefore refer the reader to some recent reviews about the subject, given by Woosley and Janka (2005), Mezzacappa (2005), and Janka et al. (2007).

36.3.4 Electron-Capture Supernovae

While electron capture plays an important role in all core-collapse supernovae, it is particularly crucial in the specific case of degenerate NeOMg cores reaching a critical density of $4.5 \times 10^9\,\mathrm{g\,cm^{-3}}$, equivalent to a mass of $1.37 M_\odot$ (Nomoto et al. 1984). These conditions are reached, according to the models by Nomoto and others, within helium cores slightly less massive than $2.5 M_\odot$ in stars of an initial mass of about $9 M_\odot$ or somewhat higher, i.e. the super-AGB stars of Sect. 34.8 (see also Fig. 34.10).

The core collapse is initiated here by the capture of electrons on ^{24}Mg and ^{20}Ne, since this is energetically preferred over keeping the electrons at high energy in the Fermi distribution. This reduces pressure, which is mainly provided by the degenerate electrons, and contraction sets in. During the ensuing collapse oxygen burning starts, but the released nuclear energy is not sufficient to stop the collapse, since the energy budget is dominated by the loss due to neutrinos emitted in the electron-capture process. The nuclear burning proceeds to nuclear statistical equilibrium, which, in the course of the collapse, first shifts to α-particles and in the final phase to neutrons and protons. The result is a neutron star of low mass ($\lesssim 1.37 M_\odot$). According to numerical simulations (Kitaura et al. 2006) the supernova explosion is driven by the neutrino heating mechanism, and comparably small amounts of metals, in particular of O, C, and Ni ($< 0.015 M_\odot$), are ejected. The overall explosion energy is of order 10^{50} erg, and therefore much lower than in type II supernovae from more massive stars. These results agree with properties of the Crab supernova remnant and pulsar, and thus this historical supernova is believed to be of the electron-capture type. It could therefore be evidence for a previous super-AGB evolution. However, the absence of hydrogen in the Crab nebula points to a previous binary star evolution.

36.3.5 Pair-Creation Instability

From Fig. 36.1 one can see that evolutionary tracks for cores of sufficient mass enter a region on the left-hand side of the diagram where also $\gamma_{\mathrm{ad}} < 4/3$ (Fowler and Hoyle 1964). In this region many photons have an energy exceeding the rest-mass energy of two electrons, $h\nu \geq 2m_e c^2$. Therefore electron-positron pairs can be spontaneously formed out of photons in the fields of nuclei. Admittedly the pairs do annihilate, creating photons again, but there is always an equilibrium number of pairs present. The *mean* energy of the photons $h\nu \approx kT$ equals the rest energy of the electron–positron pair only at a temperature of 1.2×10^{10} K, but even at 10^9 K appreciable pair creation occurs because of the high-energy photons of the Planck distribution.

For an account of the thermodynamic effects of pair creation, see, for example, Weiss et al. (2004). In many respects pair creation can be considered in analogy to ionization or dissociation (a photon being "ionized" or "dissociated" into a pair e^-, e^+). Regarding the stability of massive cores, the crucial point is that the pair creation reduces γ_{ad}, as incomplete ionization or photodisintegration does. Indeed, if the gas is compressed, not all the energy is used to increase the temperature, but part of it is used to create pairs. Other reductions of γ_{ad} are due to high radiation pressure according to (13.7), (13.12) and (13.15) and to relativistic electrons. All these effects bring γ_{ad} below the critical value $4/3$ for dynamical instability.

The total number of electrons consists of those from pairs and those from normal ionization of atoms. With increasing ϱ the Fermi energy rises. This diminishes the possibility for pair creation, since newly created electrons now need an energy exceeding the Fermi energy. Correspondingly the instability region in Fig. 36.1 is limited to the right at a density of $5 \times 10^5\,\mathrm{g\,cm^{-3}}$.

The pairs created are not relativistic, having $\gamma_{ad} = 5/3$ (Note that a photon with $h\nu = m_e c^2$ can only create a pair with zero kinetic energy!). For higher temperatures there are so many pairs that they dominate and bring γ_{ad} of the whole gas–radiation mixture slightly above $4/3$, which limits the instability region towards high temperatures. In summary, the three effects discussed in the preceding paragraphs explain the island nature of the pair-creation instability in Fig. 36.1.

For the evolution of cores into the region of pair instability, radiation pressure is important, and therefore one cannot use our simple formulae of Sect. 36.1. Furthermore, for a core instability, it is not sufficient that the evolutionary track of the star's *centre* moves through the area with $\gamma_{ad} < 4/3$. Since in reality a mean value of γ_{ad} over the whole core decides upon its dynamical stability (Sect. 40.1), an appreciable fraction of the core mass must lie in that density–temperature range. According to numerical results this happens to cores of masses of $40M_{crit}$ and more, where M_{crit} is defined in (36.5). The corresponding main-sequence masses depend on the uncertain mass loss, but a realistic guess seems to be that stars initially with $M > 80 - 100M_\odot$ later develop pair-unstable cores. This, however, assumes that no appreciable mass is lost due to radiation-driven stellar winds during the main-sequence phase. In addition, violent radial pulsations by the ϵ-mechanism (Sects. 41.1 and 41.5) may lead to a significant mass reduction for stars with $M \gtrsim 60M_\odot$. Both effects depend on metallicity. Therefore for solar metallicity models predict a maximum helium core mass of about $10M_\odot$, while for metal-free Pop. III stars, they may exceed the critical value for pair instability. It may be that this kind of supernova explosions may be restricted to the very early universe.

The final fate of stars this massive is rather uncertain. Numerical calculations indicate that, in a collapsing core of this type, oxygen is ignited explosively and the core runs into the (unstable) region of photodisintegration, which may cause a total disruption of the star. There is also the possibility of violent pulsations caused by the instability, which lead to explosive mass loss, but no total disruption. The star may thus (for increasing main-sequence mass) end in a black hole after having expelled large parts of its hydrogen/helium layers, be totally disrupted, or collapse directly into a black hole. The situation and the different outcomes have been summarized by Heger et al. (2003).

We also mention that rotation is playing a crucial role also for the final phases of massive star evolution, although the basic effects as discussed in this chapter remain the same. Details can be found in the textbook by Maeder (2009).

36.4 The Supernova-Gamma-Ray-Burst Connection

Gamma-ray bursts (GRBs) are short flashes of γ-radiation (energies in the range of 100 keV), which reach us from all directions in the sky and cosmological distances. They typically last for several ten seconds, but the total duration varies between fractions of a second to minutes. Repetitive events were never reported. The burst results from matter accelerated to highly relativistic speeds. The energy of this collimated matter jet is converted into radiation by an as yet not fully understood mechanism. The energy of the GRB is of the order of 10^{51} erg, which is the same order of magnitude as the kinetical energy of a core collapse supernova.

GRBs were detected in the 1970s by the military *Vela*-satellite, and most extensively investigated scientifically by the Batse detector on board of the *Compton Gamma Ray Observatory*. The event frequency is a few per day. While the shortest GRBs are thought to be the result of the merging of two neutron stars (or that of a neutron star with a black hole), the longer lasting type (longer than ≈ 2 s) has been associated with the core collapse of massive stars, mainly due to the association of GRBs with star-forming regions, and the coincidence, both in time and place, with SNe of type Ib and Ic. These are core collapse supernovae, which lack hydrogen in their spectra. For a review about this evidence, and more details about the connection, see Woosley and Bloom (2006).

Why do a few massive stars create highly collimated jets of matter, being ejected at more than 99.9 % of the speed of light, while the majority eject their envelopes more or less as a spherical shell? The answer is believed to lie in an exceptionally high rotation of the precursor's core. For very massive stars ($M > 30 M_{\odot}$), the core collapses into a fast-rotating black hole and infalling matter assembles in an accretion disk around it (the *collapsar* model). There are several mechanisms under discussion, how the binding energy of the disk or the rotation energy of the black hole can be converted into the collimated relativistic outflow. Alternatively, GRBs may originate from highly magnetized ($B \approx 1 \times 10^{15}$ G), fast-rotating neutron stars with rotation periods of milliseconds, thus rotating almost at breakup speed (the *magnetar* model). The rotation energy would be of order 10^{52} erg, and the spin-down luminosity would be of the right order for a GRB.

The fact that hydrogen is absent in the spectra requires high mass loss rates as for Wolf-Rayet stars, but these should not be so high as to reduce the mass too much. Since mass loss scales somehow with metallicity, it is expected that GRBs should be found mainly in low-metallicity regions, and occur more frequently in the early epochs of the universe.

Although many details are still not understood, it seems to be evident that long-duration GRBs are core collapse supernovae with very massive progenitors that have extremely fast-rotating cores.

Part VIII
Compact Objects

Stellar evolution can lead to somewhat extreme final stages. We have seen in Chaps. 33 and 35 that the evolution tends to produce central regions of very high density. On the other hand it is known that stellar matter can be ejected (see Chap. 34). The mechanisms are only partly (if at all) understood, but they do exist according to observations (normal mass loss, planetary nebulae, explosions). It may be that in certain cases the whole star explodes without any remnant left (see Chap. 36). Often enough, however, only the widely expanded envelope is removed, leaving the condensed core as a *compact object.* Relative to "normal stars" these objects are characterized by small radii, high densities, and strong surface gravity.

There are three types of compact objects, distinguished by the "degree of compactness": white dwarfs (WD), neutron stars (NS), and black holes (BH). Typical values for WD are $R \approx 10^{-2} R_{\odot}, \varrho \approx 10^6\,\mathrm{g\,cm^{-3}}$, escape velocity $v_{\mathrm{E}} \approx 0.02c$; their configuration is supported against the large gravity by the pressure of highly degenerate electrons (instead of the "thermal pressure", which dominates in the case of normal stars). For NS one has typically $R \approx 10\,\mathrm{km}$, $\varrho \approx 10^{14}\,\mathrm{g\,cm^{-3}}$, $v_{\mathrm{E}} \approx c/3$; their pressure support is provided by densely packed, partially degenerate neutrons. This is the dominant species of particles since normal nuclei do not exist above a certain density. Indeed a NS represents very roughly a huge "nucleus" of 10^{57} baryons.

As a simple illustration, suppose that in both cases (WD and NS) ideal, non-relativistic degenerate fermions (of mass m_{e} or m_{n}) provide the pressure balancing the gravity. The stars then are polytropes of index $n = 3/2$. With a mass-radius relation (19.28), where the constant of proportionality can be seen to be $\sim K \sim 1/m_{\mathrm{fermion}}$, we have $R \sim 1/m_{\mathrm{fermion}}$. The ratio of m_{n} to m_{e} then provides the ratio of typical radii for WD and NS of the same mass. The pressure–gravity balance by degenerate neutrons can only be maintained up to limiting masses corresponding to about 2×10^{57} fermions.

Clearly for objects with gravity fields like those in NS general relativity becomes important. It will be the dominant feature for the last group of compact objects, namely BH with $R \approx 1\,\mathrm{km}$ and $v_{\mathrm{E}} = c$.

The first WD was detected long before theoreticians were able to explain it, whereas NS were predicted theoretically before they were, accidentally, discovered in the sky. Today, also the existence of BH is proven beyond doubt.

The physics of compact objects is interesting and complex enough to fill special textbooks (e.g., Shapiro and Teukolsky 1983). We refer to these for details and limit ourselves to indicating a few main characteristics.

Chapter 37
White Dwarfs

It is characteristic for configurations involving degenerate matter that mechanical and thermal properties are more or less decoupled from each other. Correspondingly we will discuss these two aspects separately. When dealing with the mechanical problem (including the P and ϱ stratification, the M–R relation, etc.) one may even go to the limit $T \to 0$. Of course, such cold matter cannot radiate at all and it is more appropriate to denote these objects as "black dwarfs". The thermal properties, on the other hand, are responsible for the radiation and the further evolution of white dwarfs. The evolution indeed leads from a white dwarf (WD) to a black dwarf, since it is–roughly speaking–the consumption of fossil heat stored in the WD which we see at present (Concerning the evolution *to* the white-dwarf stage see Chaps. 34–36, and for a much deeper and more detailed review of properties of white dwarfs, Althaus et al. 2010b).

37.1 Chandrasekhar's Theory

This theory treats the mechanical structure of WD under the following assumptions. The pressure is produced only by the ideal (non-interacting) degenerate electrons, while the non-degenerate ions provide the mass. The electrons are supposed to be fully degenerate, but they may have an arbitrary degree of relativity $x = p_F/m_e c$, which varies as $\varrho^{1/3}$. Therefore we no longer have a polytrope as we had in the limiting cases $x \to 0$ and $x \to \infty$. The equation of state can be written as

$$P = C_1 f(x) \ , \quad \varrho = C_2 x^3 \ ; \quad x = p_F/m_e c \ , \tag{37.1}$$

according to (15.13) and (15.15), which also define the constants C_1 and C_2, while (15.14) gives $f(x)$.

In order to describe hydrostatic stratification we start with Poisson's equation (19.2), in which we eliminate $d\Phi/dr$ by (19.1) and substitute P and ϱ from (37.1) obtaining

R. Kippenhahn et al., *Stellar Structure and Evolution*, Astronomy and Astrophysics Library, DOI 10.1007/978-3-642-30304-3_37, © Springer-Verlag Berlin Heidelberg 2012

$$\frac{C_1}{C_2}\frac{1}{r^2}\frac{d}{dr}\left(\frac{r^2}{x^3}\frac{df(x)}{dr}\right) = -4\pi G C_2 x^3 . \tag{37.2}$$

Differentiating the left-hand side of (15.12) with respect to x, one obtains an expression for $df(x)/dx$ which shows that

$$\frac{1}{x^3}\frac{df(x)}{dr} = 8\frac{d}{dr}[(x^2+1)^{1/2}] = 8\frac{dz}{dr} , \tag{37.3}$$

with

$$z^2 := x^2 + 1 . \tag{37.4}$$

Therefore (37.2) becomes

$$\frac{1}{r^2}\frac{d}{dr}\left(r^2\frac{dz}{dr}\right) = -\frac{\pi G C_2^2}{2C_1}(z^2-1)^{3/2} , \tag{37.5}$$

and as in Sect. 19.2 we replace r and z by dimensionless variables ζ and φ:

$$\zeta := \frac{r}{\alpha} , \quad \alpha = \sqrt{\frac{2C_1}{\pi G}}\frac{1}{C_2 z_c} ,$$
$$\varphi := \frac{z}{z_c} , \tag{37.6}$$

where z_c is the central value of z, characterizing the central density. Then from (37.5)

$$\frac{1}{\zeta^2}\frac{d}{d\zeta}\left(\zeta^2\frac{d\varphi}{d\zeta}\right) = -\left(\varphi^2 - \frac{1}{z_c^2}\right)^{3/2} . \tag{37.7}$$

This is Chandrasekhar's differential equation for the structure of WD. We write it in the form

$$\frac{d^2\varphi}{d\zeta^2} + \frac{2}{\zeta}\frac{d\varphi}{d\zeta} + \left(\varphi^2 - \frac{1}{z_c^2}\right)^{3/2} = 0 \tag{37.8}$$

and see that it is very similar (differing only in the parenthesis) to the Emden equation (19.10) for polytropes. In fact (37.8) becomes the Emden equation for indices $n = 3$ and $n = 3/2$ if we go to the limits $z \to \infty$ (i.e. $x \to \infty$) and $z \to 1$ (i.e. $x \to 0$), respectively. The central conditions are now

$$\zeta = 0 : \varphi = 1 , \quad \varphi' = 0 . \tag{37.9}$$

Starting with these values, (37.8) can be integrated outwards for any given value of z_c. The density stratification is found if μ_e (which enters via C_2) is also specified:

$$\varrho = C_2 x^3 = C_2(z^2-1)^{3/2} = C_2 z_c^3\left(\varphi^2 - \frac{1}{z_c^2}\right)^{3/2} . \tag{37.10}$$

Table 37.1 Numerical results of Chandrasekhar's theory of white dwarfs

$1/z_c^2$	x_c	ζ_1	$(-\zeta^2 d\varphi/d\zeta)_1$	ϱ_c/μ_e (g cm^{-3})	$\mu_e^2 M$ ($M_\odot$)	$\mu_e R$ (km)
0	∞	6.8968	2.0182	∞	5.84	0
0.01	9.95	5.3571	1.9321	9.48×10^8	5.60	4.170
0.02	7	4.9857	1.8652	3.31×10^8	5.41	5.500
0.05	4.36	4.4601	1.7096	7.98×10^7	4.95	7.760
0.1	3	4.0690	1.5186	2.59×10^7	4.40	10.000
0.2	2	3.7271	1.2430	7.70×10^6	3.60	13.000
0.3	1.53	3.5803	1.0337	3.43×10^6	2.99	16.000
0.5	1	3.5330	0.7070	9.63×10^5	2.04	19.500
0.8	0.5	4.0446	0.3091	1.21×10^5	0.89	28.200
1.0	0	∞	0	0	0	∞

Subscripts c and 1 refer to centre and surface, respectively (After Cox and Giuli 1968, vol. II, Chap. 25)

The surface is reached at $\zeta = \zeta_1$, where ϱ becomes zero, i.e. after (37.1), (37.4) and (37.6)

$$\zeta = \zeta_1 \; : \; x_1 = 0\,, \quad z_1 = 1\,, \quad \varphi_1 = 1/z_c\,. \tag{37.11}$$

The value of R is

$$R = \alpha\zeta_1 = \sqrt{\frac{2C_1}{\pi G}\frac{1}{C_2 z_c}}\,\zeta_1\,, \tag{37.12}$$

and M can be found if we replace r and ϱ by (37.6) and (37.10):

$$\begin{aligned} M &= \int_0^R 4\pi r^2 \varrho dr \\ &= 4\pi\alpha^3 C_2 z_c^3 \int_0^{\varphi_1} \zeta^2 \left(\varphi^2 - \frac{1}{z_c^2}\right)^{3/2} d\zeta \\ &= 4\pi\alpha^3 C_2 z_c^3 \left(-\zeta^2 \frac{d\varphi}{d\zeta}\right)_1 \\ &= \frac{4\pi}{C_2^2}\left(\frac{2C_1}{\pi G}\right)^{3/2} \left(-\zeta^2 \frac{d\varphi}{d\zeta}\right)_1\,. \end{aligned} \tag{37.13}$$

The integrand in the second equation (37.13) was simply replaced by the derivative on the left-hand side of (37.7).

Table 37.1 gives the results of integrations for different values of z_c from ∞ to 1, i.e. from $x_c = \infty$ (fully relativistic) to $x_c = 0$ (non-relativistic), with the resulting M–R relation being plotted in Fig. 37.1. As in the simple case of polytropes (Sect. 19.6), we find an M–R relation with $dR/dM < 0$, but the exponent of M is no longer constant as it is in (19.28). The stellar mass M cannot exceed the Chandrasekhar limit M_{Ch} as given by (19.30),

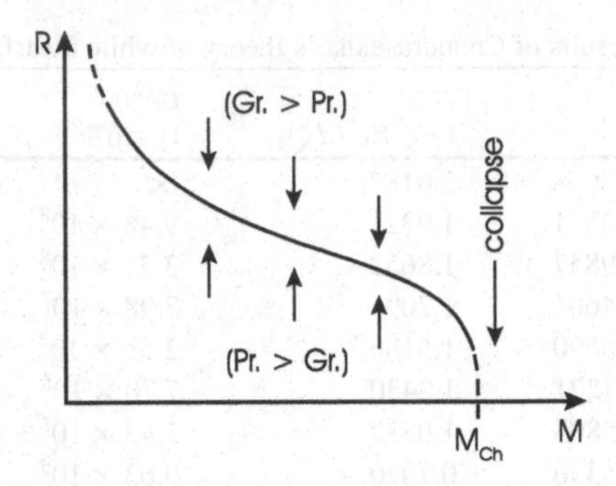

Fig. 37.1 Sketch of the classical mass–radius relation of white dwarfs according to Chandrasekhar's theory (assuming that the pressure is provided only by an ideal, degenerate electron gas). The *arrows* indicate the direction into which a non-equilibrium configuration is pushed if the gravitational force ("Gr.") is larger or smaller than the pressure gradient ("Pr."). Corrections are necessary at both ends of the curve *(dashed)*

$$M_{\mathrm{Ch}} = \left(\frac{2}{\mu_{\mathrm{e}}}\right)^2 \times 1.459 M_{\odot} , \tag{37.14}$$

since this limit case ($z_c \to \infty$) coincides with a polytropic structure of index $n = 3$.

These characteristics certainly call for a simple explanation, since they contradict the everyday experience that spheres of given material (say iron) become larger with increasing mass. This experience is not only obtained by handling small iron spheres but also by measurements of planets.

Let us consider rough averages (taken over the whole star) of the basic equation of hydrostatic equilibrium (10.2). Replacing there the absolute value of dP/dm by P/M and m/r^4 by M/R^4, we obtain

$$\frac{P}{M} \approx \frac{GM}{4\pi R^4}, \tag{37.15}$$

where P is some average value. We replace it by the average density $\varrho \sim M/R^3$, using a degenerate equation of state,

$$P \sim \varrho^\gamma \sim \left(\frac{M}{R^3}\right)^\gamma . \tag{37.16}$$

The pressure term f_{p}, i.e. the left-hand side of (37.15), and the gravity term f_{g}, on the right-hand side, are then

$$f_{\mathrm{p}} \sim \frac{M^{\gamma-1}}{R^{3\gamma}} ; \quad f_{\mathrm{g}} \sim \frac{M}{R^4} . \tag{37.17}$$

Their ratio f must be unity for hydrostatic equilibrium:

$$f := \frac{f_g}{f_P} \sim M^{2-\gamma} R^{3\gamma-4} = \begin{cases} M^{1/3} R\,, & \text{for } \gamma = 5/3 \\ M^{2/3}\,, & \text{for } \gamma = 4/3\,. \end{cases} \tag{37.18}$$

These equations describe the mass–radius relation for white dwarfs in the limiting cases of Chandrasekhar's theory. Since the pressure of a fully degenerate, non-relativistic gas with $\gamma = 5/3$ also depends on $\mu_e^{-5/3}$ (15.23), so does f_P. For He and CO white dwarfs, $\mu_e = 2$, but for ^{56}Fe it is 2.15, and therefore the mass–radius relation is shifted to smaller radii for the same mass, as the term $\mu_e^{5/3}$ appears on the right-hand side of (37.18). This shift is visible in Fig. 37.3.

Suppose we have a given stellar mass $M < M_{Ch}$ and non-relativistic electrons with $\gamma = 5/3$. Then the star can easily find an equilibrium by adjusting R such that $f = 1$. If we now slightly increase M, then $f > 1$ (gravity exceeds the pressure force), and R must *decrease* in order to regain equilibrium ($f = 1$). This explains the structure of the R–M relation (cf. Fig. 37.1).

However, if the electrons are relativistic ($\gamma = 4/3$), then f is independent of R. Equilibrium can be achieved only by adjusting M to a certain value M_{Ch}. If $M < M_{Ch}$, then $f < 1$, i.e. the dominant pressure term makes the star expand. This makes the electrons less relativistic and increases γ above the critical value 4/3. For $M > M_{Ch}$, $f > 1$, and the dominant gravity term makes the star contract; but this does not help either, and the star must collapse without finding an equilibrium. So M_{Ch} is quite obviously a mass limit for these equilibrium configurations. This mass limit again depends on μ_e, with a power of $4/3$ [cf. (15.26) and (19.30)].

37.2 The Corrected Mechanical Structure

The admirable lucidity of the theory of Sect. 37.1 is based completely on the simplicity of the equation of state for an ideal, fully degenerate electron gas used there [cf. (37.1)]. It certainly requires corrections near both ends of the mass range. For cold (or nearly cold) configurations of $M \to 0$ we should get the behaviour $R \to 0, \varrho \approx$ constant as for planets (or even smaller spheres) instead of $R \to \infty, \varrho \to 0$ (as we have already explained above). At least there should be the possibility for a smooth transition to the planets, which in this connection can well be considered cold bodies. The corrections to be applied here are due to the electrostatic interaction. Near the limiting mass, on the other hand, we have encountered very high densities, with the simple theory yielding $\varrho \to \infty$ for $M \to M_{Ch}$. In this domain we have to allow for effects of the weak interaction (inverse β decay) and the possibility of pycnonuclear reactions. Some influences on the equation of state have already been indicated in Chaps. 4 and 5.

37.2.1 Crystallization

Let us first treat the main effects of *electrostatic interaction* in a cold plasma with nuclei of type (Z, A) and electrons of density n_e. We have seen in Sect. 16.4 that matter in WD can be crystallized, and we will come back later to the condition for this. Let us suppose that the ions form a regular lattice and the electrons are evenly distributed. For the density encountered in WD the Wigner–Seitz approximation is not too bad, and so we divide the lattice into neutral Wigner–Seitz spheres of radius $R' = Z^{1/3} r_e a_0$ (r_e = average separation of the electrons in units of the Bohr radius a_0). Each sphere contains one ion (point charge $+Z$ in the centre) and Z electrons (a uniformly distributed charge $-Z$). In order to find the Coulomb energy ZE_C of the sphere we take concentric shells of radius y and charge $-3Zey^2 dy/R'^3$ and remove them to infinity, thereby overcoming the potential difference $Ze(1 - y^3/R'^3)$. An integration over the whole sphere gives the energy per electron as

$$-E_C = \frac{9}{10}\frac{Ze^2}{R'} = \frac{9}{10}\frac{Z^{2/3}e^2}{r_e a_0} \approx 2\frac{Z}{A^{1/3}}\varrho_6^{1/3}\,\text{keV}, \tag{37.19}$$

with $\varrho_6 = \varrho/10^6\,\text{g cm}^{-3}$. Even for $T \to 0$ the ions cannot sit at rest precisely on their points in the lattice. Instead, the ions of mass $m_0 = Am_u$ and density n_0 oscillate around their positions with some ion plasma frequency ω_E (with $\omega_E^2 \sim Z^2e^2 n_0/m_0$) such that the *zero-point energy* is $ZE_{zp} = 3\hbar\omega_E/2$ per ion. With $\varrho = n_0 A m_u$ we have per electron

$$E_{zp} = \frac{3}{2}\sqrt{\frac{4\pi}{3}}\frac{\hbar e}{Am_u}\varrho^{1/2} \approx \frac{0.6}{A}\varrho_6^{1/2}\,\text{keV}. \tag{37.20}$$

For ^{12}C $(Z = 6, A = 12)$ and $\varrho = 10^6\,\text{g cm}^{-3}$, the energies are $-E_C \approx 5.2\,\text{keV}$ and $E_{zp} \approx 0.05\,\text{keV} \ll -E_C$. The ratio $-E_C/E_{zp} \sim ZA^{2/3}\varrho^{-1/6}$ varies only very little with ϱ and increases towards heavier elements.

Therefore cold configurations ("black dwarfs") are crystallized. The ions form a regular lattice which minimizes the energy; they perform low-energy oscillations around their average positions, where they are kept by mutual repulsive forces.

The energy per electron is now

$$E = E_0 + E_C + E_{zp} \approx E_0 + E_C < E_0, \tag{37.21}$$

where E_0 is the mean energy of an electron in an ideal Fermi gas. The influence of E_C on the pressure is seen from

$$P \equiv -\frac{\partial E}{\partial(1/n)} \approx -\frac{\partial E_0}{\partial(1/n)} - \frac{\partial E_C}{\partial(1/n)} < P_0, \tag{37.22}$$

Table 37.2 Values of P/P_0, where P includes the Coulomb interaction and P_0 is for an ideal Fermi gas

x	$\varrho \cdot 2/\mu_e$	P/P_0 ($Z = 2$)	P/P_0 ($Z = 6$)	P/P_0 ($Z = 26$)
0.05	2.44×10^2	0.760	0.564	−0.063
0.1	1.95×10^3	0.880	0.782	0.467
1	1.95×10^6	0.988	0.975	0.933

x is the relativity parameter; ϱ is in g cm^{-3} (After Salpeter 1961)

where the derivatives are taken for constant entropy, and P_0 is the pressure of the ideal Fermi gas. The lowering of E and P due to $E_C < 0$ comes from the concentration of all positive charges into the nucleus, while the negative charges are much more uniformly distributed. The average electron–electron distance is thus larger than the average electron–nucleus distance, and the repulsion is smaller than the attraction. A few calculated values of the ratio P/P_0 for different Z and relativity parameter x are given in Table 37.2. As expected the reduction of P increases with the charge Z and with decreasing ϱ (decreasing Fermi energy). It will therefore be the dominant correction at small M, providing there the described reduction of R. The above approximation breaks down, of course, when it yields $P \lesssim 0$.

Apart from modifying the pressure, crystallization has an additional effect, which changes the chemical structure and therefore the run of pressure and density, too. This effect is *chemical* or *phase separation* and is due to the fact that different elements cannot coexist in arbitrary amounts in the solid phase. After the previous evolution most low- and intermediate-mass stars end as CO white dwarfs, with an almost flat chemical profile in the centre (Fig. 37.2). When crystallization occurs, the phase transition for the first (the lighter) element requires a lower abundance of this element in the solid phase than in the liquid (or gas) phase. The excess amount of this element (carbon in our example) will flow up to the liquid phase, above the crystallization boundary, creating there a local density inversion. This is the case, because even further away from the solid core there is the original abundance of the lighter element, and since this is now lower than in the region just above the crystallized core, the average molecular weight is higher. This induces convection, redistributing the elements in the liquid phase. Of course, with the local changes in the lighter element abundance connected is one in the abundance of the heavier element (oxygen). This effect is now continuing with decreasing temperature and a growing solid core (It can be understood in greater detail with the help of phase diagrams, as done in the textbook by Salaris and Cassisi (2005, Chap. 7)). The result is a modification of the chemical and therefore mechanical structure of the white dwarf, and a rather smooth, monotonic run of element abundances, as can be seen in Fig. 37.2, which shows the abundances for carbon and oxygen in a white dwarf before and after crystallization.

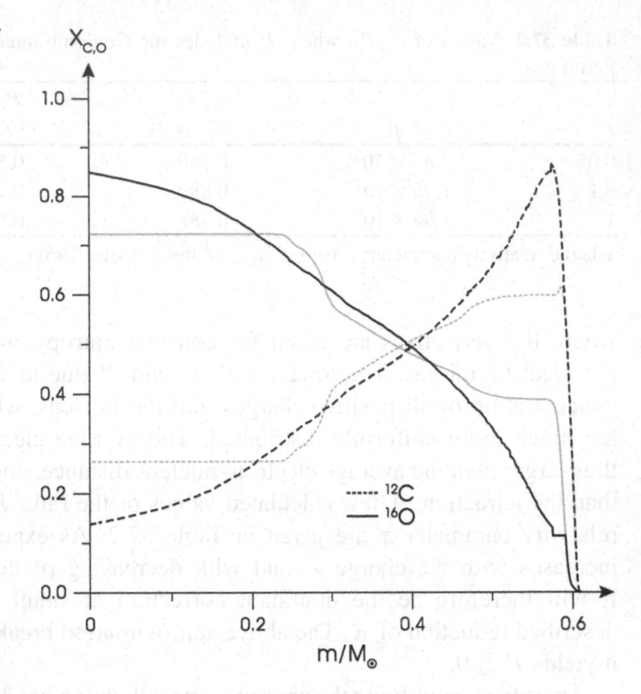

Fig. 37.2 The abundances of carbon and oxygen within the core of a white dwarf of $0.609M_\odot$, which resulted from a full evolutionary calculation (Althaus et al. 2010a) of a star of initially $2M_\odot$. The *grey lines* show the chemical profile before and the *black* ones after crystallization and phase separation have taken place (plot courtesy L. Althaus and A. Serenelli)

37.2.2 *Pycnonuclear Reactions*

In this and the following (Sect. 37.2.3) we will encounter two types of nuclear reactions which occur at very high densities and which lead to a change of the chemical composition without thermonuclear reactions. Though they are believed to be quite irrelevant for most white dwarfs, except maybe for the densest and coolest ones, they are of principal interest, as they may lead–in the extreme–to a composition defined by nuclear equilibrium (Sect. 37.2.4). Such effects will be important for neutron stars (Sect. 38).

For the very high densities occurring near the upper end of the mass range, *pycnonuclear reactions* have to be considered (cf. Sect. 18.4). These were defined as nuclear reactions which depend mainly on ϱ (instead of T, as in the case of thermonuclear reactions). They can occur even at $T \rightarrow 0$ as a consequence of the small oscillations of the nuclei in the lattice with energy E_{zp}, combined with the tunnel effect. Reactions set in rather abruptly at a certain density limit ϱ_{pyc} and use up all fuel within a short time (say 10^5 years) once $\varrho \gtrsim \varrho_{\text{pyc}}$. The limits ϱ_{pyc} for the different reactions are not well known, since the relevant cross sections are very uncertain. The values of ϱ_{pyc} increase towards heavier elements; the orders of magnitude are $\varrho_{\text{pyc}} \approx 10^6$, 10^9, and $10^{10}\,\text{g cm}^{-3}$ for burning of ^{1}H, ^{4}He, and ^{12}C, respectively. Central densities of white dwarfs may reach values up to $\varrho \approx 10^6 - 10^9\,\text{g cm}^{-3}$. However, hydrogen white dwarfs do not exist due to the previous stellar evolution, and He and CO white dwarfs reach the critical densities for the respective burning by pycnonuclear reactions only in very extreme cases. In general, therefore, the composition and structure of white dwarfs is not affected by this kind of nuclear reactions.

37.2.3 Inverse β Decays

Inverse β decay also becomes important at high densities. Consider a nucleus $(Z-1, A)$ which is β-unstable and decays under normal conditions to the stable nucleus $(Z, A) + e^- + \bar{\nu}$ (we always drop the subscript "e" for the neutrinos), the decay energy being E_d. If (Z, A) is surrounded by a degenerate electron gas with a kinetic energy at the Fermi border

$$E_F = m_e c^2[(1+x^2)^{1/2} - 1] , \tag{37.23}$$

such that $E_F > E_d$, then (Z, A) becomes unstable against electron capture, i.e. we have the inverse β decay

$$(Z, A) + e^- \rightarrow (Z-1, A) + \nu . \tag{37.24}$$

In general, we have to deal with the particularly stable even–even nuclei (Z, A) and then $E_d(Z-1, A) < E_d(Z, A)$. If $E_F > E_d(Z, A)$, then also $E_F > E_d(Z-1, A)$, and the inverse β decay proceeds further to $(Z-2, A)$. The new nuclei are now stabilized by the Fermi sea, i.e. they cannot eject an electron with $E_d (< E_F)$, since it would not find a free place in phase space. E_F increases with ϱ. Therefore for each type of nucleus (Z, A) there is a threshold ϱ_n of the density above which neutronization occurs. For ^{1}H and ^{4}He ($\varrho_n = 1.2\times 10^7$ and 1.4×10^{11} g cm^{-3}) this is of no interest, since clearly $\varrho_n \gg \varrho_{pyc}$ such that pycnonuclear burning would set in before neutronization can occur. And even for the decay $^{12}C \rightarrow {}^{12}B \rightarrow {}^{12}Be$ one still has $\varrho_n = 3.9 \times 10^{10}$ g cm^{-3} $> \varrho_{pyc}$, but the order of critical densities is reversed for heavy nuclei. The decay $^{56}Fe \rightarrow^{56} Mn \rightarrow^{56}Cr$, for example, has a threshold $\varrho_n = 1.14 \times 10^9$ g cm^{-3} $< \varrho_{pyc}$. Although neither inverse β decays nor pycnonuclear reactions are important processes that might change a white dwarf's chemical composition, which is the result of the evolution of its stellar predecessor, they demonstrate that for high densities, the outcome of the thermonuclear processes occurring during a star's life might be changed in its remnant and that the structure might be determined by the density and the energy of the Fermi sea of degenerate electrons.

37.2.4 Nuclear Equilibrium

In "normal" stars we were used to imposing the chemical composition as an arbitrary free parameter. This was reasonable, since the usual transformation of the elements by thermonuclear reactions takes a sufficiently long time, and configurations with a momentary (non-equilibrium) composition are astronomically relevant. This may be different for very high densities, at which processes such as pycnonuclear reactions or inverse β decay can transform the nuclei in relatively short timescales. The other extreme, then, is to impose only the baryon number per volume and ask for the corresponding *equilibrium composition*. In reality the approach to nuclear equilibrium may be too slow to be accomplished. But one can imagine having reached it after an artificial acceleration by suitable catalysts, leading to the expression "cold catalysed matter". Because of their history, WD will scarcely have reached that stage of equilibrium (they usually consist of ^{4}He, or ^{12}C and ^{16}O, instead of ^{56}Fe, etc.). But in order to see the connection between different types of objects, we briefly describe a few characteristics of equilibrium matter.

The equilibrium composition can be found by starting with a certain type of nucleus (Z, A) and varying Z and A until the minimum of energy is obtained. For isolated nuclei the counteraction of attracting nuclear and repelling Coulomb forces gives a maximum binding of the nucleons at ^{56}Fe (cf. Sect. 18.1). Therefore ^{56}Fe will be the equilibrium composition for small ϱ ($<8\times10^6\,\mathrm{g\,cm^{-3}}$). With increasing ϱ this balance is shifted to heavier and neutron-enriched nuclei, since replacing a proton by a neutron decreases the repulsive Coulomb force inside the nucleus; and the β decay, which would then result in isolated nuclei, is here prohibited by the filled Fermi sea of the surrounding electrons. Another influence comes from the lattice energy (37.19), which gives only a small correction to P at high ϱ, but reduces the Coulomb energy at the surface of the nucleus. The sequence of equilibrium nuclei is (the maximum density in $\mathrm{g\,cm^{-3}}$ is shown in parenthesis): ^{56}Fe(8×10^6), ^{62}Ni(2.8×10^8), ^{64}Ni(1.3×10^9), ..., ^{120}Sr(3.6×10^{11}), ^{122}Sr(3.8×10^{11}), ^{118}Kr(4.4×10^{11}). For $\varrho > 4\times10^{11}\,\mathrm{g\,cm^{-3}}$ it is energetically more favourable that further neutrons are free rather than bound in the nucleus: the "neutron drip" sets in. The composition consists of two phases: the lattice of nuclei (with sufficient electrons for neutrality) plus free neutrons. Their number increases with ϱ, and at $\varrho \approx 4\times10^{12}\,\mathrm{g\,cm^{-3}}$ their pressure P_n even exceeds P_e. At $2\times10^{14}\,\mathrm{g\,cm^{-3}}$, the nuclei are dissolved, leaving a degenerate neutron gas with a small admixture of protons and electrons (see Sect. 38.1). The $P-\varrho$ relation can be calculated, giving the equation of state as shown in Fig. 16.2.

Once an equation of state is given, one can easily integrate the mechanical equations outwards, starting from a variety of values for the central pressure which leads to a pair of values M, R. The M–R relations obtained in this way by Hamada and Salpeter (1961) are plotted as solid curves in Fig. 37.3 for different compositions (He, C, Mg, Fe, and equilibrium composition). For comparison the relations for an ideal Fermi gas (Chandrasekhar's theory) are plotted for $\mu_\mathrm{e} = 2$ (e.g. ^{4}He, ^{12}C, ^{24}Mg) and $\mu_\mathrm{e} = 2.15$ (^{56}Fe); in the latter case, the mass limit is already lowered to $M_\mathrm{Ch} \approx 1.25M_\odot$. Relative to these classical models there is a clear reduction of R, particularly at small M, owing to the Coulomb interaction reducing P. This effect increases with Z. The curve for ^{56}Fe shows a maximum of R beyond which it decreases for $M \to 0$. In fact such a maximum of R ($\approx 0.02, 0.05, 0.12_\odot$ for Fe, He, H, respectively) occurs for all compositions at values of M between a few 10^{-3} to $10^{-2}M_\odot$. In this regime the equation of state is not well known; it is certainly completely dominated by Coulomb effects, and the inhomogeneous distribution of

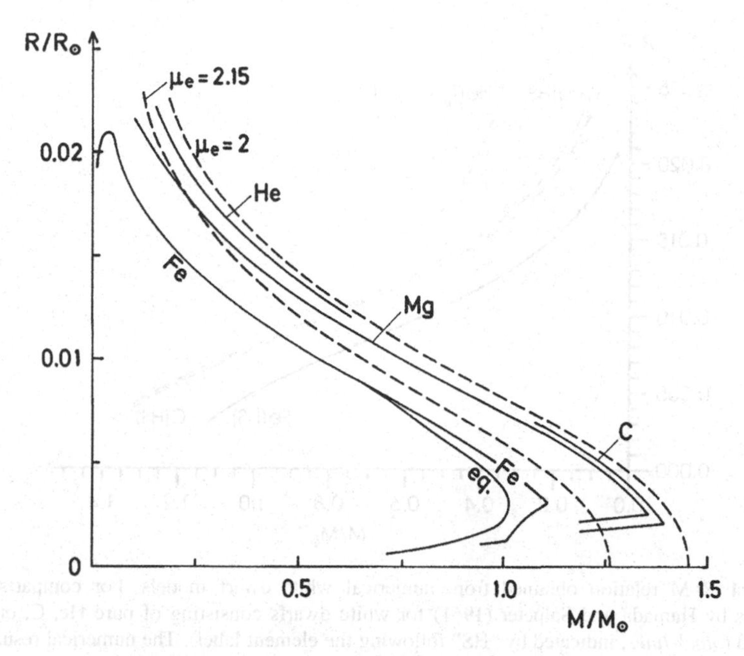

Fig. 37.3 R–M relation for white dwarfs. *Dashed lines* indicate solutions of Chandrasekhar's equation for non-interacting gases with $\mu_e = 2$ (^{4}He, ^{12}C, ^{16}O ...) and $\mu_e = 2.15$ (^{56}Fe). The other curves are for He, C, Mg, Fe, and equilibrium composition; they include interaction of the nuclei (After Hamada and Salpeter 1961)

the electrons has to be considered. In any case, we find here the natural transition between WD ($dR/dM < 0$) and planets ($dR/dM > 0$) (Note that Jupiter with $R \approx 0.1 R_\odot$ and $M \approx 10^{-3} M_\odot$ is not far from this border; in fact its radius is far above R_{max} for He and close to that of H, so that it must consist essentially of H.).

Towards large M the curves for C, Mg, and Fe show kinks at the mass limit. These are due to a phase transition in the centre, since ϱ_c reaches one of the limits described above. For ^{12}C we find here $\varrho_c = \varrho_{pyc}$, and pycnonuclear reactions then transform $^{12}\mathrm{C} \rightarrow {}^{24}\mathrm{Mg}$, which by inverse β decay becomes ^{24}Ne. Models on the lower branch beyond the kink consist of Ne cores and C envelopes. The curve for ^{24}Mg reaches M_{max} when $\varrho_c = \varrho_n$, and inverse β decay gives central cores of ^{24}Ne. For ^{56}Fe we see the result of the inverse β decay to ^{56}Cr at M_{max} and to ^{56}Ti at the following second kink (beyond which the models consist of ^{56}Ti cores, ^{56}Cr shells, and ^{56}Fe envelopes). The curve for equilibrium composition, which coincides with ^{56}Fe for $\varrho \lesssim 8 \times 10^6\,\mathrm{g\,cm^{-3}}$, is below and to the left of all other curves; it always has the largest average μ_e. At the maximum $M (\approx 1.0 M_\odot)$ one finds $\varrho_c \approx 2 \times 10^9\,\mathrm{g\,cm^{-3}}$, with ^{66}Ni nuclei giving a relatively large μ_e. Towards the end of the plotted equilibrium curve, ^{120}Kr is reached and the first neutrons are freed (From here follows the sequence of equilibrium configurations which leads to neutron stars, see Chap. 38.). The whole curve appears fairly smooth, since the change of the composition here proceeds in small steps via neighbouring nuclei,

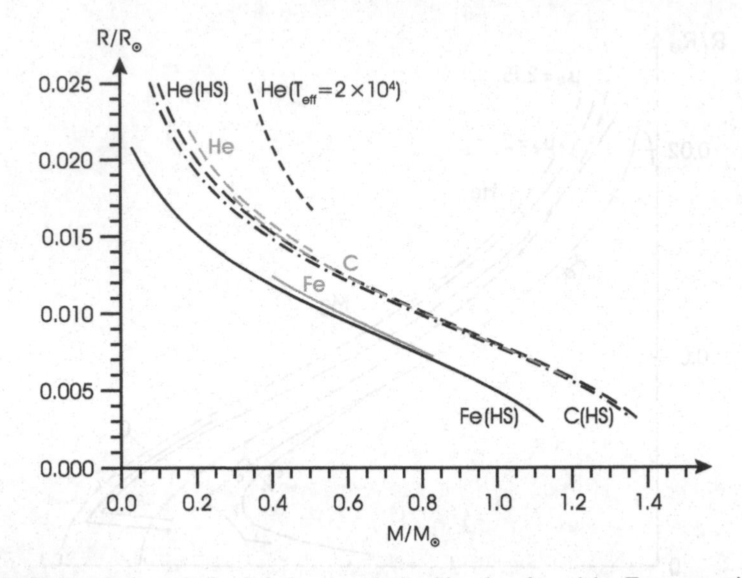

Fig. 37.4 R–M relation obtained from numerical white dwarf models. For comparison, the relations by Hamada and Salpeter (1961) for white dwarfs consisting of pure He, C, or Fe are repeated (*black lines*, indicated by "HS" following the element label). The numerical results were obtained for white dwarfs with an effective temperature of 4,000–5,000 K. These are the *shorter grey lines* with the same line style as the HS-curves. The computations were restricted to a narrow mass range resulting from realistic previous stellar evolution models. The *dashed black line* with the label "He ($T_{\mathrm{eff}} = 2 \times 10^4$)" is for a white dwarf with this effective temperature (data courtesy L. Althaus, after Panei et al. 2000)

while the transit of a non-catalysed composition to equilibrium is first delayed by large thresholds and then occurs in a big jump.

Concerning inhomogeneous models of WD with non-equilibrium composition, we briefly mention the case of a low-mass envelope of light elements (particularly ^{1}H) being placed on a WD of ^{4}He or ^{12}C and ^{16}O. This may happen by mass exchange in close binary systems. Aside from possible instabilities during the onset of nuclear burning (which can lead to the ejection of a nova shell), there is a strong influence on the equilibrium radius described by $d \lg R / d \lg M_{\mathrm{H}}$ of the order $10 \ldots 10^2$. This means that the addition of a ^{1}H envelope of only 1 % of M increases R by about 50 % and more. In fact the white dwarf will scarcely be recognizable as such. Although WD originating purely from single star evolution will also have envelopes of lighter elements, these are typically at least two orders of magnitude less massive, and the radius increase is correspondingly less drastic.

The M–R relation by Hamada and Salpeter (1961) shown in Fig. 37.3 is in fact a very good approximation for old white dwarfs with very cold (effective) temperatures, as Fig. 37.4 demonstrates. Here we compare the Hamada and Salpter relations with those obtained from full numerical models for white dwarfs with pure He, C, or Fe cores, and He–He envelopes, except for the He-WDs, which have a pure H envelope. The envelope masses range from $3 \times 10^{-4} M_{\odot}$ for the He-WDs to $10^{-2} M_\odot$ for the others. Another He-WD sequence obtained from models with a much higher $T_{\rm eff}$ of 2×10^4 K demonstrates how white dwarfs initially have much larger radii but approach the cold configurations during the cooling process. Based on mass and radius determinations of observed objects, the existence of white dwarfs with iron cores is sometimes claimed. So far, however, corrections to the radius shifted all objects back to the CO- or He-WD sequence. Such objects would indeed be a challenge to stellar evolution theory (cf. Chaps. 34 and 35).

Once L and $T_{\rm eff}$ of a white dwarf have been determined accurately enough, the Stefan-Boltzmann law (11.14) yields the radius. The mass–radius relation then delivers the mass of the white dwarf. This is the basis for the initial–final mass relation (Fig. 34.7). However, it depends on the assumption about the core composition of the white dwarf. In most cases, one can safely assume a CO-WD, as this is, at the present age of the universe, the most likely.

The connection with other types of configurations is seen in Fig. 38.3, which gives the $M-R$ relation for cold catalysed matter (equilibrium composition). When going along the curve in the direction of increasing ϱ_c, one encounters extrema of M (open circles) in which the stability properties change. An example is the point at $M = M_{\rm max}$ for the white-dwarf sequence, beyond which a branch of unstable models follows (see the discussion of Sect. 38.2).

37.3 Thermal Properties and Evolution of White Dwarfs

In the very interior of a WD, the degenerate electrons provide a high thermal conductivity. This, together with the small L, does not allow large temperature gradients. The situation is different when going to the outermost layers. With decreasing ϱ the matter is less and less degenerate, and the dominant heat transfer becomes that by radiation (or convection), which is much less effective. Therefore we expect to find a non-degenerate outer layer in which T can drop appreciably and which isolates the degenerate, isothermal interior from outer space.

We simplify matters by assuming a discontinuous transition from degeneracy to non-degeneracy (ideal gas) at a certain point (subscript 0). For the envelope we use the radiative solution (11.25) for a Kramers opacity ($\kappa = \kappa_0 P T^{-4.5}$) and a zero constant of integration:

$$T^{8.5} = BP^2 \;; \quad B = 4.25 \frac{3\kappa_0}{16\pi acG} \frac{L}{M} \,. \tag{37.25}$$

Replacing P by $\Re \varrho T/\mu$ and solving for ϱ here, we have

$$\varrho = B^{-1/2} \frac{\mu}{\Re} T^{3.25} \,. \tag{37.26}$$

The transition point is assumed to be where the degenerate electron pressure equals the pressure of an ideal gas, i.e. according to (16.6)

$$\varrho_0 = C_1^{-3/2} T_0^{3/2} \ ; \quad C_1 = 1.207 \times 10^5 \frac{\mu}{\mu_e^{5/3}} \text{ cgs} . \tag{37.27}$$

This density ϱ_0 is reached according to (37.26) at a temperature $T = T_0$ given by

$$T_0^{3.5} = \frac{B}{C_1^3} \left(\frac{\Re}{\mu} \right)^2 = \vartheta \frac{L/L_\odot}{M/M_\odot} , \tag{37.28}$$

where all factors are comprised in ϑ. For typical compositions and values of κ_0, one has roughly

$$T_0 \approx \vartheta^{2/7} \left(\frac{L/L_\odot}{M/M_\odot} \right)^{2/7} \approx \left(\frac{L/L_\odot}{M/M_\odot} \right)^{2/7} 5.9 \times 10^7 \text{ K} . \tag{37.29}$$

This simple relation between L and T_0 will turn out to be essential for deriving the cooling time of a white dwarf, (37.42). For $M = M_\odot$ and the range $L/L_\odot = 10^{-4} \ldots 10^{-2}$, (37.29) yields $T \approx 4.2 \ldots 16 \times 10^6$ K, which is, by assumption, also the temperature in the whole (isothermal) interior. Typical values for the density at the transition point are then, according to (37.27), of the order of $\varrho_0 \approx 10^3 \text{ g cm}^{-3}$ (i.e. $\ll \varrho_c$).

An idea of the radial extension $R - r_0$ of the non-degenerate envelope is easily obtained from (11.34). We neglect $T_{\text{eff}} (\lesssim 10^4 \text{ K})$ against T_0 and get

$$\frac{R - r_0}{r_0} \approx \frac{\Re T_0}{\mu \nabla} \frac{R}{GM} \approx 0.82 \frac{R/R_\odot}{M/M_\odot} \frac{T_0}{10^7 \text{ K}} . \tag{37.30}$$

(The numerical factor is given for $\mu = 4/3, \nabla = 0.4$.) The relative radial extension of the non-degenerate envelope then is typically 1 % or less, i.e. a few 10 km, since $R \approx 10^{-2} R_\odot$ and $M \approx 1 M_\odot$. This means that the radius of a WD is well approximated by the integrations which assume complete degeneracy throughout, as long as T_{eff} can indeed be neglected. As we saw in Fig. 37.4, WD models with more massive envelopes and high temperatures might be up to 50 % larger than the cold configurations.

The rather high internal temperatures of $10^6 \cdots 10^7$ K set a limit to the possible hydrogen content in the interior. If hydrogen were present with a mass concentration X_H, we would expect hydrogen burning via the *pp* chain. For average values $T = 5 \times 10^6$ K, $\varrho = 10^6 \text{ g cm}^{-3}$, (18.63) gives $\varepsilon_{pp} \approx 5 \times 10^4 X_H^2 \text{ erg g}^{-1} \text{ s}^{-1}$, and the luminosity for $M = 1 M_\odot$ would be

$$L/L_\odot \approx \frac{M_\odot}{L_\odot} \varepsilon_{pp} \approx 2.5 \times 10^4 X_H^2 , \tag{37.31}$$

such that the observed $L \le 10^{-3} L_\odot$ allows only $X_H \lesssim 2 \times 10^{-4}$. Stability considerations (Sect. 25.3.5) indeed rule out that the luminosity of normal WD is generated by thermonuclear reactions, which was first pointed out by Mestel (1952). A stable burning could only be expected in nearly cold configurations that produce their extremely small L ("black" or "brown" dwarfs) by pycnonuclear reactions near $T = 0$.

If there are no thermonuclear reactions, then which reservoirs of energy are involved when a normal WD loses energy by radiation? The means for obtaining the answer are provided in Sect. 3.1. For a configuration in hydrostatic equilibrium the virial theorem (3.9) requires $\zeta \dot{E}_i + \dot{E}_g = 0$.

The potential energy in the gravitational field $E_g (< 0)$ is given by (3.3). The total internal energy of the star $E_i = E_e + E_{ion}$ consists of the contributions from electrons and ions. By ζ we mean an average of the quantity ζ', defined by the relation

$$\zeta' u = 3 \frac{P}{\varrho} , \tag{37.32}$$

where u is the internal energy per unit mass. For highly degenerate electrons, ζ' varies from $\zeta' = 2$ (non-relativistic) to $\zeta' = 1$ (relativistic case). For the ions, $\zeta' = 2$ if they are an ideal gas [cf. (3.5)]. If there is crystallization, the contributions u_C of Coulomb energy and u_p of lattice oscillations (phonons) have to be considered. For the static Coulomb part we note that $u_C = n_e E_C/\varrho$, with $E_C \sim \varrho^{1/3}$ according to (37.19). Then one finds from (37.22) that $P_C/\varrho = u_C/3$, i.e. $\zeta' = 1$. The situation is more difficult with u_p, but this contributes relatively little.

Summing up all effects, the average over the whole WD will obviously be somewhere in the range $1 < \zeta < 2$. As in "normal" stars we have a simple relation between E_i and E_g, the absolute values of both being of the same order.

The total energy is $W = E_i + E_g$. The energy equation requires $L = -\dot{W}$, which together with the virial theorem [compare with (3.12)]

$$L = -\dot{W} = -\frac{\zeta - 1}{\zeta} \dot{E}_g = (\zeta - 1) \dot{E}_i . \tag{37.33}$$

Therefore $L > 0$ requires a contraction ($\dot{E}_g < 0$) and an increase of the internal energy ($\dot{E}_i > 0$). So far, it is the same as with normal, non-degenerate stars. The crucial question is how E_i is distributed between electrons (E_e) and ions (E_{ion}).

We recall the situation for a normal star with both electrons and ions being non-degenerate. Then there is equipartition with $E_{ion} \sim E_e \sim T$, such that also $E_i = E_{ion} + E_e \sim T$; $\dot{E}_i > 0$ means $\dot{T} > 0$. Thus the *loss of* energy ($L > 0$) leads to a heating ($\dot{T} > 0$). This was expressed in Sect. 25.3.4 by saying that the star has negative gravothermal specific heat, $c^* < 0$.

For demonstrating the behaviour of a WD, let us simply assume that the electrons are non-relativistic degenerate and the ions form an ideal gas. Then $\zeta = 2$ and $L = -\dot{E}_g/2$, i.e. the star must contract, releasing twice the energy lost by radiation. Since $-E_g \sim 1/R \sim \varrho^{1/3}$, we have $\dot{E}_g/E_g = (1/3)\dot{\varrho}/\varrho$ (Here ϱ is some average value.). The compression, however, increases the Fermi energy E_F of the electrons. Their internal energy is $E_e \approx E_F \sim p_F^2 \sim \varrho^{2/3}$, such that $\dot{E}_e/E_e = (2/3)\dot{\varrho}/\varrho$. So we have a simple relation between $\dot{E}_g$ and $\dot{E}_e$:

$$\dot{E}_e \approx 2\frac{E_e}{E_g}\dot{E}_g = -\frac{E_e}{E_i}\dot{E}_g . \tag{37.34}$$

Here E_i is introduced via the virial theorem in the form $E_g = -2E_i$.

If the WD is already cool, then $E_{ion} \ll E_e$ and $E_i = E_{ion} + E_e \approx E_e$. This means $\dot{E}_e \approx -\dot{E}_g = 2L$, and nearly as much energy as released by contraction is used up by raising the Fermi energy of the electrons. With $\dot{E}_e \approx -\dot{E}_g$, the energy balance $L = -\dot{E}_{ion} - \dot{E}_e - \dot{E}_g$ becomes

$$L \approx -\dot{E}_{ion} \sim -\dot{T} . \tag{37.35}$$

Therefore, the ions release about as much energy by cooling as the WD loses by radiation. The contraction is then seen to be the consequence of the decreasing ion pressure (even though P_{ion} is only a small part of P). In spite of the decreasing ion energy, the whole internal energy rises, since $\dot{E}_{ion} + \dot{E}_e \approx L$. This evolution tends finally to a cold black dwarf; then the contraction has stopped and all of the internal energy is in the form of Fermi energy.

Of course, the relations just derived should have somewhat different numerical factors, since ζ is not exactly 2 (a certain degree of relativity in the central part, the ion gas not being ideal, etc.). But the essence of the story remains the same.

The foregoing discussion opens the possibility of arriving at a very simple *theory of the cooling of WD*. We start with the energy equation (4.48), setting there

$$\varepsilon_g = -c_v\dot{T} + \frac{T}{\varrho^2}\left(\frac{\partial P}{\partial T}\right)_v \dot{\varrho} , \tag{37.36}$$

which follows from the first equation (4.17). We now integrate (4.48) over the whole star, taking not only $\varepsilon_n = \varepsilon_\nu = 0$, but also neglecting the compression term in (37.36),

$$-L \approx \int_0^M c_v\dot{T}\,dm \approx c_v\dot{T}_0 M , \tag{37.37}$$

where an isothermal interior is assumed with $T = T_0$. If the ions are an ideal gas, then

$$c_v^{ion} = \frac{3}{2}\frac{k}{Am_u} . \tag{37.38}$$

For the specific heat of the degenerate electrons one can derive (Chandrasekhar 1939, p. 394)

$$
\begin{aligned}
c_v^{\text{el}} &= \frac{\pi^2 k^2}{m_{\text{e}} c^2} \frac{Z}{A m_{\text{u}}} \frac{\sqrt{1+x^2}}{x^2} T \qquad [x = p_{\text{F}}/m_{\text{e}} c] \\
&\approx \frac{\pi^2 k}{2} \frac{Z}{A m_{\text{u}}} \frac{kT}{E_{\text{F}}} , \qquad \text{for } x \ll 1 .
\end{aligned} \tag{37.39}
$$

The ratio (for $x \ll 1$)

$$
\frac{c_v^{\text{el}}}{c_v^{\text{ion}}} = \frac{\pi^2}{3} Z \frac{kT}{E_{\text{F}}} \tag{37.40}
$$

is small for small kT/E_{F} and not too large Z. In the numerical examples below we will take $c_v = c_v^{\text{ion}}$. Then (37.37) describes L as given by the change of the internal energy of the ions.

In (37.28) we eliminate L with (37.37) and obtain a differential equation for T (where we drop the subscript 0 for the interior):

$$
\dot{T} = -\frac{L_\odot}{M_\odot} \frac{1}{c_v \vartheta} T^{7/2} . \tag{37.41}
$$

This can be rewritten with (37.29) as $-\dot{L} \sim L^{12/7}$, which together with $R \approx$ constant describes the motion in the HR diagram. Equation (37.41) is easily integrated from $t = 0$ when the temperature was much larger than it is now, to the present time $t = \tau$. The result gives the *cooling time*

$$
\begin{aligned}
\tau &= \frac{2}{5} \frac{M_\odot}{L_\odot} c_v \vartheta T^{-5/2} = \frac{2}{5} c_v \frac{MT}{L} \\
&= \frac{2}{5} \left(\frac{M_\odot}{L_\odot} \vartheta \right)^{2/7} c_v \left(\frac{M}{L} \right)^{5/7} \approx \frac{4.7 \times 10^7 \text{years}}{A} \left(\frac{M/M_\odot}{L/L_\odot} \right)^{5/7} .
\end{aligned} \tag{37.42}
$$

Here we have used (37.28) and (37.29). For $A = 4$, $M = M_\odot$ and $L/L_\odot = 10^{-3}$ one has $\tau \approx 10^9$ years. Equation (37.42) is the result of Mestel's model for the evolution of white dwarfs (Mestel 1952). For CO-WDs, $A \approx 14$, and the cooling time of a WD with $L = 10^{-4} L_\odot$ is estimated to be of the order of 2 Gyrs, indicating already that some very cool WDs are remnants of the earliest phases of galactic evolution. However, they are very dim, and difficult to observe except for the closest ones. Note that the more massive a WD is, and the lighter its main elements (He, CO, or ONe, depending on its initial mass and previous evolution), the longer will be cooling time.

The specific heat c_v is obviously very important. Larger values of c_v give a slower cooling ($\dot{T} \sim 1/c_v$), i.e. a larger cooling time ($\tau \sim c_v$). The simplest assumption would be $c_v = c_v^{\text{ion}} = 3k/(2Am_{\text{u}})$, but this requires several corrections. For small M (i.e. moderate $\overline{\varrho}$) and larger T and Z, one cannot neglect the contribution of the electrons. From (37.40) we have $c_v^{\text{el}} \approx 0.25 c_v^{\text{ion}}$ for $T = 10^7$ K, $M = 0.5 M_\odot$ and a carbon–oxygen mixture.

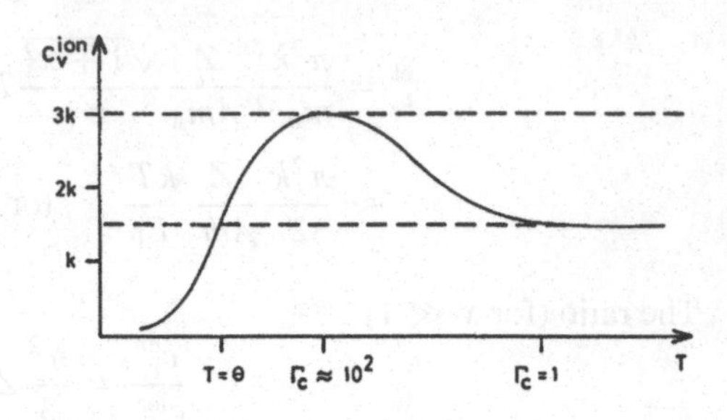

Fig. 37.5 Schematic variation of the specific heat per ion with the temperature T in white-dwarf matter

For small T the ions dominate completely: $c_v = c_v^{\text{ion}}$, but their specific heat is influenced by crystallization. We indicate only a few aspects of the rather involved theory for these processes (e.g. Mestel and Ruderman 1967; Shaviv and Kovetz 1976; Isern et al. 1997).

The properties of the ions depend critically on two dimensionless quantities, Γ_c and T/Θ. The ratio Γ_c of Coulomb energy to kinetic energy of the ions is defined in (16.25). For $\Gamma_c \approx 10^2$ a heated crystal will melt (or a cooling plasma will crystallize), which determines the melting temperature T_m given in (16.26). For $\Gamma_c < 1$ the thermal motion does not allow any correlation between the positions of the ions, no lattice is possible, and the ions behave as a gas.

The other ratio, T/Θ, contains a characteristic temperature Θ which is essentially the Debye temperature and is defined by

$$k\Theta = \hbar\Omega_p\,, \quad \Omega_p = \frac{2Ze}{Am_u}(\pi\varrho)^{1/2}\,, \tag{37.43}$$

with Ω_p being the ion plasma frequency [cf. the zero-point energy (37.20) where we used $\omega_E = \Omega_p/3$]. This gives

$$\Theta = \frac{he}{km_u\sqrt{\pi}}\frac{Z}{A}\varrho^{1/2} \approx 7.8\times10^3\,\text{K}\cdot\frac{Z}{A}\varrho^{1/2} \tag{37.44}$$

(ϱ in g cm^{-3}). $k\Theta$ is a characteristic energy of the lattice oscillations, which cannot be excited for $T/\Theta < 1$. For typical WD composed of C, O, or heavier elements, one has $\Theta < T_m$.

Figure 37.5 shows how the specific heat C_v per ion changes with T. Starting at very large $T(\Gamma_c \ll 1)$, the ions form an ideal gas. Each degree of freedom contributes $kT/2$ to the energy (i.e. $k/2$ to C_v), and $C_v = 3k/2$. With decreasing T one finds an increasing correlation of the ion positions owing to the growing importance of Coulomb forces in the range $\Gamma_c \approx 1\ldots10$. This gives additional degrees of freedom, since energy can go into lattice oscillations, and C_v increases above $3k/2$, with the maximum of $C_v = 3k$ being reached when the plasma crystallizes at $T = T_m$. With further decreasing T gradually fewer oscillations are excited, and the specific ion heat C_v even drops below $3k/2$ around $T = \Theta$. For $T \to 0$ finally, $C_v \sim T^3$.

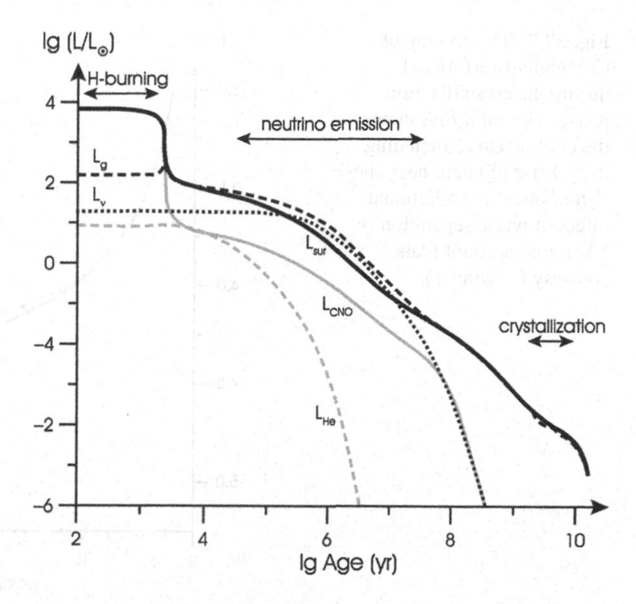

Fig. 37.6 The cooling of a CO white-dwarf model of $0.609M_\odot$, calculated from a full evolutionary sequence. The evolution has three dominant phases: initially, shell hydrogen burning (L_{CNO}) is still important, after which neutrino emission (L_ν) is the dominant channel of energy loss. At very large ages thermal energies (L_g) provide the white dwarf's luminosity (L_{sur}; thick black line) in the phase of crystallization. In this model helium burning (L_{He}) never is an important energy source (Data courtesy L. Althaus)

These large variations of C_v (increase by a factor 2, then decrease to zero) of course influence the cooling times [cf. (37.42)]. In addition there is the release of the latent heat of about kT per ion when the material crystallizes, which delays cooling. Additional energy is delivered by the phase separation (Sect. 37.2.1), which is due to the (so far neglected) dependence of the specific entropy s on chemical composition (4.7). Compared to the approximative Mestel cooling law (37.42), cooling times can be longer by up to several 10 % by these effects, which are more important when crystallization sets in at lower luminosity, because the delay is simply the extra energy divided by the luminosity. This is the case for less massive WDs, because they are less dense and reach the critical $\Gamma_c \approx 170$ only at lower T, thus lower L.

Further improvements in the theory of WD cooling result from a realistic and more accurate treatment of the equation of state, and the transport of energy within the degenerate interior and through the non-degenerate envelope by conduction, radiation, and convection. Obviously, the mass of the envelope and its composition have an influence on the cooling time. Both depend on the pre-WD evolution, in particular, therefore, on the physics of mass loss and mixing on and after the AGB phase. Generally, envelope masses between 10^{-4} and $0.01M_\odot$ are obtained for the beginning of the WD cooling phase, but later, empirical evidence from the seismology of pulsating WDs (Sect. 42.4), indicates envelope masses well below these numbers, with an average around $10^{-7}M_\odot$. The less massive the envelope, and the less hydrogen it contains, the more transparent it is for radiative losses, allowing a faster cooling of the WD.

Convection in the envelope may eventually reach into the core, and thereby the convergence of the temperature profile of the envelope to that of the outer core boundary (Sect. 11.3.3) is lost (The solution (37.25) assumes purely radiative transfer.). Instead the core depends on the conditions in the atmosphere, and

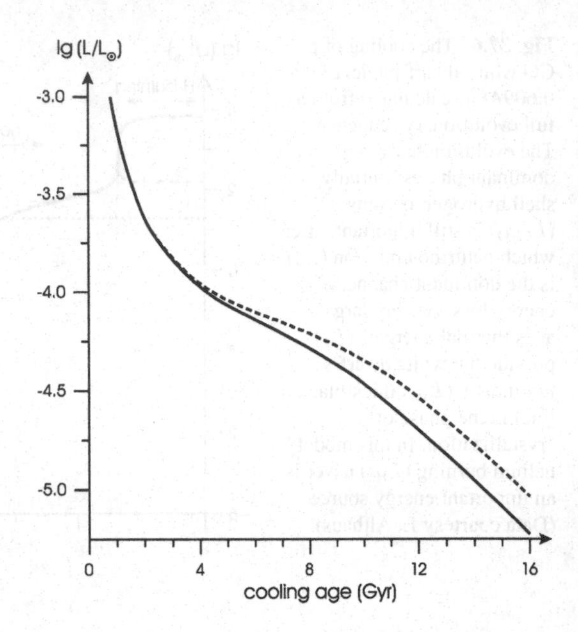

Fig. 37.7 The cooling of a CO white-dwarf model during the crystallization phase. The *solid line* shows the cooling curve including the release of latent heat, the *dotted* one if the additional effect of phase separation is taken into account (data courtesy L. Althaus)

the connection between core temperature and luminosity–or the cooling rate–is modified (D'Antona and Mazzitelli 1989). Finally, the various sources and sinks of energy–nuclear reactions, neutrino emission, gravothermal energy, and crystallization–have to be included, which are important at various phases of the WD cooling. We conclude this discussion with Fig. 37.6, which shows the cooling function of a $0.61 M_\odot$ CO white dwarf, resulting from a full evolutionary sequence, starting on the main sequence with an initial mass of $2 M_\odot$ and a metallicity of $Z = 0.01$. The figure shows the various contributions of energy sources and sinks.

In the earliest phase, lasting only a few thousand years, the total photon luminosity L emerging from the surface is provided completely by hydrogen burning from the CNO-cycles (*pp*-chains are at all times almost irrelevant). Gravothermal energies (L_g) balance the loss by neutrinos (L_ν). Then, the hydrogen shell is extinguishing rapidly, and the WD begins to cool and to get fainter. After a few 10^5 years neutrino losses become more important than photon emission for the cooling, and the energy results almost completely from gravothermal energies, being gained from the internal energy of the core. At $t \approx 10^7$ years, the neutrino sink fades away, and the WD is now in the phase where the Mestel theory applies, and the thermal energy of the ions is the only energy source. Crystallization and convection in the envelope have not yet set it. This happens only after almost 10^9 years. This final phase of cooling is the one shown in Fig. 37.7.

It is possible to connect the cooling times with the observed number of WD as a function of L. Since τ is steadily increasing with time, the evolution therefore slowing down, one expects to observe an increasing number of white dwarfs for fainter luminosities. The end of this *luminosity function* is reached when no star had enough time to cool to an even cooler temperature, respectively reach an even

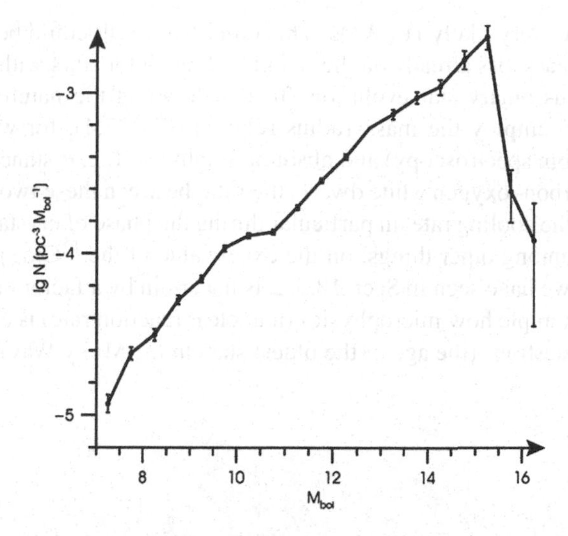

Fig. 37.8 Luminosity function of white dwarfs obtained from 6,000 objects in the *Sloan Digital Sky Survey*, data release 3 (Abazajian et al. 2005), including error bars of the star countings, which are significant after the peak of the luminosity function, due to the extreme faintness of the objects. The last few data points at and after the break of the luminosity function also depend somewhat on the assumption about the interior carbon and oxygen abundances. The break at $M_{\mathrm{bol}} = 15.3$, or equivalently $\log L/L_{\odot} = -4.3$, could correspond to an age of the coolest WDs of 8 Gyr, or be a few Gyr higher (Kilic et al. 2010, and Salaris, private communication). Data for this figure were taken from Harris et al. 2006; Fig. 7

lower luminosity. Then, the luminosity function (as the one in Fig. 37.8) should show a cut-off, and the maximum age reached indicates the age of the oldest white dwarfs and thus the oldest stars in the observed sample. With this method one could first determine the age of the galactic disk in the solar neighbourhood (found to be around 8 Gyr), and more recently that of old stellar clusters. The WD cooling curves provide an alternative way to find out about the oldest objects in our galaxy.

As discussed above, such age determinations depend on the white dwarf models used. In particular, and apart from the physics of cooling, it requires an assumption about the internal chemical composition of the WD. In case of WDs originating from single stars, it will be safe to assume a CO-WD with a He/H envelope. The arguments for this assumption are that ONeMg-WDs are very rare–if they are the result of single star evolution at all–because of the narrow mass range of super-AGB stars (Sect. 34.8; Fig. 34.10), and that the main-sequence mass of stars that develop into He-WDs by avoiding the core helium flash is so low ($\lesssim 0.50 M_{\odot}$; Sect. 35.2 and Fig. 35.2) that even the age of the universe of about 14 Gyr has not been long enough to allow them to evolve off the main sequence (an estimate for the main-sequence lifetime from (30.2) would give 75 Gyrs; a numerical model yields almost twice this number). At the present time, the oldest globular clusters produce CO-WDs of $\approx 0.53\, M_{\odot}$ (Kalirai 2009). According to Liebert et al. (2005), however, about 10 % of all white dwarfs in the solar neighbourhood have indeed a mass below $0.45 M_{\odot}$

and therefore are very likely He-WDs. This empirical result could be an indication for significant mass loss already on the red giant branch for stars with $M \gtrsim 0.8M_{\odot}$ or for a previous binary star evolution. To decide about the nature of WDs one therefore has to employ the mass–radius relation (Fig. 37.4), for which effective temperature (from spectroscopy) and absolute brightness (i.e. distance) are needed. But even for carbon–oxygen white dwarfs the ratio between these two elements will be decisive for the cooling rate, in particular during the phase of crystallization. This ratio depends, among other things, on the exact value of the $^{12}C(\alpha,\gamma)^{16}O$ reaction rate, which, as we have seen in Sect. 18.5.2, is uncertain by a factor of 2 and 3. This is a beautiful example how microphysics (a nuclear reaction rate) is connected with cosmological questions (the age of the oldest stars in the Milky Way).

Chapter 38
Neutron Stars

As early as 1934 Baade and Zwicky correctly predicted the birth of the strange objects neutron stars in supernova explosions (Baade and Zwicky 1934). The first models were calculated by Oppenheimer and Volkoff (1939), and the stage was then left for the next 28 years to particle physicists who struggled with the problem of matter at extreme densities (a struggle not yet finished). Radio astronomers accidentally found the first *pulsar* in 1967; it was interpreted soon after as a rapidly rotating neutron star (Gold 1968), emitting synchroton radiation in a narrow beacon along the magnetic axis. In addition, neutron stars were identified as sources of energetic X-ray emission, resulting from accretion in binary systems. By now, the existence of neutron stars is well established. The number of detected pulsars in the Galaxy already amounts to more than 1,800 (Lorimer 2008). These known neutron stars constitute only a tiny fraction of a population as large as a few hundred million. In some cases, their masses could be determined quite accurately (Fig. 38.2) because they are members of binary systems or from relativistic effects in their extreme gravitational potential. Everything is extreme with neutron stars, their interior state (simulating a huge nucleus), the velocity of sound (not far from c), their rotation (frequencies $1 \ldots 1{,}000$ Hz), and their magnetic fields (from 10^9 to 10^{15} gauss). One is far from really understanding them. So we content ourselves here with a few remarks on the state of matter and the resulting models. For more detailed and complete information about neutron stars, we recommend one of the many existing textbooks on compact objects (e.g., Glendenning 1997; Camenzind 2007; Haensel et al. 2007).

38.1 Cold Matter Beyond Neutron Drip

Neutron stars (NS) are born hot ($T > 10^{10}$ K) in the collapse of a highly evolved star (see Chap. 36). But the interior temperature drops rapidly because of neutrino emission: after a day, temperatures of 10^9 K are reached; after 100 years, maybe 10^8 K. And this ($kT \approx 10$ keV) can be considered cold in view of the degenerate

R. Kippenhahn et al., *Stellar Structure and Evolution*, Astronomy and Astrophysics Library, DOI 10.1007/978-3-642-30304-3_38, © Springer-Verlag Berlin Heidelberg 2012

nearly relativistic neutrons ($E_F \approx 1,000\,\text{MeV}$). The equation of state is essentially the same as for $T \approx 0$. We refer to the descriptions of high-density matter in Sect. 37.2 and of the equation of state in Chap. 16.

With increasing density the rising Fermi energy of the electrons provides an increasing neutronization by electron captures. The neutron-rich equilibrium nuclei (such as ^{118}Kr) begin to release free neutrons at $\varrho_{dr} \approx 4.3 \times 10^{11}\,\text{g cm}^{-3}$. This is called the *neutron drip.* The matter consists of nuclei (usually arranged in a lattice) plus sufficient electrons for charge neutrality, and free neutrons. Their number n_n increases with ϱ, and so does their pressure P_n. While $P \approx P_e \gg P_n$ still at $\varrho = \varrho_{dr}$, we have $P_n = P/2$ at $\varrho \approx 4 \times 10^{12}\,\text{g cm}^{-3}$ (here the Coulomb lattice is dissolving) and $P_n > 0.8P$ for $\varrho \gtrsim 1.5 \times 10^{13}\,\text{g cm}^{-3}$, and finally $P_n \approx P$. The neutrons are increasingly degenerate, but still non-relativistic, as their Fermi energy is much smaller than their rest-mass energy. Note that all characteristic densities quoted here and in the following depend in general on the model assumed for the particles and their interaction. The higher the values of ϱ, the more uncertain are the details (see below).

With progressing neutron drip the number of nuclei is diminished by fusion. The nuclei more or less touch each other at the *nuclear density* of $\varrho_{nuc} \approx 2.7 \times 10^{14}\,\text{g cm}^{-3}$, and hence they merge and dissolve, leaving a degenerate gas (or liquid) of neutrons plus a small admixture of e^- and p. The concentrations of these particles can be calculated as an equilibrium between back and forth exchanges in the reaction $n \rightleftharpoons p + e^-$ (The neutrinos leave the system immediately and can be left out of the considerations.). The conditions are that the Fermi energies fulfil $E_F^n = E_F^p + E_F^e$, and that $n_e = n_p$ for neutrality. This gives that n_p is about 1 % (or less) of n_n for a wide range of ϱ up to ϱ_{nuc}. At $\varrho \approx 6 \times 10^{15}\,\text{g cm}^{-3}$ the neutrons are relativistically degenerate. With increasing relativity of the neutrons the fraction of protons raises slowly, until at an infinite relativity parameter one finds the limiting ratio $n_n : n_p : n_e = 8 : 1 : 1$. When ϱ exceeds $10^{15}\,\text{g cm}^{-3}$, the Fermi energy of the neutrons, $E_F = [(p_F c)^2 + (m_n c^2)^2]^{1/2}$, will gradually exceed the rest masses of the hyperons of lowest mass (such as $\Lambda, \Sigma, \Delta, \ldots$). These particles will then appear, i.e. a "hyperonization" begins. Finally even free quarks can occur. Obviously, at these densities, nuclear forces, the interaction between elementary particles, and the masses of hadron states are determining the exact composition of neutron star matter. Astrophysics meets quantum chromodynamics!

We now come to the *equation of state,* in particular the dependence of P on ϱ. For ϱ up to ϱ_{drip}, the pressure is dominated by the relativistic, degenerate electrons, and $P \approx P_e \sim \varrho^{4/3}$ [cf. (15.26)].

The onset of the neutron drip ($\varrho = \varrho_{drip}$) has severe consequences for the equation of state. An increase $d\varrho$ mainly increases n_n at the expense of n_e (which yields the pressure), such that the increment dP is small (see Fig. 16.2). Therefore the gas becomes more compressible, which is described as a "softening" of the equation of state (in the opposite case one speaks of "stiffening"). In other terms the adiabatic index $\gamma_{ad} = (d\ln P/d\ln\varrho)_{ad}$ drops appreciably below the critical value 4/3 (cf. Sect. 25.3.2), and only when P_n contributes sufficiently to P will γ_{ad} again rise above 4/3 at $\varrho \approx 7 \times 10^{12}\,\text{g cm}^{-3}$.

When the neutron pressure P_n dominates one may tentatively consider the approximation that the gas consists *of ideal* (non-interacting), *fully degenerate* neutrons. These are fermions like the electrons, and they obey the same statistics, so that the same relations hold as derived in Sect. 15.2, if there m_e is replaced by m_n and μ_e by 1 (since we now have one nucleon per fermion). Instead of (15.23) and (15.26) we can write

$$P_n = K_{\gamma'} \varrho_0^{\gamma'} \tag{38.1}$$

with the non-relativistic and relativistic limit cases (for $\varrho_0 \ll 6 \times 10^{15}$ and $\varrho_0 \gg 6 \times 10^{15}\,\mathrm{g\,cm^{-3}}$ respectively)

$$\begin{aligned} \gamma' &= \frac{5}{3}, \quad K_{5/3} = \frac{1}{20}\left(\frac{3}{\pi}\right)^{2/3} \frac{h^2}{m_n^{8/3}}, \\ \gamma' &= \frac{4}{3}, \quad K_{4/3} = \frac{1}{8}\left(\frac{3}{\pi}\right)^{1/3} \frac{hc}{m_n^{4/3}}, \end{aligned} \tag{38.2}$$

with $m_u \approx m_n$. In (38.1) we have used the rest-mass density $\varrho_0 = n_n m_n$. For relativistic configurations instead of ϱ_0 one has to use the total mass-energy density $\varrho = \varrho_0 + u/c^2$. This distinction was not necessary for the electron gas, where ϱ_0 (coming mainly from the non-degenerate nucleons) was always large compared with the energy density u/c^2 coming from the degenerate electron gas. Now both ϱ_0 and u/c^2 are provided by the degenerate neutrons. For non-relativistic neutrons, $\varrho_0 \gg u/c^2$ and $\varrho \approx \varrho_0$; for relativistic neutrons, $\varrho_0 \ll u/c^2$ and $\varrho \approx u/c^2$. For relativistic particles, however, we know that $P = u/3$, i.e. $P = \varrho c^2/3$. So we can write

$$\begin{aligned} P_n &\sim \varrho^\kappa, \\ \kappa &= 5/3 \quad (\text{non-relativistic}), \\ \kappa &= 1 \quad (\text{relativistic}). \end{aligned} \tag{38.3}$$

The distinction between ϱ and ϱ_0 will be seen to be important for NS models. The relation $P = \varrho c^2/3$ also yields the velocity of sound directly as $v_s^2 = (dP/d\varrho)_{ad} = c^2/3$, i.e. $v_s = 0.577c$.

Of course, with the densities considered here, the *interaction between nucleons* is far from being negligible. It dominates the behaviour long before the limit $6 \times 10^{15}\,\mathrm{g\,cm^{-3}}$, where $p_F = m_n c$, is reached. In order to calculate its influence on the equation of state, one faces two problems. The first is the determination of a reasonable potential. In the absence of a rigorous theory and of experiments at such high densities, one has to use a model of the interacting particles that meets the results of low-energy scattering, the properties of saturation of nuclear forces, etc. It is not surprising that such models yield large uncertainties when extrapolated and applied to the densities found in NS. The qualitative influence of some effects on the equation of state is quite obvious. For example, the interaction between two nucleons depends (aside from spin and isospin properties) on their distance. When approaching each other they first feel an attraction, which turns to repulsion below a critical distance (in the extreme: at an inner hard core). Attraction (dominant at not too high ϱ) reduces P and gives a softer equation of state. Repulsion (dominant at very high ϱ and small average particle distances) increases P and thus stiffens the equation of state. Obviously details of the potential can shift the border appreciably between these two regimes.

Other uncertainties are connected with the appearance of new particles when ϱ increases. For example, if hyperons of some type occur in sufficient number, they contribute to ϱ, but scarcely to P, since their creation lowers the Fermi sea of the neutrons. Therefore "hyperonization" makes the gas more compressible. At ultra-high densities (say $\approx 10\varrho_{\mathrm{nuc}}$) so many new resonances appear that, in the extreme, attempts have been made to describe their number in a certain energy range only by statistics (which leads, e.g., to the rather soft Hagedorn equation of state). But if the nucleons almost touch each other, one might have to consider something like quark interaction. The question was even discussed whether this might lead to quark matter and possibly to *quark stars*.[1] Finally, in case that the absolute ground state of strong interactions is that of quark matter in a deconfined state, in which up, down, and strange quarks are present in about equal number, neutron stars will consist of this so-called *strange matter* and would be *strange stars* (For a discussion of quark and strange stars, see, e.g., the corresponding chapters in the book by Glendenning.).

As early as $\varrho \lesssim 2\varrho_{\mathrm{nuc}}$ the possibility of the reaction $n \rightarrow p + \pi^-$ (if $E_{\mathrm{n}} \geq E_{\mathrm{p}} + E_{\pi^-}$) gives the possibility of having a Bose–Einstein condensate of the cold π^- bosons in momentum space with zero momentum, i.e. no contribution to P but to ϱ.

The second quite general problem for determining the equation of state is that, even if the potential were known exactly, one would not know how to solve convincingly the many-body problem. Several attempts use different assumptions and yield different results.

To resume, we must stress that the equation of state is highly uncertain for at least two independent reasons (concerning the potential and the many-body problem), but there are still more open questions concerning possible effects of superfluidity and superconductivity, which might influence the evolution of neutron stars, in particular their rotation and magnetic fields. In fact particle physics cannot yet decide which of the available equations of state is correct, but the softest ones now seem to be ruled out by observation of neutron stars (see below). In Fig. 16.2 just one of them is plotted, which should resemble the general properties, but will not be exact in the details.

[1] The full beauty of this term can be savoured only in German, where the term "quark" means either a popular, soft white cheese or, in slang, complete nonsense.

38.2 Models of Neutron Stars

For a given equation of state of the form $P = P(\varrho)$ it is easy to obtain the corresponding hydrostatic models of NS. One has only to integrate the relativistic equation of hydrostatic equilibrium (2.31) (the Tolman–Oppenheimer–Volkoff equation) together with (2.30), starting at $r = 0$ with a chosen central density ϱ_c. Since the equation of state is independent of T, these two equations suffice for obtaining the mechanical structure. This is seen after replacing P by ϱ in (2.31), so that there are two equations for the variables ϱ and m. When the integration comes to $\varrho = P = 0$, the surface is reached, i.e. we have found $R = r$ and $M = m(R)$ (We do not have to worry about the obvious failure of the equation of state for $P \to 0$. The transition region to the non-degenerate atmosphere, and even the whole atmosphere, are negligibly thin so that the error made is small.).

Repeating this integration for a variety of starting values ϱ_c, one can produce a sequence of models for the chosen equation of state. They give, in particular, the relations $M = M(\varrho_c)$, $R = R(\varrho_c)$, and by elimination of ϱ_c also $R = R(M)$ (cf. Fig. 38.1).

The resulting relations $M(\varrho_c)$ and $R(M)$ change considerably if we replace the equation of state by another one, as can be seen in Fig. 38.1 for $M(R)$, where the results are plotted for several equations of state. The persisting common feature is that all relations $M(\varrho_c)$ show a minimum and a maximum of M, although at quite different values. One can easily understand the qualitative changes which occur when a soft equation of state is replaced by a stiffer one. The matter is then less compressible; for given M one expects a larger R and a smaller ϱ_c. For given ϱ_c one can put more mass on top until reaching the surface with $\varrho = 0$. This lowers the gravity inside the model, and M_{max} is higher. A particularly soft equation of state is that for the ideal degenerate neutron gas in (38.3), since the repulsive forces at small particle distances are completely neglected. Correspondingly Oppenheimer and Volkoff (1939) obtained for this equation of state a maximum mass of only $M_{max} \approx 0.72 M_\odot$. Normally the maxima range roughly between $1 M_\odot$ and $3 M_\odot$, but Fig. 38.1 also demonstrates that a particularly stiff equation of state, obtained by including interactions into the Oppenheimer–Volkoff equation, may lead to maximum masses above $3 M_\odot$. We have stressed in Sect. 38.1 that particle physics cannot yet supply the correct equation of state. All the more interesting are objects like the binary pulsar PSR B1913+16 (also called *the Hulse–Taylor pulsar*), for which the masses could be determined very accurately when details of the orbital motion were interpreted as general relativistic effects.[2] The result for the NS is $M = 1.442 M_\odot$ with a vanishing small uncertainty, which rules out all equations of state

[2]These effects include a shrinking of the orbit–and therefore a decrease of the orbital period (of the order of 60 ms)–due to the loss of gravitational waves. The observations agree perfectly with the predictions of Einstein's theory of general relativity and are considered as indirect proof for the existence of gravitational waves. J.H. Taylor and R. Hulse were awarded with the Nobel Prize for this in 1993.

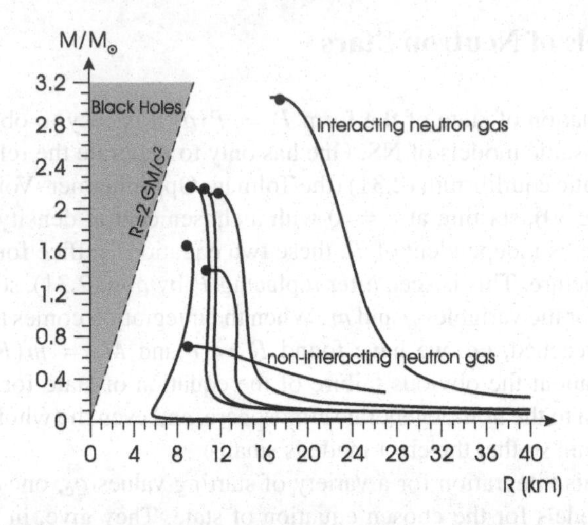

Fig. 38.1 The relation M against R of neutron-star models calculated using seven different equations of state. The maximum mass is indicated in each case by the *solid dot*. Two extreme assumptions are specifically indicated: a non-interacting neutron gas (Oppenheimer and Volkoff 1939), which leads to an extremely soft equation of state and a very low maximum mass, and one where interactions are included and which leads to a maximum neutron star mass above $3M_\odot$ (After Fig. 10.3 in Weber et al. 2009, where also details about the other five equations of state can be found)

so soft that their $M_{\max}$ is below $1.44M_\odot$. Very recently, another binary millisecond pulsar–J1614-2230–was analysed by Demorest et al. (2010), who determine a pulsar mass of $1.97\pm0.04M_\odot$, using the so-called *Shapiro delay* of the pulsar signal, which is caused by the fact that light signals do not travel a straight line, but follow null-geodesics which are bent, and therefore longer, by the gravitational potential. This result rules out at least two more equations of state of Fig. 38.1, among them one for "strange stars" (in the figure, this corresponds to the left-most line). Here seems to be one of the cases where astrophysical measurements set a discriminating limit to particle physics. A collection of accurately determined neutron-star masses is given in Fig. 38.2. They were obtained by different methods, on which we do not comment further, but refer to the respective textbooks.

The *maximum mass* for NS is very important, not only in connection with evolutionary considerations, but also in the attempt to identify compact objects with $M > M_{\max}$ as black holes. If our ignorance of the equation of state does not yet allow the determination of $M_{\max}$ to better than the interval $2\ldots3M_\odot$, we should at least understand that such a maximum mass (well below $5M_\odot$) must exist.

In order to make this plausible, we neglect effects of *general* relativity, i.e. consider the usual equation of hydrostatic equilibrium but keep those of *special* relativity as allowed for in (38.3). Let us consider some averages of P and ϱ over the whole star. As in (37.15) the normal hydrostatic equation then yields the estimate $P \sim M^2/R^4$. Here we eliminate R by $\varrho \sim M/R^3$ and obtain $P \sim M^{2/3}\varrho^{4/3}$,

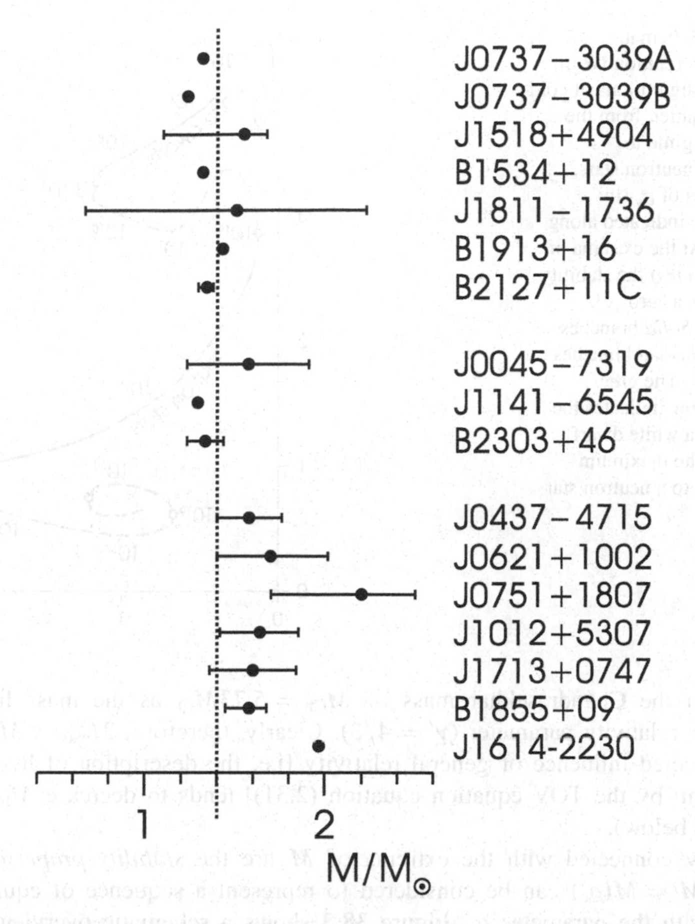

Fig. 38.2 An overview of empirically determined masses and their errors of neutron stars (identifier to the right of the data point). Except for the last object, J1614-2230 (Demorest et al. 2010), the data are taken from Fig. 6.31 in Camenzind (2007). The *vertical line* corresponds to the generic mass of $1.4 M_\odot$

introduce $\varrho \sim P^{1/\kappa}$ from the equation of state (38.3), and then solve for M and find

$$M \sim \varrho^{3(\kappa-4/3)/2}. \tag{38.4}$$

In the non-relativistic limit, $\kappa = 5/3$, giving $M \sim \varrho^{1/2}$ and $dM/d\varrho > 0$. The extreme relativistic case requires $\kappa = 1$, which gives $M \sim \varrho^{-1/2}$ and $dM/d\varrho < 0$. Somewhere on the border between the two regimes we expect $dM/d\varrho = 0$, i.e. the maximum mass(The average ϱ treated here will be a sufficient measure for ϱ_c too.). Therefore the maximum of M must occur when the neutrons *start* to become relativistic and the energy density u/c^2 begins to overtake the rest-mass density ϱ_0. Only by neglecting u/c^2 in ϱ [taking (38.1) instead of (38.3)] could

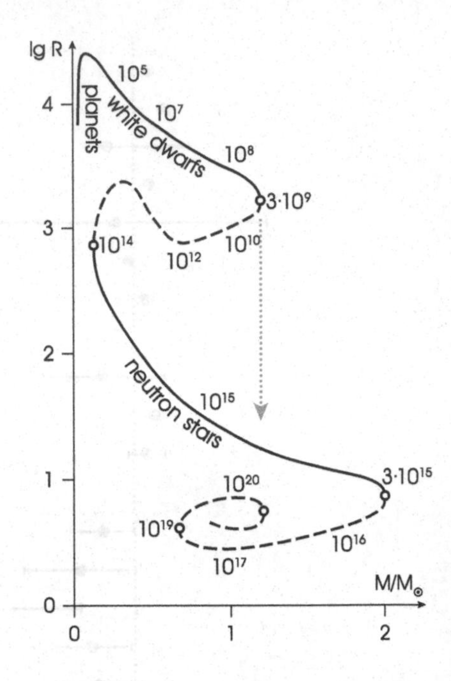

Fig. 38.3 Schematic mass–radius relation (R in km) for configurations of cold catalysed matter, from the planetary regime to ultra-dense neutron stars. Some values of ϱ_c (in g cm^{-3}) are indicated along the curve. At the extrema of M (*open circles*) the stability problem has a zero eigenvalue. *Solid* branches are stable, *dashed* branches are unstable. The *grey, vertical arrow* indicates the collapse of a white dwarf exceeding the maximum stable mass to a neutron star

we obtain the Chandrasekhar mass of $M_{\text{Ch}} = 5.73 M_\odot$ as the mass limit for an *infinite* relativity parameter ($\gamma' = 4/3$). Clearly, therefore, $M_{\text{max}} < M_{\text{Ch}}$. The here neglected influence of general relativity [i.e. the description of hydrostatic equilibrium by the TOV equation equation (2.31)] tends to decrease M_{max} even more (see below).

Closely connected with the extrema of M are the *stability properties*. The relation $M = M(\varrho_c)$ can be considered to represent a sequence of equilibrium models with the parameter ϱ_c. Figure 38.3 shows a schematic overview of the resulting $M - R$ relation for cold catalysed matter from the regime of planets to that of ultra-dense NS. Starting from planets, ϱ_c increases monotonically along the curve (compare with typical values of ϱ_c indicated in Fig. 38.3). There are extrema of R which may be interesting in other connections but are not important for the sequence $M(\varrho_c)$. However, one also encounters extrema of M (open circles). The most important are M_{min} and M_{max} for NS, as well as the maximum M for white dwarfs. These are critical points at which a detailed stability analysis shows that the stability of the equilibrium models changes. The stable parts of the curve are those with $dM/d\varrho_c > 0$, i.e. the branch of NS with $M_{\text{min}} < M < M_{\text{max}}$ (and the white-dwarf and planetary branch with M < maximum mass for white dwarfs). When further increasing ϱ_c beyond the point at which $M = M_{\text{max}}$ there follows an infinite number of maxima and minima of M. Correspondingly the curve $R = R(M)$ spirals into a limiting point, which is reached for $\varrho_c \to \infty$. All of these branches are in fact unstable. The stability analysis can also be made for general relativistic configurations. In the Newtonian limit one has the well-known result that an average of the exponent γ_{ad} of $\gamma_{\text{cr}} = 4/3$ is equivalent to marginal stability (see Sect. 25.3.2), and in addition it can be shown (see Shapiro and Teukolsky 1983) that small effects of general relativity ($GM/Rc^2 \ll 1$) change the critical value from 4/3 to

$$\gamma_{cr} = \frac{4}{3} + \Lambda \frac{GM}{Rc^2}, \tag{38.5}$$

where Λ is a positive quantity of the order of unity. Therefore general relativity increases γ_{cr}, making the star more unstable, since stability requires $\bar{\gamma}_{ad} > \gamma_{cr}$. For $M = 1M_\odot$, $R = 10\,\text{km}$ the correction term in (38.5) is about 0.15, i.e. far from being negligible. γ_{cr} can be raised well above 5/3 (even above 2 for certain models near M_{max}) such that all but the stiffest equations of state would give instability. This increase of γ_{cr} is an important factor in determining the value of M_{max} (together with the lowering of $\bar{\gamma}_{ad}$).

A very stiff equation of state, for example, gives $M_{max} = 2.7M_\odot$, with $R = 13.5\,\text{km}$ and $\varrho_c = 1.5 \times 10^{15}\,\text{g cm}^{-3}$, while a softer one yields $M_{max} = 2M_\odot$, with $R = 9\,\text{km}$ and $\varrho_c = 3.3 \times 10^{15}\,\text{g cm}^{-3}$. At present there is no equation of state that can be considered realistic and that would give M_{max} well above $3M_\odot$. This includes calculations that take into account general relativity.

The model is also marginally stable at the minimum mass M_{min}, where the curve in Fig. 38.3 begins leading to the white dwarfs. This instability is essentially caused by the lowering of γ' in connection with the neutron drip (see Sect. 38.1). We have seen that the release of free neutrons from nuclei results in $\gamma' \lesssim 4/3$ in the range $\varrho \approx 4 \times 10^{11} \dots 7 \times 10^{12}\,\text{g cm}^{-3}$. Typical models for the minimum mass of stable neutron stars give $M_{min} \approx 0.09M_\odot$, $R \approx 160\,\text{km}$, $\varrho_c \approx 1.5 \times 10^{14}\,\text{g cm}^{-3}$. The average density is, of course, much smaller ($\approx 10^{10}\,\text{g cm}^{-3}$), and the averaged γ_{ad} becomes just equal to γ_{cr} (which is here close to 4/3).

Let us dwell briefly on the meaning of the mass values quoted for NS. The stellar mass M is here always the "gravitational mass", which is the value measurable for an outside observer [cf. the comments in Sect. 2.6 after (2.29)]. M differs from the proper mass $M_0 = Nm_0$, given by the total number N of nucleons with a rest mass m_0, since in relativity, the total binding energy W of the configuration appears as a mass $\Delta M = W/c^2$, such that

$$M = M_0 + \frac{W}{c^2} = M_0 + \Delta M. \tag{38.6}$$

In the Newtonian limit (for weak fields) we were used to identifying particularly the internal energy E_i (from motion and interaction of particles) and the potential energy E_g in the gravitational field. Then for a static, stable configuration, $W = E_i + E_g < 0$, since $E_g < 0$ and $-E_g > E_i$ (In the Newtonian limit E_g and E_i were related by the virial theorem, cf. Chap. 3.). Correspondingly we may now say that the mass of a NS is increased by the internal energy and decreased by the (negative) potential energy, and the latter term wins. Therefore $W < 0$, and we have a mass defect $\Delta M < 0$. Depending on the precise model, $|\Delta M|$ can go up to 10…25 % of M near M_{max}. Formally M is given as an integral over $4\pi r^2 \varrho dr$, where ϱ is the total mass-energy density ($\varrho_0 + u/c^2$) and $4\pi r^2 dr$ is *not* the volume element. This is rather given by $dV = 4\pi r^2 e^{\lambda/2} dr$ with $e^{\lambda/2}$ being a component of the metric tensor (cf. Sect. 2.6). Then simply

$$\Delta M \equiv M - M_0 = \int_0^R (4\pi r^2 \varrho dr - \varrho_0 dV)$$
$$= \int_0^R 4\pi r^2 \varrho \left(1 - e^{\lambda/2}\frac{\varrho_0}{\varrho}\right) dr. \tag{38.7}$$

Here $\varrho_0/\varrho < 1$, but $e^{\lambda/2} > 1$, and the product of both is >1, such that $\Delta M < 0$. So if we find an NS with mass M, we know that it started off as a more massive configuration. The mass defect $|\Delta M|$ was radiated away in the course of evolution by photons, neutrinos, or gravitational radiation. In that sense the original Kelvin–Helmholtz hypothesis that contraction supplies the radiated energy has turned out to be correct. The mass defect reaches a maximum at $M = M_{max}$ and then decreases again towards models with still larger ϱ_c.

The maximum mass for NS is scarcely influenced by rotation. Except for the very few most rapidly spinning pulsars, centrifugal forces play practically no role in NS, since the overwhelming gravitational forces dominate completely. This is at least true for rigidly rotating NS stars. However, *differential rotation* may stabilize neutron stars and will lead to higher maximum masses. In the case of simple polytrope models, differential rotation can raise M_{max} by up to 50 % (see Baumgarte and Shapiro 2010, Chap. 14).

Now we turn to describe the *stratification of matter inside an NS model.* At the very outer part there must be an atmosphere of "normal" non-degenerate matter. Going inwards, we come to gradually larger densities and encounter all characteristic changes of high-density matter as described in Sect. 38.1.

The *atmosphere* of an NS is very hot and incredibly compressed. Typical temperatures are of the order of 10^6 K (see below). The extension is very small owing to the high surface gravity $g_0 \approx 1.3 \times 10^{14}$ cm s^{-2} (For comparison, $g_0 = 2.7 \times 10^4$ cm s^{-2} for the Sun and $\approx 10^8$ cm s^{-2} for white dwarfs.). This gives a pressure scale height of the order of 1 cm only. In the surface layers (say $\varrho \lesssim 10^6$ g cm^{-3}) the behaviour of the matter is still influenced by the temperature and also by strong magnetic fields.

Not far below the surface, the densities will be in and above the range typical for the interior of white dwarfs ($\gtrsim 10^6$ g cm^{-3}). As an example we discuss the model for an NS of $M = 1.4 M_\odot$ (see Fig. 38.4), calculated by using an equation of state of moderate stiffness which gives $M_{max} \approx 2 M_\odot$. The radius of the $1.4 M_\odot$ model is 10.6 km. Although there are newer models, the present one is still a good representation of the typical structure of neutron stars.

Below the surface there is a solid *crust* ($10^6 \lesssim \varrho \lesssim 2.4 \times 10^{14}$ g cm^{-3}) of thickness $\Delta r \approx 0.9$ km. The matter in the crust contains nuclei, which are mainly Fe near the surface (cf. the equilibrium composition as a function of ϱ described in Sect. 37.2). These nuclei will form a lattice, thus minimizing the energy of Coulomb interaction as in crystallized white dwarfs. The *outer crust* consists only of

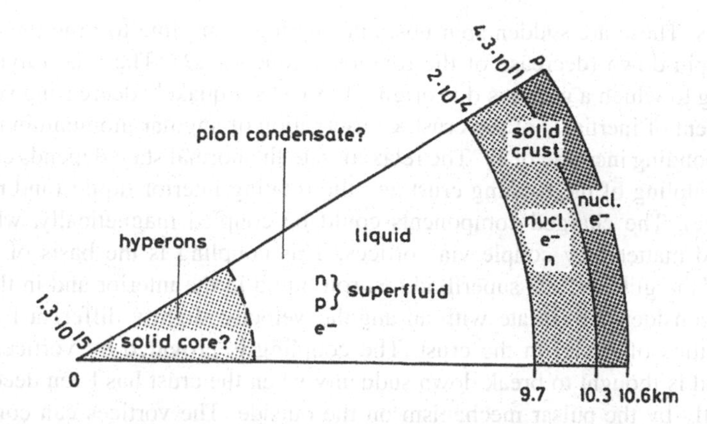

Fig. 38.4 Illustration of the interior structure of a neutron-star model with $M = 1.4M_\odot$ calculated with an equation of state similar to the intermediate ones in Fig. 38.1. A few characteristic values of the density (in g cm^{-3}) are indicated along the upper radius (After Pines 1980)

these nuclei plus a relativistically degenerate electron gas, though this changes over a depth of $\Delta r \approx 0.3$ km to where the neutron drip density $\varrho_{\mathrm{dr}} \approx 4 \times 10^{11}$ g cm^{-3} is reached. In the subsequent *inner crust* ($4 \times 10^{11} \lesssim \varrho \lesssim 2 \times 10^{14}$ g cm^{-3}), a liquid of free neutrons exists in addition to the nuclei (still arranged in a lattice, and becoming increasingly neutron-rich) and the electrons. With decreasing r the free neutrons become more and more abundant at the expense of the nuclei, and the lattice disappears with the nuclei, until all nuclei are dissolved into homogeneous, neutron-rich nuclear matter at $\varrho = \varrho_{\mathrm{nuc}} \approx 2.4 \times 10^{14}$ g cm^{-3}, which therefore defines the lower boundary of the solid crust, at a depth of 0.9 km. The equation of state throughout the crust is relatively well known; this is the reason why our aged neutron-star model is still valid.

Below the crust there is the *interior neutron liquid* ($\varrho \gtrsim 2.4 \times 10^{14}$ g cm^{-3}) consisting mainly of neutrons in equilibrium with a few protons, electrons, and muons. All constituents are strongly degenerate and the hadrons are interacting by nuclear forces. The neutrons will be superfluid, the protons superconductive. The equation of state begins not to be well-known in this density regime, and from here on the structure and composition of the inner core depends on the equation of state used.

It is unclear whether there is finally a central *solid core* in which the neutrons form a solid owing to their repulsive forces at small particle distances. The central density of our model is $\varrho_c \approx 1.3 \times 10^{15}$ g cm^{-3}. The inner core may also consist of baryon condensates (π^-, K^-), of a mixture hyperons and baryon resonances (Σ, Λ, Ξ, Δ), or deconfined quark matter. We refer the reader to Fig. 10.1 in Weber et al. (2009) for a graphical representation of various possibilities. In that figure, our model would be one of the *traditional neutron stars*.

The superfluidity of the neutron and proton liquids and the solid parts (crust and possible core) play a role in the attempts to explain the observed "glitches" of pulsars. These are sudden spin-ups, interrupting from time to time the normal, regular spin-down (decrease of the rotation frequency Ω). There is a hypothesis according to which a glitch is due originally to a "starquake", decreasing suddenly the moment of inertia I_c of the crust. Conservation of angular momentum requires a corresponding increase of Ω. The relaxation to the normal state depends critically on the coupling of the rotating crust and the rotating interior liquid (and possible solid core). The charged components could be coupled magnetically, while the superfluid matter may couple via vortices. This coupling is the basis of another model of the glitches: the superfluid neutron liquid in the interior and in the inner crust is considered to rotate with an angular velocity slightly different from that of the lattice of nuclei in the crust. The coupling is provided by vortices in the liquid and is thought to break down suddenly when the crust has been decelerated sufficiently by the pulsar mechanism on the outside. The vortices can contain an appreciable fraction of the star's angular momentum, and their distortion induces immediate changes of the observed rotation.

The thermal properties (except for the earliest stages) in principle follow once the mechanical models are given. One can then calculate the thermal conductivity, which, together with a given outward flux of energy, determines the T gradient at any point. It turns out that like white dwarfs (Sect. 37.3) the NS have a nearly isothermal interior because of the high thermal conductivity. Only in the outermost layers does T drop, by typically a factor of 10^2, to the surface temperature. Particularly in the first, hot phases, the cooling will be very rapid because of strong neutrino losses.

In this chapter we have completely ignored the strong magnetic fields of neutron stars. While they are of only minor importance for the structure and the maximum mass, they are crucial for many phenomena which allow the observation of neutron stars. Most notably this is the pulsar phenomenon, which is due to the emission of synchrotron radiation along the axis of the magnetic dipole, being inclined with respect to the rotation axis. The typical field strength of pulsar magnetic fields are of order 10^{11}–10^{13} G. Some NS possess even stronger magnetic fields, up to 10^{15} G, which is the highest known level in the universe. They are called *magnetars* and are the source for the *soft gamma repeaters* (Thompson and Duncan 1995), a class of gamma-ray bursts that exceed the Eddington luminosity by far, but are characterized by a comparably soft gamma spectrum. In a magnetar, the decaying magnetic field is the source of free energy (rather than rotation, as in pulsars). As we said at the beginning, everything is extreme in neutron stars.

Chapter 39
Black Holes

Black holes (BH) represent the ultimate degree of compactness to which a stellar configuration can evolve. Having already called the neutron star a strange object, one cannot help labelling BH as weird. From the many fascinating aspects that are accessible via the full mathematical procedure (cf. Misner et al. 1973; Shapiro and Teukolsky 1983; Chandrasekhar 1983) we will indicate only a few points, showing that this is really a final stage of evolution, not just another late phase. We limit the description to non-rotating BH without charge.

The theoretical description to be applied is that of general relativity (see, e.g., Landau and Lifshitz 1976, vol. 2). We consider the gravitational field surrounding a very condensed mass concentration M with spherical symmetry. The vacuum solution of Einstein's field equations (2.24) for this case was found as early as 1916 by K. Schwarzschild. It gives the line element ds, i.e. the distance between neighbouring events in 4-dimensional space–time as

$$\begin{aligned} ds^2 &= g_{ij}\,dx^i\,dx^j \\ &= \left(1-\frac{r_\mathrm{s}}{r}\right)c^2dt^2 - \left(1-\frac{r_\mathrm{s}}{r}\right)^{-1} dr^2 - r^2 d\vartheta^2 - r^2\sin^2\vartheta\, d\varphi^2 \\ &= \left(1-\frac{r_\mathrm{s}}{r}\right)c^2dt^2 - d\sigma^2\,, \end{aligned} \tag{39.1}$$

where one has to sum from 0 to 3 over the indices i and j, and where the usual spherical coordinates r, ϑ, φ are taken as the spatial coordinates x^1, x^2, x^3, and $x^0 = ct$. The critical parameter r_s in (39.1) is the *Schwarzschild radius*

$$r_\mathrm{s} = \frac{2GM}{c^2}\,, \tag{39.2}$$

which has the value $r_\mathrm{s} = 2.95\,\mathrm{km}$ for $M = M_\odot$. The second component of the metric tensor g_{ij}, $(1 - r_\mathrm{s}/r)^{-1}$, becomes singular at $r = r_\mathrm{s}$, but one can show that this is a non-physical singularity disappearing when other suitable coordinates are used.

R. Kippenhahn et al., *Stellar Structure and Evolution*, Astronomy and Astrophysics Library, DOI 10.1007/978-3-642-30304-3_39, © Springer-Verlag Berlin Heidelberg 2012

The proper time τ, as measured by an observer carrying a standard clock, is related to the line element ds along his world line by

$$d\tau = \frac{1}{c}ds \,. \tag{39.3}$$

For a stationary observer ($dr = d\vartheta = d\varphi = 0$) at infinity ($r \to \infty$) the proper time τ_∞ coincides with t according to (39.1). Consider two stationary observers, one at r, ϑ, φ and the other at infinity. Their proper times τ and τ_∞ are related to each other by

$$\frac{d\tau}{d\tau_\infty} = \left(1 - \frac{r_s}{r}\right)^{1/2} \,. \tag{39.4}$$

Suppose that the first of them operates a light source emitting signals at regular intervals $d\tau$, for example, an atom emitting with the frequency $\nu_0 = 1/d\tau$. The other one receives the signals and measures the intervals in his own proper time as $d\tau_\infty$, i.e. he measures another frequency $\nu = 1/d\tau_\infty$. The resulting red shift due to the gravitational field is therefore

$$z \equiv \frac{\nu_0 - \nu}{\nu} = \frac{\nu_0}{\nu} - 1 = \frac{d\tau_\infty}{d\tau} - 1 = \left(1 - \frac{r_s}{r}\right)^{-1/2} - 1 \,, \tag{39.5}$$

which gives $z \to \infty$ for $r \to r_s$.

The metric components in (39.1) show that the 4-dimensional space–time $(x^0, \dots, x^3)$ is curved, and this holds also for the 3-dimensional space (x^1, x^2, x^3). At the surface of a mass configuration of mass M and radius R, the Gaussian curvature K of position space can be written as

$$K = -\frac{GM}{c^2R^3} = -\frac{1}{2}\frac{r_s}{R}\frac{1}{R^2} \,. \tag{39.6}$$

This is usually very small compared with the curvature R^{-2} of the 2-dimensional surface. For example, $-K \approx 2\times10^{-6}R^{-2}$ at the surface of the Sun. But one already has $-K \approx 0.15R^{-2}$ for a neutron star, and the two curvatures are comparable at the surface of a BH with $R = r_s$.

Consider a test particle small enough for the gravitational field not to be disturbed which moves freely in the field from point A to B. Its world line in 4-dimensional space–time is then a *geodesic*, i.e. the length s_{AB} is an extremum. This is to say, any infinitesimal variation does not change the length:

$$\delta s_{AB} \equiv \delta \int_A^B ds = 0 \tag{39.7}$$

If the test particle moves locally with a velocity v over a spatial distance $d\sigma$, then the proper time interval $d\tau$ will be the smaller, the larger v. It becomes [cf. (39.1)]

$$d\tau = ds = 0, \quad \text{for } v = c \,, \tag{39.8}$$

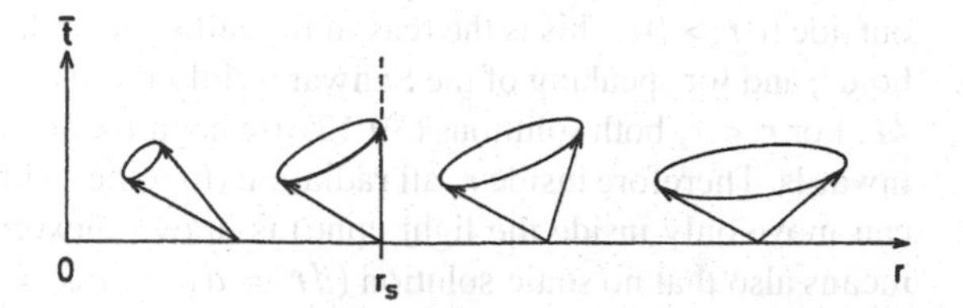

Fig. 39.1 Illustration of light cones at different distances r from the central singularity, inside and outside the Schwarzschild radius r_s

i.e. for photons or other particles of zero rest mass: they move along null geodesics. For material particles the requirement $v < c$ of special relativity (which is locally valid) means $d\tau^2$ and $ds^2 > 0$. Such separations are called *time-like*. World lines of material particles must be time-like. Separations with $ds^2 < 0$ (or $d\tau^2 < 0$) would require $v > c$; they are called *space-like*. For example, the distance between two simultaneous events ($dt = 0$) is space-like.

The null geodesies ($ds^2 = 0$), giving the propagation of photons, describe hypercones in space–time which are called *light cones*. In order to also see their properties near $r = r_s$, we introduce a new time coordinate $\bar{t}$ given by

$$\bar{t} = t + \frac{r_s}{c}\ln\left|\frac{r}{r_s} - 1\right| , \tag{39.9}$$

which transforms (39.1) to

$$\begin{aligned} ds^2 &= \left(1 - \frac{r_s}{r}\right) c^2 d\bar{t}^2 - 2\frac{r_s}{r} c\, dr\, d\bar{t} \\ &\quad - \left(1 + \frac{r_s}{r}\right) dr^2 - r^2 d\vartheta^2 - r^2 \sin^2\vartheta d\varphi^2 , \end{aligned} \tag{39.10}$$

which is non-singular at $r = r_s$. We consider only the radial boundaries of the light cones, i.e. the path of radially ($d\vartheta = d\varphi = 0$) emitted photons. Then (39.10) yields for $ds^2 = 0$, after division by $c^2 dr^2$, the quadratic equation

$$\left(1 - \frac{r_s}{r}\right)\left(\frac{d\bar{t}}{dr}\right)^2 - \frac{2r_s}{cr}\frac{d\bar{t}}{dr} - \frac{1}{c^2}\left(1 + \frac{r_s}{r}\right) = 0 , \tag{39.11}$$

which has the solutions

$$\left(\frac{d\bar{t}}{dr}\right)_1 = -\frac{1}{c} , \left(\frac{d\bar{t}}{dr}\right)_2 = \frac{1}{c}\frac{1 + r_s/r}{1 - r_s/r} \tag{39.12}$$

These derivatives are inclinations of the two radial boundaries of the light cone in an $r - \bar{t}$ plane (see Fig. 39.1). The first always corresponds to an inward motion with the same velocity c. The second derivative changes sign at $r = r_s$, being positive for $r > r_s$, where photons can be emitted outwards ($dr > 0$). With decreasing r, $(d\bar{t}/dr)_2$ becomes larger so that the light cone narrows and its axis turns to the left in Fig. 39.1. At $r = r_s$ the light cone is such that no photon can be emitted to the outside ($dr > 0$). This is the reason for calling a configuration with $R = r_s$ a "black hole", and for speaking of the Schwarzschild radius r_s as the radius of a BH of mass M. For $r < r_s$ both solutions (39.12) are negative and the whole light cone is turned inwards. Therefore inside r_s all radiation (together with all material particles, which can move only inside the light cone) is drawn inexorably towards the centre. This means also that no static solution ($dr = d\vartheta = d\varphi = 0$) is possible inside r_s, since it would require a motion vertically upwards in Fig. 39.1, i.e. outside the light cone.

In order to describe the motion of a material particle, we consider all variables to depend on the parameter τ, the proper time, varying monotonically along the world line: $d\tau = ds/c$. Dots denote derivatives with respect to τ. For example, $\dot{x}^\alpha = dx^\alpha/d\tau$ is the α component of a 4-velocity. Introducing $dx^\alpha = \dot{x}^\alpha d\tau$ into (39.1) gives the useful identity

$$c^2 = g_{ij}\dot{x}^i x^j = c^2\left(1 - \frac{r_s}{r}\right)\dot{t}^2 - \left(1 - \frac{r_s}{r}\right)^{-1}\dot{r}^2 - r^2(\dot{\vartheta}^2 + \sin^2\vartheta\,\dot{\varphi}^2). \tag{39.13}$$

The condition that the world line be a geodesic means that the variation $\delta s = \delta\tau = 0$, which yields the Euler–Lagrange equations

$$\frac{d}{d\tau}\left(\frac{\partial L}{\partial \dot{x}^\alpha}\right) - \frac{\partial L}{\partial x^\alpha} = 0\,, \tag{39.14}$$

with the Lagrangian L given by

$$2cL = [g_{ij}\dot{x}^i x^j]^{1/2} = \left[c^2\left(1 - \frac{r_s}{r}\right)\dot{t}^2 - \left(1 - \frac{r_s}{r}\right)^{-1}\dot{r}^2 - r^2\left(\dot{\vartheta}^2 + \sin^2\vartheta\dot{\varphi}^2\right)\right]^{1/2}. \tag{39.15}$$

From (39.13) and (39.15) follows the value $L = 1/2$. For $x^0 = ct$, (39.14) becomes simply

$$\frac{d}{d\tau}\left[\left(1 - \frac{r_s}{r}\right)\dot{t}\right] = 0\,, \quad \left(1 - \frac{r_s}{r}\right)\dot{t} = \text{constant} \equiv A\,. \tag{39.16}$$

We confine ourselves to the discussion of a radial infall ($\dot{\vartheta} = \dot{\varphi} = 0$) starting at $\tau = 0$ with zero velocity at the distance r_0. Instead of also deriving the equation of motion for $x^1 = r$ from (39.14), we simply introduce the second equation (39.16) into (39.13) and solve it for $\dot{r}$:

$$\dot{r} = c\left[A^2 - 1 + \frac{r_s}{r}\right]^{1/2}. \tag{39.17}$$

For our purposes we set $A^2 - 1 = -r_s/r_0$. According to (39.17) this means that the particle starts with zero velocity at $r = r_0$. The integration of (39.17) then yields

$$\tau = \frac{1}{2}\frac{r_0}{c}\sqrt{\frac{r_0}{r_s}}(\sin\eta + \eta) , \tag{39.18}$$

with the parameter $\eta = \arccos(2r/r_0 - 1)$, as can be verified by differentiation. This function $\tau = \tau(r)$ is shown in Fig. 39.2 for $r_0 = 5r_s$. Again, nothing special happens in the proper time when the particle reaches $r = r_s$. The total proper time for reaching $r = 0$ is

$$\tau_0 = \frac{\pi}{2}\frac{r_s}{c}\left(\frac{r_0}{r_s}\right)^{3/2} . \tag{39.19}$$

For $r_0 = 10\, r_s$ and $5\, r_s$ we have $\tau_0 = 49.67\, r_s/c$ and $17.56\, r_s/c$, respectively. These are very short times indeed, since for $M = M_\odot$ the characteristic time is only $r_s/c = 9.84 \times 10^{-6}$ s.

The motion in terms of the coordinate time t of an observer at infinity is quite different. The relation between t and τ is given by (39.16) as $d\tau/dt = (1 - r_s/r)/A$, which goes to zero when $r \to r_s$. By this relation and (39.17) one obtains a differential equation for $t(r)$, which is integrated to give

$$\frac{t}{r_s/c} = \ln\left|\frac{\xi + \operatorname{tg}\eta/2}{\xi - \operatorname{tg}\eta/2}\right| + \xi\left[\eta + \frac{r_0}{2r_s}(\eta + \sin\eta)\right] , \tag{39.20}$$

with η as in (39.18) and $\xi = (r_0/r_s - 1)^{1/2}$. The curve $t = t(r)$ is also shown in Fig. 39.2 for $r_0 = 5r_s$. The fact that the observer sees the τ clock of the particle slowing down completely for $r \to r_s$ has the result that $t = t(r)$ approaches $r = r_s$ only asymptotically for $t \to \infty$. Events inside $r = r_s$ are completely shielded for the distant observer by the coordinate singularity at the Schwarzschild radius acting as an "event horizon".

These few considerations may suffice to illustrate some important properties of configurations which collapse into a BH [Note that the Schwarzschild metric (39.1) is a vacuum solution, which is not valid inside the mass configuration, but holds from the surface outwards.].

As observed from the infalling surface (proper time τ) the collapse proceeds fairly rapidly and in particular quite smoothly through the Schwarzschild radius $r = r_s$. Once the surface is inside r_s a static configuration is no longer possible, and the final collapse into the central singularity within a very short time is unavoidable. This is shown by the fact that material particles have world lines only inside the local light cone, and this is open only towards $r = 0$ (even radiation falls to $r = 0$). Note that it would not help to invoke an extreme pressure exerted by unknown physical effects, since the pressure would also contribute to the gravitating energy. The singularity at $r = 0$ is an essential one (as opposed to the mere coordinate singularity at $r = r_s$) with infinite gravity, though the physical conditions there are

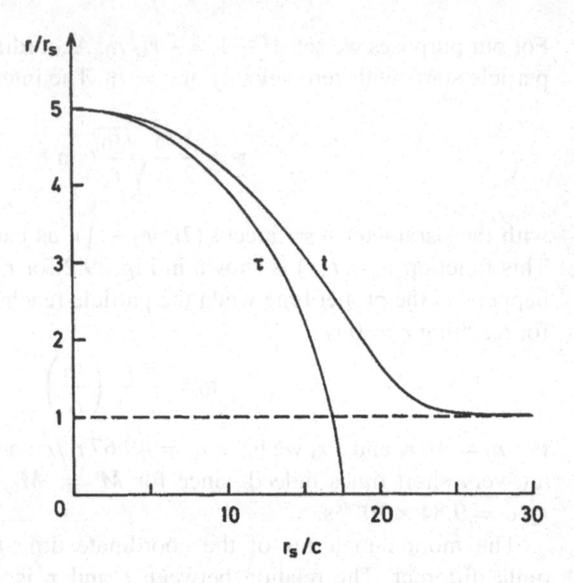

Fig. 39.2 The radial infall into a black hole for a test particle starting at a distance $5r_s$ with zero velocity. The motion is shown in terms of the particle's proper time τ, and in terms of the coordinate time t of an observer at infinity

not yet clear. Quantum effects should be included and one can speculate whether they might remove the singularity.

The collapse of a star will present itself quite differently for an astronomer who is (we hope) very far away. In his coordinate time t he will see that the collapse of the stellar surface slows down more and more, the closer it comes to r_s. In fact he will find that this critical point is not reached within finite time t; for him the collapsing surface seems to become stationary there. Of course, the approach of the surface to r_s strongly affects the light received by the distant observer. He receives photons in ever increasing intervals and with ever decreasing energy, due to the red shift $z \to \infty$ according to (39.5). Thus the collapsing star will finally "go out" for the distant observer. Only a strong gravitational field is left.

It should be mentioned that aside from the Schwarzschild solution for non-rotating, uncharged BH, there exist solutions which describe a rotating BH (Kerr metric) and a charged BH (Newman metric), the combination of these covering the full generality of possible properties of a BH: it is fully defined by mass, angular momentum, and charge. This surprising scantiness of properties left after the final collapse was summarized by Wheeler: "a black hole has no hair".

From the foregoing it is clear that black holes cannot be observed directly. However, they can be detected through their enormous influence on their surroundings. For a long time, however, BHs remained a theoretical possibility without proof of their reality. This has changed during the last few decades, and by now, their existence in two completely different mass ranges has been confirmed.

The first type of BHs are of galactic scale, sit in the centre of many galaxies, and have masses of $10^6 \lesssim M_{\mathrm{BH}}/M_\odot \lesssim 10^{10}$. They truly deserve the name *supermassive black holes*. They are detected by the analysis of the dynamics of stars in their vicinity. The overall rotation velocity (e.g. in the disks of spiral galaxies), or the velocity dispersion (in elliptical galaxies) allows to determine the total mass interior to the galactocentric radius at which it is measured. With increasing spatial resolution of the telescopes, most notably the *Hubble Space Telescope*, the central mass could be restricted to smaller and smaller central regions, until finally only a supermassive black hole could explain the dynamics. The determined masses agree well with the estimates for the mass of the central engine in quasars and other active galaxies, needed to power the energetic of these objects.

A particularly convincing case is the Seyfert galaxy NGC 4258, where microwave emission from gas orbiting the centre has been observed. For such long wavelengths the resolution is even higher and it was found that the maser clouds orbit a central mass of forty million solar masses on orbits of only half a light year! Note that the Schwarzschild radius of such a BH is 1.2×10^{13} cm, which is still only 10^{-5} of that distance. Nevertheless, no stellar cluster with that mass could be accommodated in this volume.

Our own Milky Way is hosting a supermassive BH as well (Genzel et al. 2010). From near-infrared observations of its centre, it was found that the radio source Sgr A* coincides with a supermassive BH of about 3.6 million $M_\odot$. The proof was brought about from accurate determinations of the position and movement of a cluster of about 20 stars over a decade and longer (Eckart and Genzel 1996). As before, the stars orbit with velocities of several hundred km/s around a mass of that size which is occupying a volume with a radius smaller than 0.001 pc, and this extremely high mass concentration can only be explained with a black hole.

The origin and growth of supermassive BHs is not understood, but we have good models for the creation of the so-called *stellar black holes*. Their masses are in the range $2.5 \lesssim M_{\rm BH} \lesssim 50$, and they are thought to have been created either directly in core collapse supernova explosions (see Chap. 36, and Fig. 34.10), or by the merging of binary neutron stars. They have been detected by using the fact that in binary systems mass from a companion may flow onto the black hole, and in doing so, accumulates in an accretion disk because of the conservation of angular momentum. Due to the extremely deep gravitational potential well around the BH, the energy of the infalling material is so high that any dissipative process in the disk releases X-ray photons. X-ray binary systems are therefore the ideal place to look for proof of stellar BHs. The method is rather straightforward: one measures the orbital period Π, and the maximum line-of-sight velocity $K_{\rm comp} = v \sin i$ of the visible companion using the Doppler effect. The inclination angle is not known, but can be estimated from other information or treated with a probability approach. These two quantities are used to compute the so-called *mass function*

$$f(M_{\rm BH}, i) = \frac{K_{\rm comp}^3 \Pi}{2\pi G} = \frac{M_{\rm BH} \sin^3 i}{(1 + M_{\rm BH}/M_{\rm comp})^2}, \tag{39.21}$$

where $M_{\rm comp}$ is the mass of the companion, which has to be estimated or determined using other quantities, such as the spectral type. While there are obviously a number of uncertainties in the determination of $M_{\rm BH}$ from (39.21), there are enough X-ray

Table 39.1 Some stellar black holes (as of 2008) in X-ray binaries

Object	Π	Spect.cl.	K_{comp}	i	M_{BH}	M_{comp}
V1487 Aql	33.5	K–M III	140	70	$10 - 18$	$1.0 - 1.4$
V1334 Aql	13.08211	A3–7 I	58.2		4.3 ± 0.8	12.3 ± 3.3
V404 Cyg	6.4714	K0 III–V	208.5	6.08	$10.06 - 13.38$	$0.5 - 0.8$
Cyg X–1	5.59983	O9.7 Iab	74.9	0.244	14.8 ± 1.0	$12 - 27$
LMC X–1	3.90917	O9–7 III	71.61		10.91 ± 1.54	31.79 ± 3.67
LS 5039	3.9060	ON6.5 V	25.2	0.0053	$2.7 - 5.0$	$20.0 - 26.3$
M33 X–7	3.453014	O7-8 III	108.9	0.777	15.65 ± 1.45	70.0 ± 6.9
V4641 Sgr	2.81730	B9 III	220.5	3.13	$6.82 - 7.42$	$2.92 - 3.26$
V1033 Sco	2.6219	F6 IV	215.5	2.73	$6.03 - 6.57$	$2.25 - 2.75$
BW Cir	2.54448	G0–5 III	279	7.34	$> 7.83(50)$	$> 1.02(6)$
LMC X–3	1.70479	B5 V	256.7	2.29	$9.5 - 13.6$	$3.0 - 8.3$
V381 Nor	1.5435	G8 IV–K3 II	349	6.86	$8.36 - 10.76$	< 0.9
IC 10 X–1	1.455	WR	370	7.64	$> 32.7 \pm 2.6$	35
IL Lup	1.116407	A2 V	129	0.25	$8.45 - 10.39$	$2.3 - 3.2$
V2107 Oph	0.521	K5 V	448	4.86	$6.64 - 8.30$	> 0.3
GU Mus	0.432606	K3–4 V	408	3.01	$6.47 - 8.18$	$0.7 - 1.7$
V406 Vul	0.382	G5	570	7.4	$7.6 - 12.0$	
QZ Vul	0.344092	K3–6 V	519.9	5.01	$7.15 - 7.78$	$0.25 - 0.41$
V616 Mon	0.323016	K4 V	433	2.72	$8.70 - 12.86$	$0.48 - 0.97$
MM Vel	0.285206	K7–M0 V	475.4	3.17	$3.64 - 4.74$	$0.45 - 0.75$
V518 Per	0.212160	M4–5 V	378	1.19	$3.66 - 4.97$	$0.28 - 1.55$
KV UMa	0.169930	K7 V–M0 V	701	6.1	$6.48 - 7.19$	$0.22 - 0.32$

The orbital period Π is in days, the maximumg line-of-sight velocity K_{comp} in km/s, the inclination angle i in degrees, and masses in solar units
"Spect.Cl." is the spectral class of the companion, M_{comp} its mass
Errors in K_{comp} and i have been omitted, but enter into M_{BH} (collection courtesy of H. Ritter)

binary systems to allow a quite reliable analysis in some cases. The final argument why these central masses must be BHs is that it must be a compact object (in contrast to an ordinary star which should be visible) and that its mass is beyond the maximum allowed mass for a neutron star (Fig. 38.1). In addition, from the energy of the X-ray emission, and from the timescale of its variation one can deduce the geometric scale of the hot accretion disk. This puts further constraints on possible objects. Up to now, more than 20 BH masses have been determined (Table 39.1).

There are indications for *intermediate-mass black holes* ($50 \lesssim M_{\mathrm{BH}}/M_{\odot} \lesssim 10^5$), but both their existence and their origin are still a matter of discussion. They may be created either by the merging of stellar black holes, or by the collision of massive stars in massive stellar clusters. They have been postulated to explain ultra-luminous X-ray sources. For further reading on BHs and the related physics we refer to the textbook by Camenzind (2007).

Part IX
Pulsating Stars

Throughout this book we have repeatedly considered the stability of stellar layers. A very important aspect of stellar stability is the occurrence of pulsations. Since their periods are determined by the dynamical timescale they are much easier to observe than evolutionary changes of stars, and the periods are very often determined with high precision. Since the recognition that the brightness of *Mira* (o Cet) and other stars is not constant, but varying (semi-)regularly, the interest in stellar pulsations has constantly grown, because it was realized that we can learn about the stellar interior and about the speed of stellar evolution from these pulsations. It has culminated in the field of *helioseismology*, and more recently in its generalization, *asteroseismology*.

In the following chapters we discuss only briefly the basic concepts of the theory of stellar pulsations, which is essentially the problem of solving equations that describe perturbations of a star from its hydrostatic equilibrium on dynamical timescales. The whole field has become so extended and specialised that it requires a separate textbook. We recommend the classical books by Unno et al. (1979) and Cox (1980), but in particular the very recent one by Aerts et al. (2010).

Chapter 40
Adiabatic Spherical Pulsations

40.1 The Eigenvalue Problem

The functions $P_0(m)$, $r_0(m)$, and $\varrho_0(m)$ are supposed to belong to a solution of the stellar-structure equations (10.1)–(10.4) for the case of complete equilibrium. Let us assume that we perturb the hydrostatic equilibrium, say by compressing the star slightly and releasing it again suddenly. It will expand and owing to inertia overshoot the equilibrium state: the star starts to oscillate. The analogy to the oscillating piston model (see Sect. 6.6) is obvious. More precisely we assume the initial displacement of the mass elements to be only radially directed ($d\vartheta = d\varphi = 0$) and of constant absolute value on concentric spheres. This leads to purely *radial oscillations* (or radial pulsations) during which the star remains spherically symmetric all time. For the perturbed variables at time t we write

$$P(m,t) = P_0(m) + P_1(m,t) = P_0(m)\left[1 + p(m)\mathrm{e}^{\mathrm{i}\omega t}\right],$$

$$r(m,t) = r_0(m) + r_1(m,t) = r_0(m)\left[1 + x(m)\mathrm{e}^{\mathrm{i}\omega t}\right],$$

$$\varrho(m,t) = \varrho_0(m) + \varrho_1(m,t) = \varrho_0(m)\left[1 + d(m)\mathrm{e}^{\mathrm{i}\omega t}\right], \tag{40.1}$$

where the subscript 1 indicates the perturbations for which we have made a separation ansatz with an exponential time dependence [as in (25.17)]. The relative perturbations p, x, d are assumed to be $\ll 1$.

We now insert these expressions into the equation of motion (10.2), linearize, and use the fact that P_0, r_0 obey the hydrostatic equation (10.2). Then with $g_0 = Gm/r_0^2$ we obtain

$$\frac{\partial}{\partial m}(P_0 p) = (4g_0 + r_0\omega^2)\frac{x}{4\pi r_0^2}\,. \tag{40.2}$$

R. Kippenhahn et al., *Stellar Structure and Evolution*, Astronomy and Astrophysics Library, DOI 10.1007/978-3-642-30304-3_40, © Springer-Verlag Berlin Heidelberg 2012

Using (10.2) again for $\partial P_0/\partial m$ and the relation

$$\frac{\partial}{\partial r_0} = 4\pi r_0^2 \varrho_0 \frac{\partial}{\partial m} , \tag{40.3}$$

we find

$$\frac{P_0}{\varrho_0}\frac{\partial p}{\partial r_0} = \omega^2 r_0 x + g_0(p + 4x) . \tag{40.4}$$

Quite similarly (40.1) introduced into (10.1) yields with (40.3)

$$r_0 \frac{\partial x}{\partial r_0} = -3x - d . \tag{40.5}$$

Note that the transformation (40.3) does not mean that we go back to an Eulerian description. The partial derivative $\partial/\partial t$ describes time variations at constant r_0. But since $r_0 = r_0(m)$ is given by the equilibrium solution, $\partial/\partial t$ also refers to a fixed value of m.

We know already that perturbations of hydrostatic equilibrium proceed on a timescale $\tau_{\text{hydr}} \ll \tau_{\text{adj}}$. We therefore assume here that the oscillations are adiabatic, which means that

$$p = \gamma_{\text{ad}} d . \tag{40.6}$$

This shows again the advantage of using Lagrangian variables: the adiabatic condition has the simple form (40.6) only if p and d are considered functions of m [or of $r_0 = r_0(m)$] and therefore give the variations in the *co-moving frame.* For the sake of simplicity we now assume that γ_{ad} is constant in space and time. From (40.5) and (40.6) we obtain by differentiation with respect to r_0

$$\frac{\partial x}{\partial r_0} + r_0 \frac{\partial^2 x}{\partial r_0^2} = -3\frac{\partial x}{\partial r_0} - \frac{1}{\gamma_{\text{ad}}}\frac{\partial p}{\partial r_0} . \tag{40.7}$$

Eliminating $\partial_p/\partial r_0$, p, and d from (40.4)–(40.7) gives

$$x'' + \left(\frac{4}{r_0} - \frac{\varrho_0 g_0}{P_0}\right) x' + \frac{\varrho_0}{\gamma_{\text{ad}} P_0}\left[\omega^2 + (4 - 3\gamma_{\text{ad}})\frac{g_0}{r_0}\right] x = 0 , \tag{40.8}$$

where a prime denotes a derivative with respect to r_0.

This second-order differential equation describes the relative amplitude $x(r_0)$ as function of depth for an adiabatic oscillation of frequency ω. In addition one has to fulfil boundary conditions, one at the centre and one at the surface. At the centre the coefficient of x' in (40.8) is singular, while the coefficient of x remains regular since $g_0 \sim m/r_0^2 \sim r_0$. Because one has to demand that x is regular there, this gives the central boundary condition $x' = 0$.

With a simple expansion into powers of r_0 of the form $x = a_0 + a_1 r_0 + a_2 r_0^2 + \ldots$, one finds that the regular solution starts from the centre outwards with $a_1 = 0$ and

$$a_2 = -\frac{1}{10}\frac{\varrho_c}{\gamma_{ad} P_c}\left[\omega^2 + (4 - 3\gamma_{ad})\frac{4\pi}{3} G\varrho_c\right] a_0 , \tag{40.9}$$

where the subscript c indicates central values of the unperturbed solution.

For the surface the simple condition $P_1 \equiv p\, P_0 = 0$ is often used. However, one can find a slightly more realistic boundary condition. We simplify the atmosphere by assuming its mass m_a to be comprised in a thin layer at $r = R(t)$, which follows the changing R during the oscillations and provides the outer boundary condition at each moment by its weight. We neglect, however, its inertia. Then at the bottom of the "atmosphere" we have

$$4\pi R^2 P - \frac{G m_a M}{R^2} = 0 , \tag{40.10}$$

and in the equilibrium state we have

$$4\pi R_0^2 P_0 = \frac{G m_a M}{R_0^2} . \tag{40.11}$$

Using this and (40.1), we find from (40.10) that after linearization

$$p + 4x = 0 . \tag{40.12}$$

We can rewrite this condition in terms of x and x'. If we replace p in (40.12) by (40.6) and then d by (40.5), the outer boundary condition at $r_0 = R_0$ becomes

$$\gamma_{ad} R_0 x' - (4 - 3\gamma_{ad})x = 0. \tag{40.13}$$

The interior boundary condition at $r_0 = 0$ was

$$x' = 0 \tag{40.14}$$

If we multiply the differential equation (40.8) by $r_0^4 P_0$, we can write it in the form

$$(r_0^4 P_0 x')' + \frac{r_0^4 \varrho_0}{\gamma_{ad}}\left[\omega^2 + (4 - 3\gamma_{ad})\frac{g_0}{r_0}\right] x = 0 . \tag{40.15}$$

Together with the (linear, homogeneous) boundary conditions (40.13) and (40.14) this defines a classical *Sturm–Liouville problem* with all its consequences.

From the theory of eigenvalue problems of the Sturm–Liouville type, a series of theorems immediately follows that we shall here list without proofs (which can be found in standard textbooks):

1. There is an infinite number of eigenvalues ω_n^2.
2. The ω_n^2 are real and can be placed in the order $\omega_0^2 < \omega_1^2 < \ldots$, with $\omega_n^2 \to \infty$ for $n \to \infty$.
3. The eigenfunction x_0 of the lowest eigenvalue ω_0 has no node in the interval $0 < r_0 < R_0$("fundamental"). For $n > 0$, the eigenfunction x_n has n nodes in the above interval ("nth overtone").
4. The normalized eigenfunctions x_n are complete and obey the orthogonality relation

$$\int_0^{R_0} r_0^4 \varrho_0 \, x_m \, x_n dr_0 = \delta_{mn} \; , \tag{40.16}$$

where δ_{mn} is the Kronecker symbol.

The eigenfunctions permit the investigation of the evolution in time of any arbitrary initial perturbation described by $x_m = x_m(r_0), \dot{x}_m = \dot{x}_m(r_0)$ at $t = 0$. Indeed if one writes down the expansion of the initial perturbations in terms of the eigenfunctions,

$$x_m(r_0) = \sum_{n=0}^{\infty} c_n x_n(r_0) \; , \quad \dot{x}_{\rm m}(r_0) = \sum_{n=0}^{\infty} d_n x_n(r_0) \; , \tag{40.17}$$

where the c_n, d_n are real, then

$$x(r_0, t) = \mathrm{Re}\left[\sum_{n=0}^{\infty} (a_n \mathrm{e}^{\mathrm{i}w_n t} + b_n \mathrm{e}^{-\mathrm{i}w_n t}) x_n(r_0)\right] ,$$

$$\dot{x}(r_0, t) = \mathrm{Re}\left[\sum_{n=0}^{\infty} \mathrm{i}w_n (a_n \mathrm{e}^{\mathrm{i}w_n t} - b_n \mathrm{e}^{-\mathrm{i}w_n t}) x_n(r_0)\right] \tag{40.18}$$

with complex coefficients a_n, b_n, fulfil the time-dependent equation of motion (40.15) with the initial conditions (40.17) at $t = 0$ if a_n, b_n satisfy

$$a_n + b_n = c_n \; , \quad \mathrm{Re}[\mathrm{i}w_n(a_n - b_n)] = d_n . \tag{40.19}$$

Now we come to the question of stability. Since the perturbations are assumed to be adiabatic, it is dynamical stability we are asking for. We have seen that ω_n^2 is real, so that if $\omega_n^2 > 0$, then $\pm\omega_n$ is real, and the perturbations according to (40.1) are purely oscillatory (with constant amplitude): the equilibrium is dynamically stable. If $\omega_n^2 < 0$ then $\pm\omega_n$ is purely imaginary, say $\pm\omega_n = \pm i\chi$ with real χ. The general time-dependent solution for this model is a sum of expressions of the form

$$A x_n \mathrm{e}^{-\chi t} + B x_n \mathrm{e}^{\chi t} \; , \tag{40.20}$$

where A, B are complex constants. Hence at least one of the two terms describes an amplitude growing exponentially in time. This term will necessarily show up in the expansion (40.18) of an arbitrary perturbation and dominate after sufficient time: the equilibrium is dynamically unstable.

The two regimes are separated by the case of marginal stability with $\omega_0^2 = 0$, which according to earlier considerations (Sect. 25.3.2) is expected to occur for $\gamma_{ad} = 4/3$. We now show that this in fact follows from the rather general formalism used here. For simplicity let us assume that $P_0 \to 0$ at the outer boundary.

Integration of (40.15) over the whole star for the fundamental mode ($n = 0$) gives

$$\left[r_0^4 P_0 x_0'\right]_0^{R_0} + \frac{\omega_0^2}{\gamma_{ad}} \int_0^{R_0} r_0^4 \varrho_0 x_0 d r_0 + \frac{4 - 3\gamma_{ad}}{\gamma_{ad}} \int_0^{R_0} r_0^3 \varrho_0 g_0 x_0 d r_0 = 0 . \tag{40.21}$$

The boundary term on the left vanishes and we find

$$\omega_0^2 = (3\gamma_{ad} - 4) \frac{\int_0^{R_0} r_0^3 \varrho_0 g_0 x_0 d r_0}{\int_0^{R_0} r_0^4 \varrho_0 g_0 x_0 d r_0} . \tag{40.22}$$

Since x_0, as eigenfunction of the fundamental, does not change sign in the interval, we have sign $\omega_0^2 = \text{sign}(3\gamma_{ad} - 4)$. Therefore $\gamma_{ad} > 4/3$ gives $\omega_0^2 > 0$, and the equilibrium is dynamically stable, because all $\omega_n^2 > \omega_2^0$ for $n > 0$ (see above). If $\gamma_{ad} < 4/3$, then for the fundamental (and possibly for a finite number of overtones), $\omega_n^2 < 0$, and the equilibrium is dynamically unstable.

Here we have assumed that γ_{ad} is constant throughout the stellar model, though the main result is unchanged if γ_{ad} varies; in order to guarantee dynamical stability, then, a mean value of γ_{ad} has to be $> 4/3$.

Of course, we could have carried through the whole procedure using m as independent variable instead of r_0. Then (40.4) and (40.5) would have had to be replaced by the equivalent equations (25.19) and (25.20).

40.2 The Homogeneous Sphere

To illustrate the procedure of Sect. 40.1 we apply it to the simplest, but very instructive, case of a gaseous sphere of constant density, where we have an easy analytical access to the eigenvalues and eigenfunctions.

If ϱ is constant in space, then

$$r_0 = \left(\frac{3m}{4\pi\varrho_0}\right)^{1/3} , \quad g_0 = \frac{Gm}{r_0^2} = \frac{4\pi}{3} G r_0 \varrho_0 , \tag{40.23}$$

and from integration of the equation of hydrostatic equilibrium (2.3) we find

$$P_0(r_0) = \frac{2\pi}{3} G\varrho_0^2 \left(R_0^2 - r_0^2\right) , \tag{40.24}$$

where R_0 is the surface radius in hydrostatic equilibrium.

If we introduce the dimensionless variable $\xi = r_0/R_0$ and define

$$\tilde{A} := \frac{3\omega^2}{2\pi G\varrho_0\gamma_{\text{ad}}} + \frac{2(4-3\gamma_{\text{ad}})}{\gamma_{\text{ad}}} , \tag{40.25}$$

then instead of (40.8) we can write

$$\frac{d^2x}{d\xi^2} + \left(\frac{4}{\xi} - \frac{2\xi}{1-\xi^2}\right)\frac{dx}{d\xi} + \frac{\tilde{A}}{1-\xi^2} x = 0 . \tag{40.26}$$

This differential equation has singularities at the centre and at the surface and we look for solutions which are regular at both ends.

The simplest such solution of (40.26) is obvious: $x = x_0 =$ constant is an eigenfunction for $\tilde{A} = 0$. The corresponding eigenfrequency follows from (40.25):

$$\omega_0^2 = \frac{4\pi}{3} G\varrho_0 (3\gamma_{\text{ad}} - 4) . \tag{40.27}$$

This represents the fundamental, since the eigenfunction $x =$ constant has no node. The expression (40.27) for the eigenvalue follows immediately from (40.22) for $x_0 =$ constant, $\varrho_0 =$ constant. Note that (40.27) shows the famous period–density relation for pulsating stars: $\omega_0^2/\varrho_0 =$ constant.

For the overtones we try polynomials in r_0. Indeed if for the first overtone we take $x = 1 + b\xi^2$ with constant b, then (40.26) can be solved with $b = -7/5$ and $\tilde{A} = 14$. The corresponding eigenvalue is obtained from (40.25), (40.27) and we have

$$\omega_1^2 = \omega_0^2 \left(1 + \frac{7\gamma_{\text{ad}}}{3\gamma_{\text{ad}} - 4}\right) ; \quad x_1 = 1 - \frac{7}{5}\xi^2 . \tag{40.28}$$

The eigenfunction has one node at $\xi = (5/7)^{1/2}$, i.e. at $r_0 = 0.845 R_0$. For $\gamma_{\text{ad}} = 5/3$ the ratio of the frequencies of first overtone and fundamental is $\omega_1/\omega_0 = 3.56$.

One can now try higher polynomials with free coefficients in order to find the higher overtones. But we leave this to the reader, the first three eigenfunctions being illustrated in Fig. 40.1.

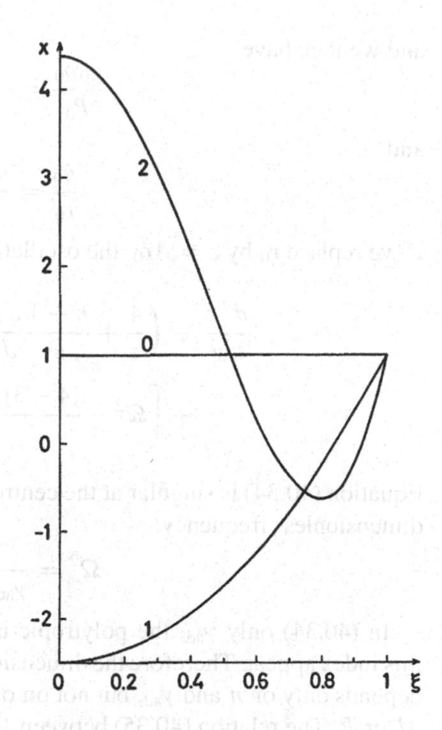

Fig. 40.1 The first three eigenfunctions for radial adiabatic pulsations of the homogeneous sphere

40.3 Pulsating Polytropes

Let us now investigate the (spherically symmetric) radial oscillations of polytropic models of index n as discussed in Chap. 19. We therefore express the quantities of the unperturbed model which appear in the coefficients of (40.8),

$$r_0\,,\quad \varrho_0 g_0/P_0\,,\quad \varrho_0/P_0\,,\quad \varrho_0 g_0/(P_0 r_0)\,,$$

by the Lane–Emden function $w(z)$ and by its dimensionless argument z. From (19.9) we have

$$g_0 = \frac{\partial \Phi_0}{\partial r_0} = A\Phi_c \frac{dw}{dz}\;;\; A^2 = \frac{4\pi G}{[(n+1)K]^n}(-\Phi_c)^{n-1}\,, \tag{40.29}$$

while (19.7) yields

$$\varrho_0 = \left[\frac{-\Phi_c w}{(n+1)K}\right]^n, \tag{40.30}$$

the subscript c denoting central values in the unperturbed model. If we use the polytropic relation (19.3), we find

$$\frac{\varrho_0}{P_0} = \frac{1}{K}\varrho^{-1/n} = -\frac{n+1}{\Phi_c w}\,, \tag{40.31}$$

and we then have

$$\frac{g_0\varrho_0}{P_0} = -A\frac{n+1}{w}\frac{dw}{dz} \tag{40.32}$$

and

$$\frac{g_0}{r_0} = \frac{\Phi_c A^2}{z}\frac{dw}{dz}. \tag{40.33}$$

If we replace r_0 by $z = Ar_0$, the oscillation equation (40.8) becomes

$$\frac{d^2x}{dz^2} + \left(\frac{4}{z} + \frac{n+1}{w}\frac{dw}{dz}\right)\frac{dx}{dz} + \left[\Omega^2 - \frac{(4-3\gamma_{ad})(n+1)}{\gamma_{ad}}\frac{1}{z}\frac{dw}{dz}\right]\frac{x}{w} = 0. \tag{40.34}$$

Equation (40.34) is singular at the centre ($z = 0$) and at the surface ($w = 0$). Ω is a dimensionless frequency:

$$\Omega^2 = \frac{n+1}{\gamma_{ad}(-\Phi_c)A^2}\omega^2 \tag{40.35}$$

In (40.34) only γ_{ad}, the polytropic index n, and the Lane–Emden function for this index appear. Therefore the dimensionless eigenvalue Ω^2 obtained from (40.34) depends only on n and γ_{ad}, but not on other properties of the polytropic model, say M or R. The relation (40.35) between Ω and ω can be expressed differently. Using (40.30) for the centre ($w = 1$) and (40.29) we have

$$\omega^2 = \frac{\gamma_{ad}(-\Phi_c)A^2}{n+1}\Omega^2 = \frac{4\pi G\gamma_{ad}\varrho_c}{n+1}\Omega^2. \tag{40.36}$$

Since for a given n the central density ϱ_c and the mean density $\bar{\varrho}$ of the whole unperturbed model differ only by a constant factor, one finds from (40.36) $w^2 =$ constant $\cdot\,\bar{\varrho}$, or with the period $\Pi = 2\pi/\omega$

$$\Pi\sqrt{\bar{\varrho}} = \left[\frac{(n+1)\pi}{\gamma_{ad}G\Omega^2}\left(\frac{\bar{\varrho}}{\varrho_c}\right)_n\right]^{1/2}. \tag{40.37}$$

For a given mode, say the fundamental, the right-hand side depends only on the polytropic index n and on γ_{ad}. This is the famous *period–density relation*. It is also approximately fulfilled for more realistic stellar models.

If one assumes for a δ Cephei star that $M = 7M_\odot$ and $R = 80R_\odot$, its mean density is $\approx 2\times10^{-5}\,\mathrm{g\,cm^{-3}}$. If the period is 11^d, then $\Pi(\bar{\varrho})^{1/2} \approx 0.049$ (Π in days, $\bar{\varrho}$ in $\mathrm{g\,cm^{-3}}$). This constant gives a period of about 220 days for a supergiant with $\bar{\varrho} = 5\times10^{-8}\,\mathrm{g\,cm^{-3}}$, while for a white dwarf (with $\bar{\varrho} \approx 10^6\,\mathrm{g\,cm^{-3}}$), it gives a period of 4 s. Indeed the supergiant period is of the order of those observed for Mira stars, while very short periods are observed for white dwarfs.

The dimensionless equation (40.34) depends on n and γ_{ad}, where the polytropic index n is a measure of the density concentration, say of $\varrho_c/\bar{\varrho}$, while γ_{ad} is a measure of the stiffness of the configuration. If $\gamma_{ad} = 4/3$, then $\Omega = 0$ is an eigenvalue and x = constant the corresponding eigenfunction, as can be seen from (40.34); the model is then marginally stable and after compression does not go back to its original size. The larger the γ_{ad}, the better the stability, since the compressed model will expand more violently after being released. This can be understood with the help of the considerations in Sect. 25.3.2.

Numerical solutions of the eigenvalue problem show how variations in n and γ_{ad} modify the solutions. Because of the singularities of (40.34) at both ends of the interval $0 < z < z_n$ (z_n is the value of z for which the Lane–Emden function of index n vanishes) the numerical solution is not straightforward. The simplest way is to choose a trial value $\Omega = \Omega^*$ and to start two integrations with power series regular at $z = 0$ and at $z = z_n$. The outward and inward integrations are continued to a common point somewhere, say at $z^* = z_n/2$. There the two solutions will have neither the same value $x(z^*)$ nor the same derivative $(dx/dz)^*$. Since the differential equation is linear and homogeneous, we can multiply one of the solutions by a constant factor such that both get the same value at z^*. But then they probably still disagree in $(dx/dz)^*$. Agreement in the derivatives can be achieved by gradually improving Ω, carrying out new integrations, and so on. By such iterations a solution for the whole interval can be obtained.

Whether by such a procedure one arrives at the fundamental or at an overtone depends in general on the trial Ω^*. If it is near the fundamental, we will end up with the fundamental eigenvalue and eigenfunction. In any case the number of nodes will reveal which mode has been found.

Since (40.34) is linear and homogeneous, the solution may be multiplied by an arbitrary constant factor, in which way we can normalize the solution such that at the surface $x(z_n) = 1$. For the polytrope $n = 3$ the eigenfunctions of different modes for $\gamma_{ad} = 5/3$ are shown in Fig. 40.2 and the eigenfunction of the fundamental for different values of γ_{ad} is displayed in Fig. 40.3.

The variation of γ_{ad} is indeed important. To see this, we assume an ideal monatomic gas with radiation pressure as discussed in Chap. 13. From (13.7), (13.12) and (13.15) we find after some algebra that

$$\gamma_{ad} = \frac{1}{\alpha - \delta \nabla_{ad}} = \frac{32 - 24\beta - 3\beta^2}{24 - 21\beta} . \tag{40.38}$$

For the limit cases $\beta = 1$ ($P_{rad} = 0$) and $\beta = 0$ ($P_{gas} = 0$) the adiabatic exponent γ_{ad} takes the values 5/3 and 4/3, respectively. We see that our assumption γ_{ad} = constant throughout the model holds only as long as β = constant. Fortunately this is the case for the polytrope $n = 3$, since $1 - \beta \sim T^4/P$ and $T \sim \omega$, $P \sim \omega^{n+1}$. In (40.34) the radiation pressure only appears in the quantity

$$\varphi := -\frac{4 - 3\gamma_{ad}}{\gamma_{ad}} = 3 - \frac{4}{\gamma_{ad}} . \tag{40.39}$$

For vanishing and dominating radiation pressure, φ takes the values 0.6 and 0, respectively.

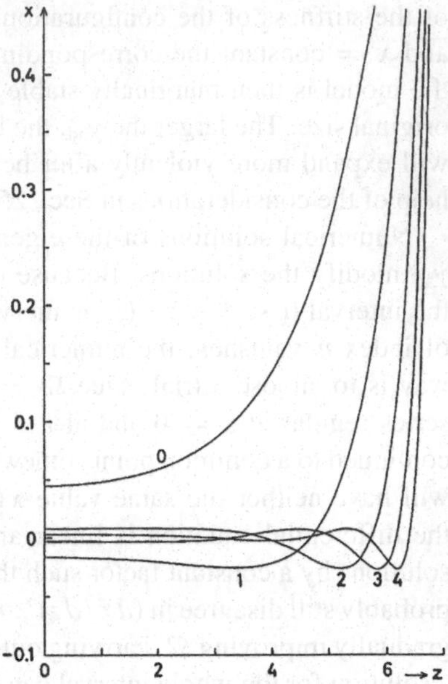

Fig. 40.2 Eigenfunctions for radial adiabatic pulsations of the polytrope $n = 3$ for $\varphi = 0.6$ (After Schwarzschild 1941)

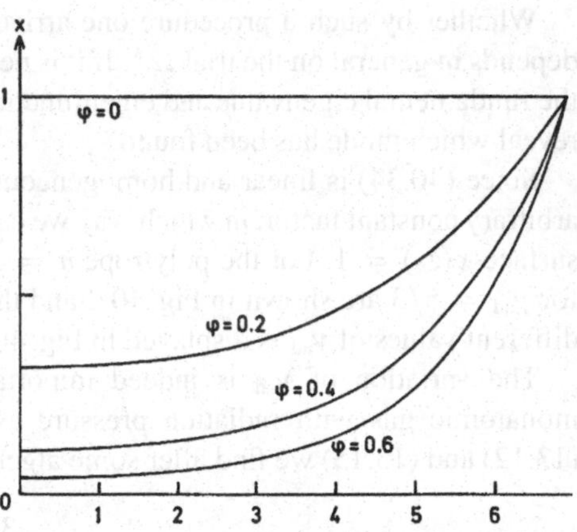

Fig. 40.3 The fundamental eigenfunction for radial adiabatic pulsations of the polytrope $n = 3$ for different values of φ. Radiation pressure diminishes the ratio of the amplitude at the surface to that of the centre. If the radiation pressure dominates the gas pressure completely ($\varphi = 0$) the relative amplitude x is constant

Fundamental and overtone solutions of (40.34) for $n = 3$ and for different values of φ have been found numerically by Schwarzschild (1941). For $\varphi = 0.6$ ($\gamma_{ad} = 5/3$) the (dimensionless) eigenfrequency for the fundamental and the first overtones are $\Omega_0^2 = 0.1367$, $\Omega_1^2 = 0.2509$, $\Omega_2^2 = 0.4209$, $\Omega_3^2 = 0.6420$, $\Omega_4^2 = 0.9117$. The corresponding eigenfunctions are shown in Fig. 40.2.

The influence of β on the fundamental eigenfunction can be seen in Fig. 40.3. With increasing radiation pressure (φ decreasing) the relative amplitude x drops less and less steeply from the surface to the centre. The ratio $x_{\text{surface}}/x_{\text{centre}}$ is 22.4 for $\varphi = 0.6$ and 9.1 for $\varphi = 0.4$. In the limit $\varphi \to 0$ (pure radiation pressure) x even becomes constant. Indeed, for $\gamma_{ad} = 4/3$ and for the eigenvalue $\Omega = 0$, $x =$ constant is a solution as we know already.

Chapter 41
Non-adiabatic Spherical Pulsations

When a star oscillates, its mass elements will generally not change their properties adiabatically. The outward-going heat flow, as well as the nuclear energy production, is modulated by the rhythm of the pulsation, and both effects cause deviations from adiabaticity. However, since the pulsation takes place on the hydrostatic timescale, which is short compared to τ_{KH}, the deviations from adiabaticity should be small in most parts of the stellar interior. In order to demonstrate the main effects of the non-adiabatic terms on the equation of motion, we discuss them at first for the simple piston model.

41.1 Vibrational Instability of the Piston Model

We go back to the description of Sect. 25.2.2. Equation (25.14) gives three eigenvalues σ for non-adiabatic oscillations of the piston model. The adiabatic period $\sigma = \pm\sigma_{\mathrm{ad}} = \pm\mathrm{i}\omega_{\mathrm{ad}}$ (with $\omega_{\mathrm{ad}}^2 > 0$) would be obtained for $e_P = e_T = 0$. For small non-adiabatic terms e_P and e_T we now write $\sigma = \sigma_{\mathrm{r}} \pm \sigma_{\mathrm{ad}}$ as in (25.15) and assume that the real part is small, $|\sigma_{\mathrm{r}}| \ll \omega_{\mathrm{ad}}$. Then, neglecting terms of the order $\sigma_{\mathrm{r}}^3, \sigma_{\mathrm{r}}^2, e_P\,\sigma_{\mathrm{r}}, e_T\,\sigma_{\mathrm{r}}$ and introducing γ_{ad} instead of 5/3, we find from (25.14) that

$$\frac{h_0 u_0}{g_0}\left(3\sigma_{\mathrm{ad}}^2\sigma_{\mathrm{r}} \pm \sigma_{\mathrm{ad}}^3\right) - \frac{h_0}{g_0}(e_P + e_T)\sigma_{\mathrm{ad}}^2 + \gamma_{\mathrm{ad}} u_0(\sigma_{\mathrm{r}} \pm \sigma_{\mathrm{ad}}) - e_T = 0. \quad (41.1)$$

Since σ_{ad} has to obey the adiabatic equation [cf. (25.15)]

$$\frac{h_0}{g_0}\sigma_{\mathrm{ad}}^2 + \gamma_{\mathrm{ad}} = 0, \quad (41.2)$$

R. Kippenhahn et al., *Stellar Structure and Evolution*, Astronomy and Astrophysics Library, DOI 10.1007/978-3-642-30304-3_41, © Springer-Verlag Berlin Heidelberg 2012

(41.1) becomes

$$2u_0\sigma_r = \nabla_{ad} e_T + e_P, \tag{41.3}$$

where we have introduced $\nabla_{ad} := (\gamma_{ad} - 1)/\gamma_{ad}$.

We now assume $\varepsilon_0 = \kappa_0 = 0$, then $e_P = 0$, $e_T = -\chi T_0/m^*$ [see (25.13)], and we find that

$$2u_0\sigma_r = -\nabla_{ad} \frac{\chi T_0}{m^*}. \tag{41.4}$$

Therefore, since $\nabla_{ad} > 0$, one has $\sigma_r < 0$, meaning that the oscillation is damped. During each cycle heat leaves and enters the gas in the container by way of the leak, kinetic energy of the piston is lost and added to the surroundings as heat.

Similarly in a star the flow of heat modulated by the oscillation can damp the motion. Since the deviation from adiabaticity is more pronounced in the outer regions, the damping time is determined by the Kelvin–Helmholtz timescale of the outer layers. In his classic book, Eddington (1926) estimated that the damping time of δ Cephei stars would be of the order of 8,000 years and concluded that there must exist a mechanism which maintains their pulsations. He actually discussed two possible mechanisms which can be easily demonstrated with the piston model.

The first is called the *κ mechanism*, since here it is the modulated absorption of radiation which can yield vibrational instability.

If for the sake of simplicity we assume that $\chi = \varepsilon_0 = 0$, then according to (25.13) one has $e_P = \kappa_0 F \kappa_P$, $e_T = \kappa_0 F \kappa_T$, and therefore (41.3) becomes

$$2u_0\sigma_r = \kappa_0 F(\nabla_{ad}\kappa_T + \kappa_P). \tag{41.5}$$

The model is vibrationally unstable ($\sigma_r > 0$) if $(\nabla_{ad}\kappa_T + \kappa_P) > 0$. This means that the instability occurs if during adiabatic compression ($d \ln P > 0$) the absorption coefficient increases: $d \ln \kappa = (\nabla_{ad}\kappa_T + \kappa_P) d \ln P > 0$. Then in the compressed state more energy is absorbed than in equilibrium and the ensuing expansion is slightly enhanced. For analogous reasons the state of maximum expansion is followed by an enhanced compression.

In stars the outgoing radiative flux can similarly cause an instability if the stellar opacity increases/decreases during the phase of contraction/expansion. As we shall see (Sect. 41.4), this is the mechanism which indeed drives the δ Cephei stars.

In the so-called *ε mechanism* the possible cause for an instability is the modulated nuclear energy generation. In order to discuss a simple case, we assume $\chi = \kappa_0 = 0$ and find from (41.3) with (25.13) that

$$2u_0\sigma_r = \varepsilon_0(\nabla_{ad}\varepsilon_T + \varepsilon_P). \tag{41.6}$$

This model is vibrationally unstable for any nuclear burning ($\varepsilon_0 > 0$), since all terms on the right-hand side are > 0. For example, the CNO cycle has typically $\varepsilon_T \gtrsim 10$, $\varepsilon_P = 1$ while $\nabla_{ad} \approx 0.4$.

In the two cases discussed above, the piston model in a certain sense mimics the stability behaviour of different layers in a star. Since $\tau_{KH} \gg 1/\omega_{ad}$, the non-adiabatic effects in a pulsating star are small, and as in the piston model one can expect that the oscillations are almost adiabatic, as described in Chap. 40. But the non-adiabatic effects will cause a small deviation of the eigenfrequency from the adiabatic value. Indeed, since the temperature variations are different in different regions of the star, these regions exchange an additional heat which–like the heat flow through the leak–causes a damping (radiative damping). A destabilizing effect on the star is caused by those regions where the opacity increases during contraction (κ mechanism) as well as those with a nuclear burning where ε increases during contraction (ε mechanism).

41.2 The Quasi-adiabatic Approximation

In order to determine the vibrational stability behaviour of a star, one has to solve the four ordinary differential equations (25.19)–(25.22) for the perturbations p, x, λ, ϑ together with homogeneous boundary conditions at the centre and at the surface. In addition to the "mechanical" boundary conditions (40.13) and (40.14) one has at the centre

$$l_0 \lambda = 0 \quad \text{at} \quad m = 0. \tag{41.7}$$

As a rough outer boundary condition one can assume that at the surface the relation $L = 4\pi R^2 \sigma T^4$ holds throughout the oscillation period, yielding

$$l = 2x + 4\vartheta. \tag{41.8}$$

This relation is not exactly true, since the photosphere (where $T = T_{\text{eff}}$) does not always belong to the same mass shell during the oscillation. With a more detailed theory of the behaviour of the atmosphere during the oscillations one can replace (41.8) by another, but also linear and homogeneous, outer boundary condition.

The homogeneous linear equations (25.19)–(25.22) and boundary conditions (40.13), (40.14) and (41.7), (41.8) define an eigenvalue problem for the eigenvalue ω.

Here we will restrict ourselves to a simplified treatment, the *quasi-adiabatic approximation*. For the given unperturbed equilibrium model we first solve the adiabatic problem described in Chap. 40, thereby obtaining a set of adiabatic eigenvalues $\omega_{\text{ad}}^{(n)}$ with the eigenfunctions $p_{\text{ad}}^{(n)}, x_{\text{ad}}^{(n)}, \vartheta_{\text{ad}}^{(n)} = \nabla_{\text{ad}} p_{\text{ad}}^{(n)}$, where the upper index n labels the different eigenvalues. In the following we will drop n, though keeping in mind that the procedure described here and in Sect. 41.3 can be carried out for each of the adiabatic eigenvalues. Of course, the real oscillations will not proceed exactly adiabatically, which, for example, is shown in luminosity perturbations. To determine an approximation to the relative luminosity perturbation λ we differentiate ϑ_{ad} with respect to m and find from (25.22)

$$\lambda = \frac{P_0}{\nabla_{\text{ad}} P_0'} \vartheta_{\text{ad}}' + 4x_{\text{ad}} - \kappa_P p_{\text{ad}} + (4 - \kappa_T)\vartheta_{\text{ad}}. \tag{41.9}$$

In this quasi-adiabatic approximation, therefore, the non-adiabatic effects determining λ are calculated from adiabatic eigenfunctions. The correct procedure would require the use of non-adiabatic eigenfunctions on the right-hand side of (41.9), while in a strictly adiabatic case we would expect $\lambda = \lambda_{\mathrm{ad}} = 0$. One can use the non-adiabatic variation λ of the local luminosity in order to estimate the change of ω due to non-adiabatic effects.

For this, one assumes the star to be forced into a periodic oscillation. If non-adiabatic processes are taken into account, periodicity can only be maintained if, during each cycle, energy is added to or removed from the whole star. If energy has to be added to maintain a periodic oscillation, the star is damped; if energy has to be removed, it is excited. In order to determine the energy necessary for maintaining a periodic pulsation one defines the energy integral.

41.3 The Energy Integral

Suppose we want to make a star undergo periodic radial pulsations. If it is vibrationally unstable, then during each cycle a certain amount W of energy has to be taken out to maintain periodicity. If the star is vibrationally stable, the energy $-W$ has to be fed into the star during each period to avoid a damping of the amplitude. In both cases W is the energy to be taken out to overcome excitation or damping. Therefore, if the star is left alone, $W > 0$ *gives amplitudes increasing in time (excitation) while for* $W < 0$ *the oscillation is damped.*

To determine W we consider a shell of mass dm which gains the energy dq/dt per units of mass and time. The energy gained per unit mass per cycle is the integral of $(dq/dt)dt$ taken over one cycle. Therefore the energy

$$dW = dm \oint \frac{dq}{dt} dt \tag{41.10}$$

has to be taken out of the mass shell to maintain periodicity. If we replace dq/dt by

$$\frac{dq}{dt} = -\cos \omega t \frac{d(l_0 \lambda)}{dm}, \tag{41.11}$$

and if we integrate over all mass shells, we have

$$W = -\int_0^M dm \oint \cos \omega t \frac{d(l_0 \lambda)}{dm} dt. \tag{41.12}$$

It is obvious that this integral vanishes: in the linear approximation there is neither damping nor excitation.

However, owing to a trick invented by Eddington it is still possible to determine the second-order quantity W with the help of solutions of the first-order theory. Since in the adiabatic case the eigenvalues are real, the time dependence of x, p, ϑ, and according to (41.9) that of λ, can be expressed by the factor $\cos \omega t$.

We first prove that

$$\oint \frac{dq}{dt} dt = \oint \vartheta \frac{dq}{dt} \cos \omega t \, dt, \tag{41.13}$$

up to second order. Indeed, since the specific entropy s is a state variable, the integral of ds over one cycle vanishes exactly. We now write $ds = dq/T$. Since we use only solutions of the adiabatic case, we can consider the variation of T as real and can write $T = T_0(1 + \vartheta_{\rm ad} \cos \omega t)$, which is correct in the first order. With the (real) adiabatic solutions x_{ad}, $p_{\rm ad}$, and $\vartheta_{\rm ad}$ according to (41.9), λ also is real, and therefore dq/dt is real, too, as can be seen from (41.12). Therefore

$$\begin{aligned} 0 = \oint \frac{ds}{dt} dt &= \oint \frac{1}{T_0} (1 - \vartheta_{\rm ad} \cos \omega t) \frac{dq}{dt} dt \\ &= \frac{1}{T_0} \oint \frac{dq}{dt} - \frac{1}{T_0} \oint \vartheta_{\rm ad} \cos \omega t \frac{dq}{dt} dt. \end{aligned} \tag{41.14}$$

This equation is exact in the second order. It therefore proves (41.13). Should the integral on the left of (41.13) vanish in the first order, its value in the second order is given by the integral on the right of (41.13), which does not vanish. We can therefore write from (41.10) by using (41.11)

$$\begin{aligned} W &= \int_0^M dm \oint \vartheta_{\rm ad} \cos \omega t \frac{dq}{dt} dt = -\int_0^M dm \oint \vartheta_{\rm ad} \frac{d(l_0 \lambda)}{dm} \cos^2 \omega t \, dt \\ &= -\int_0^M dm \oint \left[\vartheta_{\rm ad} \lambda \frac{dl_0}{dm} + l_0 \vartheta_{\rm ad} \frac{d\lambda}{dm} \right] \cos^2 \omega t \, dt. \end{aligned} \tag{41.15}$$

The time dependence of the real part is $\cos^2 \omega t$, which integrated over 2π gives π/ω. With $dl_0/dm = \varepsilon_0$ we therefore obtain

$$W = -\frac{\pi}{\omega} \left[\int_0^M \vartheta_{\rm ad} \lambda \varepsilon_0 \, dm + \int_0^M l_0 \vartheta_{\rm ad} \frac{d\lambda}{dm} dm \right]. \tag{41.16}$$

In fact we see that only second-order terms ($\sim \vartheta_{\rm ad} \lambda$ and $\sim \vartheta_{\rm ad} d\lambda/dm$) appear in the expression for W. We can now solve the adiabatic equations, insert the resulting $\vartheta_{\rm ad}$, differentiate λ given in (41.9), and determine W from (41.16).

41.3.1 The κ Mechanism

We consider here regions of the star in which no energy generation takes place ($\varepsilon_0 = 0$) and therefore in which $l_0 =$ constant. Since the adiabatic equations for the determination of x, p, ϑ are linear and homogeneous, the solutions are determined only up to a common factor. We choose it here such that $x_{\text{ad}} = 1$ at the surface. We further choose the initial point of time such that the maximal expansion of the surface is at $t = 0$. Then the first equation of (25.17) can be written $r = r_0(1 + x_{\text{ad}} \cos \omega t)$, and for $x > 0$ (expansion) the variations ϑ_{ad} and p_{ad} are certainly < 0 there. Since, for the fundamental, $\vartheta_{\text{ad}}(< 0)$ does not change sign throughout the star, one can immediately see from (41.16) that a region where λ increases outwards ($d\lambda/dm > 0$) gives a positive contribution to W: such a region has an excitational effect on the oscillation, while regions with $d\lambda/dm < 0$ have a damping influence. The last two terms on the right of (41.9) together with $\vartheta_{\text{ad}} = \nabla_{\text{ad}} p_{\text{ad}}$ can be written as

$$4\nabla_{\text{ad}} p_{\text{ad}} - (\kappa_P + \nabla_{\text{ad}}\kappa_T) p_{\text{ad}}. \tag{41.17}$$

Note that the term in parenthesis is identical with a term we encountered in (41.5) for the piston model. If for the sake of simplicity we assume κ_P, κ_T, ∇_{ad} to be constant and observe that, for the fundamental, $p_{\text{ad}} < 0$ increases inwards, then for $\kappa_P + \nabla_{\text{ad}}\kappa_T > 0$ the term $-(\kappa_P + \nabla_{\text{ad}}\kappa_T) p_{\text{ad}} > 0$ gives a contribution that helps to increase λ in an inward direction. This has a stabilizing effect. The term $4\nabla_{\text{ad}} p_{\text{ad}} < 0$ in (41.17) decreases with p_{ad} in an outward direction and has a damping effect independently of κ. This damping corresponds to the effect of the leak in the piston model.

The κ mechanism is responsible for several groups of variable stars. Before we discuss its effect on real stars we shall first deal with the other mechanism that can maintain stellar pulsations.

41.3.2 The ε Mechanism

The terms in the energy integral discussed in Sect. 41.3.1 appear everywhere in a star where radiative energy transport occurs. However, there, we have excluded nuclear energy generation, which can also be modulated by the oscillations. To investigate its influence we now concentrate on the terms which come from ε. If we put $l_0(d\lambda)/dm$ equal to the perturbation of the energy generation rate $\varepsilon : \varepsilon_0(\varepsilon_P p_{\text{ad}} + \varepsilon_T \vartheta_{\text{ad}}) = \varepsilon_0(\varepsilon_P + \nabla_{\text{ad}}\varepsilon_T) p_{\text{ad}}$, we find from (41.16) that

$$\begin{aligned} W_\varepsilon &= -\frac{\pi}{\omega}\left[\int_0^M \vartheta_{\text{ad}}\lambda\varepsilon_0 dm + \int_0^M \vartheta_{\text{ad}}\varepsilon_0(\varepsilon_P + \nabla_{\text{ad}}\varepsilon_T) p_{\text{ad}}\, dm\right] \\ &= -\frac{\pi}{\omega}\int_0^M \vartheta_{\text{ad}}[\lambda + (\varepsilon_P + \nabla_{\text{ad}}\varepsilon_T) p_{\text{ad}}]\varepsilon_0\, dm. \end{aligned} \tag{41.18}$$

Here we again find the excitation mechanism working if $\varepsilon_P + \nabla_{ad}\varepsilon_T > 0$, which is already known to us from the piston model of Sect. 41.1. All terms in the integral (41.18) contribute to the energy integral only in the very interior, where $\varepsilon_0 \neq 0$. Since the amplitudes of the eigenfunctions there are normally small compared to their values in the outer regions, one often ignores the contribution of the energy generation and instead of W computes $W_\kappa = W - W_\varepsilon \approx W$. We come back to the case where W_ε becomes important in Sect. 41.5.

41.4 Stars Driven by the κ Mechanism: The Instability Strip

If one has determined the adiabatic amplitudes for a given stellar model, one can derive λ from (41.9) and evaluate W according to (41.16). We shall first describe the influence of different layers.

In the outer layers, where deviations from adiabaticity are biggest, the κ mechanism and the damping term $4\nabla_{ad}p_{ad}$ in (41.17) become important and the sign of $(\kappa_P + \nabla_{ad}\kappa_T)$ determines whether the κ mechanism acts to damp or to excite. To illustrate this it is useful not only to plot on a lg P–lg T diagram lines of constant opacity but also to indicate at each point the slope given by $\nabla_{ad} = (d\ \lg T/d \lg P)_{ad}$ as in Fig. 41.1. The κ mechanism provides excitation if one comes to higher opacities when going along the slope towards higher pressure. For a monatomic gas one has $\nabla_{ad} = 0.4$. However, ionization reduces ∇_{ad} appreciably (see Fig. 14.1b), which according to Fig. 41.1 favours instability. This is easily seen for a simple Kramers opacity with $\kappa_P = 1$ and $\kappa_T = -4.5$: then the decisive term $(\kappa_P + \nabla_{ad}\kappa_T)$ is -0.8 for $\nabla_{ad} = 0.4$, while it is ≥ 0 for $\nabla_{ad} \leq 0.222$.

In the near-surface layers of a star with an effective temperature of about 5,000 K, there are two regions where ionization, together with a suitable form of the function $\kappa = \kappa(P, T)$, acts in the direction of instability. The outer one is quite close to the surface, where hydrogen is partially ionized, followed immediately by the first ionization of helium (see Fig. 14.1, which is plotted for the Sun). Below this ionization zone, ∇_{ad} goes back to its standard value of 0.4. But still deeper another region of excitation occurs caused by the second ionization of helium. This turns out to be the region which contributes most to instability. In still deeper layers the κ mechanism has a damping effect, but their influence is very small, since the oscillations become more adiabatic the deeper one penetrates into the star. For an estimate of the right depth of the HeII ionization zone, see Cox (1967) and Sect. 27.7 of Cox and Giuli (1968).

In Fig. 41.2 the exciting and damping regions of the outer layers of a δ Cephei star of $7M_\odot$ are shown. For a star right in the middle of the Cepheid strip the "local" energy integral

$$w(m) = -\int_m^M dm \oint \cos\omega t \frac{d\, l_0\lambda}{dm} dt \tag{41.19}$$

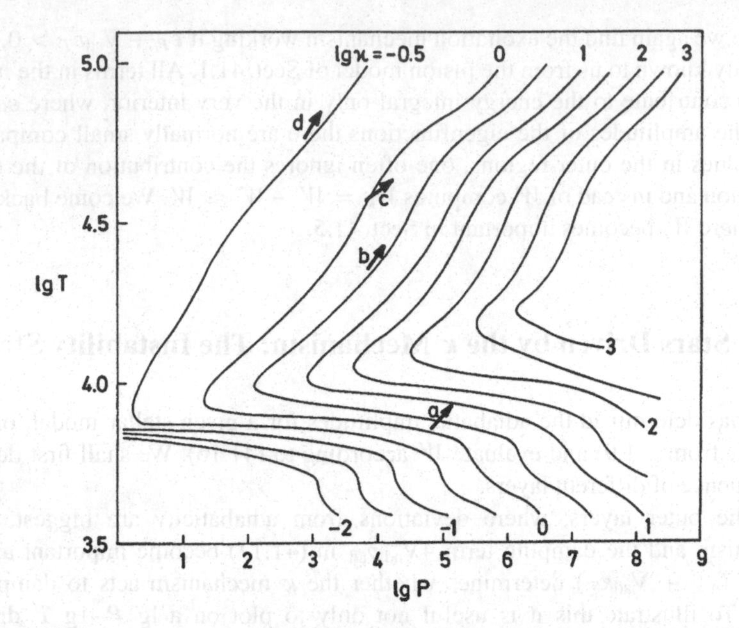

Fig. 41.1 Lines of constant opacity κ in the $\lg P$–$\lg T$ plane (all values in cgs). Four *arrows* are shown that indicate the direction in which a mass element moves during adiabatic compression. For the *arrows* labelled a, b, and d, the direction is given by $\nabla_{\text{ad}} = 0.4$. In case a the *arrow* points in the direction of increasing κ, i.e. the κ mechanism has a "driving" effect on pulsations. In cases b and d the *arrows* point in the direction of decreasing κ, indicating a "damping" (or almost neutral) effect on pulsation. In case c the direction of the *arrow* is different from that of the other ones, since ∇_{ad} is here reduced by the second ionization of helium. Because of this reduction, the *arrow* points in the direction of increasing κ, and this ionization region can contribute considerably to the excitation of pulsations in Cepheids

is plotted as a function of depth in Fig. 41.3, where lg P has been used as a measure of the depth. There one can see which regions excite the oscillations ($dw/d \lg P > 0$) and which have a damping effect ($dw/d \lg P < 0$). According to (41.12) $\omega(0) = W$.

In order that excitation wins over damping it is necessary that the zones of ionization, which provide the excitation, contain a sufficient part of the mass of the star. This means that these zones have to be situated at suitable depths, and since ionization is mainly a function of temperature, we can conclude that it is essentially a question of the surface temperature that decides whether a star is vibrationally stable or unstable via the κ mechanism.

Let us compare stellar models of the same mass (say in the range 5–10$M_\odot$), of roughly the same luminosity, and consider values for the effective temperature which range from the main sequence to the Hayashi line. At the main sequence and in some range to the right of it, the outer layers of the stars are too hot: hydrogen is fully ionized far up into the atmosphere, and even the second ionization of helium is almost complete up to the photosphere. Therefore the κ mechanism due to ionization

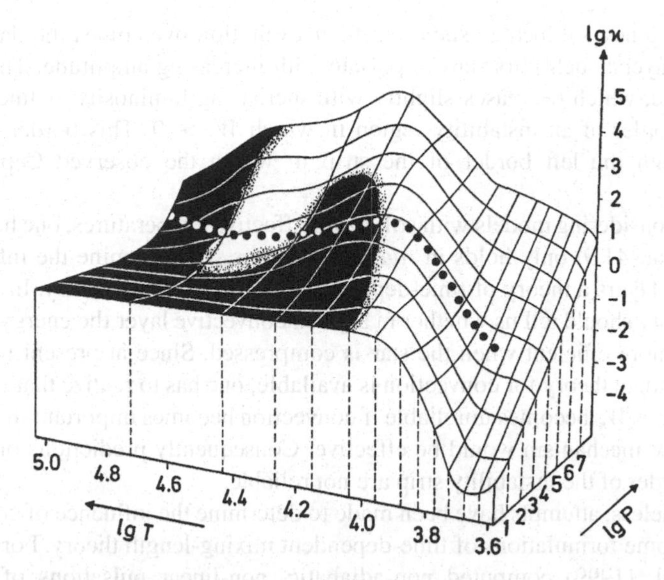

Fig. 41.2 An opacity surface ("κ mountain") for the outer layers of a star as in Fig. 17.6. But this time the dependence with respect to P (in dyn cm^{-2}) and T (in K) is shown. The *dotted line* corresponds to the stratification inside a Cepheid of $7M_\odot$. The *white areas* of the "mountain" indicate regions which excite the pulsation and the *black* ones those which damp it. The excitation in the region of $\lg T \approx 4.6$ is due to the second ionization of helium

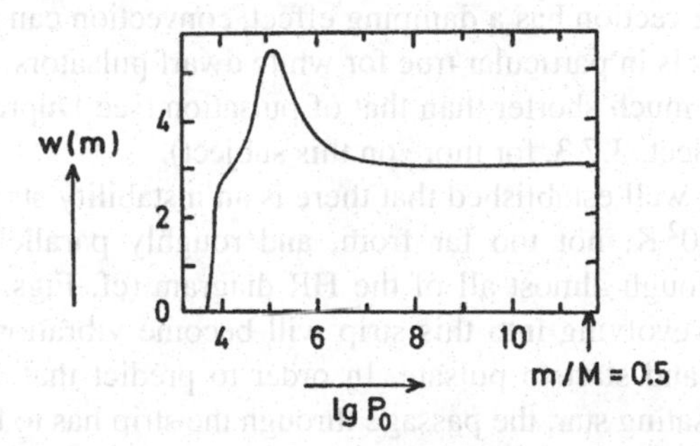

Fig. 41.3 The "local" energy integral $w(m)$ (in arbitrary units) as defined in (41.19) for a star of $7M_\odot$ and $T_{\text{eff}} = 5,300$ K as a function of the unperturbed pressure P_0 (in dyn cm^{-2}). $w(m)$ increases in regions which excite the pulsation, and falls in those regions which damp the pulsation (After Baker and Kippenhahn 1965)

as discussed in Sect. 41.3.1 does not provide much excitation. The main contribution to W comes from the layers which are in the region of the $\lg P$–$\lg T$ plane of Fig. 41.1 where the κ mechanism has a damping effect. Therefore the pulsation of such hot stars is damped. But the smaller the effective temperature, i.e. the further to the right in the HR diagram, the deeper inwards are the zones of partial ionization of H and He. Then a higher percentage of the stellar matter lies in the regions of excitation shown in Fig. 41.2. At effective temperatures below about 6,300 K, the ionization zones are located such that their excitation overcomes the damping of the other layers: such stars start to pulsate with increasing amplitude. This critical temperature, which decreases slightly with increasing luminosity, defines the left ("blue") border of an instability region in which $W > 0$. This border coincides roughly with the left border of the strip in which the observed Cepheids are located.

When considering models with still lower effective temperatures, one has to keep in mind that (41.9) only holds in *radiative* regions. To determine the influence of convective layers a theory of time-dependent convection is necessary. In particular such a theory should tell us whether in a given convective layer the energy transport is less or more efficient when the star is compressed. Since at present no reliable time-dependent theory for convection is available, one has to realize that the energy integral $W \approx W_\kappa$ becomes unreliable if convection becomes important in the layers where the κ mechanism would be effective. Consequently predictions of the right ("red") border of the instability strip are not reliable.

Nevertheless, attempts have been made to determine the influence of convection, by using some formulations of time-dependent mixing-length theory. For example, Bono et al. (1999) computed non-adiabatic, non-linear pulsations of classical Cepheid models, using the treatment of time-dependent convection described in an earlier paper (Bono and Stellingwerf 1994) about RR Lyrae stars. They could indeed determine a more realistic red edge of the instability strip than the one found when only the radiative energy transport is considered. While for these classical radial pulsators convection has a damping effect, convection can also contribute to the excitation. This is in particular true for white dwarf pulsators, where convective time scales can be much shorter than that of pulsation (see Dupret et al. 2008, and Aerts et al. 2010, Sect. 3.7.3, for more on this subject).

In any case it is well established that there is an instability strip with a probable width of a few 10^2 K, not too far from, and roughly parallel to, the Hayashi line, extending through almost all of the HR diagram (cf. Figs. 31.4 and 33.11). All stellar models evolving into this strip will become vibrationally unstable via the κ mechanism and start to pulsate. In order to predict that we can observe a corresponding pulsating star, the passage through the strip has to be slow enough.

This is fulfilled for models of typically $5 \ldots 10 M_\odot$, which during the phase of helium burning loop away from and back to the Hayashi line, thereby passing through the instability strip at least twice. These passages, in which models represent the classical Cepheids, are discussed in detail in Sect. 31.3. Depending on M, the passages occur at quite different luminosities: the larger M, the higher L. Using the adiabatic approximation one can easily determine the periods of the fundamental for models of very different L inside the instability strip. In this way one obtains a theoretical *period–luminosity relation* that is in satisfying agreement with the observed one. It is interesting to note that the passages through the instability strip do not follow lines of $R =$ constant. Since the radius and therefore the mean density changes, the period–density relations predict a certain amount of change (in both directions) of the period of a Cepheid. This period change, of the order of seconds per year, has been determined from long-term monitoring of Cepheids and is in reasonable agreement with theoretical predictions (e.g. Pietrukowicz 2001, and references therein).

Of much smaller mass are the helium-burning stars located on the horizontal branches of the HR diagrams of globular clusters. Where these branches intersect the downward continuation of the instability strip, at an effective temperature between 6,000 and 7,500 K, one finds the RR Lyrae stars (Sect. 33.7). Like the classical Cepheids these are pulsating stars driven by the κ mechanism. For a monograph see Smith (1995).

Even further down in the HR diagram, in the region of the main sequence, the instability strip is marked by another group of observed pulsating stars, the so-called δ Scuti stars or dwarf Cepheids.

Above the location of the RR Lyrae stars in the HR diagram of globular clusters one sometimes finds stars which lie in the instability strip and are therefore pulsating: the BL Herculis and W Virginis stars (see footnote on page 414). In contrast to the classical Cepheids, which belong to population I, these stars are of population II (and are called *Type II Cepheids*). It is not surprising that they do not obey the same period–luminosity relation as Cepheids. According to the evolutionary considerations of Sect. 33.7 they are low-mass stars in an evolutionary stage later than that of the horizontal branch. They obviously have lower masses than the Cepheids, which have travelled more or less horizontally from the main sequence into the instability strip. Let us assume that at the same point inside the instability strip there are two stars, a population I star of, say, $7M_\odot$ and a population II star of, say, $0.8M_\odot$. The κ mechanism will make both of them pulsate. Being at the same point in the HR diagram, the two stars have the same radii. Therefore the population II star has the lower mean density and according to the period–density relation a longer period than the population I star, although their luminosities are the same. Since the luminosity increases with the period, it follows that pulsating population I stars have a higher luminosity than pulsating population II stars of the same period. In the history of astronomy the clarification of this difference between the two period–luminosity relations caused the revision of the cosmic distance scale by W. Baade in 1944. This increase of the cosmic distance scale amounted to no less than a factor of 2, which caused the comment "The Lord made the universe–but Baade doubled it".

Up to now we have based our considerations on a linear quasi-adiabatic approximation. In the linear theory the amplitude of the solution is not determined and the time dependence is given by almost sinusoidal oscillations with amplitudes growing or decreasing very slowly in time. In reality a vibrationally unstable star would start to oscillate with increasing amplitudes until the oscillations had grown so much that they could not be described by a linear theory any more. Once the non-linear terms in the equations have become important, they have the effect of limiting the increase of amplitudes and causing a time dependence of the solutions which differs considerably from sinusoidal behaviour. Indeed the light curves of most of the observed pulsating stars have constant amplitude and are far from being sinusoidal.

Attempts have been made to reproduce the observed light curves of Cepheids by solving the non-linear equations numerically with varied parameters. A special goal was to determine the masses of Cepheids by comparing their observed light curves with computed ones (see Christy 1975; Bono et al. 2000, and related work). Such comparisons seem to indicate lower masses for Cepheids than expected from evolution theory, but as we have seen in Sect. 31.3, overshooting and/or mass loss help to solve this problem.

Besides the linearization of the equations, we have additionally simplified the problem of pulsations by applying the quasi-adiabatic approximation. With some more effort, however, one can also solve the full set of linear *non-adiabatic* equations. These four equations demand four linear boundary conditions. If they are properly chosen, one obtains one complex eigenvalue ω. Since the time dependence is given by $\exp(i\omega t)$, the imaginary part ω_I of ω determines vibrational stability. The energy integral (41.12), computed with the function λ obtained from (41.9), is connected to ω_I when one is close to the adiabatic case (Baker and Kippenhahn 1962). In most cases the quasi-adiabatic approximation seems to be sufficient. If, however, pure helium stars cross the instability strip, the oscillations are far from being adiabatic, and therefore the quasi-adiabatic approximation becomes very unreliable. This can become important, for instance, if the oscillations of stars of the type R Coronae Borealis are being investigated (Weiss 1987).

As the κ-mechanism depends on the detailed shape of the "opacity mountain", it is important to have accurate opacity data. Norman Simon in 1982 wrote a paper entitled *A plea for reaexamining heavy element opacities in stars* (Simon 1982), in which he showed that an increase in opacities by a factor of 2 and 3 in the temperature region between 10^5 and 2×10^6 K would solve two long-standing problems: the mismatch between predicted and observed period ratios for the fundamental and first overtone pulsations of classical Cepheids (so-called "double mode Cepheids"), and that between second overtone and fundamental for the "bump Cepheids". He suggested that the opacities of metals were underestimated in the then existing opacity tables. As a consequence new Rosseland mean opacities were computed (which we discussed in Sect. 17.8), and indeed the problem was solved (Moskalik et al. 1992)! Additionally, the long-sought excitation mechanism for the so-called β Cepheids (pulsating main-sequence stars of 7–10$M_\odot$ and spectral type B) was found (Moskalik and Dziembowski 1992; Kiriakidis et al. 1992): as the reason the bump around $\lg T \approx 5.3$ (see Fig. 17.6) was identified, which is due to absorption of photons by abundant metals such as C, N, Ne, and Fe. This "metal" or "iron bump" was missing in earlier opacity tables. The new opacities, required in order to understand stellar pulsations, had far-reaching consequences: they helped to compute very accurate solar models as well as improved isochrones for globular clusters.

41.5 Stars Driven by the ε Mechanism

In most stars the ε mechanism discussed in Sects. 41.1 and 41.3.2 cannot overcome the damping, the reason being that it only works in the central regions of the stars where nuclear energy is released. But there the amplitudes of the oscillations are usually very small compared to the amplitudes in the near-surface regions, which–if the star is not in the instability strip–damp the oscillations by way of the κ mechanism.

Figure 40.3 shows that for polytropes for which the radiation pressure can be neglected, the amplitude ratio $x_{\text{centre}}/x_{\text{surface}}$ is small, while it increases with decreasing φ until the ratio becomes 1 for $\varphi = 0$ (negligible gas pressure). Since the integrand of the energy integral is quadratic in the amplitudes of the oscillations, we can expect that the ε mechanism becomes more important the larger the fraction of the radiation pressure.

This is of importance at the upper end of the hydrogen main sequence (Sect. 22.4), because for such stars, the ratio of radiation pressure to gas pressure strongly increases with M. Numerical calculations with realistic stellar models instead of polytropes indicate that the ε mechanism makes the main-sequence stars pulsate if their mass exceeds a critical value of about $60M_{\odot}$ (Schwarzschild and Härm 1959); this value depends on the chemical composition. Baraffe et al. (2001) found that metal-free (Pop. III) stars may even be as massive as a few hundred $M_{\odot}$ without losing substantial amounts of material during their main-sequence lifetime. The reason is that these stars burn hydrogen at much higher central temperatures, where the H-burning reactions have a lower dependency on temperature (see Fig. 18.8), a fact which reduces $\varepsilon_P + \nabla_{\text{ad}}\varepsilon_T$ and thus stabilizes the star.

Why, then, do we not see pulsating stars in the extension of the main sequence towards higher luminosities? Non-linear pulsation calculations (Appenzeller 1970; Ziebarth 1970) indicate that the amplitudes would grow until, with each cycle, a thin mass shell is thrown into space. This would continue until the total mass is reduced to the critical mass of, say, $60M_{\odot}$. Then the pulsation would stop. However, the growth rates of the pulsations are probably longer than the main-sequence lifetime. It is therefore unclear whether the ε-mechanism really sets an upper limit to the mass range on the main sequence.

Similarly the onset of a vibrational instability due to the ε mechanism may also limit the helium main sequence towards large M (see Sect. 23.1). The critical upper mass for helium stars depends on the content of heavier elements and may lie between 7 and $8M_{\odot}$ (Boury and Ledoux 1965).

Chapter 42
Non-radial Stellar Oscillations

We use spherical coordinates r, ϑ, φ and describe the velocity of a mass element by a vector $\boldsymbol{v}$ having the components $v_r, v_\vartheta, v_\varphi$. For the radial pulsations treated in the foregoing sections, the velocity has only one non-vanishing component, v_r, which depends only on r. This is so specialized a motion that one might wonder why a star should prefer to oscillate this way at all. In fact it is easier to imagine the occurrence of perturbations that are *not* spherically symmetric, for example, those connected with turbulent motions or local temperature fluctuations. They can lead to non-radial oscillations, i.e. oscillatory motions having in general non-vanishing components $v_r, v_\vartheta, v_\varphi$, all of which can depend on r, ϑ, and φ. It is obvious that the treatment of the more general non-radial oscillations is much more involved than that of the radial case, but they certainly play a role in observed phenomena (see Sect. 42.4). We will limit ourselves to indicating a few properties of the simplest case: small (linear), adiabatic, poloidal-mode oscillations. A detailed monograph about this subject, which is the basis of the field of *asteroseismology*, was written by Aerts et al. (2010).

42.1 Perturbations of the Equilibrium Model

The unperturbed model (subscript 0) is assumed to be spherically symmetric, in hydrostatic equilibrium ($\varrho_0 \nabla \Phi_0 + \nabla P_0 = 0$) and at rest (velocity $\boldsymbol{v}_0 = 0$). We now consider perturbations which shift the mass elements over very small distances. For any mass element at r, ϑ, φ, the displacement relative to its equilibrium position is described by the vector $\boldsymbol{\xi}$ with the components $\xi_r, \xi_\vartheta, \xi_\varphi$, which, in general, depend on r, ϑ, φ, t. Owing to this displacement, such variables as pressure, density, or

R. Kippenhahn et al., *Stellar Structure and Evolution*, Astronomy and Astrophysics Library, DOI 10.1007/978-3-642-30304-3_42, © Springer-Verlag Berlin Heidelberg 2012

gravitational potential will change. This can be described either in a Lagrangian form (changes inside the displaced element) denoted by

$$P = P_0 + DP, \quad \varrho = \varrho_0 + D\varrho, \quad \Phi = \Phi_0 + D\Phi, \quad \boldsymbol{v} = d\boldsymbol{\xi}/dt \tag{42.1}$$

or as Eulerian perturbations (local changes), which we write as

$$P = P_0 + P', \quad \varrho = \varrho_0 + \varrho', \quad \Phi = \Phi_0 + \Phi', \quad \boldsymbol{v} = \partial\boldsymbol{\xi}/\partial t \tag{42.2}$$

and which are preferred in the following. The linearized connection between the two types of perturbations of any quantity q is

$$Dq = q' + \boldsymbol{\xi} \cdot \nabla q_0 = q' + \xi_r \frac{\partial q_0}{\partial r}. \tag{42.3}$$

(The last equality holds since ∇q_0 is a purely radial vector.) Together with $\boldsymbol{\xi}$, all perturbations are functions of r, ϑ, φ, and t. We have to perturb the Poisson equation and the equations of motion and continuity.

The acceleration due to gravity,

$$\boldsymbol{g} = -\nabla\Phi, \tag{42.4}$$

and its perturbations $D\boldsymbol{g}$ or $\boldsymbol{g}'$ are given by the potential Φ. Poisson's equation (2.23), together with (42.2), yields after linearization

$$\nabla^2\Phi' = 4\pi G\varrho'. \tag{42.5}$$

The equation of motion for the moving mass element is

$$\varrho\frac{d\boldsymbol{v}}{dt} = \varrho\boldsymbol{g} - \nabla P. \tag{42.6}$$

With (42.1) this gives the linearized equation

$$\varrho_0\frac{d^2\boldsymbol{\xi}}{dt^2} = \boldsymbol{g}_0 D\varrho + \varrho_0 D\boldsymbol{g} - \nabla(DP), \tag{42.7}$$

where the forces on the right-hand side are measured relative to equilibrium. From (42.7) and (42.3), the Eulerian equation of motion follows:

$$\varrho_0\frac{\partial^2\boldsymbol{\xi}}{\partial t^2} = -\varrho_0\nabla\Phi' - \varrho'\nabla\Phi_0 - \nabla P'. \tag{42.8}$$

On the right-hand side of this expression, the restoring force is represented by three terms, the last of which is due to pressure variations, while the others are gravitational terms. The first stems from the changed gravitational acceleration and is usually small compared with the second, which is essentially a buoyancy term.

The equation of continuity, $\partial\varrho/\partial t + \nabla(\varrho \boldsymbol{v}) = 0$, after insertion of (42.1) and linearization, takes the form

$$D\varrho + \varrho_0 \nabla \cdot \boldsymbol{\xi} = 0, \tag{42.9}$$

which together with (42.3) is transformed to

$$\varrho' + \boldsymbol{\xi} \cdot \nabla \varrho_0 + \varrho_0 \nabla \cdot \boldsymbol{\xi} = 0. \tag{42.10}$$

We do not have to consider the equations of energy and energy transfer, since we assume the changes to be adiabatic. The condition for adiabaticity in Lagrangian form is simply [cf. (40.6)]

$$\frac{DP}{P_0} = \gamma_{\text{ad}} \frac{D\varrho}{\varrho_0}, \tag{42.11}$$

which is transformed by (42.3) to the Eulerian condition

$$P' + \boldsymbol{\xi} \cdot \nabla P_0 = \frac{P_0}{\varrho_0} \gamma_{\text{ad}} (\varrho' + \boldsymbol{\xi} \cdot \nabla \varrho_0). \tag{42.12}$$

We shall see below that the equations derived for the perturbations constitute a fourth-order system. So we need in addition four boundary conditions.

At the surface, we require continuity of the Lagrangian variation of $\nabla\Phi$ through the surface, and a vanishing pressure perturbation, $DP = 0$, such that no forces are transmitted to the outside. These outer boundary conditions are then written as

$$\left(\frac{\partial \Phi'}{\partial r} + \boldsymbol{\xi} \cdot \nabla \Phi_0 \right)_{\text{in}} = \left(\frac{\partial \Phi'}{\partial r} + \boldsymbol{\xi} \cdot \nabla \Phi_0 \right)_{\text{out}}, \quad P' + \boldsymbol{\xi} \cdot \nabla P_0 = 0. \tag{42.13}$$

At the centre, the perturbations are required to be regular, which also yields two boundary conditions, say,

$$P' = 0, \quad \Phi' = 0. \tag{42.14}$$

42.2 Normal Modes and Dimensionless Variables

The perturbations are to be determined from (42.5), (42.8), (42.10) and (42.12)–(42.14). Aside from the perturbations $\boldsymbol{\xi}, \Phi', P', \varrho'$, these equations contain only quantities of the unperturbed equilibrium model, for which we now drop the subscript 0.

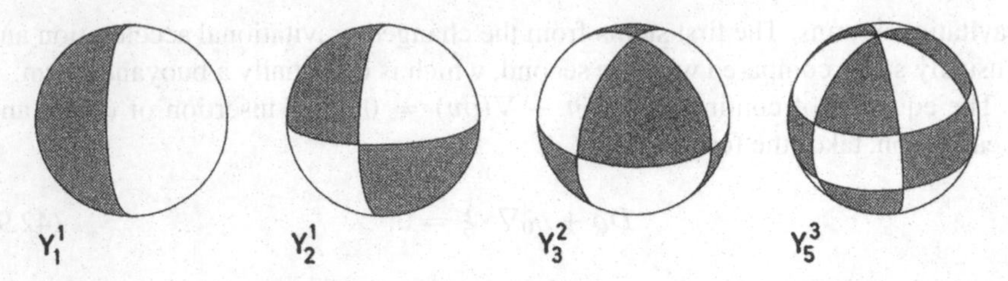

Fig. 42.1 Node lines of some spherical harmonics Y_l^m. Corresponding oscillations would show, for example, outward motion in the *shaded areas* and inward motion in the other parts of the sphere

We specify the perturbations $q(r, \vartheta, \varphi, t)$ in the usual way, assuming that all of them depend on the variables as factorized in the following separation ansatz:

$$q(r, \vartheta, \varphi, t) = \tilde{q}(r) Y_l^m(\vartheta, \varphi) \mathrm{e}^{\mathrm{i}\omega t}. \tag{42.15}$$

The perturbations are supposed to vary on all concentric spheres like the well-known spherical harmonics $Y_l^m(\vartheta, \varphi)$ of degree l and order m (see, for instance, Korn and Korn 1968). In time they vary periodically with frequency ω. The dependence on r is comprised in the function $\tilde{q}(r)$. The Y_l^m are solutions of

$$\frac{\partial^2 Y_l^m}{\partial \vartheta^2} + \mathrm{ctg}\vartheta \frac{\partial Y_l^m}{\partial \vartheta} + \frac{1}{\sin^2 \vartheta} \frac{\partial^2 Y_l^m}{\partial \varphi^2} + l(l+1) Y_l^m = 0, \tag{42.16}$$

and can be written as

$$Y_l^m = K(l, m) P_l^m(\cos \vartheta) \cos m\varphi, \tag{42.17}$$

where K is a coefficient depending on l and m, and $P_l^m(x)$ are the associated Legendre functions. Degree and order are specified by choosing the integers

$$l > 0, \quad m = -l, \ldots, +l. \tag{42.18}$$

A change of l, m changes the angular variation on concentric spheres. A few examples are illustrated in Fig. 42.1. Generally speaking, the larger l, the more node lines ($Y = 0$) are present, and the smaller are the enclosed areas in which the matter moves in the same radial direction (e.g. outwards). For example, $l = 2$ is a quadrupole oscillation, $l = 1$ a dipole oscillation, and $l = 0$ the special case of the earlier discussed radial pulsations.

We shall discuss here only perturbations of the form (42.15). The resulting oscillations of that form are called *poloidal modes.* It should be mentioned that there exists the additional class of *toroidal modes,* which do not have the form (42.15); they are independent of time and have purely transverse displacements (without radial components).

In order to get an overview of the problem, it is convenient to introduce dimensionless variables, for example,

$$\eta_1 = \frac{1}{r}\xi_r \,; \quad \eta_2 = \frac{1}{gr}\left(\frac{P'}{\varrho} + \Phi'\right) \,; \quad \eta_3 = \frac{1}{gr}\Phi' \,; \quad \eta_4 = \frac{1}{g}\frac{\partial \Phi'}{\partial r}. \tag{42.19}$$

Since they are proportional to P', ϱ', Φ', we have according to (42.15)

$$\eta_j = \tilde{\eta}_j(r) Y_l^m(\vartheta, \varphi) e^{i\omega t}, \quad j = 1, 2, 3, 4. \tag{42.20}$$

The density perturbation, which does not appear in (42.19), will always be replaced by terms in P' (and then in $\eta_2 - \eta_3$) via (42.12).

The equation of motion (42.8), together with (42.19), becomes after some algebra:

$$-\frac{\omega^2}{g}\boldsymbol{\xi} = [W(\eta_1 - \eta_2 + \eta_3) + (1 - U)\eta_2]\boldsymbol{e}_r - r\nabla\eta_2, \tag{42.21}$$

where $\boldsymbol{e}_r$ is a unit vector in the r direction. The dimensionless quantities

$$\begin{aligned} U &:= \frac{r}{m}\frac{\partial m}{\partial r} = \frac{4\pi r^3 \varrho}{m}, \\ V &:= -\frac{r}{P}\frac{\partial P}{\partial r} = \frac{g\varrho r}{P}, \\ W &:= \frac{r}{\varrho}\frac{\partial \varrho}{\partial r} - \frac{r}{P\gamma_{\mathrm{ad}}}\frac{\partial P}{\partial r} \end{aligned} \tag{42.22}$$

are to be taken from the equilibrium model. Equation (42.21) is easily verified. Its radial component will be treated later, while the tangential components

$$\frac{\omega^2}{g}\xi_\vartheta = \frac{\partial \eta_2}{\partial \vartheta}, \quad \frac{\omega^2}{g}\xi_\varphi = \frac{1}{\sin\vartheta}\frac{\partial \eta_2}{\partial \varphi} \tag{42.23}$$

are used immediately in the equation of continuity. But first we replace ω by a dimensionless frequency σ, setting

$$\frac{\omega^2 r}{g} = C\sigma^2, \quad C = \left(\frac{r}{R}\right)^3 \frac{M}{m}, \quad \sigma^2 = \omega^2 \frac{R^3}{GM}. \tag{42.24}$$

This frequency is scaled by a time of the order of the hydrostatic adjustment time or of the period of the radial fundamental.

When transforming the equation of continuity (42.10), we evaluate the term $\nabla \cdot \boldsymbol{\xi}$ by using (42.23), introduce (42.20), and eliminate all derivatives of Y_l^m with respect to ϑ and φ with the help of (42.16). Then all terms are proportional to $Y_l^m \exp(\mathrm{i}\omega t)$, which can thus be dropped. One finally obtains

$$r\frac{\partial \tilde{\eta}_1}{\partial r} = \left(3 - \frac{V}{\gamma_{\mathrm{ad}}}\right)\tilde{\eta}_1 + \left[\frac{l(l+1)}{C\sigma^2} + \frac{V}{\gamma_{\mathrm{ad}}}\right]\tilde{\eta}_2 - \frac{V}{\gamma_{\mathrm{ad}}}\tilde{\eta}_3. \qquad (42.25)$$

Similarly one finds from the radial component of the equation of motion (42.21)

$$r\frac{\partial \tilde{\eta}_2}{\partial r} = (W + C\sigma^2)\tilde{\eta}_1 + (1 - U - W)\tilde{\eta}_2 + W\tilde{\eta}_3. \qquad (42.26)$$

The next equation is simply obtained by differentiating the definition of η_3 in (42.19) with respect to r, which gives

$$r\frac{\partial \tilde{\eta}_3}{\partial r} = (1 - U)\tilde{\eta}_3 + \tilde{\eta}_4. \qquad (42.27)$$

In the Poisson equation (42.5), after elimination of ϱ' by (42.12), we introduce (42.19) and again use (42.16), arriving at

$$r\frac{\partial \tilde{\eta}_4}{\partial r} = -UW\tilde{\eta}_1 + \frac{UV}{\gamma_{\mathrm{ad}}}\tilde{\eta}_2 + \left[l(l+1) - \frac{UV}{\gamma_{\mathrm{ad}}}\right]\tilde{\eta}_3 - U\tilde{\eta}_4. \qquad (42.28)$$

With (42.25)–(42.28) we have obtained four ordinary, linear differential equations with real coefficients (given by the equilibrium model) for the four dimensionless variables $\tilde{\eta}_1 \ldots, \tilde{\eta}_4$. In addition there are four algebraic equations arising from the boundary conditions. This constitutes an eigenvalue problem with the eigenvalue σ^2.

Note that it is the assumption of adiabaticity which has reduced the problem to 4th order in the spatial variables. For the full non-adiabatic case one additionally has to consider the perturbations of the temperature and of the energy-flux vector. The perturbed energy equation contains *first* derivatives with respect to time, which according to (42.15) give terms multiplied by iω. Therefore the equations become complex and the non-adiabatic problem is of order 12 in real variables. On the other hand, for $l = 0$, one obtains the adiabatic radial oscillations, for which the problem is reduced to second order.

42.3 The Eigenspectra

For adiabatic non-radial oscillations we have obtained an eigenvalue problem of 4th order in the spatial variables and non-linear in the eigenvalue ω^2 (or the dimensionless σ^2). The problem can be shown to be self-adjoint, so that the eigenfunctions are orthogonal to one another. They have been found to form a complete set if complemented by the toroidal modes.

The eigenvalues obey an extremal principle. The self-adjointness assures that all eigenvalues are real. This means that the motion is either purely periodic ($\omega^2 > 0$, ω real: dynamical stability) or purely aperiodic ($\omega^2 < 0$, ω imaginary: dynamical instability).

Neither the equations (42.25)–(42.28) nor the boundary conditions contain explicitly the order m of the spherical harmonics. Therefore to each eigenvalue of a given l correspond $2l + 1$ solutions (for the different m values $-l, \ldots 0, \ldots, +l$). This degeneracy can be removed, for example, by centrifugal or tidal forces.

The general discussion is very much complicated by the fact that the eigenvalue $\lambda = \sigma^2$ appears non-linearly in the set (42.25)–(42.28). In order to see the typical properties of the eigenspectra, we use an approximation introduced by Cowling, assuming that the perturbation of the gravitational potential can be neglected. We then do not need (42.27), (42.28) and are left with a second-order problem. This approximation becomes the better, the more the oscillation is limited to the outer layers (e.g. high overtones of acoustic modes with sufficiently large l). The second-order problem still contains terms proportional to σ^2 [from (42.26)] and terms proportional to $1/\sigma^2$ [from (42.25)]. In order to simplify this we consider two asymptotic cases ($\sigma^2 \to \infty$ and $\sigma^2 \to 0$), in both of which the problem becomes of the classical Sturm–Liouville type.

For large σ^2 we neglect the terms proportional to $1/\sigma^2$. The only coefficient containing σ then is σ^2/c_s^2, with the velocity of sound given by $c_s^2 = \gamma_{ad} P/\varrho$. This problem has an infinite series of discretc eigenvalues $\lambda_k = \sigma_k^2$, with an accumulation point at infinity. Such oscillations are produced by acoustic waves propagating with c_s. They are dominated by pressure variations and are therefore called *p modes.* For sufficiently simple stellar models, they are easily ordered as $p_1, p_2, \ldots, p_k$ where k is the number of nodes of their eigenfunction ξ_r between centre and surface. They are analogous to the radial oscillations ($l = 0$), except for the dynamical stability: while the radial fundamental is unstable for $\gamma_{ad} < 4/3$, the p modes are all stable under reasonable conditions.

For small σ^2 we neglect the terms proportional to σ^2. The only coefficient containing σ is now $\omega_{ad}^2 l(l+1)/(\sigma^2 r^2)$, where ω_{ad} is the Brunt–Väisälä frequency as introduced in Sect. 6.2. This problem has an infinite series of eigenvalues $\lambda_k = 1/\sigma_k^2$, with an accumulation point at $\lambda = \infty$, i.e. at $\sigma^2 = 0$. The motions are dominated by gravitational forces and are therefore called *g modes* (again ordered as $g_1, g_2, \ldots, g_k$ according to the number k of nodes).

The stability of the g modes depends essentially on W, defined in (42.23). This quantity is connected with the problem of convective stability discussed in Chap. 6. One can easily verify from (6.18) that the Brunt–Väisälä frequency of an adiabatically oscillating mass element is given by

$$\omega_{ad}^2 = -grW. \tag{42.29}$$

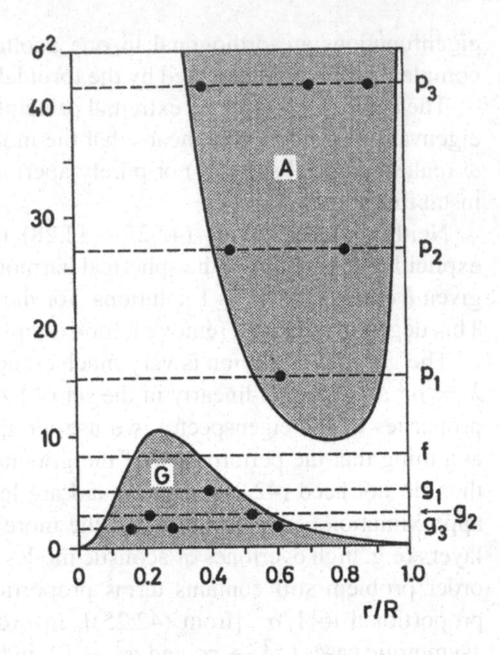

Fig. 42.2 Propagation diagram for oscillations with degree $l = 2$ in a polytropic star with index $n = 3$. The square of the dimensionless frequency σ is plotted against the distance from the centre. Propagation of acoustic and gravity waves is possible in the *shaded regions A* and *G*, respectively. For the lowest modes the eigenvalues (*broken lines*) and the positions of the nodes of the eigenfunction ξ_r (*dots*) are indicated (After Smeyers 1984)

And $rW > 0$ is just the criterion (6.4) for convective instability against adiabatically displaced elements. If in the whole star $W < 0$ (convective stability everywhere), then all g modes are stable ($\sigma^2 > 0, \sigma$ real). Such modes are also called g^+ modes and are produced by propagating gravity waves. If the star contains a region where $W > 0$ (convective instability), then unstable g^- modes also exist ($\sigma^2 < 0$, σ imaginary). So we see that convective stability (instability) coincides with dynamical stability (instability) of non-radial g modes; the onset of convection appears as the manifestation of unstable g modes.

The non-linearity in $\lambda = \sigma^2$ of the full set (42.25)–(42.28) implies that the eigenspectrum of stars is a combination of the above-described partial spectra: it contains high-frequency p modes as well as low-frequency g modes, which can be split up into the stable g^+ and the unstable g^-. Between the p and g modes of relatively simple stars there is another one, called the *f mode*, since it has no node between centre and surface (like the radial fundamental).

As mentioned above, the stable modes are produced by propagating waves. From the appropriate dispersion relations with horizontal wave numbers $[l(l+1)]^{1/2}/r$ one finds that for propagating acoustic waves $\omega \geq \omega_0 := \frac{1}{2}c_s(d \ln \varrho/dr)$, and for propagating gravity waves $\omega \leq \omega_{\text{ad}}$, where at any place $\omega_0 > \omega_{\text{ad}}$. These conditions define two main regions (G and A) of propagation inside a star: one in the deep interior for gravity waves the other in the envelope for acoustic waves (see Fig. 42.2). These regions act like cavities or resonators, inside which modes can be "trapped". At certain frequencies (the eigenvalues), the propagating waves produce standing waves by reflections at the borders such that they come back in phase with themselves. The simple polytropic model demonstrated in Fig. 42.2 is typical for

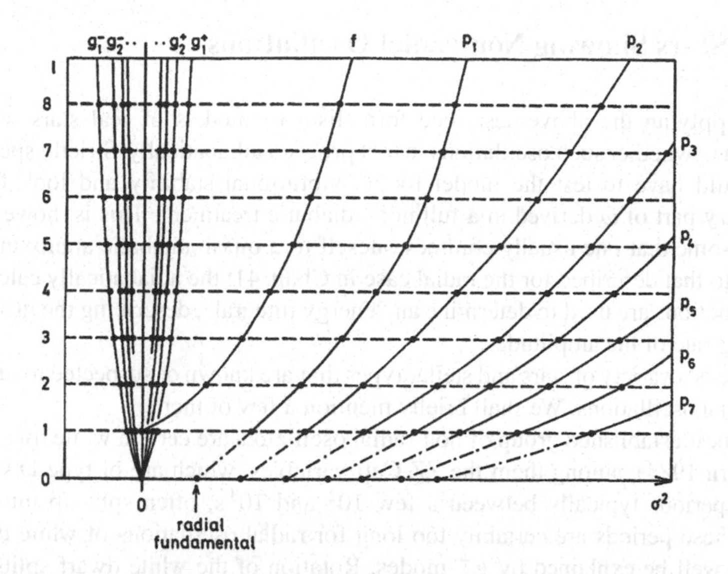

Fig. 42.3 In this scheme the *dots* indicate the eigenvalues σ^2 (plotted as abscissa) for a few modes of non-radial adiabatic oscillations with different orders l of the spherical harmonics (plotted as ordinate). Eigenvalues for the same type of mode are connected by a *solid line*. *Dot-dashed lines* give the connexion to the corresponding radial modes with $l = 0$ (p_1 to the radial fundamental, p_2 to the first radial overtone, etc.). For $l = 1$ the f mode has $\sigma^2 = 0$ (no oscillatory motion, see text)

the situation with homogeneous main-sequence stars. When during the evolution the central concentration of the model increases and a chemical inhomogeneity is built up, the maximum of the G region near the core increases far above the minimum of the A region in the envelope. Then the g_1 mode can move above the p_1 mode, etc. When they are close to each other, resonance effects provide that they exchange their properties and avoid an exact coincidence of the eigenvalues (avoided level crossing, as known, say, from quantum mechanics). So the eigenspectra can be rather involved, particularly for evolved stars.

Figure 42.3 illustrates the eigenspectra for different values of l (degree of the spherical harmonics) for the case of a rather simple star. The radial oscillations are found at $l = 0$. For dipole oscillations ($l = 1$) the f mode must have $\sigma = 0$, since otherwise it would result in an oscillatory motion of the centre of gravity, which is not possible without external forces. However, oscillations having nodes outside the centre are possible for $l = 1$, since then, for example, the core always moves in the opposite direction to the envelope such that the centre of gravity remains at rest. For higher l values the eigenspectra are generally shifted to higher frequencies. The connection between the different p modes and the radial modes as shown in the figure is based on physical considerations, as well as on solutions of (42.25)–(42.28) for continuously varying l (where of course only those for integer l have a physical meaning).

42.4 Stars Showing Non-radial Oscillations

When applying the above-described formalism to models of real stars, a basic question is whether such oscillations in fact proceed adiabatically. Strictly speaking, one would have to test the model for its vibrational stability and look for the imaginary part of ω derived in a full non-adiabatic treatment. This is, however, so cumbersome that one usually confines oneself to a quasi-adiabatic approximation, similar to that described for the radial case in Chap. 41: the adiabatically calculated eigenfunctions are used to determine an "energy integral", describing the growth or damping rate of the amplitude.

There is a variety of stars and stellar types that are known or suspected to undergo non-radial oscillations. We shall briefly mention a few of them.

The best established group of non-radial oscillators are certain white dwarfs (cf. Van Horn 1984), among them the ZZ Ceti variables, which are of type DA. They exhibit periods typically between a few 10^2 and 10^3 s, often split up into close pairs. These periods are certainly too long for radial oscillations of white dwarfs, but can well be explained by g^+ modes. Rotation of the white dwarf splits them up into oscillations with different order m. The corresponding gravity waves are "trapped" in a superficial hydrogen layer which, according to its thickness, acts as a resonator for certain modes. They are excited by the κ mechanism in zones of partial ionization. Other groups of oscillating white dwarfs, of type DB and very hot ones, have also been found.

The β Cephei stars, which are situated somewhat above the upper main-sequence, are both radial and non-radial oscillators. Some of them also seem to show the effect of rotational mode splitting. We already mentioned them in Sect. 41.4. Non-radial oscillations are also found among the δ Scuti stars and some types of supergiants.

The most prominent example of observed non-radial oscillations is our Sun (compare the early work by Christensen-Dalsgaard 1984; Deubner and Gough 1984). Detailed spectral investigations of the solar surface have shown that, again and again, areas roughly 10^5 km across start oscillating in phase for some time. These oscillations are excited by the ongoing convective motions in the solar envelope, and although they are damped, they are constantly reappearing. Their lifetime is of order a few months. The first detected and best-known oscillations have periods around 5 min. They represent standing acoustic waves trapped mainly in a region from somewhere below the photosphere down into the upper convective zone. Power spectra with ω plotted against the horizontal wave number show clearly that the phenomena contain mode oscillations with very many modes (many degrees l and radial orders k).

Meanwhile, thanks to constant monitoring of the solar surface by global networks of solar telescopes, continuous observations from the south pole during the antarctic summer, and from space, have allowed to identify tens of thousand different modes, with l being as high as 1,500. Rotational splitting due to rotation, i.e. different m-modes, has been detected as well as possibly some g-modes.

These and low-degree p-modes have allowed a detailed analysis of the solar interior (known as *helioseismology*), of solar rotation, and more recently even local helioseismology where subsurface motions in the convective envelope are measured. Helioseismology has provided us with a detailed view into the solar interior. In Sect. 29.4, we already presented those results of helioseismology of relevance for the comparison with hydrostatic, spherically symmetric solar models. But the richness of helioseismology extends far beyond this. We are already learning about convective motions, magnetic fields, active regions, and more, thanks to the analysis of non-radial oscillations.

In the future *asteroseismology* will allow similar views into the interior of other stars. Non-radial oscillations have already been found in many main-sequence stars (of which we mentioned a few classical classes above), but also in evolved stars, even in red giants. Asteroseismology will offer a unique opportunity to test stellar evolution theory.

Part X
Stellar Rotation

Rotation may influence the evolution of stars in two major aspects: it may, if sufficiently fast, affect the internal structure through an effective reduction of the gravitational pull towards the center. As a result, pressure and temperature will be different throughout the star. Second, it leads to additional mass flows, which may transport material between regions of stars that otherwise would not be connected. In particular the latter effect leads to observable modification of the surface composition of stars. This has led to a strong interest in rotation in stars. So far, full three-dimensional models of rotating stars are not available, but one-dimensional, simplified models do exist and seem to result in quite realistic models.

As in the case of stellar pulsations, the following chapters about rotating stars provide only the basic concepts and some idealized cases. Maeder (2009) has written a full textbook on all aspects of rotating stars, which reflects the large progress made in this field, and covers the present sophisticated modelling.

Chapter 43
The Mechanics of Rotating Stellar Models

The theory of rotating bodies with constant densities (liquid bodies) has been investigated thoroughly by McLaurin, Jacobi, Poincaré, and Karl Schwarzschild. We first start with a summary of their results without deriving them.

Most of the results have been obtained for solid-body rotation, i.e. for constant angular velocity ω of the self-gravitating liquid body. In this case the centrifugal acceleration c has a potential, say $c = -\nabla V$ with $V = -s^2\omega^2/2$, where s is the distance from the axis of rotation. If Φ is the gravitational potential, then according to the hydrostatic equation, the total potential $\Psi := \Phi + V$ must be constant on the surface. The main difficulty in determining the surface of a rotating liquid body lies with the gravitational potential, which in turn depends on the form of the surface.

43.1 Uniformly Rotating Liquid Bodies

For sufficiently slow rotation with constant angular velocity, the rotating liquid bodies are spheroids (i.e. axisymmetric ellipsoids) called *McLaurin spheroids*.

In order to examine the behaviour of rotating liquid masses, we define their gravitational energy E_g

$$E_g := \frac{1}{2}\int \varrho\Phi\, dV\,, \tag{43.1}$$

where Φ is the gravitational potential vanishing at infinity and dV is the volume element. The expression (43.1) is the generalization for non-spherical bodies of the definition (3.3).

Indeed in the spherical case with

$$\frac{d\Phi}{dr} = \frac{Gm}{r^2}, \tag{43.2}$$

we have from (3.3)

R. Kippenhahn et al., *Stellar Structure and Evolution*, Astronomy and Astrophysics Library, DOI 10.1007/978-3-642-30304-3_43, © Springer-Verlag Berlin Heidelberg 2012

$$
\begin{aligned}
E_{\mathrm{g}} &= -G\int_0^R \frac{m\,dm}{r} = -\frac{1}{2}G\int_0^R \frac{d(m^2)}{dm}\frac{1}{r}dm \\
&= -\frac{1}{2}G\frac{M^2}{R} - \frac{1}{2}G\int_0^R \frac{m^2 dr}{r^2} \\
&= -\frac{1}{2}\frac{GM^2}{R} - \frac{1}{2}\int_0^R \frac{d\Phi}{dr} m\,dr = \frac{1}{2}\int_0^R \Phi\,dm\,,
\end{aligned} \tag{43.3}
$$

in agreement with the definition (43.1) for the more general (non-spherical) case.

The kinetic energy

$$
T := \frac{1}{2}\int v^2\,dm \tag{43.4}
$$

is supposed to contain only the energy due to the macroscopic rotational motion, but not that due to the thermal motion of the molecules. Let us further define the dimensionless quantity

$$
\chi := \frac{\omega^2}{2\pi G\varrho}\,. \tag{43.5}
$$

It is of the order of the ratio of centrifugal acceleration to gravity at the equator and is a measure of the "strength" of rotation.

We now describe some results on the equilibrium configurations and their stability. The derivations and some details of the configurations can be found in the classic book by Jeans (1928) and in that of Lyttleton (1953).

The shape of McLaurin spheroids is described by the eccentricity e of the meridional cross section,

$$
e^2 = \frac{a^2 - c^2}{a^2}\,, \tag{43.6}
$$

where a, c are the major and the minor half axes of the meridional cross section. A sequence of increasing e leads from the sphere ($e = 0$) to the plane parallel layer ($e = 1$), and one can label each of these configurations by its value of χ. But the correspondence between e and χ is not unique. For each value of $\chi < 0.2247$ there exist two configurations with different values of e. For example, in the limit case of zero rotation with $\chi = 0$, the sphere as well as the infinite plane parallel layer are two possible equilibria, the latter of which obviously is not stable. Along the series of increasing eccentricity e, neither χ nor T is monotonic, but one can show that the angular momentum and E_{g} vary monotonically. Furthermore, ω does not vary monotonically with the total angular momentum: if we start with a liquid self-gravitating sphere ($e = 0$) and feed in angular momentum, the angular velocity, and with it the eccentricity, increases. But once the eccentricity exceeds the value of 0.9299, the angular velocity decreases again, even with further increasing angular momentum. The reason for this is that the momentum of inertia increases faster than the angular momentum, and therefore ω must decrease again.

But long before this, namely at $e = 0.8127$ or at $\chi = 0.1868$, the McLaurin spheroids become unstable. At this point the sequence of configurations shows a

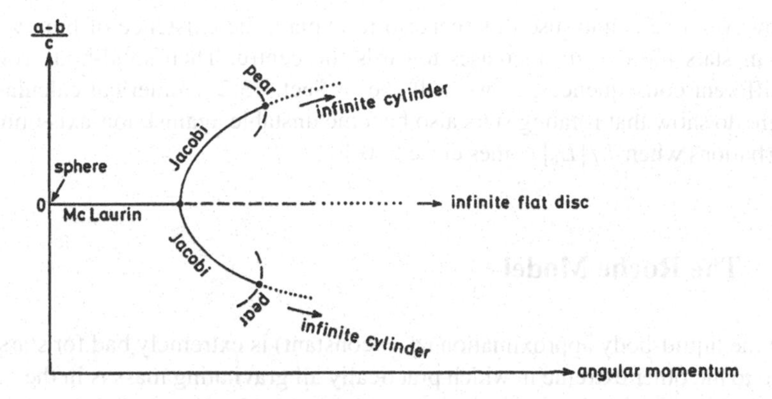

Fig. 43.1 Sequences of the McLaurin and Jacobian equilibrium configurations of a rotating incompressible fluid. In this schematic representation, each configuration is characterized by its angular momentum and its value of $(a-b)/c$, where a, b, c are the three axes of an ellipsoid. *Solid lines* indicate dynamically and secularly stable configurations, *broken lines* secularly unstable, and *dotted lines* dynamically unstable models. The branches of pear-shaped configurations are also indicated, although they cannot be plotted in a diagram with that ordinate. For more details see Ledoux (1958)

bifurcation (Fig. 43.1): another branch of stable models occurs which have a quite different shape. They are triaxial ellipsoids, the so-called *Jacobi ellipsoids.* Beyond the point of bifurcation, a McLaurin spheroid is unstable, the Jacobi ellipsoid of the same mass and angular momentum having a lower total (macroscopic kinetic plus gravitational) energy. Therefore, if there is a mechanism like friction which can use up macroscopic energy and transform it into heat, the spheroids become ellipsoids. The transition takes place on the timescale of friction as defined in Chap. 45. In analogy to the case of a blob of excess molecular weight (see Sect. 6.5) in hydrostatic equilibrium with its surroundings, the motion is controlled by a dissipative process (there heat flow, here friction). One therefore calls the instability of the McLaurin spheroids also *secular.* Instead of the oblateness, one often uses the ratio $\xi := T/|E_g|$, which reaches the value 0.1376 at the point of bifurcation. Stability analysis shows that if ξ exceeds another critical value (of about 0.16), the triaxial ellipsoids also become unstable and then assume a pear-shaped form (see Fig. 43.1).

It should be noted that here we have interpreted sequences of varying dimensionless parameters e, χ, ξ as sequences of models with increasing angular momentum, while mass and density were assumed to be constant. Models with the same dimensionless parameters can also be obtained by a sequence of increasing density, while mass and angular momentum are kept constant. In this way one can conclude from the foregoing discussion that a freely rotating body (mass and angular momentum constant) that contracts (density increasing) can start with slow rotation as a McLaurin spheroid, and can then become triaxial and finally pear-shaped. Indeed, before the Jacobi ellipsoids become long cigars they become dynamically unstable. An ensuing fission may then split the body in two.

However, one cannot use this scenario to explain the existence of binary stars, since in stars the density increases towards the centre. Then solid-body rotation has different consequences, as we will see in Sect. 43.2. Numerical calculations, though, do show that rotating stars also become unstable against non-axisymmetric perturbations when $T/|E_g|$ comes close to 0.14.

43.2 The Roche Model

Since the liquid-body approximation ($\varrho =$ constant) is extremely bad for stars, one can go to the other extreme in which practically all gravitating mass is in the centre. In Roche's approximation one assumes that the gravitational potential Φ is the same as if the total mass of the star were concentrated at the centre. Then Φ is spherically symmetric:

$$\Phi = -\frac{GM}{r} . \tag{43.7}$$

For solid-body rotation, the centrifugal acceleration can again be derived from the potential

$$V = -\frac{1}{2} s^2 \omega^2 , \tag{43.8}$$

where s is the distance from the axis of rotation. If z is the distance from the equatorial plane, then $r^2 = s^2 + z^2$, and the total potential is

$$\Psi = \Phi + V = -\frac{GM}{(s^2 + z^2)^{1/2}} - \frac{1}{2} s^2 \omega^2 . \tag{43.9}$$

The acceleration $-\nabla\Psi$ in the co-rotating frame is the sum of gravitational and centrifugal accelerations. A set of surfaces $\Psi =$ constant is plotted in Fig. 43.2. The advantage of the Roche approximation is that the gravitational field is given independently of the rotation. Eccentricity does not affect gravity. In order to investigate the rotating Roche configurations, we consider the surfaces of constant total potential Ψ:

$$\frac{GM}{(s^2 + z^2)^{1/2}} + \frac{\omega^2 s^2}{2} = \text{constant} = \frac{GM}{r_p} , \tag{43.10}$$

where r_p, the polar radius, is the distance from the centre to the point where the surface intersects the axis of rotation (i.e. the value of z for $s = 0$). With the abbreviations

$$a = \frac{1}{r_p}, \quad b = \frac{\omega^2}{2GM} , \tag{43.11}$$

we find for the equipotential surfaces

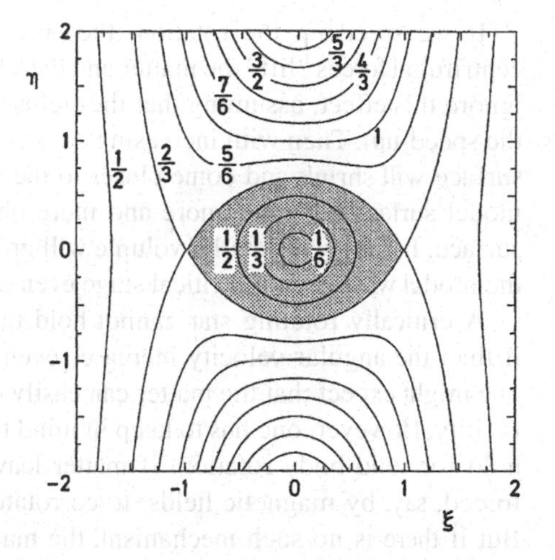

Fig. 43.2 The lines of constant total potential Ψ for the Roche model in the meridional plane. They are labelled by their values of r_p/s_{cr}. The coordinates are $\xi = s/s_{cr}$, $\eta = z/s_{cr}$. The *shaded area* is inside the critical surface

$$z^2 = \frac{1}{(a - bs^2)^2} - s^2 . \tag{43.12}$$

In the equatorial plane $z = 0$, at the circle $s = s_{cr}$ with

$$s_{cr}^3 = \frac{GM}{\omega^2}, \tag{43.13}$$

the gradient of Ψ vanishes. The corresponding critical surface intersects the axis of rotation at $z = \pm 2/3 s_{cr}$ and separates closed surfaces from those going to infinity (Fig. 43.2). In the equatorial plane $z = 0$, gravity dominates inside the critical circle, while outside, the centrifugal acceleration dominates. Both compensate each other exactly at the critical circle. Numerical integration for the volume inside the critical surface gives

$$V_{cr} = 0.1804 \times 4\pi\, s_{cr}^3 . \tag{43.14}$$

Let us now assume that a stellar model just fills its critical volume: $\bar{\varrho} = M/V_{cr}$. We redefine the dimensionless quantity χ by

$$\chi := \frac{\omega^2}{2\pi G\bar{\varrho}}, \tag{43.15}$$

which is of the order of centrifugal acceleration over gravity at the equator. The model fills its critical volume if $\chi = \chi_{cr} = 0.36075$, as can be obtained from the condition of the balance of centrifugal and gravitational acceleration together with (43.14) and (43.15). Rotating models which do not fill their critical volume have $\chi < \chi_{cr}$.

In order to see the rotational behaviour of the Roche model, let us start with very slow rotation so that the stellar surface lies safely within the critical equipotential.

If we speed up the rotation, the volume of the model star will grow, since centrifugal forces "lift" the matter and therefore reduce the effective gravity. We first ignore this effect, assuming that the stellar volume remains unchanged (in spite of the speed-up). Then with increasing ω, according to (43.13) and (43.14), the critical surface will shrink and come closer to the surface of the model. Consequently the model surface becomes more and more oblate until it coincides with the critical surface. In reality the stellar volume will grow as the angular velocity speeds up and the model will reach its critical stage even earlier.

A critically rotating star cannot hold the matter at the equator. What happens if then the angular velocity increases even more? From a first glance at Fig. 43.2 one might expect that the matter can easily escape along equipotential surfaces into infinity. However, one has to keep in mind that the equipotentials plotted there only hold for solid-body rotation. If matter leaving the star at the equator were to be forced, say, by magnetic fields, to co-rotate, it would indeed be swept into space. But if there is no such mechanism, the matter would have to conserve its angular momentum and remain in the neighbourhood of the star. If ω = constant, the centrifugal acceleration ($\sim s$) dominates over gravity ($\sim s^{-2}$) for large values of s. But in the case of constant specific angular momentum ($\omega \sim s^{-2}$), the centrifugal acceleration ($\omega^2 s \sim s^{-3}$) drops more steeply with s than gravity.

We have here considered the case of a star with increasing angular velocity and constant (or increasing) volume. A more realistic case would be that a slowly rotating star contracts. If then its radius decreases, the angular velocity increases like R^{-2} while its critical surface shrinks proportionally to $s_{cr} \sim \omega^{-2/3} \sim R^{4/3}$. The critical surface therefore shrinks faster than the star, which will become more and more oblate until its surface is critical. Then the centrifugal force balances the gravitational one at the equator. With further shrinking, the star loses mass, which is left behind as a rotating disk in the equatorial plane. This is similar to Laplace's scenario of the pre-planetary nebula.

43.3 Slowly Rotating Polytropes

In a homogeneous gaseous sphere there is no density concentration towards the centre, while for the Roche model, the assumed density concentration is too extreme compared to that of real stars. Polytropes approximate real stars better, at least with respect to their density distribution. For slowly rotating polytropes (small values of χ), equilibrium solutions have been found by solving ordinary differential equations for solid-body rotation.

As in the case of the non-rotating polytropes (see Chap. 19), one has to solve the Poisson equation for the gravitational potential. But since the centrifugal acceleration according to (43.8) can be derived from the potential V, we combine Φ and V to obtain the total potential Ψ as in (43.9). Then instead of (19.7), we have in the co-rotating frame

$$\varrho = \left[\frac{-\Psi}{(n+1)K} \right]^n , \qquad (43.16)$$

and since $\Delta \Phi = 4\pi G \varrho, \Delta V = -2\omega^2$, we find

$$\Delta \Psi = 4\pi G \varrho - 2\omega^2 , \qquad (43.17)$$

and with (43.16)

$$\Delta \Psi = 4\pi G \left[\frac{-\Psi}{(n+1)K} \right]^n - 2\omega^2 . \qquad (43.18)$$

If we now replace r in the Laplace operator by the dimensionless variable $y = Ar$, where A is defined as in (19.9), we obtain for $\omega : \Phi/\Phi_c$ with the help of (43.16)

$$\Delta_y \omega = \omega^n - \frac{\omega^2}{2\pi G \varrho_c} , \qquad (43.19)$$

with $\Delta_y = A^2 \Delta$, where Δ is the Laplace operator. In spherical coordinates, for the case of axial symmetry,

$$\Delta_y \equiv \frac{1}{y^2 \sin \vartheta} \left[\frac{\partial}{\partial y} \left(y^2 \sin \vartheta \frac{\partial}{\partial y} \right) + \frac{\partial}{\partial \vartheta} \left(\sin \vartheta \frac{\partial}{\partial y} \right) \right] . \qquad (43.20)$$

The last term on the right-hand side of (43.19) is a measure of the strength of rotation. We therefore now define for polytropes

$$\chi := \frac{\omega^2}{2\pi G \varrho_c} , \qquad (43.21)$$

and we can write (43.19) in the form

$$\Delta_y \omega = \omega^n - \chi . \qquad (43.22)$$

This partial differential equation corresponds to the Emden equation (19.10), which indeed is obtained for $\omega \to 0$. Equation (43.22) holds in the interior of the polytrope, while outside, the solution has to obey the Laplace equation, which here is $\Delta_y \omega = 0$, and has to be regular at infinity. For $\chi \ll 1$ one can approximate the solution $\omega(y, \vartheta)$ by an expansion in Legendre polynomials $L_i(\vartheta)$ with even i :

$$\omega = \omega_0(y) + \chi\omega_1(y) + \chi\omega_2(y)L_2(\cos\vartheta) + \dots , \qquad (43.23)$$

where $\omega_0(y)$ is the solution of the Lane–Emden equation. The perturbation of first order in χ is split into a spherically symmetric term and a non-spherical one, which vanishes if averaged over a sphere. The terms of higher order in χ are not explicitly written down. If the expansion (43.23) is introduced into (43.22), then the terms of the same dependence on ϑ and of the same order in χ give ordinary differential equations in y. Similarly the Laplace equation for the outside can be reduced to a set of ordinary differential equations by the expansion (43.23).

Numerical calculations by Chandrasekhar (1933) show that the oblateness of the surface defined by $(r_{\text{equ}} - r_{\text{pole}})/r_{\text{equ}}$ is 3.75χ, 5.79χ, 9.82χ, 41.81χ, 468.07χ for the polytropes of index $n = 1, 1.5, 2, 3, 4$, respectively.

Chapter 44
The Thermodynamics of Rotating Stellar Models

The theory of the structure of rotating stars becomes relatively simple if the centrifugal acceleration can be derived from a potential V:

$$\omega^2 s \boldsymbol{e}_s = -\nabla V, \tag{44.1}$$

where $\boldsymbol{e}_s$ is a unit vector perpendicular to the axis of rotation (pointing outwards) and s is the distance from this axis. One can easily see that a sufficient and necessary condition for the existence of such a potential is that in the system of cylindrical coordinates s, φ, z, the angular velocity depends on s only: $\partial\omega/\partial z = \partial\omega/\partial\varphi = 0$, i.e. ω is constant on cylinders. We call such an angular-velocity distribution (to which the case of solid-body rotation also belongs) *conservative.*

44.1 Conservative Rotation

In this case the potential V is

$$V = -\int_0^s \omega^2 s\, ds. \tag{44.2}$$

We again combine gravitational and centrifugal potentials to form the total potential

$$\Psi := \Phi + V. \tag{44.3}$$

If we now include centrifugal acceleration in the equation of hydrostatic equilibrium [compare with (2.20)], we obtain

$$\nabla P = -\varrho\nabla\Psi. \tag{44.4}$$

R. Kippenhahn et al., *Stellar Structure and Evolution*, Astronomy and Astrophysics Library, DOI 10.1007/978-3-642-30304-3_44, © Springer-Verlag Berlin Heidelberg 2012

Equation (44.4) indicates that the vectors ∇P and $-\nabla\Psi$ are parallel. In other words, the equipotential surfaces Ψ = constant coincide with the surfaces of constant pressure, which means that the pressure is a function of Ψ: $P = P(\Psi)$. It then follows that $\varrho = -dP/d\Psi$ is also a function of Ψ only. If we now have an ideal gas, then $T/\mu = P/(\varrho\Re)$ is a function of Ψ. In a chemically homogeneous star, therefore, $T = T(\Psi)$, i.e. the temperature is constant on equipotential surfaces.

Since not T but T/μ is constant on equipotentials, the temperature varies proportionally to μ on these surfaces if the chemical composition is not homogeneous. We have already encountered this case in Sect. 6.5, where we dealt with a blob of material with a higher molecular weight than that in the surroundings. In the blob the temperature was higher.

Note that this is a consequence of hydrostatic equilibrium: even small deviations from hydrostatic equilibrium can cause considerable temperature variations on equipotential surfaces, which can be seen in the case with negligible rotation. Then from (44.4) one can conclude that P, ϱ, and T/μ are constant on the equipotential surfaces of the gravitational field, say, of the earth. We know that if we light a match, the air on the horizontal equipotential planes intersecting the flame will not have the high temperature of the fire. The reason is that with the flame a circulation system is set up. With this motion, inertia terms disturb the equation of hydrostatic equilibrium. Although they cause only small perturbations, the inertia terms are sufficient to allow lower temperatures outside the flame.

In the following we discuss only the case of strict hydrostatic equilibrium for a chemically homogeneous ideal gas and therefore have $P = P(\Psi), \varrho = \varrho(\Psi)$, $T = T(\Psi)$.

Note that the coincidence of P and ϱ surfaces only holds if the rotation is conservative. Otherwise they are inclined to each other (see Sect. 45.2).

44.2 Von Zeipel's Theorem

We now investigate radiative energy transport in a homogeneous, hydrostatic star with conservative rotation. The equation for radiative transport (5.8) in vector form

$$\boldsymbol{F} = -\frac{4ac}{3\kappa\varrho}T^3\nabla T, \tag{44.5}$$

where $\boldsymbol{F}$ is the vector of the radiative energy flux. With $T = T(\Psi)$ and with $-\nabla\Psi = \boldsymbol{g}_{\text{eff}}$, the effective gravitational acceleration consisting of gravitational *and* centrifugal acceleration, one finds

$$\boldsymbol{F} = -\frac{4ac}{3\kappa\varrho}T^3\frac{dT}{d\Psi}\boldsymbol{g}_{\text{eff}} = -k(\Psi)\boldsymbol{g}_{\text{eff}}, \tag{44.6}$$

since also $\kappa(\varrho, T) = \kappa(\Psi)$. In the non-rotating case this equation is equivalent to (5.9). We now look for the equation of energy conservation and restrict ourselves to stationary states with complete equilibrium. Then, instead of (4.43), we have from (44.6)

$$\nabla \cdot \boldsymbol{F} = -\frac{dk}{d\Psi}(\nabla\Psi)^2 - k(\Psi)\Delta\Psi$$

$$= -\frac{dk}{d\Psi}(\nabla\Psi)^2 - k(\Psi)\left(4\pi G\varrho - \frac{1}{s}\frac{d(s^2\omega^2)}{ds}\right) = \varepsilon\varrho, \tag{44.7}$$

where we have made use of $\Delta\Phi = 4\pi G\varrho$ and of (44.2) (Δ is the Laplace operator.). One can easily see that this equation cannot be fulfilled. We consider a chemically homogeneous star; then P, ϱ, and T are constant on the equipotential surfaces $\Psi =$ constant. Therefore the terms $\varepsilon\varrho$ as well as $4\pi G\varrho k(\Psi)$ are constant on equipotential surfaces, but in general the remaining two terms on the left are not, and they do not cancel each other. This can be easily seen in the case of solid-body rotation, for which $(s^{-1})d(s^2w^2)/ds$ is a constant, while $(\nabla\Psi)^2$ always varies on equipotential surfaces, the effective gravity at the equator being smaller than at the poles.

The fact that radiative transport and the simple equation of energy conservation cannot be fulfilled simultaneously was first pointed out by Von Zeipel (1924) and is known as von Zeipel's theorem. The solution of the problem was independently found by Eddington (1925) and Vogt (1925).

44.3 Meridional Circulation

What is to be expected if (44.7) cannot be fulfilled? Then there must be regions in the star which would cool off, since radiation carries more energy out of a mass element than is generated by thermonuclear reactions. In other regions the mass elements would heat up. But cooling and heating cause buoyancy forces, and meridional motions occur in addition to rotation. In order to maintain a stationary state as assumed, one has to demand that meridional motions contribute to the energy transport. They carry away energy from regions where radiation cannot transport all the energy generated and they bring energy to regions which otherwise would cool off.

In order to derive the velocity field of the circulation, we write the first law of thermodynamics in the co-moving frame:

$$\nabla \cdot \boldsymbol{F} = \varepsilon\varrho - \varrho T\frac{d\sigma}{dt}. \tag{44.8}$$

We here denote the specific entropy by σ (instead of s) to avoid confusion with the distance from the axis. With $d\sigma = dq/T$, and with (4.18), one has

$$T\frac{d\sigma}{dt} = c_P\frac{dT}{dt} - \frac{\delta}{\varrho}\frac{dP}{dt}. \tag{44.9}$$

If we replace the derivatives in the co-moving frame by those in a coordinate system at rest with respect to the stellar centre, i.e. $d/dt = \partial/\partial t + \boldsymbol{v}\cdot\nabla$, we find

$$\nabla\cdot\boldsymbol{F} = \varepsilon\varrho - c_P\varrho\frac{\partial T}{\partial t} + \delta\frac{\partial P}{\partial t} - \boldsymbol{v}[c_P\varrho\nabla T - \delta\nabla P], \tag{44.10}$$

and for thermal equilibrium

$$\nabla\cdot\boldsymbol{F} = \varepsilon\varrho - c_P\varrho T\boldsymbol{v}\left[\frac{1}{T}\nabla T - \frac{\delta}{c_P\varrho T}\nabla P\right]. \tag{44.11}$$

With $\nabla T = \nabla\Psi(dT/d\Psi)$ and $\nabla P = \nabla\Psi(dP/d\Psi)$, the usual abbreviation $\nabla = d\ln T/d\ln P$, and (4.21), we can write

$$\nabla\cdot\boldsymbol{F} = \varepsilon\varrho - \frac{c_P\varrho T}{P}(\nabla - \nabla_{\text{ad}})(\boldsymbol{v}\cdot\nabla P). \tag{44.12}$$

The components of the meridional velocity field have to fulfil this equation together with the continuity equation, which in the stationary case becomes $\nabla\cdot(\varrho\boldsymbol{v}) = 0$.

We can simplify (44.12) if we assume χ, as defined in (43.5), to be small and ignore higher-order terms in χ. Since $\boldsymbol{v}$ is of first order in χ, the last term in (44.12) can be replaced by $[c_P\varrho T(\nabla - \nabla_{\text{ad}})/P]_0\nabla P_0\boldsymbol{v}$, where the subscript 0 indicates the values of the corresponding non-rotating model. Since $\nabla P_0 = -\varrho_0\boldsymbol{g}_0$ and $\boldsymbol{g}_0$ has only a radial component given by $-|\boldsymbol{g}_0| = -g_0$, we have, instead of (44.12),

$$\nabla\cdot\boldsymbol{F} = \varepsilon\varrho + \left[\frac{c_P\varrho^2 T}{P}(\nabla - \nabla_{\text{ad}})g\right]_0 v_r. \tag{44.13}$$

Comparing the non-rotating case, we have now introduced a new unknown variable v_r, which in spherical coordinates r, φ, ϑ together with the velocity component in the ϑ direction has to fulfil the continuity equation

$$\frac{1}{r^2}\frac{\partial(\varrho r^2 v_r)}{\partial r} + \frac{1}{r\sin\vartheta}\frac{\partial(\varrho v_\vartheta\sin\vartheta)}{\partial\vartheta} = 0. \tag{44.14}$$

Equations (44.13) and (44.14) are the necessary conditions for determining also the velocity field.

44.4 The Non-conservative Case

Above we have shown the existence of meridional circulation only for a conservative angular-velocity distribution. We now discuss the situation in a non-conservative case. For this we choose $\omega = \omega(r)$, but restrict ourselves to slow rotation. The equations to be solved are

$$\nabla P = -\varrho \nabla \Phi + \boldsymbol{c}, \tag{44.15}$$

$$\nabla \cdot \boldsymbol{F} = \varepsilon \varrho + \left[\frac{c_P \varrho^2 T}{P} (\nabla - \nabla_{\mathrm{ad}}) g \right]_0 v_r. \tag{44.16}$$

$$\boldsymbol{F} = -\frac{4ac}{3\kappa\varrho} T^3 \nabla T, \tag{44.17}$$

$$\Delta \Phi = 4\pi G \varrho, \tag{44.18}$$

where the functions $\varrho, \varepsilon, \kappa$ are assumed to be known functions of P and T. Without rotation the solutions are spherically symmetric, but rotation produces deviations from that symmetry. The centrifugal acceleration $\boldsymbol{c}$ appearing in (44.15) has the components

$$c_r = \omega^2 r \sin^2 \vartheta = \frac{2}{3} \omega^2 r (1 - L_2), \tag{44.19}$$

$$c_\vartheta = \omega^2 r \sin \vartheta \cos \vartheta = \frac{1}{3} \omega^2 r \frac{\partial L_2}{\partial \vartheta}, \tag{44.20}$$

where we have introduced the second Legendre polynomial $L_2(\vartheta) = (3 \cos^2 \vartheta - 1)/2$.

In order to solve the system (44.15)–(44.18), we split all the scalar functions into a spherically symmetric part (subscript 0) and one which is proportional to $L_2(\vartheta)$:

$$P(r, \vartheta) = P_0(r) + P_2(r) L_2(\vartheta), \quad T = T_0 + T_2 L_2, \quad \Phi = \Phi_0 + \Phi_2 L_2, \tag{44.21}$$

with $|P_2| \ll P_0, |T_2| \ll T_0$. For the vectors $\boldsymbol{F}$ and $\boldsymbol{v}$ we write

$$F_r = F_{r0}(r) + F_{r2}(r) L_2, \quad F_\vartheta = F_{\vartheta 2}(r) \frac{dL_2(\vartheta)}{d\vartheta},$$

$$v_r = 0 + v_{r2}(r) L_2(\vartheta), \quad v_\vartheta = v_{\vartheta 2}(r) \frac{dL_2(\vartheta)}{d\vartheta}, \tag{44.22}$$

with $|F_{r2}|$ and $|F_{\vartheta 2}|$ being small compared to $|F_{r0}|$. It should be noted that in this notation the quantities $P_0, T_0, \ldots$ are not identical with the corresponding functions of the non-rotating star, since in the centrifugal acceleration there is also a spherically symmetric component, as can be seen from (44.19).

We now ignore second-order effects and count the number of equations for the four "spherical" functions P_0, T_0, Φ_0 and F_{r0} and for the five "non-spherical" functions P_2, T_2, Φ_2, F_{r2}, and $F_{\vartheta 2}$. These are all variables appearing in (44.15)–(44.18) together with (44.21) and (44.22), if for the moment we ignore circulation ($v_r = 0$). It is obvious that each of the two scalar equations (44.16) and (44.18) gives two equations, a spherical one and a non-spherical one, though in the case of the vector equations (44.15) and (44.17) it is different. We explain this in the case of (44.15). The r component gives a "spherical" equation [compare (44.19)]

$$\frac{dP_0}{dr} = -\varrho_0 \frac{d\Phi_0}{dr} + \frac{2}{3}\varrho_0 \omega^2 r \tag{44.23}$$

and a "non-spherical" one

$$\frac{dP_2}{dr} = -\varrho_0 \frac{d\Phi_2}{dr} - \varrho_2 \frac{d\Phi_0}{dr} - \frac{2}{3}\varrho_0 \omega^2 r, \tag{44.24}$$

while the ϑ component gives [compare (44.20)]

$$P_2 = -\varrho_0 \Phi_2 + \frac{1}{3}\varrho_0 \omega^2 r. \tag{44.25}$$

Therefore the vector equation (44.15) yields the "spherical" equation (44.23) and two "non-spherical" equations (44.24) and (44.25). The same holds for the vector equation (44.17). Altogether we have four equations for the four "spherical" functions but six equations for the five "non-spherical" functions. Obviously with $v_r = 0$ the problem is overdetermined. In general it can only be solved if meridional circulations are present; then the v_r appearing in (44.16) is the sixth unknown "non-spherical" variable and the problem is no longer overdetermined. If v_r is known, the continuity equation (44.14) together with (44.21) and (44.22) gives v_ϑ.

44.5 The Eddington–Sweet Timescale

To obtain an estimate of the velocity of the circulation, we restrict ourselves to slow rotation and to the conservative case. The estimate for the non-conservative case is more complicated, but the results are very similar. We also assume $\varepsilon = 0$, which holds for the outer layers. Therefore $l =$ constant.

We now can split each function $A(r, \vartheta)$ of the model uniquely into two terms:

$$A(r, \vartheta) = \bar{A}(\Psi) + A^*(r, \vartheta), \tag{44.26}$$

where $\bar{A}(\Psi)$ is the mean value of $A(r, \vartheta)$ over the surface $\Psi =$ constant, while the integral of A^* over each Ψ surface vanishes:

$$\int_\Psi A^*(r, \vartheta) dS = 0, \tag{44.27}$$

where dS is the surface element of the Ψ surface. Then according to (44.6), $k(\Psi) = \bar{F}/\bar{g}_{\text{eff}}$, where F and g_{eff} are the absolute values of $\boldsymbol{F}$ and $\boldsymbol{g}_{\text{eff}}$, and (44.7) can be written as

$$\nabla \cdot \boldsymbol{F} = -\frac{d}{d\Psi}\left(\frac{\bar{F}}{\bar{g}}\right) g^2 - \frac{\bar{F}}{\bar{g}}\left[4\pi G\varrho - \frac{1}{s}\frac{d}{ds}(s^2\omega^2)\right], \tag{44.28}$$

where we have omitted the subscript eff in the symbols g and $\bar{g}$. We now split the terms of (44.28) according to (44.26). $\overline{\nabla \cdot \boldsymbol{F}}$ has to be zero in the steady state in regions where there is no nuclear energy generation (otherwise it has to be equal to $\varepsilon\varrho$, a function which is also constant on Ψ surfaces). But the term $(\nabla \cdot \boldsymbol{F})^*$ can only be compensated by circulation. Indeed the circulation term in (44.13) is $[c_P \varrho T(\nabla - \nabla_{\text{ad}})/P]\nabla P_0 \boldsymbol{v}$. The integral of this term over equipotential surfaces vanishes because of mass conservation, as does $(\nabla \cdot \boldsymbol{F})^*$.

We now estimate $(\nabla \cdot \boldsymbol{F})^*$ for slow rotation and take $\bar{F}/\bar{g}$ from the non-rotating model, an approximation which introduces only errors of order χ^2, since in the expression for $(\nabla \cdot \boldsymbol{F})^*$ the function $\bar{F}/\bar{g}$ appears multiplied only by terms of order χ. Then

$$\frac{\bar{F}}{\bar{g}} = \frac{L}{4\pi G m}, \tag{44.29}$$

$$\frac{d}{d\Psi}\left(\frac{\bar{F}}{\bar{g}}\right) = \frac{d}{dr}\left(\frac{\bar{F}}{\bar{g}}\right)\frac{dr}{d\Psi} = \frac{d}{dr}\left(\frac{L}{4\pi G m}\right)\frac{1}{g} = -\frac{L\varrho}{m}\left(\frac{r^2}{Gm}\right)^2, \tag{44.30}$$

and therefore

$$(\nabla \cdot \boldsymbol{F})^* = -\frac{L\varrho}{m}\left(\frac{r^2}{Gm}\right)^2 (g^2)^* - \frac{L}{4\pi G m}\left[\frac{1}{s}\frac{d(s^2\omega^2)}{ds}\right]^*. \tag{44.31}$$

Now (44.12) yields

$$\frac{\delta\varrho\bar{g}}{\nabla_{\text{ad}}}(\nabla_{\text{ad}} - \nabla)v_r = -\frac{L\varrho}{\bar{g}^2 m}(g^2)^* - \frac{L}{4\pi G m}\left[\frac{1}{s}\frac{d(s^2\omega^2)}{ds}\right]^*, \tag{44.32}$$

where in the circulation term we have made use of (4.21).

For angular velocities of the form $\omega^2 = c_1 + c_2/s^2$ the expression in the last bracket is constant and the last term vanishes for these special angular velocity distributions which include solid-body rotation ($c_2 = 0$). We first restrict ourselves to these special rotation laws. As a rough estimate, we can say that $(g^2)^*/\bar{g}^2$ is of the order of χ. Indeed g^*, the variation of g over an equipotential, is due to the difference of centrifugal acceleration between equator and poles, and therefore $g^*/g \approx \chi$, and also $(g^2)^*/g^2 \approx \chi$. We then find with $(\nabla_{\text{ad}} - \nabla)/\nabla_{\text{ad}}$ and δ of the order of 1,

$$v_r \approx \frac{L}{\bar{g}m}\chi \approx \frac{LR^2}{GM^2}\chi, \tag{44.33}$$

where we have replaced m and g by their surface values M and GM/R^2 (Replacing them by some mean values over the star would not change the order of magnitude.). The time it takes a mass element to move over the stellar radius, then, is the circulation timescale τ_{circ}, first derived by Sweet (1950):

$$\tau_{\text{circ}} \approx \frac{R}{v_r} \approx \frac{GM^2}{LR}\frac{1}{\chi} \approx \frac{\tau_{\text{KH}}}{\chi}, \tag{44.34}$$

where we have made use of (3.19), ignoring a factor 2. For the Sun one has $\chi \approx 10^{-5}$, $\tau_{\text{KH}} \approx 10^7$ years, and therefore $\tau_{\text{circ}} \approx 10^{12}$ years, which exceeds the lifetime of the Sun.

This estimate has been made ignoring the last term in (44.32). If ω is not of the special form given above, the term in the bracket will be of the order of ω^2, and since ω^* is constant on cylinders but not on equipotential surfaces, ω will be of the order of $\bar{\omega}$ and the term in question will be of the order of

$$\frac{L\omega^2}{4\pi GM} \approx \frac{L}{4\pi R^3}\chi, \tag{44.35}$$

where we have replaced $\omega^2 R/g = \omega^2 R^3/(GM)$ by χ. We estimated that the first term on the right of (44.32) is of the order of $L\varrho\chi/M$. Therefore as long as we are not too close to the surface we can replace ϱ by the mean density $\bar{\varrho} = 3M/(4\pi R^3)$, so that the two terms on the right of (44.32) are of the same order and our estimates (44.33) and (44.34) also hold for rotation laws which are not of the special form $c_1 + c_2/s^2$. But near the surface the first term on the right of (44.32) becomes small owing to the factor ϱ, and the second becomes the leading term. Then near the surface, (44.33) has to be replaced by

$$v_r \approx \frac{\bar{\varrho}}{\varrho}\frac{LR^2}{GM^2}\chi \approx \frac{L}{G\varrho RM}\chi, \tag{44.36}$$

where again we have neglected factors of the order of one. The circulation can therefore become rather fast near the surface.

The same is true at the interfaces between radiative and convective regions where $\nabla = \nabla_{\text{ad}}$, which we have excluded in our rough estimate of the left-hand side of (44.32). At these singularities the circulation speed would become so large that its inertia terms are important and (44.4) would no longer be valid.

Another more serious restriction of our estimates of v_r is the assumption of a certain time-independent angular-velocity distribution. If one starts, say, with $\omega =$ constant, then circulation will occur, and by conservation of angular momentum, it will immediately change the angular-velocity distribution, which in turn demands another circulation pattern.

The "proof" of the existence of meridional circulation in the theory of first order in χ as given in Sect. 44.4 rested on counting the number of linear equations and the number of variables. We showed that without circulation the problem is overdetermined. This, however, is only true if the linear equations are independent. But if ω is considered as a free function, it can be chosen in such a way that the equations become linearly dependent and in the first-order theory no circulation is necessary to fulfil the equations. In the (unrealistic) case ε = constant, κ = constant, the stellar-structure equations for radiative energy transport lead to a polytrope of index $n = 3$. If ε = constant, then l/m = constant and one has a very special "standard model" as discussed in Sect. 19.5. It has been shown by Schwarzschild (1942) that, for this model, solid-body rotation does not demand circulation in the first-order theory. For other, more realistic stellar models, there are also angular-velocity distributions for which there is no meridional circulation in the first-order theory (Kippenhahn 1963).

The linear dependence of the equations can also be achieved if for a given rotation law w, the molecular weight is considered a free function and chosen in an appropriate way. We will come to this problem in the next section.

44.6 Meridional Circulation in Inhomogeneous Stars

We have already estimated that for the Sun that $\tau_{circ}/\tau_{nucl} \approx 10^2$. But for more massive main-sequence stars the situation changes. According to (44.34)

$$\frac{\tau_{circ}}{\tau_{nucl}} \approx \frac{\tau_{KH}}{\tau_{nucl}} \frac{1}{\chi} \sim \frac{M^{1-\alpha}}{\chi} \approx \frac{M^{0.4}}{\chi}, \tag{44.37}$$

where we have assumed a mass-radius relation $R \sim M^\alpha$ and $\tau_{KH} \sim M^2/(RL)$, as can be derived from (3.19), and $\tau_{nucl} \sim M/L$, and we have put $\alpha = 0.6$ for the upper end of the main sequence (Sect. 22.1). Therefore, if we go from the Sun to higher masses, say, to $20M_\odot$, then the ratio τ_{circ}/τ_{nucl} (which for the Sun is about 1/100) increases by a factor 3.3. Observations of rotating B stars show that χ is larger by a factor 10^5 than for the Sun. Therefore, τ_{circ}/τ_{nucl} drops below unity towards the upper end of the main sequence, so that the circulation is rapid enough to mix the star. As a consequence one should expect that the fuel is not only used up in the central region and the star should remain chemically homogeneous. But then the stars, while converting hydrogen into helium, should move in the HR diagram from the main sequence straight towards the helium main sequence [compare (20.20) and (20.21) for $M = M'$]. But we know from observation that the stars leave the main sequence moving towards the region of the red stars and not towards the region of the (blue) helium main sequence. This indicates that they do not mix, and the explanation was found by Mestel (1953). Before the circulation can transport the material out of the burning region, the moving matter will have been enriched in helium. It therefore has a higher molecular weight than the surrounding into which it has been lifted. But then the effect discussed in connection with a blob of material of higher molecular weight μ in a gas of lower μ becomes important (Sect. 6.4). Let us

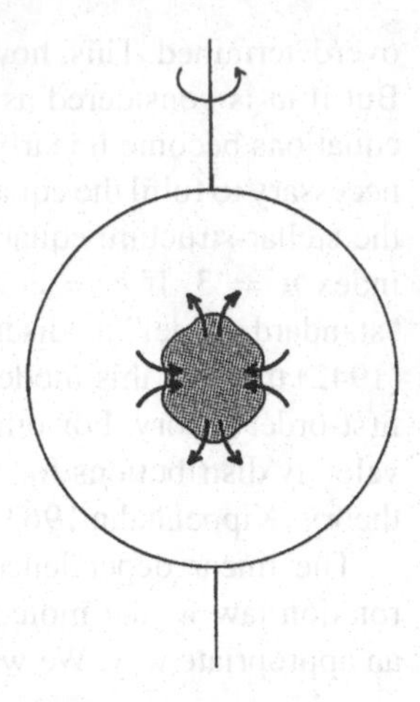

Fig. 44.1 Material of higher molecular weight in the central region of a rotating star (*grey area*) under the influence of meridional circulation

assume that the circulation lifts helium-enriched material as indicated in Fig. 44.1. Then, since in hydrostatic equilibrium T/μ must be constant on Ψ surfaces, the lifted matter has a higher temperature than the matter on the same Ψ surface which is not lifted. There is no buoyancy force acting on the lifted matter, since the higher molecular weight is compensated by the higher temperature. But as the lifted material adjusts thermally, it sinks back. This additional motion ("μ currents") acts against the circulation, and the star can only be mixed if circulation is sufficiently fast. But even in rapidly rotating main-sequence stars, the circulation is not sufficient to mix the helium formed during hydrogen burning. Obviously layers in which the molecular weight increases in an inward direction cannot easily be penetrated by meridional circulation. One therefore often speaks of *μ barriers*.

Note that μ barriers in which no circulation occurs are not in contradiction to our "proof" of the existence of meridional circulation in rotating stars. According to our considerations in Sect. 44.4, which also hold for inhomogeneous stars as long as μ is spherically symmetric, circulation would set in. But after a short time the circulation has modified the μ distribution, and the original spherically symmetric function $\mu(r)$ has become distorted and may be of the form $\mu_0(r) + \mu_2(r)L_2(\vartheta)$. Then by counting the equations and variables as was done in Sect. 44.4, we would not find the problem to be overdetermined, since $\mu_2(r)$ is an additional unknown function. It can be determined instead of v_r by the "non-spherical" equations.

The foregoing arguments do not apply to homogeneous layers in outer regions of stars, where no nuclear processes have changed the molecular weight substantially. Indeed, meridional circulation streams and other hydrodynamical matter flows due to rotation do lead to surface abundance changes, in particular of CNO elements. This is observed in massive stars and well explained by current models including rotation (see Maeder 2009).

Chapter 45
The Angular-Velocity Distribution in Stars

Stars formed out of an interstellar cloud contain a certain amount of angular momentum, which is distributed over the stellar mass. Suppose there were no transport of angular momentum between the mass elements during the formation and evolution of the stars; one would then have local conservation of angular momentum,

$$\frac{d(s^2\omega)}{dt} \equiv s^2 \frac{\partial \omega}{\partial t} + \boldsymbol{v} \cdot \nabla(s^2\omega) = 0, \tag{45.1}$$

where $\boldsymbol{v}$ is the large-scale velocity in the star. Then the angular velocity $\omega(s, \vartheta)$ would be determined by the angular momenta of the mass elements in the original cloud. However, the motion of atoms, the flow of photons through matter, and instabilities that cause small-scale motions can transport angular momentum (An example of the last of these is the convective motion in regions of dynamical instability.). We now discuss these transport mechanisms in detail.

45.1 Viscosity

Viscosity due to microscopic motion, like that of the molecules in a liquid, is given by the viscosity coefficient

$$\eta \approx \varrho \ell v_{\rm th}, \tag{45.2}$$

where ℓ is the mean free path of the particles and $v_{\rm th}$ their mean velocity. In an ionized gas the viscosity is determined by the collisions between the ions. Therefore their mean free path and their thermal velocities have to be inserted in (45.2), and one normally obtains values for η which in cgs units are of the order of 1.

In order to see whether viscosity is important in a star, one has to estimate the timescale required for viscosity to influence a given angular-velocity distribution.

R. Kippenhahn et al., *Stellar Structure and Evolution*, Astronomy and Astrophysics Library, DOI 10.1007/978-3-642-30304-3_45, © Springer-Verlag Berlin Heidelberg 2012

This can be done with the φ component of the Navier–Stokes equations of motion, which for constant viscosity can be written in the form

$$\varrho \frac{\partial \omega}{\partial t} = \eta \Delta \omega, \tag{45.3}$$

where Δ is the Laplace operator. This equation is of the form of the equation of heat transfer (5.31). In analogy to (5.32), we can estimate the viscosity timescale:

$$\tau_{\rm visc} \approx \frac{d^2 \varrho}{\eta}, \tag{45.4}$$

where d is the characteristic length on which ω varies. If for d one takes the radius of a star, say 10^{11} cm, then with $\varrho \approx 1\,{\rm g\,cm^{-3}}$, one finds $\tau_{\rm visc} \approx 10^{22}$ s, a timescale much longer than the cosmological time. In stars one can therefore neglect the viscosity due to the collisions between the ions.

In a star, photons can also cause viscosity, since they transport momentum. If they are absorbed after a mean free path $\ell_{\rm ph}$, they transfer their momentum to the absorbing particle. A rough estimate of this *radiative viscosity* $\eta_{\rm rad}$ is obtained if in (45.2) ϱ is replaced by the mass density of the radiation field $\varrho_{\rm rad} = aT^4/c^2$, $v_{\rm th}$ is replaced by c, and ℓ by $\ell_{\rm ph} \approx 1/\kappa\varrho$, the mean free path of a photon:

$$\eta_{\rm rad} \approx \frac{aT^4}{c\kappa\varrho}. \tag{45.5}$$

The characteristic timescale according to (45.4) is

$$\tau_{\rm visc} \approx \frac{d^2 \varrho}{\eta} \approx \frac{d^2 \varrho^2 c \kappa}{aT^4}. \tag{45.6}$$

With $d = 10^{11}\,{\rm cm}^2$, $\varrho = 1\,{\rm g\,cm^{-3}}$, $\kappa = 1\,{\rm cm^2\,g^{-1}}$, $T = 10^7$ K, we find the characteristic time of radiative viscosity in a star to be 10^{18} s, again a timescale long compared to the lifetime of a star. One therefore can neglect the effects of viscosity not only caused by the atomic motion but also those caused by radiation: the stellar gas moves like a frictionless fluid.

It should be noted that the radiation causes a kind of viscosity similar to that of the atomic motion only in an isotropic radiation field. For a non-isotropic field the radiative viscosity is not a scalar but a tensor.

The expression (45.2) for viscosity can also be used in convective regions, where rising and falling mass elements not only transport energy as discussed in Chap. 7 but also momentum. In the picture of the mixing-length theory, one can consider the convection elements as "particles" which are created at some place, move one mixing length $\ell_{\rm m}$, and dissolve. The corresponding "turbulent viscosity" $\eta_{\rm t}$ in analogy to (45.2) is

$$\eta_{\rm t} \approx \varrho \ell_{\rm m} v_{\rm t}, \tag{45.7}$$

where v_t is the convective velocity. In the case of the convective envelope of the Sun, we assume v_t to be 1 % of the speed of sound (as indicated in Fig. 29.5c). With $\ell_m \approx H_P \approx 10^8$ cm, $\varrho \approx 10^{-4}\,\mathrm{g\,cm^{-3}}$, a sound velocity of $v_s \approx 2\times10^6\,\mathrm{cm\,s^{-1}}$ corresponding to a temperature of 3×10^4 K, and with $v_t \approx 0.01 v_s \approx 2\times10^4\,\mathrm{cm\,s^{-1}}$, we find $\eta_t \approx 2\times10^8$ cgs and the corresponding timescale $\tau_{visc} \approx 5\times10^9\,\mathrm{s} \approx$ 160 years! One can therefore assume that the angular-velocity distribution in the convective zone of the Sun, for instance, has reached a steady state in which the initial angular-momentum distribution is smeared out by viscosity.

However, the analogy between friction caused by molecules and that by convective blobs has its limits. While the statistical motion of molecules is isotropic to a high degree, there is no reason to suppose that convection in a stellar convective zone can be described by elements with isotropic random motion. Convection is maintained in a star by the radially outgoing energy flux. The motion is caused by buoyancy forces which are antiparallel to the (radial) gravity vector. One therefore can expect that the exchange of momentum by the turbulent elements is different in the radial direction from that in other directions. The viscosity is no longer isotropic, i.e. it is a tensor.

The macroscopic behaviour of a fluid with anisotropic viscosity is peculiar. We know that in the case of isotropic viscosity, a self-gravitating sphere which initially starts out with differential rotation approaches solid-body rotation after a viscous timescale. This is not true any more for non-isotropic viscosity (Biermann 1951). One can expect that non-isotropic turbulent viscosity causes differential rotation and should therefore not be surprised that the surface of the Sun does not rotate uniformly.

In this connection it should be noted that in a large part of the solar convective zone, the layers are adiabatic (with constant ∇_{ad}) and surfaces of constant pressure and of constant density coincide (since $d\ln\varrho/d\ln P$ = constant). As in the barotropic case any angular-velocity distribution for which ω varies on cylinders of s = constant will cause dynamically driven meridional circulation which by itself changes the angular-velocity distribution.

Helioseismology has allowed to determine the rotation profile of the Sun. Schou et al. (1998) have demonstrated that the radiative interior of the Sun is rotating with nearly constant angular velocity, while the convective envelope shows differential rotation, which changes gradually with depth from the surface rotation. At 30° latitude angular velocity is nearly constant with depth, while at higher latitude, where the surface rotation velocity is about 20–30 % lower than at the poles, it rises mainly near the bottom of the convective envelope, at a relative radius of about $r/R_\odot \approx 0.28$.

45.2 Dynamical Stability

The behaviour of incompressible homogeneous rotating fluids has been thoroughly investigated (see, e.g. Chandrasekhar 1981). But in many respects compressible gases behave differently. For instance, pure rotation (without meridional motions) in the case ϱ = constant can only take place if ω is constant on cylinders of s = constant (compare Chap. 44). Otherwise the curl of the centrifugal acceleration $\omega^2 s \boldsymbol{e}_s$ would not vanish. But in the case of pure rotation the equation of motion in the meridional plane is

$$\frac{1}{\varrho}\nabla P + \nabla \Phi = \omega^2 s \boldsymbol{e}_s . \tag{45.8}$$

As long as ϱ = constant, the curl of the left-hand side vanishes. For $\partial\omega/\partial z \neq 0$ one has curl $(\omega^2 s \boldsymbol{e}_s) \neq 0$. Then the meridional components of the equation of motion can only be fulfilled if meridional motions occur, and with them additional terms appear in (45.8). This is also the case if the equation of state is barotropic (as for complete degeneracy), since for $P = P(\varrho)$, the curl of $(\nabla P/\varrho)$ also vanishes. The same holds if the equation of state is not barotropic, but if some other mechanism ensures that the surfaces of constant pressure and constant density coincide. One example is convection zones in their adiabatic regime. From the condition $\nabla = \nabla_{\mathrm{ad}}$ (where ∇_{ad} is constant or is a function of P and T) it follows that the surfaces of constant pressure and density coincide. If the convective region is chemically homogeneous, then the equation of state (say for an ideal gas) assures that also the pressure and density surfaces coincide. Therefore $\nabla \times (\nabla P/\varrho)$ vanishes and meridional flow occurs if $\partial\omega/\partial z \neq 0$.

But in a rotating star the pressure and density surfaces are normally inclined:

$$\nabla \times \left(\frac{1}{\varrho}\nabla P\right) = -\frac{1}{\varrho^2}\nabla\varrho \times +\nabla P \neq 0. \tag{45.9}$$

Here the right-hand side is obviously proportional to the sine of the angle of inclination. The vector $\nabla P/\varrho$ is no longer a gradient; it can therefore cancel the non-conservative part of $\omega^2 s \boldsymbol{e}_s$ and (45.8) can be fulfilled without any meridional velocity components.

The different behaviour of a compressible non-barotropic gas compared to that of an incompressible fluid also affects the stability behaviour.

It is well known that the shear motion of fluids can become turbulent. Then kinetic energy of the shear flow goes into the kinetic energy of the "turbulent" elements. If friction is strong, it can prevent this transition.

In an incompressible viscous fluid, therefore, the *Reynolds number Re* decides whether the flow is turbulent or laminar (Landau and Lifshitz 1987, vol. 6):

$$Re = \frac{\varrho v d}{\eta}, \tag{45.10}$$

where v is a characteristic velocity difference and d is a characteristic length. For high Reynolds numbers (say $Re \gg 3{,}000$) kinetic energy of the differential motion becomes kinetic energy of the turbulent elements and the energy which is necessarily lost because of friction is small: the flow is turbulent. If, on the other hand, *Re* is small, much more energy would have to be used up to overcome the friction of the turbulent elements than is available from the reservoir of differential motion: the flow is laminar. For a rotating star with $\varrho \approx 1\,\mathrm{g\,cm^{-3}}$, $d \approx R \approx 10^{11}\,\mathrm{cm}$, $v \approx 10^5\,\mathrm{cm\,s^{-1}}$, and $\eta \approx 1$ cgs (molecular or radiative viscosity), we find $Re \approx 10^{16}$, which means that the flow should be highly turbulent.

But the stellar gas is not incompressible and in most cases not barotropic. Therefore, for a transition from laminar to turbulent motion, the energy due to the shear motion not only has to go into kinetic energy of the turbulent elements (and via friction into heat) but also into work against the buoyancy forces. Another critical dimensionless number, the *Richardson number Ri,* can be used to decide whether shear motion becomes turbulent despite the stabilizing effect of buoyancy. In the case of a plane parallel flow $v(z)$, it is defined by

$$Ri = \frac{g}{H_P} \frac{|\nabla_{\mathrm{ad}} - \nabla|}{(\partial v/\partial z)^2}. \tag{45.11}$$

One can show that $Ri < 1/4$ is a sufficient condition for stability of the laminar motion. In the case of a layer in the deep interior of a star we may estimate $|\partial v/\partial z| \approx \omega R/R = \omega$, $|\nabla_{\mathrm{ad}} - \nabla| \approx 1$, $H_P \approx 10^9\,\mathrm{cm}$, $g \approx 10^5\,\mathrm{cm\,s^{-2}}$ and find that the rotation is laminar as long as $\omega < 2 \times 10^{-2}\,\mathrm{s^{-1}}$ or the rotation period is longer than five minutes. Only neutron stars rotate faster.

Equation (45.11) has been derived under the assumption that the turbulent elements undergo adiabatic changes during their motion. This is not necessarily always the case, not even in the very deep stellar interior. For the sake of simplicity we discuss it in the plane parallel approximation. Let us define a characteristic timescale for a turbulent element in the case of shear instability of a plane parallel flow by $\tau_\ell = |dz/dv|$. This timescale can be considered as the "lifetime" of the element. If its excess velocity over the mean velocity of its origin is $\Delta v = \ell |dv/dz|$, where ℓ is its mean free path, then it takes the time τ_ℓ to move over the distance ℓ. The motion will only be adiabatic if $\tau_\ell \ll \tau_{\mathrm{adj}}$, where τ_{adj} is the thermal adjustment time of the element. With (6.25) one finds as the condition for adiabatic changes of the turbulent elements of diameter d (as assumed in the Richardson criterion),

$$1 \gg \frac{\tau_\ell}{\tau_{\mathrm{adj}}} \approx \left|\frac{dz}{dv}\right| \frac{16acT^3}{\kappa \varrho^2 c_P d^2}. \tag{45.12}$$

One can see that this condition is violated for very small shear ($|dv/dz| \to 0$) as well as for small elements ($d \to 0$). Small elements always have time to adjust thermally to their surroundings while they are moving. Then the stabilizing effect of the temperature stratification disappears. The instability which then occurs for small turbulent elements can become important. But one has to keep in mind that extremely small turbulent elements cannot exist, since for them, even the low molecular or radiative viscosity brakes their motion. One way of estimating the lower limit would be to assume that the smallest elements are those for which τ_ℓ (which is normally short compared to the viscosity timescale of the elements)

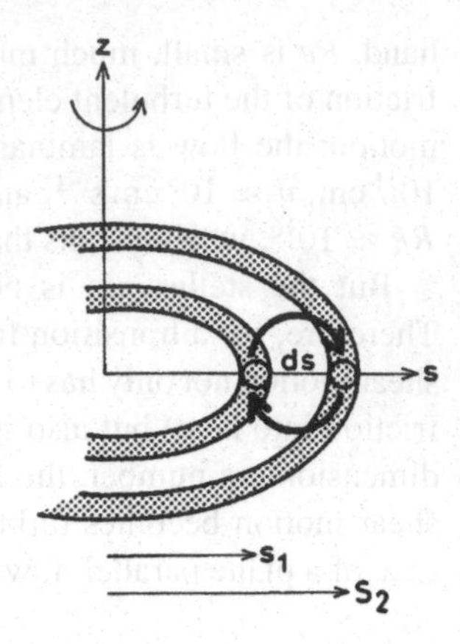

Fig. 45.1 Two tori of radii s_1 and s_2 are exchanged in order to determine the work against centrifugal forces

becomes comparable to τ_{visc}. This would mean that the critical size d of the turbulent element is given by

$$d^2 \approx \left|\frac{dz}{dv}\right| \frac{\eta}{\varrho}, \tag{45.13}$$

while for smaller elements, viscosity overcomes the instability. Since the thermal adjustment time of turbulent elements is shorter than their lifetime, however, the stabilizing effect of buoyancy is reduced and a flow can be turbulent even if $Ri < 1/4$.

There are other dynamical instabilities which are typical of rotational motion. If they occurred in a star, the flows would become turbulent and the turbulent viscosity would immediately change the original angular-velocity distribution. The simplest case of such an instability can be studied by the example of an incompressible or barotropic liquid rotating, say, in a cylindrical container. The angular velocity ω may depend on s only, making pure rotation possible (see Sect. 45.3). As "mass elements," we consider the matter within two neighbouring thin tori as indicated in Fig. 45.1. Their main radii are s_1 and $s_2 = s_1 + ds$. Their thicknesses shall be such that their mass contents dm are equal. We now try to exchange the masses of the two tori by expanding the smaller one and contracting the other without changing their angular momentum and calculate the work necessary to make the exchange against the centrifugal force. The kinetic energy of a torus is $E = \omega^2 s^2 dm/2$, which for a given mass is a function of s. If we expand (or contract) one of the rings, then conservation of angular momentum demands $\omega \sim s^{-2}$ and therefore $E \sim s^{-2}$. At their original position (s_1 and s_2), the two tori shall have the energies E_1 and E_2, respectively. Owing to the expansion $s_1 \rightarrow s_2$, the energy of the first torus changes by an amount

$$dE_1 = \frac{E_1 s_1^2}{(s_1 + ds)^2} - E_1 = -2\frac{E_1 ds}{s_1} + 3\frac{E_1 ds^2}{s_1^2} - \cdots, \tag{45.14}$$

while for the contraction $s_2 \to s_1$ of the other one, we find

$$dE_2 = 2\frac{E_2 ds}{s_2} + 3\frac{E_2 ds^2}{s_2^2} + \cdots . \tag{45.15}$$

Then the total energy required for the exchange of the two tori is

$$\begin{aligned} dE &= dE_1 + dE_2 = 2\left[\frac{E_2}{s_2} - \frac{E_1}{s_1}\right] ds + 6\frac{E_1 ds^2}{s_1^2} + \cdots \\ &= 2\frac{d}{ds}\left(\frac{E}{s}\right) ds^2 + 6\frac{E}{s^2} ds^2 + \cdots , \end{aligned} \tag{45.16}$$

where in the last term of (45.15), we have replaced E_2/s_2 by E_1/s_1, which only introduces third-order errors in ds/s_1. In the last equation (45.16), E means, for instance, a value between E_1 and E_2. With $E/s = s\omega^2 dm/2$, we find

$$dE = 2\omega^2 dm \left[\frac{d\ln\omega}{d\ln s} + 2\right] ds^2. \tag{45.17}$$

Since dE is the energy which has to be supplied for the exchange, $dE > 0$ indicates stability, while $dE < 0$ gives instability (energy is gained). We therefore find the condition for stability,

$$\frac{d\ln\omega}{d\ln s} > -2. \tag{45.18}$$

This is the *Rayleigh criterion,* which we have derived here in a heuristic way. It says that if the specific angular momentum $s^2\omega$ decreases with distance from the axis of rotation, the flow will be turbulent. We have to keep in mind that it has been derived by assuming axisymmetric perturbations only. Since additional non-axisymmetric instabilities may exist, (45.18) is only a necessary condition for stability. Experiments with rotating incompressible fluids between coaxial cylinders indicate that the transition from laminar to turbulent flow occurs when the left-hand side of (45.18) becomes equal to −2. But a liquid between a slowly rotating inner cylinder and a very rapidly rotating outer one can become turbulent even though condition (45.18) is fulfilled.

In the derivation of the Rayleigh criterion we have assumed that the gas is incompressible or at least barotropic. But in all other cases buoyancy forces become important and the work against them has to be taken into account. In the case of gas rotating with $\omega = \omega(s)$ and with gravity pointing towards the axis of rotation (as it is in the equatorial plane of a star), instead of (45.18) one has as stability condition

$$\frac{1}{s^3}\frac{\partial s^4\omega^2}{\partial s} - g_s\frac{\partial \ln P}{\partial s}(\nabla - \nabla_{\rm ad}) > 0, \tag{45.19}$$

where $g_s (< 0)$ is the component of gravity in the s direction.

If the second term on the left is neglected, the Rayleigh criterion is recovered. Without rotation (45.19) gives the Schwarzschild criterion (6.13) for stability.

As in the case of the Rayleigh criterion the derivation of (45.19) assumes that the exchange of toroidal mass elements takes place only in the s direction. If in a star the directions of gravity and of exchange do not coincide, then the *Solberg-Høiland* criterion decides whether the flow is stable or not. We introduce the specific entropy σ:

$$\sigma = c_P \ln(\varrho P^{-1/\gamma_{\rm ad}}) + \text{constant}. \tag{45.20}$$

As long as the equipotential surfaces are not too far from being spherical we can write approximately that

$$\boldsymbol{g} \cdot \nabla\sigma = \frac{|\boldsymbol{g}|}{H_P}(\nabla_{\rm ad} - \nabla). \tag{45.21}$$

With the specific angular momentum $j = s^2\omega$, the Solberg–Høiland criterion (Tassoul 1978; Zahn 1974) requires for stability

$$\frac{1}{s^2}\frac{\partial j^2}{\partial s} - \frac{|\boldsymbol{g}|}{H_P}c_P(\nabla - \nabla_{\rm ad}) > 0, \tag{45.22}$$

$$g_z\left[\frac{\partial j^2}{\partial s}\frac{\partial\sigma}{\partial z} - \frac{\partial j^2}{\partial z}\frac{\partial\sigma}{\partial z}\right] < 0, \tag{45.23}$$

$$g_z\frac{\partial\sigma}{\partial z} > 0. \tag{45.24}$$

All three conditions have to be fulfilled in order to obtain stability; otherwise, the flow is unstable. They are necessary and sufficient for stability as long as only axisymmetric perturbations are allowed. They are also necessary for stability if non-axisymmetric perturbations are permitted.

One immediately sees that (45.22) is identical to (45.19) and gives stability for exchange in the s direction. Condition (45.23) is fulfilled as long as j increases on surfaces of σ = constant on the way from the pole to the equator. Exchange on such surfaces does not imply buoyancy forces, and therefore it reproduces our old condition (45.18). Condition (45.24) says that the Schwarzschild criterion has to be fulfilled for exchange in directions parallel to the axis of rotation in which there is no centrifugal acceleration.

For the problem of dynamical stability in the more general case $\omega = \omega(z, s)$, we refer to Tassoul (1978) and Zahn (1974).

45.3 Secular Stability

We have seen that buoyancy forces can stabilize angular-velocity distributions which otherwise are dynamically unstable. In the case of non-conservative rotation of a barotropic fluid, there can be no hydrostatic equilibrium between centrifugal, gravitational, and pressure accelerations. Therefore circulation currents are necessary to fulfil the equation of motion in the meridional plane. If buoyancy forces are present, equilibrium can exist for any rotation law $\omega = \omega(s, z)$ as long as gravity overcomes the centrifugal force.

However, buoyancy forces are not as reliable as, for instance, gravity. Let us consider the axisymmetric case of a fluid between two rotating cylinders and let us assume the Rayleigh criterion (45.18) to be violated, while the Solberg–Høiland criterion (45.22)–(45.24) gives stability. We then know that if a toroidal mass element is exchanged with another one further outwards in the s direction, energy is gained from centrifugal forces, but the work which goes into buoyancy is larger. Therefore, if kicked outwards, it will go back and, in the pure adiabatic case, starts to oscillate around its original position. This reminds us of the oscillating blob discussed in Chap. 6. But we have seen there that a blob with an excess of molecular weight will sink while adjusting thermally. The situation is very similar in the case of a rotating star in which buoyancy forces guarantee dynamical stability.

Let us discuss the case of non-conservative rotation. It is called "baroclinic", since the P and ϱ are inclined against each other. Then centrifugal acceleration is not curl-free and cannot be balanced by the (conservative) gravity. We now consider a closed line in one quadrant of the meridional plane (Fig. 45.2). The vector of a line element is $\boldsymbol{dl}$. Then the integral of the centrifugal acceleration taken along the line is

$$\oint \boldsymbol{c} \cdot \boldsymbol{dl} \neq 0. \tag{45.25}$$

This means that the centrifugal acceleration produces a torque on the matter along this line. In a barotropic (or incompressible) fluid this torque would cause a meridional flow. In the more general case, $\nabla P/\varrho$ can balance this torque. But the matter will follow the torque within the timescale during which heat can leak out.

The matter will also flow if the Rayleigh criterion (45.18) is violated, but the Richardson number (45.11) gives stability. This is analogous to the case of the salt-finger experiment (see Sect. 6.5). If we then exchange two coaxial tori adiabatically as indicated in Fig. 45.1, buoyancy will bring them back to their old position. But since it takes a finite time to return to the initial state, heat will leak out of, or go into, the two tori and they will never come back exactly to the old position. As the blobs in the salt-finger experiment exchange chemical species, here a meridional flow will exchange angular momentum. This flow is again controlled by the time during which heat can leak away from the matter.

What is the timescale of such a thermally controlled flow? Let us go back to the baroclinic case and the example indicated in Fig. 45.2. Along each closed meridional line there is a torque. The heat exchange can take place most effectively if the thickness d is small, just as the thinnest salt-finger moves fastest, as can be seen from (6.25) and (6.29). One would therefore expect that the smallest elements move fastest. Indeed, with decreasing thickness, the velocity increases like $v \sim d^{-2}$. Certainly for small mass elements friction becomes important, but since the molecular (or radiative) viscosity is low, the elements slowed down by friction

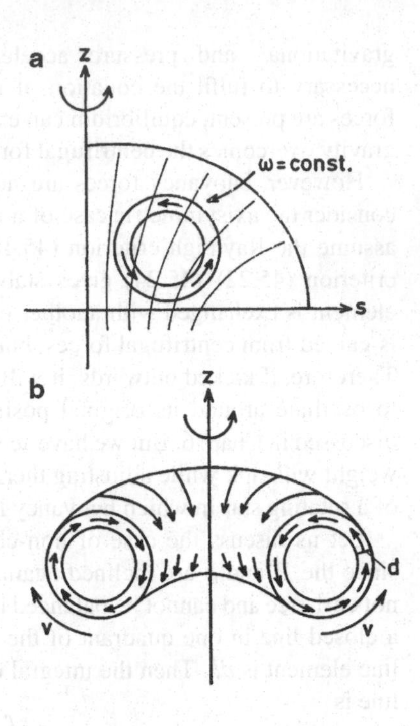

Fig. 45.2 (a) The meridional plane of a rotating star with $d\omega/dz \neq 0$. *Thin lines* give $\omega =$ constant. Along each closed line, the integral over the centrifugal acceleration as defined in (45.25) does not vanish, giving rise to a torque which causes meridional motions as indicated in (b)

are rather small. Estimates indicate that they are of the order of some metres in the radiative interior of the Sun.

Here we have discussed the instabilities by rather heuristic arguments. A mathematically more satisfying treatment of this problem has been carried out by Goldreich and Schubert (1967) and Fricke (1968). They find as conditions necessary for secular stability

$$\frac{\partial \ln \omega}{\partial \ln s} > -2, \qquad \frac{\partial \omega}{\partial z} = 0. \tag{45.26}$$

Although the first condition is identical with (45.18) we have to keep in mind that there we discussed dynamical stability in the barotropic (or incompressible) case, while here we deal with *secular stability*. The second condition of (45.26) does not correspond to a stability condition in the barotropic case. If in this case it is violated, there is no equilibrium. Only buoyancy forces can establish equilibrium in the non-barotropic case, but this equilibrium is thermally unstable.

Several estimates have been made of the timescale by which the thermal instabilities change the overall angular-velocity distribution, violating conditions (45.26). There is no definite answer, but it may well be that it is the Eddington–Sweet timescale (44.34) (Kippenhahn et al. 1980b).

What kind of angular-velocity distribution really does occur in radiative regions of stars? Let us start with a conservative angular-velocity distribution, $\omega = \omega(s)$, say with $\omega =$ constant. Then meridional motions will start. Since they are due to the thermal imbalance between polar and equatorial regions, their characteristic length scale should be of the order of the stellar radius. They will change the angular-velocity distribution, and ω will become a function of z too. But then the Goldreich–Schubert–Fricke criterion (45.26) is violated and instabilities will occur, which grow fastest for small-scale perturbations. Therefore one again expects eddies of the size of metres. At the present time, a complete picture of the transport of angular momentum does not exist. Multidimensional hydrodynamical simulations as well as advanced theories are used to understand the physics determining the properties of rotation in stars. Seismology may yield observational evidence, too. The present situation is summarized in more detail in the textbook by Maeder (2009). Rotation is one of the big unsolved and important questions of stellar structure theory. It will require much more physical insight and many more numerical efforts to understand it.

References

Abazajian, K., et al. (2005): Astron. J., **129**, 1755

Adelberger, E.G., García, A., Robertson, R.G, Snover, K.A., Balantekin, A.B., et al. (2011): Rev. Mod. Phys. **83**, 195

Aerts, C., Christensen-Dalsgaard, J., Kurtz, D.W. (2010): *Asteroseismology*, (Springer, Dordrecht)

Aizenman, M.L., Perdang, J. (1971): Astron. Astrophys. **12**, 232

Alexander, D.R., Ferguson, J.W. (1994): Astrophys. J. **437**, 879

Althaus, L.G., Miller Bertolami, M.M., Córsico, A.H, García/Berro, E., Gil-Pons, P. (2005): Astrophys. J. **440**, L1

Althaus, L.G., García-Berro, E., Renedo, I., Isern, J., Córsico, A.H., Rohrmann, R.D. (2010a): Astrophys. J. **719**, 612

Althaus, L.G., Córsico, A.H., Isern, J., García-Berro, E. (2010b): Astron. Astrophys. Rev. **18**, 471

Andersen, J. (1991): Astron. Astrophys. Rev. **3**, 91

Angulo, C., Arnould, M., Rayet, M., Descouvemont, P., Baye, D., et al. (1999): Nucl. Phys. A, **656**, 3

Appenzeller, I. (1970): Astron. Astrophys. **5**, 355

Appenzeller, I., Tscharnuter, W. (1974): Astron. Astrophys. **30**, 423

Appenzeller, I., Tscharnuter, W. (1975a): Astron. Astrophys. **40**, 397

Appenzeller, I., Tscharnuter, W. (1975b): private communication

Arnett, W.D. (1969): Astrophys. Space Sci. **5**, 180

Arnould, M., Takahashi, K. (1999): Rep. Prog. Phys. **62**, 395

Asplund, M., Grevesse, N., Sauval, A.J. (2005): In *Cosmic Abundances as Records of Stellar Evolution and Nucleosynthesis (ASP Conf. Ser. 336)*, ed. by T.G. Barnes III, F.N. Bash (Astron. Soc. Pacific, Los Angeles), p. 25

Asplund, M., Grevesse, N., Sauval, A.J., Scott, P. (2009): Ann. Rev. Astron. Astrophys. **47**, 481

Baade, W., Zwicky, F. (1934): Phys. Rev. **45**, 138

Bahcall, J.N., Davies, R., Jr. (2000): Pub. Astron. Soc. Pac. **112**, 429

Baker, N., Kippenhahn, R. (1962): Z. Astrophys. **54**, 114

Baker, N., Kippenhahn, R. (1965): Astrophys. J. **142**, 868

Banerjee, R., Pudritz, R.E. (2007): Astrophys. J. **660**, 479

Baraffe, I., Heger, A., Woosley, S.E. (2001): Astrophys. J., **550**, 890

Baraffe, I., Chabrier, G., Barman, T., Allard, F., Hauschildt, P. (2003): Astron. Astrophys. **402**, 701

Baumgarte, Th.W., Shapiro, S.L. (2010): *Numerical Relativity*, (Cambridge University Press, Cambridge)

Biermann, L. (1951): Z. Astrophys. **28**, 304

Blinnikov, S.I., Dunina-Barkovskaya, N.V., Nadyozhin, D.K. (1996): Astrophys. J. Suppl. Ser. **106**, 171

Blöcker, T. (1995): Astron. Astrophys. **297**, 727

R. Kippenhahn et al., *Stellar Structure and Evolution*, Astronomy and Astrophysics Library, DOI 10.1007/978-3-642-30304-3, © Springer-Verlag Berlin Heidelberg 2012

Böhm-Vitense, E. (1958): Z. Astrophys. **46**, 108
Bonner, W.B. (1956): Mon. Not. R. Astron. Soc. **116**, 351
Bono, G., Stellingwerf, R.F. (1994): Astrophys. JSupp. **93**, 233
Bono, G., Marconi, M., Stellingwerf, R.F. (1999): Astrophys. JSupp. **122**, 167
Bono, G., Marconi, M., Stellingwerf, R.F. (2000): Astron. Astrophys. **360**, 245
Boury, A., Ledoux, P. (1965): Ann. d'Astrophys. **28**, 353
Bromm, V., Larson, R.B. (2004): Ann. Rev. Astron. Astrophys. **42**, 79
Burbidge, E.M., Burbidge, G.R., Fowler, W.A., Hoyle, F. (1957): Rev. Mod. Phys. **29**, 547
Burgers, J.M. (1969): *Flow equations for composite gases, Applied Mathematics and Mechanics, vol. 11*, (New York: Academic Press)
Busso, M., Gallino, R., Wasserburg, G.J. (1999): Ann. Rev. Astron. Astrophys. **37**, 239
Camenzind, M. (2007): *Compact Objects in Astrophysics*, (Springer, Berlin Heidelberg)
Canuto, V.M. (2008): In *Interdisciplinary aspects of turbulence*, ed. by W. Hillebrandt, F. Kupka (Springer, Berlin Heidelberg), p. 107
Canuto, V.M., Mazzitelli, I. (1991): Astrophys. J. **370**, 295
Cassisi, S., Schlattl, H., Salaris, M., Weiss, A. (2003): Astrophys. J. Letters **582**, L4
Castellani, V., Giannone, P., Renzini, A. (1971): Astrophys. Space Sci. **10**, 340
Caughlan, G.R., Fowler, W.A. (1988): Atomic Data and Nuclear Data Tables, **40**, 283
Chabrier, G., Baraffe, I. (2000): Ann. Rev. Astron. Astrophys. **38**, 337
Chandrasekhar, S. (1933): Mon. Not. R. Astron. Soc. **93**, 390
Chandrasekhar, S. (1939): *An Introduction to the Study of Stellar Structure* (University of Chicago Press, Chicago)
Chandrasekhar, S. (1981): *Hydrodynamic and Hydromagnetic Stability* (Dover, Oxford, New York)
Chandrasekhar, S. (1983): *The Mathematical Theory of Black Holes* (Clarendon Press, Oxford)
Chapman, S., Cowling, T.G. (1970): *The Mathematical Theory of Non-uniform Gases*, 3rd ed. (Cambridge University Press, Cambridge)
Choudhuri, A.R. (1998): *The physics of fluids and plasmas* (Cambridge: Cambridge University Press)
Christensen-Dalsgaard, J. (1984): In *Theoretical Problems in Stellar Stability and Oscillations*, ed. by A. Noels, M. Gabriel, 25th Liège Intern. Astrophys. Coll., p. 155
Christy, R.F. (1975): In *Problèmes d'Hydrodynamique Stellaire*, 19th Liège Intern. Astrophys. Coll., p. 173
Cignoni, M., Sabbi, E., Nota, A., Tosi, M., Degl'Innocenti, S., Prada Moroni, P.G., Angeretti, L., Carlson, L.R., Gallagher, J., Meixner, M., Sirianni, M., Smith, L.J. (2009): Astron. J. **137**, 3668
Clayton, D.B. (1968): *Principles of Stellar Evolution and Nucleosynthesis* (McGraw-Hill, New York)
Courant, R., Friedrichs, K.O. (1976): *Supersonic Flow and Shock Waves* (Springer, New York)
Cowling, T.G. (1935): Mon. Not. R. Astron. Soc. **96**, 42 (Appendix)
Cox, J.P. (1967): In *Aerodynamic Phenomena in Stellar Atmospheres*, ed. by R.N. Thomas, IAU Symp. 28 (Academic Press, London), p. 3
Cox, J.P. (1980): *Theory of Stellar Pulsation* (Princeton University Press, Princeton)
Cox, A.N. (1980): Ann. Rev. Astron. Astrophys. **18**, 15
Cox, A.N. (2000): *Allen's Astrophysical Quantities*, 4th edition (Springer, New York)
Cox, J.P., Giuli, R.T. (1968): *Principles of Stellar Structure*, Vols. I, II (Gordon and Breech, New York)
D'Antona, F., Mazzitelli, I. (1989): Astrophys. J. **347**, 934
Demorest, P.B., Pennucci, T., Ransom, S.M., Roberts, M.S.E., Hessels, J.W.T (2010): Nature, **467**, 1081
Deubner, F.L., Gough, D. (1984): Ann. Rev. Astron. Astrophys. **22**, 593
Dupret, M.A., Quirion, P.O., Fontaine, G., Brassard, P., Grigahcéne, A. (2008): Journal of Physics, Conf. Ser., **118**, 012051(1–6)
Ebert, R. (1955): Z. Astrophy. **37**, 217
Eckart, A., Genzel, R. (1996): Nature **383**, 415
Eddington, A.S. (1925): Observatory **48**, 73

Eddington, A.S. (1926): The Internal Constitution of Stars (Cambrigde University Press, Cambridge)
Emden, R. (1907): Gaskugeln (Teubner, Leipzig, Berlin)
Evans, N.R., Böhm-Vitense, E., Carpenter, K., Beck-Winchatz, B., Robinson, R. (1998): Astrophys. J. **494**, 768
Faulkner, J. (1966): Astrophys. J. **144**, 978
Ferguson, J.W., Alexander, D.R., Allard, F., Barman, T., Bodnarik, J.G., Hauschildt, P.H., Heffner-Wong, A., Tamanai, A. (2005): Astrophys. J. **623**, 585
Fowler, W.A., Hoyle, F. (1964): Astrophys. J. Suppl. **9**, 201
Fricke, K.J. (1968): Z. Astrophys. **68**, 317
Fricke, K.J., Strittmatter, P.A. (1972): Mon. Not. R. Astron. Soc. **156**, 129
Fujimoto, M.Y., Iben, I., Jr., Hollowell, D. (1990): Astrophys. J. **349**, 580
García-Berro, E., Iben, I.Jr. (1994): Astrophys. J. **434**, 306
Genzel, R., Eisenhauer, F., Gillesen, S. (2010): Rev. Mod. Phys. **82**, 3121
Giannone, P., Kohl, K., Weigert, A. (1968): Z. Astrophys. **68**, 107
Glendenning, N.K. (1997): *Compact Stars*, (Springer, New York)
Gold, T. (1968): Nature **218**, 731
Goldreich, P., Schubert, G. (1967): Astrophys. J. **150**, 571
Goldreich, P., Weber, S.V. (1980): Astrophys. J. **238**, 991
Gong, Z., Zejda, L., Däppen, W., Aparicio, J.M. (2001): Comp. Phys. Comm. **136**, 294
Gough, D.O. (1986): In *Seismology of the Sun and distant stars*, ed. by D.O. Gough (Reidel, Dordrecht), p. 125
Grevesse, N., Noels, A. (1993): Phys. Scr. **T47**, 133
Grevesse, N., Sauval, A.J. (1998): Space Sci. Rev. **65**, 161
Habing, H.J., Olofsson, H. (2003): *Asymptotic Giant Branch Stars* (Springer New York)
Haensel, P., Potekhin, A.Y., Yakolev, D.G. (2007): *Neutron Stars I* (Springer, New York)
Haft, M., Raffelt, G., Weiss, A. (1994): Astrophys. J. **425**, 222
Hamada, T., Salpeter, E.E. (1961): Astrophys. J. **134**, 683
Härm, R., Schwarzschild, M. (1972): Astrophys. J. **172**, 403
Harris, M.J., Fowler, W.A., Caughlan, G.R., Zimmerman, B.A. (1983): Ann. Rev. Astron. Astrophys. **21**, 165
Harris, H.C., Munn, J.A., Kilic, M., Liebert, J., Williams, K.A., von Hippel, T., Levine, S.E., Monet, D.G., Eisenstein, D.J., Kleinman, S.J, Metcalfe, T.S., Nitta, A., Winget, D.E., Brinkmann, J., Fukugita, M., Knapp, G.R., Lupton, R.H., Smith, J.A., Schneider, D.P. (2006): Astron. J., **131**, 571
Hayashi, C. (1961): Publ. Astron. Soc. Japan **13**, 450
Heger, A., Fryer, C.L., Woosley, S.E., Langer, N., Hartmann, D.H. (2003): Astrophys. J. **591**, 288
Henyey, L.G., Forbes, J.E., Gould, N.L. (1964): Astrophys. J. **139**, 306
Henyey, L.G., Vardya, M.S., Bodenheimer, P.L. (1965): Astrophys. J. **142**, 841
Herwig, F. (2005): Ann. Rev. Astron. Astrophys. **43**, 435
Hillebrandt, W. (1991): In *Proceedings of the International School of Physics, "Enrico Fermi", Course CXIII*, ed. by S. Eliezer, R.A. Ricci (North Holland, Amsterdam), p. 399
Hillebrandt, W., Niemeyer, J.C. (2000): Ann. Rev. Astr. Astrophys. **38**, 191
Hirschi, R., Meynet, G., Maeder, A. (2004): Astron. Astrophys. **425**, 649
Hubbard, W.B., Lampe, M. (1969): Astrophys. J. Suppl. **18**, 297
Iben, I., Jr. (1975): Astrophys. J. **196**, 549,
Iben, I., Jr., Renzini, A. (1983): Ann. Rev. Astron. Astrophys. **21**, 271
Iglesias, C.A., Rogers, F.J. (1996): Astrophys. J. **464**, 943
Iliadis, C. (2007): Nuclear Physics of Stars, (Wiley, Wenheim, Germany)
Isern, J., Mochkovitch, R., García-Berro, E., Hernanz, M. (1997): Astrophys. J. **485**, 308
Itoh, N., Mitake, S., Iyetomi, H., Ichimaru, S. (1983): Astrophys. J. **273**, 774
Itoh, N., Hayashi, H., Nishikawa, A., Kohyama, Y. (1996): Astrophys. J. Suppl. Ser. **102**, 411
Ivanova, L.N., Imshennik, V.S., Chechetkin V.M. (1977): Sov. Astron. **21**, 5

Janka, H.-Th., Langanke, K., Marek, A., Martínez-Pinedo, G., Müller, B. (2007): Phys. Rep. **442**, 38
Jeans, J. (1928): *Astronomy and Cosmogony* (Cambridge University Press, Cambridge), republished 1961 (Dover, New York)
Junker, M., D'Alessandro, A., Zavatarelli, S., Arpesella, C., Bellotti, E., et al. (1998): Phys. Rev. C **57**, 2700
Kähler, H. (1972): Astron. Astrophys. **20**, 105
Kähler, H. (1975): Astron. Astrophys. **43**, 443
Kähler, H. (1978): In *The HR Diagram,* ed. by A.G.D. Philip and D.S. Hayes, IAU Symp. **80** (Reidel, Dordtrecht), p. 303
Kalirai, J.S., Saul Davis, D., Richer, H.B., Bergeron, P., Catelan, M., Hansen, B.M.S., Richer, R.M. (2009): Astrophys. J., **705**, 408
Karakas, A.I. (2010): Mon. Not. R. Astron. Soc. **403**, 1413
Keller, S.C. (2008): Astrophys. J. **677**, 483
Kilic, M., Leggett, S.K., Tremblay, P.-E., von Hippel, T., Bergeron, P., Harris, H.C., Munn, J.A., Williams, K.A., Gates, E., Farihi, J. (2010): Astrophys. J. Supp. **190**, 77
Kippenhahn, R. (1963): In *Star Evolution,* Proc. International School of Physics "Enrico Fermi", Course XXVIII, ed. by L. Gratton (Academic Press, New York), p. 330
Kippenhahn, R., Thomas, H.-C. (1964): Z. Astrophys. **60**, 19
Kippenhahn, R., Weigert, A., Hofmeister, E. (1967): Meth. Comp. Phys. **7**, 129
Kippenhahn, R., Thomas, H.-C., Weigert, A. (1968): Z. Astrophys. **69**, 265
Kippenhahn, R., Ruschenplatt, G., Thomas, H.-C. (1980a): Astron. Astrophys. **91**, 175
Kippenhahn, R., Ruschenplatt, G., Thomas, H.-C. (1980b): Astron. Astrophys. **91**, 181
Kiriakidis, M., El Eid, M.F., Glatzel, W. (1992): Mon. Not. R. Astron. Soc. **255**, 1p
Kitaura, F.S., Janka, H.-Th., Hillebrandt, W. (2006): Astron. Astrophys. **450**, 345
Korn, G.A., Korn, T.M. (1968): *Mathematical Handbook for Scientists and Engineers,* 2nd ed. (McGraw-Hill, New York)
Kudritzki, R.-P., Puls, J. (2000): Ann. Rev. Astron. Astrophys. **38**, 613
Kunz, R., Fey, M., Jaeger, M., Mayer, A., Hammer, J.W., Staudt, G., Harissopulos, S., Paradellis, T. (2002): Astrophys. J. **567**, 643
Kupka, F. (2008): In *Interdisciplinary aspects of turbulence*, ed. by W. Hillebrandt, F. Kupka (Springer, Berlin Heidelberg), p. 49
La Salle, J., Lefschetz, S. (1961): *Stability by Liapunov's Direct Method with Applications* (Academic Press, New York)
Lamers, H.J.G.L.M. (1981): Astrophys. J. **245**, 593
Landau, L.D., Lifshitz, E.M. (1976): *The Classical Theory of Fields,* Vol. 2 of Course of Theoretical Physics, 4th edition reprinted 2003 (Butterworth-Heinemann, Amsterdam)
Landau, L.D., Lifshitz, E.M. (1980): *Statistical Physics, part 1,* Vol. 5 of Course of Theoretical Physics, 3rd edition reprinted 2003 (Butterworth-Heinemann, Amsterdam)
Landau, L.D., Lifshitz, E.M. (1987): *Fluid Mechanics,* Vol. 6 of Course of Theoretical Physics, 2nd edition reprinted 2004 (Butterworth-Heinemann, Amsterdam)
Langer, N. (1989): In *R*ev. Mod. Astron, ed. G. Klare (Springer, Berlin Heidelberg), vol. **2**, p. 306
Langer, N. (1991): Astron. Astrophys. **252**, 669
Langer, N., El Eid, M.F., Fricke, KJ. (1984): In *Liege International Astrophysical Colloquia,* **25**, 120
Langer, N., El Eid, M.F., Fricke, KJ. (1985): Astron. Astrophys. **145**, 169
Larson, R.B. (1969): Mon. Not. R. Astron. Soc. **145**, 271
Lauterborn, D., Refsdal, S., Weigert, A. (1971a): Astron. Astrophys. **10**, 97
Lauterborn, D., Refsdal, S., Roth, M.L. (1971b): Astron. Astrophys. **13**, 119
Ledoux, P. (1958): In *Handbuch der Physik,* ed. by S. Flügge (Springer, Berlin, Heidelberg), Vol. LI, p. 605
Liebert, J., Bergeron, P., Holberg, J.B. (2005): Astrophys. JṠupp. **156**, 47
Limongi, M., Chieffi, A. (2006): Astrophys. J. **647**, 483
Lorimer, D.R. (2008): Living Reviews in Relativity **11**, 8

Low, C., Lynden-Bell, D. (1976): Mon. Not. R. Astron. Soc. **176**, 367
Ludwig, H.-G., Freytag, B., Steffen, M. (1999): Astron. Astrophys. **346**, 111
Lyttleton, R.A. (1953): *The Stability of Rotating Liquid Masses* (Cambridge University Press, Cambridge)
Maeder, A. (1975): Astron. Astrophys. **40**, 303
Maeder, A. (2009) *Physics, formation and evolution of rotating stars* (Springer, Berlin Heidelberg)
Maeder, A., Meynet, G. (1987): Astron. Astrophys. **182**, 243
Maeder, A., Meynet, G. (1991): Astron. Astrophys. Suppl. Ser. **89**, 451
Malkov, O.Y., Oblak, E., Snegireva, E.A., Torra, J. (2006): Astron. Astrophys. **446**, 785
Marigo, P., Bressan, A., Chiosi, C. (1996): Astron. Astrophys. **313**, 545
Masunaga, H., Inutsuka, S.-I. (2000): Astrophys. J. **531**, 350
Masunaga, H., Miyama, S.M., Inutsuka, S.-I. (1998): Astrophys. J. **495**, 346
Matraka, B., Wassermann, C., Weigert, A. (1982): Astron. Astrophys. **107**, 283
Mazzitelli, I., D'Antona, F. (1986): Astrophys. J. **308**, 706
McKee, Ch.F., Ostriker, E.C. (2007): Ann. Rev. Astr. Astrophys. **45**, 565
Mac Low, M.-M., Klessen, R.S. (2004): Rev. Mod. Phys. **76**, 125
Mendoza, C., Seaton, M.J., Buerger, P., Bellorn, A., Melndez, M., Gonzlez, J., Rodrguez, L.S., Delahaye, F., Palacios, E., Pradhan, A.K., Zeippen, C.J. (2007): Mon. Not. R. Astron. Soc. **378**, 1031
Merrill, S.P.W. (1952): Astrophys. J. **116**, 21
Mestel, L. (1952): Mon. Not. R. Astron. Soc. **112**, 583 and 598
Mestel, L. (1953): Mon. Not. R. Astron. Soc. **113**, 716
Mestel, L., Ruderman, M.A. (1967): Mon. Not. R. Astron. Soc. **136**, 27
Meyer, B.S. (1994): Ann. Rev. Astron. Astrophys. **32**, 153
Mezzacappa, A. (2005): Ann. Rev. Nuc. Part. Sci. **55**, 467
Mihalas, D., Hummer, D.G., Däppen, W. (1988): Astrophys. J. **331**, 815
Misner, C.W., Thorne, K.S., Wheeler, J.A. (1973): *Gravitation* (Freeman, San Francisco)
Mocák, M., Müller, E., Weiss, A., Kifonidis, K. (2008): Astron. Astrophys. **490**, 265
Mocák, M., Müller, E., Weiss, A., Kifonidis, K. (2009): Astron. Astrophys. **501**, 659
Moskalik, P., Dziembowski, W.A. (1992): Astron. Astrophys. **256**, L5
Moskalik, P., Buchler, J.R., Marom, A. (1992): Astrophys. J. **385**, 685
Müller, E. (1997): In Computational Methods for Astrophysical Fluid Flow, Saas Fee Advanced Course 27, ed. by LeVeque, R.J., Mihalas, D., Dorfi, E.A., Müller, E. (Springer, Berlin Heidelberg), p. 343
Musella, I., Ripepi, V., Brocato, E., Castellani, V., Caputo, F., Del Principe, M., Marconi, M., Piersimoni, A.M., Raimondo, G., Stetson, P.B., Walker, A.R. (2006): Mem. Soc. Astron. Italiana **77**, 291
Nomoto, K. (1984): Astrophys. J **277**, 791
Nomoto, K., Sugimoto, D., Neo, S. (1976): Astrophys. Space Sci. **39**, L37
Nomoto, K., Thielemann, F.-K., Yokoi, K. (1984): Astrophys. J. **286**, 644
Ogino, S., Tomisaka, K., Nakamura, F. (1999): Publ. Astron. Soc. Japan **51**, 637
Oppenheimer, J.R., Volkoff, G.M. (1939): Phys. Rev. **55**, 374
Paczyński, B. (1970): Acta Astron. **20**, 47
Paczyński, B. (1971): Acta Astron. **21**, 271
Paczyński, B. (1972): Acta Astron. **22**, 163
Paczyński, B. (1975): Astrophys. J. **202**, 558
Panei, J.A., Althaus, L.G.,, Benvenuto, O.G. (2000): Astron. Astrophys. **353**, 970
Paquette, C., Pelletier, C., Fontaine, G., Michaud, G. (1986): Astrophys. J. Suppl. **61**, 177
Parker, P.D., Bahcall, J.N., Fowler, W.A. (1964): Astrophys. J. **139**, 602
Pethick, C.J., Ravenhall, D.G. (1991): In *An Introduction to Matter at Subnuclear Densities, NATO ASIC Proc. 344: Neutron Stars*, ed. by J. Ventura, D. Pines (Kluwer Academic Publishers, Dordrecht), p. 3
Pichon, B. (1989): Comp. Phys. Comm. **55**, 127
Pietrukowicz, P. (2001): Acta Astron. **51**, 247

Pietrinferni, A., Cassisi, S., Salaris, M., Castelli, F. (2004): Astrophys. J. **612**, 168
Pines, D. (1980): Journal de Physique **41**, Coll. C2, suppl. au no. 3, pp. C2–111
Potekhin, A.Y., Baiko, D.A., Haensel, P., Yakolev, D.G. (1999): Astron. Astrophys. **346**, 345
Rees, M.J. (1976): Mon. Not. R. Astron. Soc. **176**, 483
Refsdal, S., Weigert, A. (1970): Astron. Astrophys. **6**, 426
Reimers, D. (1975): Mémoires de la Société Royale des Sciences de Liège **8**, 369
Renzini, A. (1981): In *Physical Processes in Red Giants*, ed. by I. Iben, Jr., A. Renzini (Reidel, Dordrecht), p. 431
Renzini, A. (1987): Astron. Astrophys. **188**, 49
Renzini, A., Voli, M. (1981): Astron. Astrophys. **94**, 175
Richer, J., Michaud, G., Rogers, F., Iglesias, C., Turcotte, S., LeBlanc, F. (1998): Astrophys. J. **492**, 833
Richtmyer, R.D., Morton, K.W. (1967): *Difference Methods for Initial-Value Problems,* 2nd ed. (Interscience, New York)
Robertson, J.W. (1971): Astrophys. J. **164**, L105
Roepke, F.K. (2008): In Supernovae: lights in the darkness, Proc. of Science (SISSA, Trieste), 024
Rogers, F.J., Iglesias, C.A. (1992): Astrophys. J. Suppl. Ser. **79**, 507
Rogers, F.J., Swenson, F.J., Iglesias, C.A. (1996): Astrophys. J. **456**, 902
Roth, M.L. (1973): Dissertation, University of Hamburg
Salaris, M., Cassisi, S. (2005): *Evolution of Stars and Stellar Populations* (Wiley, Chichester, UK)
Salaris, M., Serenelli, A., Weiss, A., Miller Bertolami, M. (2009): Astrophys. J. **692**, 1013
Salpeter, E.E. (1961): Astrophys. J. **134**, 669
Saslaw, W.C., Schwarzschild, M. (1965): Astrophys. J. **142**, 1468
Saumon, D., Chabrier, G., van Horn, H.M. (1995): Astrophys. J. Suppl. Ser. **99**, 713
Schaerer, D. (1996): Astron. Astrophys. **309**, 129
Schlattl, H. (1999): *The Sun - a Laboratory for Neutrino- and Astrophysics*, PhD thesis, TechnÜniv. Munich
Schlattl, H., Salaris, M. (2003): Astron. Astrophys. **402**, 29
Schlattl, H., Weiss, A., Ludwig, H.-G. (1997): Astron. Astrophys. **322**, 646
Schlattl, H., Cassisi, S., Salaris, M., Weiss, A. (2001): Astrophys. J. **559**, 1082
Schönberg, M., Chandrasekhar, S. (1942): Astrophys. J. **96**, 161
Schönberner, D. (1979): Astron. Astrophys. **79**, 108
Schönberner, D. (2008): In *Hydrogen-Deficient Stars*, ed. by K. Werner, T. Rauch (Astr. Soc. of the Pacific, San Francisco), ASP Conf. Ser. **391**, p. 139
Schou, J., Antia, H.M., Basu, S., Bogart, R.S., Bush, R.I., Chitre, S.M., Christensen-Dalsgaard, J., Di Mauro, M.P., Dziembowski, W.A., Eff-Darwich, A., Gough, D.O., Haber, D.A., Hoeksma, J.T., Howe, R., Korzennik, S.G., Kosovichev, A.G., Larsen, R.M., Pijpers, F.P., Scherrer, P.H., Sekii, T., Tarbell, T.D., Title, A.M., Thompson, M.J., Toomre, J. (1998): Astrophys. J. **505**, 390
Schröder, K.-P., Cuntz, M. (2005): Astrophys. J. Letters **630**, L73
Schwarzschild, M. (1941): Astrophys. J. **94**, 245
Schwarzschild, M. (1942): Astrophys. J. **95**, 441
Schwarzschild, M. (1946): Astrophys. J. **104**, 203
Schwarzschild, M. (1958): *Structure and Evolution of the Stars* (Princeton University Press, Princeton)
Schwarzschild, M., Härm, R. (1959): Astrophys. J. **129**, 637
Schwarzschild, M., Härm, R. (1965): Astrophys. J. **142**, 855
Sedlmayer, E., Dominik, C. (1994): Space Sci. Rev. **73**, 211
Serenelli, A., Weiss, A. (2005): Astron. Astrophys. **442**, 1041
Shapiro, S.L., Teukolsky, S.A. (1983): *Black Holes, White Dwarfs, and Neutron Stars. The Physics of Compact Objects* (Wiley, New York)
Shaviv, G., Kovetz, A. (1976): Astron. Astrophys. **51**, 383
Shaviv, G., Salpeter, E.E. (1973): Astrophys. J. **184**, 191
Siess, L. (2006a): Astron. Astrophys. **448**, 717

Siess, L. (2006b): In *Stars and Nuclei: A Tribute to Manuel Forestini*, ed. by T. Montmerle, C. Kahane (EDP Science, Les Ulis, France), EAS Pub. Ser. **19**, p. 103
Simon, N.R. (1982): Astrophys. J. Letters **260**, L87
Smartt, S.J. (2009): Ann. Rev. Astr. Astrophys. **47**, 63
Smeyers, P. (1984): In *Theoretical Problems in Stellar Stability and Oscillations,* ed. by A. Noels, M. Gabriel, Proc. 25th Liège Intern. Coll., p. 68
Smith, H.A. (1995): *R*R Lyrae Stars, Cambridge Astrophysics Series, **vol. 27** (Cambridge University Press, Cambridge, UK)
Spitzer, L., Jr. (1968): *Diffuse Matter in Space* (Wiley, New York)
Stothers, R.B., Chin, C.W. 1992: Astrophys. J. **390**, 136
Sweet, P.A. (1950): Mon. Not. R. Astron. Soc. **110**, 548
Sweigart, A.V., Gross, P.G. (1978): Astrophys. J. Suppl. **36**, 405
Tassoul, J.-L. (1978): *Theory of Rotating Stars* (Princeton University Press, Princeton)
Thomas, H.-C. (1967): Z. Astrophys. **67**, 420
Thompson, C., Duncan, R. (1995): Mon. Not. R. Astron. Soc. **275**, 255
Thoul, A.A., Bahcall, J.N., Loeb, A. (1994): Astrophys. J. **421**, 828
Truran, J.W., Iben, I., Jr. (1977): Astrophys. J. **216**, 797
Turner, D.G., Abdel Sabour Abdel Latif, M., Berdnikov, L.N. (2006): Pub. Astron. Soc. Pac. **118**, 410
Unno, W., Osaki, Y., Ando, H., Shibahashi, H. (1979): *Nonradial Oscillations of Stars* (University of Tokyo Press, Tokyo)
Van Horn, H.M. (1984): In *Theoretical Problems in Stellar Stability and Oscillations,* ed. by A. Noels, M. Gabriel, 25th Liège Intern. Astrophys. Coll., p. 307
Van Horn, H.M. (1986): Mitt. Astron. Ges. **67**, 63
Van Loon, J., Cioni, M.-R., Zijlstra, A., Loop, C. (2005): Astron. Astrophys. **438**, 273
VandenBerg, D.A., Edvardsson, B., Eriksson, K., Gustafsson, B. (2008): Astrophys. J. **675**, 746
Vink, J.S., de Koter, A., Lamers, H.J.G.L.M. (2001): Astron. Astrophys. **369**, 574
Vogt, H. (1925): Astron. Nachr. **223**, 229
Von Zeipel, H. (1924): In *Probleme der Astronomie,* Festschrift für H. v. Seeliger, ed. by H. Kienle (Springer, Berlin) p. 144
Wachter, A., Schröder, K.-P., Winters, J.M., Arndt, T.U., Sedlmayr, E. (2002): Astron. Astrophys. **384**, 452
Wagenhuber, J. (1996): PhD thesis, Tech. Univ. München
Wagenhuber, J., Groenewegen, M.A.T. (1998): Astron. Astrophys. **340**, 183
Wagenhuber, J., Weiss, A. (1994): Astron. Astrophys. **286**, 121
Weber, F., Negreiros, R., Rosenfield, P. (2009): In *N*eutron Stars and Pulsars, ed. by W. Becker (Springer, Berlin Heidelberg), pp. 213
Weidemann, V. (1977): Astron. Astrophys. **59**, 411
Weidemann, V. (2000): Astron. Astrophys. **363**, 647
Weigert, A. (1966): Z. Astrophys. **64**, 395
Weiss, A. (1987): Astron. Astrophys. **185**, 178
Weiss, A., Ferguson, J.W. (2009): Astron. Astrophys. **508**, 1343
Weiss, A., Hillebrandt, W., Thomas, H.-C., Ritter, H. (2004) *Cox & Giuli's Principles of Stellar Structure, second edition* (Cambridge Scientific Publishers, Cambridge)
Willson, L.A. (2000): Ann. Rev. Astron. Astrophys. **38**, 573
Woosley, S.E., Bloom, J.S. (2006): Ann. Rev. Astr. Astrophys. **44**, 507
Woosley, S., Janka, H.-Th. (2006): Nature Physics **1**, 147
Woosley, S.E., Pinto, P.A., Ensman, L. (1988): Astrophys. J. **324**, 466
Wrubel, M.H. (1958): In *Handbuch der Physik,* ed. by S. Flügge (Springer, Berlin, Heidelberg), Vol. LI, p. 1
Wuchterl, G., Tscharnuter, W.M. (2003): Astron. Astrophys. **398**, 1081
Yahil, A., Lattimer, J.M. (1982): In *S*upernovae: A survey of current research, ed. by Rees, M.J., Stoneham, R.J. (Reidel, Dordrecht), p. 53
Yoon, S.-C., Langer, N., van der Sluys, M. (2004): Astron. Astrophys. **425**, 207

Zahn, J.-P. (1974): In *Stellar Instability and Evolution,* ed. by P. Ledoux, A. Noels and A.W. Rodgers, IAU Symp. **59** (Reidel, Dordrecht), p. 185
Zeldovich, Ya. B., Novikov, I.D. (1971): *Relativistic Astrophysics,* Vol. I "Stars and Relativity" (University of Chicago Press, Chicago)
Ziebarth, K. (1970): Astrophys. J. **162**, 947
Zinnecker, H., Yorke, H.W. (2007): Ann. Rev. Astron. Astrophys. **45**, 481
Zoccali, M., Cassisi, S., Piotto, G., Bono, M., Salaris, M. (1999): Astrophys. J. Letters **518**, L49

Index

R. Kippenhahn et al., *Stellar Structure and Evolution*, Astronomy and Astrophysics Library, DOI 10.1007/978-3-642-30304-3, © Springer-Verlag Berlin Heidelberg 2012